Green Energy and Technology

For further volumes:
http://www.springer.com/series/8059

Michael Nosonovsky · Bharat Bhushan
Editors

Green Tribology

Biomimetics, Energy Conservation and Sustainability

Springer

Prof. Michael Nosonovsky
Department of Mechanical Engineering
University of Wisconsin–Milwaukee
North Cramer Street 3200
Milwaukee
WI 53211-0413
USA
e-mail: nosonovs@uwm.edu

Prof. Bharat Bhushan
Nanoprobe Laboratory for
Bio- and Nanotechnology
and Biomimetics (NLB^2)
Ohio State University
201 West 19th Avenue
Columbus
OH 43210-1142
USA
e-mail: bhushan.2@osu.edu

ISSN 1865-3529
ISBN 978-3-642-43525-6
DOI 10.1007/978-3-642-23681-5
Springer Heidelberg Dordrecht London New York

e-ISSN 1865-3537
ISBN 978-3-642-23681-5 (eBook)

Printed on acid-free paper

Springer is part of Springer Science+Business Media (www.springer.com)

Preface

Tribology (from the Greek word τρίβω "tribo" meaning "to rub") is the interdisciplinary area of science and technology that involves the study of the interaction of solid surfaces in relative motion. Typical tribological studies cover friction, wear, lubrication, and adhesion. These studies involve the efforts of mechanical engineers, material scientists, chemists, and physicists. The word "tribology" was coined in the 1960s when it was realized that it may be beneficial for engineers and scientists studying friction, lubrication, and wear to collaborate in the framework of the new interdisciplinary area. Since then, many new areas of tribological studies have been suggested, which are at the interface of various scientific disciplines. These areas include nanotribology, biotribology, the tribology of magnetic storage devices, and micro/nanoelectromechanical systems. The research in these areas is driven mostly by the advent of new technologies and new experimental techniques for surfaces characterization.

Green tribology is a new, separate research area that is emerging, and it is defined as the science and technology of the tribological aspects of ecological balance and of environmental and biological impacts. There are a number of tribological problems that can be put under the umbrella of green tribology, and they are of mutual benefit to one another. These problems include tribological technology that mimics living nature (biomimetic surfaces) and thus is expected to be environment-friendly, the control of friction and wear that is of importance for energy conservation and conversion, environmental aspects of lubrication and surface modification techniques, and tribological aspects of green applications such as wind-power turbines, tidal turbines, or solar panels.

Since the 2000s, there have been several publications dealing with the economic and social implications of the ecological aspects of tribology. Most of these papers were prepared by economists and people involved in the strategic planning of research. The first scientific volume completely devoted to green tribology, which emphasized scientific rather than societal and economic aspects, appeared in 2010, and it was the theme issue of the Philosophical Transactions of the Royal Society, Series A (Volume 368, Number 1929) edited by M. Nosonovsky and B. Bhushan. In that volume, three areas of green tribology were identified:

biomimetic tribology, eco-friendly lubrication and materials, and tribological aspects of sustainable energy applications. The assumption was that combining these three areas, rather than focusing on narrow issues such as biodegradable lubrication, would mutually enhance them and establish new connections. Several workshops, conference sections, and symposia took place after that, which confirmed this inclusive approach, as well as the interest in green tribology in general.

The present publication in Springer to a certain degree extends that work: whereas some authors who participated in that volume also submitted their new results into the present volume, new authors participated as well. Prominent experts in various areas were invited that fit the definition of Green Tribology. The international group of authors include tribologists from the U.S., the U.K., Austria, Australia, Canada, India, South Africa, China, Israel, and Malaysia. Some of the authors are from academic institutions, while others are practical engineers from the industry. At University of Wisconsin–Milwaukee (UWM) a big group of tribologists has worked since 2009 on various aspects related to green tribology, and the results of their efforts are presented in the current volume. The biomimetic surfaces, including those using the Lotus, rose petal, gecko, and shark skin effects as well as tribology of human skin and hair were studied actively at the Ohio State University (OSU) in the past decade.

After a review of the current state of green tribology and its history, the main content of this book is divided into three parts. First, biomimetics in tribology is discussed, including biomimetic surfaces, materials, and methods. Biomimetic approaches follow the ways found in living nature and thus are expected to be eco-friendly. This includes non-adhesive surfaces mimicking flower (e.g., Lotus and rose) leaves, wetting transitions on these surfaces, biomimetic adhesion control for antifouling, polymeric and metal-based composite materials, and surfaces capable of friction-induced self-organization (self-lubrication, self-cleaning, and self-healing) as well as biomimetics in nanotribology. Second, green and sustainable materials and lubricants are reviewed. This involves water, ice, and natural oil-based lubrication, eco-friendly products for tribological applications involving natural fiber reinforced composites, fly ash, cements, and lubricant additives. The third part includes tribology of eco-friendly applications, such as wind turbines, biorefinaries, and marine wave energy collectors. Some of the chapters emphasize the review of the current state of the area, while others stress the research conducted by the investigators.

We would like to thank our colleagues, the authors, who responded to our invitations and contributed to this edited book. In addition, we would like to acknowledge help in preparation of the manuscripts of Ms. Caterina Runyon-Spears (OSU) and Mr. Mehdi Mortazavi (UWM).

July, 2011

Michael Nosonovsky
Bharat Bhushan

Contents

Part I Introduction

1 **Green Tribology, its History, Challenges, and Perspectives** 3
Michael Nosonovsky and Bharat Bhushan

Part II Biomimetics Surfaces, Materials and Methods

2 **Lotus Versus Rose: Biomimetic Surface Effects** 25
Michael Nosonovsky and Bharat Bhushan

3 **Self-Organization at the Frictional Interface** 41
Michael Nosonovsky and Vahid Mortazavi

4 **Advanced Nanostructured Surfaces for the Control of Biofouling: Cell Adhesions to Three-Dimensional Nanostructures** 79
Chang-Hwan Choi and Chang-Jin Kim

5 **Green Nanotribology and Sustainable Nanotribology in the Frame of the Global Challenges for Humankind** 105
I. C. Gebeshuber

6 **Towards Understanding Wetting Transitions on Biomimetic Surfaces: Scaling Arguments and Physical Mechanisms** 127
Edward Bormashenko and Gene Whyman

7 **Wear-Resistant and Oleophobic Biomimetic Composite Materials** 149
Vahid Hejazi and Michael Nosonovsky

8 **Polymer Adhesion and Biomimetic Surfaces for Green Tribology** 173
Mehdi Mortazavi and Michael Nosonovsky

Part III Green and Sustainable Lubricants and Materials

9 **A Biomimetic Approach to Ice Friction** 223
Anne-Marie Kietzig

10 **Green Lubricants: Role of Additive Size** 265
Pradeep L. Menezes, Michael R. Lovell, M. A. Kabir, C. Fred Higgs III and Pradeep K. Rohatgi

11 **Natural Oil-Based Lubricants** 287
K. R. Sathwik Chatra, N. H. Jayadas and Satish V. Kailas

12 **Studies on the Tribological Behavior of Natural Fiber Reinforced Polymer Composite** 329
Pradeep L. Menezes, Pradeep K. Rohatgi and Michael R. Lovell

13 **Water Lubricated Bearings** 347
R. Pai and D. J. Hargreaves

14 **Green Tribological Behavior of Waxes, Adhesives and Lubricants** 393
W. Li, X. H. Kong, M. Ruan, F. M. Ma, X. H. Zuo and Y. Chen

15 **Tribo-Chemical Activation of Green Eco-Cements** 413
Konstantin Sobolev

16 **Tribological Properties of Fly Ash-Based Green Friction Products** 429
Pradeep K. Rohatgi, Pradeep L. Menezes and Michael R. Lovell

17 **Self-Lubricating Behavior of Graphite Reinforced Metal Matrix Composites** 445
Pradeep L. Menezes, Pradeep K. Rohatgi and Michael R. Lovell

Part IV Applications

18 **Wind Turbine Tribology** 483
Elon J. Terrell, William M. Needelman and Jonathan P. Kyle

19 Ecological Aspects of Water Desalination Improving Surface Properties of Reverse Osmosis Membranes 531
Tyler G. Hurd, Saman Beyhaghi and Michael Nosonovsky

20 Biorefining: A Green Tribological Perspective 565
P. L. de Vaal, L. F. Barker, E. du Plessis and D. Crous

21 Tribological and Mechanical Design Considerations for Wave Energy Collecting Devices 607
Andrew Fronek, Michael Nosonovsky, Ben Barger and Ilya Avdeev

Editor's Biography 621

Index 625

Contributors

Dr. Ilya V. Avdeev University of Wisconsin–Milwaukee, Milwaukee, WI 53211-0413, USA, e-mail: avdeev@uwm.edu

Benjamin David Barger Department of Mechanical Engineering, University of Wisconsin–Milwaukee, North Cramer Street 3200, Milwaukee, WI 53211-0413, USA, e-mail: bdbarger@uwm.edu

L. F. Barker Eskom Generation Business Engineering, Johannesburg, South Africa, e-mail: Leslie.barker@eskom.co.za

Saman Beyhaghi Department of Mechanical Engineering, University of Wisconsin–Milwaukee, North Cramer Street 3200, Milwaukee, WI 53211-0413, USA, e-mail: beyhagh2@uwm.edu

Prof. Bharat Bhushan Nanoprobe Laboratory for Bio- and Nano-technology and Biomimetics (NLB^2), Ohio State University, 201 West 19th Avenue, Columbus, OH 43210-1142, USA, e-mail: bhushan.2@osu.edu

Prof. Edward Bormashenko Laboratory of Polymers, Ariel University Center of Samaria, P.O.B. 3, Ariel 40700, Israel, e-mail: edward@ariel.ac.il

Mr. Sathwik Chathra Department of Mechanical Engineering, Indian Institute of Science, Bangalore 560 012, India, e-mail: sathwikchathra@gmail.com

Y. Chen Hubei Key Laboratory of Environmental Pollution Control, Resources Reuse, School of Chemical and Materials Engineering, Huangshi Institute of Technology, Huangshi 435003, People's Republic of China, e-mail: ychen21@163.com

Chang-Hwan Choi Ph.D. Department of Mechanical Engineering, Castle Point on Hudson, Stevens Institute of Technology, Hoboken, NJ 07030, USA, e-mail: Chang-Hwan.Choi@stevens.edu

Duncan Crous Department of Chemical Engineering, University of Pretoria, Pretoria, South Africa, e-mail: i_dc_i@yahoo.com

P. L. de Vaal Department of Chemical Engineering University of Pretoria, Pretoria, South Africa, e-mail: pdv@up.ac.za

E. du Plessis Producut Lubrication Technologies (Pty) Ltd., P.O. Box 2309, Silverton 0127, South Africa, e-mail: producut@icon.co.za

Andrew Fronek Department of Mechanical Engineering, University of Wisconsin–Milwaukee, North Cramer Street 3200, Milwaukee, WI 53211-0413, USA, e-mail: ajfronek@gmail.com

Prof. Dr. Ille C. Gebeshuber Institute of Microengineering and Nanoelectronics (IMEN), Universiti Kebangsaan Malaysia, 43600 UKM, Bangi, Selangor, Malaysia, e-mail: ille.gebeshuber@mac.com; ille.gebeshuber@ukm.my

Douglas John Hargreaves School of Engineering Systems, Queensland University of Technology, P.O. Box 2434, Brisbane 4001, Australia, e-mail: d.hargreaves@qut.edu.au

Vahid Hejazi Department of Mechanical Engineering University of Wisconsin–Milwaukee, North Cramer Street 3200, Milwaukee, WI 53211-0413, USA, e-mail: vhejazi@uwm.edu

C. Fred Higgs III Department of Mechanical Engineering, Carnegie Mellon University, 5000 Forbes Avenue, Pittsburgh, PA 15213-3890, Australia, e-mail: higgs@andrew.cmu.edu

Tyler Gregory Hurd Department of Mechanical Engineering, University of Wisconsin–Milwaukee, North Cramer Street 3200, Milwaukee, WI 53211-0413, USA, e-mail: tghurd@uwm.edu

N. H. Jayadas Department of Mechanical Engineering, Cochin University of Science and Technology, Ernakulam 682 022, India, e-mail: jayadasnh@cusat.ac.in

Dr. Aslam Kabir Research Associate DOE - NETL, Pittsburgh, PA 15236, USA, e-mail: mohd.kabir@netl.doe.gov

Satish V. Kailas Ph.D. Department of Mechanical Engineering, Indian Institute of Science, Bangalore 560 012, India, e-mail: satish.kailas@gmail.com; satvk@mecheng.iisc.ernet.in

Anne Kietzig Ph.D. Department of Chemical Engineering, McGill University, 3610 University Street, Montreal, Quebec H3A 2B2, Canada, e-mail: anne.kietzig@mcgill.ca

Prof. Chang-Jin Kim Mechanical and Aerospace Engineering Department, University of California, 3610 University Street, Los Angeles, CA 90095, USA, e-mail: cjkim@seas.ucla.edu

X. H. Kong Hubei Key Laboratory of Environmental Pollution Control, Resources Reuse, School of Chemical and Materials Engineering, Huangshi Institute of Technology, Huangshi 435003, People's Republic of China, e-mail: xkong@ualberta.ca

Jonathan P. Kyle Mechanical Engineering Department, Columbia University, S.W. Mudd, Room 220B 500 West 120th St., New York, NY 10027, USA, e-mail: jpk2128@columbia.edu

Dr. Wen Li Hubei Key Laboratory of Environmental Pollution Control, Resources Reuse, School of Chemical and Materials Engineering, Huangshi Institute of Technology, Huangshi 435003, People's Republic of China, e-mail: wenl@ualberta.ca

Prof. Michael Lovell University of Wisconsin–Milwaukee, Milwaukee, WI 53201, USA, e-mail: mlovell@uwm.edu

F. M. Ma Hubei Key Laboratory of Environmental Pollution Control, Resources Reuse, School of Chemical and Materials Engineering, Huangshi Institute of Technology, Huangshi 435003, People's Republic of China, e-mail: fmma1@yahoo.cn

Dr. Pradeep Lancy Menezes Department of Materials, University of Wisconsin–Milwaukee, Milwaukee, WI 53211-0413, USA, e-mail: menezesp@uwm.edu

Mehdi Mortazavi Department of Mechanical Engineering University of Wisconsin–Milwaukee, North Cramer Street 3200, Milwaukee, WI 53211-0413, USA, e-mail: mortaza3@uwm.edu

Vahid Mortazavi Department of Mechanical Engineering, University of Wisconsin–Milwaukee, North Cramer Street 3200, Milwaukee, WI 53211-0413, USA, e-mail: mortaza2@uwm.edu

William M. Needelman Donaldson Company, Inc., 8 Hillside Court, Huntington, NY 11743, USA, e-mail: bill.needelman@donaldson.com

Michael Nosonovsky Ph.D. Department of Mechanical Engineering, University of Wisconsin–Milwaukee, North Cramer Street 3200, Milwaukee, WI 53211-0413, USA, e-mail: nosonovs@uwm.edu

Prof. Raghuvir Pai Department of Mechanical and Manufacturing Engineering, Manipal Institute of Technology, Manipal University, 5000 Forbes Avenue, Manipal 576104, India, e-mail: raghuvir.pai@manipal.edu

Dr. Pradeep K. Rohatgi Advanced Materials Manufacture, UWM Centers for Composites, University of Wisconsin–Milwaukee, P.O. Box 784, Milwaukee, WI 53201, USA, e-mail: prohatgi@uwm.edu

M. Ruan Hubei Key Laboratory of Environmental Pollution Control, Resources Reuse, School of Chemical and Materials Engineering, Huangshi Institute of Technology, Huangshi 435003, People's Republic of China, e-mail: rm@126.com

Dr. Konstantin Sobolev Department of Civil Engineering, University of Wisconsin–Milwaukee, EMS 939 3200 North Cramer Street, P.O. Box 784, Milwaukee, WI 53211, USA, e-mail: sobolev@uwm.edu

Elon J. Terrell Mechanical Engineering Department, Columbia University, S.W. Mudd, Room 220C 500 West 120th St., New York, NY 10027, USA, e-mail: eterrell@columbia.edu

Dr. Gene Whyman Laboratory of Polymers, Ariel University Center of Samaria, P.O.B. 3, Ariel 40700, Israel, e-mail: gevaiman@ariel.ac.il

X. H. Zuo Hubei Key Laboratory of Environmental Pollution Control, Resources Reuse, School of Chemical and Materials Engineering, Huangshi Institute of Technology, Huangshi 435003, People's Republic of China, e-mail: xhzuo@126.com

Part I
Introduction

Chapter 1
Green Tribology, its History, Challenges, and Perspectives

Michael Nosonovsky and Bharat Bhushan

Abstract In this chapter the concept of green tribology and its relation to other areas of tribology is discussed as well as other "green" disciplines, namely, green engineering and green chemistry. The twelve principles of green tribology are formulated: the minimization of (1) friction and (2) wear, (3) the reduction or complete elimination of lubrication, including self-lubrication, (4) natural and (5) biodegradable lubrication, (6) using sustainable chemistry and engineering principles, (7) biomimetic approaches, (8) surface texturing, (9) environmental implications of coatings, (10) real-time monitoring, (11) design for degradation, and (12) sustainable energy applications. Three areas of green tribology are further defined: (1) biomimetics for tribological applications, (2) environment-friendly lubrication, and (3) the tribology of renewable energy application. The integration of these areas remains a primary challenge for this novel area of research. The challenges of green tribology and future directions of research are also discussed.

1.1 Introduction

Tribology (from the Greek word τρίβω "tribo" meaning "to rub") is defined by the Oxford dictionary as "the branch of science and technology concerned with interacting surfaces in relative motion and with associated matters (as friction,

M. Nosonovsky
College of Engineering and Applied Science, University of Wisconsin,
Milwaukee, WI 53201, USA

B. Bhushan (✉)
Nanoprobe Laboratory for Bio- and Nanotechnology and Biomimetics (NLB2),
The Ohio State University, 201 W. 19th Avenue,
Columbus, OH 43210-1142, USA
e-mail: bhushan.2@osu.edu

M. Nosonovsky and B. Bhushan (eds.), *Green Tribology*,
Green Energy and Technology, DOI: 10.1007/978-3-642-23681-5_1,

wear, lubrication, and the design of bearings—Oxford English Dictionary)". The word "tribophysics" was used already in the 1940s by David Tabor and Philip Bowden for the name of their laboratory in Melbourne University, Australia [36]. The term "tribology" was officially introduced in 1966 by Prof. H. Peter Jost, then the chairman of a working group of lubrication engineers, in his published report for the U.K. Department of Education and Science. It was reported that huge sums of money had been lost in the UK annually due to the consequences of friction, wear, and corrosion. Therefore, it was strongly recommended to unite multidisciplinary scientific and engineering efforts in these areas, so that they could benefit from one another. As a result, several centers for tribology were created in many countries. Since then the term has diffused into the international engineering field; various tribological organizations and societies have been established, such as the Society of Tribologists and Lubrication Engineers (STLE), and many specialists now claim to be tribologists.

Typical tribological studies cover friction, wear, lubrication, and adhesion and involve the efforts of mechanical engineers, material scientists, chemists, and physicists [9, 11, 12]. Since the emergence of the word tribology almost 50 years ago, many new areas of tribological studies have developed which are at the interface of various scientific disciplines, and various aspects of interacting surfaces in relative motion have been the focus of tribology. These areas include, for example, nanotribology, biotribology, the tribology of magnetic storage devices and micro/nanoelectromechanical systems (MEMS/NEMS), and adhesive contact [8–12, 15–17]. The research in these areas is driven mostly by the advent of new technologies and new experimental techniques for surface characterization.

Few researchers have reported the need for ecological or "green" tribology. Bartz [5] stated that, "Savings of resources of energy and reducing the impact on the environment are the most important aspects of ecotribology. In the course of relevant practices savings of basic resources and materials, optimum design, optimum operation, reduced energy consumption and the protection of the environment have to be covered... Using environmentally acceptable lubricants is the key factor for this." He suggested a diagram which he called "ECO-Balance-Sheet-Tree" which summarized all of these aspects. Sasaki [67] has emphasized the need for ecological tribology, which he considered a response to the world economic and financial crisis of 2008 as well as the global worming crisis as reflected by the "Kyoto Protocol." The new concept of "green tribology" has been defined as "the science and technology of the tribological aspects of ecological balance and of environmental and biological impacts." Jost [38] elaborated on the need for green tribology and has mentioned that the influence of economic, market, and financial triumphalisms have retarded tribology and could retard 'green tribology' from being accepted as a not-unimportant factor in its field... Therefore, by highlighting the economic benefits of tribology, tribology societies, groups and committees are likely to have a far greater impact on the makers of policies and the providers of funding than by only preaching the scientific logic... Tribology societies should highlight to the utmost the economic advantage of tribology. It is the language financial oriented policy makers and markets, as well as governments, understand.

These earlier mentions by researchers called scientists to pay attention to green tribology; however, they did not define the field in a rigorous scientific or academic manner. The first scientific volume on green tribology emerged in 2010, when *Philosophical Transaction of the Royal Society A*, the oldest (published since 1666) and one of the most prestigious research journals in the world, decided to devote a theme issue to "green tribology," edited by M. Nosonovsky and B. Bhushan. In that volume, the editors defined green tribology in quite a broad way, so that it encompassed biomimetic tribology (which follows the ways of living nature to solve engineering problems, eco-friendly lubrication, and clean and sustainable energy applications [58]. The presumption was that the combination of these areas under the umbrella of green tribology could enhance them all and help to benefit from one another by establishing new links.

Ten chapters were published in the theme issue. Nosonovsky and Bhushan [58] suggested twelve principles and three areas of green tribology. Several chapters were devoted to biomimetic surfaces. Bormashenko [22] reviewed wetting transitions on biomimetic superhydrophobic surfaces, while Bhushan and Nosonovsky [20] discussed various wetting regimes with emphasis on the lotus effect and rose petal effect regimes. Shark-skin effect surfaces were discussed by Dean and Bhushan [27], whereas antifouling biomimetic surfaces were investigated by Salta et al. [66]. Nosonovsky [49] discussed the principles of friction-induced self-organization. In the application area, Kotzalas and Doll [42] reviewed the tribological aspects of wind power turbines, whereas Wood et al. [75] discussed tribological constraints of marine renewable energy systems. Lovell et al. [46] discussed the effect of boric acid additives on the performance of green lubricants, and Li et al. [44] studied green waxes, adhesives, and lubricants.

Several workshops, conference sections, and symposia took place after that, which confirmed the volume's inclusive approach, as well as the interest in green tribology in general. Green tribology topics have been covered at a number of conferences [50, 51, 59].

The specific field of green or environment-friendly tribology emphasizes the aspects of interacting surfaces in relative motion, which are of importance for energy or environmental sustainability or which have impact upon today's environment. This includes tribological technology that mimics living nature (biomimetic surfaces) and thus is expected to be environment friendly, the control of friction and wear that is of importance for energy conservation and conversion, environmental aspects of lubrication and surface modification techniques, and tribological aspects of green applications, such as wind-power turbines, tidal turbines, or solar panels (Fig. 1.1). It is clear that a number of tribological problems could be put under the umbrella of "green tribology" and are of mutual benefit to one another.

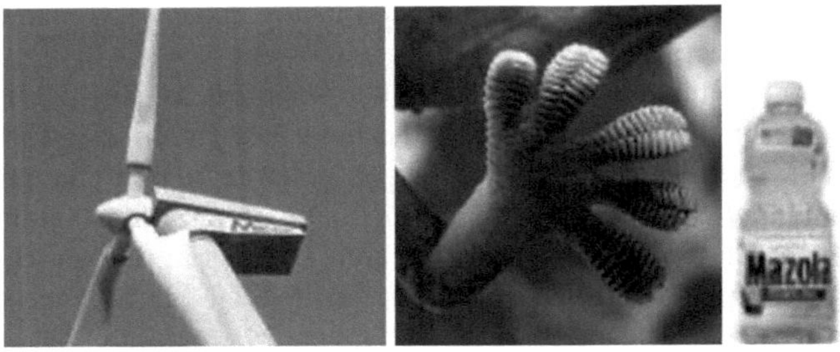

Fig. 1.1 The paradigm of green tribology: renewable energy (represented by a wind turbine), biomimetic surfaces (represented by the gecko foot), and biodegradable lubrication (represented by natural vegetable oil)

1.2 Green Tribology and Green Chemistry and Engineering

Green tribology can be viewed in the broader context of two other 'green' areas: green engineering and green chemistry. The US Environmental Protection Agency (EPA) defined green engineering as "the design, commercialization and use of processes and products that are technically and economically feasible while minimizing (1) generation of pollution at the source (2) risk to human health and the environment" [3]. Besides that, the three tiers of green engineering assessment in design involve: (1) process research and development, (2) conceptual/preliminary design, and (3) detailed design pollution prevention; process heat/energy integration; process mass integration [1].

Another related area is green chemistry, also known as sustainable chemistry, which is defined as "the design of chemical products and processes that reduce or eliminate the use or generation of hazardous substances" [3]. The focus of green chemistry is on minimizing the hazard and maximizing the efficiency of any chemical choice. It is distinct from environmental chemistry which focuses on chemical phenomena in the environment. While environmental chemistry studies the natural environment as well as pollutant chemicals in nature, green chemistry seeks to reduce and prevent pollution at its source. Green chemistry technologies provide a number of benefits, including reduced waste, eliminating costly end-of-the-pipe treatments, safer products, reduced use of energy and resources, and improved competitiveness of chemical manufacturers and their customers. Green chemistry consists of chemicals and chemical processes designed to reduce or eliminate negative environmental impacts. The use and production of these chemicals may involve reduced waste products, non-toxic components, and improved efficiency.

Anastas and Warner [2] formulated The Twelve Principles of Green Chemistry which provided a road map for chemists to implement green chemistry:

1. Prevention of waste is better than cleaning up.
2. Maximum incorporation into the final product of all materials used in the process.
3. Chemical synthesis should incorporate less hazardous or toxic materials, when possible.

4. Chemical products should be designed to reduce toxicity.
5. Auxiliary substances, such as solvents, should be safe whenever used.
6. Energy efficiency requirements should be recognized. Synthetic methods should be conducted at ambient temperature and pressure, whenever possible.
7. A raw material or feedstock should be renewable, whenever possible.
8. Reduce unnecessary derivatives.
9. Catalytic reagents are superior to stoichiometric reagents.
10. Chemical products should be degradable at the end of their function.
11. Real-time analysis, monitoring and control should be implemented to prevent the formation of hazardous substances.
12. Substances and their use in the chemical process should be chosen to minimize the risk of accidents and prevent fires, explosions, spills, etc.

A number of green chemistry metrics have been suggested to quantify the environmental efficiency of a chemical process. These metrics include the environmental factor ("E-factor"), which is equal to the total mass of waste divided by the mass of the product [69], the atom economy [71], the effective mass yield [37], the carbon efficiency, and reaction mass efficiency [26], etc.

Attempts are made not only to quantify the "greenness" of a chemical process but also to factors in other parameters such as chemical yield, the price of reaction components, safety in handling chemicals, hardware demands, energy profile, and ease of product workup and purification. Green chemistry is increasingly seen as a powerful tool that researchers must use to evaluate the environmental impact of nanotechnology. As nanomaterials are developed, the environmental and human health impacts of both the products themselves and the processes to make them must be considered to ensure their long-term economic viability. While many scientists use the term "green chemistry," there are also critics who argue that green chemistry is no more than a public relations label, since some chemists use the term green chemistry without relating it to the green chemistry principles proposed by Anastas and Warner [2], as it was pointed out by Linthorst [45]. Green tribology may have to deal with the same problem.

Since tribology is an interdisciplinary area which involves, among other fields, chemical engineering and materials science, the principles of green chemistry are applicable to green tribology as well. However, since tribology involves not only the chemistry of surfaces but also other aspects related to the mechanics and physics of surfaces, there is a need to modify these principles. The principles of green tribology will be formulated in the following section.

1.3 Twelve Principles of Green Tribology

Nosonovsky and Bhushan [58] suggested that twelve principles of green tribology can be formulated, similar to the principles of green chemistry. Some principles are related to the design and manufacturing of tribological applications (3–10), while others belong to their operation (1–2, 11–12).

1. *Minimization of heat and energy dissipation*. Friction is the primary source of energy dissipation. According to some estimates, about one-third of the energy consumption in the US is spent to overcome friction. Most energy dissipated by friction is converted into heat and leads to heat pollution of the atmosphere and the environment. The control of friction and friction minimization, which leads to both energy conservation and the prevention of damage to the environment due to the heat pollution, is a primary task of tribology. It is recognized that for certain tribological applications (e.g., car brakes and clutches) high friction is required; however, ways of effective use of energy for these applications should be sought as well.
2. *Minimization of wear* is the second most important task of tribology which has relevance to green tribology. In most industrial applications wear is undesirable. It limits the lifetime of components and therefore creates the problem of their recycling. Wear can lead also to catastrophic failure. In addition, wear creates debris and particles which contaminate the environment and can be hazardous for humans in certain situations. For example, wear debris generated after human joint replacement surgery is the primary source of long-term complications in patients.
3. *Reduction or complete elimination of lubrication and self-lubrication.* Lubrication is a focus of tribology since it leads to the reduction of friction and wear. However, lubrication can also lead to environmental hazards. It is desirable to reduce lubrication or achieve the self-lubricating regime, when no external supply of lubrication is required. Tribological systems in living nature often operate in the self-lubricating regime. For example, joints form essentially a closed self-sustainable system.
4. *Natural lubrication* (e.g., vegetable oil-based) should be used in cases when possible, since it is usually environmentally friendly.
5. *Biodegradable lubrication* should also be used when possible to avoid environmental contamination. In particular, water lubrication is an area which attracted researchers in recent years. Natural oil (such as canola) lubrication is another option, especially discussed in the developing countries.
6. *Sustainable chemistry and green engineering principles* should be used for the manufacturing of new components for tribological applications, coatings, and lubricants.
7. *Biomimetic approach* should be used whenever possible. This includes biomimetic surfaces, materials, and other biomimetic and bio-inspired approaches, since they tend to be more ecologically friendly.
8. *Surface texturing* should be applied to control surface properties. Conventional engineered surfaces have random roughness, and the randomness is the factor which makes it extremely difficult to overcome friction and wear. On the other hand, many biological functional surfaces have complex structures with hierarchical roughness, which defines their properties. Surface texturing provides a way to control many surface properties relevant to making tribosystems more ecologically friendly.

9. *Environmental implications of coatings* and other methods of surface modification (texturing, depositions, etc.) should be investigated and taken into consideration.
10. *Design for degradation* of surfaces, coatings, and tribological components. Similar to green chemistry applications, the ultimate degradation/utilization should be taken into consideration during design.
11. *Real-time monitoring*, analysis, and control of tribological systems during their operation should be implemented to prevent the formation of hazardous substances.
12. *Sustainable energy applications* should become the priority of the tribological design as well as engineering design in general.

1.4 Areas of Green Tribology

The following three focus areas of tribology have the greatest impact on environmental issues, and therefore, they are of importance for green tribology: (1) biomimetic and self-lubricating materials/surfaces; (2) biodegradable and environment-friendly lubricants, coatings, and materials; and (3) tribology of renewable and/or sustainable sources of energy. Below, the current state of these areas and their relevance for the novel field of green tribology are briefly discussed.

1.4.1 Biomimetic Surfaces

Biomimetics (also referred to as bionics or biomimicry) is the application of biological methods and systems found in nature to the study and design of engineering systems and modern technology. It is estimated that the 100 largest biomimetic products generated approximately US $1.5 billion over the years 2005–2008 and the annual sales are expected to continue to increase dramatically [14]. Many biological materials have remarkable properties which can hardly be achieved by conventional engineering methods. For example, a spider can produce huge amounts (compared with the linear size of his body) of silk fiber which is stronger than steel without any access to the high temperatures and pressures which would be required to produce such materials as steel using conventional human technology. These properties of biomimetic materials are achieved due to their composite structure and hierarchical multiscale organization [30]. The hierarchical organization provides biological systems with the flexibility needed to adapt to the changing environment. As opposed to the traditional engineering approach, biological materials are grown without the final design specifications, but by using the recipes and recursive algorithms contained in their genetic code. The difference of natural versus engineering design is the difference of growth versus fabrication

[29, 54, 55]. Hierarchical organization and the ability of biological systems to grow and adapt also provides a natural mechanism for the repair or healing of minor damage in the material.

The remarkable properties of the biological materials serve as a source of inspiration for materials scientists and engineers indicating that such performance can be achieved if the paradigm of materials design is changed. While in most cases it is not possible to directly borrow solutions from living nature and to apply them in engineering, it is often possible to take biological systems as a starting point and a source of inspiration for engineering design. Molecular scale devices, superhydrophobicity, self-cleaning, drag reduction in fluid flow, energy conversion and conservation, high adhesion, reversible adhesion, aerodynamic lift, materials and fibres with high mechanical strength, biological self-assembly, antireflection, structural coloration, thermal insulation, self-healing, and sensory-aid mechanisms are some of the examples found in nature that are of commercial interest.

Biomimetic materials are also usually environmentally friendly in a natural way, since they are a natural part of the ecosystem. For this reason, the biomimetic approach in tribology is particularly promising. In the area of biomimetic surfaces, a number of ideas have been suggested [4, 14, 28, 34, 54, 68].

1. The lotus effect based non-adhesive surfaces. The term "lotus effect" stands for surface roughness-induced superhydrophobicity and self-cleaning. Superhydrophobicity is defined as the ability to have a large (>150°) water contact angle and, at the same time, low contact angle hysteresis. The lotus flower is famous for its ability to emerge clean from dirty water and to repel water from its leaves. This is due to a special structure of the leaf surface (multiscale roughness) combined with hydrophobic coatings, Fig. 1.2 [52, 53]. These surfaces have been fabricated in the lab with comparable performance [19, 21].

Adhesion is a general term for several types of attractive forces that act between solid surfaces, including the van der Waals force, electrostatic force, chemical bonding, and the capillary force due to the condensation of water at the surface. Adhesion is a relatively short-range force, and its effect (which is often undesirable) is significant for microsystems which have contacting surfaces. The adhesion force strongly affects friction, mechanical contact, and tribological performance of such a system's surface, leading, for example, to "stiction" (combination of adhesion and static friction [8, 16]), which precludes microelectromechanical switches and actuators from proper functioning. It is therefore desirable to produce non-adhesive surfaces, and applying surface microstructure mimicking the lotus effect has been successfully used for the design of non-adhesive surfaces, which are important for many tribological applications. In some applications, high adhesion surfaces are of interest. High adhesion surfaces have been produced using the so-called "Petal effect," Fig. 1.3 [18, 20].

Fig. 1.2 **a** SEM micrographs (shown at three magnifications) of lotus (*Nelumbo nucifera*) leaf surface, which consists of microstructure formed by papillose epidermal cells covered with epicuticular wax tubules on the surface, which create nanostructure; **b** image of water droplet sitting on the lotus leaf [21]

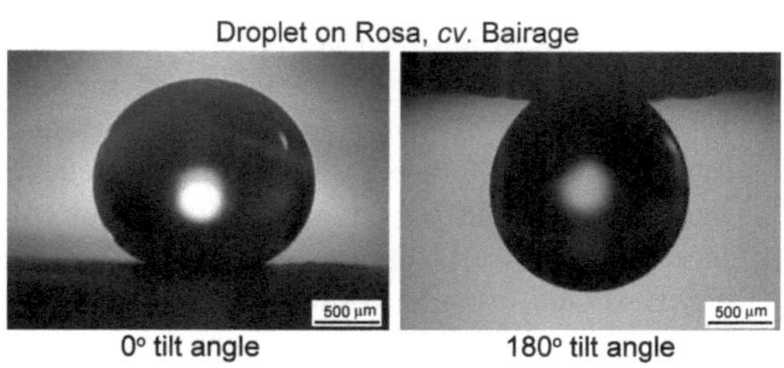

Fig. 1.3 Optical micrographs of water droplets on Rosa, cv. Bairage at 0 and 180° tilt angles. Droplet is still suspended when the petal is turned upside down [18]

2. The Gecko effect, which stands for the ability of specially structured hierarchical surfaces to exhibit controlled adhesion. Geckos are known for their ability to climb vertical walls due to a strong adhesion between their toes and a number of various surfaces. They can also detach easily from a surface when needed (Fig. 1.4). This is due to a complex hierarchical structure of gecko feet surface. The Gecko effect is used for applications when strong adhesion is needed (e.g., adhesive tapes) or for reversible adhesion (e.g., climbing robot) [13, 54].

Fig. 1.4 Tokay gecko has the ability to climb walls and detach from surfaces easily at will

3. Microstructured surfaces for underwater applications, including easy flow due to boundary slip, the suppression of turbulence (the shark-skin effect, Fig. 1.5), and anti-biofouling (the fish-scale effect). Biofouling and biofilming are the undesirable accumulation of microorganisms, plants, and algae on structures which are immersed in water. Conventional antifouling coatings for ship hulls are often toxic and environmentally hazardous. On the other hand, in living nature there are ecological coatings (e.g., fish scale), so a biomimetic approach is sought [25, 27, 32, 40, 56].
4. Oleophobic surfaces capable of repelling organic liquids. The principle can be similar to superhydrophobicity, but it is much more difficult to produce an oleophobic surface, because surface energies of organic liquids are low, and they tend to wet most surfaces [48, 56, 72, 73]. Underwater oleophobicity can be used also to design self-cleaning and antifouling surfaces, Fig. 1.6 [39].
5. Microstructured surfaces for various optical applications, including non-reflective (the Moth-eye effect), highly reflective, colored (in some cases, including the ability to dynamically control coloration), and transparent surfaces. Optical surfaces are sensitive to contamination, so the self-cleaning ability should often be combined with optical properties [14, 33, 54].
6. Microtextured surfaces for de-icing and anti-icing (Fig. 1.7). De-icing (the removal of frozen contaminant from a surface) and anti-icing (protecting against the formation of frozen contaminant) are significant problems for many applications that have to operate below the water freezing temperature: aircrafts, machinery, road and runway pavements, traffic signs and traffic lights, etc. The traditional approaches to de-icing include mechanical methods, heating, the deposition of dry or liquid chemicals that lower the freezing point of water. Anti-icing is accomplished by applying a protective layer of a viscous anti-ice fluid. All anti-ice fluids offer only limited protection, dependent upon frozen contaminant type and precipitation rate, and it fails when it can no longer absorb the contaminant. In addition to limited efficiency, these de-icing fluids, such as propylene glycol or ethylene glycol, can be toxic and raise

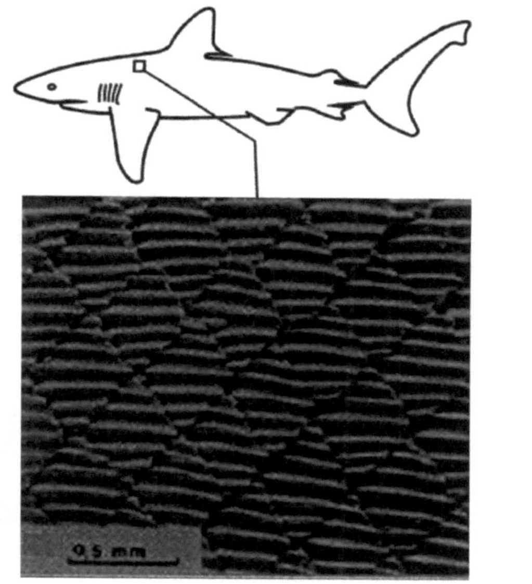

Fig. 1.5 Scale structure on a Galapagos shark (*Carcharhinus galapagensis*; [64])

environmental concerns. Anti-icing on roadways is used to prevent ice and snow from adhering to the pavement, allowing easier removal by mechanical methods.

Ice formation occurs due to the condensation of vapor phase water and further freezing of liquid water. For example, droplets of supercooled water that exist in stratiform and cumulus clouds crystallize into ice when they are struck by the wings of passing airplanes. Ice formation on other surfaces, such as pavements or traffic signs also occurs via the liquid phase. It is therefore suggested that a water repellent surface can also have de-icing properties [24]. When a super-hydrophobic surface is wetted by water, an air layer or air pockets are usually kept between the solid and the water droplets. After freezing, ice will not adhere to solid due to the presence of air pockets and will be easily washed or blown away.

7. Microelectromechanical system (MEMS)-based dynamically tunable surfaces for the control of liquid/matter flow and/or coloration (for example, mimicking the coloration control in cephalopods), used for displays and other applications, the so-called "origami" [23, 70].
8. Various biomimetic microtextured surfaces to control friction, wear and lubrication [14, 15, 74].
9. Self-lubricating surfaces, using various principles, including the ability for friction-induced self-organization [57].

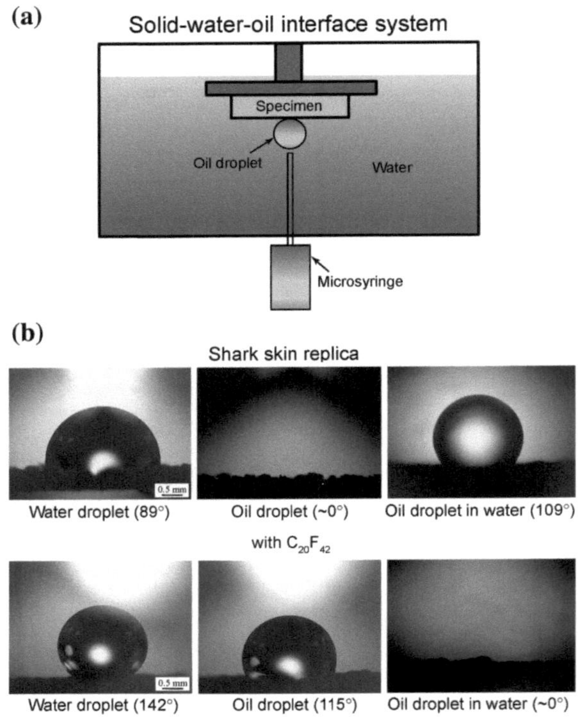

Fig. 1.6 **a** Schematics of a solid-water-oil interface system. A specimen is first immersed in water, and then an oil droplet is gently deposited using a microsyringe, and the static contact angle is measured; **b** opticalmicrographs of droplets at three-different-phase interfaces on a micropatterned surface (shark skin replica) without and with $C_{20}F_{42}$ [39]

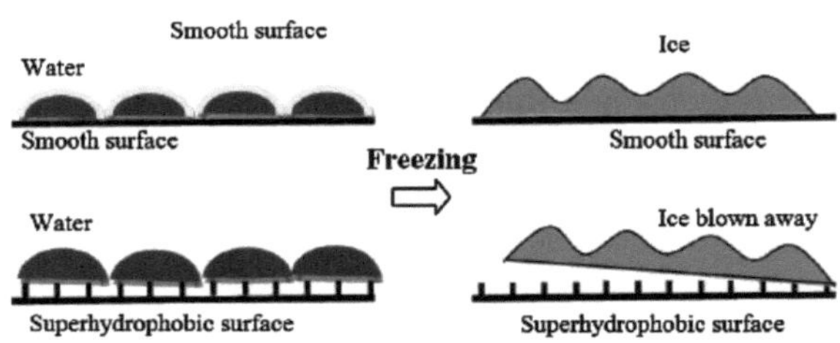

Fig. 1.7 The principle of applying of surface microstructure for de-icing

10. Self-repairing surfaces and materials, which are able to heal minor damage (cracks, voids), [57, 58].
11. Various surfaces with alternate (and dynamically controlled) wetting properties for micro/nanofluidic applications, including the Darkling beetle effect, e.g., the ability of a desert beetle to collect water on its back using the hydrophilic spots on the otherwise hydrophobic surface of its back [54, 61, 63], Fig. 1.8.
12. Water strider effect mimicking the ability of insects to walk on water using the capillary forces. The hierarchical organization of the water strider leg surface plays a role in its ability to remain dry on water surface, Fig. 1.9 [31].

Fig. 1.8 The water-capturing surface of the fused overwings (elytra) of the desert beetle *Stenocara* sp. **a** Adult female, dorsal view; peaks and valleys are evident on the surface of the elytra; **b** SEM image of the textured surface of the depressed areas [61]

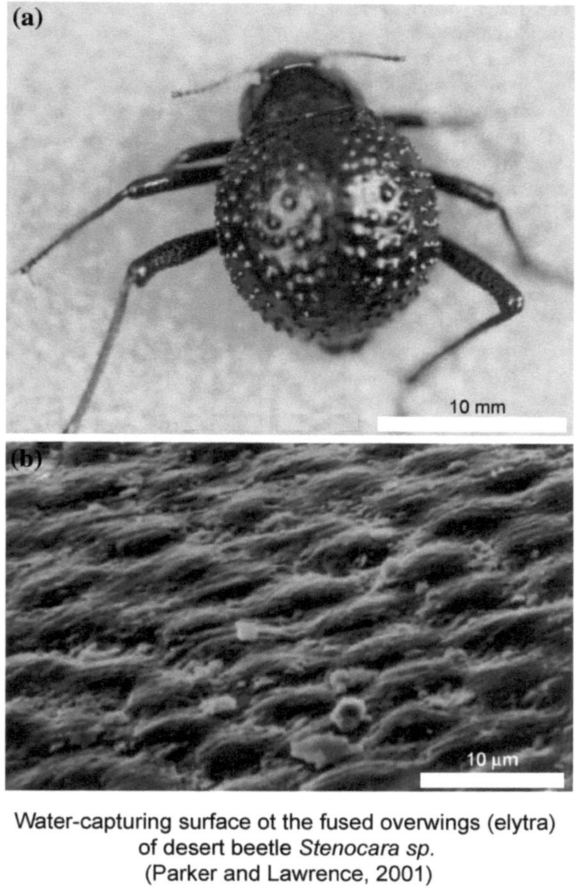

13. The "sand fish" lizard effect, able to dive and "swim" in loose sand due to special electromechanical properties of its scale [54, 63].
14. Composite and nanocomposite materials tailored in such way that they can produce required surface properties, such as self-cleaning, self-lubrication, and self-healing. Metal-matrix composites, and polymeric composites as well as ceramics (including concrete) have been recently used for this purpose. Natural fiber-reinforced composites are among these materials. The difference between microstructured surfaces and composite materials is that the latter have hydrophobic reinforcement in the bulk and thus can be much more wear-resistant than microstructured surfaces, which are vulnerable even to moderate wear rates.
15. Green biomimetic nanotribology, including cell adhesion, nanoornamentics, and biochemistry is another new area associated with green tribology.

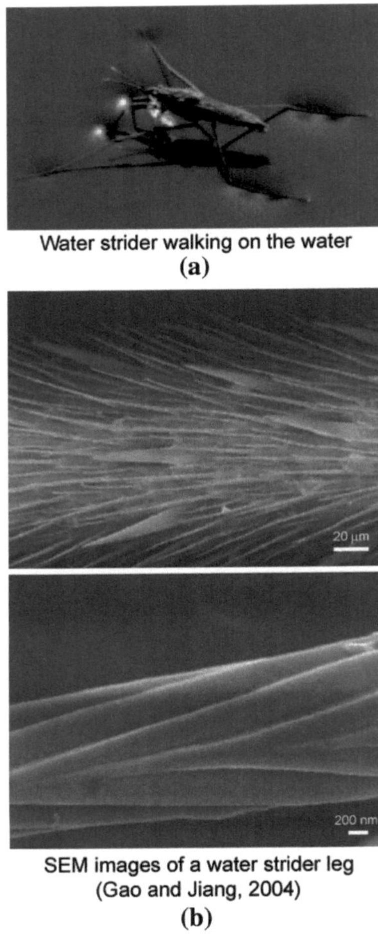

Fig. 1.9 **a** Water strider (Pond skater, *G. remigis*) walking on water; **b** SEM images of a pond skater leg showing (*top*) numerous oriented microscale setae and (*bottom*) nanoscale grooved structures on a seta [31]

Environmental engineers have only just started paying attention to biomimetic surfaces. Raibeck et al. [62] investigated the potential environmental benefits and burdens associated with using the lotus effect based self-cleaning surfaces. They found that while the use phase benefits are apparent, production burdens can outweigh them when compared with other cleaning methods, so a more thoughtful and deliberate use of bio-inspiration in sustainable engineering is needed. Clearly, more studies are likely to emerge in the near future.

1.4.2 Biodegradable Lubrication, Coatings, and Materials

In the area of environment-friendly and biodegradable lubrication several ideas have been suggested:

1. The use of natural (e.g., vegetable-oil based or animal-fat based) biodegradable lubricants. This involves oils that are used for engines, hydraulic applications, and metal cutting applications. In particular, corn, soybean, coconut oils have been used so far (the latter is of particular interest in tropical countries such as India). These lubricants are potentially biodegradable, although in some cases chemical modification or additives for best performance are required. Vegetable oils can have excellent lubricity, far superior than that of mineral oil. In addition, they have a very high viscosity index and high flash/fire points. However, natural oils often lack sufficient oxidative stability, which means that the oil will oxidize rather quickly during use, becoming thick and polymerizing to a plastic-like consistency. Chemical modification of vegetable oils and/or the use of antioxidants can address this problem [47].
2. Ionic liquids for green molecular lubrication. Common industrial lubricants include natural and synthetic hydrocarbons and perfluoropolyethers (PFPEs), where the latter is widely used in commercial applications requiring extreme operating conditions due to their high temperature stability and extremely low vapor pressure. However, PFPEs exhibit low electrical conductivity, making them undesirable in some nanotechnology applications. Ionic liquids (ILs) have been explored as lubricants for various device applications due to their excellent electrical conductivity as well as good thermal conductivity, where the latter allows frictional heating dissipation [60]. Since they do not emit volatile organic compounds, they are regarded as "green" lubricants.
3. Powder lubricants and,in particular, boric acid lubricants. In general, these tend to be much more ecologically friendly than the traditional liquid lubricants [41]. Boric acid and MoS_2 powder can also be used as an additive to the natural oil. Friction and wear experiments show that the nanoscale (20 nm) particle boric acid additive lubricants significantly outperformed all of the other lubricants with respect to frictional and wear performance. In fact, the nanoscale boric acid powder based lubricants exhibited a wear rate more than an order of magnitude lower than the MoS_2 and larger sized boric acid additive-based lubricants [46].
4. Self-replenishing lubrication that uses oil-free environmentally benign powders for lubrication of critical components such as bearings used in fuel cell compressors and expanders [76].
5. Water lubrication of bearings and other tribological components. Recently, a lot of attention has been paid to water lubrication which is considered an eco-friendly method.
6. New eco-friendly coating materials for tribological applications.
7. Environmental effect of wear particles. It has been suggested that environmental aspects should become an integral part of brake design [77]. Preliminary data obtained with animal experiments revealed that the inhaled metallic particles remain deposited in the lungs of rats 6 month after the exposure. The presence of inhaled particles had a negative impact on health and led to emphysema (destroyed alveoli), inflammatory response, and morphological changes of the lung tissue.

1.4.3 Renewable Energy

The tribology of renewable sources of energy is a relatively new field of tribology. Today, there are meetings and sessions devoted to the tribology of wind turbines at almost every tribology conference, and they cover certain issues specific for these applications. Unlike in the case of the biomimetic approach and environment-friendly lubrication, it is not the manufacturing or operation, but the very application of the tribological system which involves "green" issues, namely, environmentally friendly energy production. The following issues can be mentioned.

1. Wind power turbines have a number of specific problems related to their tribology, and constitute a well-established area of tribological research. These issues include water contamination, electric arcing on generator bearings, issues related to the wear of the mainshaft and gearbox bearings and gears, the erosion of blades (solid particles, cavitation, rain, hail stones), etc. [43].
2. Tidal power turbines are another important way of producing renewable energy, which involves certain tribiological problems. Tidal power turbines are especially popular in Europe (particularly, in the U.K.), which remains the leader in this area, although several potential sites in North America have been suggested. There are several specific tribological issues related to tidal power turbines, such as their lubrication (seawater, oils, and greases), erosion, corrosion, and biofouling, as well as the interaction between these modes of damage [6].
3. Besides tidal, the ocean water flow and wave energy and river flow energy (without dams) can be used with the application of special turbines, such as the Gorlov helical turbine [35], which provides the same direction of rotation independent of the direction of the current flow. These applications also involve specific tribological issues.
4. Geothermal energy plants are used in the US (in particular, at the Pacific coast and Alaska); however, their use is limited to the geographical areas at the edges of tectonic plates [65]. In 2007, they produced 2.7 GW of energy in the US, with Philippines (2.0 GW) and Indonesia (1.0 GW) in the second and third places [7]. There are several issues related to the tribology of geothermal energy sources which are discussed in the literature.

1.5 Challenges

In the preceding sections the need for green tribology, its principles, and primary areas of research have been outlined. As a new field, green tribology has a number of challenges. One apparent challenge is the development of the above-mentioned fields in such a manner that they could benefit from each other. Only where such synthesis is performed is it possible to see green tribology as a coherent and self-sustained field of science and technology, rather than a collection of several topics of research in tribology and surface engineering. There is potential synergy in the use of biomimetic approach, microstructuring, biodegradable lubrication, self-lubrication, and other novel approaches as well as in developing methods of their applications to sustainable engineering and energy production. Clearly, more research should be performed for the integration of these fields. Some ideas could be borrowed from the related field of green chemistry, for example, developing quantitative metrics to assess the environmental impact of tribological technologies.

The creator of the periodic table of elements, chemist Mendeleev used to say that science starts when the measurement begins. This saying should apply to green tribology as well. It is important to develop quantitative measures and metrics which would allow comparison of which tribological material, technology, or application is "greener," i.e., produces smaller carbon footprint, less chemical, or thermal pollution of the environment.

Green tribology should be integrated into world science and make its impact on the solutions for worldwide problems, such as the change of climate and the shortage of food and drinking water. Jost [38] mentioned the economical potential of the new discipline: "the application of tribological principles alone will, of course, not solve these world-wide problems. Only major scientific achievements are likely to be the key to their solution, of which I rate Energy as one of the most important ones. For such tasks to be achieved, the application of Tribology, and especially of green tribology can provide a breathing space which would enable scientists and technologists to find solutions to these, mankind's crucial problems and allow time for them to be implemented by governments, organizations and indeed everyone operating in this important field. Consequently, this important—albeit limited—breathing space may be extremely valuable to all working for the survival of life as we know it. However, the ultimate key is science and its application. Tribology—especially green tribology can and—I am confident—will play its part to assist and give time for science to achieve the required solutions and for policy makers to implement them."

1.6 Conclusions

Green tribology is a novel area of science and technology. It is related to other areas of tribology as well as other "green" disciplines, namely, green engineering and green chemistry. The twelve principles of green tribology are formulated, and three areas of tribological studies most relevant to green tribology are defined. The integration of these areas remains the primary challenge of green tribology and defines the future directions of research.

References

1. D.T. Allen, D.R. Shonnard, *Green Engineering: Environmentally Conscious Design Of Chemical Processes* (Prentice Hall, Upper Saddle River, 2001)
2. P.T. Anastas, J.C. Warner, *Green Chemistry: Theory and Practice* (Oxford University Press, Oxford, 1998)
3. Anonymous, *Green Engineering* (2010) http://www.epa.gov/oppt/greenengineering/
4. Y. Bar-Cohen, *Biomimetics: Nature-Based Innovation* (CRC Press, Boca Raton, 2011)
5. W.J. Bartz, Ecotribology: Environmentally Acceptable Tribological Practices. Tribol. Int. **39**, 728–733 (2006)
6. W.M.J. Batten, A.S. Bahaj, A.F. Molland, J.R. Chaplin, The Prediction of the Hydrodynamic Performance of Marine Current Turbines. Renew. Energ. **33**, 1085–1096 (2008)
7. E.R. Bertani, World Geothermal Generation in 2007. Geo Heat Cent. Bull. **28**, 8–19 (2007)
8. B. Bhushan, *Tribology and Mechanics of Magnetic Storage Devices*, 2nd edn. (Springer, New York, 1996)
9. B. Bhushan, *Principles and Applications of Tribology* (Wiley, New York, 1999)
10. B. Bhushan, *Mechanics and Reliability of Flexible Magnetic Media*, 2nd edn. (Springer, New York, 2000)
11. B. Bhushan, *Modern Tribology Handbook, Vol. 1–Principles of Tribology; Vol. 2–Materials, Coatings, and Industrial Applications* (CRC Press, Boca Raton, 2001)
12. B. Bhushan, *Introduction to Tribology* (Wiley, New York, 2002)
13. B. Bhushan, Adhesion of Multi-level Hierarchical Attachment Systems in Gecko feet. J. Adhesion Sci. Technol. **21**, 1213–1258 (2007, invited)
14. B. Bhushan, Biomimetics: Lessons from Nature—an overview. Phil. Trans. R. Soc. A **367**, 1445–1486 (2009)
15. B. Bhushan, *Springer Handbook of Nanotechnology*, 3rd edn. (Springer, Heidelberg, 2010)
16. B. Bhushan, *Nanotribology and Nanomechanics I-Measurement Techniques and Nanomechanics, II-Nanotribology, Biomimetics, and Industrial Applications*, 3rd edn. (Springer, Heidelberg, 2011)
17. B. Bhushan, B.K. Gupta, *Handbook of Tribology: Materials, Coatings, and Surface Treatments* (McGraw-Hill, New York, 1991)
18. B. Bhushan, E.K. Her, Fabrication of Superhydrophobic Surfaces with high and low Adhesion Inspired from rose petal. Langmuir **26**, 8207–8217 (2010)
19. B. Bhushan, Y.C. Jung, Natural and biomimetic artificial surfaces for superhydrophobicity, self-cleaning, low adhesion, and drag reduction. Prog. Mater. Sci. **56**, 1–108 (2011)
20. B. Bhushan, M. Nosonovsky, The rose petal effect and the modes of superhydrophobicity. Phil. Trans. Royal. Soc. A. **368**, 4713–4728 (2010)
21. B. Bhushan, Y.C. Jung, K. Koch, Micro-, nano- and hierarchical structures for superhydrophobicity, self-cleaning and low adhesion. Phil. Trans. R. Soc. A **367**, 1631–1672 (2009)
22. E. Bormashenko, Wetting transitions on biomimetic surfaces. Phil. Trans. Royal. Soc. A. **368**, 4695–4712 (2010)
23. M.A. Bucaro, P.R. Kolodner, A. Taylor, A. Sidorenko, J. Aizenberg, T.N. Krupenkin, Tunable liquid optics: electrowetting-controlled liquid mirrors based on self-assembled janus tiles. Langmuir **25**, 3876–3879 (2009)
24. L. Cao, A.K. Jones, V.K. Sikka, J. Wu, D. Gao, Anti-icing superhydrophobic coatings. Langmuir **25**, 12444–12448 (2009)
25. L.D. Chambers, K.R. Stokes, F.C. Walsh, R.J.K. Wood, Modern approaches to marine antifouling coatings. Surf. Coat. Tech. **201**, 3642–3652 (2006)
26. D.J.C. Constable, A.D. Curzons, L.M. Freitas dos Santos, G.R. Geen, R.E. Hannah, J.D. Hayler, J. Kitteringham, M.A. McGuire, J.E. Richardson, P. Smith, R.L. Webb, M. Yu, Green chemistry measures for process research and development. Green Chem. **3**, 7–9 (2001)

27. B. Dean, B. Bhushan, Shark-skin surfaces for fluid drag reduction in turbulent flow. Phil. Trans. Royal. Soc. A. **368**, 4775–4806 (2010)
28. E. Favret, N.O. Fuentas, *Functional Properties of Bio-inspired Surfaces: Characterization and Technological Applications* (World Scientific, Hackensack, 2009)
29. P. Fratzl, Biomimetic materials research: what can we really learn from nature's structural materials? J. R. Soc. Interf. **4**, 637–642 (2007)
30. P. Fratzl, R. Weinkamer, Nature's hierarchical materials. Prog. Mater. Sci. **52**, 1263–1334 (2007)
31. X.F. Gao, L. Jiang, Biophysics: water-repellent legs of water striders. Nature **432**, 36 (2004)
32. J. Genzer, K. Efimenko, Recent developments in superhydrophobic surfaces and their relevance to marine fouling: a review. Biofouling **22**, 339–360 (2006)
33. A. Gombert, B. Blasi, The Moth-Eye-Effect—from Fundamentals to Commercial Exploitation, in *Functional Properties of Bio-inspired Surfaces: Characterization and Technological Applications*, ed. by E. Favret, N.O. Fuentas (World Scientific Publishing Co., Hackensack, 2009), pp. 79–102
34. S. Gorb, Functional Surfaces in Biology: Mechanisms and Applications, in *Biomimetics: Biologically Inspired Technologies*, ed. by Y. Bar-Cohen (Taylor and Francis, Boca Raton, 2006), pp. 381–397
35. A.N. Gorban', A.M. Gorlov, V.M. Silantyev, Limits of the turbine efficiency for free fluid flow. ASME J. Energ. Resour. Technol. **123**, 311–317 (2001)
36. N.N. Greenwood, J.A. Spink, An antipodean laboratory of remarkable distinction. Notes Rec. R. Soc. Lond. **57**, 85–105 (2003)
37. T. Hudlicky, D.A. Frey, L. Koroniak, C.D. Claeboe, L.E. Brammer, Toward a reagent-free synthesis. Green Chem. **1**, 57–59 (1999)
38. H.P. Jost, Green Tribology—A Footprint where Economics and Environment Meet, presented at 4th World Tribology Congress, Kyoto, 6–11 September 2009 (unpublished)
39. Y.C. Jung, B. Bhushan, Wetting behavior of water and oil droplets in three-phase interfaces for hydrophobicity/philicity and oleophobicity/philicity. Langmuir **25**, 14165–14173 (2009)
40. Y.C. Jung, B. Bhushan, Biomimetic structures for fluid drag reduction in laminar and turbulent flows. J. Phys. Cond. Matt. **22**, 035104 (2010)
41. M.A. Kabir, C.F. Higgs III, M.R. Lovell, A pin-on disk experimental study on a green particulate-fluid lubricant. J. Tribol. **130**(4), 041801 (6 pp) (2008)
42. M.N. Kotzalas, G.L. Doll, Tribological advancements for reliable wind turbine performance. Phil. Trans. Royal. Soc. A. **368**, 4829–4850 (2010)
43. M. Kotzalas, D. Lucas, Comparison of bearing fatigue life predictions with test data, in *Proceedings AWEA Wind Power*, Los Angeles, 3–6 June 2007 [CD-ROM] (2007)
44. W. Li, X.H. Kong, M. Ruan, F.M. Ma, Y.F. Jiang, M.Z. Liu, Y. Chen, X.H. Zuo, Green waxes, adhesives and lubricants. Phil. Trans. Royal. Soc. A. **368**, 4869–4890 (2010)
45. J.A. Linthorst, An overview: origins and development of green chemistry. Found. Chem. **12**, 55–68 (2010)
46. M.R. Lovell, M.A. Kabir, P.L. Menzes, C.F. Higgs III, Influence of boric acid additive size on green lubricant performance. Phil. Trans. Royal. Soc. A. **368**, 4851–4868 (2010)
47. J.K. Mannekote, and S.V. Kailas, Performance evaluation of vegetable oils as lubricant in a four stroke engine, in *Proceedings Fourth World Tribology Congress*, Kyoto, 6–11 Sept 2009, p. 331
48. M. Nosonovsky, Multiscale roughness and stability of superhydrophobic biomimetic interfaces. Langmuir **23**, 3157–3161 (2007)
49. M. Nosonovsky, Self-organization at the frictional interface for green tribology. Phil. Trans. Royal. Soc. A. **368**, 4755–4774 (2010)
50. M. Nosonovsky, Towards the 'green tribology': biomimetic surfaces, biodegradable lubrication, and renewable energy, in *Proceedings 1st International Brazilian Conference on Tribology TriboBr*-2010-17172, 24–26 Nov 2010, Rio de Janeiro (2010b)
51. M. Nosonovsky, Towards 'green tribology': self-organization at the sliding interface for biomimetic surfaces, in *Proceedings. ASME 10th Biennial Conference on Engineering Systems Design and Analysis*, 12–14 July 2010, Istanbul, Turkey, ESDA2010-25047 (2010c)

52. M. Nosonovsky, B. Bhushan, Biomimetic superhydrophobic surfaces: multiscale approach. Nano. Lett. **7**, 2633–2637 (2007)
53. M. Nosonovsky, B. Bhushan, Multiscale friction mechanisms and hierarchical surfaces in nano- and bio-tribology. Mater. Sci. Eng. R **58**, 162–193 (2007)
54. M. Nosonovsky, B. Bhushan, *Multiscale Dissipative Mechanisms and Hierarchical Surfaces: Friction, Superhydrophobicity, and Biomimetics* (Springer, Heidelberg, 2008)
55. M. Nosonovsky, B. Bhushan, Multiscale effects and capillary interactions in functional biomimetic surfaces for energy conversion and green engineering. Phil. Trans. R. Soc. A **367**, 1511–1539 (2009)
56. M. Nosonovsky, B. Bhushan, Superhydrophobic surfaces and emerging applications: non-adhesion, energy, green engineering. Curr. Opin. Colloid Interface Sci. **14**, 270–280 (2009)
57. M. Nosonovsky, B. Bhushan, Thermodynamics of surface degradation, self-organization, and self-healing for biomimetic surfaces. Phil. Trans. R. Soc. A **367**, 1607–1627 (2009)
58. M. Nosonovsky, B. Bhushan, Green tribology: principles, research areas and challenges. Phil. Trans. Royal. Soc. A. **368**, 4677–4694 (2010)
59. M. Nosonovsky, and B. Bhushan, Towards the 'Green Tribology:' Biomimetic Surfaces, Biodegradable Lubrication, and Renewable Energy, in *Proceedings STLE/ASME* 2010 *International Joint Tribology Conference*, 17–20 Oct 2010, San Francisco, IJTC2010-41157 (2010b)
60. P. Palacio, B. Bhushan, A review of ionic liquids for green molecular lubrication in nanotechnology. Tribol. Lett. **40**, 247–268 (2010)
61. A.R. Parker, C.R. Lawrence, Water capture by a desert beetle. Nature **414**, 33–34 (2001)
62. L. Raibeck, J. Reap, B. Bras, Investigating environmental burdens and benefits of biologically inspired self-cleaning surfaces. CIRP J. Manufact. Sci. Technol. **1**, 230–236 (2009)
63. I. Rechenberg, A.R. El Khyeri, *The sandfish of the Sahara. A model for friction and wear reduction.* (Department of Bionics and Evolution Techniques, Technical University of Berlin, Berlin), See http://www.bionik.tu-berlin.de/institut/safiengl.htm
64. W.E. Reif, Squamation and ecology of sharks. Cour. Forschung. Senck. **78**, 1–255 (1985)
65. L. Rybach, Geothermal sustainability. Geo Heat Cent. Q. Bull. **28**, 2–7 (2007)
66. M. Salta, J.A. Wharton, P. Stoodley, S.P. Dennington, L.R. Goodes, S. Werwinski, U. Mart, R.J.K. Wood, K.R. Stokes, Designing biomimetic antifouling surfaces. Phil. Trans. Royal. Soc. A. **368**, 4729–4754 (2010)
67. S. Sasaki, Environmentally friendly tribology (eco-tribology). J. Mech. Sci. Technol. **24**, 67–71 (2010)
68. M. Scherge, S. Gorb, *Biological Micro- and Nanotribology: Nature's Solutions* (Springer, Heidelberg, 2001)
69. R.A. Sheldon, Organic synthesis: past, present and future. Chem. Ind. **23**, 903–906 (1992)
70. A. Sidorenko, T. Krupenkin, A. Taylor, P. Fratzl, J. Aizenberg, Reversible switching of hydrogel-sctuated nanostructures into complex micropatterns. Science **315**, 487–490 (2007)
71. B.M. Trost, The atom economy—a search for synthetic efficiency. Science **254**, 1471–1477 (1991)
72. A. Tuteja, W. Choi, M. Ma, J.M. Mabry, S.A. Mazzella, G.C. Rutledge, G.H. McKinley, R.E. Cohen, Designing superoleophobic surfaces. Science **318**, 1618–1622 (2007)
73. A. Tuteja, W. Choi, J.M. Mabri, G.H. McKinley, R.E. Cohen, Robust omniphobic surfaces. Proc. Nat. Acad. Sci. U. S. A. **105**, 18200–18205 (2008)
74. M. Varenberg, S. Gorb, Hexagonal surface micropattern for dry and wet friction. Adv. Mater. **21**, 483–486 (2009)
75. R.J.K. Wood, A.S. Bahaj, S.R. Turnock, L. Wang, M. Evans, Tribological design constraints of marine renewable energy systems. Phil. Trans. Royal. Soc. A. **368**, 4807–4828 (2010)
76. E.Y.A. Wornyoh, V.K. Jasti, C.F. Higgs III, A review of dry particulate lubrication: powder and granular materials. ASME J. Tribol. **129**, 438–449 (2007)
77. R. Yun, Y. Lu, P. Filip, Application of extension evaluation method in development of novel eco-friendly brake materials. SAE Int. J. Mater. Manuf. **2**, 1–7 (2010)

Part II
Biomimetics Surfaces, Materials and Methods

Chapter 2
Lotus Versus Rose: Biomimetic Surface Effects

Michael Nosonovsky and Bharat Bhushan

Abstract The Lotus and rose petal effects have become a subject of active investigation by scientists, as they involve different modes of the interaction of wetting with roughness. The contact angle (CA) and CA hysteresis are two parameters, which characterize the hydrophobicity/philicity of a solid surface. Lotus-effect surfaces have a high CA and low CA hysteresis. However, it was found recently that a high CA can coexist with strong adhesion between water and a solid surface (and high CA hysteresis) in the case of the so-called "rose petal effect." It is clear now that wetting cannot be characterized by only the CA, since several modes or regimes of wetting of a rough surface can exist, including the Wenzel, Cassie, Lotus, and Petal regimes. This is due to the hierarchical structure of rough surfaces built of micro- and nanoscale roughness, so that a composite interface can exist at the microscale, while a homogeneous interface can exist at the nanoscale or vice versa. The understanding of the wetting of rough surfaces is important in order to design non-adhesive surfaces for various applications, including environmental.

M. Nosonovsky
College of Engineering and Applied Science, University of Wisconsin,
Milwaukee, WI 53201, USA

B. Bhushan (✉)
Nanoprobe Laboratory for Bio- and Nanotechnology and Biomimetics (NLB2),
The Ohio State University, 201 W. 19th Avenue, Columbus,
OH 43210-1142, USA
e-mail: bhushan.2@osu.edu

M. Nosonovsky and B. Bhushan (eds.), *Green Tribology*,
Green Energy and Technology, DOI: 10.1007/978-3-642-23681-5_2,

2.1 Introduction

Wetting of rough surfaces is a complex problem which continues to attract scientists, in particular due to the emergence of new materials with controlled surface micro-, nano-, and hierarchical structure. During the past decade, the so-called "Lotus effect," or surface roughness-induced superhydrophobicity and self-cleaning, became a subject of active investigation. According to early Wenzel [37] and Cassie and Baxter [9] models, there are two regimes of wetting of a rough surface: a homogeneous regime with a two-phase solid–water interface and a non-homogeneous or composite regime with a three phase solid–water–air interface (air pockets are trapped between the solid surface and water). Both models predict that surface roughness affects the water CA and can easily bring it to the extreme values close to 180° (superhydrophobicity) or close to 0° (superhydrophilicity). The studies of wetting of microstructured surfaces have concentrated on the investigation of the two regimes and the factors which affect the transition between the regimes [3, 5, 7, 16, 26–32].

Recent experimental findings and theoretical analyses made it clear that the early Wenzel [37] and Cassie and Baxter [9] models do not explain the complexity of interactions during wetting of a rough surface, which can follow several different scenarios [4, 5, 12, 14, 15, 22, 33, 36, 39]. As a result, there are several modes of wetting of a rough surface, and therefore, wetting cannot be characterized by a single number, such as the CA.

The concept of surface (or interface) energy is central for the analysis of wetting phenomena. Atoms or molecules at the surface of a solid or liquid have fewer bonds with neighboring atoms than those in the bulk. Energy is spent for breaking the bonds when a surface is created. As a result, the atoms at the surface have higher energy. This excess surface energy or surface tension, γ, is measured in N/m, and it is equal to the energy needed to create a surface with the unit area. If a liquid droplet is placed on a solid surface, the liquid and solid surfaces come together under equilibrium at a characteristic angle called the static CA, θ_0, given by the Young equation [1, 2],

$$\cos\theta_0 = \frac{\gamma_{SA} - \gamma_{SL}}{\gamma_{LA}} \tag{2.1}$$

γ_{SL}, γ_{SA}, and γ_{LA} are the surface energies of the solid–liquid, solid–air, and liquid–air interfaces, respectively. For a large number of combinations of materials and liquids, $\gamma_{SA} + \gamma_{LA} > \gamma_{SL}$, which means that it is energetically profitable for a liquid to wet the solid surface rather than to have an air film separating the solid and liquid. On the other hand, for many material combinations, $\gamma_{SL} + \gamma_{LA} > \gamma_{SA}$, which means that it is energetically profitable for a solid to be in contact with air, rather than to be covered by a thin liquid film. As a result, in most situations $-1 < (\gamma_{SA} - \gamma_{SL})/\gamma_{LA}) < 1$, and there exists a value of the CA given by Eq. 2.1. The CA is the angle under which the liquid–air interface comes in contact with the solid surface locally, and it does not depend on the shape of the body of water.

If water CA $0° < \theta_0 < 90°$, then the surface is usually called "hydrophilic," whereas a surface with water CA $90° < \theta_0 < 180°$ is usually called "hydrophobic."

In the ideal situation of a perfectly smooth and homogeneous surface, the static CA is a single number which corresponds to the unique equilibrium position of the solid–liquid–air contact line (the triple line). However, when the contact takes place with a rough surface, there may be multiple equilibrium positions which result in an entire spectrum of possible values of the CA. In addition, the value of the surface energy itself exhibits so-called "adhesion hysteresis" and can depend on whether it is measured during the approach of the two bodies or when they are taken apart. As the result, there is always the minimum value of the CA called the receding CA, θ_{rec}, and the maximum value of the CA called the advancing CA, θ_{adv}. The difference between the advancing and receding CA is called CA hysteresis (Fig. 2.1a).

Consider now a rough solid surface with the roughness factor $R_f > 1$ equal to the ratio of the solid surface area to its flat projected area. When water comes in contact with such a surface, the effective values of the solid–liquid and solid–air surface energies become $R_f\,\gamma_{SL}$ and $R_f\,\gamma_{SA}$ (Fig. 2.1b). This leads to the Wenzel equation for the CA with a rough surface [37]

$$\cos\theta = R_f \cos\theta_0 \tag{2.2}$$

If some air is trapped between the rough solid surface and the liquid, then only the fraction $0 < f_{SL} < 1$ constitutes the solid–liquid contact interface (Fig. 2.1c). The area of the solid–liquid interface is now $R_f f_{SL}$ per unit area, and in addition, there is $(1 - f_{SL})$ of the liquid–air interface under the droplet. The effective values of the solid–liquid and solid–air surface energies become $R_f f_{SL}\gamma_{SL}$ and $R_f f_{SL}\gamma_{SA} + (1 - f_{SL})\gamma_{LA}$. The CA is then given by the Cassie and Baxter [9] equation

$$\cos\theta = R_f f_{SL} \cos\theta_0 - 1 + f_{SL} \tag{2.3}$$

If a surface is covered by holes filled (or impregnated) with water, the contact angle is given by

$$\cos\theta = 1 + f_{SL}(\cos\theta_0 - 1) \tag{2.4}$$

This is the so-called "impregnating" Cassie wetting regime [32] (Fig. 2.1d).

The CA is a macroscale parameter characterizing wetting. However, hydrophobicity/philicity is dependent upon the adhesion of water molecules to the solid. On the one hand, a high CA is a sign of low liquid–solid adhesion. On the other hand, low CA hysteresis is a sign of low liquid–solid adhesion as well. There is an argument in the literature as to whether superhydrophobicity is adequately characterized only by a high CA and whether a surface can have a high CA but at the same time strong adhesion. It is now widely believed that a surface can be superhydrophobic and at the same time strongly adhesive to water (e.g., [15]). The so-called "petal effect" is exhibited by a surface that has a high CA, but also a large CA hysteresis and strong adhesion to water. The phenomenon of the large

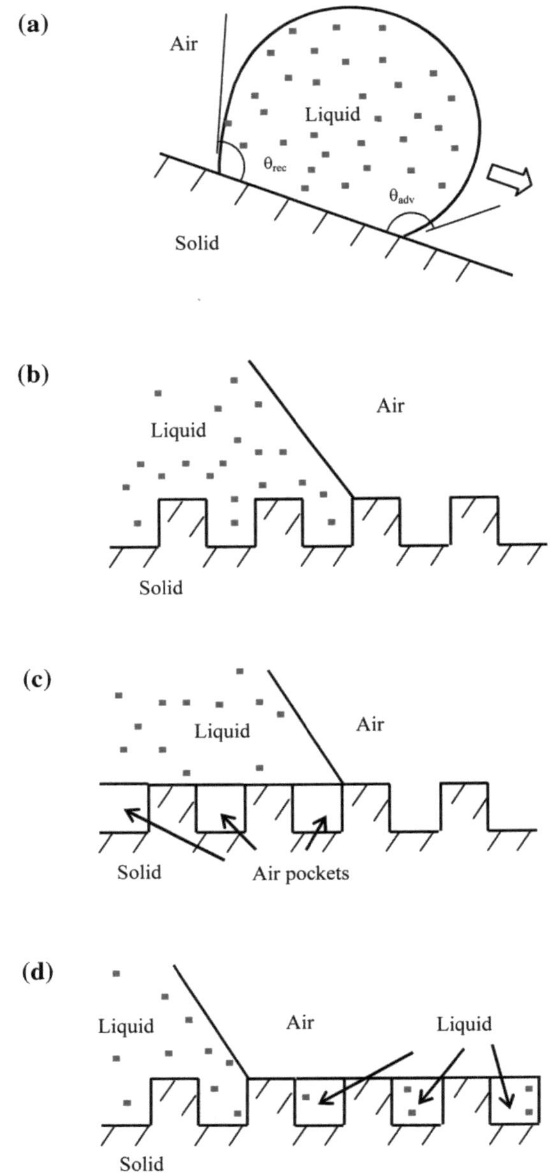

Fig. 2.1 **a** Schematics of a droplet on a tilted substrate showing advancing (θ_{adv}) and receding (θ_{rec}) contact angles. The difference between these angles constitutes the contact angle hysteresis. Configurations described by **b** the Wenzel equation for the homogeneous interface, **c** Cassie–Baxter equation for the composite interface with air pockets, and **d** the Cassie equation for the homogeneous interface

CA hysteresis and high water adhesion to rose petals (and similar surfaces), as opposed to small CA hysteresis and low adhesion to Lotus leaf, was observed by several research groups [4, 8, 10]. Bormashenko et al. [8] reported a transition between wetting regimes, e.g., the penetration of liquid into the micro/nanostructures.

Li and Amirfazli [19] argued that since "superhydrophobicity" means a strong fear of water or lacking affinity to water, "the claim that a superhydrophobic surface also has a high adhesive force to water is contradictory." Gao and McCarthy [14] pointed out that the terms "hydrophobic/phillic" should be defined in a more accurate way. They suggested several experiments showing that even Teflon®, which is usually considered very hydrophobic, can be, under certain conditions, considered hydrophilic, i.e., has affinity to water. They argued that the concepts of "shear and tensile hydrophobicity" should be used, so that the wettability of a surface is characterized by two numbers, advancing and receding CAs, and "the words hydrophobic, hydrophilic, and their derivatives can and should only be considered qualitative or relative terms." Instead, "shear and tensile hydrophobicity" should be investigated, which makes wetting ("solid–liquid friction") similar to the friction force, as it has been pointed out in the literature earlier [23]. McHale [22] noted that all solid materials, including Teflon®, are hydrophobic to some extent, if they have Young CA $<180°$. Therefore, it is energetically profitable for them to have contact with solid, at least to some extent. Wang and Jiang [36] suggested five superhydrophobic states (Wenzel's state, Cassie's state, so-called "Lotus" and "Gecko" states, and a transitional state between Wenzel's and Cassie's states). It may be useful also to see the transition between the Wenzel, Cassie, and dry states as a phase transition and to add the ability of a surface to bounce off a water droplet to the definition of the superhydrophobicity [31]. In addition, there is an argument on how various definitions of the CA hysteresis are related to each other [7, 8, 10, 17, 39]. A number of wetting regimes and transitions between them have been studied since 2010 [6, 11, 13, 34]. Modern research has concentrated on the ability to switch between the wetting states by tuning the surface energy [20, 21].

The Lotus effect has been comprehensively discussed in earlier publications. The objective of this paper is to discuss various wetting modes of rough surfaces, beyond the classical Wenzel [37] and Cassie and Baxter [9] regimes in light of recent experimental data on the petal effect and strong adhesion with superhydrophobic surfaces referred to as the "rose petal effect."

2.2 Modeling CA Hysteresis

Predicting CA hysteresis for a rough surface with a given topography is a difficult task. One approach is a numerical simulation; however, in most cases the simulations are limited to two-dimensional (2D) topography. Kusumaatmaja and Yeomans [18] showed that contact angle hysteresis is sensitive to the details of the surface patterning. Despite that, certain conclusions about the relation of the contact angle hysteresis to roughness can be made. It is known that the energy gained for surfaces during contact is greater than the work of adhesion for separating the surfaces, due to so-called adhesion hysteresis. Factors that affect contact angle hysteresis include adhesion hysteresis, surface roughness, and heterogeneity. Nosonovsky and Bhushan [26–29] assumed that contact angle hysteresis is equal to the adhesion hysteresis term and the term corresponding to the pinning effect of roughness, H_r. They further noted that the adhesion hysteresis can be assumed to be proportional to the fractional solid–liquid area $(1 - f_{LA})$. Using Eq. 2.3, the difference of cosines of the advancing and receding angles is related to the difference of those for a nominally smooth surface, θ_{adv0} and θ_{rec0}, as

$$\cos\theta_{adv} - \cos\theta_{rec} = R_f(1 - f_{LA})(\cos\theta_{adv0} - \cos\theta_{rec0}) + H_r \tag{2.5}$$

The first term in the right-hand part of the equation, which corresponds to the inherent contact angle hysteresis of a smooth surface, is proportional to the fraction of the solid–liquid contact area, $1 - f_{LA}$. The second term H_r is the effect of surface roughness, which is proportional to the length of the triple line. Thus Eq. 2.5 involves both the term proportional to the solid–liquid interface area and to the triple line length. It is observed from Eqs. 2.4 and 2.5 that increasing $f_{LA} \to 1$ results in increasing the contact angle ($\cos\theta \to -1$, $\theta \to \pi$) and decreasing the contact angle hysteresis ($\cos\theta_{adv} - \cos\theta_{rec} \to 0$). In the limiting case of a very small solid–liquid fractional contact area under the droplet, when the contact angle is large ($\cos\theta \approx -1 + (\pi - \theta)^2/2$, $\sin\theta \approx \theta - \pi$) and where the contact angle hysteresis is small ($\theta_{adv} \approx \theta \approx \theta_{rec}$), based on Eq. 2.5 [28],

$$\pi - \theta = \sqrt{2(1 - f_{LA})(R_f\cos\theta_0 + 1)} \tag{2.6}$$

$$\theta_{adv} - \theta_{rec} = (1 - f_{LA})R_f\frac{\cos\theta_{a0} - \cos\theta_{r0}}{-\sin\theta} = (\sqrt{1 - f_{LA}})R_f\frac{\cos\theta_{r0} - \cos\theta_{a0}}{\sqrt{2(R_f\cos\theta_0 + 1)}} \tag{2.7}$$

For the homogeneous interface, $f_{LA} = 0$, whereas for the composite interface f_{LA} is a non-zero number. It is observed from Eqs. 2.6–2.7 that for a homogeneous interface, increasing roughness (high R_f) leads to increasing the contact angle hysteresis (high values of $\theta_{adv} - \theta_{rec}$), while for a composite interface, an approach to unity of f_{LA} provides both high contact angle and small contact angle hysteresis [16, 26–28]. Therefore, the composite interface is desirable for self-cleaning.

A different semi-phenomenological model of the contact angle hysteresis has been proposed recently by Whyman et al. [38]. According to their model, the contact angle hysteresis is given by the equation $\theta_{adv} - \theta_{rec} = \left(\frac{8U}{\gamma R_0}\right)^{1/2} h(\theta*)$, where U is the height of the potential barrier connected with the motion of the triple line along a substrate, R_0 is the initial radius of the spherical drop before deposition on the substrate, and $h(\theta^*)$ is the dimensionless function of the apparent contact angle θ^*.

Vedantam and Panchagunula [35] suggested a semi-empirical phase field method to calculate the CA hysteresis. In this method, the order-parameter $\eta(x, y)$ is selected in such a manner that $\eta = 0$ for the non-wetted regions of the surface and $\eta = 1$ for wetted regions, whereas $0 < \eta < 1$ for partially wetted regions. After that, the energy function $f(\eta)$ is constructed, and its minima correspond to the equilibrium states of the system (e.g., the Wenzel and Cassie states). After that, the energy functional is written as

$$L = \int_A \left\{ f(\eta) + \frac{\lambda}{2} |\nabla f(\eta)|^2 \right\} \mathrm{d}A \tag{2.8}$$

where λ is the gradient coefficient. The functional that should be minimized involves the free energy and the gradient of the free energy. The latter term is needed to account for the fact that creating an interface between two phases is energetically unprofitable. The kinetic equation is given in the form

$$\beta \dot{\eta} = -\frac{\mathrm{d}L}{\mathrm{d}\eta} = \lambda \nabla^2 \eta - \frac{\partial f}{\partial \eta} \tag{2.9}$$

where $\beta > 0$ is the kinetic coefficient. Vedantam and Panchagunula [35] showed that in the case of $\beta = \text{const}$ for an axisymmetric drop flowing with the velocity V, Eq. 2.9 leads to

$$\cos \theta_{\text{adc}} - \cos \theta_{\text{rec}} = 2\alpha\beta V \tag{2.10}$$

In other words, assuming that the kinetic coefficient is constant, the contact angle hysteresis is expected to be proportional to the flow velocity. A more complicated form of the kinetic coefficient may lead to a more realistic dependence of the contact angle hysteresis on the velocity.

There is an asymmetry between the wetting and dewetting processes, since less energy is released during wetting than the amount required for dewetting due to adhesion hysteresis. Adhesion hysteresis is one of the reasons that leads to contact angle hysteresis, and it also results in the hysteresis of the Wenzel-Cassie state transition. The Cassie-Wenzel transition and CA hysteresis both may be considered as different manifestations of the same wetting–dewetting cycle behavior. Both the CA hysteresis and Cassie-Wenzel transition cannot be determined from the macroscale equations and are governed by micro- and nanoscale phenomena.

Note that the size of the surface roughness details is an important factor. It is generally assumed that the roughness factor R_f as well the fractional area of contact f_{SL} can be determined by averaging the surface roughness over some area, which is itself small relative to the size of the liquid droplet. For R_f and f_{SL} fractional areas changing with a spatial coordinate, special generalized Wenzel and Cassie equations, proposed by Nosonovsky [24], should be used. The size of the surface roughness also affects the ability of the interface to pin the triple line and thus affects the CA hysteresis. It could be claimed that CA hysteresis is a "second order" effect which is expected to vanish with the decreasing ratio of the size of

Fig. 2.2 **a** Optical images, **b** Scanning Microscope micrographs, and **c** Atomic Force Microscope roughness maps of petals of two roses [Rosa Hybrid Tea, cv. Bairage (Rosa, cv. Bairage), and Rosa Hybrid Tea, cv. Showtime (Rosa, cv. Showtime)] (adapted from [4])

the surface roughness and heterogeneity details to the droplet radius. This, however, does not happen since surface roughness and heterogeneity is an inherent property of any surface. There is a deep similarity between the dry friction and the wetting of a solid surface [23]. In the ideal situation of absolutely homogeneous and smooth surfaces there would be no friction and no CA hysteresis due to the absence of energy dissipation. However, in the real situations, surfaces are not ideal, and this leads to both dry friction and CA hysteresis. The development of quantitative relationships between the degrees of surface non-ideality (e.g., Shannon entropy of a rough surface) and CA hysteresis, remains an interesting task similar to the same task for friction [25].

2.3 Investigation of the Petal Effect

Plant leaves and petals provide an example of surfaces with high CA and high and low CA hystereses. Bhushan and Her [4] studied two kinds of superhydrophobic rose petals: (1) Rosa Hybrid Tea, cv. Bairage and (2) Rosa Hybrid Tea,

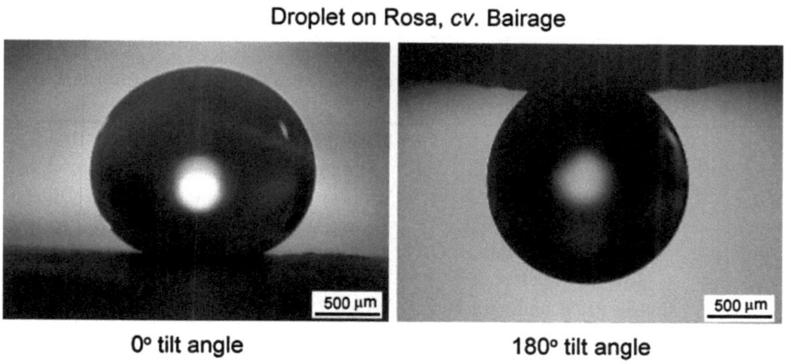

Fig. 2.3 Optical micrographs of water droplets on Rosa, cv. Bairage at 0° and 180° tilt angles. Droplet is still suspended when the petal is turned upside down [4]

Table 2.1 Surface roughness statistics for the two rose petals [4]

	Peak-to-base height (μm)	Midwidth (μm)	Peak radius (μm)	Bump density (1/10,000 μm^2)
Rosa, cv. Bairage (high adhesion)	6.8	16.7	5.8	23
Rosa, cv. Showtime (low adhesion)	8.4	15.3	4.8	34

Table 2.2 Wetting regimes of a surface with a single level of hierarchy of roughness

State	Cassie–Baxter	Wenzel	Impregnating cassie
Cavities	Air	Water under droplet	Water everywhere
CA	High	High	High
CA hysteresis	Low	Can be high	Low

cv. Showtime, referred to as Rosa, cv. Bairage and Rosa, cv. Showtime, respectively. Figure 2.2 shows optical micrographs and scanning electron microscopy (SEM) images and atomic force microscope (AFM) surface height maps of two rose petals. Figure 2.3 shows a sessile and a suspending water droplet on Rosa, cv. Bairage demonstrating that it can simultaneously have high CA and high adhesion and high CA hysteresis.

The surface roughness of the two rose petals was measured with the AFM, and the results for the peak-to-base height of bumps, the midwidth, peak radius, and bump density are summarized in Table 2.1. The data indicates that the low adhesion specimen (Rosa, cv. Showtime) has a higher density and height of the bumps, indicating that the penetration of water between the micro-bumps is less likely. Wetting of a rough surface with a single level of hierarchy of roughness details can follow several scenarios (Table 2.2).

Table 2.3 Different regimes of wetting of a surface with dual roughness

	Air in microstructure	Water under droplet in microstructure	Water impregnating microstructure
Air in nanostructure	Lotus, high CA, low CA hysteresis	Rose, high CA, high CA hysteresis	Rose filled microstructure
Water under droplet in nanostructure	Cassie (air-filled microstructure, water in nanostructure), high CA, low CA hysteresis	Wenzel (water in micro- and nanostructure), high CA, high or low CA hysteresis	Wenzel filled microstructure
Water impregnating nanostructure	Cassie filled nanostructure	Wenzel filled nanostructure	Wenzel filled micro and nanostructure

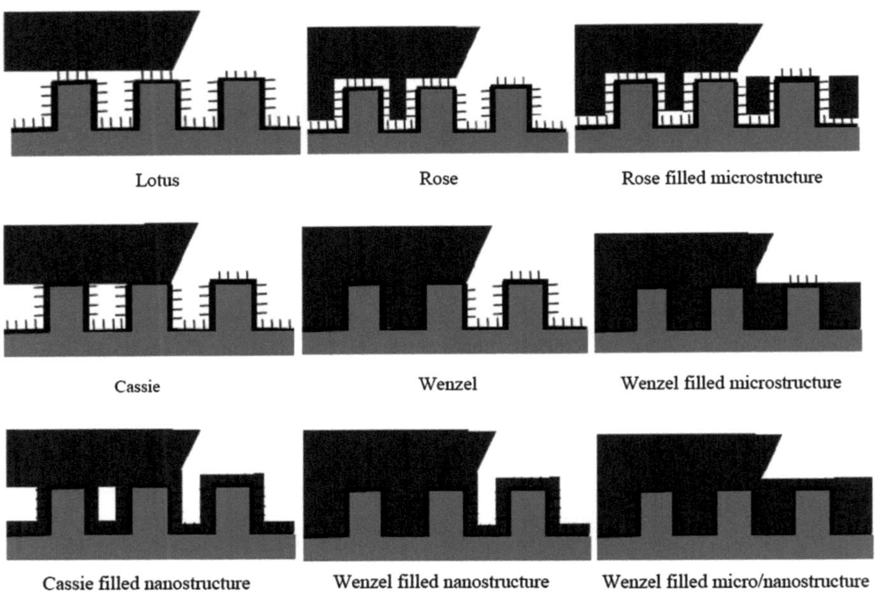

Fig. 2.4 Schematics of nine wetting scenarios for a surface with hierarchical roughness

For a hierarchical structure with small bumps on top of the larger bumps, a larger number of scenarios are available, and they are summarized in Table 2.3 and Fig. 2.4. Water can penetrate either in the micro- or nanostructure, or into both. In addition, the micro- or nanostructure can be impregnated by water or air. The regimes with water penetrating into the microstructure can have high solid–water adhesion and therefore high CA hysteresis.

Bhushan and Her [4] conducted a series of carefully designed experiments to decouple the effects of the micro- and nanostructures. They synthesized micro-structured surfaces with pillars out of epoxy resin. The epoxy surfaces were

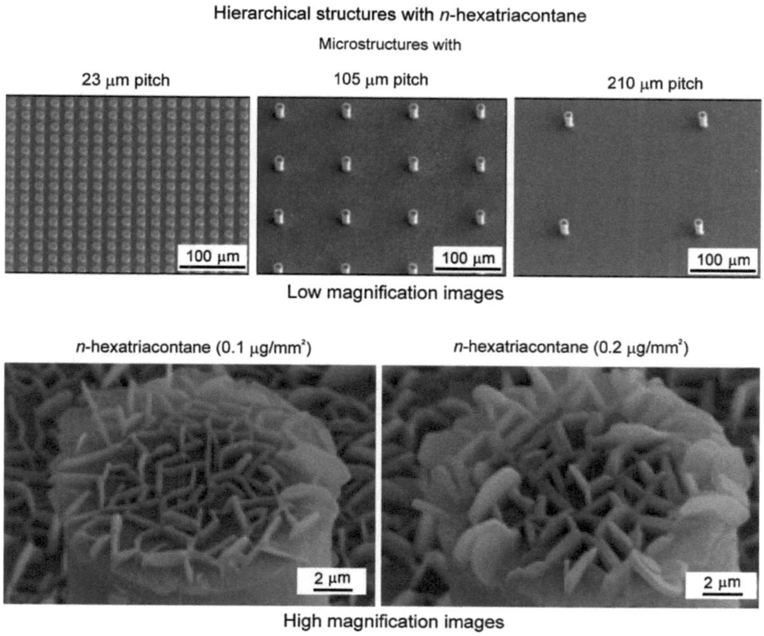

Fig. 2.5 SEM micrographs of the microstructures and nanostructures fabricated with two different masses of n-hexatriacontane for hierarchical structure. All images were taken at 45° tilt angle. All samples are positive replicas, obtained from negative replica with dental wax and Si micropatterned master template (14 μm diameter and 30 μm height) fabricated with epoxy resin coated with n-hexatriacontane [4]

reproduced from model Si templates and were created by a two-step molding process producing a dual replica (first a negative replica and then a positive replica of the original Si template). Surfaces with a pitch (the periodicity of the structure of the pillars) of 23, 105, and 210 μm and with the same diameter (14 μm) and height (30 μm) of the pillars were produced. After that, nanostructures were created on the microstructured sample by self-assembly of the alkane n-hexatriacontane ($CH_3(CH_2)_{34}CH_3$) deposited by a thermal evaporation method. Alkanes of varying chain lengths are common hydrophobic compounds of plant waxes. On smooth surfaces, alkanes can cause a large contact angle and a small contact angle hysteresis for water droplets. To fabricate the nanostructure, various masses of n-hexatriacontane were coated on a microstructure. The nanostructure is formed by three-dimensional platelets of n-hexatriacontane. Platelets are flat crystals, grown perpendicular to the surface. They are randomly distributed on the surface, and their shapes and sizes show some variation. Figure 2.5 shows selected images. When different masses of wax are applied, the density of the nanostructure is changed.

For surfaces with a small pitch of 23 mm, while the mass of n-hexatriacontane is changed, there are only small changes in the static contact angle and contact

Table 2.4 CA and CA hysteresis for surfaces with various micro- and nanoroughness (based on [4])

Mass of n-hexatriacontane (μg/mm^2)	Pitch					
	23 μm		105 μm		210 μm	
	CA	CA hysteresis	CA	CA hysteresis	CA	CA hysteresis
0.1	164°	3°	152	87	135	45
0.12	165°	3°	153	20	135	42
0.16	166°	3°	160	5	150	12
0.2	167°	3°	168	4	166	3

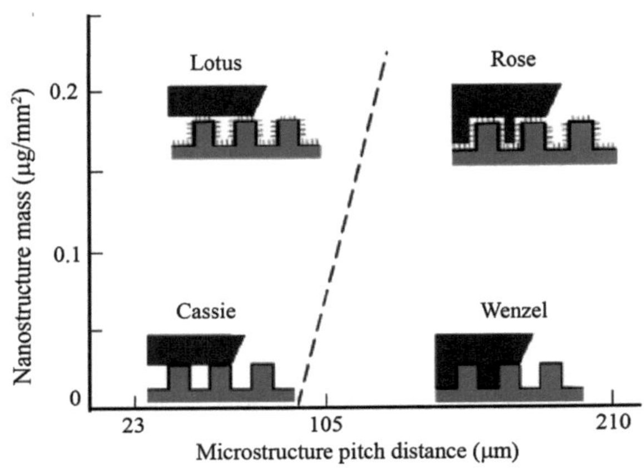

Fig. 2.6 Schematic of a wetting regime map as a function of microstructure pitch and the mass of nanostructure material. The mass of nanostructure material equal to zero corresponds to microstructure only (with the Wenzel and Cassie regimes). Higher mass of the nanostructure material corresponds to higher values of pitch, at which the transition occurs

angle hysteresis values, which means that they are always in the "Lotus" wetting regime. On the surface with a 210 μm pitch value, as the mass of n-hexatriacontane is increased, the static contact angle is increased, and the reverse trend was found for the contact angle hysteresis. This was interpreted as evidence that the nanostructure is responsible for the CA hysteresis and low adhesion between water and the solid surface. The results are summarized in Table 2.4. The wetting regimes are shown schematically in Fig. 2.6 as a function of the pitch of the microstructure and the mass of n-hexatriacontane. A small mass of the nanostructure material corresponds to the Cassie and Wenzel regimes, whereas a high mass of nanostructure corresponds to the Lotus and rose regimes. The Lotus regime is more likely for larger masses of the nanostructure material. Figure 2.7 shows a droplet on a horizontal surface of a hierarchical structure with 23 and 105 μm pitch and n-hexatriacontane (0.1 μg/mm^2). Air pockets are observed in the first case and not observed in the second case, indicating the difference between the two regimes [4].

To further verify the effect of wetting states on the surfaces, evaporation experiments with a droplet on a hierarchical structure coated with two different

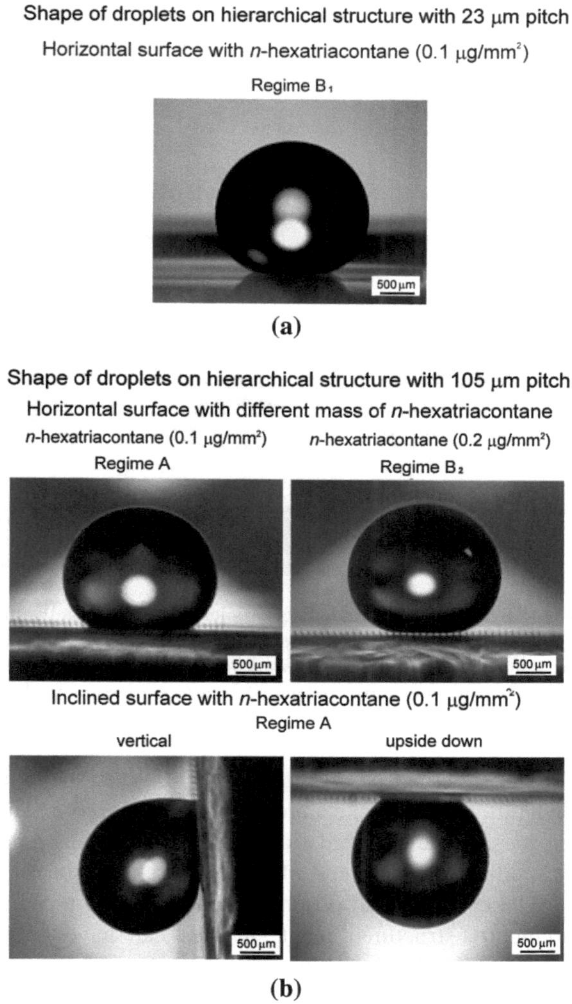

Fig. 2.7 **a** Droplet on a horizontal surface of hierarchical structure with 23 μm pitch and n-hexatriacontane (0.1 μg/mm^2) showing air pocket formation and **b** droplet on a hierarchical structure with 105 μm pitch and n-hexatriacontane (0.1 μg/mm^2) and 0.2 μg/mm^2 showing no air pocket and air pocket formation, respectively. Also shown is the image taken on the inclined surface with hierarchical structure with 0.1 μg/mm^2 showing that droplet is still suspended [4]

amounts of n-hexatriacontane were performed. Figure 2.8 shows the optical micrographs of a droplet evaporating on two different hierarchical structured surfaces. On the n-hexatriacontane (0.1 μg/mm^2) coated surface, an air pocket was not visible at the bottom area of the droplet. However, the droplet on the surface has a high static contact angle (152°) since the droplet still cannot completely impregnate the nanostructure. The footprint size of the droplet on the surface has only small changes from 1820 to 1791 μm. During evaporation, the initial contact area between the droplet and hierarchical structured surface does not decrease until the droplet evaporates completely, which means complete wetting between droplet and microstructures. For the n-hexatriacontane (0.2 μg/mm^2) coated surface, the light passes below the droplet, and air pockets can be seen, so to start with the

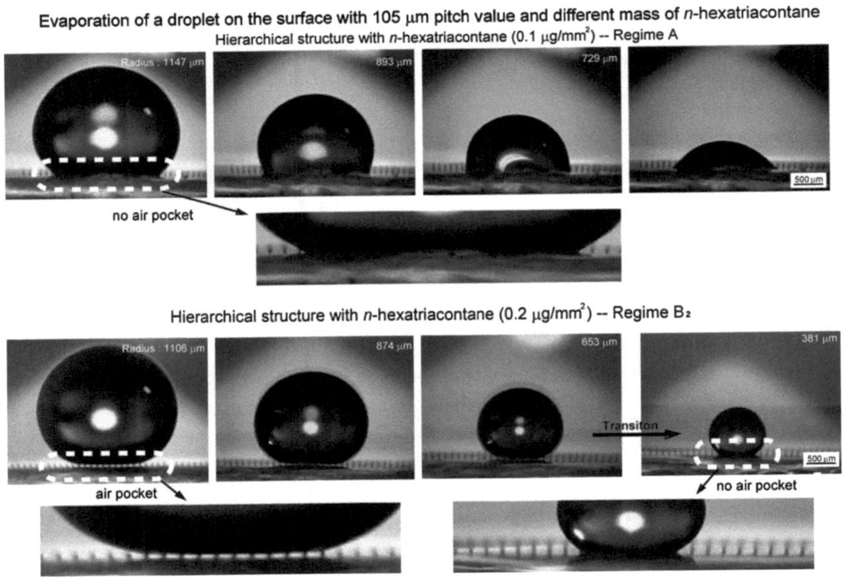

Fig. 2.8 Optical micrographs of droplet evaporation on the hierarchical structured surfaces with 105 μm pitch value. n-Hexatriacontane (0.1 μg/mm^2) coated sample has no air pocket formed between the pillars in the entire contact area until evaporation was completed. Hierarchical structure with n-hexatriacontane (0.2 μg/mm^2) has air pocket, and then the transition from the "Lotus" regime to the "Rose petal" regime occurred [4]

droplet is in the Cassie–Baxter regime. When the radius of the droplet decreased to 381 μm, the air pockets are not visible anymore. The footprint size of the droplet on the surface changed from 1177 to 641 μm, since droplet remained on only a few pillars until the end of the evaporation process.

The experimental observations of the two types of rose petals show that hierarchically structured plant surfaces can have both adhesive and non-adhesive properties at the same time with high CA. This is due to the existence of various modes of wetting of a hierarchical surface, so that water can penetrate either into macro- or nanoroughness, or into both. Water penetration into the microroughness tends to result in high adhesion with the solid surface, whereas the presence of the nanoroughness still provides high CA. As a result, two distinct modes of wetting are observed, one can be called the "Lotus" mode (with low adhesion) and the other is the "rose" mode with high adhesion.

2.4 Conclusions

In this work several modes of wetting of rough surfaces were investigated. Rose petals have different hierarchically organized surface micro- and nanostructures, and can exhibit high and low adhesion to water. The pitch spacing and height of the microstructures controls the wetting regime, since it controls the penetration of water into the microstructure. The microstructure controls the CA hysteresis, whereas the nanostructure provides high CA. As a result, the rose petal can exhibit typical "Lotus effect" properties (high CA and low CA hysteresis) or "Petal effect" properties (high CA high CA hysteresis). Artificial surfaces which mimic rose petals were investigated and similar behavior found. Various wetting regimes are possible, depending on air and water penetration into the micro- and nano-structures. The understanding of the wetting of rough surfaces is important in order to design non-adhesive surfaces for various applications including green tribology.

Acknowledgment Michael Nosonovsky acknowledges the support of the UWM Research Growth Initiative grant.

References

1. B. Bhushan, *Principles and Applications of Tribology* (Wiley, New York, 1999)
2. B. Bhushan, *Introduction to Tribology* (Wiley, New York, 2002)
3. B. Bhushan, *Springer Handbook of Nanotechnology*, 3rd edn. (Springer, Heidelberg, 2010)
4. B. Bhushan, E.K. Her, Fabrication of superhydrophobic surfaces with high and low adhesion inspired from rose petal. Langmuir **26**, 8207–8217 (2010)
5. B. Bhushan, Y.C. Jung, Natural and biomimetic artificial surfaces for superhydrophobicity, self-cleaning, low adhesion, and drag reduction. Prog. Mater. Sci. **56**, 1–108 (2011)
6. B. Bhushan, M. Nosonovsky, The rose petal effect and the modes of superhydrophobicity. Phil. Trans R. Soc. A **368**, 4713–4728 (2010)
7. E. Bormashenko, Y. Bormashenko, T. Stein, G. Whyman, R. Pogreb, Z. Barkay, Environmental scanning electron microscope study of the fine structure of the triple line and cassie-wenzel wetting transition for sessile drops deposited on rough polymer substrates. Langmuir **23**, 4378–4382 (2007)
8. E. Bormashenko, T. Stein, R. Pogreb, D. Aurbach, "Petal effect" on surfaces based on lycopodium: high-stick surfaces demonstrating high apparent contact angles. J. Phys. Chem. C **113**, 5568–5572 (2009)
9. A. Cassie, S. Baxter, Wettability of porous surfaces. Trans. Faraday Soc. **40**, 546–551 (1944)
10. F.M. Chang, S.J. Hong, Y.J. Sheng, H.K. Tsao, High contact angle hysteresis of superhydrophobic surfaces: hydrophobic defects. Appl. Phys. Lett. **95**, 064102 (2009)
11. M.K. Dawood, H. Zheng, T.H. Liew, K.C. Leong, Y.L. Foo, R. Rajagopalan, S.A. Khan, W.K. Choi, Mimicking both petal and lotus effects on a single silicon substrate by tuning the wettability of nanostructured surfaces. Langmuir **27**, 4126–4133 (2011)
12. L. Feng, Y. Zhang, J. Xi, Y. Zhu, N. Wang, F. Xia, L. Jiang, Petal effect: a superhydrophobic state with high adhesive force. Langmuir **24**, 4114 (2008)
13. L. Feng, Y.A. Zhang, Y.Z. Cao, X.X. Ye, L. Jiang, The effect of surface microstructures and surface compositions on the wettabilities of flower petals. Soft Matter **7**, 2977–2980 (2011)
14. L. Gao, T.J. McCarthy, Teflon is hydrophilic. Comments on definitions of hydrophobic, shear versus tensile hydrophobicity, and wettability characterization. Langmuir **24**, 9184–9188 (2008)
15. M.H. Jin, X.L. Feng, L. Feng, T.L. Sun, J. Zhai, T.J. Li, L. Jiang, Superhydrophobic aligned polystyrene nanotube films with high adhesive force. Adv. Mater. **17**, 1977–1981 (2005)
16. Y.C. Jung, B. Bhushan, Contact angle, adhesion, and friction properties of micro- and nanopatterned polymers for superhydrophobicity. Nanotechnology **17**, 4970–4980 (2006)

17. B. Krasovitski, A. Marmur, Drops down the hill: theoretical study of limiting contact angles and the hysteresis range on a tilted plane. Langmuir **21**, 3881–3885 (2004)
18. H. Kusumaatmaja, J.M. Yeomans, Modeling contact angle hysteresis on chemically patterned and superhydrophobic surfaces. Langmuir **23**, 6019–6032 (2007)
19. W. Li, A. Amirfazli, Superhydrophobic surfaces: adhesive strongly to water? Adv. Mater. **19**, 3421–3422 (2007)
20. M.J. Liu, L. Jiang, Switchable adhesion on liquid/solid interfaces. Adv. Func. Mater. **20**, 3753–3764 (2010)
21. M.J. Liu, Y.M. Zheng, J. Zhai, L. Jiang, Bioinspired super-antiwetting interfaces with special liquid–solid adhesion. Acc. Chem. Res. **43**, 368–377 (2010)
22. G. McHale, All solids, including Teflon, are hydrophilic (to some extent), but some have roughness induced hydrophobic tendencies. Langmuir **25**, 7185–7187 (2009)
23. M. Nosonovsky, Model for solid–liquid and solid–solid friction for rough surfaces with adhesion hysteresis. J. Chem. Phys. **126**, 224701 (2007)
24. M. Nosonovsky, On the range of applicability of the wenzel and cassie equations. Langmuir **23**, 9919–9920 (2007)
25. M. Nosonovsky, Entropy in tribology: in search of applications. Entropy **12**, 1345–1390 (2010)
26. M. Nosonovsky, B. Bhushan, Biomimetic superhydrophobic surfaces: multiscale approach. Nano Lett. **7**, 2633–2637 (2007)
27. M. Nosonovsky, B. Bhushan, Multiscale friction mechanisms and hierarchical surfaces in nano- and bio-tribology. Mater. Sci. Eng. R **58**, 162–193 (2007)
28. M. Nosonovsky, B. Bhushan, Hierarchical roughness makes superhydrophobic surfaces stable. Microelectron. Eng. **84**, 382–386 (2007)
29. M. Nosonovsky, B. Bhushan, Hierarchical roughness optimization for biomimetic superhydrophobic surfaces. Ultramicroscopy **107**, 969–979 (2007)
30. M. Nosonovsky, B. Bhushan, Biologically-inspired surfaces: broadening the scope of roughness. Adv. Func. Mater. **18**, 843–855 (2008)
31. M. Nosonovsky, B. Bhushan, Energy transitions in superhydrophobicity: low adhesion, easy flow and bouncing. J. Phys. Condens. Matter **20**, 395005 (2008)
32. M. Nosonovsky, B. Bhushan, *Multiscale Dissipative Mechanisms and Hierarchical Surfaces: Friction, Superhydrophobicity, and Biomimetics* (Springer, Heidelberg, 2008)
33. M. Nosonovsky, B. Bhushan, Superhydrophobic surfaces and emerging applications: non-adhesion, energy, green engineering. Curr. Opin. Colloid Interface Sci. **14**, 270–280 (2010)
34. A. Tonosaki, T. Nishide, Novel petal effect of hafnia films prepared in an aqueous solution and containing hydroxy acids. Appl. Phys. Express **3**, 125801 (2010)
35. S. Vedantam, M.V. Panchagnula, Phase field modeling of hysteresis in sessile drops. Phys. Rev. Lett. **99**, 176102 (2007)
36. S. Wang, L. Jiang, Definition of superhydrophobic states. Adv. Mater. **19**, 3423–3424 (2007)
37. R.N. Wenzel, Resistance of solid surfaces to wetting by water. Indust. Eng. Chem. **28**, 988–994 (1936)
38. G. Whyman, E. Bormashenko, T. Stein, The rigorous derivation of young, cassie–baxter and wenzel equations and the analysis of the contact angle hysteresis phenomenon. Chem. Phys. Lett. **450**, 355–359 (2008)
39. F. Xia, L. Jiang, Bio-inspired, smart, multiscale interfacial materials. Adv. Mater. **20**, 2842–2858 (2008)

Chapter 3
Self-Organization at the Frictional Interface

Michael Nosonovsky and Vahid Mortazavi

Abstract Despite the fact that self-organization during friction has received relatively little attention of the tribologists so far, it has a potential for the creation of self-healing and self-lubricating materials, which are of importance for the green or environment-friendly tribology. The principles of the thermodynamics of irreversible processes and of the nonlinear theory of dynamical systems are used to investigate the formation of spatial and temporal structures during friction. The transition to the self-organized state with low friction and wear occurs through the destabilization of the steady-state (stationary) sliding. The criterion for the destabilization is discussed and examples like formation of a protective film and slip waves are discussed. Some cases like running-in stage, elastic structures, and Turing pattern formation as evidences of self-organization are studied. A special self-healing mechanism may be embedded into material by coupling corresponding required forces. The analysis provides a structure–property relationship which can be applied for the design optimization of composite self-lubricating and self-healing materials for various ecologically friendly applications and green tribology.

3.1 Introduction

The emergence of the green tribology makes it important to reconsider some conventional approaches to friction and wear. Friction and wear are usually viewed as irreversible processes, which lead to energy dissipation (friction) and

---

M. Nosonovsky (✉) · V. Mortazavi
College of Engineering and Applied Science, University of Wisconsin,
Milwaukee, WI 53201, USA
e-mail: nosonovs@uwm.edu

V. Mortazavi
e-mail: mortaza2@uwm.edu

M. Nosonovsky and B. Bhushan (eds.), *Green Tribology*,
Green Energy and Technology, DOI: 10.1007/978-3-642-23681-5_3,

material deterioration (wear). On the other hand, it is known that under certain circumstances frictional sliding can result in the formation (self-organization) of spatial and temporal patterns. Friction-induced self-organization has a significant potential for the development of self-lubricating, self-healing, and self-cleaning materials, which can reduce environmental contamination, and thus it is relevant to green tribology.

Historically, the first attempts to investigate friction-induced self-organization were made in Russia during the 1970s. Several groups can be mentioned. First, Kostetsky and Bershadsky [8, 9] in Kiev investigated the formation of the so-called self-organized "secondary structures". According to Bershadsky, friction and wear are two sides of the same phenomenon and they represent the tendency of energy and matter to achieve the most disordered state. However, the synergy of various mechanisms can lead to the self-organization of the secondary structures, which are "nonstoichiometric and metastable phases," whereas "the friction force is also a reaction on the informational (entropic) excitations, analogous to the elastic properties of a polymer, which are related mostly to the change of entropy and have the magnitude of the order of the elasticity of a gas" [9].

These ideas were influenced by the theory of self-organization developed by Ilya Prigogine, a Belgian physical chemist of Russian origin and the winner of the 1977 Nobel Prize in Chemistry. Prigogine [58] used the ideas of non-equilibrium thermodynamics to describe the processes of self-organization (Nicolis and Prigogine [50]). Later, Prigogine wrote several popular books about the scientific and philosophical importance of self-organization [59, 60]. At the same time, at the end of the 1970s, the concept of "Synergetics" was suggested by Haken [31] as an interdisciplinary science investigating the formation and self-organization of patterns and structures in open systems, which are far from thermodynamic equilibrium. The word "synergetics" was apparently coined by the architect and philosopher R.B. Fuller (the same person, after whom the C_{60} molecule was later called "fullerene"). The books by Prigogine and Haken were published in the USSR in Russian translations and became very popular in the 1980s.

The second group to mention is Bushe and Gershman in Moscow. Their results were summarized in the edited book by Fox-Rabinovich and Totten [25]. In 2002, Bushe won the most prestigious tribology award, the Tribology Gold Medal, for his studies on tribological compatibility and other related effects. The third group is Garkunov [28] and co-workers, who claimed the discovery of the synergetic "non-deterioration effect" also called the "selective transfer." Garkunov's mostly experimental research also received an international recognition when he was awarded the 2005 Tribology Gold Medal "for his achievements in tribology, especially in the fields of selective transfer."

In the English language literature, the works by Klamecki [35–38] were the first to use the concepts of non-equilibrium thermodynamics to describe friction and wear. His work was extended by Zmitrowicz [67], Dai et al. [15], Doelling et al. [17] and others. An important entropic study of the thermodynamics of wear was conducted by Bryant et al. [12], who introduced a degradation function and formulated the Degradation-Entropy Generation theorem in their approach intended to study the friction and wear in complex. They note that friction and wear, which are often treated as unrelated processes, are in fact manifestations of the same dissipative physical processes occurring at sliding interfaces. The possibility of the reduction of friction between two elastic bodies due to a pattern of propagating slip waves was investigated by Adams [3] and Nosonovsky and Adams [49], who used the approach of the theory of elasticity.

A completely different approach to friction-induced self-organization is related to the theory of dynamical systems and involves the investigation of friction test results as time-series. Since the 1980s, it has been suggested that a specific type of self-organization, called "self-organized criticality (SOC)," plays a role in diverse "avalanche-like" processes, such as the stick–slip phenomenon during dry friction. The researches of Zypman [69] and Ferrante [4, 16, 24] and others deal with this topic.

A different approach to describe some mechanisms resulting in self-organization was based on the reaction–diffusion (RD) systems and their important class the "Turing systems" [34]. Mortazavi and Nosonovsky [47] investigated whether RD systems can describe certain types of friction-induced pattern formation involving heat transfer and diffusion-like mass transfer due to wear. In somehow similar trend, Nosonovsky and Bushan [55, 56] and Nosonovsky et al. [57] suggested treating self-lubrication and surface-healing as a manifestation of self-organization. They noted that the orderliness at the interface can increase (and, therefore, the entropy is decreased) at the expense of the entropy either in the bulk of the body or at the microscale. They also suggested that self-organized spatial patterns (such as interface slip waves) can be studied by the methods of the theory of self-organization.

The reduction of friction and wear due to the self-organization at the sliding frictional interface can lead to self-lubrication, i.e., the ability to sustain low friction and wear without the external supply of a lubricant. Since lubricants constitute environmental hazards, while friction and wear often lead to heat and chemical contamination of environment, self-lubrication has potential for the green tribology. Self-lubrication is also common in the living nature, and therefore, is of interest to scientists and engineers who are looking for the biomimetic approach, which also has potential for green tribology. In this chapter, we review some different researches done on self-organization including modeling of formation of self-organized patterns during friction through destabilization of the steady state, the methods of investigating of such patterns, and their relation to self-lubrication and environment-friendly tribology.

3.2 Thermodynamic Methods of Study of Self-Organized Tribological Systems

Tribology is defined as the science and technology of interacting surfaces in relative motion, or, in other words, the study of friction, wear, and lubrication. The concept of entropy often seems difficult and confusing for non-physicists.

The reason is that, unlike in the case of temperature and energy, there is no direct way of measuring entropy, so our everyday intuition does not work well when we have to deal with entropy. This is apparently the main reason why entropy, being the most important quantitative measure of disorder and dissipation, has not yet become the major tool for analyzing such dissipative processes as friction and wear.

While the net entropy grows in most systems in accordance with the Second law of thermodynamics, some thermodynamic systems may lead to an increasing orderliness and self-organisation [58]. These are thermodynamically open systems that operate far from thermodynamic equilibrium and can exchange energy, matter, and entropy with the environment. Many of these self-organizing systems (such as the Bénard cells in boiling liquid and oscillating chemical reactions) were known a long time ago; however, as described in the previous section, the universality and generality of the processes involved in these systems was understood only with the works by Prigogine [55, 56] and Haken [31]. It is believed that this ability for self-organization of physical systems led to the formation of complex hierarchical chemical and biological systems.

The flow of heat, entropy, and material away from the interface during the dry friction and wear can lead to self-organization when so-called "secondary structures" form. The secondary structures are either patterns that form at the interface (e.g., stick and slip zones) or those formed as a result of mutual adjustment of the bodies in contact. The entropy production rate reaches its minimum in the self-organized state. Therefore, the self-organization is usually beneficial for the tribolological system, as it leads to the reduction of friction and wear [55].

3.2.1 Qualitative Studies

The earlier works on self-organization are characterized by qualitative, rather than quantitative analysis of self-organization mechanisms. For example, it is noted in many publications that typical tribosystems possess the qualities which, according to Prigogine, are required (but not sufficient) for the self-organization, for example, the system should be thermodynamically open, nonlinear, and it should operate far from the equilibrium.

Bershadsky [8] suggested a classification of various friction-induced self-organization effects. In search of self-organization during friction he investigates quite a diverse range of processes and phenomena—auto-hydrodynamic effects, the evolution of micro topography, the formation of chemical and convective patterns, and the oscillation of various parameters measured experimentally during friction. Some of these phenomena had well-investigated principles and mechanisms, while others were studied in a phenomenological manner, so it was not possible to approach them all in a uniform manner. He, therefore, suggested that the state and evolution of a self-organized tribosystem might be described using different methods, including the equations of motion, statistical description, measurement of a certain parameter, etc. Depending on the method of description, different features of self-organization ("synergism") were observed, but in most situations a self-regulated parameter existed and the governing principle or target function could be identified. Table 3.1 summarizes the features of "synergism" in various tribological phenomena and the corresponding governing principles and target function, based on Bershadsky [8].

An important entropic study of the thermodynamics of wear was conducted by Bryant et al. [12], who introduced a degradation function and formulated the Degradation-Entropy Generation theorem in their approach intended to study the friction and wear in complex. They note that friction and wear, which are often treated as unrelated processes, are in fact manifestations of the same dissipative physical processes occurring at sliding interfaces. The possibility of the reduction of friction between two elastic bodies due to a pattern of propagating slip waves was investigated by Adams [3] and Nosonovsky and Adams [49], who used the approach of the theory of elasticity.

Nosonovsky [51, 52], Nosonovsky and Bhushan [55], and Nosonovsky et al. [57] suggested entropic criteria for friction-induced self-organization on the basis of the multiscale structure of the material (when self-organization at the macroscale occurs at the expense of the deterioration at the microscale) and coupling of the healing and degradation thermodynamic forces. Table 3.2 summarizes their interpretation of various tribological phenomena, which can be interpreted as self-organization. In addition, self-organization is often a consequence of coupling of friction and wear with other processes, which creates a feedback in the tribosystem.

In addition, self-organization is often a consequence of coupling of friction and wear with other processes, which create a feedback in the tribosystem. These "other processes" may include radiation, electricity, ultrasound, electric field etc. Following Haken and Prigogine, Bershadsky considered self-organization as a general property of matter, which is complimentary to wear and degradation. These ideas, while interesting from the philosophical point of view, caused the criticism of the synergetics as not being a sufficiently "scientific" field in terms of quantitative analysis. It took several decades until the investigation of spatial and temporal pattern formation during friction found a foundation in the thermodynamics and the theory of dynamical systems.

3.2.2 Entropy During Friction and Dissipation

Before going through quantitative investigation of self-organization phenomena, we will discuss entropic methods of the description of friction and wear. Consider a rigid body sliding upon a flat solid surface with the sliding velocity $V = dx/dt$ (Fig. 3.1). The normal load W is applied to the body and the friction force $F = \mu W$ is generated [55]. The work of the friction force is equal to the dissipated energy, and, therefore, we will assume for now that all dissipated energy is converted into heat

Table 3.1 Self-organization effects in tribosystems [52]

Effect	Description of the state or evolution	Features of synergism	Self-regulated parameter	Target function and/or governing principle
Auto-hydrodynamic effects (wedges, gaps, canyons)	Equations of motion, competing processes for entropy and negentropy production	Bifurcation; self-excited vibrations and waves; feedback and target functions	Gap thickness, temperature, and microtopography distributions	Minimum friction
Self-reproducing micro-topography, waviness	Equations of motion or kinetics	Bifurcations; self-excited vibrations and waves	Rough surface microtopography	Minimum energy dissipation; pressure or heat flow distribution
Steady state microtopography of worn surfaces ("natural wear shape")	Competing processes for entropy and negentropy (information) production	Feedback and target function	Shape of the profile	Minimum energy dissipation
Self-excited vibrations of wear, electric resistance, stresses, etc.	Measurements of a parameter of the system (friction force, electrical resistance, wear rate, etc.)	Instabilities and self-excited oscillations of the measured parameter	Corresponding parameter	Minimum entropy production
Spatial or periodic chemical pattern	Molecular, atomic, or dislocation structure	Large-scale ordered structures	Secondary heterogeneity at the surface	Dissipative principles
Periodic or concentric structures, such as Bénard cells	Molecular, atomic, or dislocation structure; Entropy is measured	Large-scale order structures; a sudden decrease in entropy production	–	Minimum entropy production
Decrease in macrofluctuation of temperature, particle size and other parameters	Order-parameter dependent on generalized coordinate, Measurements of a parameter of the system	Microfluctuations; phase transitions; instabilities and self-excited vibrations of the measured parameter	–	Sub-minimal friction

Table 3.2 Self-organization effects in tribosystems [51]

Effect	Mechanism/ driving force	Condition to initiate	Final configuration
Stationary microtopography distribution after running in	Feedback due to coupling of friction and wear	Wear affects microtopography until it reaches the stationary value	Minimum friction and wear at the stationary microtopography
In situ tribofilm formation	Chemical reaction leads to the film growth	Wear decreases with increasing film thickness	Minimum friction and wear at the stationary film thickness
Slip waves	Dynamic instability	Unstable sliding	Reduced friction
Self-lubrication	Embedded self-lubrication mechanism	Thermodynamic criteria	Reduced friction and wear
Surface-healing	Embedded self-healing mechanism	Proper coupling of degradation and healing	Reduced wear

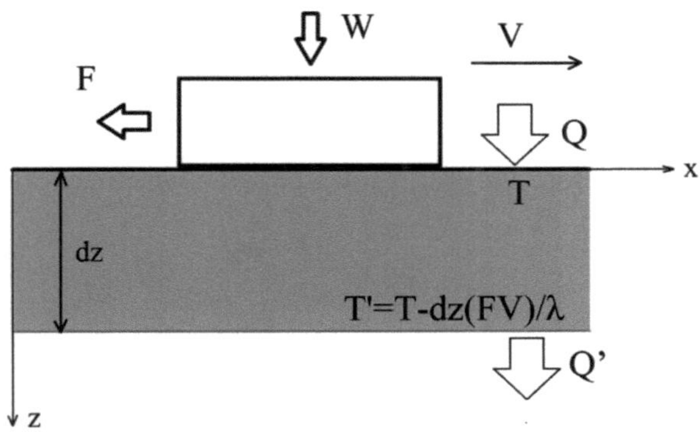

Fig. 3.1 Heat flow away from the frictional interface [51]

$$dQ = \mu W dx. \tag{3.1}$$

The rate of entropy generation during friction is given by

$$\frac{dS}{dt} = \frac{\mu WV}{T} \tag{3.2}$$

It is noted that friction is a non-equilibrium process. When a non-equilibrium process, which can be characterized by a parameter q (a so-called generalized coordinate), occurs, a generalized thermodynamic force X that drives the process can be introduced in such a manner that the work of the force is equal to $dQ = X dq$. The flux (or flow rate) $J = \dot{q}$ is associated with the generalized coordinate. For many linear processes the flow rate is linearly proportional to Y. For sliding friction, the flow rate $J = V$, and the thermodynamic force $X = \mu W/T$. Note that, for the Coulombian friction, J is not proportional to Y, which is the case for the viscous friction. Nosonovsky [52] discussed in detail the problem of bringing the linear friction in compliance with the linear thermodynamics.

The net entropy growth rate for frictional sliding of rigid bodies is given by Eq. 3.2. However, if instead of the net entropy, the entropy per surface area at the frictional interface is considered, the rate equation becomes more complicated. We consider the 1D flow of entropy near the infinite interface in the steady-state situation, and suppose the flow is equal to entropy generation. Heat dQ is generated at the interface in accordance with Eq. 3.1. For simplicity, we assume that all generated heat is dissipated in one of the two contacting bodies and ignore the division of heat between the two bodies. The heat is flown away from the interface in accordance with the heat conduction equation

$$\lambda \frac{\partial T}{\partial z} = \mu WV \tag{3.3}$$

where z is the vertical coordinate (distance from the interface), and λ is the heat conductivity. Consider a thin layer near the interface with the thickness dz. The temperature drop across the layer is $\mathrm{d}T = (\mu WV/\lambda)\mathrm{d}z$. The ratio of the heat released at the interface, dQ, to that radiated at the bottom of the layer, dQ', is equal to the ratio of the temperatures at the top and at the bottom of the layers

$$\frac{\mathrm{d}Q'}{\mathrm{d}Q} = \frac{T - \mu WV\mathrm{d}z/\lambda}{T} \tag{3.4}$$

Therefore, the energy released at the subsurface layer of depth dz is given by [55]

$$\mathrm{d}q = \frac{\mathrm{d}Q - \mathrm{d}Q'}{\mathrm{d}z} = \mathrm{d}Q\frac{\mu WV}{\lambda T} = \frac{(\mu WV)^2}{\lambda T} \tag{3.5}$$

Thus the entropy in the subsurface layer, dS/dt = dq/T, is given by

$$\frac{\mathrm{d}S}{\mathrm{d}t} = \frac{(\mu WV)^2}{\lambda T^2} \tag{3.6}$$

Note that S in Eq. 3.6 is entropy per unit surface area and thus it is measured in $\mathrm{JK^{-1}m^{-2}}$, unlike the total entropy Eq. 3.2, which is measured in $\mathrm{JK^{-1}}$ [55, 56].

The difference between Eqs. 3.2 and 3.6 is that the latter takes into account the thermal conductivity and that in Eq. 3.2 gives the net entropy rate, while Eq. 3.6 gives the rate of entropy in the subsurface layer. Note that the form for the thermodynamic flow is now $J = \mu WV$, and the thermodynamic force is $X = \mu WV/(\lambda T^2)$.

3.2.3 Thermally Activated Self-Organization

With describing the total entropy rates and surface entropy rate during friction (i.e. Eqs. 3.2 and 3.6), their application to the tribosystems will be discussed in this section. Frictional sliding and wear are irreversible processes, since they are inhomogeneous and often non-stationary. The transition from the steady-state (stationary) sliding regime to the regime with self-organized structures occurs through the destabilization of the steady-state regime. At the steady state, the rate of entropy production is at minimum. The stability condition for the thermodynamic system is given in the variational form by

$$\frac{1}{2}\delta^2\dot{S} = \sum_k \delta X_k \delta J_k \geq 0 \tag{3.7}$$

where $\delta^2\dot{S}$ is the second variation of entropy production rate [55] and k is the number of the generalized forces and flows. Equation 3.7 states that the energy dissipation per unit time at the steady state should be at its minimum or the variations of the flow and the force should be of the same sign. Otherwise, the steady-state regime becomes unstable and the transition to the self-organized regime with patterns can occur. Equation 3.7 is valid for a wide range of interactions, including mechanical, thermal, and chemical, however, the corresponding terms in the entropy production rate should be considered. When Eq. 3.7 is not satisfied, the system is driven away from the equilibrium, which creates the possibility for self-organization.

In the situation when only mechanical interactions are significant, and the change of temperature T has a negligible effect on friction, the entropy is proportional to the dissipated energy divided by temperature $dS = dQ/T$. Consider first the situation when the production of entropy depends on the sliding velocity V. Considering that the rate of entropy production is given by Eq. 3.2, the stability condition (Eq. 3.7) now yields

$$\frac{1}{2}\delta^2\dot{S} = \delta\left(\frac{\mu W}{T}\right)\delta(V) = \frac{W}{T}\frac{\partial\mu}{\partial V}(\delta V)^2 \geq 0 \tag{3.8}$$

If the slope of the $\mu(V)$ curve (the partial derivative $\mu'_V \equiv \frac{\partial\mu}{\partial V}$) is negative, then the steady-state sliding becomes unstable. And understandably so, since decreasing friction leads to increasing sliding velocity and to further increasing friction, and thus to the positive feedback loop.

Suppose that one contacting material has microstructure characterized by a certain parameter ψ, such as, for example, the size of reinforcement particles in a composite material. Such values of ψ that $\mu'_V(\psi) > 0$ correspond to steady-state sliding. However, $\mu'_V(\psi) = 0$ corresponds to the destabilization of the steady-state solution. As a result, new equilibrium position will be found with a lower value of μ. Suppose now that the coefficient of friction depends also on a microstructure parameter ϕ, such as the thickness of the interface film (Fig. 3.2). The difference between ψ and ϕ is that the parameter ψ is constant (the composition of the material does not change during the friction), whereas the parameter ϕ can change during friction (the film can grow or decrease due to a friction-induced chemical reaction or wear). The stability condition is now given by

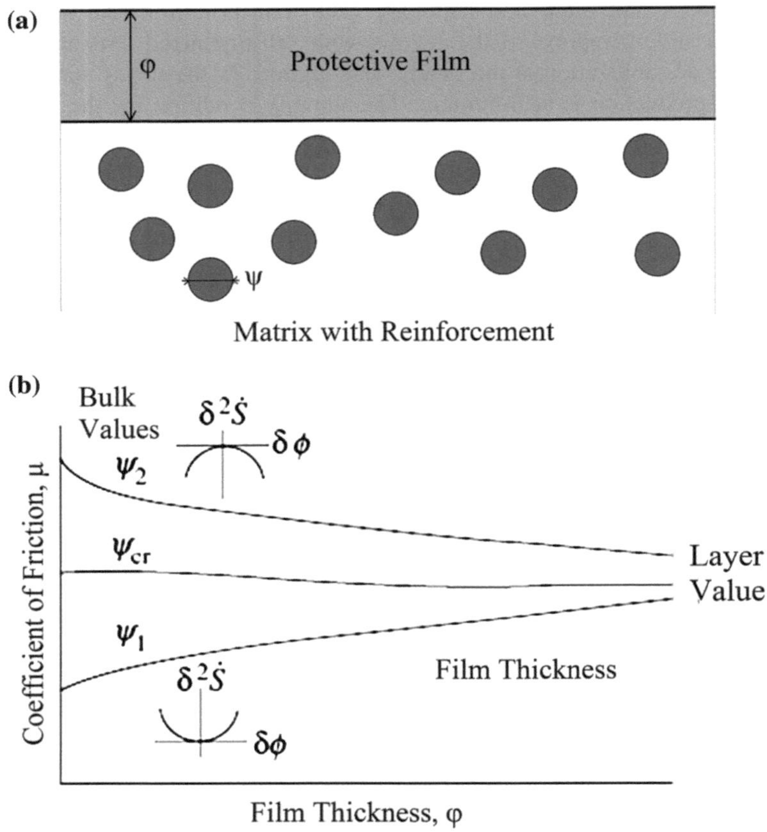

Fig. 3.2 **a** Self-organized protective film at the interface of a composite material. **b** The coefficient of friction as a function of film thickness for various values of the microstructure parameter ψ. Sub-critical values of $\psi < \psi_{cr}$ result in the positive slope (no layer formed), whereas $\psi > \psi_{cr}$ results in the instability and self-organization of the protective layer. The slope depends on the ratio of the bulk and layer values of μ, which allows to find composite microstructure providing the self-organization of the layer [51]

$$\frac{1}{2}\delta^2\dot{S} = \delta\left(\frac{\mu W}{T}\right)\delta(V) = \frac{W}{T}\frac{\partial\mu}{\partial\phi}\frac{\partial V}{\partial\phi}(\delta\phi)^2 \geq 0 \tag{3.9}$$

If the stability condition is violated for a certain value of ϕ, then further growth of the film will result in decreasing friction and wear, which will facilitate the further growth of the film. The destabilization occurs at $\mu'_\varphi(\psi, \phi) = 0$. Note that Eq. 3.9 becomes Eq. 3.8 if $\phi = V$. At this point, we are not discussing the question of which particular thermodynamic force is responsible for the growth of the film.

Since we are interested in the conditions of the formation of such a protective film, consider now the limit of the thin film ($\phi \to 0$). With increasing film

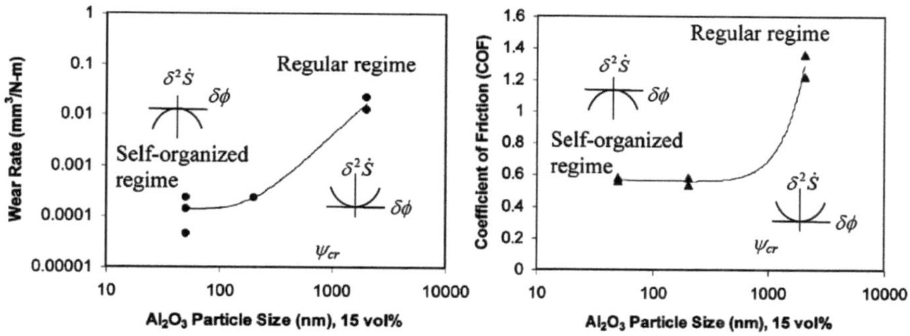

Fig. 3.3 A significant wear and friction reduction with decreasing particle size in Al-Al_2O_3 nanocomposite (based on [33] can be attributed to surface self-organization [51])

thickness the value of μ changes from that of the bulk composite material to that of the film material. On the other hand, the value for the bulk composite material depends also on its microstructure ψ (Fig. 3.2). The critical value, ψ_{cr}, corresponds to $\mu'_{\phi}(\psi, 0) = 0$. For the size of reinforcement particles finer than ψ_{cr}, the bulk (no film, $\phi = 0$) values of the coefficient of friction are lower than the values of the film. That can lead to a sudden destabilization (formation of the film with thickness ϕ_0) and reduction of friction to the value of $\mu(\psi, \phi_0)$ as well as wear reduction. Here, we do not investigate the question of why the film would form and how its material is related to the material of the contacting bodies. However, it is known that such a reaction occurs in a number of situations when a soft phase is present in a hard matrix, including Al-Sn and Cu-Sn-based alloys [14].

An experimental example of such sudden decrease of friction and wear with a gradual decrease of the size of reinforcement particles, which could be attributed to the destabilization, is presented in Fig. 3.3 for Al_2O_3 reinforced Al matrix nanocomposite friction and wear tests (steel ball-on-disk in ambient air) based on Jun et al. [33]. The abrupt decrease of friction and wear occurs for reinforcement particles smaller than $\psi_{cr} = 1$ μm in size and can be attributed to the changing sign of the derivative $\mu'_{\phi}(\psi_{cr}, 0) = 0$. The decrease is sudden and dramatic, so it can be explained by the loss of stability (cf. Eq. 3.7) rather than by a gradual change of properties; although additional study is required to prove it.

For the entropy production governed by Eq. 3.6, the stability condition of Eq. 3.7 yields

$$\frac{\partial}{2\partial t}(\delta^2 S) = \frac{1}{2}\delta^2\left(\frac{(\mu WV)^2}{\lambda T^2}\right) = \delta X \delta J = \delta(\mu WV)\delta\left(\frac{\mu WV}{\lambda T^2}\right) \geq 0 \tag{3.10}$$

The coefficient of friction and the thermal conductivity depend upon material's microstructure, ϕ, so that

$$\mu = \mu(\psi, \phi) \tag{3.11}$$

$$\lambda = \lambda(\psi, \phi)$$

The stability condition given by Eq. 3.20 takes the form of

$$\frac{1}{2}\delta^2 \dot{S} = \frac{V^2 W^2}{T^2}\frac{\partial \mu}{\partial \phi}\left(\frac{1}{\lambda}\frac{\partial \mu}{\partial \phi} - \frac{\mu}{\lambda^2}\frac{\partial \lambda}{\partial \phi}\right)(\delta\phi)^2 \geq 0 \tag{3.12}$$

The stability condition can be violated if

$$\frac{\partial \mu}{\partial \phi}\frac{\partial \lambda}{\partial \phi} < 0 \tag{3.13}$$

It is known from non-equilibrium thermodynamics that when the secondary structure is formed, the rate of entropy production reduces [26]. Therefore, if Eq. 3.13 is satisfied, the frictional force and wear can reduce. By selecting appropriate values of ψ (e.g., the density of a micro pattern), the condition of Eq. 3.13 can be satisfied. Note, that the wear rate is related to the rate of surface entropy production

$$\frac{\mathrm{d}w}{\mathrm{d}t} = B\frac{\mathrm{d}S}{\mathrm{d}t} = YJ \tag{3.14}$$

It is suggested to use the theory presented in this section to optimize the microstructure of a composite material in order to ensure that the self-organized regime occurs. To that end, the dependencies Eq. 3.11 should be investigated experimentally and their derivatives obtained. Following that, the value of ψ should be selected, which provides the best chances for the transition to the self-organized regime [51].

A particular field where this approach has been applied is the electrical contact, involving the current collection (e.g., for a railroad locomotive [29]). It is noted that when Tribology, which is today considered the science and technology of friction, wear, and lubrication, emerged in the 1960s, it was meant to include the fourth component, namely, the electrical contact. The electrical current is important in many applications involving the mechanical contact, such as the microelectromechanical systems (MEMS). Besides that, electromechanical contact is a typical example of a system, where two coupled processes (friction and electrical current) take place simultaneously which creates a potential for feedback, destabilization, and pattern formation. Frictional self-organization during electromechanical contact was investigated by Gershman [30]. Other areas where this approach has been successfully applied include cutting tools and the theory of "tribological compatibility" of materials [14].

3.2.4 The Concept of "Selective Transfer"

The concept of the "selective transfer" or the "non-deterioration effect" was developed by Garkunov [27]. According to his definition, the selective transfer is a type of friction which is characterized by the formation at the contact interface of a thin non-oxidized metallic film with a low shear resistance which is not able to accumulate dislocations. According to the selective transfer scheme, a selective dissolution of a component of a copper alloy occurs, followed by the transfer of the component to a contacting body (a steel shaft). The standard example of the selective transfer is the formation of a copper layer in a bronze–steel system lubricated by glycerin.

Kragelsky and Garkunov studied in the 1950s the state of airplane chassis which included a bronze–steel lubricated frictional system and found that a protective copper film can form, which reduces the wear to very small values. The copper film is formed due to the anodic dissolution of bronze (an alloy of copper and tin with additive elements). The additives, such as iron, zinc, and aluminum, as well as tin, dissolved in the lubricant, while copper forms a film on the surfaces of the contacting materials. The film is in a dynamic equilibrium, while contacting layers are worn and destroyed, new layers of copper are formed, resulting in virtual absence of wear and the friction force reduction by an order of magnitude. A similar effect can be achieved by the diffusion of copper ion dissolved in a lubricant. The authors called the effect the "selective transfer" of copper ions, and the protective effect "servovite films" (or "serfing-films") [28], although a more common term in the modern literature is "in situ formation of self-lubricating films."

While the "non-deterioration" effect was presented as an unique and universal type of self-organization in tribosystems and as a very promising way of wear reduction, the concept also caused criticism of scientists who did not see any generality in the suggested effect. For example, Bershadsky [8] wrote: "Selective transfer shows neither new mechanisms, nor new phenomena that change our concepts. The concept of self-organization is used here in an artificial manner without any convincing proof."

3.2.5 The Concept of "Tribofatigue"

The concept of tribofatigue (or "tribofatika") was suggested and developed since the early 1990s by Dr. L. Sosnovsky from Gomel, Byelorussia. The concept implies the coupling between the wear damage and fatigue damage. It stresses the fact that in engineering calculations of lifetime of components it may not be sufficient to consider the two modes separately. Sosnovsky and Sherbakov [64] provide several examples when "the normative documents and methods of calculation" could not predict a catastrophic failure in a machine, such as a gigantic turbine of the 1.2 GW power plant, which he calls a ticking "tribofatigue bomb." Apparently, the phenomenon is a manifestation (or even just another mane) of the fretting fatigue.

The concepts of entropy and information play a certain role in the "tribofatogue" analysis. For the tribofatigue damage, ω_Σ, the change of energy in the volume Wp, where degradation occurs, the effective energy and the tribofatigue entropy production are given by

$$dU_\Sigma = \gamma_1 \omega_\Sigma dW_p \tag{3.15}$$

$$dS_{TF} = \frac{\gamma_1 \omega_\Sigma}{T} dW_p$$

where γ_1 is a coefficient characterizing the proportionality.

Sosnovsky and Sherbakov [64] further consider the rate of change of the damaged volume dW_p/dt as the generalized flux and $\gamma_1\omega_\Sigma/T$ as the corresponding generalized force. The authors claim that, on the basis of Prigogine's approach, the state of thermodynamic equilibrium is characterized either by the minimum entropy generation (in the self-organized state) or by the maximum entropy. They also distinguish several states of objects depending on the value of the tribofatigue damage parameter: (a) $\omega_\Sigma = 0$, (undamaged), (b) $0 < \omega_\Sigma < 1$ (damaged) (c) $\omega_\Sigma = 1$ (critical), (d) $1 < \omega_\Sigma < \infty$ (supercritical), (e) $\omega_\Sigma = \infty$ (decomposition). Further investigation of the structure–property relationship on the basis of this model can be promising.

3.3 Friction-Induced Self-Organization

The entropic methods are especially promising for the analysis of friction-induced self-organization, especially self-lubrication, surface-healing, and self-cleaning. According to the preliminary studies, there are two key features in these processes. First, the hierarchical (multiscale) organization of the material, which allows implementing healing (as observed at the macroscale) at the expense of microscale deterioration. Second, the presence of positive and/or negative feedbacks that lead to dynamic destabilization and friction-induced vibrations. Hierarchical organization is the key feature of biological systems, which have capacity for self-organization, self-healing, and self-lubrication, so it is not surprising that hierarchy plays a central role also in artificial materials capable of self-organization. In most cases, the self-organization is induced by bringing the system out of thermodynamic equilibrium, so that the restoring force is coupled with the degradation in such a manner that healing (decreasing degradation) occurs during the return to the thermodynamic equilibrium. In this section, we will discuss the phenomenological observations of friction-induced self-organization, and will review several examples.

3.3.1 Running-In

When sliding between two contacting, solid bodies is initiated, friction and wear often remain high during a certain initial transient period, referred to as the running-in period [11]. After running-in, friction and wear tend to decrease to their steady-state values. Numerous phenomenological explanations of the running-in phenomenon deal with various mechanisms, such as creep and viscoplastic deformation of asperities. However, a general thermodynamic description of the running-in, which would explain its underlying physical principles, is missing.

During running-in, the surface topography evolves until it reaches a certain stationary state referred to as the equilibrium roughness distribution [11]. The surfaces adjust to each other, leading to a more ordered state, resulting in lower energy dissipation rates, demonstrating friction-induced self-organization [25]. Friction and wear are two sides of the same irreversible process leading to energy dissipation and material deterioration. It is therefore natural to use entropy as a measure of irreversibility to describe them.

Consider a two-dimensional surface height profile $y_n = y(x_n)$, (where $n = 1, N$) measured with some resolution so that every value of y_n belongs to a certain "bin". The Shannon entropy S of a rough surface profile is a measure of its randomness [24], where

$$S = -\sum_{j=1}^{B} p_j \ln(p_j) \tag{3.16}$$

and $p_j = N_j/N$ is the probability of appearance of some height in the bin j, N_j is the number of appearances in the bin j, and B is the total number of bins. A surface profile having a lower Shannon entropy is considered to be "more ordered" (or "less random") than one with a higher Shannon entropy. Thus, a decrease in Shannon entropy during the transient process is an indication of self-organization.

Note that the profile entropy as a measure of randomness is not the same as the surface roughness in a more traditional sense, such as the root-mean-square (RMS) roughness. Two profiles can have the same degree of randomness at different values of the RMS roughness and conversely, different RMS roughness for the same level of randomness.

Two types of events can occur during the evolution of the surface profile: (1) a smoothening event, which is typical of deformation-driven friction and wear; high asperities fracture due to deformation (abrasive wear) resulting in a smoothening of the profile; and (2) a roughening event, which is typical for the adhesive-driven friction and wear, which is higher for smoother surfaces.

When a particle is removed from the interface due to abrasive or adhesive wear, a data point $y(x)$ is driven from a higher bin j into a lower bin i, so the Shannon entropy would change by:

$$\Delta S = \frac{N_j - 1}{N} \ln \frac{N_j - 1}{N} - \frac{N_j}{N} \ln \frac{N_j}{N} + \frac{N_{j-1} + 1}{N} \ln \frac{N_{j-1} + 1}{N} - \frac{N_{j-1}}{N} \ln \frac{N_{j-1}}{N} \tag{3.17}$$

Using $\ln(x + \Delta x) - \ln(x) \approx \Delta x/x$ yields

$$\Delta S = \frac{1}{N} \ln \frac{N_i}{N_j} \tag{3.18}$$

Smoothening occurs when a particle moves from a less populated bin j into a more populated bin i. Here the change in entropy is negative, whereas in the opposite case of roughening, the change of entropy change is positive. Furthermore, for a very rough surface, smoothening is significant whereas for a very smooth surface roughening is significant, so that a certain value of equilibrium roughness can exist, corresponding to $S = S_0$. Therefore, the rate of entropy change is given by

$$\dot{S} = f(S) = f_S(S) + f_R(S) \tag{3.19}$$

where $f_S(S)$, designating the rate due to smoothening, is a negative monotonically decreasing function such that $f_S(0) = 0$, and $f_R(S)$, designating the rate due to roughening, is a positive monotonically decreasing function. The value S_0 corresponds to the stationary state at $f(S_0) = 0$, which is stable if the derivative is negative, $f'(S_0) < 0$.

Sliding friction is associated with a number of phenomena. The intensity of some of these phenomena (such as the deformation of asperities) grows with roughness, whereas the intensity of others (such as adhesion) decreases, so that the coefficient of friction μ is given by $\mu = \mu_{\text{def}}(S) + \mu_{\text{adh}}(S)$, where $\mu_{\text{def}}(S)$ is a positive increasing function and $\mu_{\text{adh}}(S)$ is a positive decreasing function. The minimum value of $\mu = \mu(S_m)$ is obtained from $\mu'_{\text{def}}(S_m) + \mu'_{\text{adh}}(S_m) = 0$. The coefficient of friction will always decrease with decreasing roughness only if $S_0 = S_m$. This occurs when $f_S(S) \propto \mu_{\text{def}}(S)$ and $f_R|S| \propto \mu_{\text{adh}}(S)$. Physically, such proportionality implies that the mechanisms affecting surface roughness (e.g., abrasive and adhesive wear) are proportional to friction mechanisms (e.g., asperity deformation and adhesion).

To investigate these observations, Mortazavi and Nosonovsky [48] conducted an experiment. The coefficient of friction between a tungsten carbide (WC) pin sliding over a copper (Cu) substrate was measured experimentally in an ultrahigh vacuum (UHV) tribometer, operating at the pressure $\sim 3.5 \times 10^{-10}$ Torr. A pure 99.99% Cu sample (about 20 × 10 × 1 mm) was vacuum cleaned, and mounted to a sample manipulator, which was oriented horizontally and on the opposite side of the chamber to the tribometer. A WC tribopin, mounted to the end of an arm, was brought into contact with the sample and rubbed its surface during each scan. The tribometer was under computer control, so that the normal loads, scan speed, scan area, and scan pattern could be selected. More details of the apparatus are presented in Ref. [66].

The aim was to produce wear scars after different number of passes of the tribopin over the surface, each one representing different periods of the transient running-in process. Measurements were made using a single pass in the same direction with the sliding speed of 4 mm/s using a normal load of 0.9 N. The experiment was conducted at room temperature, and five different wear scars were created, each with a length of 4 mm collected for different rubbing times. The experiment was carried out by increasing the number of scans of the tribopin on the surface of sample from 1, 4, 8, 16, and 32 scans. Two series of wear scars were produced.

The coefficient of friction during the running-in process was calculated as the ratio of the lateral to the normal forces, and is plotted in Fig. 3.4. The coefficient of

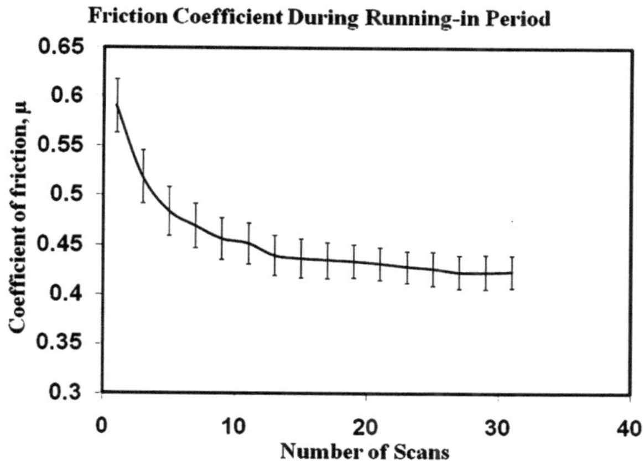

Fig. 3.4 Evolution of the coefficient of friction μ plotted as a function of the number of scans during the running-in process [48]

friction decreases monotonically during run in until it reached a steady-state value of $\mu = 0.43 \pm 0.02$.

Images of surface topography of the sample at different positions for each of the wear tracks were obtained using a Pacific Nanotechnology Nano-R^{TM} Atomic Force Microscopy (AFM) operating in contact mode. The images were collected using a conical Si cantilever coated with Si_3N_4 with the tip radius of about 10 nm. Scans (512×215 pixels) were collected at a minimum of three different locations on each wear track. The images were processed using a standard software package (SPIP from Image Metrology) to obtain statistical roughness parameters such as the RMS, Skewness (SK), and Kurtosis (K), as shown in Fig. 3.5.

Both the RMS roughness and surface SK decreased during the running-in process since the surfaces were smoothened steadily during the running-in period until an equilibrium roughness was achieved. However, the K increases from $K = 4.39 \pm 0.05$ to 6.52 ± 0.05 as the number of scans increases. Increasing K implies an increasing degree of pointedness of surface profile.

To compute the Shannon entropy during running-in stage we proceeded as follows. We obtained the height probability distribution for each location using the above-mentioned software. We took a fixed range of heights in this profile and divided this interval into B bins. We numerically found that for number of bins more than 1,000 the amount of error in calculation of Shannon entropy was less than 1%, thus we subsequently chose B = 1,000. Then using Eq. 3.16, the results were plotted as shown in Fig. 3.6. It is observed that the entropy tended to decrease indicating that self-organization had occurred during the run-in period.

This indicates that self-organization occurred during the transient running-in process due to mutual adjustment of surface roughness topography. The corresponding decrease in the friction force can be explained by postulating that the same mechanisms (deformation and adhesion) affect both the roughness evolution and the change in friction, which therefore have stationary values at close values of roughness.

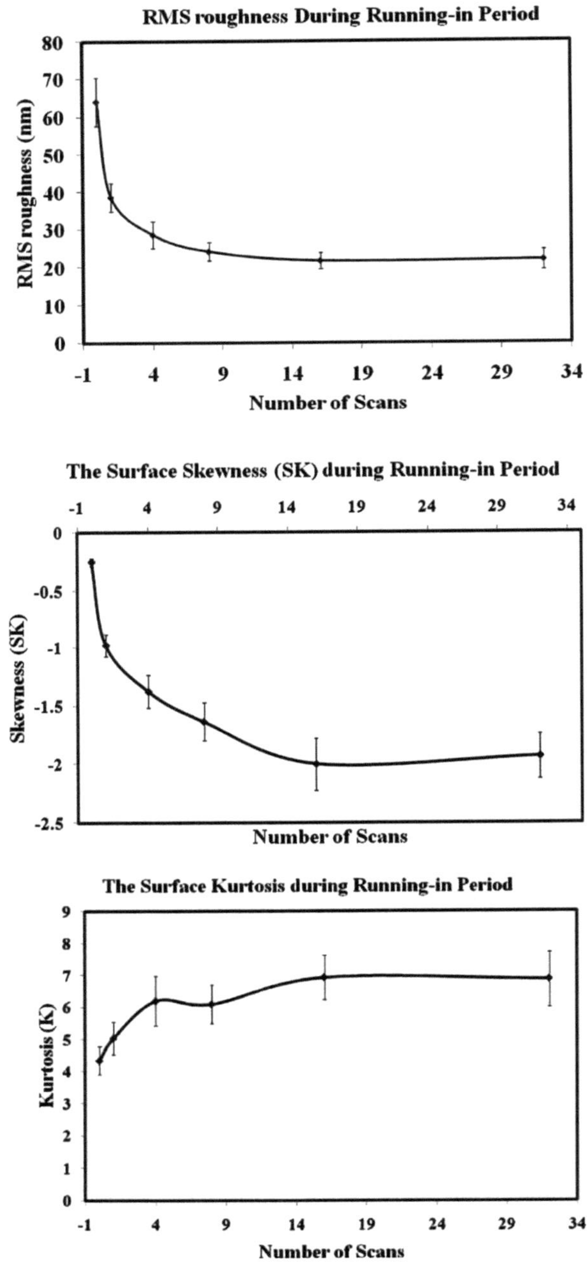

Fig. 3.5 Evolution of **a** the root-mean square roughness, **b** SK and, **c** K measured within the wear track as a function of the number of scans during running-in

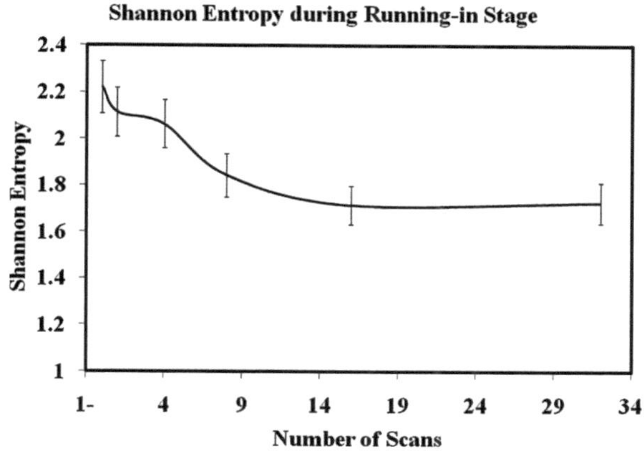

Fig. 3.6 Evolution of Shannon entropy during running-in [48]

3.3.2 Self-Organized Elastic Structures

In the preceding sections we discussed the self-organization of interface films and micro topography evolution. Another type of self-organized microstructures is elastic slip waves. The mathematical formulation of quasi-static sliding of two elastic bodies (half-spaces) with a flat surface and a frictional interface is a classical contact mechanics problem. Interestingly, the stability of such sliding was not investigated until the 1990s, when, separately, Adams [2] and Martins [44] showed that the steady sliding of two elastic half-spaces is dynamically unstable, even at low sliding speeds. Steady-state sliding was shown to give rise to a dynamic instability [Adams-Matrins instabilities (AMI)] in the form of self-excited oscillations with exponentially growing amplitudes. These oscillations confine to a region near the sliding interface and can eventually lead to either partial loss of contact or to propagating regions of stick–slip motion (slip waves). The existence of AMI instabilities depends upon the elastic properties of the surfaces, however, it does not depend upon the friction coefficient, nor does it require a nonlinear contact model. The same effect was predicted theoretically by Nosonovsky and Adams [54] for the contact of rough periodic elastic surfaces.

It is well known that two types of elastic waves can propagate in an elastic medium: the shear and dilatational waves. In addition, surface elastic waves (Rayleigh waves) may exist, and their amplitude decreases exponentially with the distance from the surface. For two slightly dissimilar elastic materials in contact, the generalized Rayleigh waves (GRW) may exist at the interface zone. The instability mechanism described above is essentially one of GRW destabilization, that is, when friction is introduced, the amplitude of the GRW is not constant anymore, but exponentially grows with time.

The stability analysis involves the following scheme. First, a steady-state solution should be obtained. Second, a small arbitrary perturbation of the steady state solution is considered. Third, the small arbitrary perturbation is presented as a superposition of modes, which correspond to certain eigenvalues (frequencies). Fourth, the equations of the elasticity (Navier equations) with the boundary conditions are formulated for the modes, and solved for the eigenvalues. Positive real parts of the eigenvalues show that the solution is unstable.

For the GRW, the 2D displacement field at the interface is given by

$$\begin{aligned} u &= \sum_k u_k(x, y, t), \\ v &= \sum_k v_k(x, y, t) \\ u_k(x, 0, t) &= \mathrm{Re}\left[A_k e^{ikx/l} e^{\Lambda t}\right], \\ v_k(x, 0, t) &= \mathrm{Re}\left[B_k e^{ikx/l} e^{\Lambda t}\right] \end{aligned} \tag{3.20}$$

where k stands for the x- or y-component of the displacement field at the interface ($y = 0$), A_k is the complex amplitude, k/l is the wavenumber, l is the wavelength of the lowest wavenumber, and Λ is the complex frequency. It can be shown that for the frictionless case $\Lambda = \pm\, i\lambda$ is purely imaginary and thus, for real A_k, the displacement is a propagating generalized Rayleigh wave, $u_k(x, 0, t) = A_k \cos(kx \pm \lambda t)$. It can be shown also that if small friction with the coefficient μ is present, then $\Lambda = \pm(i\lambda + \alpha\mu)$, where α is a real number, and thus one root of Λ always has a positive real component, leading to the instability [2, 61]. As a result, the amplitude of the interface waves grows with time. In a real system, of course, the growth is limited by the limits of applicability of the linear elasticity and linear vibration theory. This type of friction-induced vibration may be, at least partially, responsible for noise (such as car brake squeal) and other effects during friction, which are often undesirable [54].

Whereas the GRW occurs for slightly dissimilar (in the sense of their elastics properties) materials, for very dissimilar materials, waves would be radiated along the interfaces, providing a different mechanism of pumping the energy away from the interface [49]. These waves can form a rectangular train of slip pulses propagates, two bodies shift relative to each other in a “caterpillar” or “carpet-like” motion (Fig. 3.7). This microslip can lead to a significant reduction of the observed coefficient of friction, as the slip is initiated at a shear load much smaller than μW [10, 53].

The motion is observed as the reduction of the coefficient of friction (in comparison with the physical coefficient of friction μ) to the apparent value of $\mu_{\mathrm{app}} = q/p$, where q is applied shear force per unit area and p is the normal pressure. The slip pulses can be treated as “secondary structures” self-organized at the interface, which result in the reduction of the observed coefficient of friction. Note that the analysis in this case remains linear and it just shows that the equations of elasticity with friction are consistent with the existence of such waves. The amplitude of the slip waves cannot be determined from this analysis, since they are dependent on the initial and boundary conditions. In order to investigate whether the slip waves will actually occur, it is important to ask the question whether it is energetically profitable for them to exist. To that end, the

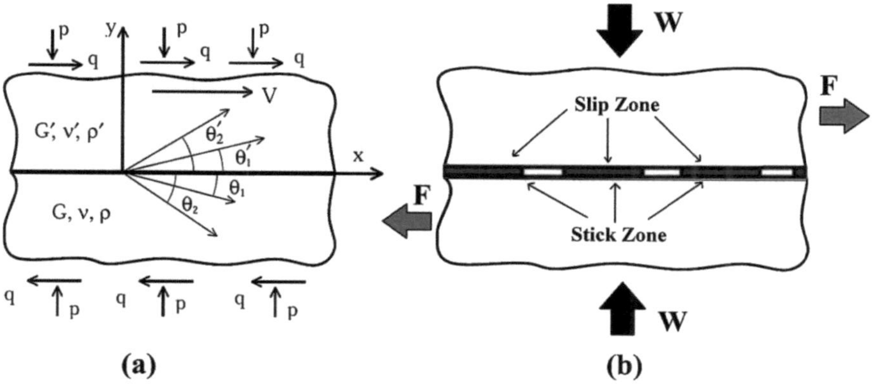

Fig. 3.7 **a** Elastic waves radiated from the frictional interface. **b** Friction reduction due to propagating stick–slip zones [49]

energy balance should be calculated of the work of the friction force and the energy dissipated at the interface and radiated away from the interface. A stability criterion based on Eq. 3.8 can be used.

3.3.3 SOC and Avalanche Dynamics

The methods in the preceding sections of this paper describe the onset of the instability, which can lead to the self-organization of spatial and temporal structures. The linear stability analysis is independent of the amplitudes of small vibrations and in the case of the instability it predicts an exponential growth of the amplitudes until they become so large that the linear analysis cannot apply anymore. It is much more difficult to perform the nonlinear analysis to find the amplitudes and actual finite motion. Several approaches have been suggested.

One is that a specific type of self-organization, known as SOC plays a role in frictional systems. SOC is a concept in the theory of dynamic systems that was introduced in the 1980s [6]. The best studied example of SOC is the "sandpile model," representing grains of sand randomly placed into a pile until the slope exceeds a threshold value, transferring sand into the adjacent sites and increasing their slope in turn. There are typical external signs of a SOC system, such as the power-law behavior (the magnitude distribution of the avalanches) and the "one-over-frequency" noise distribution. The concept has been applied to such diverse fields as physics, cellular automata theory, biology, economics, sociology, linguistics, and others [6].

In the case of dry frictional sliding, it has been suggested that a transition between the stick and slip phases during dry friction may be associated with the SOC, since the slip is triggered in a similar manner to the sandpile avalanches and earthquake slides. Zypman et al. [69] showed that in a traditional pin-on-disk experiment, the probability distribution of slip zone sizes follows the power law. In a later work, the same group found nanoscale SOC-like behavior during AFM studies of at least some materials [13, 69]. Thus "stick" and "slip" are two phases, and the system tends to achieve the critical state between them: in the stick state elastic energy is accumulated until slip is initiated, whereas energy release during slip leads, again, to the stick state. The entropic methods of analysis of friction of the systems with SOC were used by Zypman and co-workers [24], who calculated Shannon entropy of the surface profile and showed that the entropy of the profile decreased indicating self-organization.

3.3.4 Pattern Formation and Turing Systems

As discussed in the previous sections, under certain circumstances friction can also result in self-organization or formation of patterns and structures at the frictional interface. These self-organized patterns or "secondary structures" [25] can include a broad range of phenomena, such as in-situ formed tribofilms [5], patterns of surface topography, and other interfacial patterns including propagating trains of stick and slip zones formed due to dynamic sliding instabilities [39, 49]. Moreover, experimental observations demonstrate various structures formed by friction and wear, such as the Schallamach [42] and slip waves [69], wave-like topography patterns [46], honeycomb-like and other structures [40].

Frictional sliding generates heat, and heat tends to lead to the thermal expansion of materials. This can result in so-called thermoelastic instabilities (TEI), which have been studied extensively in the past 30 years [7]. The effect of TEI on sliding is similar to that of AMI discussed in Sect. 3.3.2. Besides heat generation, friction can be coupled with other effects, such as wear, which can lead to similar instabilities. In addition, the coefficient of friction can depend upon the sliding velocity, load, and interface temperature, which can also lead to complex dynamic behavior [52].

The so-called RD systems and their important class called the "Turing systems" constitute a different type of self-organization mechanism [33, 47]. These systems can describe certain types of friction-induced pattern formation involving heat transfer and diffusion-like mass transfer due to wear. While AMI and TEI involve wave propagation (hyperbolic) partial differential equations (PDE), which describes dynamic behavior of elastic media, the RD system involves parabolic PDE, typical for diffusion and heat propagation problems. The RD systems describe evolution of concentrations of reagents in space and time due to local chemical reactions and the diffusion of the product of reactions [39]. The RD system of PDEs is given by

$$\frac{\partial w_j}{\partial t} = f_i(w_j) + d_{ij}\Delta w_j \tag{3.21}$$

where w_j is the vector of concentrations, f_i represent the reaction kinetics and d_{ij} and Δ are a diagonal matrix of diffusion coefficients and Laplace operator, respectively. Alan Turing [65] showed that a state that is stable in the local system can become unstable in the presence of diffusion, which is counterintuitive, since diffusion is commonly associated with a stabilizing effect. The solutions of RD equations demonstrate a wide range of behaviors including the traveling waves and self-organized patterns, such as stripes, hexagons, spirals, etc.

While parabolic RD equations cannot describe elastic deformation, they may be appropriate for other processes, such as viscoplastic deformation and interface film growth. For a system of two components, u and v, Eq. 3.21 has the following form:

$$\frac{\partial u}{\partial t} = f(u, v) + d_u \left(\frac{\partial^2 u}{\partial x^2} + \frac{\partial^2 u}{\partial y^2} \right) \tag{3.22}$$

$$\frac{\partial v}{\partial t} = g(u, v) + d_v \left(\frac{\partial^2 v}{\partial x^2} + \frac{\partial^2 v}{\partial y^2} \right) \tag{3.23}$$

where f and g are the reaction kinetics functions and d_{ij} is a diagonal matrix $(d_{11} = d_u$ and $d_{22} = d_v)$. Suppose u represents the non-dimensional temperature at the sliding interface and v is the local slip velocity, also non-dimensional. The non-dimensional values of u, v, x, t, and other parameters are obtained from the dimensional values by division of the latter by the corresponding scale parameters. Then Eq. 3.22 is interpreted as the description of heat transfer along the interface, and Eq. 3.23 describes the flow of viscous material along the interface. In this paper, we discuss several types of kinetic functions which can lead to the formation of periodic patterns.

One of the standard forms for functions $f(u, v)$ and $g(u, v)$ based on transferring original RD equations and proper scaling was proposed to be [19, 20]:

$$f(u, v) = \gamma(a - u + u^2 v) \tag{3.24}$$

$$g(u, v) = \gamma(b - u^2 v) \tag{3.25}$$

In order to investigate the possibility of pattern formation in a Turing system, a stability analysis should be performed [47]. A RD system exhibits diffusion-driven instability or Turing instability if the homogeneous steady state is stable to small perturbations in the absence of diffusion; however, it is unstable to small spatial perturbations when diffusion is introduced [49]. Diffusion results in the spatially inhomogeneous instability and determines the spatial patterns that evolve.

From the linear stability analysis one can show when a solution with pattern formation (Turing pattern) can exist [19, 47, 49]. The results of the stability analysis shown depend on the parameters of Eqs. 3.22 and 3.23 (i.e., d_v, d_u, a, b and γ); three regions of stability could be identified. In region 1, the steady state is stable to any perturbation. In region 2, the steady state exhibits an oscillating instability. In region 3, the steady state is destabilized by inhomogeneous perturbations, which is "Turing space".

In this section we discuss several exemplary cases of parameter values and initial conditions which can lead to patterns. Cases 1 and 2 are the classical examples of Turing systems, whereas case 3 is justified by the frictional mechanism of heat generation and mass transfer.

Examples of Turing patterns for two different cases with the following values of the parameters in Eqs. 3.22 and 3.23 are shown in Figs. 3.8 and 3.9. For case 1:

$$\gamma = 10{,}000,\ d_v = 20,\ d_u = 1,\ a = 0.02 \text{ and } b = 1.77 \tag{3.26}$$

and considering a random distribution function as the initial conditions for both values of u and v. For case 2:

$$\gamma = 10{,}000,\ d_v = 20,\ d_u = 1,\ a = 0.07 \text{ and } b = 1.61 \tag{3.27}$$

For initial conditions, we considered the following harmonic functions

$$u(x,y) = 0.919145 + 0.0016\cos(2\pi(x+y)) \tag{3.28}$$

$$v(x,y) = 0.937903 + 0.0016\cos(2\pi(x+y)) \tag{3.29}$$

The results in Figs. 3.8 and 3.9 present five consecutive snapshots of systems presented by Eqs. 3.22 and 3.23 corresponding to different time steps (t = 0, 0.001, 0.005, 0.01, and 0.1). It is observed in both cases that the initially random pattern ($t = 0$) evolves finally, into a so-called hexagonal-like (or honeycomb) pattern ($t = 0.1$), which indicates that the pattern formation occurs.

Whereas the two first cases showed that the model is capable of capturing the self-organized patterns in Turing systems, in the third case Mortazavi and Nosonovsky [47] tried to use more specific functions of reaction kinetics, which were expected to characterize friction-induced reaction mechanisms. They assumed functions $f(u, v)$ and $g(u, v)$ to be in the following forms:

$$f(u,v) = w_0(\mu_0 + \alpha_1 u + \beta_1 v)u \tag{3.30}$$

$$g(u,v) = (\alpha_2 u + \beta_2 v) \tag{3.31}$$

or:

$$g(u,v) = (\alpha_2 u + \beta_2 v)u \tag{3.32}$$

where $\mu_0 = \alpha_1 u + \beta_1 v$ is the coefficient of friction dependent on the temperature u and local slip velocity v, and the non-dimensional coefficients α_1 and β_1 and w_0 are constant. The function $g(u, v)$ characterizes the rheological properties of the material and depends on its viscous and plastic properties.

Such interpretation of Eqs. 3.13–3.22 can be used if the growth of a tribo-film (a thin interfacial layer activated by friction) is considered. Whereas u still represents the interfacial temperature, v can be interpreted as the non-dimensional thickness of the tribofilm formed at the interface. The tribofilm can grow due to the material transfer to the interface via diffusion activated by friction, due to precipitation of a certain component (e.g., a softer one) in an alloy or composite material, due to a chemical reaction, temperature gradient, etc. For example, during the contact of bronze versus steel, a protective Cu tribofilm can form at the

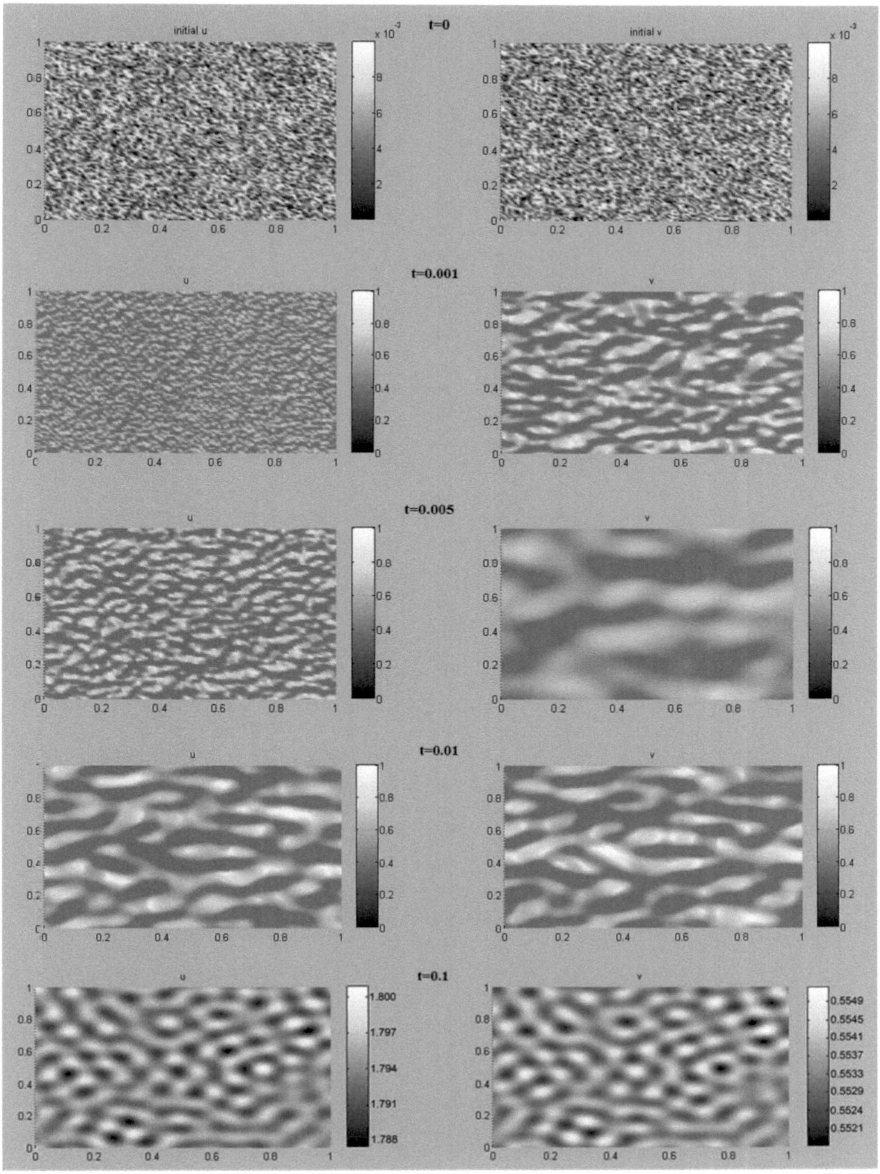

Fig. 3.8 Time evolution of amounts of u and v in different time steps, at $t = 0$, 0.001, 0.005, 0.01, and 0.1 (First case) [47]

interface, which significantly reduces the wear. Such in situ tribofilm has protective properties for the interface since it is formed dynamically and compensates the effect of wear. Furthermore, if wear is a decreasing function of the tribofilm

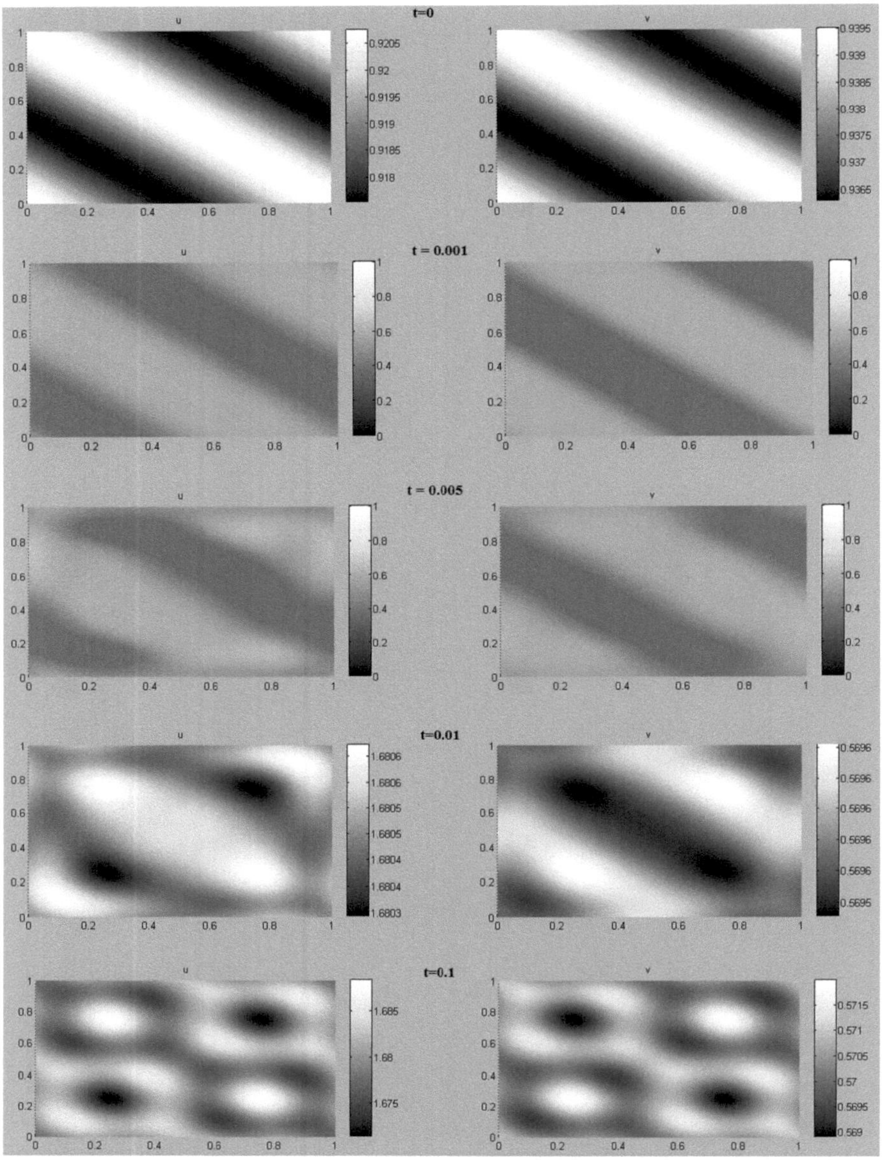

Fig. 3.9 Time evolution of u and v values in different time steps, $t = 0, 0.001, 0.005, 0.01, 0.05$ and 0.1 (Second case) [47]

thickness, it is energetically profitable for the film to grow, since a growing film reduces wear and further stimulates its growth forming a feedback loop, until a certain equilibrium thickness is attained. Therefore, such tribofilms can be used for

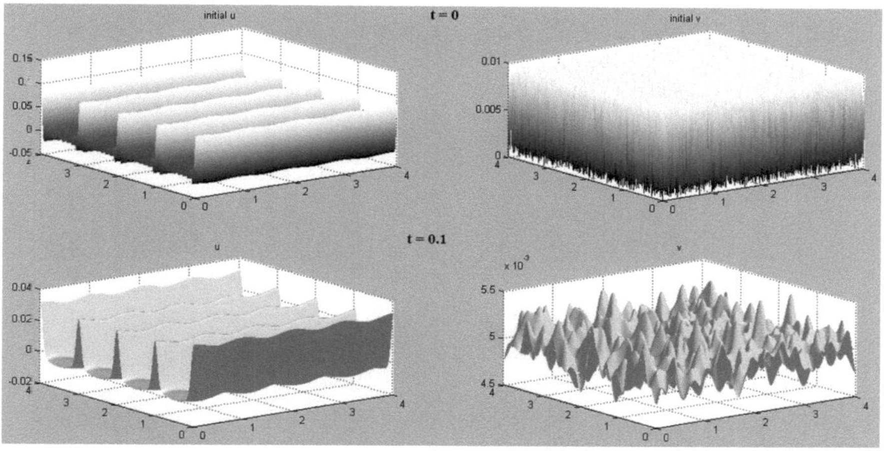

Fig. 3.10 Time evolution of u and v values in two different time steps (Third case, using Eqs. 3.30 and 3.31) [47]

machine tool protection and other applications, as discussed in the literature [5]. The growth of the film is governed by interfacial diffusion and by a local kinetic function $g(u, v)$ dependent upon the temperature and local film thickness.

Use the following equation as a temperature distribution for initial conditions

$$u(x,y) = 0.002\cos(2\pi(x+y)) + 0.01\sum_{j=1}^{8}\cos(2\pi jx) \tag{3.33}$$

and random distribution function as a initial roughness for $v(x, y)$ and, moreover, consider the following values for parameters of Eqs. 3.22 and 3.23

$$w_0 = 10_2, \alpha_1 = 10^{-4}, \beta_1 = 10^{-4}, \mu_0 = 5\times 10^{-1},$$
$$\alpha_2 = 10^{-4}, \beta_2 = 10^{-4} \tag{3.34}$$

The obtained results for different time steps are shown in Figs. 3.10 and 3.11 considering Eqs. 3.31 and 3.32 as $g(u, v)$, respectively. While the patterns found in Figs. 3.10 and 3.11 are not the same as the patterns found in the two previous cases, similar trends in the evolution of tribofilm thickness (v) from an initially random distribution of roughness to a more organized and more patterned distribution could be observed. However, the investigation of pattern formation based on solving complete three-dimensional heat and mass transfer equations in tribofilm is needed to show a more realistic picture of how and under what conditions such patterns could occur.

The modeling analysis in the preceding sections shows that properly selected functions $f(u, v)$ and $g(u, v)$ can lead to pattern formation. The question remains as to whether any experimental data can be interpreted as friction-induced patterns formed

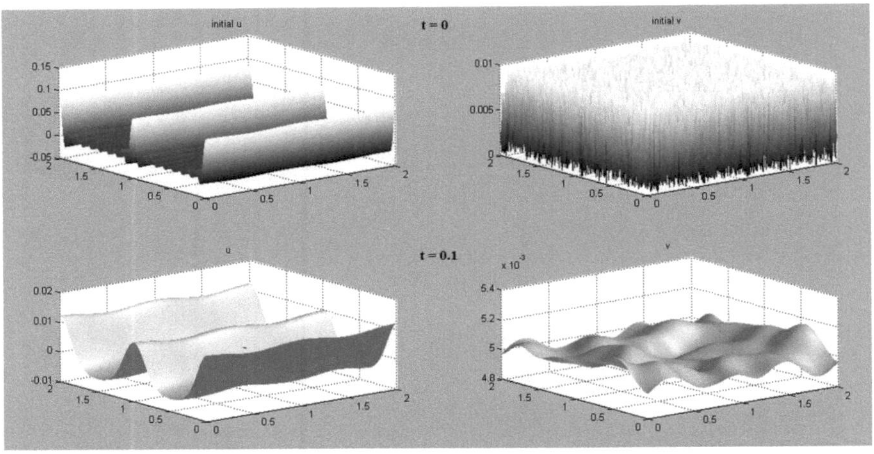

Fig. 3.11 Time evolution of u and v values in different time steps (Third case, using Eqs. 3.30 and 3.32) [47]

by the RD mechanism? It would be appropriate to look for this type of self-organization in processes involving viscoplastic contact or diffusion-dominated effects, e.g., in-situ tribofilms. There are several effects which can be interpreted in this way. The so-called "secondary structures" can form at the frictional interface due to the self-organization [25], and some of these structures can have a spatial pattern. The so-called "selective transfer" discussed in Sect. 3.2.4 could be another example.

The selective layer (an in situ tribofilm due to the selective transfer) formed during friction between steel and a copper alloy (bronze) was investigated experimentally by Ilie and Tita [32]. In the presence of glycerin or a similar lubricant, the ions of copper were selectively transferred from bronze to the frictional interface forming the copper tribofilm. This copper was different in its structure from the copper that falls out through normal electrolytic procedures. Ilia and Tita [32] investigated the selective layer using the AFM and found that the layer formed a micro-island pattern with the size on the order of 1 μm, rather than a uniform film of constant thickness. A schematic figure of such layer can be seen in Fig. 3.12.

Another important example of patterns formation reported in the literature is related to the self-adaptive mechanisms improving the frictional properties of hard coatings, e.g., during dry cutting, by tailoring their oxidation behavior [21, 22, 41]. Thus, boric acid formation on boron carbide is a potential mechanism for reaching ultra-low friction. Such mechanism uses the reaction of the boric oxide (B_2O_3) with ambient humidity (H_2O) to form a thin boric acid (H_3BO_4) film. The low friction coefficient of boric acid is associated with its layered triclinic crystal structure [23, 63]. The layers consist of closely packed and strongly bonded boron, oxygen, and hydrogen atoms, but the layers are widely separated and attracted by van der Waals forces only. During sliding, these atomic layers can align themselves parallel to the direction of relative motion and slide easily over one another [22].

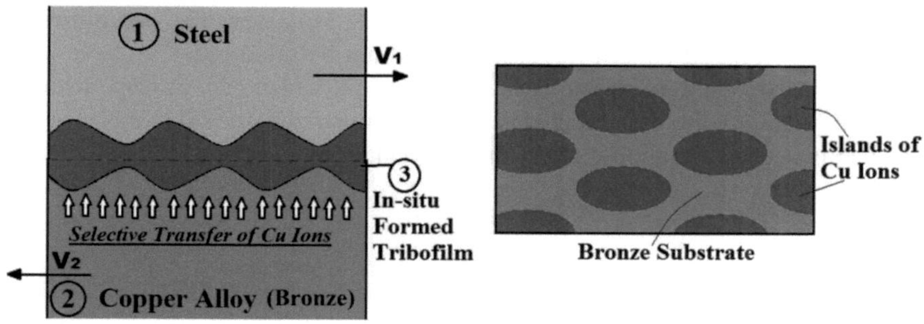

Fig. 3.12 Schematic presentation of the selective layer obtained by friction between a surface of steel and a surface of copper alloy (bronze) [47]

The tribological behavior of protective coatings formed by both ex-situ and in situ transfer films were studied by Singer et al. [63]. Coatings that exhibit long life in sliding contact often do so because the so-called "third body" forms and resides in the sliding interface. The concept of the "third body" as a separate entity, different from the two contacting bodies, is very similar to the concept of the tribofilm. Ex-situ surface analytical studies identified the composition and structure of third bodies and provided possible scenarios for their role in accommodating sliding and controlling friction. In situ Raman spectroscopy clearly identified the third bodies controlling frictional behavior during sliding contact between a transparent hemisphere and three solid lubricants: the amorphous Pb–Mo–S coating was lubricated by an MoS_2 transfer film; the diamond-like carbon/nanocomposite (DLC/DLN) coating by a graphite-like transfer film; and the annealed boron carbide by H_3BO_3 and/or carbon couples. In situ optical investigations identified third body processes with certain patterns responsible for the frictional behavior [63].

TiB_2 thin films are well known for their high hardness which makes them useful for wear-resistant applications. Mayrhofer et al. [43] showed that overstoichiometric $TiB_{2.4}$ layers have a complex self-organized columnar nanostructure precursor. Selected area electron diffraction (SAED) pattern from a $TiB_{2.4}$ layer showed a texture near the film/substrate interface with increased preferred orientation near the film surface [43]. The film has a dense columnar structure with an average column diameter of $\sim$20 nm and a smooth surface with an average RMS roughness essentially equal to that of the polished substrate surface, $\sim$15 nm.

Aizawa et al. [5] investigated in situ TiN and TiC ceramic coating films utilized as a protective coating for dies and cutting tools. They found that chlorine ion implantation assists these lubricious oxide (TiO and Ti_nO_{2n-1}) films to be in situ formed during wearing. They also performed the microscopic analysis and observed worn surfaces and wear debris and found microscale patterns.

Lin and Chu [40] described Bénard cell-like surface structures found from the observation of the Transmission electron microscopy (TEM) images of the scuffed worn surface as a result of lubricated steel versus steel contact. They attributed the cells to high temperatures (800°C) and very strong fluid convection or even evaporation occurring inside the scuffed surface. However, the possibility of diffusion-driven based pattern formation should not be ruled out.

The experimental evidences of pattern formation, which can, at least theoretically attributed to the RD mechanism, are summarized in Table 3.3. Further evidence is needed to rule out alternative explanations.

3.4 Self-Lubrication

The term 'self-lubrication' implies the ability of details and components to operate without lubrication and refers to several methods and effects in addition to the above mentioned self-organization that reduce friction or wear. Among these methods are the deposition of self-lubricating coatings that are either hard (to reduce wear) or with low surface energy (to reduce adhesion and friction). Besides coatings, self-lubrication refers to the development of metal-, polymer-, or ceramic-based composite self-lubricating materials, often with a matrix that provides structural integrity and a reinforcement material that provides low friction and wear. The nanocompisites have become a focus of this research, as well as numerous attempts to include carbon nanotubes and fullerene C_{60} molecules [62]. Simple models assume that these large molecules and nanosized particles serve as "rolling bearings" that reduce friction; however, it is obvious now that the mechanism can be more complicated and involve self-organization.

Dynamic self-organization is thought to be responsible for self-lubrication in the atomic force microscopy experiments with the atomic resolution. A protective layer can be formed also due to a chemical reaction of oxidation or a reaction with water vapor. For example a self-lubricating layer of the boric acid (H_3BO_3) is formed as a result of a reaction of water molecules with B_2O_3 coating [18]. Another type of self-lubricating material involves lubricant embedded into the matrix, e.g., inside microcapsules that rupture during wear and release the lubricant. Surface micro texturing that provides holes and dimples which serve as reservoirs for lubricant can be viewed as another method of providing self-lubrication. In addition, we should mention that self-lubrication is observed also in many biological systems (e.g., human joints) and that the term "self-lubrication" is used also in geophysics where it refers to animally low friction between tectonic plates observed during some earthquakes.

3.5 Surface Healing

Self-healing or surface-healing materials constitute a special class of novel smart materials, usually composites, which are capable of repairing minor damage caused by wear and deterioration [68]. Self-healing is a non-equilibrium process. When a system exchanges mass and energy with its surroundings, various irreversible processes inside the system may interact with each other. These

Table 3.3 Summary of different experimental pattern formation evidences discussed in literature

Materials in contact	Tribofilm material	Mechanism	Pattern	Reference
1. Steel versus copper alloy (bronze)	Copper	The selective transfer	Cu islands ($\sim$1 μm)	Ilie and Tita [32]
2. Boric oxide (B_2O_3)	Thin boric acid (H_3BO_4) film	Oxidation	Layered pattern	Erdemir et al. [2, 21, 22] and Lovell et al. [41]
3. Glass versus Pb–Mo–S	MoS_2 films	Transfer film	MoS_2 islands (nanoscale)	Singer et al. [63]
4. Sapphire against a DLC/DLN coating	Silicone-formed DLC/DLN coating	Transfer film	Silicone islands (nanoscale)	Singer et al. [63]
5. TiN versus AlN	TiB2 thin films	Oxidation	Columnar structure ($\sim$20 nm)	Mayrhofer et al. [43]
6. Steel versus steel	TiN or TiC films of 1 μm	Oxidation	Microscale patterns	Aizawa et al. [5]
7. Steel versus steel	α-Fe inside the substrate of the unscuffed surface	Wear	Bénard cell like patterns	Lin and Chu [40]

interactions are called thermodynamic couplings and they provide a mechanism for a process without its primary driving force or may even move the process in a direction opposite to the one imposed by its own driving force. For example, in thermodiffusion a substance diffuses because of a temperature gradient rather than a concentration gradient. When a substance flows from a low to a high concentration region, it must be coupled with a compensating process. The principles of thermodynamics allow the progress of a process without or against its primary driving force X_n only if it is coupled with another process. In the widely accepted linear approximation, the thermodynamic flows are related to the forces by

$$J_k = \sum_n L_{\mathrm{kn}} X_n \tag{3.35}$$

where L_{ki} are Onsager coefficients [16]. Many linear empirical laws of physics, such as the Ohm's law of electrical resistance, the Fourier law of heat conduction, or Fick's law of diffusion, which relate thermodynamic flows J_k with thermodynamic forces X_n, are consequences of the equations of the linear thermodynamics given in Eq. 3.35. The heat production per unit time is given by

$$\frac{\mathrm{d}Q}{\mathrm{d}t} = \sum_n J_n X_n \tag{3.36}$$

and the rate of entropy production is

$$\dot{S} = \frac{\mathrm{d}Q}{T\mathrm{d}t} = \frac{1}{T} \sum_n J_n X_{ni} \tag{3.37}$$

which is a linear function of thermodynamic flows.

In most self-healing schemes, the self-healing material is driven away from the thermodynamic equilibrium either by the deterioration process itself or by an external intervention, such as heating. After that, the composite slowly restores thermodynamic equilibrium, and this process of equilibrium restoration drives the healing.

In order to characterize degradation, it is convenient to introduce a so-called "degradation parameter" ξ, to represent, for example, the wear volume or the total volume of the cracks. When a self-healing mechanism is embedded in the system, another generalized coordinate, the healing parameter, ζ, can also be introduced, for example, the volume of released healing agent. The corresponding thermodynamic flows are linearly related to the thermodynamic forces (Eq. 3.35) and entropy rates (Eq. 3.37). The generalized degradation and healing forces are external forces that are applied to the system, and flows are related to the forces by the governing equations

$$\begin{aligned} J^{\circ} &= LX^{\circ} + MX^{\mathrm{heal}} \\ J^{\mathrm{heal}} &= NX^{\circ} + HX^{\mathrm{heal}} \end{aligned} \tag{3.38}$$

where L, M, N, H are corresponding Onsager coefficients (Fig. 3.13). The degradation force X° is an externally applied thermodynamic force that causes the

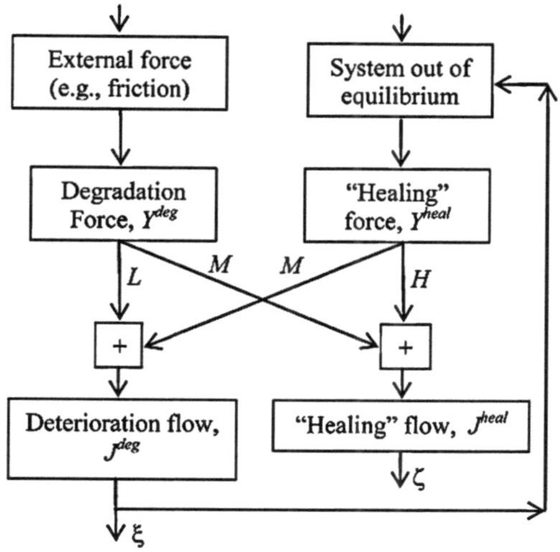

Fig. 3.13 Block diagram of the healing process. Deterioration is caused by an external force. The deteriorated system is brought out of equilibrium so that the restoring ("healing") force is created, which is coupled with the degradation flow through the parameter *M* [55]

degradation. In most self-healing mechanisms the system is placed out of equilibrium and the restoring force X^{heal} emerges. Since the restoring force is coupled with the degradation parameter ξ by the negative coefficients $N = M$, it also causes degradation decrease or healing.

Surface is the most vulnerable part of a material sample, and not surprisingly, deterioration often occurs at the surface (wear, fretting, etc.) and is induced by friction. The empirical Coulomb (or Amontons-Coulomb) law of friction states that the dry friction force F is linearly proportional to the normal load force W

$$F = \mu W \tag{3.39}$$

where μ is the coefficient of friction, which is independent of load, sliding velocity and the nominal area of contact. Unlike many other linear empirical laws, the Coulomb law cannot be directly deduced from the linear equations of the non-equilibrium thermodynamics, such as Eq. 3.35. Indeed, in the case of dry or lubricated friction, the sliding velocity is the thermodynamic flow, $V = J$, which, in accordance with Eq. 3.7 should be proportional to the friction force $F = X$ (as it is the case for the viscous friction), so that the energy dissipation rate is given by the product of the thermodynamic flow and force

$$\dot{\psi} = JX = VF \tag{3.40}$$

However, the Coulomb friction force is independent of sliding velocity.

To overcome this difficulty, we consider the normal degree of freedom y, in addition to the sliding coordinate x (Fig. 3.14). Introducing the normal degree of freedom is a standard procedure in the study of the dynamic friction, where normal vibrations are often coupled with the in-plane vibrations. We further define the generalized flows as $J_1 = \dot{x}$, $J_2 = \dot{y}$ and forces as $X_1 = F$, $X_2 = W$ [57].

Fig. 3.14 Normal (y) and tangential (x) degrees of freedom during friction [51]

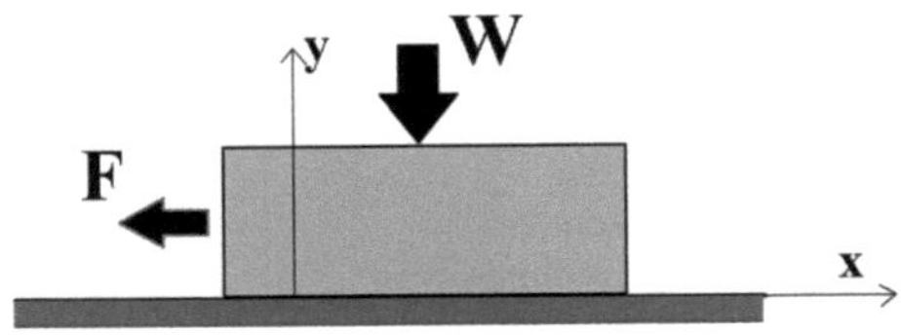

The thermodynamic equations of motion Eq. 3.35 immediately yield the law of viscous friction in the form of

$$\begin{pmatrix} \dot{x} \\ \dot{y} \end{pmatrix} = \begin{bmatrix} L_{11} & L_{12} \\ L_{21} & L_{22} \end{bmatrix} \begin{pmatrix} F \\ W \end{pmatrix} \tag{3.41}$$

Note that Eq. 3.41 is valid, in a general case, for the bulk of 3D deformable medium and not necessarily for the interface between two solids.

The interface between sliding bodies has highly anisotropic properties, because a small force in the direction of the interface causes large displacements, whereas a small force in the normal direction causes only small displacements. To compensate for this anisotropy, we substitute coordinates using a small parameter ε as $(x, y) \rightarrow (\varepsilon x, y)$. The force–displacement relationships are now given by

$$\begin{pmatrix} \dot{x} \\ \dot{y} \end{pmatrix} = \begin{bmatrix} L_{11}/\varepsilon & L_{12}/\varepsilon \\ L_{21} & L_{22} \end{bmatrix} \begin{pmatrix} F \\ W \end{pmatrix} \tag{3.42}$$

In the limit of $\varepsilon \rightarrow 0$, Eq. 3.42 yields

$$F = -\frac{L_{12}}{L_{11}} W \tag{3.43}$$

$$\dot{x} = (L_{11}F + L_{12}W)/\varepsilon = 0/0$$

$$\dot{y} = \left(L_{22} - \frac{L_{12}L_{21}}{L_{11}} \right) W$$

Any velocity $\dot{x}$ satisfies Eq. 3.43, provided $L_{11}F + L_{12}W = 0$, which is exactly the case of Coulomb friction if $\mu = -L_{12}/L_{11}$ [57].

Thus the friction can be included into the linear framework of non-equlibrium thermodynamics as the force driving the degradation process (e.g., the wear). The force which leads to healing should be related to the material microstructure in order to design an optimized material with the highest potential for self-healing and self-lubrication.

3.6 Conclusions

Despite the fact that self-organization during friction has received relatively little attention in the tribological community so far, it has a potential for the creation of self-healing and self-lubricating materials, which are of importance for the green or environment-friendly tribology. The principles of the thermodynamics of irreversible processes and of the non-linear theory of dynamical systems are used to investigate the formation of spatial and temporal structures during friction. These structures lead to friction and wear reduction (self-lubrication). The self-organization of these structures occurs through the destabilization of the stationary sliding. The stability criteria involve minimum entropy production. Some evidences of friction-induced self-organization like running-in stage, elastic structures, and Turing pattern formation were studied. A self-healing mechanism, which provides the coupling of the degradation and healing, can be embedded into the material. Structure–property relationships were formulated and can be used for optimized design of self-lubricating and self-healing materials and surfaces for various ecologically friendly applications and green tribology.

Acknowledgements The authors acknowledge the support of the University of Wisconsin-Milwaukee (UWM) Research Growth Initiative (RGI) and UWM Research Foundation Bradley Catalyst grants.

References

1. H. A. Abdel-Aal, Wear and irreversible entropy generation in dry sliding, Annals of University “Dunarea De Jos” Galati, Fascicle VIII, Tribology. 34–45 (2006) ISSN 1221–4590
2. G.G. Adams, Self-excited oscillations of the two elastic half-spaces sliding with a constant coefficient of friction. ASME J. Appl. Mech. **62**, 867–872 (1995). doi:10.1115/1.2896013
3. G.G. Adams, Steady sliding of two elastic half-spaces with friction reduction due to interface stick-slip. ASME J. Tribol. **65**, 470–475 (1998)
4. M. Adler, J. Ferrante, A. Schilowitz, D. Yablon, F. Zypman, Self-organized criticality in nano tribology. Mater. Res Soc. **782**, 111 (2004)
5. A. Aizawa, A. Mitsuo, S. Yamamoto, T. Sumitomo, S. Muraishic, Self-lubrication mechanism via the in situ formed lubricious oxide tribofilm. Wear **259**(1–6), 708–718 (2005)
6. P. Bak, *How Nature Works: The Science of Self-Organized Criticality* (Springer, New York, 1996)
7. J.R. Barber, Thermoelastic instabilities in the sliding of conforming solids. Proc. R. Soc. London A **312**, 381–394 (1969)
8. L.I. Bershadski, On the self-organization and concepts of wear-resistance in tribosystems. Trenie I Iznos (Russian Friction and Wear) **13**, 1077–1094 (1992)
9. L.I. Bershadski, B.I. Kostetski, General concept in tribology. Trenie I Iznos (Russian Friction and Wear) **14**, 6–18 (1993)
10. B. Bhushan, M. Nosonovsky, Scale effects in friction using strain gradient plasticity and dislocation-assisted sliding (microslip). Acta Mater. **51**, 4331–4340 (2003). doi:10.1016/S1359-6454(03)00261-1
11. P.J. Blau, *Friction and Wear Transitions of Materials: Break-In, Run-In, Wear-In* (Noyes Publications, New Jersey, 1989)
12. M.D. Bryant, M.M. Khonsari, F.F. Ling, On the thermodynamics of degradation. Proc. R. Soc. A **464**, 2001–2014 (2008). doi:10.1098/rspa.2007.0371
13. S.V. Buldyrev, J. Ferrante, F.R. Zypman, Dry friction avalanches: experiments and theory. Phys. Rev. E **74**, 066–110 (2006). doi:10.1103/PhysRevE.74.066110
14. N. Bushe, I.S. Gershman, Compatibility of Tribosystems, in *Self-Organization During Friction. Advanced Surface-Engineered Materials and Systems Design*, ed. by G.S. Fox-Rabinovich, G.E. Totten (CRC Taylor & Francis, Boca Raton, 2006), pp. 59–80

15. Z. Dai, S. Yang, Q. Xue, Thermodynamic model of fretting wear. J. Nanjing Univ. Aeronaut. Astronaut. **32**, 125–131 (2000)
16. S.R. De Groot, P. Mazur, *Non-Equilibrium Thermodynamics* (Interscience, New York, 1962)
17. K.L. Doelling, F.F. Ling, M.D. Bryant, B.P. Heilman, An experimental study of the correlation between wear and entropy flow in machinery components. J. Appl. Phys. **88**, 2999–3003 (2000)
18. C. Donnet, A. Erdemir, Historical developments and new trends in tribological and solid lubricant coatings. Surf. Coat. Technol. **180**, 76–84 (2004)
19. V. Dufiet, J. Boissonade, Conventional and nonconventional turing patterns. J. Chem. Phys. **96**, 664–673 (1991)
20. V. Dufiet, J. Boissonade, Numerical studies of turing patterns selection in a two-dimensional system. Physician A **188**, 158–171 (1992)
21. A. Erdemir, R.A. Erck, J. Robles, Relationship of hertzian contact pressure to friction behavior of selflubricating boric acid films. Surf. Coat. Technol. **49**, 435–438 (1991)
22. A. Erdemir, C. Bindal, C. Zuiker, E. Savrun, Tribology of naturally occurring boric acid films on boron carbides. Surf. Coat. Technol. **86–87**, 507–510 (1996)
23. A. Erdemir, in *Modern Tribology Handbook*, ed. by B. Bhushan, vol II (CRC, Boca Raton, 2001)
24. P. Fleurquin, H. Fort, M. Kornbluth, R. Sandler, M. Segall, F. Zypman, Negentropy generation and fractality in dry friction of polished surfaces. Entropy **12**, 480–489 (2010)
25. G.S. Fox-Rabinovich, G.E. Totten (eds.), *Self-Organization during Friction* (CRC Press, FL, 2006)
26. G.S. Fox-Rabinovich, S.C. Veldhuis, A.I. Kovalev, D.L. Wainstein, I.S. Gershman, S. Korshunov, L.S. Shuster, J.L. Endrino, Features of self-organization in ion modified nanocrystalline plasma vapor deposited AlTiN coatings under severe tribological conditions. J. Appl. Phys. **102**, 74–305 (2007). doi:10.1063/1.2785947
27. D.N. Garkunov, *Triboengineering (Wear and Non-Deterioration)* (Moscow Agricultural Academy Press, Moscow, 2000) (in Russian)
28. D.N. Garkunov, *Scientific Discoveries in Tribotechnology* (MSHA, Moscow, 2004) (in Russian)
29. I.S. Gershman, N. Bushe, Elements of Thermodynamics of Self-Organization during Friction, in *Self-Organization during Friction. Advanced Surface-Engineered Materials and Systems Design*, ed. by G.S. Fox-Rabinovich, G.E. Totten (CRC Taylor & Francis, Boca Raton, 2006), pp. 13–58
30. I.S. Gershman, Formation of Secondary Structures and Self-Organization Process of Tribosystems during Friction with the Collection of Electric Current, in *Self-Organization during Friction. Advanced Surface-Engineered Materials and Systems Design*, ed. by G.S. Fox-Rabinovich, G.E. Totten (CRC Taylor & Francis, Boca Raton, FL, 2006), p. 197
31. H. Haken, *Synergetics. An Introduction. Non equilibrium Phase Transitions in Physics, Chemistry and Biology*, 3rd edn. (Springer-Verlag, New York, 1983)
32. F. Ilie, C. Tita, Investigation of layers formed through selective transfer with atomic force microscopy. Paper presented in ROTRIB'07, 10th international conference on tribology, Bucharest, Romania, 2007
33. Q. Jun, A. Linan, P.J. Blau, Sliding friction and wear characteristics of Al_2O_3-Al nanocomposites. STLE/ASME international joint tribology conference, ASME, New York, 2006 (IJTC 2006, Ser. 2006)
34. E. Kagan, Turing systems, entropy, and kinetic models for self-healing surfaces. Entropy **12**, 554–569 (2010)
35. B.E. Klamecki, Wear—an entropy production model. Wear **58**, 325–330 (1980)
36. B.E. Klamecki, A thermodynamic model of friction. Wear **63**, 113–120 (1980)
37. B.E. Klamecki, Energy dissipation in sliding. Wear **77**, 115–128 (1982)
38. B.E. Klamecki, An entropy-based model of plastic deformation energy dissipation in sliding. Wear **96**, 319–329 (1984)
39. T Leppänen, Computational studies of pattern formation in turing systems, Dissertation for the degree of Doctor of science in technology, Helsinki University of Technology (2004)
40. J.F. Lin, H.Y. Chu, Analysis of the bénard cell-like worn surface type occurred during oil-lubricated sliding contact. Proceedings of the ASME/STLE 2009 international joint tribology conference, IJTC2009, Memphis, Tennessee (2009)

41. M.R. Lovell, M.A. Kabir, P.L. Menezes, C.F. Higgs III, Influence of boric acid additive size on green lubricant performance, Phil. Trans. R. Soc. A **368**(1929), 4851–4868 (2010)
42. S. Maegawa, K. Nakano, Mechanism of stick-slip associated with Schallamach waves. Wear **268**, 924–930 (2010)
43. P.H. Mayrhofera, C. Mitterer, J.G. Wen, J.E. Greene, I. Petrov, Self-organized nanocolumnar structure in superhard TiB_2 thin films. Appl. Phys. Lett. **86**, 131909 (2005)
44. J.A.C. Martins, J. Guimarães, L.O. Faria, Dynamic surface solutions in linear elasticity and viscoelasticity with frictional boundary conditions. ASME J. Vib. Acoust. **117**, 445–451 (1995)
45. P.H. Mayrhofera, C. Mitterera, L. Hultmanb, H. Clemensa, Microstructural design of hard coatings. Prog. Mater. Sci. **51**(8), 1032–1114 (2006)
46. P.L. Menezes, Kishore, S.V. Kailas, Studies on friction and transfer layer using inclined scratch. Tribol. Int. **39**(2), 175–183 (2006)
47. V. Mortazavi, M. Nosonovsky, In *Friction-induced pattern-Formation and Turing systems*. Langmuir. **27**(8), 4772–4779 (2011a)
48. V. Mortazavi, M. Nosonovsky, Wear-induced microtopography evolution and wetting properties of self-cleaning, lubricating and healing surfaces. J. Adhes. Sci. Technol. **25**(12), 1337–1359 (2011b)
49. J.D. Murray, *Mathematical Biology*, 2nd edn. (Springer Verlag, Berlin, 1989)
50. G. Nicolis, I. Prigogine, *Self-Organization in Nonequilibrium Systems* (Wiley, New York, 1977)
51. M. Nosonovsky, Self-organization at the frictional interface for green tribology. Phil. Trans. R. Soc. A **368**(1929), 4755–4774 (2010a)
52. M. Nosonovsky, Entropy in tribology: in search of applications. Entropy **12**(6), 1345–1390 (2010b), doi: 10.3390/e12061345
53. M. Nosonovsky, G.G. Adams, Dilatational and shear waves induced by the frictional sliding of two elastic half-spaces. Int. J. Eng. Sci. **39**, 1257–1269 (2001). doi:10.1016/S0020-7225(00)00085-9
54. M. Nosonovsky, G.G. Adams, Vibration and stability of frictional sliding of two elastic bodies with a wavy contact interface. ASME J. Appl. Mech. **71**, 154–300 (2004). doi:10.1115/1.1653684
55. M. Nosonovsky, B. Bhushan, Thermodynamics of surface degradation, self-organization and self-healing for biomimetic surfaces. Phil. Trans. R. Soc. A **367**, 1607–1627 (2009)
56. M. Nosonovsky, B. Bhushan, Surface self-organization: from wear to self-healing in biological and technical surfaces. Appl. Surf. Sci. **256**, 3982–3987 (2010)
57. M. Nosonovsky, R. Amano, J.M. Lucci, P.K. Rohatgi, Physical chemistry of self-organization and self-healing in metals. Phys. Chem. Chem. Phys. **11**, 9530–9536 (2009)
58. I. Prigogine, *Thermodynamics of Irreversible Processes* (Willey, New York, 1968)
59. I. Prigogine, *From Being to Becoming* (WH Freeman and Company, San Francisco, CA, 1980)
60. I. Prigogine, I. Stengers, *Order Out of Chaos* (Bantam, New York, 1984)
61. K. Ranjith, J.R. Rice, Slip dynamics at a dissimilar material interface. J. Mech. Phys. Solids **49**, 341–361 (2001)
62. L. Rapoport, N. Fleischer, R. Tenne, Fullerene-like WS2 nanoparticles: superior lubricants for harsh conditions. Adv. Mater. **15**, 651–655 (2003)
63. I.L. Singer, S.D. Dvorak, K.J. Wahl, T.W. Scharf, Role of third bodies in friction and wear of protective coatings. J. Vac. Sci. Technol. A **21**, 232–240 (2003)
64. L.A. Sosnovskiy, S.S. Sherbakov, Surprizes of Tribo-Fatigue (Magic Book, Minsk, 2009)
65. A.M. Turing, The chemical basis of morphogenesis. Phil. Trans. R. Soc. B **237**, 37–72 (1952)
66. G. Wu, F. Gao, M. Kaltchev, J. Gutow, J.K. Mowlem, W.C. Schramm, P.V. Kotvis, W.T. Tysoe, An investigation of the tribological properties of thin KCl films on iron in ultrahigh vacuum: modeling the extreme-pressure lubricating interface. Wear **252**, 595–606 (2002)
67. A. Zmitrowicz, A thermodynamical model of contact, friction, and wear. Wear **114**, 135–221 (1987)

68. S. Zwaag, van der, Self-healing behaviour in man-made engineering materials: bioinspired but taking into account their intrinsic character. Phil. Trans. R. Soc. A **367**, 1689–1704 (2009)
69. F. Zypman, J. Ferrante, M. Jansen, K. Scanlon, P. Abel, Evidence of self-organized criticality in dry sliding friction. J. Phys. Condens. Matter. **15**, L191–L196 (2003). doi:10.1088/0953-8984/15/12/101

Chapter 4
Advanced Nanostructured Surfaces for the Control of Biofouling: Cell Adhesions to Three-Dimensional Nanostructures

Chang-Hwan Choi and Chang-Jin Kim

Abstract In marine environments or industrial water systems, microorganisms are likely to adhere onto surfaces and form biofilms. Such biofouling creates significant adverse effects, e.g., increases flow friction by roughening surfaces. Previous studies demonstrated the effectiveness of surface microstructures on the prevention of biofouling, which is also closely associated with the surface energy and wettability. Unfortunately, the study of the anti-biofouling property of the micro- and nanostructured surfaces with regulated surface wettability is underperformed at present. In this paper, we report on the bio-adhesions of various cell types on nanoengineered surfaces with dense-array nanostructures whose physical and chemical properties are systematically controlled for the prevention of biofouling. Two nanopatterns (pillar and grating) with varying three-dimensionalities (e.g., structural heights are varied from 50 to 500 nm while the pattern periodicity is fixed at 230 nm) are tested in both hydrophilic and hydrophobic surface conditions. The structural tips are especially sharpened (<10 nm in tip radius) to minimize the cell contact to the substrate and potentially biofouling. The experimental results show that cells were much smaller and their proliferation significantly lower on taller nanostructures in both hydrophilic and hydrophobic surface conditions. Cells were found levitated by sharp tips and easily peeled off, i.e., their adherence to the sharp-tip tall nanostructures was relatively weak regardless of the surface wettability. The ability to control adherence and growth of cells by nanoscale surface topographies can empower the micro- and nanotechnology-based materials, devices, and systems

C.-H. Choi (✉)
Department of Mechanical Engineering, Stevens Institute of Technology, Castle Point on Hudson, Hoboken, NJ 07030, USA
e-mail: chang-hwan.choi@stevens.edu

C.-J. Kim
Mechanical and Aerospace Engineering Department, University of California, Los Angeles, CA 90095, USA
e-mail: cjkim@seas.ucla.edu

M. Nosonovsky and B. Bhushan (eds.), *Green Tribology*, Green Energy and Technology, DOI: 10.1007/978-3-642-23681-5_4,

for anti-biofouling and anti-microbial applications. The knowledge obtained through this investigation will also be useful in engineering problems that involve contact with biological substances and in the development of energy efficient surfaces for green tribology.

4.1 Introduction

Biofouling [1–4] is a serious problem for many engineering systems such as marine vehicles and industrial pipe systems due to the reduced process efficiency and the clean-up cost. For example, biofouling results in significant surface roughness for hydrodynamic systems and increases the frictional drag [5–10] such that fouled ships are required to burn 40% more fuel in order to maintain the same speed [11]. Every year, huge sums of money are spent to combat the consequences of biofouling. The global economic cost of this was estimated as $7.5 billion in 2000 [11]. The US Navy spends $500 million per annum on propulsion fuel, and even with the current antifouling technology it is estimated that approximately $30–60 million a year is still wasted in hydrodynamic drag [12, 13]. From these figures, it can be seen that the potential environmental and economic savings that could be made from the development and utilization of effective, hydrodynamically efficient, antifouling surfaces is substantial. At present, most methods used to prevent biofouling are paints or coatings, which continuously release toxic substances like copper, organotin (TBT), or organic biocides [14, 15]. They are now being prohibited due to their harmful effects on non-target species and environment. It should also be noted that some members of biofouling communities, such as *Enteromorpha*, are tolerant of biocidal antifouling paints [16]. Therefore, it is necessary to develop non-toxic alternatives for anti-biofouling [17]. One attractive approach is to use micro- or nano-textured surfaces with well-controlled surface wettability [18–23]. Studies of biofouling with systematically varied micro- and nanostructure dimensions and surface wettability will help to improve the understanding of biofouling mechanisms and to develop more efficient anti-biofouling techniques. However, the effects of the surface topography and chemistry on cell adhesions, especially for the topographical details in nanoscale, have not been studied systematically and interactively yet mostly due to the inability to create well-controlled nanostructures over a relatively large surface area.

Since cell adhesion is a nanoscale three-dimensional (3D) phenomenon, the full complement of substrate topography, molecular composition, and mechanical properties at a nanoscale are reportedly important for cells to form proper adhesion and to activate desired intracellular pathways for growth [24–26]. To know more about the reactions of cells to the nanofeatures and how to control them, it is desired to develop a proper nanostructured surface model which can provide statistically meaningful data with a large substrate area of multiple samples. A recent achievement in nanofabrication has made it possible to fabricate well-regulated nanofeatures (nano-periodic structures) uniformly over a large sample area (several cm^2) with independent controllability for the structural aspect ratio (>10) and shape (e.g., sidewall profile and tip sharpness) at a low cost and simplicity [27], allowing systematical studies of the effect of 3D nanotopography on cell adhesions [28–30]. Extending from the previous studies, this article reports on the cellular adhesion of various cell types on surfaces with well-defined nanotopographies of pillar and grating patterns of varying heights and controlled surface wettability, especially considering the future development of advanced nanostructured surfaces for the control of biofouling.

4.2 Materials and Methods

4.2.1 Fabrication of 3D Nanostructures

To fabricate the 3D nanotopography model surfaces for the study of cell adhesion, a simple and effective method to fabricate silicon nanostructures over a large substrate area with superior control of pattern regularity was developed and used for the sample preparation. The key idea was to combine laser interference lithography with deep reactive ion etch (DRIE), which allowed the fabrication of well-ordered nano-periodic array of nanostructures over a large substrate area at a wafer level with precise control of the structural three-dimensionality. While the detailed fabrication processes were reported elsewhere [27], they are summarized in the following. Interference lithography system of Lloyds-mirror configuration using the HeCd laser of the wavelength of 325 nm was first used for the nano-patterning of a photoresist layer spin-coated on a polished silicon substrate (2×2 cm^2). A single exposure creates a periodic grating pattern, while double exposure with an angular rotation of the substrate produces a periodic pillar pattern with positive-tone photoresist material. While the minimum pattern periodicity that can be defined by the lithography technique is around a half of the wavelength of the laser light, the pattern periodicity is fixed at 230 nm in this study to allow maximum pattern coverage area by the lithography [31]. The substrate was then etched by DRIE using the above patterned photoresist as an etching mask directly. The structural height and sidewall profile were determined in the DRIE by regulating the etching parameters. In this study, sharp-tip nanostructures in both pillar and grating patterns were designed to minimize the cell contact to the substrate surface for the prevention of cell adhesion and biofouling. For the sharp-tip nanostructures, nanostructures with a positively tapered profile were first created in the DRIE and their tips were next sharpened by thermal oxidation and buffered oxide etching. To test the effects of structural aspect ratios on the cell adhesion with the maximized tip sharpness, nanostructures of varying structural heights were prepared ranging from 50 to 500 nm. The nanostructured sample of 2×2 cm^2 was then cleaved into four chips of 1×1 cm^2, which was used for multiple experiments for statistical significance, to guarantee the uniformity of the pattern geometries among the chips. Planar silicon samples (1×1 cm^2) of an unpatterned surface (named 'Smooth') were also prepared as controls. The root mean square surface roughness of the Smooth control samples was measured to be <1 nm by atomic force microscopy over a 10×10 μm^2 scan area.

To test the combined effects of hydrophilic and hydrophobic surface wettability with the 3D nanotopography, surface treatments were further performed on the nanostructured samples. For a hydrophilic condition, the samples were cleaned with a Piranha solution (H_2SO_4:H_2O_2, 3:1 by volume), which would form a thin native SiO_2 layer on the silicon surface, followed by a rinse with water and a blow dry with nitrogen gas. For a hydrophobic treatment, Teflon solution (0.2% amorphous fluoropolymer AF 1600 in perfluoro-compound FC-75) was spin-coated on the sample to get a film thickness of ~10 nm, which was confirmed by scanning electron microscope (SEM) images taken before and after the coating.

4.2.2 Cell Culture

Human foreskin fibroblast and NIH-3T3 line (Swiss mouse embryonic fibroblast) were obtained from American type culture collection. Neonatal cardiac fibroblasts and smooth muscle cells were isolated from Dawley rat. All cells were cultured in Dulbecco's Modified Eagle medium supplemented with 10% fetal bovine serum, penicillin/streptomycin (100 U/ml) and kept at 37°C and 5% CO_2. The cells were seeded on each sample (1×1 cm^2), which had been placed on polystyrene culture well containing culture medium. The seeding density of cells was 1×10^4 to 2×10^4 cm^{-2}. Cells were kept in culture for periods ranging from a few days to several months.

4.2.3 Immunostaining

Cell viability was determined by means of molecular probes for fluorescence microscopy (double-staining with Calcein and Propidium Iodide). Calcein was used for the detection of cells with intact plasma membrane (live cells) and Propidium Iodide for the detection of dead cells. Calcein and Propidium Iodide were added to the culture medium. After 15 min incubation at room temperature, the viability of the cells was detected by a fluorescence microscope.

To label F-actin, the samples were fixed in 4% paraformaldehyde for 20 min, followed by washing with Dulbecco's phosphate-buffered saline (5 min three times) and permeabilization by 0.5% Triton X-100 (10 min). To reduce non-specific background staining, 1% bovine serum albumin was added to the staining solution of Rhodamine phalloidin in phosphate-buffered saline. After several washes with phosphate-buffered saline, mounting medium containing DAPI was added to the samples for nuclei staining. Digital images were captured of the stained cultures with a digital camera from a Leica DM IRB inverted microscope.

4.2.4 SEM

For SEM images, the samples containing cells were rinsed with 0.1 M sodium cacodylate buffer (pH 7.2) supplemented with 5% sucrose for 10 min, and fixed for 30 min in 2% paraformaldehye/2% glutaraldehyde in 0.1 M sodium cacodylate buffer (pH 7.2) supplemented with 5% sucrose, followed by dehydration (30, 50, 70, 80, and 95% ethanol for 10 min each, 100% ethanol for 10 min 3 times, and finally 100% ethanol for 40 min). The samples were then dried by incubating in 100% ethanol/hexamethyldisilazane (1:1) for 20 min, followed by pure hexamethyldisilazane for 20 min. Then, the pure hexamethyldisilazane solution was evaporated during 20 min by air-drying. Once dry, the samples were coated with ~10 nm thick gold/palladium (Au/Pd) by Denton Desk II sputtering system and examined with a Hitachi S-4700 field emission SEM.

4.3 Results

4.3.1 3D Sharp-Tip Nanostructures

Figure 4.1 shows SEM images of the 3D sharp-tip nanostructures fabricated on silicon substrates [28, 30]. The two different patterns, 'Nanopillar' and 'Nanograting', were used for the investigation of the dependency of cell behaviors on the anisotropy of the patterns. To differentiate nanotopographical 3D effects, only structure heights were varied from 'Low' (50–100 nm), 'Mid' (200–300 nm), to 'High' (400–500 nm), while the pattern periodicity (230 nm) and the tip sharpness (<10 nm in tip apex radius) were fixed. The silicon nanostructures were uniformly created over the top surface of the sample (with <10% deviation in structural size and shape for each sample of 1×1 cm^2).

4.3.2 Effect of Nanotopography

In order to examine the effects of nanotopography on the behaviors of cell growth and adhesion, human foreskin fibroblasts were first tested on the surface samples of varying patterns and structural aspect ratios, while their surface conditions were all hydrophilic. Figures 4.2, 4.3, 4.4, 4.5, 4.6, 4.7 and 4.8 show the results of microscopy.

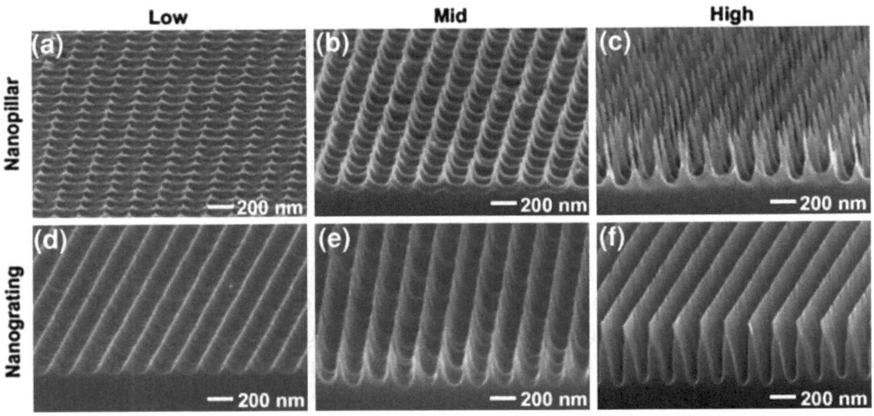

Fig. 4.1 SEM images of 3D sharp-tip nanostructures [28, 30]. Well-ordered silicon nanostructures (**a–c** 'Nanopillar', **d–f** 'Nanograting') covered the surface of the sample (1×1 cm^2) uniformly. While the pattern periodicity was maintained to be 230 nm and the structural tips were sharpened to be <10 nm in apex radius for all the samples, the structural heights were varied from 'Low' (**a**, **d** 50–100 nm), 'Mid' (**b**, **e** 200–300 nm) to 'High' (**c**, **f** 400–500 nm), in order to test the effects of nanotopographical three-dimensionalities on cell adhesions

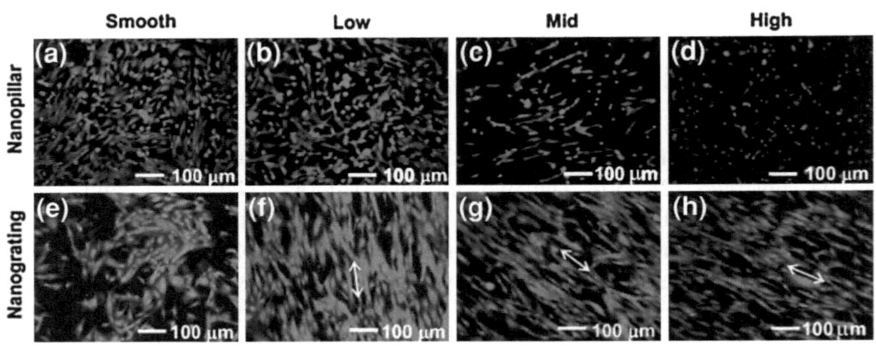

Fig. 4.2 Cell viability of human foreskin fibroblast on hydrophilic Nanopillar (**a–d**) and Nanograting (**e–h**) of varying aspect ratios. The fluorescence microscopy images were taken in 3 days for Smooth (**a**, **e**), Low (**b**, **f**), Mid (**c**, **g**), and High (**d**, **h**) samples. In the images, green color represents viable cells. In (**f–h**), the arrows (*left–right arrow*) indicate the direction of the grating patterns on the samples

Figure 4.2 shows fluorescent microscope images for cell viability, taken after 3 days. On Nanopillar, smaller cell population and cell size were observed on Nanopillar-Mid (Fig. 4.2c) and Nanopillar-High (Fig. 4.2d) than on Smooth (Fig. 4.2a). Nanopillar-Low (Fig. 4.2b) does not show any significant difference from Smooth. On Nanograting, cells were mostly viable on all the samples. Compared to Smooth (Fig. 4.2e), cells were elongated to the grating direction

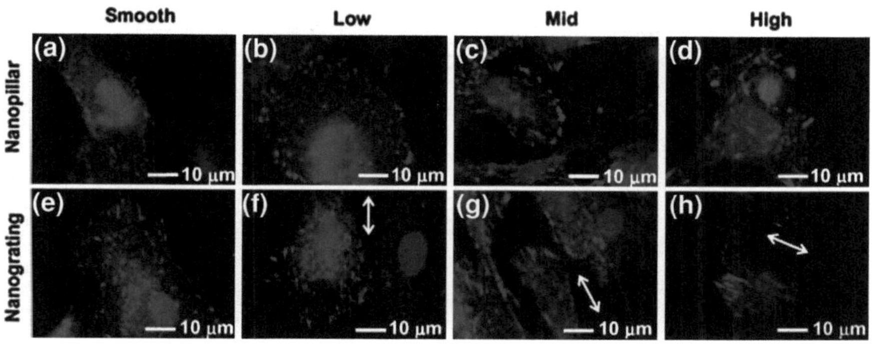

Fig. 4.3 Cell adhesions of human foreskin fibroblast on hydrophilic Nanopillar (**a–d**) and Nanograting (**e–h**) [28]. The fluorescence microscope images of immunostaining (nuclei: *blue*, pFAK: *red*, in colors) were taken in 3 days for Smooth (**a**, **e**), Low (**b**, **f**), Mid (**c**, **g**), and High (**d**, **h**) samples. Each arrow (*left–right arrow*) in (**f–h**) represents the direction of grating patterns on the samples

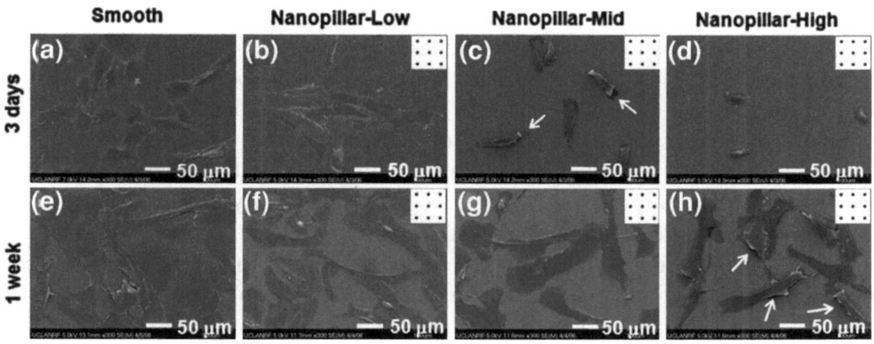

Fig. 4.4 Cell morphology of human foreskin fibroblast on hydrophilic Nanopillar topography [28, 30]. The SEM images (top view) were taken in 3 days (**a–d**) and 1 week (**e–h**) for Smooth (**a**, **e**), Low (**b**, **f**), Mid (**c**, **g**), and High (**d**, **h**) samples. Each inset in Nanopillar (**b–d**, **f–h**) represents the orientation of the pillar array on each sample. Detached cells were often observed on Nanopillar-Mid and Nanopillar-High during the sample preparation for SEM, as indicated by arrows (*right arrow*) in (**c**, **h**)

significantly, and the elongation was more pronounced on taller gratings. As reported elsewhere [28], the quantitative measurement of cellular metabolic activity by the alamarBlue™ assay also agreed with the qualitative results of Fig. 4.2.

Figure 4.3 shows the fluorescent microscope images of the immunostaining of phosphorylated focal adhesion kinase (pFAK) taken after 3 days [28]. Human foreskin fibroblasts on Smooth (Fig. 4.3a, e) and Nanopillar (Fig. 4.3b–d)

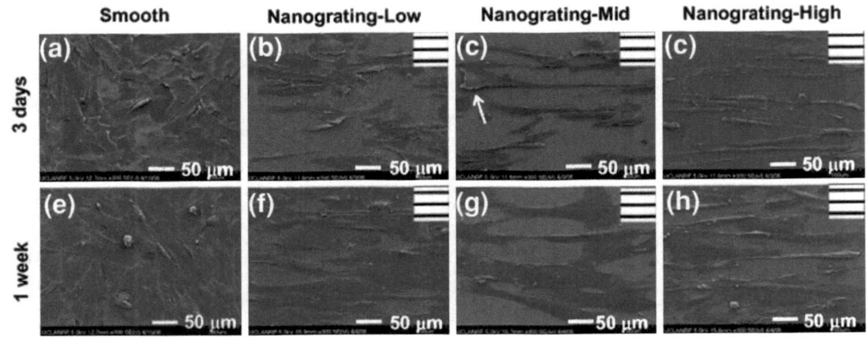

Fig. 4.5 Cell morphology of human foreskin fibroblast on hydrophilic Nanograting [28, 30]. The SEM images (top view) were taken in 3 days (**a–d**) and 1 week (**e–h**) for Smooth (**a**, **e**), Low (**b**, **f**), Mid (**c**, **g**), and High (**d**, **h**) samples. Each *inset* in Nanograting (**b–d**, **f–h**) represents the orientation of the grating array on each sample. Detached cells were often observed on the tall Nanograting during the sample preparation for SEM, as indicated by an arrow (*right arrow*) in (**c**)

exhibited punctuated adhesion complexes, whereas cells on Nanograting (Fig. 4.6f–h) demonstrated dash-like adhesion complexes clearly aligned to the grating direction. This result suggests that the 3D nanotopographies also induce distinct intracellular processes that may modify the cell's behaviors.

Figure 4.4 shows the SEM images of cell morphology on Nanopillar, taken after 3 days and 1 week [28, 30]. On day 3, cells on Smooth (Fig. 4.4a) exhibited well-spread and flattened cell morphology, which is typical for 2D planar substrates. In contrast, Nanopillar gave rise to significantly different cell morphologies. While being well-spread and flattened (similar to that for Smooth), cell morphology on Nanopillar-Low was elongated (Fig. 4.4b). On Nanopillar-Mid (Fig. 4.4c), an enhanced elongation with a slender morphology was observed, and cell size was significantly smaller. On Nanopillar-High (Fig. 4.4d), fibroblasts were not spread well and exhibited a swollen morphology with much smaller cell size, which is indicative of a poor adhesion and an apoptosis-like process. In 1 week, cell population, spreading, and size increased on all the samples. Cells on Smooth (Fig. 4.4e) and Nanopillar-Low (Fig. 4.4f) maintained the flattened cell morphology with more spreading and larger size. The elongated cell morphology was maintained on Nanopillar-Low (Fig. 4.4f). Compared to day 3, cells on Nanopillar-Mid (Fig. 4.4g) and Nanopillar-High (Fig. 4.4h) after one week were more spread, flattened, and enlarged, with an enhanced elongation and slender morphology. Also, compared to those on Smooth (Fig. 4.4e) and Nanopillar-Low (Fig. 4.4f), cells exhibited a relatively swollen morphology with a smaller cell size, especially on Nanopillar-High (Fig. 4.4h).

Figure 4.5 shows SEM images of cell morphology on Nanograting, taken after 3 days and 1 week [28, 30]. On day 3, although the cell spreading on Nanograting (Fig. 4.5b–d) was not as good as that on Smooth (Fig. 4.5a), fibroblasts spread

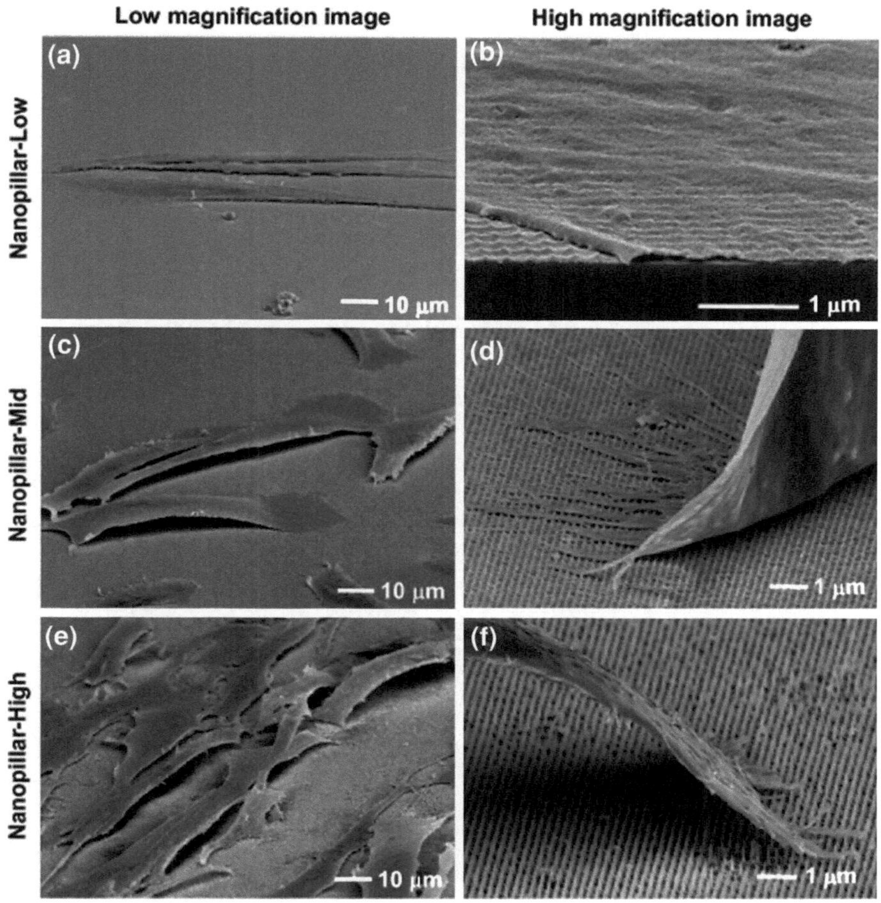

Fig. 4.6 SEM images (*tilted view*) of human foreskin fibroblast cells cultured on hydrophilic Nanopillar samples: Nanopillar-Low (**a**, **b** 3 days), Nanopillar-Mid (**c**, **d** 3 days), and Nanopillar-High (**e**, **f** 7 days) [30]. The high-magnification image (**b**) is from the sample cleaved through the silicon substrate and cell to show the cross section. While already discussed with the top-view images in Fig. 4.3, the *tilted-view* images more clearly show the detachment of cell bodies from the Nanopillar-Mid and Nanopillar-High. Cells detached during the sample preparation for SEM

relatively well on Nanograting compared to Nanopillars (Fig. 4.4b–d). Fibroblasts on Nanograting were more significantly elongated than on Nanopillar, showing clear alignment along the grating direction. The elongated cell morphology and alignment with the grating direction were more pronounced for taller gratings. Relatively swollen cell morphology was often observed on Nanograting-Mid (Fig. 4.5c) and Nanograting-High (Fig. 4.5d). In 1 week, cell population, spreading, and size increased on all samples. Cells on Nanograting, compared to

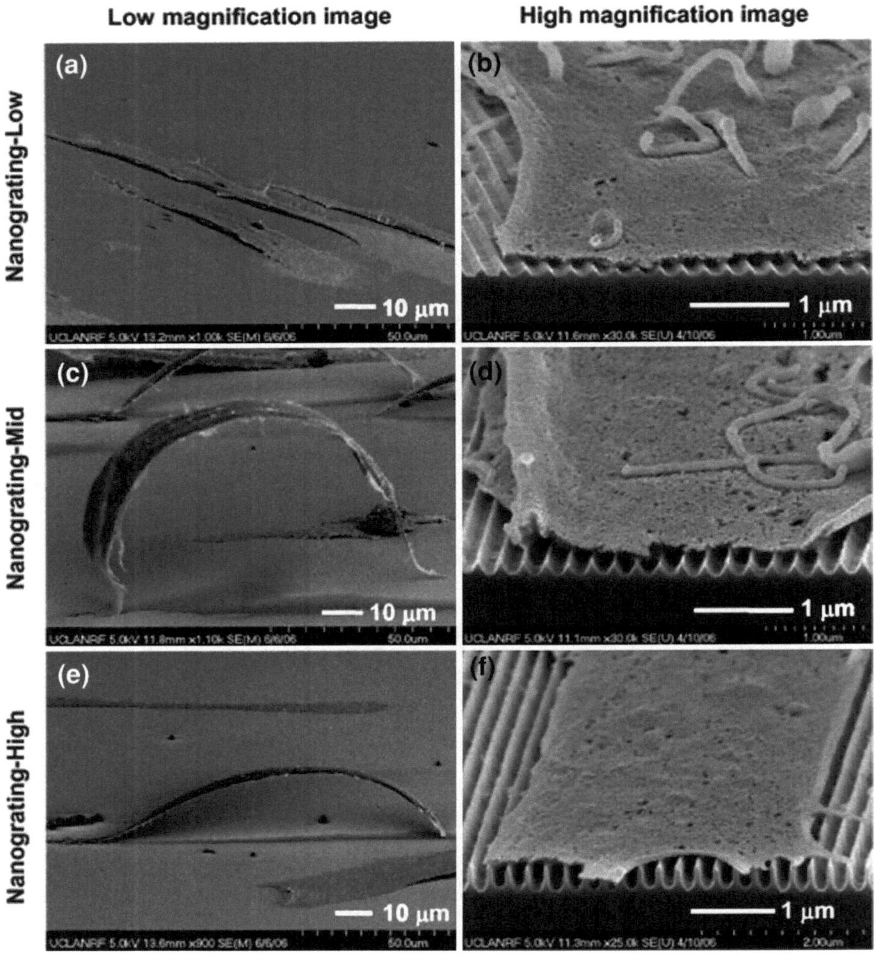

Fig. 4.7 SEM images (*tilted view*) of human foreskin fibroblast cells cultured on hydrophilic Nanograting samples: Nanograting-Low (**a**, **b** 3 days), Nanograting-Mid (**c**, **d** 3 days), and Nanograting-High (**e**, **f** 3 days) [30]. The high-magnification images on the right column (**b**, **d**, **f**) are from the sample cleaved through the silicon substrate and cell to show the cross sections. While already discussed with the top-view images in Fig. 4.5, the *tilted-view* images more clearly show the detachment of cell bodies from the Nanograting-Mid and Nanograting-High. Cells detached during the sample preparation for SEM

3 days, spread better, with flattened morphology and enlarged size, while the cell elongation and clear alignment to the grating direction was maintained. As reported elsewhere [28, 30], more quantitative data of the cell morphology obtained by image analyses further support the differences between Nanopillar and Nanograting patterns of the varying aspect ratios.

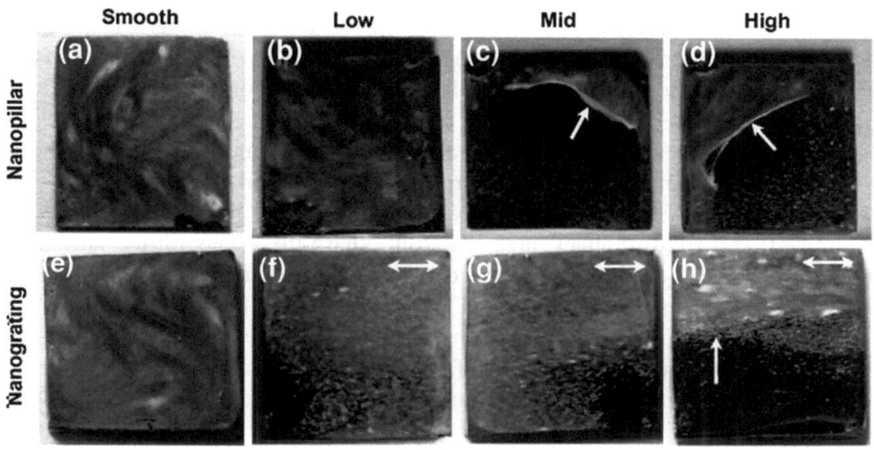

Fig. 4.8 Optical images (*top view*) of human foreskin fibroblast cells cultured for 3 weeks on the hydrophilic samples (**a** and **e** Smooth samples used as controls for Nanopillar and Nanograting, respectively; **b–d** Nanopillar-Low, -Mid, and -High, respectively; **f–h** Nanograting-Low, -Mid, and -High, respectively) [30]. Each image shows the entire surface of the sample of 1×1 cm^2. Although a cell sheet was formed on all the nanostructure samples after 3 weeks, the cell sheets formed on the tall nanostructures (**c**, **d**, **h**), as indicated by the arrows (*right arrow*), were detached from the surface and rolled up or removed away in the sample preparation for the images. In (**f–h**), the arrows (*left–right arrow*) indicate the grating directions on the samples

To examine the physical contact between the cells and the nanostructured surfaces, their interfaces were inspected by the SEM, as shown in Figs. 4.6 and 4.7 [30]. Figure 4.6 shows the SEM images of human foreskin fibroblasts on Nanopillar samples. On Nanopillar-Low (Fig. 4.6a, b), it was often shown (Fig. 4.6b) that the short Nanopillar structures were projected through the well-spread, flattened cell body, suggesting that cells accommodated the mild topography by an endocytosis-like process, which would result in a strong cell adherence to the surface. On taller Nanopillar-Mid and Nanopillar-High (Fig. 4.6c–f), as noticed in the top-view images (Fig. 4.4c, h), detached cell bodies were frequently observed after the samples were prepared for SEM. The ends of the detached cells shown in the magnified images (Fig. 4.6d, f) further suggest that the cell bodies were only supported by the Nanopillars' sharp tips, not entering into the valleys between the pillars (i.e., no endocytosis-like process). It also suggests that cell-substrate interactions occurred at the Nanopillars' sharp tips, and that cells' locomotion (e.g., spreading) continued while levitated by the tall, sharp-tip Nanopillar structures. The minimized adhesive area of cells to the surface by the levitation on the sharp tips would result in the weaker cell adherence to the surface, causing the cell detachment during the sample preparations for the SEM images.

Figure 4.7 shows the tilted-view SEM images of human foreskin fibroblasts on Nanograting samples. On Nanograting-Low (Fig. 4.7a, b), unlike on Nanopillar-Low (Fig. 4.6b), nanogratings did not project through the cells (Fig. 4.7b)

(i.e., no endocytosis-like process). As seen in the top-view images (Fig. 4.5c), detached cell bodies were frequently observed on tall Nanograting samples (Fig. 4.7c, e). Although the clear cell levitation was not observed on Nanograting-Low (Fig. 4.7b), the magnified tilted-view images of cleaved samples on the right column clearly show that on tall Nanograting samples (Fig. 4.7d, f) cell bodies were only supported by the sharp-tip ridges and did not go down into the valleys. Consistent with the case of Nanopillar samples, it is believed that the minimized cell-solid contact area by tall, sharp-tip Nanograting structures resulted in the self-detachment of the cell bodies during the sample preparation for the SEM images.

Human foreskin fibroblasts were cultured for an extended period (past the 7 days) to see whether they could proliferate enough to form a robust sheet of cell layers and gain good adhesion on the nanostructured substrates. Figure 4.8 shows the optical images of the fibroblasts cultured for 3 weeks [30]. Although the fibroblasts proliferated slower on the tall nanostructures in the beginning, a sheet of cells eventually formed on the nanostructured surfaces after 3 weeks. During the sample preparation for the image (fixation and drying), the cell sheets on Nanopillar-Mid (Fig. 4.8c) and Nanopillar-High (Fig. 4.8d) detached from one side and their edges curled up, while the cell sheets on Smooth (Fig. 4.8a) and Nanopillar-Low (Fig. 4.8b) remained attached. A part of a cell sheet on Nanograting also detached and curled away from the surface (parallel to the grating direction) in the course of the sample preparation. The detachment was more pronounced on taller Nanogratings (Fig. 4.8f–h), while a whole cell sheet on Smooth remained attached (Fig. 4.8e). In simple mechanical peel tests using tweezers and also jet shear flows, before drying (in a cell culture medium) as well as after drying, the cell sheets formed on the tall nanostructures (Nanopillar-Mid, Nanopillar-High, and Nanograting-High) were also found easier to peel off than those on the other surfaces. This result suggests that the cells on tall nanostructures, even when they formed a 3D sheet, were levitated on the sharp tips so that the cell-solid real contact area was minimized, resulting in a weak overall adherence. It should also be noted that, although less evident compared with the 1-week-old cell bodies, the formed cell sheets on Nanograting still showed the cells' directional elongation to the underlying grating structures, suggesting that the alignment was maintained over time [30].

4.3.3 Effect of Surface Wettability

Next, the effects of surface wettability on the cell behaviors in growth and adhesion were studied by testing neonatal rat cardiac fibroblasts on Nanopillar samples of varying structural aspect ratios in both hydrophilic and hydrophobic surface conditions. Figures 4.9, 4.10, 4.11, 4.12 show the results of microscopy.

Figure 4.9 shows the SEM images of neonatal rat cardiac fibroblasts cultured on hydrophilic samples for 2 days to 8 weeks. In 2 days, cells on Smooth

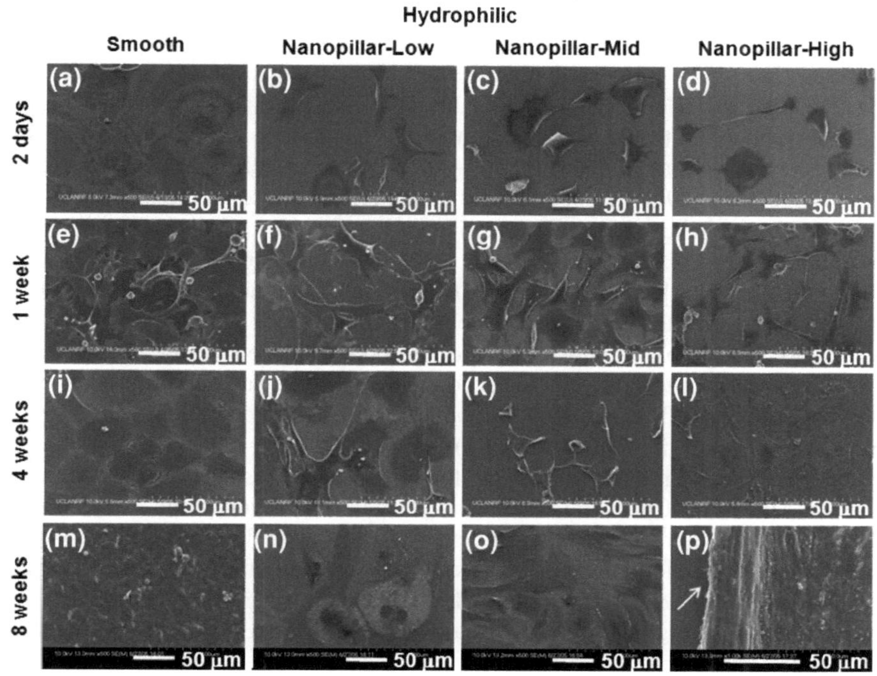

Fig. 4.9 SEM images of neonatal rat cardiac fibroblasts cultured on hydrophilic samples. Neonatal rat cardiac fibroblasts were cultured on Smooth (**a**, **e**, **i**, and **m**), Nanopillar-Low (**b**, **f**, **j**, and **n**), Nanopillar-Mid (**c**, **g**, **k**, and **o**), and Nanopillar-High (**d**, **h**, **l**, and **p**) in a hydrophilic surface condition for 2 days (**a–d**), 1 week (**e–h**), 4 weeks (**i–l**), and 8 weeks (**m–p**). In (**p**), an arrow (*right arrow*) indicates the boundary of cell sheet curled up from the surface

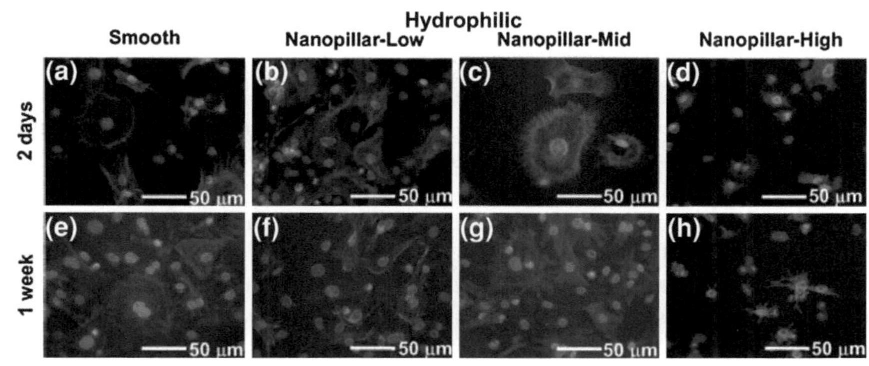

Fig. 4.10 Fluorescence microscope images of neonatal rat cardiac fibroblasts cultured on hydrophilic samples. Neonatal rat cardiac fibroblasts were cultured on Smooth (**a**, **e**), Nanopillar-Low (**b**, **f**), Nanopillar-Mid (**c**, **g**), and Nanopillar-High (**d**, **h**) in a hydrophilic surface condition for 2 days (**a–d**) and 1 week (**e–h**). *Blue* and *red* colors represent nuclei and actins, respectively

exhibited well-spread, flattened, and rounded cell morphology (Fig. 4.9a). In contrast, cells on Nanopillar showed less-spread, spindled or elongated cell morphology (Fig. 4.9b–d). In particular, the cell shape became slenderer and the cell size smaller for taller Nanopillars. Although cell population increased on all samples in 1 week, the slender and spindle-like shape with a smaller cell size was still observed on Nanopillar (Fig. 4.9f–h). In terms of cell size, cells on Nanopillar-High (Fig. 4.9h) were significantly smaller than those on the other samples. In 4 weeks, cells on Nanopillar exhibited better spreading with less slender morphology (Fig. 4.9j–l) than before, while cells on Smooth became confluent (Fig. 4.9i). In 8 weeks, a cell sheet was formed on all the samples (Fig. 4.9m–p). It was observed that the edges of the cell sheet formed on Nanopillar-High were curled up (Fig. 4.9p), while the edges on Smooth remained firmly attached. In a simple detachment test, the cell sheet formed on the tall and sharp-tip Nanopillar was easier to peel off than that on Smooth. This result was consistent with the case of human foreskin fibroblasts (Fig. 4.8).

Furthermore, Fig. 4.10 shows the fluorescence microscope images of actin filament (F-actin) of neonatal rat cardiac fibroblasts on hydrophilic samples after 2 days and 1 week. F-actins were less developed on Nanopillar-High (Fig. 4.10d, h), supporting the smaller cell size in cell morphology. Many radiations of F-actins at the periphery of cells were observed on Nanopillar-Mid (Fig. 4.10c, g) and Nanopillar-High (Fig. 4.10d, h), whereas the radiating length was much longer on Nanopillar-Mid than on Nanopillar-High.

In contrast to the hydrophilic samples, Fig. 4.11 shows the SEM images of neonatal rat cardiac fibroblasts cultured on hydrophobic samples from 2 days to 8 weeks. Nanopillar, compared to Smooth, tended to support cell proliferation in early periods (till 1 week), which was especially evident on Nanopillar-High (Fig. 4.11d, h). Cells on Nanopillar-High appear to be more viable than those on the other samples and showed stellate morphology (Fig. 4.11d, h, l). However, compared to the hydrophilic samples, cells did not proliferate much on the hydrophobic samples, and eventually few viable cells were left on the hydrophobic samples after 4 weeks. This result suggests that neonatal rat cardiac fibroblasts do not attain sufficient adhesion for proliferation on the hydrophobic surface. Although cells on the Nanopillar-High exhibited better cell proliferation than those on other samples, we speculate that it was because cells have less contact area to the hydrophobic surface due to the levitation by the tall and sharp-tip nanostructures.

Figure 4.12 also shows the fluorescence microscope images of actin filament (F-actin) of neonatal rat cardiac fibroblasts on hydrophobic samples in 2 days and 1 week. F-actins were seldom developed on hydrophobic samples, suggesting poor cell adhesion. Considerable development of F-actins was only observed on Nanopillar-High in 1 week (Fig. 4.12h). F-actins radiated at the cells' periphery (Fig. 4.12h), which was consistent with the cells' morphology.

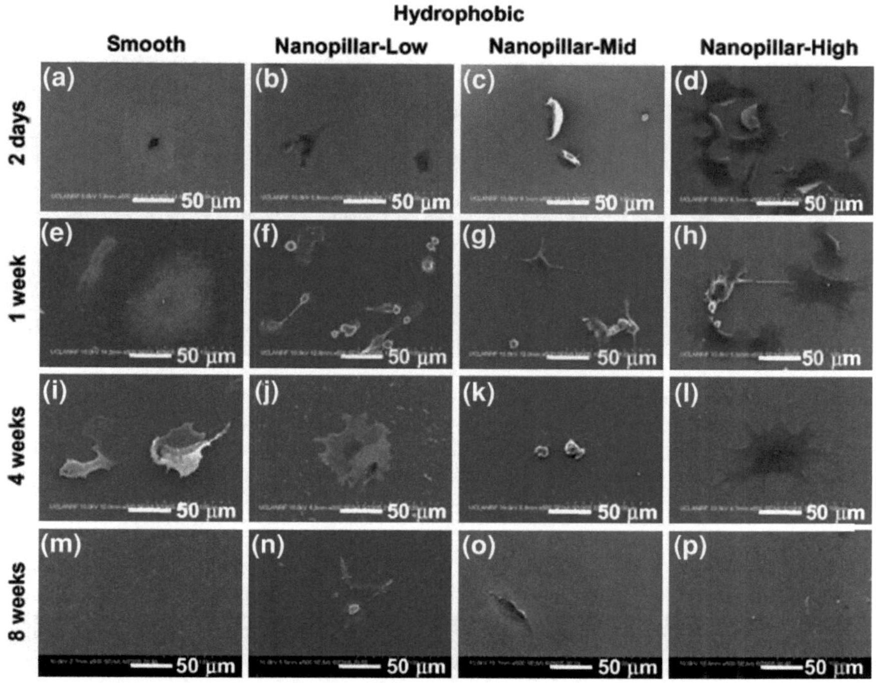

Fig. 4.11 SEM images of neonatal rat cardiac fibroblasts cultured on hydrophobic samples. Neonatal rat cardiac fibroblasts were cultured on Smooth (**a**, **e**, **i**, and **m**), Nanopillar-Low (**b**, **f**, **j**, and **n**), Nanopillar-Mid (**c**, **g**, **k**, and **o**), and Nanopillar-High (**d**, **h**, **l**, and **p**) in a hydrophobic surface condition for 2 days (**a–d**), 1 week (**e–h**), 4 weeks (**i–l**), and 8 weeks (**m–p**)

4.3.4 Effect of Cell Type

The dependency of cell behaviors on cell types were also studied by testing three different cells (human foreskin fibroblast, NIH-3T3, and smooth muscle cell) on both Nanopillar and Nanograting samples of a hydrophilic surface condition. Figures 4.13, 4.14, 4.15 show the results of microscopy. For the human foreskin fibroblast, the results shown in Figs. 4.2, 4.4, and 4.5 are presented in Figs. 4.13, 4.14, 4.15 again, respectively, to help the comparison.

Figure 4.13 shows the fluorescent microscope images of NIH-3T3 cell viability on hydrophilic Nanopillar after 3 days, compared to human foreskin fibroblast. Contrary to that of human foreskin fibroblast (HFF, Fig. 4.13a–d), no significant difference was observed with NIH-3T3 among the samples (Fig. 4.13e–h), suggesting that the viability of NIH-3T3 should not be affected considerably by the presence of the Nanopillar topographies. Figure 4.14 further shows the SEM images of NIH-3T3 cell morphology on Smooth and Nanopillar samples after 3 days (Fig. 4.14e–h), compared to human foreskin fibroblast (HFF, Fig. 4.14a–d).

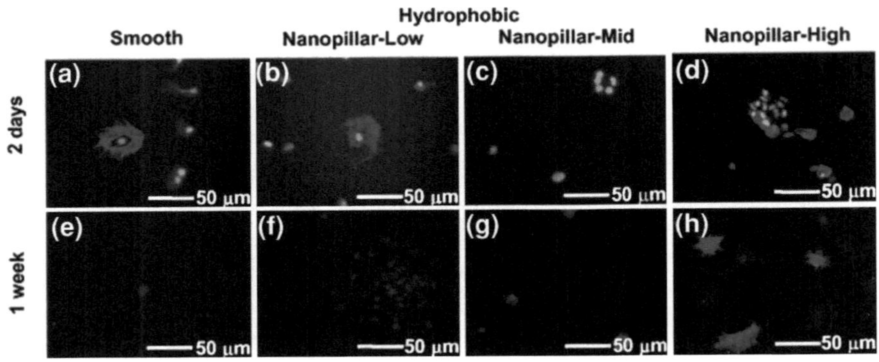

Fig. 4.12 Fluorescence microscope images of neonatal rat cardiac fibroblasts cultured on hydrophobic samples. Neonatal rat cardiac fibroblasts were cultured on Smooth (**a**, **e**), Nanopillar-Low (**b**, **f**), Nanopillar-Mid (**c**, **g**), and Nanopillar-High (**d**, **h**) in a hydrophobic surface condition for 2 days (**a**–**d**) and 1 week (**e**–**h**). *Blue* and *red* colors represent nuclei and actins, respectively

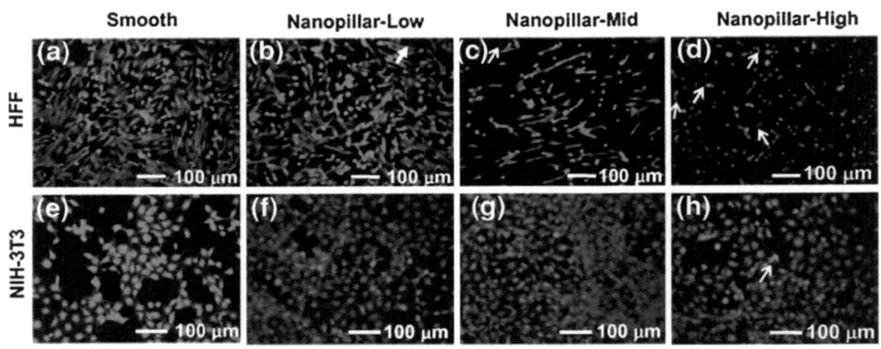

Fig. 4.13 Cell viability of NIH-3T3 (**e**–**h**) on hydrophilic Nanopillar surfaces, compared to human foreskin fibroblast (**a**–**d**). The fluorescence microscopy images were taken in 3 days for Smooth (**a**, **e**), Nanopillar-Low (**b**, **f**), Nanopillar-Mid (**c**, **g**), and Nanopillar-High (**d**, **h**). In the images, *green* color represents viable cells, while *red* cells are inactive (indicated with an *arrow*)

In terms of cell population, size, and shape, no noticeable difference was observed with NIH-3T3 among the samples. Along with the cell viability result (Fig. 4.13), this confirms that cell morphology of NIH-3T3 is not significantly affected by the existence of the Nanopillar topographies. However, NIH-3T3 cells were often detached on Nanopillar-Mid (Fig. 4.14g) and Nanopillar-High (Fig. 4.14h) during the sample preparation for SEM, similarly as human foreskin fibroblast cells were.

Figure 4.15 shows the SEM images of smooth muscle cell morphology on hydrophilic Smooth and Nanograting surfaces after 1 week (SMC, Fig. 4.15a–d), compared to human foreskin fibroblast (HFF, Fig. 4.15e–h). Compared to Smooth

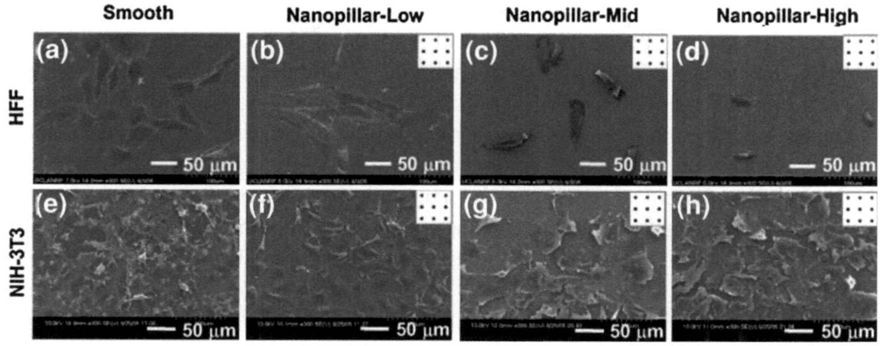

Fig. 4.14 Cell morphology of NIH 3T3 (**e–h**) on hydrophilic Nanopillar surfaces, compared to human foreskin fibroblast (**a–d**). The SEM images (*top view*) were taken in 3 days for Smooth (**a**, **e**), Nanopillar-Low (**b**, **f**), Nanopillar-Mid (**c**, **g**), and Nanopillar-High (**d**, **h**). Each inset in Nanopillar (**b–d**, **f–h**) represents the orientation of the pillar array on each sample

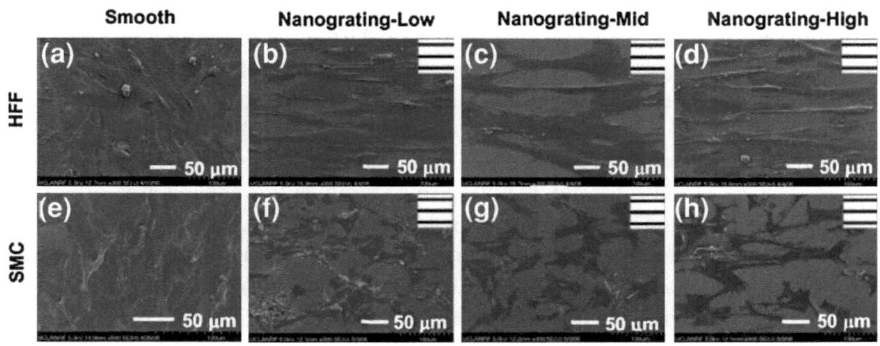

Fig. 4.15 Cell morphology of smooth muscle cell (**e–h**) on hydrophilic Nanograting surfaces, compared to human foreskin fibroblast (**a–d**). The SEM images (*top view*) were taken in 1 week for Smooth (**a**, **e**), Nanograting-Low (**b**, **f**), Nanograting-Mid (**c**, **g**), and Nanograting-High (**d**, **h**). Each *inset* in Nanograting (**b–d**, **f–h**) represents the orientation of the grating array on each sample

(Fig. 4.15e), smooth muscle cell on Nanograting exhibited noticeable elongation to the grating direction, which is more pronounced on Nanograting-High. However, the aligned elongation of smooth muscle cell on Nanograting was less significant than human foreskin fibroblast. The cell detachment of smooth muscle cell on Nanograting was also less significant than that of human foreskin fibroblast. However, compared to Smooth surfaces, the smooth muscle cell was often detached on tall Nanograting surfaces during the sample preparation for SEM.

Comparison of the cell morphologies of human foreskin fibroblast, NIH-3T3, and smooth muscle cell suggests that the influence of 3D nanotopographies on the cell morphologies is dependent on cell types. However, the NIH-3T3 and smooth muscle cell were still prone to detachment from the high aspect ratio nanostructure samples during the sample handling in liquid. Thus, the comparison of the cell detachments of human foreskin fibroblast, NIH-3T3, and smooth muscle cell suggests that the influence of 3D sharp-tip nanotopographies on the mechanical cell adhesion is independent of cell types. As long as the cell is much larger than the lateral dimension (i.e., pitch) of the nanopatterns, the contact area between the cell and the solid is reduced on the nanostructured substrates regardless of cell types, resulting in a weaker cell adherence to the sample surface.

4.4 Discussion

Our results showed that bodies of the tested cells (human foreskin fibroblast, NIH-3T3, and smooth muscle cell) were able to conform to the surface nano-topography of the short (50–100 nm) nanostructures but not of the medium (200–300 nm) and tall (400–500 nm) nanostructures. Note that all the nano-structures have the same periodicity (pitch of 230 nm) and similar tip sharpness ($\sim$10 nm). It is reasonable to interpret that the cells deformed to fill shallow valleys of the short nanostructures and contacted the bottom surface of the substrate, but could not deform enough to fill the deep valleys of tall nanostructures. Instead, they were suspended by the tall nanostructures and remained off the bottom surface of the substrate, resulting in a significantly reduced area of contact with the substrate and a much diminished overall adhesion.

Our result and interpretation not only agree well with the reports of others [32–36] but further suggest that taller (i.e., high aspect ratio) structures in a smaller spacing (i.e., small periodicity) are more effective in sustaining cells off the bottom surface and consequently obtain weak cell adherence on the substrate. It is important to note that a high curvature of a cell membrane is energetically unfavorable [37]. Considering the mechanical properties (e.g., elasticity or rigidity of a cell membrane), we expect that cells will conform to the surface nano-topography of scarcely populated or low-aspect-ratio nanostructures, which can be filled with a relatively small curvature (i.e., small deformation) of a cell membrane. If surface nanostructures are densely populated with a high aspect ratio, on the other hand, the cells of the same membrane pliability will not be able to develop a conformal contact (or endocytosis-like process).

In addition to the mechanical contact area, the adhesiveness of cells to the substrate is influenced by other factors, such as cell types and the effective amount of the adhesion complexes present at the cell-surface interfaces [38, 39]. In [38], where human corneal epithelial cells contacted only the tops of the ridge structures (220–2,200 nm in width, 400–4,000 nm in pitch, 250–600 nm in height, respectively), cell adhesion was found stronger on smaller structures (i.e., larger aspect ratio) than on larger structures (i.e., smaller aspect ratio) or even on a planar surface—a seemingly opposite trend to our results. Although the different trend may suggest the difference in the cell type simply overshadowed the adhesive responses to topographical stimuli [39], we also speculate that more surface discontinuities present on the smaller (denser) structures might have provided more available binding sites for protein conformations, resulting in the increased overall adhesion. Such speculation can be supported by a few other reports [35, 40, 41]. In [40], human gingival fibroblasts cultured on titanium-coated microscale grooves (6–10 μm in pitch, 3 μm in depth) produced a higher amount of the adhesive protein fibronectin compared with those on a smooth surface. In [41], a confined contact and a localized high adhesion of rat epitenon fibroblast cells were observed along the sharp edges of the microscale ridges (~10 μm in width, ~20 μm in pitch, 200 nm in height). In [35], similar to the observation in [41], actin molecules in HeLa cells were strongly localized only on the circumferences of the pillar head (500 nm in diameter, ~1 μm in pitch, 1 μm in height). If the concentration of adhesive proteins is higher at the sharp edges of surface structures, in [38] the adhesion may have been increased by the higher number of edges on the smaller (denser) structures more than it decreased by the reduced contact area between the cells and the substrate. Meanwhile, it should also be noted that the lateral dimensions of the focal adhesion have been reported to be 250–500 nm [42]. In [36, 39], focal adhesions formed on 70 nm wide ridges were found sporadic and significantly narrower than those on ridges of more than 400 nm wide, suggesting that there may exist a threshold lateral dimension for the proper formation of focal adhesion and associated stress fibers. The nanostructure tip size (~10 nm) tested in the present report is much smaller than the typical dimension of focal adhesion, suggesting that a highly intensified adhesion structure is not likely to occur on such slender and sharp-tip structures, in which case surface adhesiveness would be consistently low for a variety of cells.

Smaller cell populations and weaker cell adherence on the sharp-tip tall nanostructures in both hydrophilic and hydrophobic conditions indicate that such nanoengineered surfaces of well-tailored 3D nanotopography may prevent biofouling in many engineering systems and components such as lab-on-a-chip microdevices and biomedical implanting devices. In addition to the prevention of biofouling in hydrostatic conditions (e.g., no flow), our results of cell detachment in hydrodynamic conditions (e.g., shearing by a water jet flow) further suggest that such prevention will be more effective in dynamic flow conditions such as pipe flows and marine vessels. The control of bio-adhesion by only a physical means (i.e., surface nanotopographies) without any chemical modification or active mechanism will present many advantages in the designs of non-toxic and energy-efficient surfaces for green tribology. In addition to the reduction of biofouling, such nanostructured surfaces in hydrophobic condition are also considered promising for frictional drag reduction in various flows [43, 44], which is also expected to contribute to greener tribology.

To date, the effects of nano-periodic surface features on the cell behavior have been examined by using nanoscale structures, such as columns [32, 35, 45–50], dots [51–58], pits [59–62], pores [63–68], gratings [36, 39, 69–74], meshwork [75–84], nanophase grain [85–87], and random surface roughness [88–91], created by a variety of lithography and fabrication techniques, including e-beam lithography [36, 39], dip-pen lithography [53], laser irradiation [70, 72], nanoimprint [35, 59–62, 74], capillary lithography [49], X-ray lithography [71], interference lithography [69, 73], polymer demixing [46, 50], block-copolymer lithography [54], nanoparticle or colloidal lithography [32, 47, 48, 51, 52, 55–58], chemical vapor deposition of carbon nanofibers or nanotubes [75–84], black silicon method [45], electrochemical porous etching [63–68], chemical etching [88–91], and thermal sintering [85–87]. However, the serial methods such as e-beam lithography, dip-pen lithography, or direct laser irradiation are not adequate to pattern a large area (over cm^2s). X-ray lithography is a parallel process and can pattern a large area, but it is too expensive for most applications. Soft lithography-based fabrication, such as nanoimprint or capillary lithography, replicates patterns in a parallel fashion, but it still needs a master mold manufactured by e-beam or X-ray lithography. Most non-lithographic methods, such as nanotemplates made by self-assembly (e.g., copolymers, nanoparticles), direct growth of nanostructures (e.g., carbon nanofibers or nanotubes), and blanket etching or surface roughening (e.g., black silicon methods, chemical or electrochemical etching, sintering) generate poorly ordered patterns with irregular feature sizes, especially lacking the regularity over a large area. The ability to control the surface topography on the nanoscale has been very limited; the inability to independently control structural three-dimensionality and periodicity over a wide range in the nanoscale has been precluding systematic investigations of the 3D effects of nanoscale features on cell behaviors. The new nanofabrication technique combining the interference lithography and DRIE has opened the way for such investigations by realizing 3D nanostructures of well-regulated dimensions and shapes uniformly over a large substrate area (e.g., on a wafer level) with a relatively low cost. We hope the nanostructured surfaces used in this report provide a new opportunity to elucidate many nanobiological issues of the cell and open further possibilities in bioengineering application [92–94].

4.5 Conclusion

Recent nanotechnology revolutions have presented increased opportunities along with new challenges to bioscience and technology including bio-adhesion of cells. Surface topography and chemistry tailored by nanotechnology can make significant contribution to such applications by applying the technology to study how cells migrate and adhere on 3D nanofeatures with varying surface chemistry such as wettability. In this study, we demonstrated that nanostructures with systematically controlled patterns, sizes, and shapes can serve as a new means to elucidate the 3D cell-material interactions, which play a critical role in biofouling. For example, the dependency of cell behaviors on nanotopographical three-dimensionality (i.e., structural aspect ratio) was verified by using well-regulated sharp-tip nanostructures of independently controlled height and surface wettability.

Although cell-specific, for all the cell types tested, the trend was consistent: fewer cells and smaller cell size on taller nanostructures of a given pitch. Compared with 2D smooth surfaces, cells on 3D nanostructured surfaces were found more elongated to the directions of the structure patterns. The elongation phenomenon was more pronounced on grating patterns than pillar patterns; it was also more pronounced on taller gratings than shorter. The result illuminates that cell proliferation and morphological characteristics are dependent not only on the type of surface patterns but also on the topographical three-dimensionality. On the molecular level, the sharp-tip nanotopographies resulted in a marked difference in the formation and arrangement of intracellular adhesion molecules, which not only influenced the cell proliferation and morphology but also were likely to affect the intracellular activities such as gene expression and protein production. In regard to cell attachment, when the nanostructure was relatively short, cells adhered well to the entire surface including the valleys (e.g., endocytosis-like process on short pillar structures). When the nanostructure was tall, however, cells were levitated on the nanostructures whether the surface was hydrophilic or hydrophobic. As a result, the cell attachment to the substrate was weak. Not only individual cells but also the 3D cell layers formed on the tall nanostructures were easily peeled off, suggesting the possibility of using sharp-tip nanostructures of a high aspect ratio for the prevention and removal of biofoulants. The control of cell adherence and directional growth merely by the nanoscale surface topographies can open new application possibilities in biological tribology.

Acknowledgments This work was supported by the National Science Foundation Nanoscale Interdisciplinary Research Teams Grant 0103562. The authors thank Profs. Benjamin Wu and James Dunn for numerous assistance and discussions as this work evolved and Dr. Sepideh Hagvall for the help in the cell culture and data collection/interpretation.

References

1. G.G. Geesey, Z. Lewandowski, H.-C. Flemming, *Biofouling and Biocorrosion in Industrial Water Systems* (CRC Press, Boca Raton, 1994)
2. M. Fingerman, R. Nagabhushanam, M.-F. Thompson, *Recent Advances in Marine Biotechnology: Biofilms, Bioadhesion, Corrosion, and Biofouling* (Science Publishers, Enfield, 1999)
3. J. Walker, S. Surman, J. Jass, *Industrial Biofouling: Detection, Prevention and Control* (Wiley, Chichester, 2000)
4. A.I. Raikin, *Marine Biofouling: Colonization Processes and Defenses* (CRC Press, Boca Raton, 2003)
5. Z. Lewandowski, P. Stoodley, Flow induced vibrations, drag force, and pressure drop in conduits covered biofilm. Wat. Sci. Tech. **32**, 19–26 (1995)
6. P. Stoodley, Z. Lewandowski, J.D. Boyle, H.M. Lappin-Scott, Oscillation characteristics of biofilm streamers in turbulent flowing water as related to drag and pressure drop. Biotechnol. Bioeng. **57**, 536–544 (1998)
7. M.P. Schultz, G.W. Swain, The influence of biofilms on skin friction drag. Biofouling **15**, 129–139 (2000)

8. E.R. Holm, M.P. Schultz, E.G. Haslbeck, W.J. Talbott, A.J. Field, Evaluation of hydrodynamic drag on experimental fouling-release surfaces, using rotating disks. Biofouling **20**, 219–226 (2004)
9. M.P. Schultz, Effects of coating roughness and biofouling on ship resistance and powering. Biofouling **23**, 331–341 (2007)
10. A.F. Barton, M.R. Wallis, J.E. Sargison, A. Buia, G.J. Walker, Hydraulic roughness of biofouled pipes, biofilm character, and measured improvements from cleaning. J. Hydraul. Eng. **134**, 852–857 (2008)
11. R.L. Townsin, The ship hull fouling penalty. Biofouling **19**, 9–15 (2003)
12. M.E. Callow, Marine biofouling: a sticky problem. Biologist **49**, 1–5 (2002)
13. D. Howell, B. Behrends, A review of surface roughness in antifouling coatings illustrating the importance of cutoff length. Biofouling **22**, 401–410 (2006)
14. J.A. Lewis, Marine biofouling and its prevention on underwater surfaces. Mater. Forum **22**, 41–61 (1998)
15. J.A. Lewis, Antifoulings: towards and beyond the global TBT ban. Ships Ports **12**, 28 (2000)
16. M.E. Callow, Ship-fouling: the problem and methods of control. Biodeterior. Abstr **10**, 411–421 (1996)
17. A.S. Clare, Towards nontoxic antifouling. J. Mar. Biotech. **6**, 3–6 (1998)
18. P.E. Dyrynda, Defensive strategies of modular organisms. Philos. TR. Soc. B **313**, 227–243 (1986)
19. M. Andersson, K. Berntsson, P. Jonsson, P. Gatenholm, Microtextured surfaces: towards macrofouling resistant coatings. Biofouling **14**, 167–178 (1999)
20. P. Ball, Shark skin and other solutions. Nature (London) **400**, 507–508 (1999)
21. A.V. Bers, M. Wahl, The influence of natural surface microtopographies on fouling. Biofouling **20**, 43–51 (2004)
22. L. Hoipkemeier-Wilson, J.F. Schumacher, M.L. Carman, A.L. Gibson, A.W. Feinberg, M.E. Callow, J.A. Finlay, J.A. Callow, A.B. Brennan, Antifouling potential of lubricious, micro-engineered, PDMS elastomers against zoospores of the green fouling alga Ulva (Enteromorpha). Biofouling **20**, 53–63 (2004)
23. H. Zhang, R. Lamb, J. Lewis, Engineering nanoscale roughness on hydrophobic surface—Preliminary assessment of fouling behaviour. Sci. Technol. Adv. Mat. **6**, 236–239 (2005)
24. G.A. Abrams, S.L. Goodman, P.F. Nealey, M. Franco, C.J. Murphy, Nanoscale topography of the basement membrane underlying the corneal epithelium of the rhesus macaque. Cell Tissue Res. **299**, 39–46 (2000)
25. E. Cukierman, R. Pankov, D.R. Stevens, K.M. Yamada, Taking cell-matrix adhesions to the third dimension. Science **294**, 1708–1712 (2001)
26. E. Cukierman, R. Pankov, K.M. Yamada, Cell interactions with three-dimensional matrices. Curr. Opin. Cell Biol. **14**, 633–639 (2002)
27. C.-H. Choi, C.-J. Kim, Fabrication of dense array of tall nanostructures over a large sample area with sidewall profile and tip sharpness control. Nanotechnology **17**, 5326–5333 (2006)
28. C.-H. Choi, S.H. Hagvall, B.M. Wu, J.C.Y. Dunn, R.E. Beygui, C.-J. Kim, Cell interaction with three-dimensional sharp-tip nanotopography. Biomaterials **28**, 1672–1679 (2007)
29. S.H. Hagvall, C.-H. Choi, J.C.Y. Dunn, S. Heydarkhan, K. Schenke-Layland, W.R. MacLellan, R.E. Beygui, Influence of systematically varied nano-scale topography on cell morphology and adhesion. Cell Comm. Adhes. **14**, 181–194 (2007)
30. C.-H. Choi, S.H. Hagvall, B.M. Wu, J.C.Y. Dunn, R.E. Beygui, C.-J. Kim, Cell growth as a sheet on three-dimensional sharp-tip nanostructures. J. Biomed. Mater. Res. A **89**, 804–817 (2009)
31. I. Wathuthanthri, W. Mao, C.H. Choi, Two degrees-of-freedom Lloyd-mirror interferometer for superior pattern coverage area. Opt. Lett. **36**(9), 1593–1595 (2011)
32. M.J. Dalby, C.C. Berry, M.O. Riehle, D.S. Sutherland, H. Agheli, A.S.G. Curtis, Attempted endocytosis of nano-environment produced by colloidal lithography by human fibroblasts. Exp. Cell Res. **295**, 387–394 (2004)

33. A.S.G. Curtis, M.J. Dalby, N. Gadegaard, Nanoimprinting onto cells. J. R. Soc. Interface **3**, 393–398 (2006)
34. U. Gimsa, A. Iglic, S. Fiedler, M. Zwanzig, V. Kralj-Iglic, L. Jonas, J. Gimsa, Actin is not required for nanotubular protrusions of primary astrocytes grown on metal nano-lawn. Mol. Membr. Biol. **24**, 243–255 (2007)
35. S. Nomura, H. Kojima, Y. Ohyabu, K. Kuwabara, A. Miyauchi, T. Uemura, Cell culture on nanopillar sheet: study of HeLa cells on nanopillar sheet. Jap. J. Appl. Phys. **44**, L1184–L1186 (2005)
36. A.I. Teixeira, G.A. Abrams, P.J. Bertics, C.J. Murphy, P.F. Nealey, Epithelial contact guidance on well-defined micro- and nanostructured substrates. J. Cell Sci. **116**(10), 1881–1892 (2003)
37. R. Lipowsky, The conformation of membranes. Nature (London) **349**, 475–481 (1991)
38. N.W. Karuri, S. Liliensiek, A.I. Teixeria, G. Abrams, S. Campbell, P.F. Nealey, C.J. Murphy, Biological length scale topography enhanced cell-substratum adhesion of human corneal epithelial cells. J. Cell Sci. **117**, 3153–3164 (2004)
39. A.I. Teixeira, P.F. Nealey, C.J. Murphy, Responses of human keratocytes to micro- and nanostructured substrates. J. Biomed. Mater. Res. **71A**, 369–376 (2004)
40. L. Chou, J.D. Firth, V.J. Uitto, D.M. Brunette, Substratum surface topography alters cell shape and regulates fibronectin mRNA level, mRNA stability, secretion and assembly in human fibroblasts. J. Cell Sci. **108**, 1563–1573 (1995)
41. A.S.G. Curtis, B. Casey, J.O. Gallgher, D. Pasqui, M.A. Wood, C.D.W. Wilkinson, Substratum nanotopography and the adhesion of biological cells. Are symmetry or regularity of nanotopography important? Biophys. Chem. **94**, 275–283 (2001)
42. P.T. Ohara, R.C. Buck, Contact guidance in vitro: a light, transmission, and scanning electron microscopic study. Exp. Cell Res. **121**, 235–249 (1979)
43. C.-H. Choi, C.-J. Kim, Large slip of aqueous liquid flow over a nanoengineered superhydrophobic surface. Phys. Rev. Lett. **96**, 066001 (2006)
44. C.-H. Choi, U. Ulmanella, J. Kim, C.-M. Ho, C.-J. Kim, Effective slip and friction reduction in nanograted superhydrophobic microchannels. Phys. Fluids **18**, 087105 (2006)
45. S. Turner, L. Kam, M. Isaacson, H.G. Craighead, W. Shain, J. Turner, Cell attachment on silicon nanostructures. J. Vac. Sci. Technol. B **15**, 2848–2854 (1997)
46. M.J. Dalby, D. Pasqui, S. Affrossman, Cell response to nano-islands produced by polymer demixing: a brief review. IEE Proc. Nanobiotechnol. **151**, 53–61 (2004)
47. M.J. Dalby, M.O. Riehle, D.S. Sutherland, H. Agheli, A.S.G. Curtis, Fibroblast response to a controlled nanoenvironment produced by colloidal lithography. J. Biomed. Mater. Res. **69A**, 314–322 (2004)
48. M.J. Dalby, M.O. Riehle, D.S. Sutherland, H. Agheli, A.S.G. Curtis, Changes in fibroblast morphology in response to nano-columns produced by colloidal lithography. Biomaterials **25**, 5415–5422 (2004)
49. D.-H. Kim, P. Kim, I. Song, J.M. Cha, S.H. Lee, B. Kim, K.Y. Suh, Guided three-dimensional growth of functional cardiomyocytes on polyethylene glycol nanostructures. Langmuir **22**, 5419–5426 (2006)
50. C.C. Berry, M.J. Dalby, D. McCloy, S. Affrossman, The fibroblast response to tubes exhibiting internal nanotopography. Biomaterials **26**, 4985–4992 (2005)
51. M.A. Wood, D.O. Meredith, G.Rh. Owen, Steps toward a model nanotopography. IEEE Trans. Nanobios. **1**, 133–140 (2002)
52. C.C. Berry, S. Rudershausen, J. Teller, A.S.G. Curtis, The influence of elastin-coated 520-nm- and 20-nm-diameter nanoparticles on human fibroblasts in vitro. IEEE Trans. Nanobios. **1**, 105–109 (2002)
53. K.-B. Lee, S.-J. Park, C.A. Mirkin, J.C. Smith, M. Mrksich, Protein nanoarrays generated by dip-pen nanolithography. Science **295**, 1702–1705 (2002)
54. M. Arnold, E.A. Cavalcanti-Adam, R. Glass, J. Blummel, W. Eck, M. Kantlehner, H. Kessler, J.P. Spatz, Activation of integrin function by nanopatterned adhesive interfaces. Chem. Phys. Chem. **5**, 383–388 (2004)

55. M.A. Wood, C.D.W. Wilkinson, A.S.G. Curtis, The effects of colloidal nanotopography on initial fibroblast adhesion and morphology. IEEE Trans. Nanobios. **5**, 20–31 (2006)
56. J.M. Rice, J.A. Hunt, J.A. Gallagher, P. Hanarp, D.S. Sutherland, J. Gold, Quantitative assessment of the response of primary derived human osteoblasts and macrophages to a range of nanotopography surfaces in a single culture model in vitro. Biomaterials **24**, 4799–4818 (2003)
57. A.-S. Andersson, F. Backhed, A. von Euler, A. Richter-Dahlfors, D. Sutherland, B. Kasemo, Nanoscale features influence epithelial cell morphology and cytokine production. Biomaterials **24**, 3427–3436 (2003)
58. A.-S. Andersson, P. Olsson, U. Lidberg, D. Sutherland, The effects of continuous and discontinuous groove edges on cell shape and alignment. Exp. Cell Res. **288**, 177–188 (2003)
59. J.O. Gallagher, K.F. McGhee, C.D.W. Wilkinson, M.O. Riehle, Interaction of animal cells with ordered nanotopography. IEEE Trans. Nanobios. **1**, 24–28 (2002)
60. A.S.G. Curtis, N. Gadegaard, M.J. Dalby, M.O. Riehle, C.D.W. Wilkinson, G. Aitchison, Cells react to nanoscale order and symmetry in their surroundings. IEEE Trans. Nanobios. **3**, 61–65 (2004)
61. M.J. Dalby, N. Gadegaard, M.O. Riehle, C.S.W. Wilkinson, A.S.G. Curtis, Investigating filopodia sensing using arrays of defined nano-pits down to 35 nm diameter in size. Int. J. Biochem. Cell Biol. **36**, 2005–2015 (2004)
62. E. Martines, K. McGhee, C. Wilkinson, A. Curtis, A parallel-plate flow chamber to study initial cell adhesion on a nanofeatured surface. IEEE Trans. Nanobios. **3**, 90–95 (2004)
63. S.C. Bayliss, P.J. Harris, L.D. Buckberry, C. Rousseau, Phosphate and cell growth on nanostructured semiconductors. J. Mater. Sci. Lett. **16**, 737–740 (1997)
64. S.C. Bayliss, R. Heald, D.I. Fletcher, L.D. Buckberry, The culture of mammalian cells on nanostructured silicon. Adv. Mater. **11**, 318–321 (1999)
65. S.C. Bayliss, L.D. Buckberry, I. Fletcher, M.J. Tobin, The culture of neurons on silicon. Sensor Actuat. A **74**, 139–142 (1999)
66. S.C. Bayliss, L.D. Buckberry, P.J. Harris, M. Tobin, Nature of the silicon-animal cell interface. J. Porous Mat. **7**, 191–195 (2000)
67. A.H. Mayne, S.C. Bayliss, P. Barr, M. Tobin, L.D. Buckberry, Biologically interfaced porous silicon devices. Phys. Stat. Sol. A **182**, 505–513 (2000)
68. A.V. Sapelkin, S.C. Bayliss, B. Unal, A. Charalambou, Interaction of B50 rat hippocampal cells with stain-etched porous silicon. Biomaterials **27**, 842–846 (2006)
69. P. Clark, P. Connolly, A.S.G. Curtis, J.A.T. Dow, C.D.W. Wilkinson, Cell guidance by ultrafine topography in vitro. J. Cell Sci. **99**, 73–77 (1991)
70. B. Zhu, Q. Zhang, Q. Lu, Y. Xu, J. Yin, J. Hu, Z. Wang, Nanotopographical guidance of C6 glioma cell alignment and oriented growth. Biomaterials **25**, 4215–4223 (2004)
71. N.W. Karuri, S. Liliensiek, A.I. Teixeira, G. Abrams, S. Campbell, P.F. Nealey, C.J. Murphy, Biological length scale topography enhances cell-substratum adhesion of human corneal epithelial cells. J. Cell Sci. **117**, 3153–3164 (2004)
72. B. Zhu, Q. Lu, J. Yin, J. Hu, Z. Wang, Alignment of osteoblast-like cells and cell-produced collagen matrix induced by nanogrooves. Tissue Eng. **11**, 825–834 (2005)
73. H. Baac, J.-H. Lee, J.M. Seo, T.H. Park, H. Chung, S.-D. Lee, S.J. Kim, Submicron-scale topographical control of cell growth using holographic surface relief grating. Mater. Sci. Eng. C **24**, 209–212 (2004)
74. E.K.F. Yim, R.M. Reano, S.W. Pang, A.F. Yee, C.S. Chen, K.W. Leong, Nanopattern-induced changes in morphology and motility of smooth muscle cells. Biomaterials **26**, 5405–5413 (2005)
75. M.P. Mattson, R.C. Haddon, A.M. Rao, Molecular functionalization of carbon nanotubes and use as substrates for neuronal growth. J. Mol. Neurosci. **14**, 175–182 (2000)
76. K.L. Elias, R.L. Price, T.J. Webster, Enhanced functions of osteoblasts on nanometer diameter carbon fibers. Biomaterials **23**, 3279–3287 (2002)

77. T.J. Webster, M.C. Waid, J.L. McKenzie, R.L. Price, J.U. Ejiofor, Nano-biotechnology: carbon nanofibres as improved neural and orthopaedic implants. Nanotechnology **15**, 48–54 (2004)
78. J.L. McKenzie, M.C. Waid, R. Shi, T.J. Webster, Decreased functions of astrocytes on carbon nanofiber materials. Biomaterials **25**, 1309–1317 (2004)
79. R.L. Price, K. Ellison, K.M. Haberstroh, T.J. Webster, Nanometer surface roughness increases select osteoblast adhesion on carbon nanofiber compacts. J. Biomed. Mater. Res. **70A**, 129–138 (2004)
80. H. Hu, Y. Ni, V. Montana, R.C. Haddon, V. Parpura, Chemically functionalized carbon nanotubes as substrates for neuronal growth. Nano Lett. **4**, 507–511 (2004)
81. H. Hu, Y. Ni, S.K. Mandal, V. Montana, B. Zhao, R.C. Haddon, V. Parpura, Polyethyleneimine functionalized single-walled carbon nanotubes as a substrate for neuronal growth. J. Phys. Chem. B **109**, 4285–4289 (2005)
82. L. Zanello, B. Zhao, H. Hu, R.C. Haddon, Bone cell proliferation on carbon nanotubes. Nano Lett. **6**, 562–567 (2006)
83. R.A. MacDonald, B.F. Laurenzi, G. Viswanathan, P.M. Ajayan, J.P. Stegemann, Collagen-carbon nanotube composite materials as scaffolds in tissue engineering. J. Biomed. Mater. Res. **74A**, 489–496 (2005)
84. A.V. Liopo, M.P. Stewart, J. Hudson, J.M. Tour, T.C. Pappas, Biocompatibility of native and functionalized single-walled carbon nanotubes for neuronal interface. J. Nanosci. Nanotechnol. **6**, 1365–1374 (2006)
85. T.J. Webster, R.W. Siegel, R. Bizios, Osteoblast adhesion on nanophase ceramics. Biomaterials **20**, 1221–1227 (1999)
86. T.J. Webster, C. Ergun, R.H. Doremus, R.W. Siegel, R. Bizios, Specific proteins mediate enhanced osteoblast adhesion on nanophase ceramics. J. Biomed. Mater. Res. **51**, 475–483 (2000)
87. T.J. Webster, C. Ergun, R.H. Doremus, R.W. Siegel, R. Bizios, Enhanced osteoclast-like cell functions on nanophase ceramics. Biomaterials **22**, 1327–1333 (2001)
88. A. Thapa, T.J. Webster, K.M. Haberstroh, Polymers with nano-dimensional surface features enhance bladder smooth muscle cell adhesion. J. Biomed. Mater. Res. **67A**, 1374–1383 (2003)
89. A. Thapa, D.C. Miller, T.J. Webster, K.M. Haberstroh, Nano-structured polymers enhance bladder smooth muscle cell function. Biomaterials **24**, 2915–2926 (2003)
90. D.C. Miller, A. Thapa, K.M. Haberstroh, T.J. Webster, Endothelial and vascular smooth muscle cell function on poly (lactic-co-glycolic acid) with nano-structured surface features. Biomaterials **25**, 53–61 (2004)
91. Y.W. Fan, F.Z. Cui, S.P. Hou, Q.Y. Xu, L.N. Chen, I.-S. Lee, Culture of neural cells on silicon wafers with nano-scale surface topograph. J. Neurosci. Methods **120**, 17–23 (2002)
92. T.A. Desai, Micro- and nanoscale structures for tissue engineering constructs. Med. Eng. Phys. **22**, 595–606 (2000)
93. A. Curtis, C. Wilkinson, Nanotechniques and approaches in biotechnology. Trends Biotechnol. **19**, 97–101 (2001)
94. A. Curtis, M. Riehle, Tissue engineering: the biophysical background. Phys. Med. Biol. **46**, R47–R65 (2001)

Chapter 5
Green Nanotribology and Sustainable Nanotribology in the Frame of the Global Challenges for Humankind

I. C. Gebeshuber

5.1 Introduction

This chapter deals with green and sustainable nanotribology. It highlights the challenges, development and opportunities of these new, emerging fields of science and embeds them in the major frame of the most serious problems we currently face on our planet. Fifteen global challenges are annually identified by the Millennium Project, a major undertaking that was started in 1996 and that incorporates organizations of the United Nations, governments, corporations, non-governmental organizations, universities and individuals from more than 50 countries from around the world. Green Nanotribology is of specific relevance when addressing Global Challenge 13 (Energy) and Global Challenge 14 (Science and Technology). These two challenges are introduced in more detail, and the contributions of Green Nanotribology to specifically address issues that arise due to these two challenges are outlined. Subsequently, the concept of sustainable nanotribology is introduced by correlating nanotribological developments with principles of sustainability identified by the US American Biomimicry Guild. Conclusions and outlook as well as recommendations round up the chapter.

I. C. Gebeshuber (✉)
Department of Microengineering and Nanoelectronics,
Universiti Kebangsaan Malaysia, 43600 UKM Bangi, Selangor, Malaysia
e-mail: gebeshuber@iap.tuwien.ac.at; ille.gebeshuber@mac.com

I. C. Gebeshuber
AC^2T Austrian Center of Competence for Tribology, Wiener Neustadt, Austria

I. C. Gebeshuber
Institute of Applied Physics, Vienna University of Technology, 1040 Wien, Austria

M. Nosonovsky and B. Bhushan (eds.), *Green Tribology*,
Green Energy and Technology, DOI: 10.1007/978-3-642-23681-5_5,

5.1.1 The "Most Serious Problems"

In 2005, Jared Diamond published his book 'Collapse: How societies choose to fail or succeed' [17]. In this work, Diamond identifies four major issues that lead to the collapse of societies. The first issue comprises destruction and loss of natural resources (e.g. destruction of natural habitats, aquacultures, biodiversity loss, erosion and soil damage), the second ceilings on natural resources (e.g. fossil fuels, water, photosynthesis ceiling), the third harmful things that we produce and move around (e.g. toxic man-made chemicals, alien species, ozone hole) and the fourth comprises population issues (e.g. population growth, impact of population on the environment). We need to go green if we want to sustain ourselves.

5.1.2 The "15 Global Challenges"

The Millennium Project was initiated in 1996. Until now, it has comprised the work of 2500 futurists, scholars, decision makers and business planners from over 50 countries. The project has nodes in 30 countries. The Millennium Project publishes the annual State of the Future Report (SOF) [36]. The SOF identifies and deals in detail with the fifteen major global challenges for humanity (Fig. 5.1), and provides an action plan for the world.

The 15 global challenges identified by the Millennium Project are:

1. How can sustainable development be achieved for all while addressing global climate change?
2. How can everyone have sufficient clean water without conflict?
3. How can population growth and resources be brought into balance?
4. How can genuine democracy emerge from authoritarian regimes?
5. How can policymaking be made more sensitive to global long-term perspectives?
6. How can the global convergence of information and communications technologies work for everyone?
7. How can ethical market economies be encouraged to help reduce the gap between rich and poor?
8. How can the threat of new and reemerging diseases and immune microorganisms be reduced?
9. How can the capacity to decide be improved as the nature of work and institutions change?
10. How can shared values and new security strategies reduce ethnic conflicts, terrorism, and the use of weapons of mass destruction?
11. How can the changing status of women help improve the human condition?
12. How can transnational organized crime networks be stopped from becoming more powerful and sophisticated global enterprises?
13. How can growing energy demands be met safely and efficiently?

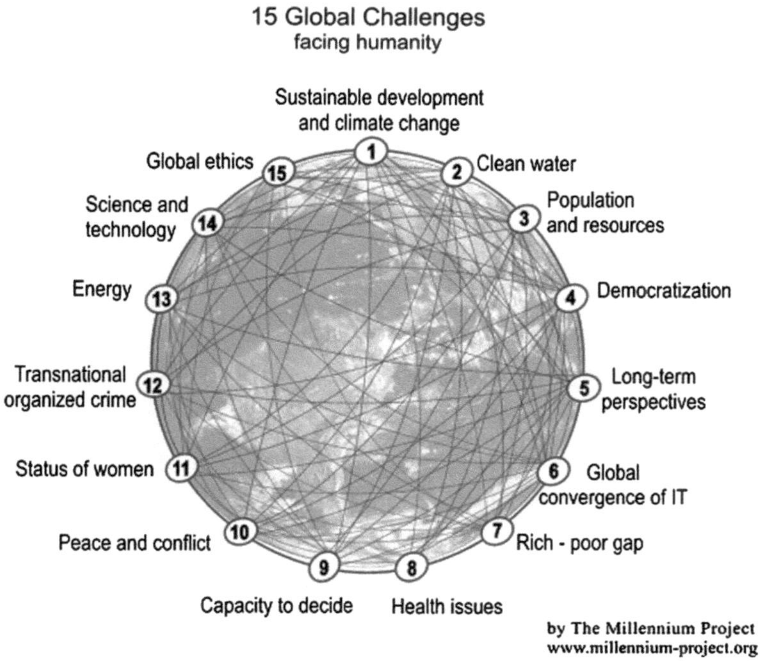

Fig. 5.1 15 Global challenges facing humanity as identified by the millennium project. Source: http://www.millennium-project.org/millennium/images/15-GC.jpg (last accessed 4 October 2011)

14. How can scientific and technological breakthroughs be accelerated to improve the human condition?
15. How can ethical considerations become more routinely incorporated into global decisions?

Green Nanotribology with all its beneficial consequences is of specific relevance for Global Challenge 13 (Energy) and Global Challenge 14 (Science and Technology).

5.2 Green (Nano-)Tribology

Si-wei Zhang, past chairman of the Chinese Tribology Institution, coined the term 'Green Tribology' and launched it as an international concept in June 2009.

> Green tribology is the science and technology of the tribological aspects of ecological balance and of environmental and biological impacts. Its main objectives are the saving of energy and materials and the enhancement of the environment and the quality of life (Peter Jost 2009, [2]).

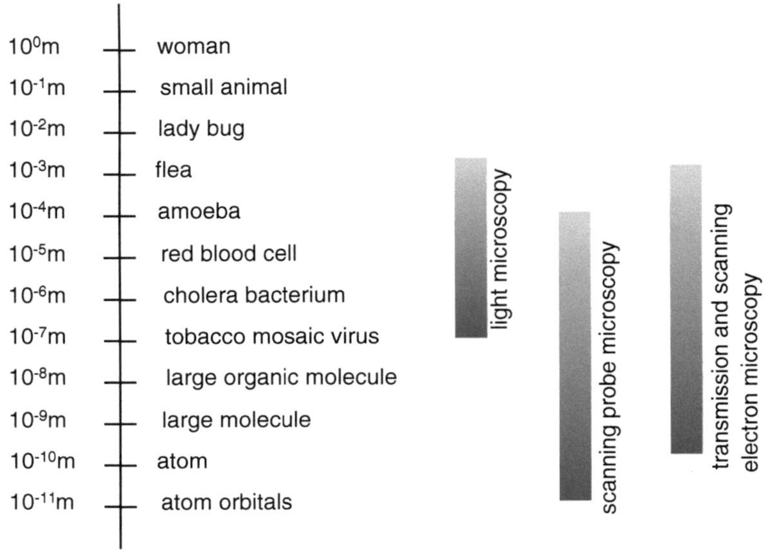

Fig. 5.2 From the macro- to the nanoscale. New types of microscopy allow for visualization and active interaction on very small scales, and therefore open up whole new domains for science and technology, e.g. nanotribology. © 2010 Springer [28]

We need green tribology because of the current pressures on energy, materials and food [58]. A focus on tribology might give breathing space while fuller solutions to environmental problems are being addressed [2]. Tribology must fall into line with the major politics of world environment and energy. Economic benefits derived from the application of tribology for the UK comprise £8–10 billion, out of which 60–70% would be energy related, all this largely from existing and applied research (innovation) [2].

Tribology covers all length scales. In this chapter, we concentrate on nanotribology (length scale some 10^{-9} nm). New types of microscopy now allow access to the nanocosmos, not just to view, but to interact (Fig. 5.2). This opens up completely new opportunities.

Nanotribology deals with nanosurfaces, nanoagents and nanoprocesses. For green nanosurfaces, points such as nanostructured surfaces, hierarchical surfaces, material selection, coated materials and monomolecular lubricant layers need to be addressed. The importance of nanosurfaces regarding Green Nanotribology is in the medium range. Of very high importance for Green Nanotribology are nanoagents. Points to address here are physical and chemical properties, the effect on the environment and biology, and the changes of the properties during the tribo-process. Regarding green nanoprocesses, the importance of points to address is in the medium to low range. Points to address comprise energy efficiency, the share between process relevant energy, destructive energy and waste and reusable energy as well as the effectiveness of reusing process energy [21].

In going green, tribology can benefit from a look at biology. Recently, biology has changed from being a highly descriptive science to a science that can be understood by engineers and researchers coming from the hard sciences, in terms of concepts, ideas, languages and approaches [24, 26]. In former times, tribology as well as biology used to be very descriptive. Inter- and transdisciplinary connections between the two fields were nearly impossible because of limited causal knowledge and limited causal relationships in both fields. This has changed. Today, we have increased causal knowledge [29] in both fields and therefore a promising area of overlap between tribology and biology [25, 30]. Causal knowledge indicates the fact that we know the relevant natural laws and can therefore construct explanations and forecasts. In biology, this is very often the case in physiology: We get cold feet when the vessels contract, *because* according to the laws of physics (fluid mechanics) less blood flows through these vessels (Drack, 2011, personal communication).

Biomimetics, the field that deals with knowledge transfer from biology to engineering and the arts, is a booming science that attracts more and more researchers, papers and attention [5, 48, 83]. Otto Schmitt [69], the inventor of the Schmitt trigger coined this field in 1982. One of the interesting aspects of such an interdisciplinary science as biomimetics is the variety of the publication channels. The author of this chapter for example has long been working in the field of bioinspired nanotribology and has published in journals as diverse as the Polish Botanical Journal (touching on the tribology of photosynthetic microorganisms with rigid parts in relative motion on the nanoscale) [78], Nano Today, elaborating on the tribology of biological hinges and interlocking devices, natural switchable adhesives and self-repairing molecules [22], the Proceedings of the Institution of Mechanical Engineers Part J: Journal of Engineering Tribology (touching on hinges and interlocking devices in microorganisms, [23]) and Tribology [25], proposing new ways of scientific publishing and accessing human knowledge inspired by transdisciplinary approaches regarding nanotribology. Biological best practice systems regarding nanotribology are functional and—in many cases—beautiful (Fig. 5.3).

Turning nanotribology green implies more than just the usage of sustainable additives [21]. Tribology is a systems science; therefore also the environment and the development with time have to be accounted for.

Nanoagents in tribology are additives, products of the additives and byproducts that appear in the system after the technological application. Reaction products (which can be harmful) have to be either chemically inert after use or are fed back to the system for further usage (waste-to-wealth concept). Not-used nanoagents need to be either inert or fed back to the reaction. Potentially harmful byproducts that have nothing to do with the initial nanoagent need to be either neutralized or re-used. Biomimetic tribological nanotechnology might help to turn nanotribology green, but it cannot be stated often enough that biomimetics does not automatically yield sustainable or even just green products [29].

Green control of friction, wear and lubrication on the nanoscale can be achieved by taking into consideration environmental aspects of nanoscale lubrication layers, environmental aspects of nanotechnological surface modification techniques and

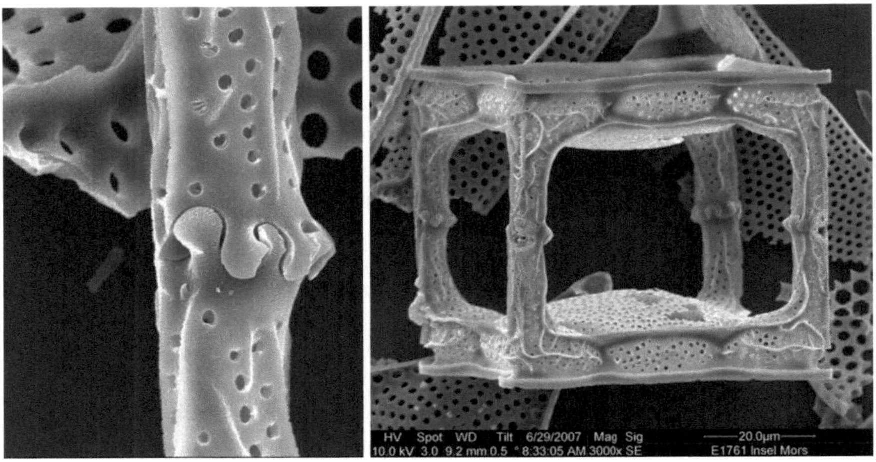

Fig. 5.3 This fossil diatom *Solium exsculptum* lived 45 millions of years ago on the island of Mors in Denmark. Scanning electron microscopy reveals unbroken diatom shells, with elaborate linking structures, and various micromechanical, incl. tribological, optimizations. © F. Hinz, Alfred Wegener Institute Bremerhaven, Germany. Image reproduced with permission

nanotribological aspects of green applications such as artificial photosynthesis. Questions that need to be addressed in turning nanotribology green are for example [21]:

- Do the processes get greener with the envisaged nanotribology (e.g. better coatings, less wear, less stiction)?
- Do the processes turn worse because of chemical reactions?
- Is the envisaged Green Nanotribology only pseudo-green, and in reality the negative impact on the environment/biology is only translated to other layers?

The usage of new technologies, materials and devices might increase advantages, but generates new problems. Exact eco-balance calculations need to be performed to prevent pseudo-green approaches. Biodiesel for example might be greener in the production than conventional products, yet still has technical concerns when used at concentrations greater than 5% [16].

5.3 The Relation of Green (Nano-)Tribology and "Global Challenge 13: Energy"

Issue: How can growing energy demand be met safely and efficiently?

The Millennium Projects annually publishes the State of the Future report (with few pages in print and vast information on CD) and has an extensive webpage (http://www.millennium-project.org). Concise information about this global challenge as well as the Executive Summary of the 2011 State of the Future report can be accessed online at http://www.millennium-project.org/millennium/Global_Challenges/chall-13.html and at http://www.millennium-project.org/millennium/SOF2011-English.pdf. The following quotations from the Millennium Project webpage on Global Challenge 13 give a glimpse on the points touched by the futurists:

> The world energy demand is expected to increase by between 40 and 50% over the next 25 years, with the vast majority of the increase being in China and India.
>
> G20 leaders pledged to phase out fossil fuel subsidies in the medium term.
>
> The World Bank estimates that countries with underperforming energy systems may lose up to 1–2% of growth potential every year, while billions of gallons of petroleum are wasted in traffic jams around the world.
>
> Massive saltwater irrigation can produce 7,600 L/ha-year of biofuels via halophyte plants and 200,000 L/ha-year via algae and cyanobacteria, instead of using less-efficient fresh-water biofuel production that has catastrophic effects on food supply and prices. Exxon announced its investment of $600 million to produce liquid transportation fuels from algae.
>
> CO_2 emissions from coal plants might be re-used to produce biofuels and perhaps carbon nanotubes. The global market value for liquid biofuel and bioenergy manufacturing is estimated at $102.5 billion in 2009 and is projected to grow to nearly $170.4 billion by 2014.
>
> Innovations are accelerating: concentrator photovoltaics that dramatically reduce costs; waste heat from power plants, human bodies, and microchips to produce electricity; genomics to create hydrogen-producing photosynthesis; buildings to produce more energy than consumed; solar energy to produce hydrogen; microbial fuel cells to generate electricity; and compact fluorescent light bulbs and light-emitting diodes to significantly conserve energy, as would nanotubes that conduct electricity. Solar farms can focus sunlight atop towers with Stirling engines and other generators. Estimates for the potential of wind energy continue to increase, but so do maintenance problems. Plastic nanotech photovoltaics printed on buildings and other surfaces could cut costs and increase efficiency. The transition to a hydrogen infrastructure may be too expensive and too late to affect climate change, while plug-in hybrids, flex-fuel, electric, and compressed air vehicles could provide alternatives to petroleum-only vehicles sooner. Unused nighttime power production could supply electric and plug-in hybrid cars. National unique all-electric car programs are being implemented in Denmark and Israel, with discussions being held in 30 other countries.

According to the Millennium Project, Global Challenge 13 will have been addressed seriously when the total energy production from environmentally benign processes surpasses other sources for five years in a row and when atmospheric CO_2 additions drop for at least five years.

Opportunities: Contribution of Green Nanotribology to meeting the energy demand safely and efficiently.

Energy is one of two global challenges identified by the Millennium Project where optimized tribology can substantially contribute. Opportunities regarding this challenge comprise renewable fuels, the use of waste energy and more efficient energy conversion systems. Regarding Green Nanotribology, energy management, wear management and self-healing coatings are of high potential.

Biofuels still have major unresolved tribological issues related to their hygroscopic properties and related absorption or adsorption of water, leading to microbiological activity, corrosion and fuel instability. These issues need to be addressed with tribology on all length scales. Furthermore, underperforming energy systems, MEMS energy harvesters and wind energy plants need to be tribologically optimized. Green Nanotribology can provide its share in all these areas.

5.4 The Relation of Green (Nano-)Tribology and "Global Challenge 14: Science and Technology"

Issue: How can scientific and technological breakthroughs be accelerated to improve the human condition?

The following quotations from the Millennium Project webpage on Global Challenge 14 gives a glimpse on the points touched by the futurists (http://www.millennium-project.org/millennium/Global_Challenges/chall-14.html):

> The acceleration of S and T innovations from improved instrumentation, communications among scientists, and synergies among nanotechnology, biotechnology, information technology, cognitive science, and quantum technology continues to fundamentally change the prospects for civilization.
>
> Nanobots the size of blood cells may one day enter the body to diagnose and provide therapies and internal VR imagery.
>
> Nanotechnology-based products have grown by 25% in the last year to over 800 items today for the release of medicine in the body, thin-film photovoltaics, super hard surfaces, and many lightweight strong objects.
>
> Despite these achievements, the risks from acceleration and globalization of S and T remain ... and give rise to future ethical issues...
>
> The environmental health impacts of nanotech are in question.
>
> ... supporting basic science is necessary to improve knowledge that applied science and technology draws on to improve the human condition.
>
> We need a global collective intelligence system to track S and T advances, forecast consequences, and document a range of views so that politicians and the public can understand the potential consequences of new S and T.

Scientific and technological breakthroughs need to be accelerated to improve the human condition. We currently have a major issue with over-information, and how to deal with it. Another issue relates to the major gaps between investors, innovators and inventors, regarding their goals and visions, dreams and approaches, reward systems and driving reasons. There is a need for new ways, concise visions and researchers who understand the big picture. There is a need for specialists who are coordinated by generalists.

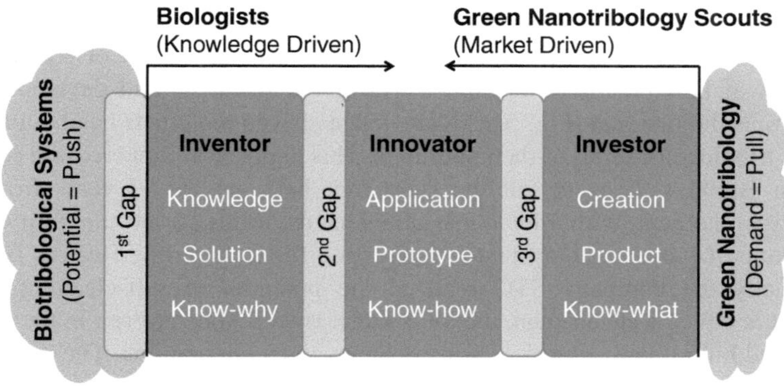

Fig. 5.4 The potential of Green Nanotribology in addressing Global Challenge 14, Science and Technology. See text

Questions that need to be addressed comprise: What good is science for if the work of scientists disappears in journals and books, and nobody applies it? What good is it if the scientists are only sustaining the publishing industry and/or serving the 'research market' and their results are hard/costly to access for the general public and the industry?

According to the Millennium Project, Global Challenge 14 will have been addressed seriously when the funding of R and D for societal needs reaches parity with funding for weapons and when an international science and technology organization is established that routinely connects world S and T knowledge for use in R and D priority setting and legislation.

Opportunities: Contribution of Green Nanotribology to accelerate scientific and technological breakthroughs to improve the human condition.

Sustainable Nanotribology ensures that there are no adverse environmental health impacts of nanotech on biology and the environment. Health impacts of nanotechnology are currently dealt with extensively [4, 12, 20, 40, 41, 62, 74–76, 79]. Green Nanotribology can, e.g. help address nanobot issues such as stiction and too high adhesion, emerging 3D MEMS tribology issues [66] and increase the quality of lab-on-a-chip devices. Micro- and nanotribological research has furthermore inspired a concept for a global collective intelligence system to track S and T advances [25] which might be extended to forecast consequences and to document a range of views, which is important regarding politics and governance.

Opportunities in science and technology are furthermore improvement of the human condition, aiming at having a context of knowledge and an equilibrium between generalists and specialists. The contribution of Green Nanotribology comprises establishment of a best practice example of a pipeline from the inventor to the innovator and the investor (Fig. 5.4). Nanobiotribological systems such as hinges and interlocking devices or self-repairing adhesives in diatoms [23, 34] or switchable adhesion exemplified by the selectin/integrin system [60] or the dry adhesives of the Gecko foot [3] are just some examples of the treasure box of best practice examples for the green nanotribologist looking for inspiration in nature. They provide the potential (push, Fig. 5.4). Between this world of solutions, of examples, of best practices, and the world of the scientists and developers, the inventors, is the first gap (Fig. 5.4). Knowledge–driven biologists head out and try to reach tribologists, offering their solutions. This happens with increasing success. Wilhelm Barthlott likes to tell the story that he went for 11 years from one company to the next, with a lotus leaf in his hands, trying to sell the purity of the sacred lotus [6] as something that is interesting for technology. He was ignored, until finally the company STO realized the potential of self-cleaning paints. Barthlott is now a well-off man, the lotus effect is well known, even in the general public, and biomimetics is a field with very positive connotations. The next gap is between the inventor and the innovator (How do the research results become prototypes? Who identifies promising designs/developments and promotes commercialization?), and the third gap is between the innovator and the investor (How do the prototypes translate into marketable products? Who 'sells' the prototypes to industry? Who does the communication between researchers and developers and industry?). From the technology side, market-driven green nanotribologists head out to screen beyond the gaps, and identify people from other fields to work with, and fulfill the demand (pull) of the market (Fig. 5.4). What we need is a pipeline from knowledge via the application to the creation, from the solution via the prototype to the product, and from the know-why via the know-how to the know-what.

5.5 Bioinspired Optimization Levers in Green Nanotribology

Scherge and Rehl [68] identified four optimization levers in tribology: breaking-in, additives, finishing and material selection. Green Nanotribology can address all these four levers [21]: (1) The key advance of biomimetic Green Nanotribology is in the area of breaking-in. Our current technological systems and devices are sequentially produced and need 'breaking-in'. Organisms that are growing from a nucleus on the other hand need 'initial consolidation'. Contrary to the process of breaking-in, the initial consolidation in organisms is equivalent to a 'peak' that establishes function and controls quality via 'hail or fail'. Biological systems mainly have soft materials and use water-based lubricants; they have tribological properties in many cases superior to the ones of technical devices and can serve as inspiration for new tribological approaches. (2) Examples for additives in biological lubricants are proteins, e.g. in the mucus of snails (which acts as adhesive and lubricant, [46]) and in fish mucus (which, amongst other functions, reduces drag, [13, 42, 51]) as well as in synovial joints (hip, knee, shoulder, [54]). Biological lubricants are water soluble. Currently, various research projects aim at the development of water-soluble lubricants for technological applications (e.g. by the V.A. Kargin Polymer Chemistry and Technology Research Institute or by the Singapore Institute of Manufacturing Technology). Environmental compliance and improved tribological properties could go hand-in-hand. (3) Nature's materials are complex, multi-functional, hierarchical and responsive and in most instances functionality on the nanoscale is combined with performance on the macroscale. Biological materials have elaborate surface textures and finishing: the finishing of biological tribosurfaces has been the focus of investigation in major research networks, and the literature is bursting with examples of soft matter with excellent finishing. Examples comprise hierarchical surface structures down to the nanoscale in the sand skink [8], in snake skin [39], in the surface of the mammal eye [14]. Optimized surfaces might render the use of lubricants obsolete, and would be a most elegant way to approach tribology issues. (4) The optimization lever 'material selection' is of utmost importance in biological nanotribology. The correlation of structure and function and structure and material need not be forgotten in this regard. In biology, few chemical elements and few different materials are used; these materials are either slightly changed to obtain added value and to fulfill various functions (cf. collagen) and/or they are structured, in some cases even hierarchically. Diatoms [65] are excellent examples from nature for tribological optimization under limited material variety conditions [23, 32].

5.6 Goals for Effective Green Nanotribology

Goals for effective Green Nanotribology are located in three main areas: Production (agents), reaction (agents; object to nanoproduct; waste agents direct effects) and nanoproduct life cycle (effects on the environment during the service period and during degradation). In the production the goal is minimum pollution, in the reaction the goal is to keep the reaction only where it needs to be and not beyond, and in the life cycle care has to be taken to ensure minimum influence on substances and improved degradation characteristics (e.g. via a decay booster).

5.7 Sustainable Nanotribology

The distinction between 'green' and 'sustainable' can best be explained with the example of wood. Wood is green, but if it comes from the rainforest and not from plantations it is not sustainable. Not everything that is green is sustainable, and not everything that is sustainable is green.

In this section we develop a concept of Sustainable Nanotribology inspired by 'Life's Principles' (Table 5.1) as introduced by the U.S. American 'Biomimicry Guild'. The Biomimicry Guild has been combining sustainability and biomimetics from their very beginnings at the end of the 1990s, stating that these two have to be inseparably connected. Other researchers state that biomimicry is a design method, and as such is independent from a value such as sustainability [29].

Table 5.1 Life's principles as established by the Biomimicry Guild (http://www.biomimicryguild.com/) can provide a guide towards rendering nanotribology green

Life's principles–design lessons from nature			
1. Earth's operating conditions:			
	1.1. Water-based		
	1.2. Subject to limits and boundaries		
	1.3. In a state of dynamic non-equilibrium		
2. Life creates conditions conducive to life			
	2.1. Optimizes rather than maximizes		
		2.1.1. Multi-functional design	
		2.1.2. Fits form to function	
		2.1.3. Recycles all materials	
	2.2. Leverages interdependence		
		2.2.1. Fosters cooperative relationships	
		2.2.2. Self-organizing	
	2.3. Benign manufacturing		
		2.3.1. Life-friendly materials	
		2.3.2. Water-based chemistry	
		2.3.3. Self-assembly	
3. Life adapts and evolves			
	3.1. Locally attuned and responsive		
		3.1.1. Resourceful and opportunistic	
			3.1.1.1. Shape rather than material
			3.1.1.2. Simple, common building blocks
			3.1.1.3. Free energy
		3.1.2. Feedback loops	
			3.1.2.1. Antennae, signal, response
			3.1.2.2. Learns and imitates
	3.2. Integrates cyclic processes		
		3.2.1. Feedback loops	
		3.2.2. Cross-pollination and mutation	
	3.3. Resilient		
		3.3.1. Diverse	
		3.3.2. Decentralized and distributed	
		3.3.3. Redundant	

The Biomimicry Guild webpage (http://www.biomimicryguild.com) gives six categories for sustainable biomimetics, each with various subcategories. In the following, we correlate this list with nanotribology, identify applications in green nanotribology and give tribological examples where these principles are already incorporated (research, prototype or device stage).

1. Evolve to survive

Principle: Sustainability can be ensured when information to ensure enduring performance is continually incorporated and embedded.

Application in Green Nanotribology: Keeping nanotribology state-of-the art. Continuous implementation of the most recent research and development results ensures continuous optimization of nanotribological materials, structures and processes.

There are three major ways to implement this principle: (1) *When strategies that work are replicated, when successful approaches are repeated.* Implications for Green Nanotribology would be biomimetic approaches, learning from nature, looking at natural biotribological model systems (best practices) and learn from them [22, 27, 30, 31]. (2) *When the unexpected is integrated.* In biology, the incorporation of mistakes in ways that can lead to new forms and functions turned out to be highly successful. Regarding Green Nanotribology, this way could be implemented by policies that allow for tribological research along unconventional paths, by responsible and environmentally concerned researchers. (3) *Exchange and alteration of information (reshuffling of information).* In biology this implementation has highly successful examples in bacteria (e.g. gene swapping, [61]). Concerning Green Nanotribology the implications would be biomimetics, learning from nature and tribology–related research and development directed by deep understanding of the underlying principles.

2. Resource efficiency regarding material and energy

Principle: Skillfully and conservatively take advantage of local resources and opportunities.

Application in Green Nanotribology: Keeping the tribosystem closed, reusing and recycling of substances and of waste energy.

There are four major ways to implement this principle: (1) *Multi-functional design ensures the meeting of multiple needs with one solution.* Comparable to the pluripotency of stem cells, for example surface textures or formulations of additives or tribological processes can be designed in a way that is sustainable. If now the textures, formulations and processes are pluripotent, and can easily be adjusted for the respective tribosystem, the advantage would be that there are already established green routes of production, usage and disposal, and no new research to make the new approaches green would have to be performed. In nature, we have very often just slight variations in the same material, structure or function, to accommodate totally different needs. One example is collagen that occurs in bones, skin, tendons and the cornea [67]. (2) *The usage of low energy processes ensures minimum energy consumption.* This can be realized, e.g. by temperature, pressure or time reduction. One fabulous example from nature is biomineralization. More than 60 different minerals are produced by organisms, at ambient conditions—ceramics in our teeth, magnets in bacteria or silica in diatoms are just some examples [9, 70]. (3) *Recycling of all materials keeps materials in a closed loop.* Food chains in biology are exquisite examples for closed-loop materials usage. Concerning Green Nanotribology, closed tribosystems, reuse of energy and material as well as recycling are of relevance. (4) *The selection for shape or pattern based on need ensures the fit of form to function.* In biology, structure–function relationships are omnipresent [71, 72, 77]. Especially in natural micro-

and nanotribological systems, e.g. hinges and interlocking devices in diatoms, this relationship is obvious, and can serve as inspiration for new and emerging man-made micro- and nanotribological systems [23, 32, 33].

3. Adaptation to changing conditions

Principle: Appropriately respond to dynamic contexts.

Application in Green Nanotribology: Multifunctional responsive nanosurfaces, nanoagents and nanoprocesses that change dependent on the environment and that are used in amounts as minimal as necessary.

There are three major ways to implement this principle: (1) *The maintenance of integrity through self-renewal ensures persistency by constantly adding energy and matter to the system*. This energy and matter is subsequently used to heal/repair and improve the system. Passive and active tribological systems such as self-repairing adhesives or anti-corrosion layers are the respective examples from tribology [1, 35, 47, 81]. (2) *Resilience through variation, redundancy and decentralization ensures the maintenance of function following disturbance*. (Resilience—as defined by http://www.wordnet.princeton.edu—denotes the physical property of a material that can return to its original shape or position after deformation that does not exceed its elastic limit). In biology this is achieved by the incorporation of a variety of duplicate forms, processes or systems that are not exclusively located together. In the concept development of Sustainable Nanotribology, implementation of this feature might prove complicated, and some additional brainstorming might be necessary to identify tribologically relevant biological best practices and the related translation to engineering. (3) *The incorporation of diversity (inclusion of multiple forms, processes or systems) to meet a functional need has proven highly successful in biology*. Examples can be found on all scales, from single biomolecules (size some nanometers) to tissues and limbs (size some centimeters) to whole organisms and ecosystems. Nanodiversity as a concept in nanotribology was just recently introduced as one of the major goals for effective Green Nanotribology [21] and should be developed further towards sustainability.

4. Integration of development with growth

Principle: Invest optimally in strategies that promote both development and growth.

Application in Green Nanotribology: New research and development in sustainable nanotribology should not just be concerned with growth and revenue—even when profits are high, a balance has to be sought between development and growth—new research results have to be implemented, even if it is not beneficial for the economy in the short run—because in the long run only sustainable approaches ensure our survival [17].

There are three major ways to implement this principle: (1) *The combination of modular and nested components progressively fits multiple units within each other, from simple to complex*. In biology, hierarchy and multifunctionality can be seen in many organisms ([49, 82, 19]). In tribology, and especially in nanotribology, we have just started to develop such elaborate approaches, e.g. nanotribological multiscale friction mechanisms and hierarchical surfaces [55]. (2) *Building from the bottom up allows the use of code, work with molecular building blocks and inclusion of functionalities*. Organisms are all built bottom–up, our current technology, however, is still in most cases using top–down techniques [11]. The development and successful application of nanotribological bottom–up approaches would be of highest interest, since one of the important levers of optimization lies in surface finishing [21]. (3) *Self-organization, the creation of globally coherent patterns from just local interactions, with no central control unit*. Realizations of this implementation can be seen on all length scales in biology, from single molecules to social behavior. In Green Nanotribology it would be beneficial if the tribosystem would locally react according to varying demands, being only as good as necessary, and not as good as possible.

5. Responsiveness and being locally attuned

Principle: Fit into and integrate with the surrounding environment.

Application in Green Nanotribology: Sometimes we overdo it with our current technology. Sometimes our systems are too good, too expensive, too anything. Responsive Green Nanotribological systems would be just as good as necessary, with additional benefits of energy saving and even greater environmental soundness.

There are four major ways to implement this principle: (1) *Usage of readily available materials and energy*. Implementation of this principle ensures building with abundant, accessible materials, while harnessing freely available energy [59, 63]. A possible realization of this would be flexible nanotribology that uses only readily available materials and energy. (2) *The cultivation of cooperative relationships finds value through win–win interactions*. This aspect is omnipresent in biology, and lacks in current technology. Again, this might be a point where brainstorming could yield some interesting new approaches. (3) *The leverage of cyclic processes takes advantage of phenomena that repeat themselves*. Also this aspect is omnipresent in biology, and lacks in current technology. Brainstorming for new approaches is necessary. (4) *The use of feedback loops engages in cyclic information flows to modify a reaction appropriately*. Reinforcement via feedback loops is important and is applied in biology and in technology [15, 37, 73]. Regarding nanoprocesses, we are just beginning to apply cyclic information flows [43, 52, 53].

6. Usage of life-friendly chemistry

Principle: Use chemistry that supports life processes.

Application in Green Nanotribology: Green chemistry is a prerequisite for Green Nanotribology [21, 56, 57].

There are three major ways to implement this principle: (1) *Building selectively with a small subset of elements (assembly of relatively few elements in elegant ways)*. This principle is strongly implemented in biology. In biological materials, just a couple of elements are used in major amounts, and even just a couple of

Table 5.2 Major goals for effective Green Nanotribology and how it can benefit from biology [21]

Green Nanotribology major goals		Importance	Nature's solutions
Optimized system energy balance	Minimizing destructive energy	Medium	Water–based lubricants; Predetermined breaking points; Responsive marerials; Structure rather than material
	Shield tribosystem against damaging consequences	High	Integration intstead of additive construction; Optimization of the whole instead of maximization of a single component future; Multi-functionality instead of mono-functionality; Energy efficiency; Development via trial-and-error processes
	Reuse energy and neutralize waste energy	Low	Organisms
Protection of the environment from process residues	Unused agents	Low	Reuse in the ecosystem
	Pollutants	Medium	Biodegradability; Confined spaces for chemical processes
	Process reliability and worst case scenario dangers	High	Highly developed over time (evolution)
Environmental cost of the process itself	Effort to produce agents	High	Optimized (on system level)
	Purer inputs with more waste in the preparation	Medium	Water-based chemistry; Cell organelles serve as nano factories; Shielding of process via membranes
	Different economy of scale	Low	Major evolutionary transitions; Mammals versus cold-blooded animals
Preservation of nanodiversity		Unknown	Additives in ultra-low concentration ensure nanodiversity; Reuse of the same base material with slight modifications for various applications

chemical compounds. Especially in small biological entities (some molecules, nanotextured surfaces, nanostructures, etc.) the structure rather than the material determines the function [18, 44, 64]. This goes as far as properties of biological entities that rely on structure alone. One impressive example in this regard are structural colors, where the coloration does not come from pigments but from minuscule structures with sizes on the order of the wavelength of visible light, generating colors by physical mechanisms such as diffraction, scattering and interference (see e.g. [10, 45, 50, 84]). One impressive example from biology for building with a small subset of elements are the diatoms, unicellular algae with a skeleton made from silica (see Fig. 5.3, [65]). The silica is structured with various functional levels of hierarchy, it is micromechanically optimized [38], has optical properties that are of high interest for nanotechnology [84] and is produced in ambient conditions which makes it interesting for nanoengineers attempting to build structures with intrinsic functions at ambient conditions and with life-friendly chemistry. (2) *Breaking down of products into benign constituents using chemistry in which decomposition results in no harmful by-products.* Nature does it this way, and we with our current technology are still far away from doing it this way—although regarding going green it would be very beneficial. In green nanotribology, we could start another attempt to develop entities that decompose into harmless end products [21]. (3) *Water is the generally used solvent in the chemistry of life.* Such water-based approaches are slowly starting to enter our current technology. In their 2004 article in Nature on the nonlinear nature of friction, Israelachvili and co-workers stress that current oil-based lubricants are by far outclassed by the water-based lubricants used in nature [80]. The hip joint with its amazingly low friction coefficient is one of the examples from nature on highly efficient tribosystems lubricated with water-based chemistry [54].

The goals of Green Nanotribology are to provide technical support to the preservation of resources and energy and to propel the society forward towards sustainability.

The goals of Green Nanotribology and how it can benefit from biology can be grouped into four categories (Table 5.2). The best practices from nature listed in Table 5.2 are highly diverse and show a multitude of different approaches that can be learnt from.

5.8 Conclusions and Outlook

Green Nanotribology and Sustainable Nanotribology have high potential in science and technology. Tribology together with the other fields of engineering will increasingly aim at providing sustainable solutions. We need to establish ethical codes of conduct for using novel technologies, to make the transition smooth and well rounded. Successful tribologists are inherently transdisciplinary thinkers—this is needed in our increasingly complex world. Tribologists will successfully contribute to address major global challenges.

The Sustainable Nanotribology concept developed in this book chapter is inspired by life's principles as introduced by the Biomimicry Guild. Open points regarding Sustainable Nanotribology touch upon adaptation to changing conditions and responsiveness and being locally attuned: Resilience through variation, redundancy and decentralization that ensure the maintenance of function following disturbance as well as the cultivation of cooperative relationships that finds value through win–win interactions as well as the leverage of cyclic processes that takes advantage of phenomena that repeat themselves need to be implemented in the concept. Including the sustainability approach as introduced by Albert Bartlett [7] can help refine the concept further, yielding Sustainable Nanotribology for society.

Acknowledgments The National University of Malaysia funded part of this work with its leading-edge research project scheme 'Arus Perdana'. Profs. F. Aumayr, H. Störi and G. Badurek from the Vienna University of Technology are acknowledged for enabling ICG three years of research in the inspiring environment in Malaysia.

References

1. D.V. Andreeva, D. Fix, H. Moehwald, D.G. Shchukin, Self-healing anticorrosion coatings based on pH-sensitive polyelectrolyte/inhibitor sandwichlike nanostructures. Adv. Mater. **20**(14), 2789–2794 (2008)
2. Anonymous, Summary: world tribology congress 2009 (WTC IV) (2010), International Tribology Council Information, No. 191, https://www.stle.org/UserFiles/File/Tribology%20 Conference/TR2009/ITC191-News%20_Jan%202010_%20WTC%20Report.pdf Accessed 4 October 2011
3. K. Autumn, Y.A. Liang, S.T. Hsieh, W. Zesch, W.P. Chan, T.W. Kenny, R. Fearing, R.J. Full, Adhesive force of a single gecko foot-hair. Nature **405**(6787), 681–685 (2000)
4. D.M. Balshaw, M. Philbert, W.A. Suk, Research strategies for safety evaluation of nanomaterials, Part III: nanoscale technologies for assessing risk and improving public health. Toxicol. Sci. **88**(2), 298–306 (2005)
5. Y. Bar-Cohen, *Biomimetics: Biologically Inspired Technologies* (CRC Press, Boca Raton, 2005)
6. W. Barthlott, C. Neinhuis, The purity of sacred lotus or escape from contamination in biological surfaces. Planta **202**, 1–8 (1997)
7. A. Bartlett, Environmental sustainability (1997), Talk given at the American Association of Physics Teachers Meeting, Denver, Colorado, 16 August 1997, http://www.AlBartlett.org/articles/art1997aug16.html last accessed 6 October 2011
8. W. Baumgartner, F. Saxe, A. Weth, D. Hajas, D. Sigumonrong, J. Emmerlich, M. Singheiser, W. Böhme, J.M. Schneider, The sandfish's skin: morphology, chemistry and reconstruction. J. Bionic. Eng. **4**(1), 1–9 (2007)
9. P. Behrens, E. Baeuerlein, *Handbook of Biomineralization: Biomimetic and Bioinspired Chemistry* (Wiley–VCH, Weinheim, 2009)
10. S. Berthier, *Iridescences: The Physical Colors of Insects* (Springer, New York, 2006)
11. B. Bhushan (ed.), *Springer Handbook of Nanotechnology*, 3rd edn. (Springer, Heidelberg, 2010)
12. P. Borm, F.C. Klaessig, T.D. Landry, B. Moudgil, J. Pauluhn, K. Thomas, R. Trottier, S. Wood, Research strategies for safety evaluation of nanomaterials, Part V: role of dissolution in biological fate and effects of nanoscale particles. Toxicol. Sci. **90**(1), 23–32 (2006)
13. T.L. Daniel, Fish mucus: in situ measurements of polymer drag reduction. Biol. Bull. **160**, 376–382 (1981)
14. J.P. Davim (ed.), in *Biomaterials and Medical Tribology: Research and Development*. Woodhead Publishing Reviews: Mechanical Engineering Series No. 4 (2011)
15. A.P. del Pobil, J. Mira, A. Moonis (eds.), Methodology and tools in knowledge-based systems. 11th International Conference on Industrial and Engineering Applications of Artificial Intelligence and Expert Systems. Notes in Artificial Intelligence, vol. 1, (Springer, Berlin, 1998)

16. A. Demirbas, *Biodiesel: A Realistic Fuel Alternative for Diesel Engines* (Springer-Verlag, London, 2008)
17. J. Diamond, *Collapse: How Societies Choose to Fail or Succeed* (Viking Books, NY, 2005)
18. H. Fan, Y. Lu, A. Stump, S.T. Reed, T. Baer, R. Schunk, V. Perez-Luna, G.P. López, C.J. Brinker, Rapid prototyping of patterned functional nanostructures. Nature **405**, 56–60 (2000)
19. P. Fratzl, R. Weinkamer, Nature's hierarchical materials. Prog. Mater. Sci. **52**, 1263–1334 (2007)
20. A. Gazsó, The Austrian experience–project nanotrust (2008), OECD Database, http://www.oecd.org/dataoecd/35/58/42326290.pdf Accessed 14 Aug 2010
21. I.C. Gebeshuber, Green nanotribology. Proc. IMechE Part C: J. Mech. Eng. Sci., Special Issue, Guest Editors: Ille C. Gebeshuber, Manish Roy, (to appear in February 2012)
22. I.C. Gebeshuber, Biotribology inspires new technologies. Nano Today **2**(5), 30–37 (2007)
23. I.C. Gebeshuber, R.M. Crawford, Micromechanics in biogenic hydrated silica: hinges and interlocking devices in diatoms. Proc. IMechE Part J: J. Eng. Tribol. **220**(8), 787–796 (2006)
24. I.C. Gebeshuber, M. Drack, An attempt to reveal synergies between biology and engineering mechanics. Proc. IMechE Part C: J. Mech. Eng. Sci. **222**(7), 1281–1287 (2008)
25. I.C. Gebeshuber, B.Y. Majlis, New ways of scientific publishing and accessing human knowledge inspired by transdisciplinary approaches. Tribol.–Surf. Mater. Interfaces **4**(3), 143–151 (2010)
26. I.C. Gebeshuber, B.Y. Majlis, 3D corporate tourism: a concept for innovation in nanomaterials engineering. Int. J. Mat. Eng. Innov. **2**(1), 38–48 (2011)
27. I.C. Gebeshuber, M. Drack, M. Scherge, Tribology in biology. Tribol.–Surf. Mater. Interfaces **2**(4), 200–212 (2008)
28. I.C. Gebeshuber, M. Drack, F. Aumayr, H.P. Winter, F. Franek, Scanning Probe Microscopy: From Living Cells to the Subatomic Range, in *Biosystems Investigated by Scanning Probe Microscopy. Fuchs*, 1st edn., ed. by H. Fuch, B. Bhushan (Springer, New York, 2010), p. 834
29. I.C. Gebeshuber, P. Gruber, M. Drack, A gaze into the crystal ball–biomimetics in the year 2059. Proc. IMechE Part C: J. Mech. Eng. Sci. **223**(12), 2899–2918 (2009)
30. I.C. Gebeshuber, B.Y. Majlis, H. Stachelberger, Tribology in biology: biomimetic studies across dimensions and across fields. Int. J. Mech. Mat. Eng. **4**(3), 321–327 (2009)
31. I.C. Gebeshuber, B.Y. Majlis, H. Stachelberger, Biomimetics in tribology, in *Biomimetics–Materials, Structures and Processes. Examples, Ideas and Case Studies*, ed. by P. Gruber, D. Bruckner, C. Hellmich, H.-B. Schmiedmayer, H. Stachelberger, I.C. Gebeshuber. Series: Biological and Medical Physics, Biomedical Engineering, Series Editor Claus Ascheron (Springer, Berlin), 25–50 (in press)
32. I.C. Gebeshuber, H. Stachelberger, M. Drack, Diatom bionanotribology–biological surfaces in relative motion: their design, friction, adhesion, lubrication and wear. J. Nanosci. Nanotechnol. **5**(1), 79–87 (2005)
33. I.C. Gebeshuber, H. Stachelberger, B.A. Ganji, D.C. Fu, J. Yunas, B.Y. Majlis, Exploring the innovational potential of biomimetics for novel 3D MEMS. Adv. Mat. Res. **74**, 265–268 (2009)
34. I.C. Gebeshuber, J.B. Thompson, Y. Del Amo, H. Stachelberger, J.H. Kindt, In vivo nanoscale atomic force microscopy investigation of diatom adhesion properties. Mat. Sci. Technol. **18**(7), 763–766 (2002)
35. S.K. Ghosh, *Self-healing Materials: Fundamentals, Design Strategies, and Applications* (Wiley-VCH, Weinheim, 2009)
36. J.C. Glenn, T.J. Gordon, E. Florescu, 2011 State of the future. (The Millennium Project, 2011)
37. P.O. Haikonen, *Robot Brains: Circuits and Systems for Conscious Machines* (Wiley-Interscience, Chichester, 2007)
38. C.E. Hamm, R. Merkel, O. Springer, P. Jurkojc, C. Maier, K. Prechtel, V. Smetacek, Architecture and material properties of diatom shells provide effective mechanical protection. Nature **421**, 841–843 (2003)

39. J. Hazel, M. Stone, M.S. Grace, V.V. Tsukruk, Nanoscale design of snake skin for reptation locomotions via friction anisotropy. J. Biomech. **32**(5), 477–484 (1999)
40. M.P. Holsapple, L.D. Lehman-McKeeman, Forum Series: research strategies for safety evaluation of nanomaterials. Toxicol. Sci. **87**(2), 315 (2005)
41. M.P. Holsapple, W.H. Farland, T.D. Landry, N.A. Monteiro-Riviere, J.M. Carter, N.J. Walker, K.V. Thomas, Research strategies for safety evaluation of nanomaterials, Part II: Toxicological and safety evaluation of nanomaterials, current challenges and data needs. Toxicol. Sci. **88**(1), 12–17 (2005)
42. S. Jakowska, Mucus secretion in fish–a note. Ann. N. Y. Acad. Sci. **160**, 458–462 (1963)
43. R.A.L. Jones, Challenges in soft nanotechnology. Faraday Discuss. **2009**(143), 9–14 (2009)
44. A. Kessel, N. Ben-Tal, *Introduction to Proteins: Structure, Function, and Motion*. Chapman and Hall/CRC Mathematical and Computational Biology (CRC Press, Boca Raton, 2010)
45. S. Kinoshita, *Structural Colors in the Realm of Nature* (World Scientific Publishing Company, Singapore, 2008)
46. A. Kobayashi, I. Yamamoto, T. Aoyama, in *Tribology of a Snail* (*terrestrial gastropod*). Proceedings 29th Leeds-Lyon Symposium on Tribology, Tribology Series, vol. 41 (Elsevier, Amsterdam, 2004), pp. 429–436
47. A. Kumar, L.D. Stephenson, J.N. Murray, Self-healing coatings for steel. Progr. Org. Coat. **55**(3), 244–253 (2006)
48. C.S.S.R. Kumar (ed.), *Biomimetic and Bioinspired Nanomaterials*, Series Nanomaterials for Life Sciences (Wiley-VCH, Weinheim, 2010)
49. R. Lakes, Materials with structural hierarchy. Nature **361**, 511–515 (1993)
50. D. Lee, *Nature's Palette: The Science of Plant Color* (University of Chicago Press, Chicago, 2007)
51. S.C. Ling, T.Y.J. Ling, Anomalous drag-reducing phenomenon at a water/fish-mucus or polymer interface. J. Fluid Mech. **65**(3), 499–512 (1974)
52. Y. Liu, E. Kim, R. Ghodssi, G.W. Rubloff, J.N. Culver, W.E. Bentley, G.F. Payne, Biofabrication to build the biology–device interface. Biofabrication **2**, 022002 (2010)
53. M.O. Macqueen, J. Mueller, C.F. Dee, I.C. Gebeshuber, GEMS: a MEMS-based way for the innervation of materials. Adv Mater Res **254**(8), 34–37 (2011)
54. J.M. Mansour, Biomechanics of Cartilage, in *Kinesiology: The Mechanics and Pathomechanics of Human Movement*, ed. by C.A. Oatis (Lippincott Williams and Wilkins, Philadelphia, 2003)
55. M. Nosonovsky, B. Bhushan, Multiscale friction mechanisms and hierarchical surfaces in nano- and bio-tribology. Mat. Sci. Eng. R **58**(3–5), 162–193 (2007)
56. M. Nosonovsky, B. Bhushan, Green tribology: principles, research areas and challenges. Phil. Trans. R. Soc. A **368**(1929), 4677–4694 (2010)
57. M. Nosonovsky, B. Bhushan, Theme issue green tribology. Phil. Trans. Roy. Soc. A **368**(1929), 4675–4890 (2010)
58. M. Nosonovsky, B. Bhushan, *Green Tribology: Biomimetics Energy Conservation and Sustainability* (Springer, Berlin, 2011). this volume
59. T. Okada, M. Kaneko (eds.), *Molecular Catalysts for Energy Conversion*, Springer Series in Materials Science (Springer, Berlin Heidelberg, 2010)
60. C.E. Orsello, D.A. Lauffenburger, D.A. Hammer, Molecular properties in cell adhesion: a physical and engineering perspective. Trends Biotechnol. **19**, 310–316 (2001)
61. E. Pennisi, Microbes, immunity, and disease: is it time to uproot the tree of life? Science **284**(5418), 1305–1307 (1999)
62. K.W. Powers, S.C. Brown, V.B. Krishna, S.C. Wasdo, B.M. Moudgil, S.M. Roberts, Research strategies for safety evaluation of nanomaterials. Part VI. Characterization of nanoscale particles for toxicological evaluation. Toxicol. Sci. **90**(2), 296–303 (2006)
63. S. Priya, D.J. Inman (eds.), *Energy Harvesting Technologies* (Springer, New York, 2010)
64. L. Richert, F. Vetrone, J.-H. Yi, S.F. Zalzal, J.D. Wuest, F. Rosei, A. Nanci, Surface nanopatterning to control cell growth. Adv. Mater. **15**, 1–5 (2008)

65. F.E. Round, R.M. Crawford, D.G. Mann, *The Diatoms: Biology and Morphology of the Genera* (Cambridge University Press, Cambridge, 1990)
66. Z. Rymuza, *Tribology of Miniature Systems Tribology Series* (Elsevier Science Ltd, Amsterdam, 1989), p. 576. ISBN 978-0444874016
67. C. Sanchez, H. Arribart, M.M. Giraud-Guille, Biomimetism and bioinspiration as tools for the design of innovative materials and systems. Nature Mater. **4**, 277–288 (2005)
68. M. Scherge, A. Rehl, Hebel der tribologischen Optimierung, presentation given at the opening of the MikroTribologie Centrum μTC, Karlsruhe,Germany, May 2010
69. O.H. Schmitt, Biomimetics in solving engineering problems (1982), Talk given on April 26, 1982. http://160.94.102.47/OttoPagesFinalForm/BiomimeticsProblem%20Solving.htm Accessed 4 October 2011
70. A. Sigel, H. Sigel, R.K.O. Sigel (eds.), *Biomineralization: From Nature to Application (Metal Ions in Life Sciènces)*, vol. 2 (Wiley, Chichester, 2008)
71. C. Starr, R. Taggart, *Animal Structure and Function*, vol. 5, Biology the Unity and Diversity of Life (Brooks Cole, Stamfordt, 2008)
72. C. Starr, R. Taggart, *Plant Structure and Function*, vol. 4, Biology: The Unity and Diversity of Life (Brooks Cole, Stamford, 2008)
73. R. Thomas, R. D'Ari, *Biological Feedback* (CRC Press, Boca Raton, 1990)
74. K. Thomas, P. Sayre, Research strategies for safety evaluation of nanomaterials, Part I: Evaluating the human health implications of exposure to nanoscale materials. Toxicol. Sci. **87**(2), 316–321 (2005)
75. K. Thomas, P. Aguar, H. Kawasaki, J. Morris, J. Nakanishi, N. Savage, Research strategies for safety evaluation of nanomaterials, Part VIII: International efforts to develop risk-based safety evaluations for nanomaterials. Toxicol. Sci. **92**(1), 23–32 (2006)
76. T. Thomas, K. Thomas, N. Sadrieh, N. Savage, P. Adair, R. Bronaugh, Research strategies for safety evaluation of nanomaterials, Part VII: Evaluating consumer exposure to nanoscale materials. Toxicol. Sci. **91**(1), 14–19 (2006)
77. D'.A.W. Thompson, *On growth and Form* (Cambridge University Press, Cambridge, 1917)
78. M.A. Tiffany, R. Gordon, I.C. Gebeshuber, *Hyalodiscopsis plana*, a sublittoral centric marine diatom, and its potential for nanotechnology as a natural zipper-like nanoclasp. Pol. Bot. J. **55**(1), 27–41 (2010)
79. J.S. Tsuji, A.D. Maynard, P.C. Howard, J.T. James, C.-W. Lam, D.B. Warheit, A.B. Santamaria, Research strategies for safety evaluation of nanomaterials, Part IV: Risk assessment of nanoparticles. Toxicol. Sci. **89**(1), 42–50 (2006)
80. M. Urbakh, J. Klafter, D. Gourdon, J. Israelachvili, The nonlinear nature of friction. Nature **430**, 525–528 (2004)
81. S. van der Zwaag (ed.), *Self Healing Materials: An Alternative Approach to 20 Centuries of Materials Science*, Springer Series in Materials Science (Springer, Dortrecht, 2007)
82. J.F.V. Vincent, Deconstructing the design of a biological material. J. Theor. Biol. **236**, 73–78 (2005)
83. J.F.V. Vincent, O.A. Bogatyreva, N.R. Bogatyrev, A. Bowyer, A. Pahl, Biomimetics—its practice and theory. J. R. Soc. **3**(9), 471–482 (2006)
84. P. Vukusic, J.R. Sambles, Photonic structures in biology. Nature **424**, 852–855 (2003)

Chapter 6
Towards Understanding Wetting Transitions on Biomimetic Surfaces: Scaling Arguments and Physical Mechanisms

Edward Bormashenko and Gene Whyman

Abstract Biomimetic hierarchical surfaces demonstrate a potential for a variety of green technologies, including energy conversion and conservation, due to their remarkable water repellence. The design of such surfaces allowing emerging green applications remains a challenging scientific and technological task. Understanding the physical mechanism of wetting transitions is crucial for design of highly stable superhydrophobic materials. The main experimental and theoretical approaches to wetting transitions are reviewed. General scaling arguments shedding light on the complicated physics of wetting transitions are supplied. Reducing the micro-structural scales is the most efficient measure needed to enlarge the threshold pressure of a wetting transition. The trends of future investigations are envisaged.

Keywords Biomimetic surfaces · Wetting states · Superhydrophobicity · Wetting transitions

6.1 Introduction

Revealing the natural surfaces demonstrating pronounced water repellence (the so-called "lotus effect" [5, 59]) stimulated extended theoretical and experimental research of wetting phenomena occurring on rough surfaces. It is already well understood that certain kinds of hierarchically rough reliefs supply extreme water repellency (superhydrophobicity) to interfaces [48, 78, 90, 91]. Superhydrophobic properties demonstrate plants, birds' wings, legs of water striders and other natural

E. Bormashenko (✉) · G. Whyman
Laboratory of Polymers, Ariel University Center of Samaria,
Post Box 3, 40700 Ariel, Israel
e-mail: edward@ariel.ac.il

M. Nosonovsky and B. Bhushan (eds.), *Green Tribology*,
Green Energy and Technology, DOI: 10.1007/978-3-642-23681-5_6,

Fig. 6.1 8 μl droplet deposited on the polymer superhydrophobic surface prepared as described by [12]

objects [9, 14, 40, 83]. Various sophisticated experimental techniques including UV-lithography and plasma etching were applied for manufacturing lotus-like biomimetic surfaces [8, 53]. Typical superhydrophobic wetting is shown in Fig. 6.1: an 8 μl droplet is deposited on the polymer water-repellent surface. The detailed recent reviews of the state of art summarized the accumulated experience in the field [77, 84]. Superhydrophobic surfaces demonstrate a potential for a variety of green technologies, including energy conversion and conservation and environment-friendly self-cleaning underwater surfaces [33, 45, 77].

The wetting of rough surfaces is characterized by the apparent contact angle (APCA), which is different from the local contact angle given by Young's relation [68]. However, a high APCA does not necessarily guarantee true superhydrophobicity and self-cleaning properties. Moreover, high-stick surfaces demonstrating large APCA were reported recently [23, 28, 39]. Low contact angle hysteresis resulting in low sliding angles and high stability of "lotus-like" wetting state supply water repellent and self-cleaning properties to the surface [41, 42, 73]. Design of such surfaces allowing emerging green applications remains a challenging scientific and technological task.

6.2 Wetting States

The wetting of atomically smooth surfaces is characterized by the equilibrium or Young angle θ_Y given by the well-known Young equation:

$$\cos\theta_Y = \frac{\gamma_{SA} - \gamma_{SL}}{\gamma}, \tag{6.1}$$

where γ, γ_{SL}, γ_{SA} are the surface tensions at the liquid/air (vapor), solid/liquid, and solid/air interfaces, respectively. The contact angle of a droplet deposited on a solid depends on external parameters, such as temperature. The temperature change may stimulate the transition from the partial to complete wetting of a solid substrate. In this case we observe a wetting transition on a smooth solid surface [10, 49].

The Young equation supplies the sole value of contact angle for a certain combination of solid, liquid, and gaseous phases. Regrettably, the experimental situation is much more complicated; even on the atomically flat surfaces the diversity of contact angles is observed. This is due to the long-range interaction between molecules forming the triple (three-phase) line of the droplet and molecules forming the solid substrate [100]. It was shown that the droplet-surface attraction is time-dependent due to re-orientation of the molecules constituting the solid substrate; this fact hinders experimental tribology investigations of a solid/liquid contact and calls for novel experimental techniques such as the recently reported centrifugal adhesion balance [95].

The maximal contact angle observed on the surface is called the advancing angle, θ_{adv}; the minimal one is called the receding angle, θ_{rec} [32, 38, 44, 47]. The advancing and receding contact angles are equilibrium (though metastable) angles [69]. The difference between advancing and receding contact angles $\theta_{adv} - \theta_{rec}$ is called the contact angle hysteresis [93, 94]. The experimental establishment of advancing and receding angles is a challenging task, and it should be mentioned that reported contact angles are sensitive to the experimental technique used for their measurement [19, 35, 60].

Chemical heterogeneities and roughness strengthen the contact angle hysteresis [4, 36]. Various models explaining the phenomenon of hysteresis were proposed [54, 74, 96, 99]. The effect was related to the pinning of the triple line by defects, which produces the potential barrier U to be surpassed by the droplet under its displacement [99]. The general expression for the contact angle hysteresis is supplied by Eq. 6.2

$$\theta_{adv} - \theta_{rec} = \left(\frac{8U}{\gamma R_0}\right)^{1/2} h(\theta), \tag{6.2}$$

where R_0 is the initial radius of the spherical drop before deposition on the substrate; the function $h(\theta)$ of the equilibrium contact angle is given in [99]. However, a general theory of the contact angle hysteresis is still not built. For experimental study of hysteresis, the manufacture of well-defined microscopically scaled defects is necessary, which is a complicated technological task [85].

The wetting of flat, chemically heterogeneous surfaces is characterized by APCA θ_C predicted by the Cassie–Baxter wetting model [29, 30]. Consider the wetting of a composite flat surface comprising several materials. Each material is characterized by its own surface tension coefficients $\gamma_{i,SL}$ and $\gamma_{i,SA}$, and by the fraction f_i in the substrate surface, $f_1 + f_2 + \cdots + f_n = 1$. The APCA in this case is supplied by the Cassie–Baxter equation:

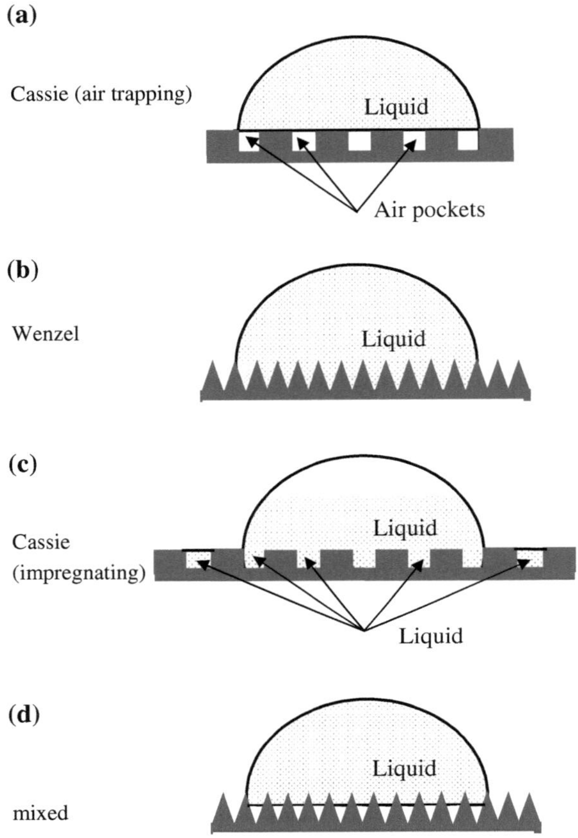

Fig. 6.2 Various wetting states occurring on rough surfaces. **a** Cassie air trapping state. **b** Wenzel state. **c** Cassie impregnating wetting state. **d** mixed wetting state

$$\cos\theta_C = \frac{1}{\gamma}\sum_{1}^{n} f_i(\gamma_{i,\mathrm{SA}} - \gamma_{i,\mathrm{SL}}) = \sum_{1}^{n} f_i \cos\theta_i \tag{6.3}$$

where θ_i are equilibrium (Young) contact angles for the ith material. The Cassie–Baxter equation can also be applied to the solid surface comprising pores (the contact angle for pores equals π and $\cos\theta = -1$, see Fig. 6.2a). In this case the Cassie–Baxter equation yields:

$$\cos\theta_C = -1 + f_S(\cos\theta_Y + 1) \tag{6.4}$$

where f_S and $1-f_S$ are the relative fractions of the solid and air fractions underneath the droplet. In the case of pillar reliefs, the Cassie air trapping wetting state is also called the "fakir state". The Cassie-like air trapping wetting results in unusual tribology of the surface providing an easy sliding of water droplets. The slip lengths as high as 200–400 μm were reported recently [63, 64].

It is noteworthy that the derivation of Eq. 6.4 from Eq. 6.3 is not straightforward, because the triple (three phase) line could not be in rest on pores [20]. Taking into account the fine structure of the triple line justifies the success of the Cassie–Baxter formula for predicting APCAs on porous surfaces [15, 20, 34, 92]. When the relief is hierarchical the Cassie–Baxter equation obtains more complicated forms taking into account the interrelation between scales constituting the topography of the relief [12, 21, 22, 48]. It should be mentioned that the dependence of the APCA on equilibrium contact angles is weak on the hierarchical surfaces when compared to those on single-scale surfaces [21, 22]. Modifications of the Cassie–Baxter equation considering the peculiarities of complicated topography of biomimetic surfaces were reported recently [34, 62, 76].

The wetting of rough chemically homogeneous surfaces is governed by the Wenzel model [98]. According to the Wenzel model, the surface roughness r defined as the ratio of the real surface in contact with liquid to its projection onto the horizontal plane, always magnifies the underlying wetting properties (see Fig. 6.2b). Both hydrophilic and hydrophobic properties are strengthened by surfaces textures. The Wenzel apparent contact angle is given by Eq. 6.5:

$$\cos\theta_W = r\cos\theta_Y. \tag{6.5}$$

Actually, pure Cassie and Wenzel wetting situations are rare in occurrence [37], and Marmur introduced a mixed wetting state [67]. In the mixed wetting state, a droplet partially wets the side surface of pores and partially sits on air pockets as described in Fig. 6.2d. The APCA is supplied in this case by Eq. 6.6:

$$\cos\theta_{\text{mix}} = r_f f\cos\theta_Y + f - 1. \tag{6.6}$$

In this equation f is the fraction of the projected area of the solid surface that is wetted by the liquid. When $f = 1$, Eq. 6.6 turns into the Wenzel Eq. 6.5.

One more wetting state has been introduced [7, 50, 52]. This is the Cassie impregnating state depicted in Fig. 6.2c. In this case liquid penetrates into grooves of the solid and the drop finds itself on a substrate viewed as a patchwork of solid and liquid (solid "islands" ahead of the drop are dry, as shown in Fig. 6.2c). The APCA of Cassie impregnating state is established as (the contact angle at pores is zero and $\cos\theta = 1$; [7]):

$$\cos\theta_{\text{CI}} = 1 - f_S + f_S\cos\theta_Y \tag{6.7}$$

The Cassie impregnating wetting is possible when the Young angle satisfies Eq. 6.8 [7]:

$$\cos\theta_Y > \frac{1 - f_S}{r - f_S} \tag{6.8}$$

It should be stressed that Eqs. 6.3–6.7 could be applied when the radius of the droplet is much larger than the characteristic scale of surface heterogeneities [75].

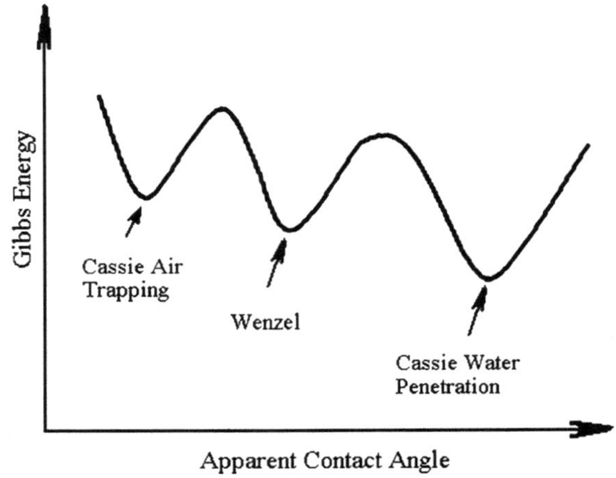

Fig. 6.3 Multiple minima of the Gibbs energy of a droplet deposited on a rough surface

The rigorous thermodynamic derivation of Eqs. 6.3–6.7 was obtained in the series of theoretical works [7, 24, 25, 46, 66, 71, 99].

Various wetting states featured by very different APCA can co-exist on the same heterogeneous surface. The diversity of APCA could be easily understood if one takes into account that the Gibbs energy curve for a droplet on a real surface is characterized by multiple minima points [22, 68]. It could be shown that the Wenzel state is energetically favorable compared to the Cassie state when [6]:

$$\cos\theta_Y > \frac{f_S - 1}{r - f_S} \tag{6.9}$$

The lowest minimum of the Gibbs energy usually corresponds to the Cassie impregnating APCA given by Eq. 6.7 (see Fig. 6.3). When APCA changes spontaneously or under external stimuli we observe the wetting transition (WT). It should be emphasized that physical mechanisms of WT on flat and rough surfaces are quite different.

6.3 Wetting Transitions: Experimental Data

Wetting transitions were observed under various experimental techniques utilizing a diversity of factors: gravity [101], pressure [61], bouncing of droplets [6, 56, 57, 76] evaporation of droplets [55, 70, 72], electric field in the electrowetting experiments [1, 2], and vibration of droplets [11, 16–18, 21, 22]. The interesting experimental technique allowing the study of an air layer responsible for the formation of the Cassie state was reported recently [88]. A superhydrophobic surface exposed to hydrostatic water pressure was irradiated by a laser beam. The jump in the reflectivity of the laser beam indicated the occurrence of WT.

Fig. 6.4 Wetting transitions observed by vibration of 15 μl water drop deposited on the micrometrically rough PDMS surface. **a** The initial Cassie state. **b** The Cassie impregnating state induced by vibrations

The reflection interference contrast microscopy was used for the study of the air–water interface on the textured PDMS surfaces [72]. ESEM technique was used successfully for the study of WT during micro-droplet evaporation [76].

It should be mentioned that various experimental methods used for the investigation of WT supplied the close values of a pressure necessary for the Cassie–Wenzel transition, which is of the order of magnitude of 100–300 Pa for 10 μl droplets deposited on micrometrically scaled rough surfaces [16, 17, 61]. It is noteworthy, that the Cassie air trapping wetting regime observed on natural objects (birds' wings) was much more stable when compared to that on artificial surfaces [21, 23]. Single and two-stage pathways of WT were observed, including Cassie (air-trapping)–Wenzel–Cassie (impregnating), Wenzel–Cassie (impregnating) and Cassie (air-trapping)–Cassie (impregnating) transitions [21, 23]. The lowest energy state corresponds to the Cassie impregnating wetting regime. Vibration-induced Cassie (air-trapping)–Cassie (impregnating) transition observed on (polydimethylsiloxane) PDMS substrate is illustrated in Fig. 6.4. The mechanisms of WT will be discussed in detail below.

6.4 Time Scaling of Wetting Transitions

As already mentioned, wetting transitions were observed under evaporation, pressing, vibration, and bouncing of droplets.

1. Pressing droplets. When a droplet deposited on a rough surface is exposed to external pressure the characteristic time of the pressure change, τ, should be introduced. Pressuring a droplet causes a subsequent change of the apparent contact angle with the same characteristic time τ (due to the phenomenon of the contact angle hysteresis) [61]. This time has to be compared with the time of transitionsτ_{tr}. Actually, τ_{tr} is the time necessary for filling the texture with liquid. It definitely depends on the topography of the relief. As was established [72] for microscopically scaled reliefs usually used for the study of wetting transitions, $\tau_{tr} \approx 2 \times 10^{-2}$ s. Two experimental situations are possible, the first of which is a "quick" transition occurring when $\tau \gg \tau_{tr}$. In this case, the transition takes place under the constant apparent contact angle. The second

case corresponds to slow transitions when $\tau_{tr} \gtrsim \tau$, and the change of the apparent contact angle in the course of transition should be taken into account.
2. Evaporation of droplets. The characteristic time of evaporation of 10 μl droplet equals dozens of minutes and it is much larger than the time of transitions τ_{tr}. Thus the transition occurring under evaporation of a droplet is a "quick" one.
3. Droplets are vibrated with the frequency ν. If the contact line is pinned under vibration, both the contact angle and the pressure in the drop vary with the characteristic time $\tau \approx (1/\nu)$ For the reported value of $\nu = 36$ Hz (see Refs. [14–16], $\tau \approx 10^{-2}$ s, and it is seen that $\tau \approx \tau_{tr}$. It is the most complicated situation, and the mechanism of the wetting transition is explained by lateral displacement of the three-phase (triple) line. The apparent contact angle changes in the course of transition.
4. Bouncing of droplets. When droplets fall on a solid substrate, the characteristic time of the pressure and contact angle equilibration τ equals the so-called contact time. It was shown that τ is independent of the velocity of the bouncing droplet and depends strongly on its radius R [86]. It was also demonstrated that for droplets with a radius in the range of 0.1–10 mm, τ varies from 0.5 to 100 ms [86]. Thus, it could be concluded that for bouncing droplets, various interrelations between τ and τ_{tr} are possible.

6.5 Mechanisms of Wetting Transitions: Statics

6.5.1 Composite Wetting State

As mentioned, the Cassie air trapping wetting state corresponds to the highest of multiple minima of Gibbs energy of the droplet deposited on the rough surface (with biological and hierarchical surfaces being exceptions). Thus, for the WT the energy barrier should be surmounted [3, 50, 79]. It was supposed that this energy barrier corresponds to the surface energy variation between the Cassie state and the hypothetical *composite* state with the almost complete filling of surface asperities by water keeping the liquid–air interface under the droplet and the contact angle constant, as shown in Fig. 6.2d. Contrastingly to the equilibrium mixed wetting state [67], the composite state is unstable for hydrophobic surfaces and corresponds to energy maximum (transition state). For the simple topography depicted in Fig. 6.5, the energy barrier could be calculated as follows [26]:

$$W_{trans} = 2\pi R^2 h(\gamma_{SL} - \gamma_{SA})/p = -2\pi R^2 h\gamma \cos\theta_Y/p \tag{6.10}$$

where h and p are the geometric parameters of the relief, shown in Fig. 6.5, and R is the radius of the contact area. The numerical estimation of the energetic barrier according to formula (6.10) with the parameters $p = h = 20$ μm, $R = 1$ mm, $\theta_Y = 105°$(corresponding to low density polyethylene (LDPE)), and

Fig. 6.5 Geometric parameters of the model relief used for the calculation of Cassie-Wenzel transition energetic barrier

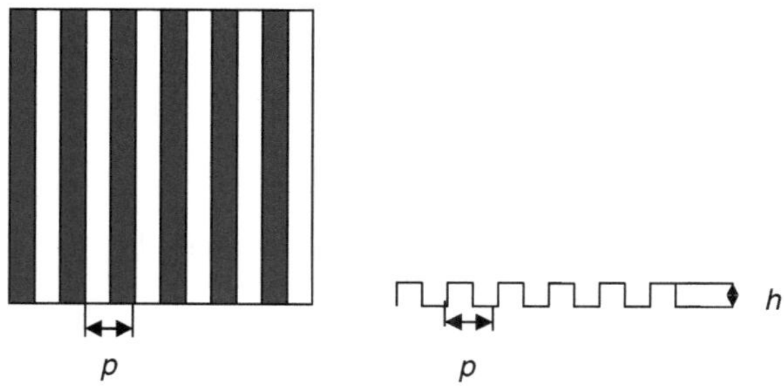

$\gamma = 72$ mJ m^{-2} gives a value of $W_{\text{trans}} = 120$ nJ. It should be stressed that according to (6.10) the energy barrier scales as $W_{\text{trans}} \sim R^2$. The validity of this assumption will be discussed below. The energetic barrier is extremely large compared to thermal fluctuations: $\frac{W_{\text{trans}}}{kT} \approx \left(\frac{R}{a}\right)^2 \gg 1$, where a is an atomic scale [50]. At the same time W_{trans} is much less than the energy of evaporation of the droplet $Q \approx (4/3)\pi R^3 \lambda$, where λ is the volumetric heat of water evaporation, $\lambda = 2 \times 10^9$ J/m^3. For a 3 μl droplet with the radius $R \approx 1$ mm it yields $Q \approx 10$ J, hence $kT \ll W_{\text{trans}} \ll Q$. Actually, this interrelation between characteristic energies makes wetting transitions possible. If it was not the case, a droplet exposed to external stimuli might evaporate before wetting transition. It is instructive to estimate the radius, at which $W_{\text{trans}} \approx Q$. Equating W_{trans} given by Eq. 6.10 to Q yields $R \approx -(3/2)\gamma \cos\theta_Y/\lambda \approx 5 \times 10^{-11}$m. It means that wetting transitions are possible for any volume of a droplet. It is noteworthy that the ratio γ/λ is practically the same for all liquids, and it is of the order of magnitude of molecular size [27]. Hence, wetting transitions are possible for any liquid and any volume.

6.5.2 Energy Barrier and Cassie, Wenzel, and Young Contact Angles

It is well established experimentally that wetting transitions are usually irreversible. This conveys the suggestion that some general reasons for such irreversibility exist. It turns out that a variety of wetting states, as well as transitions between them, may be described on the same mathematical basis that gives the possibility to elucidate their features which are independent of peculiarities of a particular substrate. Starting with the spherical model for the droplet shape, it can be shown that the surface-energy dependence $E(\theta)$ on the (non-equilibrium) apparent contact angle θ looks like [99].

$$E_i(\cos\theta) = \left[\frac{9\pi V^2}{(1-\cos\theta)(2+\cos\theta)^2}\right]^{1/3}\gamma(2-\cos\theta_i(1+\cos\theta)), \qquad (6.11)$$

and its minimum E_{0i} is expressed as

$$E_{0i} = \gamma[9\pi V^2(1-\cos\theta_i)^2(2+\cos\theta_i)]^{1/3}, \qquad (6.12)$$

where V is the droplet volume and θ_i is the equilibrium APCA in a given wetting state. In particular, Eq. 6.12 supplies equilibrium energies in the Cassie (i = C) and Wenzel (i = W) states or in the wetting state on flat surfaces (i = Y, Young's angle $\theta_i = \theta_Y$) with the corresponding angles in Eqs. 6.4, 6.5, and 6.1, respectively. Moreover, for a definite mechanism of transition, Eqs. 6.11 and 6.12 give the energies of the transition state (i = trans). For hydrophobic materials and orthogonal reliefs it can be shown that the energy in the transition state is also expressed by (6.11), (6.12) with

$$\cos\theta_{\text{trans}} = \cos\theta_W + \cos\theta_C - \cos\theta_Y \qquad (6.13)$$

The mentioned mechanism of the Cassie–Wenzel transition is described as wetting the side surfaces of hydrophobic relief [3] accompanied with the energy increase. A transition (composite) state corresponds to the almost complete filling of relief asperities. A transition barrier is overcome when liquid touches their bottoms, and the high-energy liquid–air interface under the droplet disappears.

Equations 6.11–6.13 enable one to calculate the energy barrier of transition by using the measured or calculated values of contact angles in the wetting states without entering geometrical details of a substrate relief. In this way, e.g., the results [3] of the barrier calculation can be reproduced.

As mentioned in Sect. 6.4, wetting transitions may proceed quickly or slowly. Accordingly, two types of wetting transitions may proceed in principle: adiabatic transitions with a fixed value of the contact angle, and slow non-adiabatic transitions when a droplet has time to relax and the contact angle changes in the course of liquid penetration into depressions (or going out from them). Both these types of energy barriers can be calculated on the basis of the presented model, e.g., for the transition from the Cassie state to the Wenzel one, as

$$W_{\text{adia}} = E_{\text{trans}}(\cos\theta_C) - E_{0C}, W_{\text{nonadia}} = E_{0\text{trans}} - E_{0C} \qquad (6.14)$$

The irreversibility of wetting transitions is seen from peculiarities of the dependence (6.12) of the equilibrium surface energies on the equilibrium APCAs (Fig. 6.6). The function $E_{0i}(\cos\theta_i)$ is a monotonically decreasing one, with a weak dependence for low values of $\cos\theta_i$ (~ -1) and a strong one for higher values. Furthermore, as it can be proven, $\cos\theta_{\text{trans}} < \cos\theta_C$, $\cos\theta_{\text{trans}} < \cos\theta_W$, i.e. $\cos\theta_{\text{trans}}$ is located out of the interval between $\cos\theta_C$ and $\cos\theta_W$, closer to the lower limit. Consequently, the energy barrier is very asymmetric, low from the side of

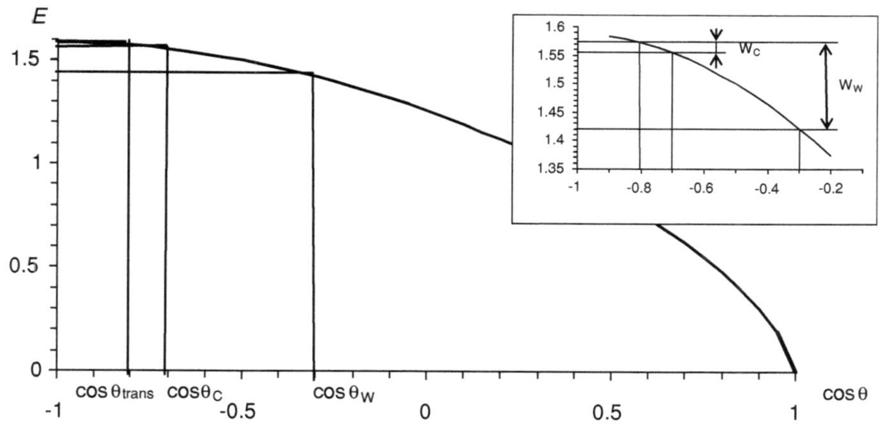

Fig. 6.6 Dependence (Eq. 6.12) of the equilibrium surface energy (in units of $\gamma(9\pi V^2)^{1/3}$) on the equilibrium APCA and barriers of wetting transitions. Numerical values of APCAs are 107.4°(Wenzel), 134.4° (Cassie), and 101.5° (Young). The transition state angle calculated according (6.13) is 143.1°. The heights of the highly asymmetric energy barrier for a water droplet of a volume of 3 μl are: from the side of the (metastable) Cassie state $W_C = 8$ nJ and from the side of the (stable) Wenzel state $W_W = 70$ nJ

the metastable (higher energy) state and high from the side of the stable state. Calculations of real transitions based on (6.12–6.14) gave the difference of almost one order of magnitude. Taking into account exponential (Arrhenius-type) dependence of the transition probability on the barrier height shows that the reverse transition is impossible.

The results of this section remain true for inherently hydrophilic substrates, where the transition mechanism is different compared to hydrophobic substrates. In this case the existence of a barrier may be due to the pinning of liquid at discontinuities, such as pore or pillar borders, that leads to the formation of a new liquid–vapor interface at additional energy expense, as shown in Fig. 6.7. The wetting transition takes place when the menisci touch the bottoms of the relief [12, 13, 15, 34]. The connection between APCAs for hydrophilic substrates is given by

$$\cos\theta_{\text{trans}} = 1 + \cos\theta_C - \frac{\cos\theta_W}{\cos\theta_Y}. \tag{6.15}$$

6.5.3 Critical Pressure Necessary for Wetting Transition

Consider a single-scale pillar-based biomimetic surface, similar to that studied by Yoshimitsu with pillar width a, and groove width b [101]. Analysis of the balance

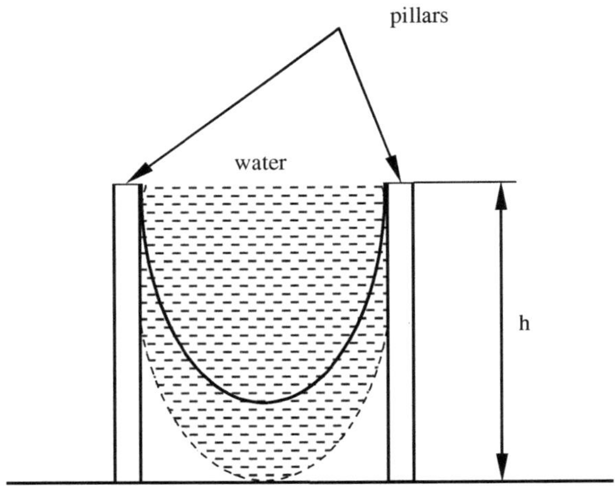

Fig. 6.7 Pressure induced displacement of the water front leading to the collapse of the Cassie air trapping wetting state

of forces at the air–liquid interface, at which the equilibrium is still possible, yields for the critical pressure p_c [102]:

$$p_c > -\frac{\gamma f_s \cos\theta_Y}{(1-f_s)\lambda} \tag{6.16}$$

where $\lambda = (A/L)$, A and L are pillar cross-sectional area and perimeter respectively. As an application of Eq. 6.16 with $\theta = 114^{\circ}$(Teflon), $a = 50$ μm, $b = 100$ μm we obtain $p_c = 296$ Pa, in excellent agreement with experimental results [101, 102]. Recalling that the dynamic pressure of rain droplets may be as high as 10^4–10^5 Pa, which is much larger than $p_c \approx 300\,\text{Pa}$, we conclude that creating biomimetic reliefs with very high critical pressure is of practical importance [102]. The concept of critical pressure leads to the conclusion that reducing the micro-structural scales (e.g., the pillars diameters and spacing) is the most efficient measure needed to enlarge the critical pressure [72, 102]. The energy barrier separating the Cassie and Wenzel states is given by an expression similar to (6.10) and scales as R^2 [102]. It is noteworthy that neither Eq. 6.10 nor Eq. 6.16 explains the existence of Cassie wetting on inherently hydrophilic surfaces [12, 13, 22, 58, 87]. Indeed, W_{trans} and p_c calculated according to Eqs. 6.10 and 6.16 are negative for $\theta < (\pi/2)$, this makes the traditional Cassie wetting inapplicable to hydrophilic surfaces, and the alternative physical reasoning explaining experimentally observed high APCAs should be involved, as discussed below.

An alternative mechanism of WT based on the concept of critical pressure was proposed [65, 81]. It was supposed that as the pressure applied to the droplet increased, the meniscus will move towards the flat substrate as shown in Fig. 6.7. The meniscus will eventually touch the substrate, this will cause the collapse of the

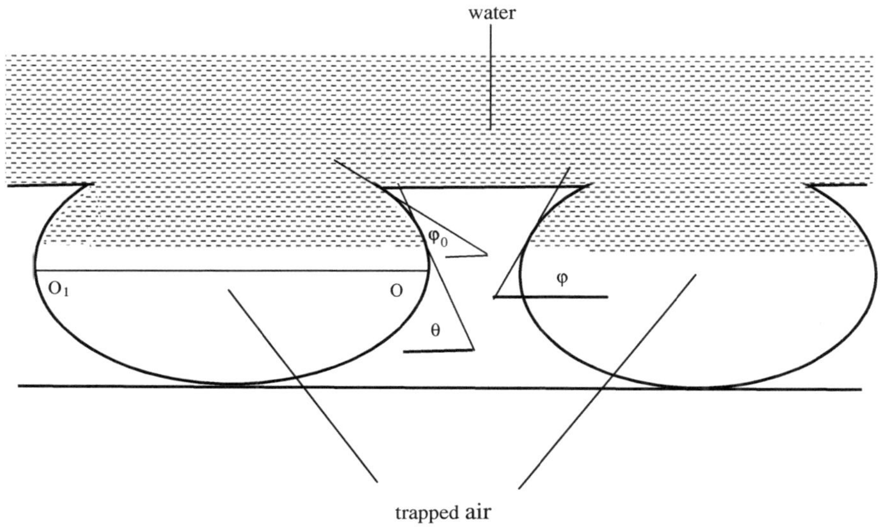

Fig. 6.8 Geometrical air trapping on hydrophilic reliefs

Cassie wetting, and consequently lead to the Cassie–Wenzel transition ([65, 81]; see also the preceding section).

6.5.4 Cassie Wetting of Inherently Hydrophilic Surfaces: Criteria for Gas Entrapping

For the explanation of the roughness-induced superhydrophobicity of inherently hydrophilic materials it was supposed that air is entrapped by cavities constituting the topography of the surface [13, 80, 97]. The simple mechanism of "geometrical" trapping could be explained as follows: consider a hydrophilic surface ($\theta_Y < 90^\circ$) comprising pores as depicted in Fig. 6.8. It is seen that air trapping is possible only if $\theta_Y > \varphi_0$, where φ_0 is the angle between the tangent in the highest point of the pattern and the horizontal symmetry axis O_1O. Indeed, when the liquid level is descending, the angle θ is growing (see Fig. 6.8), and if the condition $\theta_Y > \varphi_0$ is violated, the equilibrium $\theta = \theta_Y$ will be impossible.

In the equilibrium position, small fluctuations of the contact angle lead to the appearance of curvature on the plane air–water interface that is energetically unfavorable. Below the central plane O—O_1, where $\theta > 90^\circ$, the equilibrium is impossible in the case of $\theta_\gamma < (\pi/2)$. For a sufficiently large pore, small fluctuations of θ can lead to touching the pore bottom near its center by the curved air–liquid interface followed with filling the pore and consequent collapse of the

Cassie air trapping wetting regime. The effects of compressibility of trapped air on WT on hydrophilic surfaces have also been considered [80].

6.6 Mechanisms of Wetting Transitions: Dynamics

The experimental data related to the dynamics of WT are scanty [72, 82, 89]. The characteristic time of Cassie–Wenzel transition was established with reflection interference contrast microscopy as less than 20 ms [72]. The dynamics of WT for droplets placed on polymer micrometer-size square pillars was studied by optical microscopy in combination with high-speed imaging [82]. Two regimes of the droplet front displacement were observed: zipping and non-zipping. In the zipping regime, the velocity of front in one direction (to advance to the next row of pillars) is much smaller than the velocity in the other direction (liquid filling up one row of micro-pillars). The topography of the surface (pillar height and gap size between pillars) and water contact angle were varied. It was established that the velocity of the wetting front increases with increasing gap size, decreasing pillar height, or decreasing contact angle [82]. A velocity of the wetting front as high as 1.5 m/s was registered [82]. Balancing interfacial energy contributions with viscous dissipation yielded universal equations for the zipping and the transition dynamics [82].

6.7 Dimension of Wetting Transitions

One of the debatable problems in the field of WT is the problem of the "dimension" of transitions, or, in other words: whether all pores underneath the droplet should be filled by liquid (the "2D scenario"), or, perhaps, only the pores adjacent to the three-phase (triple) line are filled under external stimuli such as pressure, vibrations or impact (the "1D scenario"). Indeed, APCA is dictated by the area adjacent to the triple line, and not by the total area underneath the droplet [43]. 1D and 2D scenarios of WT are schematically illustrated in Fig. 6.9. The experiments carried out with vibrated drops supported the 1D scenario of wetting transitions. It has been established that the transition occurs when the condition $F_{cr} = \text{const}$ is fulfilled, where F_{cr} is the critical force acting on the unit length of the triple line, and the transition is caused by de-pinning of the triple line [17, 21, 22]. The critical value of de-pinning force F_{cr} has been established experimentally for various microscopically structured surfaces as $F_{cr} \approx 200 - 350\,\text{mN}\,\text{m}^{-1}$ [21, 22]. The energy barrier W to be surmounted for the elementary displacement of the triple line δr could be estimated as

$$W \approx 2\pi R F_{cr} \delta r, \tag{6.17}$$

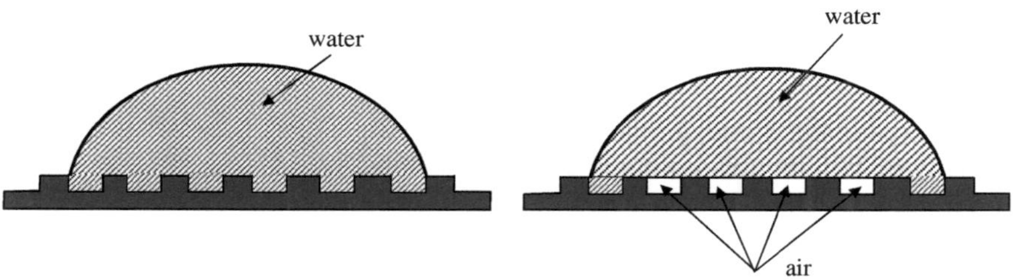

Fig. 6.9 **a** The Wenzel wetting state (2D scenario of transition is depicted; all pores underneath the droplet are filled by water). **b** The Wenzel wetting state (1D scenario of transition is depicted; only pores in the vicinity of three-phase line are filled)

which scales as R. This scaling law corresponds to the results obtained with vibrated drops, but contradicts the scaling law given by expression (6.10). The electrowetting stimulated WT also supported the 1D mechanism of transitions [2]. De-pinning of the triple line was observed directly with use of reflection interference contrast microscopy [72]. The potential barrier calculated according to expression (6.17) for a drop with a radius of $R \approx 1$ mm deposited on the LDPE relief presented in Fig. 6.5 ($F_{cr} \approx 350\,\mathrm{mN\,m^{-1}}$, the elementary displacement $\delta r \approx p/2 \sim 10^{-5}\mathrm{m}$) equals $W \approx 20\,\mathrm{nJ}$, smaller than the value predicted by formula (6.10) but still much larger than thermal fluctuations [26].

In contrast, it was suggested that the Cassie–Wenzel transition occurs via a nucleation mechanism starting from the drop center [51]. Patankar in his recent theoretical investigation supposed that both mechanisms of WT (i.e., de-pinning of the triple line and the critical pressure induced collapse of Cassie wetting depicted in Fig. 6.7) are possible [81]. Lack of trustworthy experimental data in the field should be stressed.

6.8 Trends of Future Investigations

The following trends of investigations in the field of wetting transitions could be proposed:

1. Study of wetting transitions on hydrophobic surfaces with well-defined and controlled topography. Such experiments will shed light on the impact exerted by surface topography on wetting transitions.
2. Study of wetting transitions on well-defined micro-scaled hydrophilic surfaces. These kinds of investigations will clarify the conditions of air trapping by highly developed hydrophilic reliefs.

3. Direct experimental evidencing of the Cassie impregnating state is necessary. Perhaps, this could be done with environmental scanning electron microscopy (ESEM).
4. The experimental study of the dynamics of the triple line under WT is required.
5. Study of WT for various organic liquids is called for [31].
6. Clearing up the true "dimension" of WT is necessary.
7. Development of theoretical threshold criteria of WT is required.
8. Development of surfaces demonstrating highly stable Cassie states is urgent for green applications of biomimetic surfaces such as energy conversion and conservation.
9. Investigation of the temperature dependent WT is actual.

6.9 Conclusions

The main experimental and theoretical approaches to WT are reviewed; the importance of WT for understanding of the tribology of rough liquid/solid interfaces is discussed. The threshold pressure and energy barrier necessary for WT are supplied. The energy barrier of WT is large compared to thermal fluctuations and small compared to the heat of evaporation of the droplet. General spatial and temporal considerations allowing understanding of WT are presented. The problem of dimension of WT is discussed. 1D and 2D pathways of transitions are presented. In spite of the intensive theoretical and experimental effort expended for the study of wetting transitions, the physical mechanism of these phenomena remains unclear. At the same time the profound understanding of this mechanism is vital for manufacturing "green" biomimetic surfaces holding water repellence and self-cleaning properties under external stimuli such as pressure, vibration, or impact. The trends of future investigations are envisaged.

Acknowledgments The authors are indebted to Professor Roman Pogreb for extremely useful discussions. The authors are thankful to Mrs. Ye. Bormashenko, Mrs. A. Musin and Dr. T. Stein for their help in preparing this paper.

References

1. V. Bahadur, S.V. Garimella, Electrowetting-based control of droplet transition and morphology on artificially microstructured surfaces. Langmuir **24**, 8338–8345 (2008). doi:10.1021/la800556c
2. V. Bahadur, S.V. Garimella, 1D Preventing the Cassie–Wenzel transition using surfaces with noncommunicating roughness elements. Langmuir **25**, 4815–4820 (2009). doi:10.1021/la803691m

3. L. Barbieri, E. Wagner, P. Hoffmann, Water wetting transition parameters of perfluorinated substrates with periodically distributed flat-top microscale obstacles. Langmuir **23**, 1723–1734 (2007). doi:10.1021/la0617964
4. F.E. Bartell, J.W. Shepard, Surface roughness as related to hysteresis of contact angles: I. the system paraffin-water-air. J. Phys. Chem. **57**, 211–215 (1953). doi:10.1021/j150503a017
5. W. Barthlott, C. Neinhuis, Purity of the sacred lotus, or escape from contamination in biological surfaces. Planta **202**, 1–8 (1997). doi:10.1007/s004250050096
6. D. Bartolo, F. Bouamrirene, E. Verneuil, A. Buguin, B. Silberzan, S. Moulinet, Bouncing of sticky droplets: impalement transitions on superhydrophobic micropatterned surfaces. Europhys. Lett. **74**, 299–305 (2006). doi:10.1209/epl/i2005-10522-3
7. J. Bico, U. Thiele, D. Quéré, Wetting of textured surfaces. Colloids Surf. A **206**, 41–46 (2002)
8. N. Blondiaux, E. Scolan, A.M. Popa, J. Gavillet, R. Pugin, Fabrication of superhydrophobic surfaces with controlled topography and chemistry. Appl. Surf. Sci. **256S**, S46–S53 (2009). doi:10.1016/j.apsusc.2009.04.194
9. R. Blossey, Self-cleaning surfaces—virtual realities. Nat. Mater. **2**, 301–306 (2003). doi:10.1038/nmat856
10. D. Bonn, D. Ross, Wetting transitions. Rep. Prog. Phys. **64**, 1085–1163 (2001)
11. J.B. Boreyko, Ch.-H. Chen, Restoring superhydrophobicity of lotus leaves with vibration-induced dewetting. Phys. Rev. Lett. **103**, 174502 (2009). doi:10.1103/PhysRevLett.103.174502
12. E. Bormashenko, T. Stein, G. Whyman, Ye. Bormashenko, R. Pogreb, Wetting properties of the multiscaled nanostructured polymer and metallic superhydrophobic surfaces. Langmuir **22**, 9982–9985 (2006). doi:10.1021/la061622m
13. E. Bormashenko, Ye. Bormashenko, G. Whyman, R. Pogreb, O. Stanevsky, Micrometrically scaled textured metallic hydrophobic interfaces validate the Cassie–Baxter wetting hypothesis. J. Colloid Interface Sci. **302**, 308–311 (2006). doi:10.1016/j.jcis.2006.06.016
14. E. Bormashenko, Ye. Bormashenko, T. Stein, G. Whyman, E. Bormashenko, Why do pigeon feathers repel water? Hydrophobicity of pennae, Cassie–Baxter wetting hypothesis and Cassie–Wenzel capillarity-induced wetting transition. J. Colloid Interface Sci. **311**, 212–216 (2007). doi:10.1016/j.jcis.2007.02.049
15. E. Bormashenko, Ye. Bormashenko, T. Stein, G. Whyman, R. Pogreb, Environmental scanning electron microscopy study of the fine structure of the triple line and Cassie-Wenzel wetting transition for sessile drops deposited on rough polymer substrates. Langmuir **23**, 4378–4382 (2007). doi:10.1021/la0634802
16. E. Bormashenko, R. Pogreb, G. Whyman, M. Erlich, Vibration-induced Cassie-Wenzel wetting transition on rough surfaces. Appl. Phys. Lett. **90**, 201917 (2007). doi:10.1063/1.2738364
17. E. Bormashenko, R. Pogreb, G. Whyman, M. Erlich, Cassie-Wenzel wetting transition in vibrating drops deposited on rough surfaces: is dynamic Cassie-Wenzel wetting transition a 2D or 1D affair? Langmuir **23**, 6501–6503 (2007). doi:10.1021/la700935x
18. E. Bormashenko, R. Pogreb, G. Whyman, M. Erlich, Resonance Cassie-Wenzel wetting transition for horizontally vibrated drops deposited on a rough surface. Langmuir **23**, 12217–12221 (2007). doi:10.1021/la7016374
19. E. Bormashenko, Y. Bormashenko, G. Whyman, R. Pogreb, Al. Musin, R. Jager, B. Barkay, Contact angle hysteresis on polymer substrates established with various experimental techniques, its interpretation, and quantitative characterization. Langmuir **24**, 4020–4025 (2008). doi:10.1021/la703875b
20. E. Bormashenko, Why does the Cassie-Baxter equation apply? Colloids Surf. A **324**, 47–50 (2008). doi:10.1016/j.colsurfa.2008.03.025
21. E. Bormashenko, R. Pogreb, T. Stein, G. Whyman, M. Erlich, A. Musin, V. Machavariani, D. Aurbach, Characterization of rough surfaces with vibrated drops. Phys. Chem. Chem. Phys. **27**, 4056–4406 (2008). doi:10.1039/b800091c

22. E. Bormashenko, T. Stein, G. Whyman, R. Pogreb, S. Sytovsky, Y. Danoch, Y. Shoham, Y. Bormashenko, B. Sorokov, D. Aurbach, Superhydrophobic metallic surfaces and their wetting properties. J. Adhes. Sci. Technol. **22**, 379–385 (2008). doi:10.1163/156856108X29549t
23. E. Bormashenko, T. Stein, R. Roman Pogreb, D. Aurbach, "Petal effect" on surfaces based on lycopodium: high stick surfaces demonstrating high apparent contact angles. J. Phys. Chem. C **113**, 5568–5572 (2009). doi:10.1021/jp900594k
24. E. Bormashenko, Young, Boruvka–Neumann, Wenzel and Cassie–Baxter equations as the transversality conditions for the variational problem of wetting. Colloids Surf. A **345**, 163–165 (2009). doi:10.1016/j.colsurfa.2009.04.054
25. E. Bormashenko, Wetting of flat and rough curved surfaces. J. Phys. Chem. C **113**, 17275–17277 (2009). doi:10.1021/jp905237v
26. E. Bormashenko, R. Pogreb, G. Whyman, Comment on water wetting transition parameters of perfluorinated substrates with periodically distributed flat-top microscale obstacles. Langmuir **25**, 13694–13695 (2009). doi:10.1021/la9020959
27. E. Bormashenko, Why are the values of the surface tension of most organic liquids similar? Am. J. Phys. **78**, 1309–1311 (2010). doi:10.1119/1.3471939
28. H. Caps, D. Vandormael, J. Loicq, S. Dorbolo, N. Vanderwalle, Hybrid wetting state on micro-waffle textures. Europhys. Lett **88**, 16002 (2009). doi:10.1209/0295-5075/88/16002
29. A.B.D. Cassie, S. Baxter, Wettablity of porous surfaces. Trans. Faraday Soc. **40**, 546–551 (1944). doi:10.1039/TF9444000546
30. A.B.D. Cassie, Contact angles. Discuss. Faraday Soc. **3**, 11–16 (1948). doi:10.1039/DF9480300011
31. D. Chandra, S. Yang, Capillary-force-induced clustering of micropillar arrays: Is it caused by isolated capillary bridges or by the lateral capillary meniscus interaction force? Langmuir **25**, 10430–10434 (2009). doi:10.1021/la901722g
32. E. Chibowski, Surface free energy of a solid from contact angle hysteresis. Adv. Colloid Interface Sci. **103**, 149–172 (2003)
33. Ch.-Hw. Choi, Ch.-J. Kim, Large slip of aqueous liquid flow over a nanoengineered superhydrophobic surface. Phys. Rev. Lett. **96**, 066001 (2006). doi:10.1103/Phys.Rev.Lett.96.066001
34. W. Choi, A. Tuteja, J.M. Mabry, R.E. Cohen, G.H. McKinley, A modified Cassie–Baxter relationship to explain contact angle hysteresis and anisotropy on non-wetting textured surfaces. J. Colloid Interface Sci. **339**, 208–216 (2009). doi:10.1016/j.jcis.2009.07.027
35. E.L. Decker, B. Frank, Y. Suo, S. Garoff, Physics of contact angle measurement. Colloids Surf. A **156**, 177–189 (1999)
36. R.H. Dettre, R.E. Johnson Jr., Contact angle hysteresis. IV. Contact angle measurements on heterogeneous surfaces. J. Phys. Chem. **69**, 1507–1515 (1965). doi:10.1021/j100889a012
37. H.Y. Erbil, C.E. Cansoy, Range of applicability of the Wenzel and Cassie-Baxter equations for superhydrophobic surfaces. Langmuir **25**, 14135–14145 (2009). doi:10.1021/la902098a
38. C.W. Extrand, Y. Kumagai, An experimental study of contact angle hysteresis. J. Colloid Interface Sci. **191**, 378–383 (2002)
39. L. Feng, Y. Zhang, J. Xi, Y. Zhu, N. Wang, F. Xia, L. Jiang, Petal effect: a superhydrophobic state with high adhesive force. Langmuir **24**, 4114–4119 (2008). doi:10.1021/la703821h
40. X. Gao, L. Jiang, Biophysics: water-repellent legs of water striders. Nature **432**, 36 (2004). doi:10.1038/432036a
41. L. Gao, Th.J. McCarthy, Contact angle hysteresis explained. Langmuir **22**, 6234–6237 (2006). doi:10.1021/la060254j
42. L. Gao, Th.J. McCarthy, A commercially available perfectly hydrophobic material ($\theta_A/\theta_R =$ 180°/180°). Langmuir **23**, 9125–9127 (2007). doi:10.1021/la701097k
43. L. Gao, Th.J. McCarthy, How Wenzel and Cassie were wrong? Langmuir **23**, 3762–3765 (2007). doi:10.1021/la062634a
44. P.G. de Gennes, F. Brochard-Wyart, D. Quéré, *Capillarity and wetting phenomena* (Springer, Berlin, 2003)

45. J. Genzer, A. Marmur, Biological and synthetic self-cleaning surfaces. MRS Bull. **33**, 742–746 (2008)
46. R.J. Good, A thermodynamic derivation of Wenzel's modification of Young's equation for contact angles; together with a theory of hysteresis. J. Am. Chem. Soc. **74**, 5041–5042 (1952). doi:10.1021/ja01140a014
47. V.R. Gray, Surface aspects of wetting and adhesion. Chem. Ind. **23**, 969–978 (1965)
48. S. Herminghaus, Roughness-induced non-wetting. Europhys. Lett. **52**, 165–170 (2000)
49. K.D. Humfeld, St. Garoff, Geometry-driven wetting transition. Langmuir **20**, 9223–9226 (2004). doi:10.1021/la048614v
50. C. Ishino, K. Okumura, D. Quéré, Wettingy transitionys on rough surfaces. Europhys. Lett. **68**, 419–425 (2004). doi:10.1209/epl/i2004-10206-6
51. C. Ishino, K. Okumura, Nucleation scenarios for wetting transition on textured surfaces: the effect of contact angle hysteresis. Europhys. Lett. **76**, 464–470 (2006). doi:10.1209/epl/i2006-10288-0
52. C. Ishino, K. Okumura, Wetting transitions on textured hydrophilic surfaces. Eur. Phys. J. E **25**, 415–424 (2008). doi:10.1140/epje/i2007-10308-y
53. H.E. Jeong, M.K. Kwak, I. Park Ch, K.Y. Suh, Wettability of nanoengineered dual-roughness surfaces fabricated by UV-assisted capillary force lithography. J. Colloid Interface Sci. **339**, 202–207 (2009). doi:10.1016/j.jcis.2009.07.020
54. J.F. Joanny, P.G. de Gennes, A model for contact angle hysteresis. J. Chem. Phys. **81**, 552–562 (1984)
55. Y.C. Jung, B. Bhushan, Wetting transition of water droplets on superhydrophobic patterned surfaces. Scr. Mater. **57**, 1057–1060 (2007). doi:10.1016/j.scriptamat.2007.09.004
56. Y.C. Jung, B. Bhushan, Dynamic effects of bouncing water droplets on superhydrophobic surfaces. Langmuir **24**, 6262–6269 (2008). doi:10.1021/la8003504
57. Y.C. Jung, B. Bhushan, Dynamic effects induced transition of droplets on biomimetic superhydrophobic surfaces. Langmuir **25**, 9208–9218 (2009). doi:10.1021/la900761u
58. A.-M. Kietzig, S.G. Hatzikiriakos, P. Englezos, Patterned superhydrophobic metallic surfaces. Langmuir **25**, 4821–4827 (2009). doi:10.1021/la8037582
59. K. Koch, H.F. Bohn, W. Barthlott, Hierarchically sculptured plant surfaces and superhydrophobicity. Langmuir **25**, 14116–14120 (2009). doi:10.1021/la9017322
60. D.Y. Kwok, A.W. Neumann, Contact angle interpretation: re-evaluation of existing contact angle data. Colloids Surf. A **161**, 49–62 (2000)
61. A. Lafuma, D. Quéré, Superhydrophobic states. Nat. Mater. **2**, 457–460 (2003). doi:10.1038/nmat924
62. S.T. Larsen, R. Taboryski, A Cassie-like law using triple phase boundary line fractions for faceted droplets on chemically heterogeneous surfaces. Langmuir **25**, 1282–1284 (2009). doi:10.1021/la8030045
63. C. Lee, C.-H. Choi, C.-J. Kim, Structured surfaces for a giant liquid slip. Phys. Rev. Lett. **101**, 064501 (2008). doi:10.1103/PhysRevLett.101.064501
64. C. Lee, C.-J. Kim, Maximizing the giant liquid slip on superhydrophobic microstructures by nanostructuring their sidewalls. Langmuir **25**, 12812–12818 (2009). doi:10.1021/la901824d
65. B. Liu, F. Lange, Pressure induced transition between superhydrophobic states: Configuration diagrams and effects of surface feature size. J. Colloid Interface Sci. **298**, 899–909 (2006). doi:10.1016/j.jcis.2006.01.025
66. A. Marmur, Line tension effect on contact angles: axisymmetric and cylindrical systems with rough or heterogeneous solid surfaces. Colloids Surf. A **136**, 81–88 (1998)
67. A. Marmur, Wetting on hydrophobic rough surfaces: to be heterogeneous or not to be? Langmuir **19**, 8343–8348 (2003). doi:10.1021/la0344682
68. A. Marmur, Soft contact: measurement and interpretation of contact angles. Soft Matter **2**, 12–17 (2006). doi:10.1039/b514811c
69. A. Marmur, A guide to the equilibrium contact angles maze in contact angle wettability and adhesion, vol. 6, ed by K.L. Mittal (VSP,Brill, Leiden, 2009), pp 3–18

70. G. McHale, S. Aqil, N.J. Shirtcliffe, G.M.I. Newton, H.Y. Erbil, Analysis of droplet evaporation on a superhydrophobic surface. Langmuir **21**, 11053–11060 (2005). doi:10.1021/la0518795
71. G. McHale, Cassie and Wenzel: were they really so wrong? Langmuir **23**, 8200–8205 (2007). doi:10.1021/la7011167
72. S. Moulinet, D. Bartolo, Life and death of a fakir droplet: impalement transitions on superhydrophobic surfaces. Eur. Phys. J. E **24**, 251–260 (2007). doi:10.1140/epje/i2007-10235-y
73. M. Nosonovsky, B. Bhushan, Hierarchical roughness optimization for biomimetic superhydrophobic surfaces. Ultramicroscopy **107**, 969–979 (2007). doi:10.1016/j.ultramic.2007.04.011
74. M. Nosonovsky, Model for solid-liquid and solid–solid friction of rough surfaces with adhesion hysteresis. J. Chem. Physics **126**, 224701 (2007). doi:10.1063/1.2739525
75. M. Nosonovsky, On the range of applicability of the Wenzel and Cassie equations. Langmuir **23**, 9919–9920 (2007). doi:10.1021/la701324m
76. M. Nosonovsky, B. Bhushan, Patterned nonadhesive surfaces: superhydrophobicity and wetting regime transitions. Langmuir **24**, 1525–1533 (2008). doi:10.1021/la702239w
77. M. Nosonovsky, B. Bhushan, Superhydrophobic surfaces and emerging applications: non-adhesion, energy, green engineering. Curr. Opin. Colloid Interface Sci. **14**, 270–280 (2009). doi:10.1016/j.cocis.2009.05.004
78. N.A. Patankar, Mimicking the lotus effect: influence of double roughness structures and slender pillars. Langmuir **20**, 8209–8213 (2004). doi:10.1021/la048629t
79. N.A. Patankar, Transition between superhydrophobic states on rough surfaces. Langmuir **20**, 7097–7102 (2004). doi:10.1021/la049329e
80. N.A. Patankar, Hydrophobicity of surfaces with cavities: making hydrophobic substrates from hydrophilic materials? *J.* Adhes. Sci. Technol. **23**, 413–433 (2009). doi:10.1163/156856108X370073
81. N.A. Patankar, Consolidation of hydrophobic transition criteria by using an approximate energy minimization approach. Langmuir ASAP paper, (2010). (doi:10.1021/la9047424)
82. A.M. Peters, C. Pirat, M. Sbragaglia, B.M. Borkent, M. Wessling, D. Lohse, R.G.H. Lammertink, Cassie-Baxter to Wenzel state wetting transition: scaling of the front velocity. Eur. Phys. J. **29**, 391–397 (2009). doi:10.1140/epje/i2009-10489-3
83. D. Quéré, M. Reyssat, Non-adhesive lotus and other hydrophobic materials. Phil. Trams. R. Soc. A **366**, 1539–1556 (2008). doi:10.1098/rsta.2007.2171
84. D. Quéré, Wetting and roughness. Annu. Rev. Mater. Res. **38**, 71–99 (2008). doi:10.1146/annurev.matsci.38.060407.132434
85. M. Reyssat, D. Quéré, Contact angle hysteresis generated by strong dilute defects. J. Phys. Chem. B **113**, 3906–3909 (2009). doi:10.1021/jp8066876
86. D. Richard, C. Clanet, D. Quere, Contacting time of a bouncing droplet. Nature **417**, 811–812 (2002)
87. A. Safaee, D.K. Sarkar, M. Farzaneh, Superhydrophobic properties of silver-coated films on copper surface by galvanic exchange reaction. Appl. Surf. Sci. **254**, 2493–2498 (2008). doi:10.1016/j.apsusc.2007.09.073
88. M. Sakai, T. Yanagisawa, A. Nakajima, Y. Kameshima, K. Okada, Effect of surface structure on the sustainability of an air layer on superhydrophobic coatings in a water-ethanol mixture. Langmuir **25**, 13–16 (2009). doi:10.1021/la802763h
89. M. Sbragaglia, A.M. Peters, C. Pirat, B.M. Borkent, R.G.H. Lammertink, M. Wessling, D. Lohse, Spontaneous breakdown of superhydrophobicity. Phys. Rev. Lett. **99**, 156001 (2007). doi:10.1103/PhysRevLett.99.156001
90. A. Shibuichi, T. Onda, N. Satoh, K. Tsujii, Super water-repellent surfaces resulting from fractal structure. J. Phys. Chem. **100**, 19512–19517 (1996)
91. N.J. Shirtcliffe, G. McHale, G.M.I. Newton, C.C. Perry, Wetting and wetting transitions on copper-based super-hydrophobic surfaces. Langmuir **21**, 937–943 (2005). doi:10.1021/la048630s

92. V.M. Starov, M.G. Velarde, Surface forces and wetting phenomena. J. Phys: Condens. Matter **21**, 464121 (2009). doi:10.1088/0953-8984/21/46/464121
93. R. Tadmor, Line energy and the relation between advancing, receding, and Young contact angles. Langmuir **20**, 7659–7664 (2004). doi:10.1021/la049410h
94. R. Tadmor, P.S. Yadav, As-placed contact angles for sessile drops. J. Colloid Interface Sci. **317**, 241–246 (2008). doi:10.1016/j.jcis.2007.09.029
95. R. Tadmor, Pr. Bahadur, A. Leh, H.E. N'guessan, R. Jaini, L. Dang, Measurement of lateral adhesion forces at the interface between a liquid drop and a substrate. Phys. Rev. Lett. **103**, 266101 (2009). doi:10.1103/PhysRevLett.103.266101
96. S. Vedantam, M.V. Panchagnula, Constitutive modeling of contact angle hysteresis. J. Colloid Interface Sci. **321**, 393–400 (2008). doi:10.1016/j.jcis.2008.01.056
97. J. Wang, D. Chen, Criteria for entrapped gas under a drop on an ultrahydrophobic surface. Langmuir **24**, 10174–10180 (2008). doi:10.1021/la801092y
98. R.N. Wenzel, Resistance of solid surfaces to wetting by water. Ind. Eng. Chem. **28**, 988–994 (1936). doi:10.1021/ie50320a024
99. G. Whyman, E. Bormashenko, T. Stein, The rigorous derivation of young, Cassie–Baxter and Wenzel equations and the analysis of the contact angle hysteresis phenomenon. Chem. Phys. Lett. **450**, 355–359 (2008). doi:10.1016/j.cplett.2007.11.033
100. V.V. Yaminsky, Molecular Mechanisms of Hydrophobic Transitions, in *Apparent and Microscopic Contact Angles*, ed. by J. Drelich, J.S. Laskowski, K.L. Mittal (VSP BV, Utrecht, 2000), pp. 47–95
101. Z. Yoshimitsu, A. Nakajima, T. Watanabe, K. Hashimoto, Effects of surface structure on the hydrophobicity and sliding behavior of water droplets. Langmuir **18**, 5818–5822 (2002). doi:10.1021/la020088p
102. Q.-S. Zheng, Y. Yu, Z.-H. Zhao, Effects of hydraulic pressure on the stability and transition of wetting modes of superhydrophobic surfaces. Langmuir **21**, 12207–12212 (2005). doi:10.1021/la052054y

Chapter 7
Wear-Resistant and Oleophobic Biomimetic Composite Materials

Vahid Hejazi and Michael Nosonovsky

Abstract The Lotus effect involving roughness-induced superhydrophobicity is a way to design biomimetic non-wetting, non-sticky, self-cleaning, omniphobic, icephobic, and anti-fouling surfaces, which can be applied for various purposes related to green tribology. However, such surfaces require micropatterning, which is extremely vulnerable to even small wear rates. This limits the applicability of the Lotus effects to situations, when wear is practically non-present. To design sustainable superhydrophobic surfaces, we suggest using metal matrix composites (MMC) with hydrophobic reinforcement in the bulk of the material, rather than at its surface. Such surfaces provide roughness and heterogeneity needed for superhydrophobicity. In addition, they are sustainable since when surface layer is deteriorated and removed due to wear, hydrophobic reinforcement and roughness remains. We present a model and experimental data on wetting of MMCs. We also conduct experiments with graphite-reinforced MMCs and show that the contact angle can be determined from the model. In order to decouple the effects of reinforcement and roughness, the experiments were conducted for initially smooth and etched matrix and composite materials. Micropatterned surfaces can be used for underwater oleophobicity and self-cleaning, in a manner, similar to the Lotus effect. However, wetting of a rough surface by oil (or any non-polar organic liquid) can follow more complex scenarios than just wetting of a rough surface by water, since a four-phase solid–oil–water–air interface can be involved.

V. Hejazi · M. Nosonovsky (✉)
College of Engineering and Applied Science, University of Wisconsin, Milwaukee, WI 53201, USA
e-mail: nosonovs@uwm.edu

M. Nosonovsky and B. Bhushan (eds.), *Green Tribology*,
Green Energy and Technology, DOI: 10.1007/978-3-642-23681-5_7,

7.1 Introduction

Contact angle is the main parameter that characterizes wetting of solid surfaces by liquids. Water droplets on smooth hydrophobic surfaces do not usually form contact angles with the solid surface greater than 120°. When the angle achieved exceeds 150° this is termed superhydrophobicity. Superhydrophobic surfaces also usually have low contact angle hysteresis, show self-cleaning properties, and have low drag for fluid flow. The range of actual and potential application of self-cleaning surfaces is diverse including optical (e.g. self-cleaning lenses), building and architecture (windows, exterior paints, roof tiles), textiles, solar panels, microdevices (where the reduction of adhesion is crucial), applications requiring antifouling from biological and organic contaminants, etc [1].

Two models of superhydrophobic behavior are used known as Wenzel [2] and Cassie–Baxter [3] models. The main difference between the two models is whether the liquid droplet retains contact with the solid surface at all points or whether the liquid bridges only across surface protrusions, thus, resulting in a droplet suspended on a composite solid and vapor surface [4]. In the Wenzel model (Fig. 7.1a), the solid–liquid contact occurs at all points below the droplet and the observed equilibrium contact angle with the rough surface, θ, is given by

$$\cos\theta = R_f \cos\theta_0 \tag{7.1}$$

where the roughness factor $R_f = A_{SL}/A_F \geq 1$ is the ratio of the real substrate area A_{SL} to the projected area A_F, and θ_0 is the contact angle on a smooth surface of the same material. In the Cassie–Baxter model (Fig. 7.1b) the droplet suspends itself across surface protrusions, and an average of the cosines of the angle on the solid (i.e. cos θ_0) and on the air (i.e. cos 180° $= -1$) below the drop is used. If f_{SL} is the fraction of the solid surface upon which the drop sits and $(1 - f_{SL})$ is the fraction below the drop that is air, then the Cassie–Baxter equation applies. When solid–liquid interface is rough, the roughness factor should also be included as

$$\cos\theta = f_{SL}\, R_f \cos\theta_0 - (1 - f_{SL}) \tag{7.2}$$

The overall conclusion is that two factors are needed to produce a superhydrophobic surface: roughness (providing high R_f and f_{SL}) and a certain extent of initial hydrophobicity (e.g. a coating), such that $\cos\theta_0 < 0$. Surface roughness magnifies the hydrophobicity bringing the contact area into the superhydrophobic region, $150° \leq \theta \leq 180°$. Furthermore, proper surface roughness is more critical than the initial superhydrophobicity, since under certain conditions even initially hydrophilic surface can show superhydrophobic properties [5]. There is also evidence that surfaces with dual-scale roughness (nanoroughness superimposed on microroughness) makes hydrophobic properties much more sustainable [6].

Since the 1990s, when new technologies emerged to produce microstructured surfaces, a huge amount of research work was done on design, fabrication, and characterization of superhydrophobic surfaces from various materials, ranging

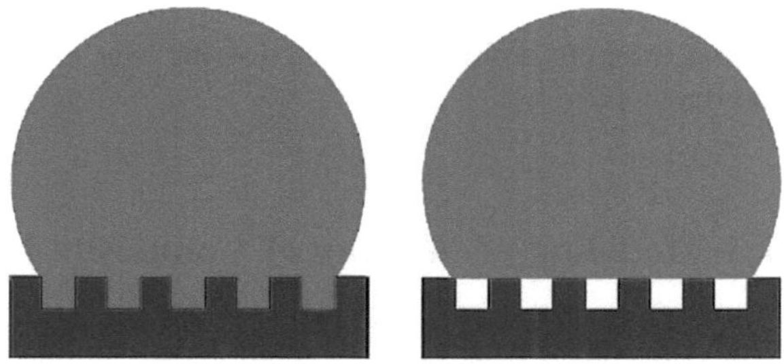

Fig. 7.1 Wetting of a microstructured surface in the (**a**) homogeneous (Wenzel, solid–liquid) and (**b**) composite (Cassie–Baxter, solid–liquid–air) regime

from polymers and ceramics to textiles, etc. A significant limitation on the practical application of the Lotus effect for self-cleaning is the sustainability of superhydrophobic microstructured coatings, which is often extremely vulnerable even to small wear rates and contamination [7].

It is much more difficult to produce a superhydrophobic metallic material than a polymer- or ceramic-based one, because metals tend to have higher surface energies [8–11]. In the area of metallic superhydrophobic materials a number of advances have been made. Yet in the 1950s, Bikerman [12] investigated wetting of stainless steel plates with different finishes with the contact angles around 90° and proposed that the surface roughness provides resistance for the sliding of water droplets. Since then, few studies of non-wetting metallic materials have been conducted. Baitai et al. [13] studied the effect of the surface roughness induced by the chemical etching on metallic composites super-hydrophobicity. They used Al, Cu, and Zn specimens immersed into an etchant (a mixture of HCl, H_2O, and HF) at room temperature for time periods from 5 to 15 s. Shirtcliffe and McHale [14] studied the wettability of Cu-base superhydrophobic surfaces. They used Cu to form the base material and a coating to hydrophobize it. The removal or addition of material roughened the surface to control wetting by combining roughness with surface patterning. Sommers and Jacobi [15] achieved anisotropic wettability on an Al surface by controlling its surface micro-topography.

Metal matrix composites (MMCs) are composite materials which have a metallic matrix and a reinforcement of another metallic or nonmetallic (ceramics, polymer, etc.) material [16]. MMCs with hydrophobic reinforcement can provide much broader opportunities than pure metals for design and fabrication of composite surfaces and readily supply the reinforcement hydrophobic fraction and surface roughness due to the reinforcement. However, superhydrophobic MMCs have not yet been explored in the literature. Furthermore, in a composite material the hydrophobic reinforcement is in the bulk of the material rather than at the surface and thus wear does not necessary lead do the deterioration of the hydrophobic coatings making these materials appropriate to the situations where traditional Lotus-effect coatings cannot be used. The use of composite materials with hydrophobic reinforcement in the bulk has already been suggested for concretes to prevent water penetration [17]. In this paper we investigate wetting of MMCs with the potential for various applications where self-cleaning sustainable surfaces are needed ranging from antifouling for the water industry to magnetic tape-head interfaces [18].

7.2 Modeling of Wetting of Composite Materials

The reinforcement particles or fibers have certain dimensions and wetting properties. Some models are presented here. First a general model is considered to predict the contact angle on MMCs surface. Second, it is considered that the matrix has smooth surface and the surface roughness is produced only by the reinforcement present at the surface. Finally, the surfaces are etched by etchants so that their roughness increases. For the two latter models, two types of the reinforcements geometry are considered: particles and fibers. It is assumed that the reinforcement particles are spherical whereas fibers are aligned beams. The wetting of the surfaces is modeled based on our assumptions using both Wenzel and Cassie–Baxter models.

7.2.1 General Model

For a composite interface built of two fractions with the fractional areas of f_1 and f_2 (so that $f_1 + f_2 = 1$) the contact angle is given by the Cassie equation

$$\cos\theta = f_1 \cos\theta_1 + f_2 \cos\theta_2 \tag{7.3}$$

where θ_1 and θ_2 are the contact angles of the fractions. If a composite material has a matrix and reinforcement with the volume fractions of f_m and f_r (so that $f_m + f_r = 1$) forming a rough surface the contact angle is then given by

$$\cos\theta = R_{\mathrm{fm}}(1 - f_r)\cos\theta_m + R_{\mathrm{fr}} f_r \cos\theta_r \tag{7.4}$$

where θ_m and θ_r are the contact angles for the matrix and reinforcement materials, and R_{fm} and R_{fr} are the corresponding roughness factors. Note that for spherical reinforcement particles the roughness factor is equal to the ratio of half of the sphere's area $2\pi R^2$ to the cross-sectional area πR^2 or $R_{\mathrm{fr}} = 2$. Solving for the reinforcement fraction yields the volume of the reinforcement fraction providing the desired contact angle θ

$$f_r = \frac{\cos\theta - R_{\mathrm{fm}}\cos\theta_m}{R_{\mathrm{fr}}\cos\theta_r - R_{\mathrm{fm}}\cos\theta_m} \tag{7.5}$$

Further assuming $R_{\mathrm{fr}} = 2$, $R_{\mathrm{fm}} = 1$ (no roughness expected from the reinforcement particles), and $\theta = 180°$ (the superhydrophobic limit) yields

$$f_r = \frac{-1 - \cos\theta_m}{2\cos\theta_r - \cos\theta_m} \tag{7.6}$$

which has a solution ($f_r < 1$) if $\theta_r > 120°$. Thus it is difficult to produce a composite interface by only using the reinforcement roughness.

If water forms partial contact with the solid (composite or Cassie–Baxter) interface with the fractional solid–liquid contact areas f_{SLm} and f_{SLr}, the contact angle is given by

$$\cos\theta = R_{\mathrm{fm}}(1-f_r)f_{\mathrm{SLm}}\cos\theta_m + R_{\mathrm{fr}}f_r f_{\mathrm{SLr}}\cos\theta_r - 1 + f_r f_{\mathrm{SLr}} + (1-f_r)f_{\mathrm{SLm}} \tag{7.7}$$

Solving for the reinforcement fraction yields the volume of the reinforcement fraction providing the desired contact angle θ

$$f_r = \frac{\cos\theta - R_{\mathrm{fm}}f_{\mathrm{SLm}}\cos\theta_m + 1 - f_{\mathrm{SLm}}}{R_{\mathrm{fr}}f_{\mathrm{SLr}}\cos\theta_r - R_{fm}f_{\mathrm{SLm}}\cos\theta_m + f_{\mathrm{SLr}} - f_{\mathrm{SLm}}} \tag{7.8}$$

Making the assumptions of $R_{\mathrm{fr}} = 2$, $R_{\mathrm{fm}} = 1$, $f_{\mathrm{SLr}} = 1$, and $\theta = 180°$ yields

$$f_r = \frac{-f_{\mathrm{SLm}} - f_{\mathrm{SLm}}\cos\theta_m}{2\cos\theta_r - f_{\mathrm{SLm}}\cos\theta_m + 1 - f_{\mathrm{SLm}}} \tag{7.9}$$

Let us apply Eqs. 7.3–7.9 to the metallic (aluminum or copper) matrix with amorphous graphite reinforcement. For that end, we need to substitute the material properties of these materials. We measured the water contact angle with graphite, aluminum, and copper, which were used to produce MMC samples at the UWM Center for Composite materials, using the ramé-hart Model 250 goniometer/tensiometer. We found the values of the contact angle equal to 140°, 47.2° and 47.7°, respectively. The results for metal matrices are close to those available in the literature [19]. To measure the water contact angle of amorphous graphite, we compressed it at first to obtain a smooth surface. The main reason for high water contact angle of graphite is that the surface still remains rough even after compressing. Since we need the water contact angle of smooth graphite, we used the result of Fowkes and Harkins [20]. They measured the contact angle of water on smooth graphite using the tilting plate method and found a value of 86°. A similar result (84°) is reported by Morcos [21]. Figure 7.2 shows the variation of the contact angle of water droplet on surface of metal matrix composite reinforced by graphite particles versus reinforcement volume fraction ($\theta_m = 47°$, $\theta_r = 86°$), as obtained from Eq. 7.8 for various values of f_{SLm}. As seen in this figure, for $f_{\mathrm{SLm}} > 0.4$, the contact angle increases with increase in the reinforcement volume fraction, whilst for $f_{\mathrm{SLm}} \leq 0.4$, it decreases with increase in the reinforcement volume fraction. Figure 7.3 shows the variation of f_{SLm} versus the reinforcement volume fraction, f_r as obtained from Eq. 7.9 for cases where θ is equal to 150° and 180°. For this case, it is assumed that the matrix is made of aluminum whereas the reinforcement is made of a material with the water contact angle of 140°. As observed from this figure,

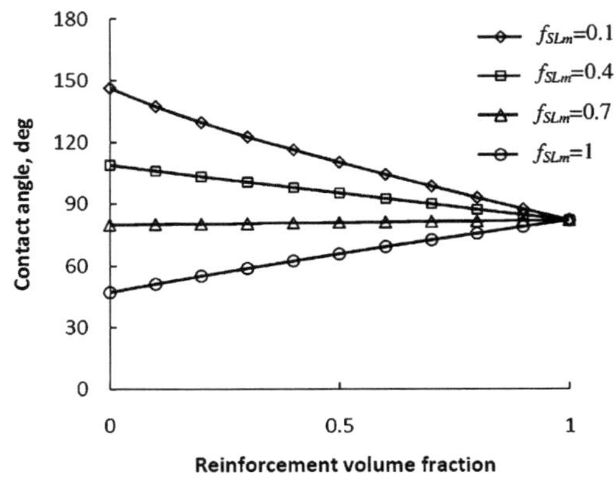

Fig. 7.2 The water contact angle as a function of graphite reinforcement volume fraction for different values of $f_{\rm SLm}$

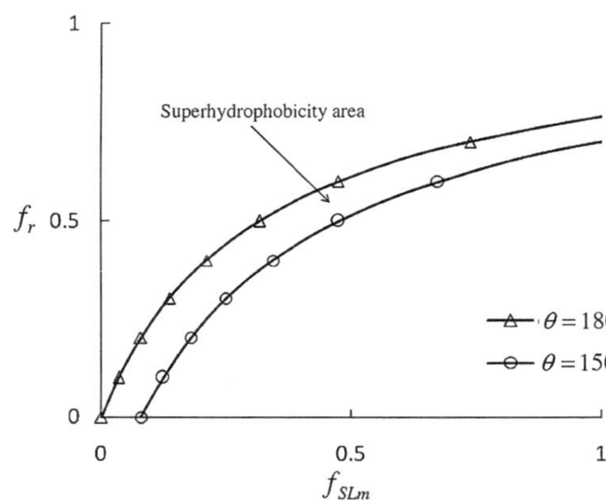

Fig. 7.3 The variation of the reinforcement volume fraction, f_r, versus $f_{\rm SLm}$

the reinforcement volume fraction, f_r, is proportional to $f_{\rm SLm}$. The area between two lines of $\theta = 150^{\rm o}$ and $\theta = 180^{\rm o}$ is called the superhydrophobicity area.

The effect of wear on a composite material is twofold. First, the matrix roughness factor, $R_{\rm fm}$, can be changed due to material removal and evolve to a certain "equilibrium value" [18]. This can affect the solid–liquid fractional area, $f_{\rm SLm}$. Second, the reinforcement particles can be removed as matrix surface layers are removed due to the deterioration. However, new particles come in contact so it is expected that the values of $R_{\rm fr}$ and $f_{\rm SLr}$ do not change significantly. To decouple the effect of reinforcement and matrix roughness we investigated experimentally wetting of composite materials with initially smooth surface and with the matrix roughness by etching, as described in the next section.

7.2.2 Underwater Oleophobicity

If an oil droplet is placed on a solid surface in water, the contact angle of an oil droplet in water, $\theta_{\rm OW}$, is given by Young's equation:

$$\cos\theta_{\mathrm{OW}} = \frac{\gamma_{\mathrm{OA}}\cos\theta_{\mathrm{O}} - \gamma_{\mathrm{WA}}\cos\theta_{w}}{\gamma_{\mathrm{OW}}} \tag{7.10}$$

where θ_O and θ_W are contact angles of oil and water with the solid surface, γ_{OA}, γ_{WA} and γ_{OW} are interfacial energies for oil–air, water–air, and oil–water interfaces. As a consequence, a superhydrophobic (in air) surface can become oleophobic when immersed in water, under certain circumstances, which are summarized in Table 7.1. An oleophobic surface repels organic liquids and thus prevent organic contaminants from accumulation and decrease the adhesion of bacteria, thus facilitating antifouling properties.

Similar to the superhydrophobic surfaces, besides the homogeneous solid–oil interface (Wenzel state), a composite solid–oil–water interface (Cassie–Baxter or Cassie state) with water pockets trapped between the solid and the oil droplet can exist. The contact angle is then given by

$$\cos\theta_{\mathrm{OW}} = \frac{\gamma_{\mathrm{OA}}\cos\theta_{\mathrm{O}} - \gamma_{\mathrm{WA}}\cos\theta_{W}}{\gamma_{\mathrm{OW}}} f_{\mathrm{SO}}R_f + 1 - f_{\mathrm{SO}} \tag{7.11}$$

Where $0 < f_{SO} < 1$ is the fractional solid–oil contact area and R_f is the roughness factor. The rules of the Cassie–Wenzel wetting regime transition are the same as in the case of superhydrophobic surfaces, as discussed in the preceding chapters. A more complex four-phase solid–oil–water–air system can form if both water and air bubbles are present at the composite interface. It is expected that with time air will dissolve so that $f_{SA} \rightarrow 0$.

A wetting regime transition, similar to the Cassie–Wenzel transition can apparently occur at the solid–oil–water interface. The evidence of this is presented in Fig. 7.4 showing solid–oil–water system with the same aluminum alloy of different surface roughness. In our experiments samples with lower roughness having Root-mean-square (RMS) $R_a = 0.2$ μm, showed low contact angles of $\theta = 43.10^{\circ}$ and $\theta = 24.96^{\circ}$, whereas samples with higher roughness (etched by an acid) $R_a = 0.8$ μm showed contact angles of $\theta = 140.78^{\circ}$ and $\theta = 141.30^{\circ}$. We attribute such an abrupt change of roughness to the fact that the Wenzel state (homogeneous solid–oil interface) was realized for the low roughness samples, whereas the Cassie–Baxter state (composite solid–oil–water interface) was realized for high roughness samples.

7.2.3 Reinforcement-Induced Roughness

In this section we investigate wetting properties of a composite material with particulate or fiber reinforcement. The reinforcement, such as graphite, modifies wetting properties of the matrix material, such as a metal, by providing heterogeneity and surface roughness.

Table 7.1 Oleophobic and oleophilic interfaces

Interface					
Solid–air–water	Hydrophobic $\gamma_{SA} > \gamma_{SW}$			Hydrophobic $\gamma_{SA} < \gamma_{SW}$	
Solid–air–oil	Oleophilic if $\gamma_{SA} > \gamma_{SO}$	Oleophobic if $\gamma_{SA} < \gamma_{SO}$		Oleophilic if $\gamma_{SA} > \gamma_{SO}$	Oleophobic if $\gamma_{SA} < \gamma_{SO}$
Solid–water–oil	Oleophilic	Oleophilic if $\gamma_{OA}\cos\theta_0 < \gamma_{WA}\cos\theta_W$	Oleophobic if $\gamma_{OA}\cos\theta_0 > \gamma_{WA}\cos\theta_W$	Oleophilic if $\gamma_{OA}\cos\theta_0 < \gamma_{WA}\cos\theta_W$	Oleophobic if $\gamma_{OA}\cos\theta_0 < \gamma_{WA}\cos\theta_W$

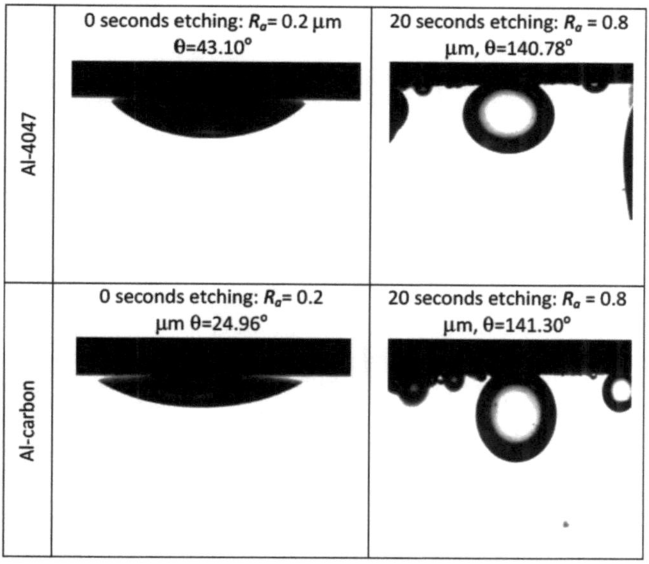

Fig. 7.4 Solid–oil–water system with low roughness (*left*) and high roughness, A dramatic change of the contact angle is attributed to the Cassie–Wenzel wetting regime change

7.2.3.1 Homogeneous (Wenzel) Solid–Liquid Interface

The Wenzel model defines the homogeneous (solid–liquid) wetting regime. According to that model, water penetrates inside all crevices and there are no air gaps left between the liquid and solid (Fig. 7.1a).

Particle Reinforcement

Now we model our cases starting with particle reinforcements. For this case, we assumed that the spherical particles make roughness on the surface. We considered a layer of matrix which contains randomly distributed particles with radii r (Fig. 7.5).

The reinforcements volume fraction, f_v, is obtained through dividing the total volume of all particles inside the layer, V_r, by volume of layer, V_{tot}

$$f_v = \frac{v_r}{v_{tot}} \tag{7.12}$$

$$V_r = \frac{4n\pi r^3}{3} \tag{7.13}$$

$$V_{tot} = A_{tot} \cdot r \tag{7.14}$$

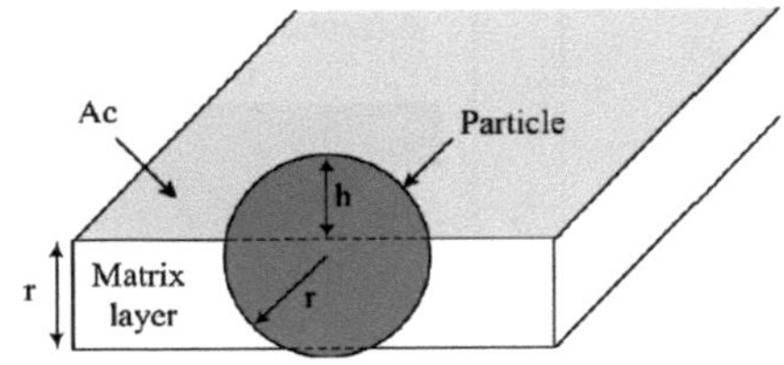

Fig. 7.5 The schematic of a matrix layer passes through spherical particles

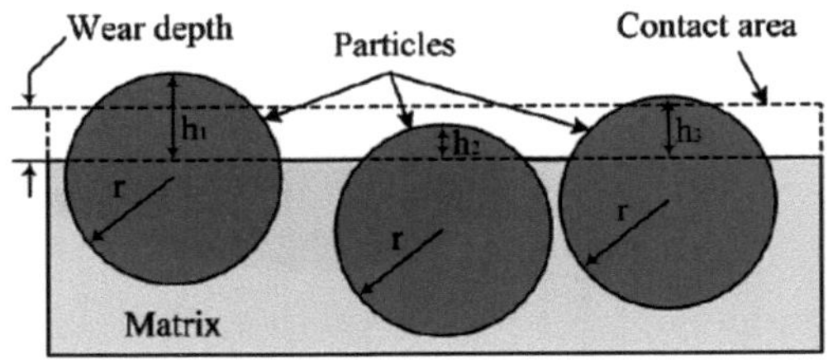

Fig. 7.6 The schematic of changing h due to wear

where A_{tot} is the top surface of the layer and n is the total number of particles inside the layer.

Substituting (7.13) and (7.14) into (7.12), one can show that

$$n = \frac{3f_v v_{\text{tot}}}{4\pi r^3} \tag{7.15}$$

The height of each particle out of matrix layer is defined by h. Since the metal matrices are usually softer than reinforcements, this height can be changed due to wear during the time. We assumed that $0 \leq h \leq r$ (Fig. 7.6).
therefore,

$$f_r = f_v \tag{7.16}$$

where the total area of reinforcements in contact area, A_c, can be obtained by

$$A_r = n\left(\frac{1}{r}\int_0^r 2\pi rh\,\mathrm{d}h\right) = n\pi r^2 \tag{7.17}$$

We can find the f_m from the following equation:

$$f_m + f_r = 1 \quad \rightarrow \quad f_m = 1 - f_v \tag{7.18}$$

We define averaged Wenzel roughness factor as follows:

$$\frac{1}{R_f} = 1 - \frac{A_r}{4A_{\text{tot}}} \quad \rightarrow \quad \frac{1}{R_f} = 1 - \frac{3}{16}f_v \tag{7.19}$$

From Wenzel and Cassie–Baxter equations:

$$Cos\theta_c = R_f \cdot f_r \cdot Cos\theta_r + f_m \cdot Cos\theta_m \tag{7.20}$$

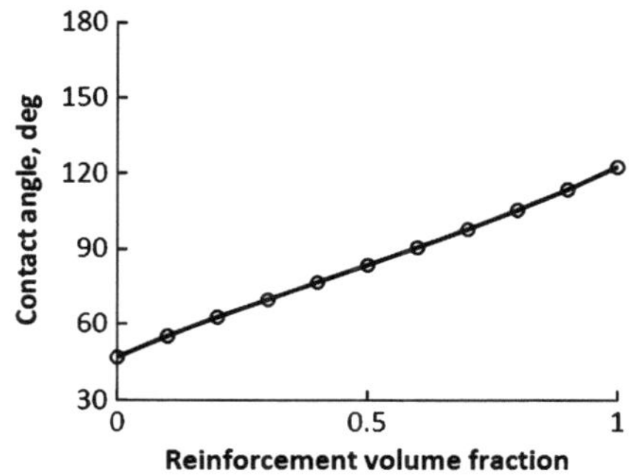

Fig. 7.7 Contact angle of water droplet versus graphite particles volume fraction in metal matrix (Wenzel model)

Substituting (7.16), (7.18), and (7.19) into (7.20) we have:

$$\theta_c = \cos^{-1}\left[\frac{12f_v}{16-3f_v}\cos\theta_r - \frac{3}{4}f_v\cos\theta_m + \cos\theta_m\right] \tag{7.21}$$

Let us apply Eq. 7.21 to the metallic (aluminum or copper) matrix with graphite reinforcement. For that end, we need to substitute the material properties of these materials. We measured the water contact angle with graphite, aluminum, and copper, which were used to produce MMC samples at the UWM Center for Composite materials, using the ramé-hart Model 250 goniometer/tensiometer. Figure 7.7 shows the variation of the contact angle of water droplet on surface of metal matrix composite reinforced by graphite particles versus reinforcement volume fraction. As seen in this figure, the contact angle increases with increase in the reinforcement volume fraction.

Fiber Reinforcement

For this case, we assumed that the aligned fibers perpendicular to the surface make roughness on it. The length of each fiber out of matrix layer divided by fiber radius is defined by a (Fig. 7.8).

The Wenzel roughness factor is defined as follows:

$$R_f = \frac{A_c + 2n\pi r l_p}{A_c} = 1 + \frac{2n\pi r l_p}{A_c}; \quad a = \frac{l_p}{r} \tag{7.22}$$

where l_p is the length of each fiber out of matrix layer. r and n are the radius and total number of fibers inside certain volume, respectively.

One can easily prove the following equations:

$$f_r = f_v \quad \rightarrow \quad f_m = 1 - f_v \tag{7.23}$$

So after calculating the following roughness factor is obtained:

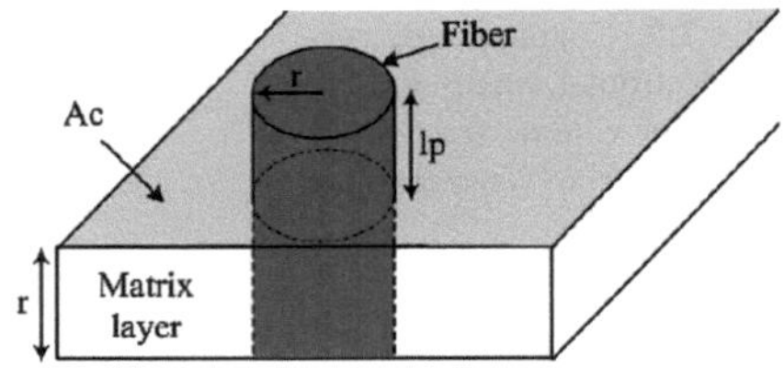

Fig. 7.8 The schematic of a certain volume passes perpendicular to the reinforcement fiber length

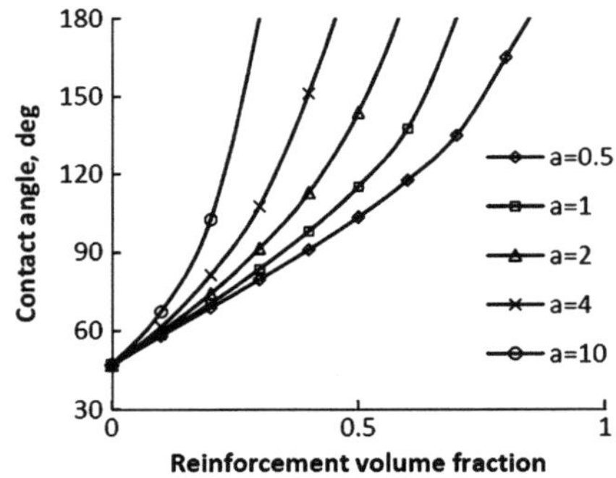

Fig. 7.9 Variation of contact angle versus graphite aligned fibers volume fraction in metal matrix (Wenzel model)

$$R_f = 1 + 2af_v \tag{7.24}$$

Using Eq. 7.20 and substituting the values of f_r, f_m and R_f into it, we have the following equation for calculating the contact angle on composite surface:

$$\theta_c = \cos^{-1}\left[2a\,\cos\theta_r f_v^2 + (\cos\theta_r - \cos\theta_m)f_v + \cos\theta_m\right] \tag{7.25}$$

Then we apply Eq. 7.25 to the metallic (aluminum or copper) matrix with graphite reinforcement. Figure 7.9 shows the contact angle of water droplet on surface of metal matrix composite reinforced by graphite fibers versus reinforcement volume fraction. As it is seen in this figure, the contact angle increases with increasing the reinforcement volume fraction and the value of "a" as well.

7.2.3.2 Composite (Cassie–Baxter) Solid–Liquid–Air Interface

The Cassie–Baxter model defines the heterogeneous wetting regime. Based on Cassie–baxter model, there are some air gaps left between liquid and solid surfaces and we deal with a three-phase system of liquid–solid–gas (Fig. 7.1b).

In order to calculate the contact angle, θ, for a rough surface in a manner similar to the previous section but for Cassie–Baxter model, the differential area of the liquid–air interface under the droplet, $f_{\mathrm{LA}}\mathrm{d}A_c$, should be subtracted from the differential of the total liquid–air area $\mathrm{d}A_{\mathrm{LA}}$, which yields

$$\cos\theta = R_f\cos\theta_0 - f_{\mathrm{LA}}(R_f\cos\theta_0 + 1) \tag{7.26}$$

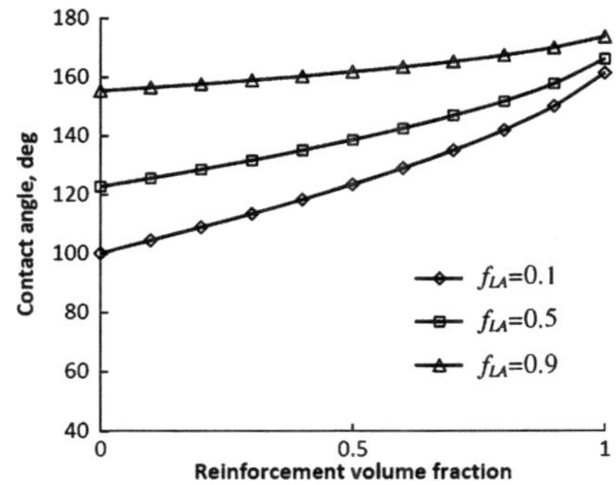

Fig. 7.10 Contact angle of water droplet versus graphite particles volume fraction in metal matrix for different fractions of liquid–air in contact (C–B model)

such that the f_{LA} is the fraction of contact area between liquid and air to total contact area and θ_0 is contact angle between the water droplet and the solid with relatively smooth surface.

Particle Reinforcement

For this case, we modeled our MMCs based on Cassie–baxter in case the spherical particles make roughness on the surface, so that the Cassie–Baxter equation is turned into

$$Cos\theta_c = R_f f_r \cos\theta_r + f_m \cos\theta_m - f_{\mathrm{LA}}(R_f f_r \cos\theta_r + f_m \cos\theta_m + 1) \qquad (7.27)$$

The values of R_f, f_m, and f_r are defined by Eqs. 7.16, 7.18, and 7.19 and after substituting into Eq. 7.26 we obtained the following contact angle:

$$\theta_c = \cos^{-1}\left[(1 - f_{\mathrm{LA}})\left(\frac{16f_v}{16 - 3f_v}\cos\theta_r + (1 - f_v)\cos\theta_m + 1\right) - 1\right] \qquad (7.28)$$

Figure 7.10 shows the contact angle of water droplet on the surface of aluminum matrix composite versus graphite particles volume fraction. As seen in this figure, the contact angle increases with increase in the reinforcement volume fraction and the value of f_{LA} as well.

Fiber Reinforcement

For this case, we modeled our MMCs based on Cassie–baxter in case the aligned fibers make roughness on the surface, so that the Cassie–Baxter equation is the same as Eq. 7.20.

The values of R_f, f_m, and f_r are defined by Eqs. 7.23 and 7.24 and after substituting into Eq. 7.27, we obtained the following contact angle:

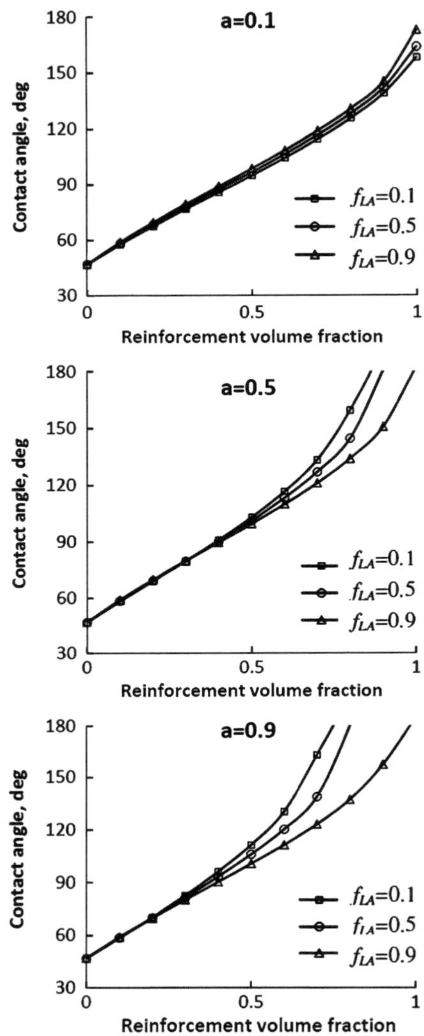

Fig. 7.11 Contact angle of water droplet versus graphite aligned fibers volume fraction in metal matrix for different aspect ratio and fractions of liquid–air in contact (C–B model)

$$\theta_c = \cos^{-1}\left[2a\cos\theta_r(1-f_{\mathrm{LA}})f_v^2 + [\cos\theta_r - \cos\theta_m - f_{\mathrm{LA}}(\cos\theta_r+1)]f_v + \cos\theta_m\right] \tag{7.29}$$

Figure 7.11 shows the variation of the contact angle of water droplet on the surface of aluminum matrix composite versus graphite fibers volume fraction inside certain volume. As seen in this figure, the contact angle increases with increase in the reinforcement volume fraction. These figures show that for $a = 0.1$, the contact angle is proportional to f_{LA}, whereas for $a = 0.5$ and $a = 0.9$, it is proportional to the inverse of f_{LA}.

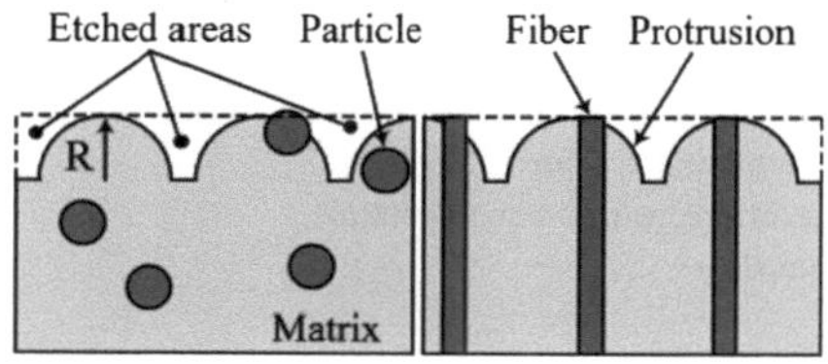

Fig. 7.12 The schematic of surface pattern after etching for particles and fibers reinforcement

7.2.4 Etching-Induced Roughness

In this section we consider wetting of composite materials after their surface roughness was enhanced by etching inducing protrusions, which made typical roughness details larger than a typical reinforcement. According to Wenzel [2] the contact angle of water droplet on a rough surface is given by Eq. 7.1. Therefore for calculating the contact angle after roughening the surface by etching, all we need is to find the roughness factor for the etched surface.

One can assume that the etching makes hemispherical protrusion patterns on a surface (Fig. 7.12). For such a rough surface, the roughness factor is given by

$$R_f = 1 + \frac{N\pi R^2}{A_c} \tag{7.30}$$

where A_c, N, and R are the flat projected area, the number of crevices, and crevice radius, respectively. We assumed that the term $\frac{N\pi R^2}{A_c}$ is equal to the average roughness of surface and showed that with R_a. Thus the roughness factor for an etched surface is given by

$$R_f = 1 + R_a \tag{7.31}$$

7.2.4.1 Homogeneous (Wenzel) Solid–Liquid Interface

Particle Reinforcement

For spherical particles, to measure contact angle of water droplet on an etched surface using Eqs. 7.1, 7.21, and 7.31 we have:

$$\theta_c = \cos^{-1}\left[(1 + R_a)\left(\frac{12f_v}{16 - 3f_v}\cos\theta_r - \frac{3}{4}f_v\cos\theta_m + \cos\theta_m\right)\right] \tag{7.32}$$

Figure 7.13 shows the variation of contact angle of water droplet on MMC surface versus particles volume fraction in case the matrix and particles are made by aluminum and graphite, respectively.

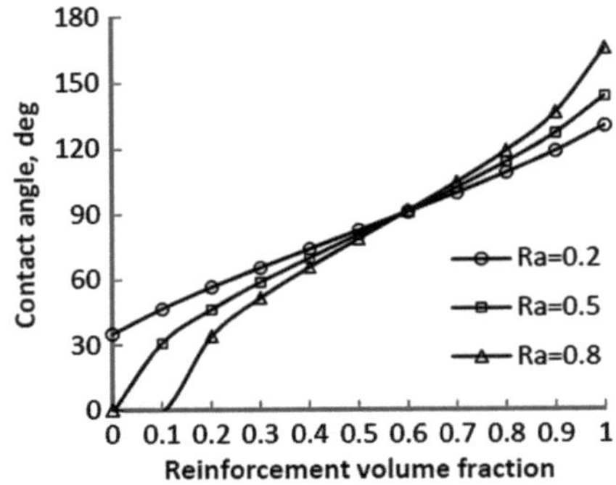

Fig. 7.13 Contact angle of water droplet versus graphite particles volume fraction in metal matrix for different average roughness (Wenzel model)

Fiber Reinforcement

For aligned fibers, we used Eq. 7.25 to find the contact angle of water droplet for etched surface, therefore

$$\theta_c = \cos^{-1}\left[(1+R_a)\left(2a\cos\theta_r f_v^2 + (\cos\theta_r - \cos\theta_m)f_v + \cos\theta_m\right)\right] \quad (7.33)$$

Figure 7.14 shows the variation of contact angle of water droplet on MMC surface versus fibers volume fraction in case the matrix and fibers are made of aluminum and graphite, respectively.

7.2.4.2 Composite (Cassie–Baxter) Solid–Liquid–Air Interface

Particle Reinforcement

For this case, using Eq. 7.28 and applying roughness factor for an etched surface in it, the contact angle of water droplet is obtained by

$$\theta_c = \cos^{-1}\left[(1+\mathrm{R}_a)\left((1-f_{LA})\left(\frac{16f_v}{16-3f_v}\cos\theta_r + (1-f_v)\cos\theta_m + 1\right) - 1\right)\right] \quad (7.34)$$

Figure 7.15 shows the variation of the contact angle of water droplet on the surface of aluminum matrix composite versus graphite particles volume fraction inside certain volume. As seen in this figure, the contact angle increases with increase in the reinforcement volume fraction and the amount of R_a as well. Comparing figures concludes that with increase in the value of f_{LA}, the contact angle decreases.

Fiber Reinforcement

To calculate the contact angle for this case, we applied the roughness factor into Eq. 7.29 so that the θ_0 is obtained by Cassie–Baxter equation, the contact angle is given by:

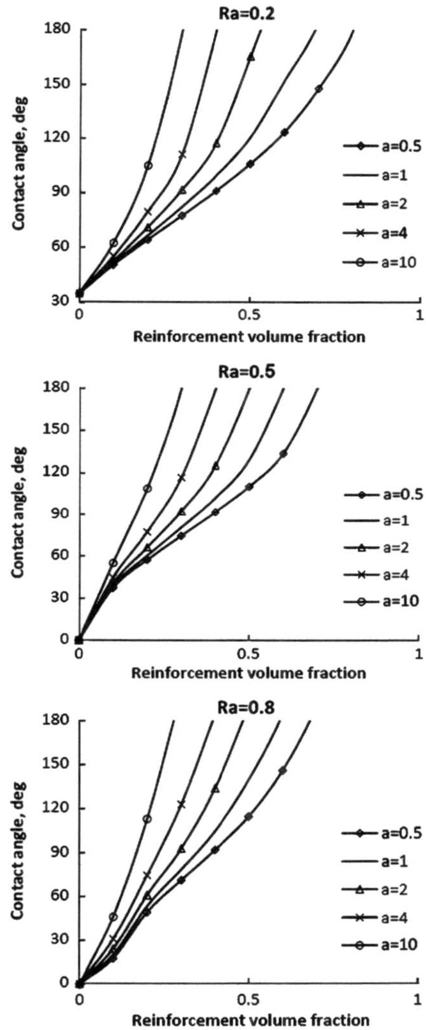

Fig. 7.14 Contact angle of water droplet versus graphite aligned fibers volume fraction in metal matrix for different average roughness (Wenzel model)

$$\theta_c = \cos^{-1}\left[(1 + R_a)\left(2a\cos\theta_r(1 - f_{\mathrm{LA}})f_v^2 + [\cos\theta_r - \cos\theta_m - f_{\mathrm{LA}}(\cos\theta_r + 1)]f_v + \cos\theta_m\right)\right] \quad (7.35)$$

Figure 7.16 shows the variation of the contact angle of water droplet on the surface of aluminum matrix composite versus graphite fibers volume fraction inside certain volume. As seen in this figure, for a constant aspect ratio, the contact angle increases with increase in the reinforcement volume fraction and the amount of R_a as well. Comparing figures concludes that with increase in the amount of f_{LA}, the contact angle decreases.

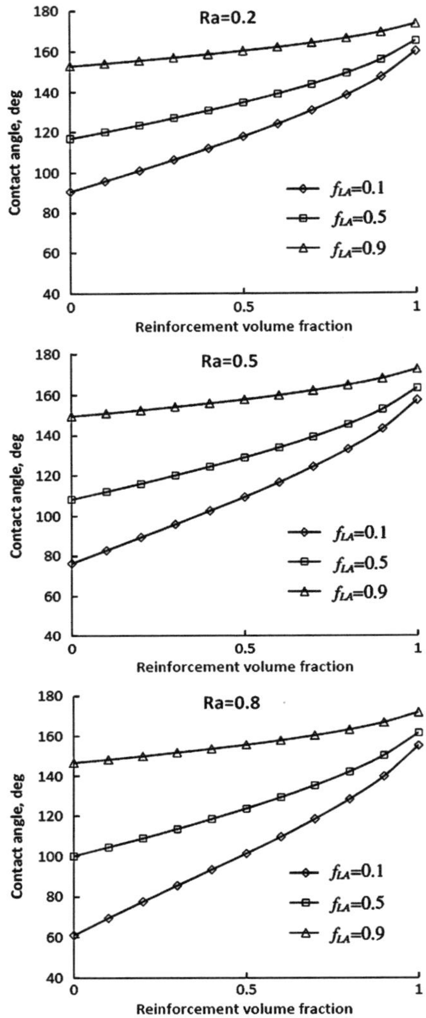

Fig. 7.15 Contact angle of water droplet versus graphite paricles volume fraction in metal matrix for different average roughness and fractions of liquid–air in contact (C–B model)

7.3 Experimental

In order to verify experimentally the models presented in the preceding section, we prepared eight samples of MMCs, four with relatively smooth surfaces and four roughened by etching, and measured their roughness and contact angles. In fact by using etching, we tried to simulate the corrosive and erosive wear on the samples which are probable especially due to environmental conditions.

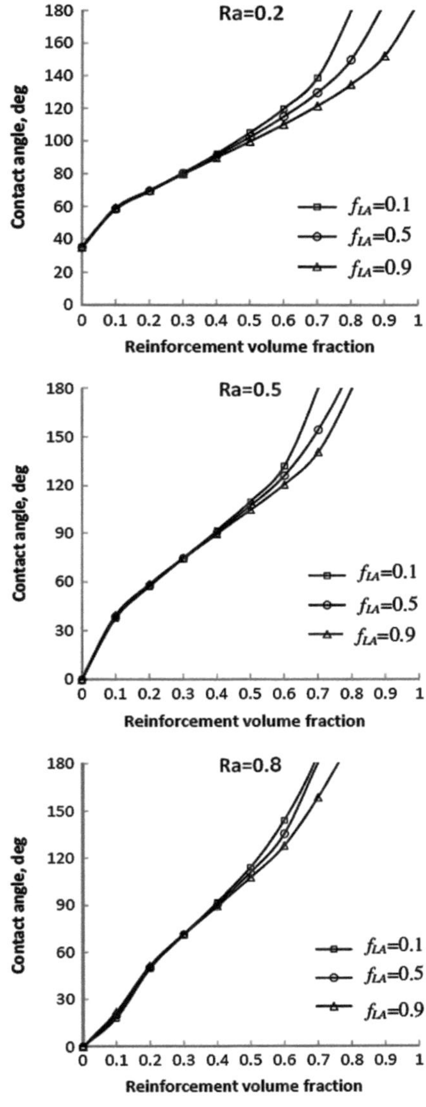

Fig. 7.16 Variation of contact angle versus graphite aligned fibers volume fraction in metal matrix for different average roughness, R_a and fractions of liquid–air in contact and $a = 0.5$ (C–B model)

7.3.1 Sample Preparation

The four samples of Al- and Cu-based alloys and their graphite composites were sectioned to 2.0 × 1.5 × 0.2 cm pieces. The diameter of graphite particles used to make MMCs is estimated to be between 10 and 15 μm. We used Al and Cu because they are the standard material used as a matrix in the literatures due to their high conductivity and ductility for making MMCs. Furthermore, Al- and

Table 7.2 Chemical composition of the samples

Sample material	Composition
Copper base alloy	Cu (79.0–82.0%), Sn (2.5–3.5%), Pb (6.3–7.7%), Zn (7.0–10%), P (0.02%), Al (0.05%), Si (0.005%)
Copper-graphite composite	Cu (81%), Ni (5%), Fe (4%), Al (9%), Mn (1%), 60% volume of graphite.
Al base alloy	Al (88%), Si (12%)
Al-graphite composite	Al (35%), Si (5%), 60% volume of graphite

Fig. 7.17 SEM images of polished samples

Cu-based samples are easy to work with and inexpensive as well. Table 7.2 presents the chemical composition of the samples.

The samples were ground and polished to create a smooth surface before the etching process. The grinding involved successive steps with 400, 600, and 1,200 grit SiC paper. Polishing was done with a soft cloth impregnated with 1 μ alumina. The Scanning electron microscope (SEM) images of the polished samples are as shown in Fig. 7.17.

After that four samples were etched. The reagents used in the selective dislocation etching are as follows [13]. For the aluminum base alloy and the aluminum–graphite composite, the etchant consisted of 40 ml of 37 wt% HCl, 12.5 ml of H_2O, and 2.0 ml of 48 wt% of HF. The etching time was set for 20 s, for a two-cycle test. On the other hand, the etchants used for the copper-based alloy and the copper–graphite composite was 0.1 molar concentration of 37 wt% HCl. The etching time was 20 h for both the sample types. Since the Al-based samples are softer than Cu-based samples, these time periods for etching, make roughness of the same order of magnitude for all samples. All samples were washed, cleaned, and dried before conducting experiments.

Table 7.3 Measured and calculated surafce roughness and contact angles

Sample	Measured			Calculated		
	Etching time	R_a (μm)	Measured contact angle	R_{fm}	Contact angle (Wenzel, $f_{SLm} = 1$)	Contact angle (Cassie–Baxter, $f_{SLm} = 0$)
Al-graphite composite	0	0.2	59.1°	1	69°	92°
Cu-graphite composite	0	0.2	86.7°	1	69°	92°
Al base alloy	20 s	16	82.5°	1.89	0°	73°
Cu base alloy	20 h	8	99.4°	1.28	29°	97°
Al-graphite composite	20 s	14	86.5°	1.72	56°	81°
Cu-graphite composite	20 h	8	87.3°	1.28	64°	88°

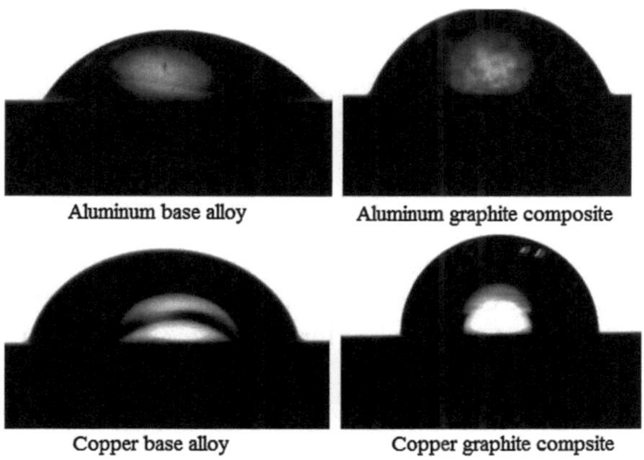

Fig. 7.18 Images of water droplet on non-etched samples

7.3.2 Characterization of Samples

The surface roughness of the samples was measured before etching process using a surface profilometer (Mitutoyo Surftest. 402). The roughness of the samples was measured again after the etching. The roughness parameter measured in the above cases is the average roughness value, R_a, defined as the arithmetic average of the absolute values of the roughness profile ordinates [22].

After that, the contact angles were measured. These measurements were done using the model 250 Rame Hart Goniometer. The measured contact angles showed that none of the surfaces were hydrophobic before etching, though surfaces of the composites had a higher contact angle with water (Table 7.3). This can be attributed to the graphite particles which have an effect on the surface roughness of the composites when compared to the base alloys. Figure 7.18 shows the contact angle profiles of these samples before etching.

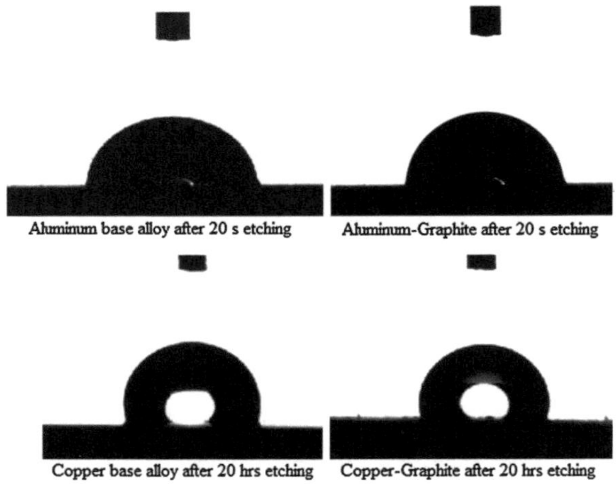

Fig. 7.19 Images of water droplet on etched samples

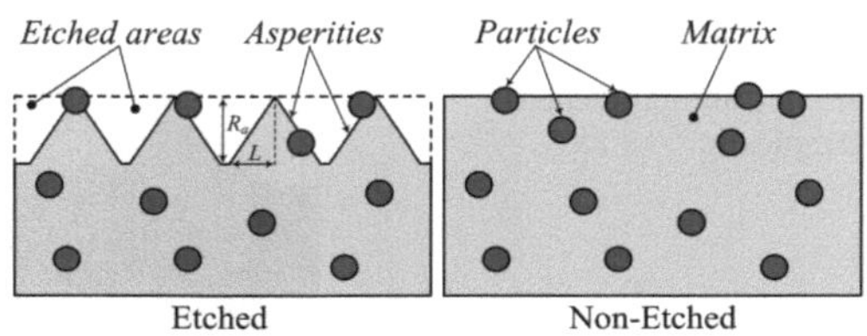

Fig. 7.20 The schematic of surface before and after etching for a matrix reinforced with particles

It is observed that all etched samples except Al base alloy became hydrophobic, i.e. have the contact angle greater than 90°. The contact angle profiles for Al- and Cu-samples that were etched for 20 s and 20 h, respectively are as shown in Fig. 7.19.

7.4 Discussion

To compare the experimental results with the model and determine whether the homogeneous or composite interface is realized, the contact angle was calculated for the Wenzel Eq. 7.4 and Cassie–Baxter wetting regimes Eq. 7.7. For that end, it was assumed that non-etched sample have smooth matrix ($R_{\mathrm{fm}} = 1$) and the surfaces roughness is caused by spherical reinforcement particles ($R_{\mathrm{fr}} = 2$). For etched samples, the matrix roughness is significant in comparison with the reinforcement roughness, although the exact values of R_{fm} cannot be calculated from the measured R_a. However, it can be estimated as

$$R_{\mathrm{fm}} = \sqrt{1 + (R_a/L)^2} \tag{7.36}$$

where L is a typical length of an asperity (Fig. 7.20). For the Cassie–Baxter model we assumed $f_{\mathrm{SLm}} = 0$ (no matrix in contact with water) and $f_{\mathrm{SLr}} = 1$ (entire reinforcement area is covered by water). Assuming $L = 10\ \mu m$, the roughness factors and the contact angles were further calculated (Table 7.3).

It is observed from Table 7.2 that for smoother samples (no etching) the Wenzel model predicts more accurate results, whereas for the etched samples the Cassie–Baxter model is better. As a matter of fact, we decoupled the effects of roughness and reinforcement. The results showed that both roughness and reinforcement are essential for having a superhydrophobic surface. The experimental data show that, for all samples, the contact angle increased with increasing roughness. The change in the contact angle of the Al-graphite sample after etching was greater than that for Cu-graphite. The main reason for this is the difference in their average roughness values, R_a. Since the average roughness values of samples are different for Al- and Cu-graphite composites, respectively, so each sample follows a different behavior.

7.5 Conclusion

The standard method of creating a superhydrophobic surface implies roughening the surface and placing some low-energy (hydrophobic) material on it. Instead, we suggested here to place low-energy reinforcement into the matrix of the composite material. We presented a model for the overall contact angle of such a composite material and for the reinforcement volume fraction needed to make the material superhydrophobic. We also conducted experiments with low-energy graphite-reinforced aluminum- and copper-based composites and showed that the contact angle can be determined from the model with certain reasonable assumptions about the roughness factors and solid–liquid fractional contact areas. In order to decouple the effects of reinforcement and roughness, the experiments were conducted for initially smooth and etched matrix and composite materials. The composite materials can be used to produce sustainable (wear-resistant) superhydrophobic surfaces.

Acknowledgments The authors acknowledge the support of the University of Wisconsin-Milwaukee (UWM) RGI, NSF I/UCRC for Water Equipment and Policy, and UWM Research Foundation Bradley Catalyst grants. The authors are also thankful to Prof. Pradip K. Rohatgi and Mr. Aniedi Nyong from the UWM Center for Composite materials for metallic samples.

References

1. M. Nosonovsky, B. Bhushan, Superhydrophobic surfaces and emerging applications: non-adhesion, energy, green engineering. Curr. Opin. Coll. Interface Sci. **14**, 270–280 (2009)
2. R. Wenzel, Resistance of solid surfaces to wetting by water. Ind. Eng. Chem. **28**, 988 (1936)
3. A.B.D. Cassie, S. Baxter, Wettability of porous surfaces. Trans. Faraday Soc. **40**, 546 (1944)
4. D. Quere, Rough ideas on wetting. Physica A **313**(1–2), 32–46 (2002)

5. M. Nosonovsky, B. Bhushan, Biologically-inspired surfaces: broadening the scope of roughness. Adv. Funct. Mater. **18**, 843–855 (2008)
6. M. Nosonovsky, Multiscale roughness and stability of superhydrophobic biomimetic interfaces. Langmuir **23**, 3157–3161 (2007)
7. T. Verho et al., Mechanically durable superhydrophobic surfaces. Adv. Mater. **23**, 673–678 (2011)
8. A.M. Kietzig, S.G. Hatzikiriakos, P. Englezos, Patterned superhydrophobic metallic surfaces. Langmuir **25**, 4821–4827 (2009)
9. R. Tadmor et al., Measurement of lateral adhesion forces at the interface between a liquid drop and a substrate. Phys. Rev. Lett. **103**, 266101 (2009)
10. E. Bormashenko et al., Wetting properties of the multiscaled nanostructured polymer and metallic superhydrophobic surfaces. Langmuir **22**, 9982–9985 (2006)
11. E. Bormashenko et al., Micrometrically scaled textured metallic hydrophobic interfaces validate the Cassie–Baxter wetting hypothesis. J. Colloid Interface Sci. **302**, 308–311 (2006)
12. J.J. Bikerman, Sliding of drops from surfaces of different roughnesses. J. Colloid Sci. **5**, 349 (1950)
13. B. Qian, Z. Shen, Fabrication of superhydrophobic surfaces by dislocation-selective chemical etching on aluminum, copper, and zinc substrates. Langmuir **21**, 9007–9009 (2005)
14. N.J. Shirtcliffe, G. McHale, M.I. Newton, C.C. Perry, Wetting and wetting transitions on copper-based super-hydrophobic surfaces. Langmuir **21**, 937–943 (2005)
15. A.D. Sommers, A.M. Jacobi, Creating micro-scale surface topology to achieve anisotropic wettability on an aluminum surface. J. Micromech. Microeng. **16**, 1571–1578 (2006)
16. P.K. Rohatgi, Metal-matrix composites. Def. Sci. J. **43**(4), 323–349 (1993)
17. K. Sobolev, M. Ferrada-Gutiérrez, How nanotechnology can change the concrete world: part 1. Am. Ceram. Soc. Bull. 84(10), 14–17 (2005)
18. V. Mortazavi, M. Nosonovsky, Wear-induced microtopography evolution and wetting properties of self-cleaning, lubricating and healing surfaces. J. Adhes. Sci.Technol. 25(12), 1337–1359 (2011)
19. V. Mortazavi, M. Nosonovsky, Friction-induced pattern-formation and Turing systems. Langmuir 27(8), 4772–4779 (2011). doi: 10.1021/la200272x
20. S. Wang, L. Fen, H. Liu, T. Sun, X. Zhang, L. Jiang, D. Zhu, Manipulation of surface wettability between superhydrophobicity and superhydrophilicity on copper films. Chem. Phys. Chem. **6**, 1475–1478 (2005)
21. F.M. Fowkes, W.D. Harkins, The state of the monolayers adsorbed at the interface solid–liquid solution. J. Am. Chem. Soc. **62**(12), 3377–3386 (1940)
22. I. Morcos, Surface tension of stress-annealed pyrolitic graphite. J. Chem. Phys. **57**(4), 1801–1802 (1972)

Chapter 8
Polymer Adhesion and Biomimetic Surfaces for Green Tribology

Mehdi Mortazavi and Michael Nosonovsky

Abstract Adhesive properties of polymeric materials and modern techniques of surface modification make polymers appropriate for Green Tribology applications, which require functional surfaces and the ability to control, and modify and surface properties, such as adhesion and wetting. Polymers, along with polymer composites, are appropriate materials for coating and various biomimetic applications, such as those utilizing the Lotus and gecko effects. In this chapter, we review polymer properties relevant to adhesion and wetting, modern methods and techniques of surface modification which are used to synthesize and produce superhydrophobic biomimetic materials as well as the methods of surface characterization.

8.1 Introduction

The emergence of Green Tribology brought attention to the need for the development of new smart materials and surfaces, which possess functionality and can minimize the impact on the environment. An important class of such materials and surfaces are micro- and nanostructured biomimetic materials. These materials are expected to be more environment friendly, since they mimic the patterns, mechanisms, and approaches found in living nature. The most typical examples of micro/nanostructured biomimetic surfaces include those capable for the

M. Mortazavi · M. Nosonovsky (✉)
Department of Mechanical Engineering, University of Wisconsin-Milwaukee, Milwaukee, WI 53211-0413, USA
e-mail: nosonovs@uwm.edu

M. Mortazavi
e-mail: mortaza3@uwm.edu

M. Nosonovsky and B. Bhushan (eds.), *Green Tribology*, Green Energy and Technology, DOI: 10.1007/978-3-642-23681-5_8,

Lotus-effect (surface roughness induces superhydrophobicity and self-cleaning), gecko-effect (smart and controlled adhesion) and other similar effects, such as shark-skin, water strider, etc.

Among the major classes of materials, such as metals, polymers, and ceramics, as well as their composites, polymers play a special role in the development of microstructured surfaces. The macromolecular structure of polymeric materials, as well as the ability of polymer chains to form crystalline structures and cross-link bounds is responsible for the observed unique chemical and mechanical properties of polymeric materials. These materials are easily deformed and can stretch significantly without rupturing. At the interface, interdiffusion of polymeric chains can occur. The electrostatic component of the adhesion force can be significant due to dissimilar polymers. Polymers are used in many applications where low adhesion, friction, and wear are desired. In this chapter, we review the modern techniques to develop functional micro/nanostuctured polymeric surfaces as well as their properties in relevance to Green Tribology.

8.2 Polymer Properties Relevant to Polymer Adhesion and Wetting

In this section, we discuss the properties of polymers which are relevant to adhesion force, surface energy, and wetting.

8.2.1 Normal Forces Between Two Surfaces

8.2.1.1 Van der Waals Forces

Van der Waals forces are relatively weak intermolecular interactions which exist between atoms, molecules, or particles. These interactions were first documented by Johannes Diderik van der Waals in 1873 for developing a theory to account for the properties of real gases. Unlike other kinds of forces, Van der Waals forces are always present. They may be small compared to other kinds of forces, but become significant when dealing with large bodies. Figure 8.1 presents the various types of Van der Waals forces between molecules as well as macroscopic solids.

Van der Waals force is believed to be responsible for observing reversible dry adhesion in natural surfaces like gecko toes. In order to investigate the principal mechanism of adhesion in gecko toes, Autumn et al. [1] measured the adhesion force between the toes of a live gecko and a highly hydrophobic substrate as well as a highly hydrophilic one. The toes of the gecko adhered equally well to both the surfaces. It is in agreement with the van der Waals hypothesis which predicts high adhesion force on polarizable surfaces, regardless of hydrophobicity.

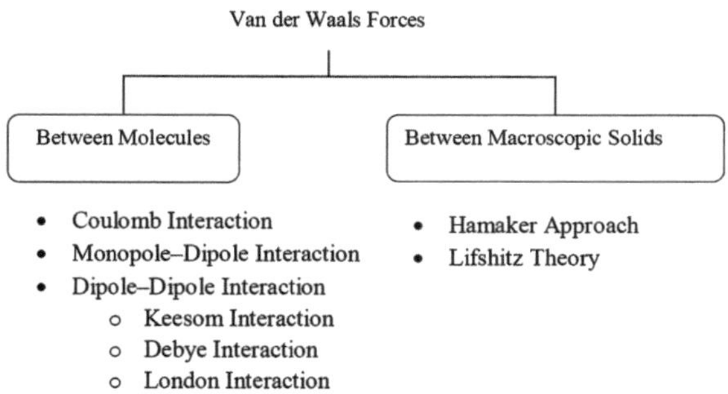

Fig. 8.1 Different types of Van der Waals interactions

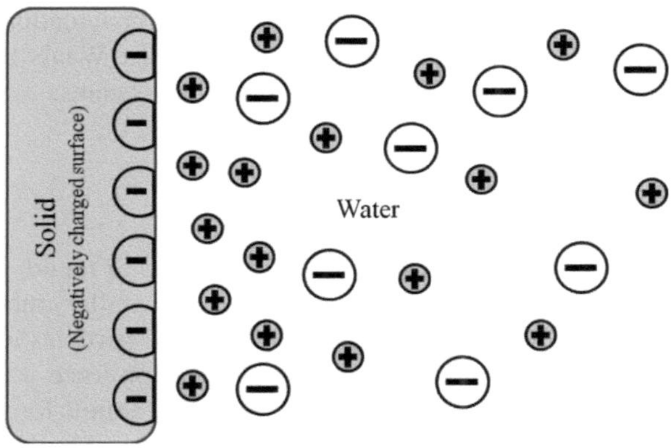

Fig. 8.2 Schematic of double layer in water at contact with a solid

Lin et al. [52] investigated the adhesion mechanism of pedal foot of a marine species, the abalone. The experimental pull-off force measured by Atomic force microscopy (AFM) was in agreement with the theoretical vand der Walls adhesion force obtained by using the Johnson–Kendall–Roberts (JKR) equation (Sect. 8.2.2). It was also shown that the pull-off force increased with increasing humidity for the hydrophilic substrate while, it remained unchanged for the hydrophobic substrate. The combination of van der Waals force and capillary force was believed to be responsible for the strong adhesion of abalone to various substrates.

8.2.1.2 Electrostatic Double-Layer Forces

An electrical double layer is a structure that forms on the surface of an object when it comes into contact with a liquid. This model visualizes the ionic environment and layer of surface charges. Charged surfaces are balanced by oppositely charged ions (counterions) in solution, so electrical neutrality is preserved. Some counterions adsorb to the surface, and others are distributed near the surface within the double layer. The region near the surface is called the electrical double layer (Fig. 8.2).

When two charged surfaces approach enough, their electrical double layers will start to overlap, and an electrostatic double layer force arises. This approach forms

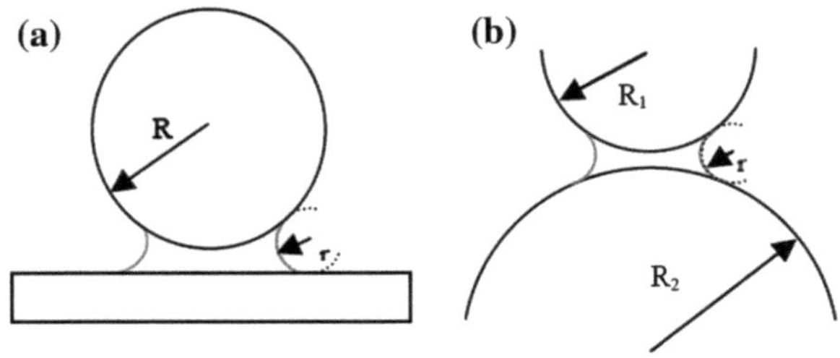

Fig. 8.3 Water meniscus formed between two surfaces. **a** sphere-plane. **b** sphere–sphere

the basis of the well-known DLVO theory of colloidal stability developed by Derjaguin, Landau, Verwey, and Overbeek [15, 84]. DLVO theory is the combined effect of two forces. This theory describes coagulation of dispersed particles by the interplay between double layer and van der Waals forces, and has been successfully applied to many aqueous colloidal systems.

8.2.1.3 Capillary Forces

Capillary forces result from the interaction of liquid, gas, and solid surfaces, at the interface between them. Rise of water in a capillary tube, and filling the pore space of soil with water and the attachment of soil particles to each other are examples of capillary force. The importance of adhesion force was first realized by Fisher and Haines in the 1920s [20, 25]. The advent of atomic force microscopy (AFM) enabled the measurements of capillary adhesion forces. The AFM provides a simple means of probing the capillary adhesion force with pN–nN sensitivity (Sect. 8.4.3.2).

When two solid bodies in contact are placed in a humid environment, water vapor condenses and forms a meniscus at the contact area of the two particles [55]. An attractive force arises from the formed meniscus. The attraction force is due to the surface tension of the liquid which pulls the particles together, as well as the negative Laplace pressure which attracts the particles toward each other. There would be no capillary force for the particles in contact in a vacuum or a dry environment. The meniscus formed between a sphere and a plane, and two spheres are presented in Fig. 8.3.

Capillary forces are often negligible on macroscopic structures, while it becomes important at micro and nano scales. A nano-scale water meniscus is formed between the AFM tip and the substrate at ambient condition. The observed capillary force usually governs the interaction between the nano-sized AFM tip and the substrate.

8.2.1.4 Solvation Forces

Two surfaces in a solvent will experience an oscillatory force as they approach each other. At separation distances in which the molecules are closely packed, the

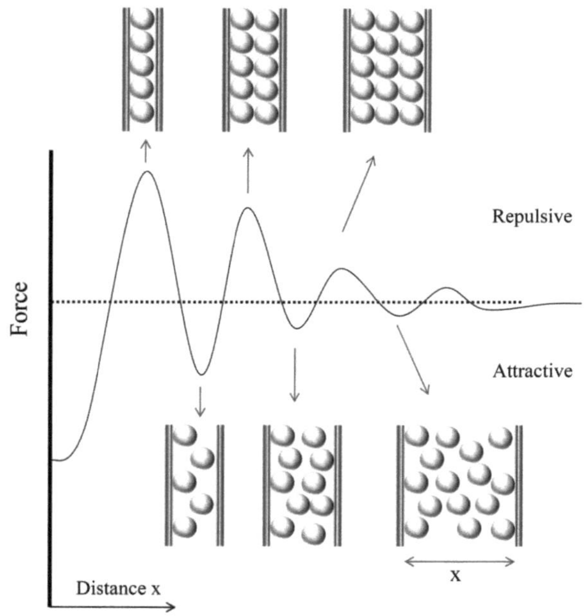

Fig. 8.4 Diagram showing molecules of a liquid confined between two parallel planes

maximum points in Fig. 8.4, the density between the two surfaces is higher compared with the bulk. This situation which is entropically unfavorable results in a repulsive force between the two surfaces. On the other hand, at separation distances in which the molecules are disordered, the minimum points in Fig. 8.4, the density is less than that of bulk. An attractive force arises due to the difference in density between two sides of the planes.

The salvation force can be calculated by applying the contact theorem [36, 75]. Solvation force is often described by an exponential decaying cosine function. For two parallel planes the salvation force is

$$f(x) = f_0 \cos \frac{2\pi x}{d_0} e^{\frac{-x}{x_0}} \tag{8.1}$$

where f is the force per unit area, f_0 is the force extrapolated to zero separation distance, x is the separation distance for the surfaces, d_0 molecular cross-section or diameter and x_0 is the characteristic decay length.

8.2.2 Contact Mechanics and Adhesion

When two bodies come into contact the interacting forces across the contact boundary result in adhesion force which is the combination of different

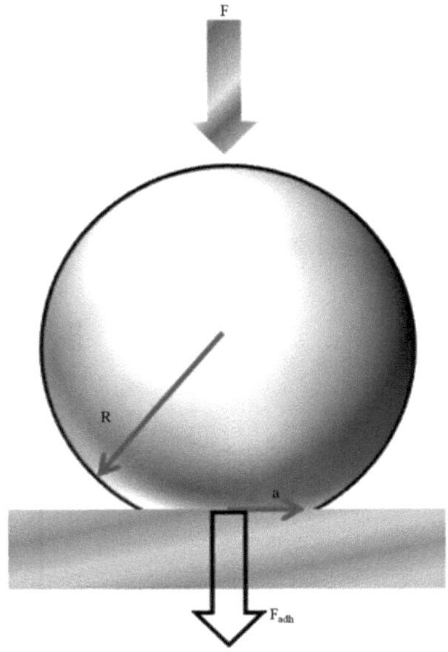

Fig. 8.5 The schematic of a sphere of radius R attached to a rigid flat surface

contributions (Van der Waals force, electrostatic force, capillary force, and others). This adhesion force depends on different parameters such as the number of contributing forces, the strength of the contributing forces, area of contact between two bodies, the wettability of the two surfaces, and the ambient conditions.

Two bodies in contact deform as a result of either external or surface forces. The original work in contact mechanics dates back to 1882 when Hertz solved the problem involving contact between a sphere and a planar surface, and between two spheres [28]. The Hertzian model of contact does not consider the adhesion between two bodies. A schematic of a contact between a sphere and a flat surface is presented in Fig. 8.5. The radios of contact a and the indentation δ based on Hertz model is given by

$$a^3 = \frac{3FR*}{4E*} \tag{8.2}$$

$$\delta = \frac{a^2}{R*} \tag{8.3}$$

where F is the external force, $E*$ the reduced Young's modulus, and $R*$ is the reduced radius.

The Johnson-Kendall-Roberts (JKR) model [40] and the Derjaguin-Muller-Toporov (DMT) models [16, 63, 64] were the first attempts to extend the Hertz theory and include the contribution of adhesion forces. The JKR model has been employed to incorporate the effect of adhesion in the Hertzian model. This model considers the adhesion forces inside the area of contact and neglects the interactions outside the area of contact. The JKR model is more suitable for soft materials with high surface energy, and large radii of curvature. The radios of contact area a it and the JKR indentation δ are obtained with the following equations:

$$a^3 = \frac{3R*}{4E*}\left(F + 3\pi\gamma R * + \sqrt{6\pi\gamma R * F + (3\pi\gamma R*)^2}\right) \tag{8.4}$$

$$\delta = \frac{a^2}{R*} - \sqrt{\frac{2\pi a\gamma}{E*}} \tag{8.5}$$

where γ is the surface energy.

The JKR model predicts that the force needed to remove the particle (the pull-off force) is given by

$$F_{\mathrm{adh}}^{\mathrm{JKR}} = \frac{3}{4}\pi\gamma R* \tag{8.6}$$

DMT model is another modified Hertzian contact theory taking into account the surface attraction forces. DMT model has been successfully applied to systems with small particles, small surface energy, and high elastic modulus. The pull-off force predicted by DMT is given by

$$F_{\mathrm{adh}}^{\mathrm{DMT}} = -2\pi\gamma R* \tag{8.7}$$

8.2.3 Wetting of Polymers

8.2.3.1 Definition of Contact Angle: Young's Equation

Wetting is the process of making an interface between a solid and a liquid. The primary parameter for a liquid wetting a solid is the angle at which the liquid–vapor interface meets the solid–liquid interface, called the static contact angle θ as seen in Fig. 8.6.

In this figure, θ is the angle between the solid surface and the tangent to the liquid–vapor interface. γ_{SG}, γ_{SL}, and γ_{LG} are solid–vapor, solid–liquid and liquid–vapor interfacial energies respectively. At the equilibrium, Young's equation relates the static contact angle to the interfacial energies

$$\cos\theta = \frac{\gamma_{\mathrm{SG}} - \gamma_{\mathrm{SL}}}{\gamma_{\mathrm{LG}}} \tag{8.8}$$

In general, if the static water contact angle is less than 90° the surface is wettable or hydrophilic, whereas a water contact angle above 90° indicates nonwetting or hydrophobic surface. Table 8.1 presents the critical surface tensions

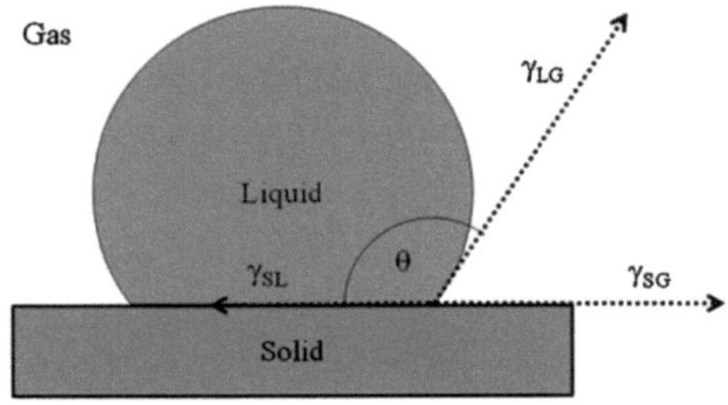

Fig. 8.6 Liquid droplet profile showing surface tensions acting at the three phase line of contact

Table 8.1 Critical surface tension and water contact angle for common polymers

Polymer	$\gamma_{SG}\left(\frac{mJ}{m^2}\right)$	$\theta(°)$
Polyvinyl acetate (PVA)	35.3	60.6
Nylon 6,6	42.2	68.3
Polyethylene terephthalate (PET)	39	72.5
Polyvinyl chloride (PVC)	37.9	85.6
Polystyrene (PS)	34	87.4
Polyethylene (PE)	31.6	96
Polydimethylsiloxane (PDMS)	20.1	107.2
Polytetrafluoroethylene (PTFE)	19.4	109.2

and the water contact angles for some polymers. Among these are Polydimethylsiloxane (PDMS) and Polytetrafluoroethylene (PTFE) which have very low surface energies and have been used for non-sticky and water repellant coatings.

8.2.3.2 Rough Surfaces: Wenzel's Model

The influence of the substrate surface roughness on wetting was first described by the well-known thermodynamic approach of Wenzel. The Wenzel equation for the contact angle on a rough surface is defined by the following equation:

$$\cos\theta = r\cos\theta_0 \tag{8.9}$$

where θ is the observed contact angle on a rough surface and θ_0 is its contact angle on a smooth surface. The roughness ratio, r is defined as the ratio of the surface area A_{SL} to its flat projected area A_F (Fig. 8.7a)

$$r = \frac{A_{SL}}{A_F} \tag{8.10}$$

This simple relationship predicts that as roughness increases, a hydrophobic surface ($\theta_0 > 90°$) becomes more hydrophobic, while a hydrophilic ($\theta_0 < 90°$) surface becomes more hydrophilic. According to Wenzel equation polymers such as Polyvinyl chloride (PVC) and Polyethylene terephthalate (PET) with $\theta_0 < 90°$,

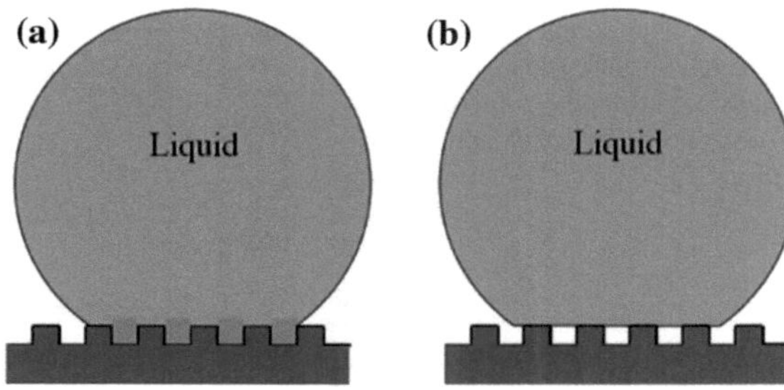

Fig. 8.7 Schematic illustration of a liquid droplet in **a** homogeneous (Wenzel), **b** heterogeneous (Cassie–Baxter) wetting regimes

should become more hydrophilic by introduction of surface roughness. However, highly hydrophobic surfaces have been fabricated from these polymers by increasing the surface roughness, indicating that the Wenzel equation fails to predict the water contact angle of these surfaces [48, 82].

8.2.3.3 Heterogeneous Surfaces: Cassie–Baxter Model

The alternative method for the heterogeneous surfaces is that of Cassie–Baxter model [10]. When the surface is composed of two different materials, the surface has a surface area fraction f_1 with a contact angle θ_1 and a surface area fraction f_2 with a contact angle θ_2. The contact angle for the heterogeneous interface can be expressed by the following equation:

$$\cos\theta = f_1 \cos\theta_1 + f_2 \cos\theta_2 \tag{8.11}$$

In the Cassie–Baxter model the liquid droplet sits on a composite surface made of solid ($f_1 = f_{SL}$) and air ($f_1 = f_{LA}$) (Fig. 8.7b). Substituting $\theta = 180°$ and $\theta_1 = \theta_0$ into Eq. 8.10 and combining it with Eq. 8.9 yields the Cassie–Baxter equation

$$\cos\theta = rf_{SL} \cos\theta_0 - 1 + f_{SL} \tag{8.12}$$

Figure 8.8 presents the contact angle values for two different surfaces with θ_0 values, close to those for common polymers presented in Table 8.1, as a function of surface roughness for different fractional solid–liquid areas. According to this figure, for polymers like Polystyrene (PS) and Polyvinyl chloride (PVC) with contact angle close to 90°, the fractional solid–liquid area should decrease to about 0.1 ($f_{SL} = 0.1$) in order to have a contact angle above 150° (superhydrophobic surface). In case of hydrophobic polymers such as Polydimethylsiloxane (PDMS) and Polytetrafluoroethylene (PTFE) in Table 8.1, the combination of surface roughness and f_{SL} should be selected in order to fabricate a superhydrophobic surface.

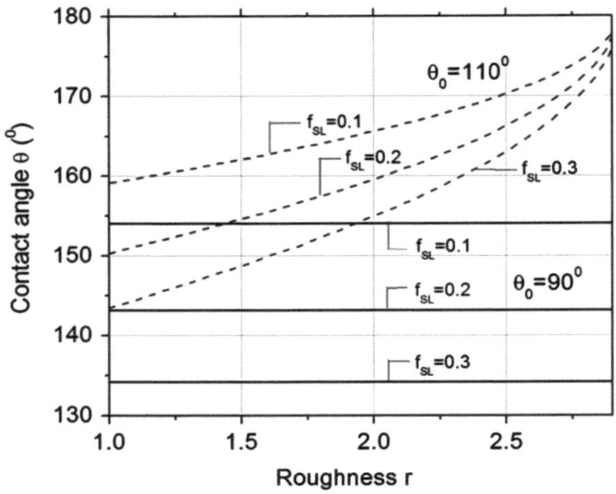

Fig. 8.8 The contact angle for rough surface (θ) as a function of roughness factor (r) for two different θ_0 and various f_{SL}

8.3 Methods for Fabrication of Superhydrophobic Materials

In this section, the methods of surface modification methods and fabrication of superhydrophobic polymeric surfaces are reviewed. Typical polymers and the corresponding techniques that have been recently used for fabrication of superhydrophobic polymeric surfaces are summarized in Table 8.2.

8.3.1 Electrospinning

Electrospinning is a process in which a high electrical field is used to fabricate fibers with diameters ranging from a few nanometers to tens of micrometers from liquid polymeric solutions. A composition of solvents can be used as well. The process is suitable for a wide variety of polymers and can be conducted at room temperature. The morphology, surface texture, and wetting properties of the fabricated fibrous mat can be controlled by changing the solution characteristics or electrospinning parameters. The solution concentration, viscosity, applied voltage, ambient relative humidity, and temperature are the most important parameters which control this process. A schematic of an electrospinning process is shown in Fig. 8.9a.

The process of electrospinning was invented in 1902 by J.F. Cooley and W.J. Morton. A number of further developments were made after that in the 1930–1960s by A. Formhals, G.I. Taylor, N.D. Rozenblum, I.V. Petryanov-Sokolov, and others. Since the early 1990s the technique was used for many polymers and the term "electrospinning" was popularized, being coined from "electrostatic spinning" and now is widely used in the literature.

Table 8.2 Typical polymers and corresponding techniques to produce micro- and nanostructured surfaces

Polymer	Technique	Contact angle (°)	Note	Source
PDMS	Nanocomposite	164	$CaCO_3/SiO_2$ particles	Yang et al. [92]
	Polymer casting		Plant leaf replica	Lee et al. [49]
	Replica molding+CVD	160		Rahmawan et al. [71]
Cotton	Assembly of CNT	>150	Special textile	Liu et al. [57]
	Nanocomposite	170	Functionalized SiO_2	Xue et al. [90]
UV curable	Nanoimprint lithography		Plant leaf replica	Lee et al. [47, 48]
	Nanoparticle+ozon treatment	160	Al_2O_3 nanoparticles	Choi et al. [14]
	Nanoimprint lithography	132	Cicada wing replica	Hong et al. [29]
Polyurethane (PU)	Nanocomposite		Plant leaf replica	Feng et al. [19]
	Electrospinning			Wu et al. [88]
	Nanocomposite		PS nanoparticles	Zhang et al. [100]
PE	Nano-micro replication + Blending			Puukilainen et al. [70]
	Nano-micro replication			Sheng and Zhang [76]
	Solution method	153	High stability	Yuan et al. [98]
	Solution method	173		Lu et al. [58]
PP	Nano-micro replication + Blending			Puukilainen et al. [70]
	Nano-micro replication		Using AAO membrane	Yoo et al. [94]
	Solution method		Anticoagulation property	Hou et al. [33]
PMMA	Plasma treatment	152	Transparent	Vourdas et al. [85]
Carbon fiber	Chemical vapor deposition chemical-wet impregnation	169	CNT decoration	Hsieh et al. [35]
			SiO_2 particles on CF	Hsieh et al. [34]

(continued)

Table 8.2 (continued)

Polymer	Technique	Contact angle (°)	Note	Source
Epoxy resin	Replication+ Self-assembly			Bhushan et al. [5]
PS	Spin coating	124	PS Micro/nano spheres	Yan et al. [91]
	Vertical deposition		PS nano spheres	Ge et al. [22]
	Spin coating		PS nanoparticles	Shiu et al. [78]
	Spray coating	160	PS-CNT nanocpmposite	Yang et al. [93]
	Reactive-ion etching (RIE)	180	SiO_2 surface treatment	Park et al. [68]
	Replication		Nanoporous templates	Lee et al. [46]
	Solution casting		PS/SiO_2 nanocomposite	Hou and Wang [32]
	Molding	>150	Alumina porous template	Cheng et al. [12]
	Electrospinning	154	DMF/PS solution	Kang et al. [42]
	Electrospinning	178	TiO_2 + graphene	Asmatulu et al. [2]
	Electrospinning	150	High tensile strength	Li et al. [55]
	Solution method	156		Yuan et al. [97]
	Replication	158		Yuan et al. [96]
PVFD	Solution casting + molding			Bormashenko et al. [8]
	Chemical vapor deposition	155		Zheng et al. [105]
	Chemical bath deposition	157	Grafting by DTS/DDS	Zheng et al. [106]
	Solution casting	155		Peng et al. [69]
PTFE	Electrospinning		Teflon sheath/PCL core	Han and steckl [26]
	Nnaocomposite	165	PTFE nanoparticles	Hong and pan [30]
	Radio frequency sputtering	164		Sarkar et al. [74]
	Molding		Filter paper as template	Hou and Wang [31]
	Physical vapor deposition	165	Surface treatment of Al	Weibel et al. [87]
	Pulse electron deposition	166		Gupta et al. [24]

(continued)

Table 8.2 (continued)

Polymer	Technique	Contact angle (°)	Note	Source
PC	Solution casting	160		Zhang et al. [103]
	Solvent induced crystallization			Zhoa et al. [104]
PET	Nanocomposite	158		Bae et al. [4]
	Plasma treatment + CVD	>150		Teshima et al. [59]
PCL	Electrospinning + CVD	175		Ma et al. [59]
	Electrospinning	172		Yoon et al. [95]
PVC	Solution method	155	Good stability	Li et al. [54]
	Solution method	154		Chen et al. [11]
PI	Nanocomposite	160	Layer-by layer assembly	Zhao et al. [104]
PLA	Solution method	150		Shi et al. [77]

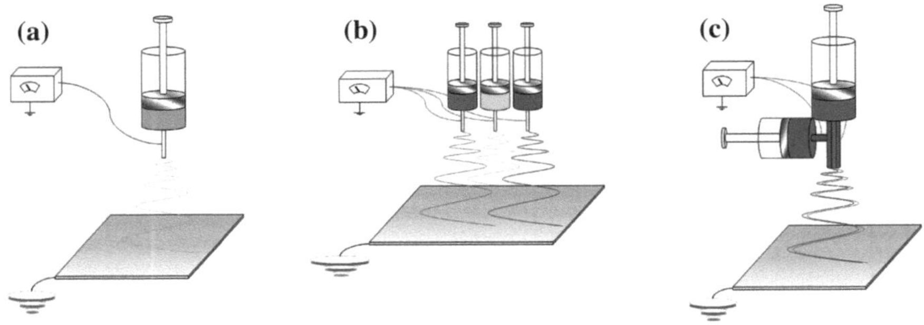

Fig. 8.9 Electrospinning process. **a** Conventional electrospinning. **b** Multi-jet electrospinning. **c** Coaxial electrospinning

Electrospinning was first used for fabrication of superhydrophobic surfaces by Jiang et al. [41]. Solutions of polystyrene (PS) in Dimethylformamide (DMF) at different polymer concentrations were prepared and the effect of concentration on the morphology of the resultant surfaces was investigated. Polystyrene (PS) is a hydrophobic polymer with a water contact angle of 95°. Nanofibrous structure with WCA of 139° was formed from a concentrated solution. Porous microparticles with WCA as high as 162°, but with low structural stability were formed from a high concentrated solution. A porous microsphere/nanofiber composite coating with WCA of 160° and good stability was obtained from a medium concentrated solution. No information about contact angle hysteresis was reported in this study.

Several attempts have been conducted to improve the wetability and mechanical properties of the electrospun mats. Modification of polymers with low surface energy segments can alter its water replant property (REFS). Ma et al. [59] investigated the wetting properties of fibrous mats prepared by the electrospinning of poly (styrene-*b*-dimethylsiloxane) block copolymer. Randomly oriented fibers with broad distribution were formed. Excess concentration of PDMS was formed on the surface of the fibers due to the lower surface energy of PDMS block compared to PS block. The resulted PS-PDMS copolymer electrospun mat had a larger WCA and much lower contact hysteresis compared to mat prepared from PS homopolymer. The enhanced superhydrophobicity of PS-PDMS copolymer was believed to be due to the combined result of macro/nano scale roughness and the excess concentration of low surface energy PDMS blocks on the surface of the fibers. Wu et al. [88] investigated the effect of surface chemistry on the wetting behavior of polyurethanes synthesized with and without low surface energy segments. Bead-thread microstructure and nanofibrous structures were prepared from low and high concentrated solutions respectively. Higher WCA values for both structures were measured for the surfaces containing low surface energy segments.

The volatility of the solvent is also an important factor which influences the surface roughness of the electrospun mats. Using a non-volatile dimethylformamide (DMF) solvent, Kang et al. [42] fabricated a rough surface with a higher water contact angle compared to the case of using high-volatile solvent like tetrahydrofuran (THF).

Simultaneous spinning of two or more polymer solution is an approach to fabricate electrospun mats with enhanced mechanical and wettability properties. Li et al. [55] fabricated modified electrospun mats by simultaneous elctrospinning of a Polystyrene (PS) and a polyamid6 (PA6) via a multi-spinneret electrospinning Process. Polyamide 6 is a polymer with higher tensile strength and more hydrophilic than PS. Electrospun mats with tensile strength three times higher than the pure PS mats were obtained, while the water contact angle remained higher than 150° (Fig. 8.9b).

Coaxial electrospinning is another modification to the ordinary electrospinning process, and is a powerful approach in cases where a polymer solution cannot be electrospun in an ordinary process, or a modification in the properties of the fibers is needed. In this process, two polymer solutions are electrospune simultaneously from a coaxial capillary to produce core–sheath structured fibers (Fig. 8.9c). Fluorinated polymers with low surface energy and inherent hydrophobicity have a great potential to be used for fabrication of superhydrophobic coatings. However, they exhibit relatively low dielectric constant which prevents them from being successfully electrospun. To solve this problem and make electrospun fluorinated polymers, Han and Steckl [26] investigated a coaxial electrospinning of a solution of a Teflon AF fluoropolymer and a poly (ε-caprolactone) (PCL) to fabricate superhydrophobic and olephobic membranes. Fibers with core-sheath structure were obtained in which, the PCL core is responsible for good mechanical properties and the Teflon AF fluoropolymer sheath is essential for providing low surface energy (Fig. 8.9c).

Nanoparticles can be incorporated into polymer solutions to modify the wetting property of electrospun membranes. Asmatulu et al. [2] investigated the effect of the incorporation of Titania nanoparticles and graphene nanoflakes on the superhydrophobicity of polystyrene (PS) and poly (vinyl chloride) (PVC) electrospun surfaces. Titania nanoparticles and graphene nanoflakes were added to the polymer solutions at different concentrations and the nanocomposite solutions were electrospun. The WCA was increased by up to 18% on the electrospun nanocomposites due to formation of nano–scale roughness. By addition of PTFE nanoparticles to a low concentrated electrospinning solution, Menini and Farzaneh [61] fabricated superhydrophobic electrospun mats with dual fiber–bead microstructures. Electrospun mats were prepared with polystyrene (PS) and a fluoropolymer with and without PTFE nanoparticles. For the PS mats, fibrous structure and Bead-fiber dual structure were obtained at higher and lower concentrations respectively. The WCA increased with the addition of PTFE nanoparticles but were still lower than 150°. For the fluoropolymer, superhydrophobic surface with WCA up to 161° was obtained after addition of PTFE nanoparticles.

Yoon et al. [95] modified the electrospinning process to obtain a superhydrophobic surface from a low concentration poly (ε-caprolactone) (PCL) solution. In this method the solution was electro sprayed into a water bath connected to the ground resulting in a hierarchical surface composed of microscale pyramid

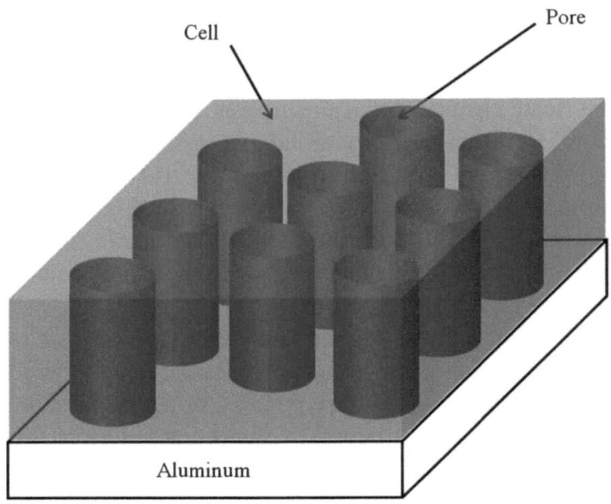

Fig. 8.10 Schematic presentation of an AAO template

structure and nanosized pores. A WCA of 172 was obtained in the modified method whereas it was only 128 in the regular method.

In summary, electrospinning is a very simple and versatile process that can be used to fabricate superhydrophobic surfaces from various polymers or a combination of polymers. Multi-spinneret and coaxial arrangements can also be used to obtain surfaces with modified chemistry and surface roughness.

8.3.2 Molding or Template-Assisted Techniques

8.3.2.1 Molding by Using Anodic Aluminum Oxide Templates

Anodic Aluminum Oxide (AAO) is a self-ordered nanostructureed material formed on aluminum surface from an electrochemical oxidation in acidic solutions. AAO is composed of uniform nanosized vertical nanopores with various pore diameter and pore depth. An alumina layer is formed along the pores which separate the pores from the aluminium bulk (Fig. 8.10). High aspect ratio and tunable pore dimensions make AAO a desirable material for fabrication of self-aliened materials and a commonly used template for nanofabrication.

Lee et al. [46] fabricated vertically aliened polymeric nanofibers on polystyrene (PS) surface with different aspect ratios and surface morphologies using AAO template. AAO templates with different pattern sizes and aspect ratios were first prepared by varying the electrochemical parameters, resulting in close-packed hexagonal nanostructures. The obtained AAO templates were then utilized in nanoimprint embossing to transfer the nanopatterns into the polymer surface. After removing the template by using a chemical wet etching, nanopatterns with

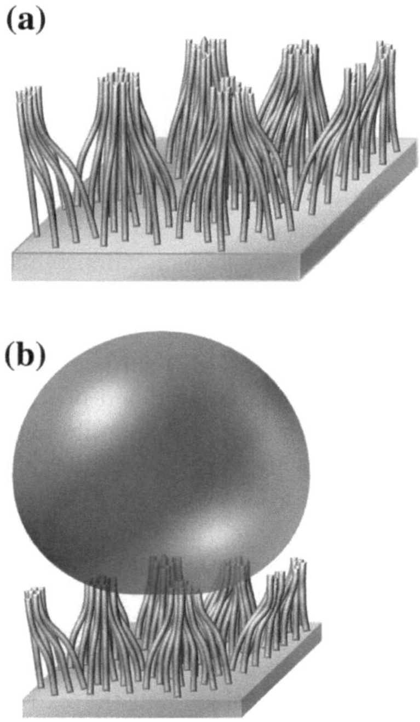

Fig. 8.11 Aliened nanofibers prepared from molding by using AAO template. **a** Nanofibers adhere and form bundles **b** Water droplet on top of the bundles

the pore diameter and length similar to those of the templates were obtained. The study of the wettability of the surfaces revealed that the surface texture had a great impact on the WCA and the contact angle hysteresis. Superhydrophobic surfaces were obtained by tailoring the pore dimensions of the templates (Fig. 8.11).

Injection molding is a widely used polymer process in which molten plastic is forced under high pressure to flow into a mold cavity where it is held until it cools and hardens into the shape of the mold. Injection molding can be used to fabricate materials with nano-patterned surfaces. AAO nanotemplates can be mounted into the mold to transfer the nanopatterns to the molten polymer. Puukilainen et al. [70] fabricated two AAO templates with various pore nanostructures by altering the voltage used in the electrochemical reaction. The lower voltage resulted in the higher order structure. The fabricated AAO templates were used to make nano-patterns on the surface of two polyethylene (PE) and one polypropylene (PP) samples by injection molding. Well-ordered nanostructures comparable to those of the AAO templates were formed. The wettability property of the surface was investigated by static contact angle measurements of the nanostructured and flat surfaces. The WCA was increased due to the enhanced surface roughness. More increase was observed for the surfaces made by using higher order templates.

Yoo et al. [94] investigated the fabrication of nano-textured surfaces by injection molding process and studied the effect of mold temperature on the ability of molten polymer to fill the nanopores of the applied AAO templates. AAO templates with the nanoholes diameter of 200 nm were used as template to make nanohair structure on a polypropylene (PP) surface. Samples were injected molded at different mold temperatures. It was shown that at low temperatures, the polymer melt did not penetrate completely into the nanopores, resulting in low aspect ratio bumps. As the mold temperature increased, the molten polymer penetrated more, and nanohairs with higher aspect ratios were obtained. The WCA measurements on the prepared surfaces indicated that the surfaces decorated with longer nanohairs, which were obtained at higher mold temperatures exhibited higher WCAs.

Sheng and Zhang [76] fabricated nanotextured hydrophobic high-density polyethylene (HDPE) surfaces by extruding molten polymer into AAO templates at specific temperature and pressure. Templates with different pore sizes resulted in the formation of nanofibrous surfaces with different fiber diameters. The nanofibers then collapsed due to self-aggregation and bundles with different surface morphologies were formed. The WCAs of the prepared surfaces were sufficiently higher than that for the smooth surface, but almost the same for different surfaces. However, significant difference was found for the sliding angles of different surfaces. The surface with the finer nanofibers exhibited the highest sliding angle which is believed to be due to the pore created by the self-assembly of bundles in the case of very fine nanofibers. It was concluded that by tuning the template pore sizes and the extrusion pressure, superhydrophobic surfaces with different surface properties can be made.

Cheng et al. [12] used AAO templates to fabricate superhydrophobic surfaces of polystyrene (PS) with different adhesion characteristics. Smooth surface of polystyrene was first prepared on substrate using a solution-casting method. The AAO nano-hole pattern was then transferred onto the underlying polystyrene surface by a replication method. By altering the replication parameters, surfaces with different nanoscale roughness and surface topography were obtained. Polystyrene nanopillars with diameter comparable to that of the used template and terminating in flat and concave tips as well as polystyrene nanotubes were resulted (Fig. 8.12). Superhydrophobic surfaces with WCAs larger than 150° were obtained for all surfaces. Small differences observed for the different topographies can be explained by the difference in water–solid contact area. The adhesion property of the fabricated surfaces was also investigated by measuring the adhesive force between water droplet and the surfaces. Different adhesives force was measured for different surfaces which can be explained by the contribution of the different forces acting on the surface (van der Waals forces and the capillary force induced by negative pressure). Different surface topographies not only provide different contact area which alter the contribution of the van der Waals forces, but also provide different negative pressures which change the contribution of the capillary force. The highest adhesive force was obtained for the nanopillar with concave tips, which was due to high capillary force, while the least adhesive force was measured for the nanotube arrays, which was believed to be due to the low van der Waals and capillary forces.

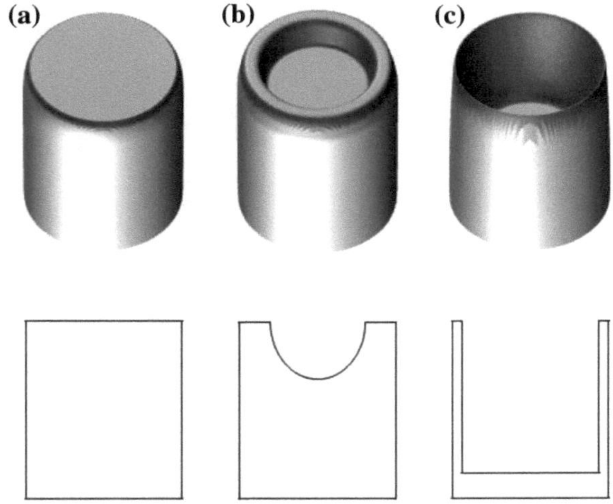

Fig. 8.12 Different nanopillars with different tip geometry. **a** Flat tip. **b** Concave tip. **c** Nanotubes

8.3.2.2 Molding by Using Silicon Templates

Patterned silicon wafer is also a suitable candidate for micro- and nano- fabrication of polymeric surfaces. However, since silicon wafer is expensive and fragile it has been used to make intermediate polymer molds which are more flexible and can be used many times. Jeong et al. [37] fabricated polymeric molds by photolithography on silicon masters. The prepared molds were then used to fabricate micro- and nanoscale roughness on a UV-curable resin by capillary molding technique. Capillary molding technique, which is based on raising the liquid resin in micropores of the mold due to capillary force is developed recently [38]. The micro-patterned polymeric mold was first placed on the UV-curable resin followed by partial curing of the resin, resulting in micropatterns on the resin. The surface of the micro-patterned resin which is cured partially remains tacky, whereas the resin beneath the surface cures completely, allowing for subsequent molding of the nano-patterns by using the nano-patterned polymeric mold on top of the macro-patterned surface. The final dual-scale hierarchical surface is composed of nanosized roughness on top of the micropatterns (Fig. 8.13). After treatment of the surfaces with a low surface energy material, wettability was studied on the dual-sized hierarchical structures as well as single-scale ones. The effect of spacing to width ratio of the micro- and nanopillars on the WCA as well as Cassie to Wnzel transition was investigated on the single micro and nano-patterned surfaces and was shown that Wenzel state is more stable than the Cassie State as the spacing to width ratio increased. On the other hand, stable Cassie state with enhanced WCA's was obtained in dual-scale hierarchical surfaces. A thermodynamic model was presented to explain the contribution of micro- and nanosized roughness on the wettability as well as Cassie to Wenzel transition (Fig. 8.13).

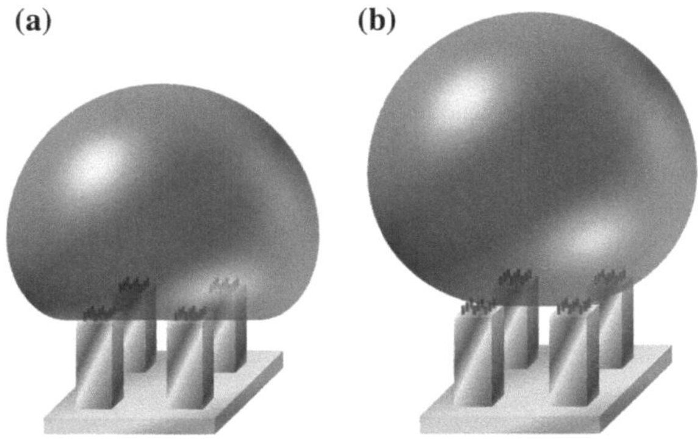

Fig. 8.13 Water droplets on dual-scale hierarchical structures. **a** Cassie–Wenzel state, **b** Cassie–Cassie state

Rahmawan et al. [71] introduced a new method to make nanosized roughness on the surface of micropillar PDMS and fabricated dual-size rough surfaces. Micropillar arrays with various aspect ratios (pillar separation distance to the pillar diameter) were replicated on the surface of PDMS using silicon masters. The micropillar PDMS surface was further treated with a thin layer of a low surface energy material via a chemical vapor deposition (CVD) technique. The deposited film was observed to form nanoscale wrinkles on the surface. The mechanism of wrinkling, which is believed to be due to the combined effect of residual stress in the low surface energy material, and the large difference in Young's modulus of the low surface energy material and the substrate was studied. It was also shown that the aspect ratio of the micropillars, which control the microscale roughness as well as the thickness of the low surface energy layer, which control the nanoscale roughness, govern the wetting properties of the created surfaces.

8.3.2.3 Other Molding Methods

Most of the molding techniques that have been used for fabricating of nanopatterns on polymer substrates are expensive and the entire process takes a significant amount of time; so development of new cost effective and fast methods are needed. Another problem with using nano-textured templates (for example AAO) is the detachment and dissolving of the attached template after the molding process. Hou and Wang [31] presented a new technique for fabrication of surface roughness by using a filter paper as a template and prepared stable superhydrophobic surface. Polytetrafluoroethylene (PTFE) sheets were molded against filter paper and then were sintered at various temperatures, resulting in micro- and nanoscale lotus-like rough surfaces. The number of microscale bumps at the surface decreased as the sintering temperatures increased, resulted in the decrease in the surface roughness, which consequently resulted in the decrease in the WCAs and the increase in the sliding angles.

Choi et al. [14] used a micromolding technique to fabricate micro- and nanoscale hierarchical surfaces on a UV-curable resin containing alumina nanoparticles. Prism- and pyramid-like micro-patterned mold were used to form microscale roughness, and the alumina nanoparticles were used for nanoscale roughness. An ultraviolet ozone treatment on the surface resulted in selective etching of the nanocomposite surface, leaving a rough surface with silica nanoparticles on the outermost surface. After further treatment with a fluoroalkylsilane monolayer, hierarchical suoperhydrophobic surface was obtained. The method provided the ability to adjust the microscale roughness by changing the micromold patterns, and the nanoscale roughness by changing the size of nanoparticles as well as the etching time during the ozone treatment.

8.3.3 Replication of Natural Surfaces

Many surfaces in nature such as lotus leaves and wings of some insects are superhydrophobic exhibiting WCA larger than 150° [67, 86]. The water repellency property is mostly considered to be the effect of micro and nanoscale roughness on these surfaces. A simple approach for fabrication of superhydrophobic surfaces is the replication of the surface structure of natural surfaces with polymeric materials using various methods.

8.3.3.1 Direct Replication of Natural Surface

In direct replication method, the micro- and nanopatterns of the natural surface are directly transferred to the polymer surface using lithography, polymer casting, or hot embossing process.

Lee et al. [47] investigated a single-step method of replicating the surface micro- and nano-textures of plant leaves in a UV curable polymer. The leaf was attached and pressed against a UV curable polymer to transfer the patterns of the leaf to the polymer. The system was then exposed to a UV light from the polymer side to fix the prepared patterns. The micro- and nano-features of the leaves were successfully transferred to the polymer surface, resulting in a remarkably large increase in the WCAs of all the replicated surfaces compared to the WCA values of the smooth polymer film.

In another study, the nano-pillar array of a cicada wing was transferred to a PVC film by a hot embossing process resulting in a negative nanopatterned PVC replica. This negative replica was then employed to transfer the nanopatterns to a UV curable polymer through a nanoimprint lithography (NIL) process. A nanopatterned hydrophobic surface with a WCA comparable to that of cicada wing (138°) was obtained [29].

The effect of combined micro and nano hierarchical structures on superhydrophobicity is discussed in many studies and it is indicated that the nanoscale roughness is important as well as microsclae roughness [66]. However, it is not possible in all cases to make nanopaterns using simple replication techniques. Various nanoparticles can be used to introduce nanoscale roughness into the replicated polymer surfaces. A nanocomposite solution was obtained by incorporation of a waterborne polyurethane (WPU) solution into antimony doped tin oxide (ATO) nanoparticles and the resulted nanocomposite solution was coated onto a PDMS negative mold, replicated from a lotus leaf. A superhydrophobic film with hierarchical micro and nanostructure roughness was obtained with the addition of ATO nanoparticles. No nanoscale structure was formed using pure WPU, demonstrating that the nanoscale structure observed on the WPU/ATO nanocomposite was due to the presence of nanoparticles [19].

8.3.3.2 Replication by Using an Intermediate Nickel Template

In this method, a nickel mold is first made via an electroforming process. In the next step, the prepared nickel mold is used to transfer the patterns to the surface of a polymer. The proposed replication method, which was based on making a nickel mold, had the potential to be used for replicating a wide variety of natural surfaces with various polymeric surfaces [48]. Typical nickel mold making and nanoimprint lithography is presented in Fig. 8.14.

Fabrication of superhydrophobic surfaces by using micro- and nano-textured mold, replicated from natural surfaces can be conducted for polymers ranging from hydrophobic to hydrophilic nature by using polymer molding techniques. Lee et al. [49] fabricated superhydrophobic surfaces from both an intrinsically hydrophobic polymer (PDMS) and an intrinsically hydrophilic UV curable polymer. The micro- and nanopatterns transferred to the UV curable polymer more precisely than to the PDMS surface. However, the PDMS surface displayed better superhydrophobicity due to the intrinsic hydrophobicity of the PDMS.

Biological surfaces of animals can be transferred to polymer surfaces with the same process. Using a negative nickel master fabricated through electroforming upon a surface of dung beetles, Nagaraja et al. [65] prepared a hydrophobic surface from an intrinsically hydrophilic thermoplastic acrylonitrile– butadiene–styrene (ABS) copolymer. The obtained rough surface mimicked the microstructure of the surface of dung beetles.

Biomimetic superhydrophobic plant-leaf-like polymer surface was fabricated by using PDMS negative replica instead of nickel [96]. In this study, negative replica was obtained by casting a liquid PDMS prepolymer on a taro leaf. After solidification, the PDMS negative mold in which, the leaf patterns were precisely replicated, was peeled off the taro leaf. A taro-leaf like PS surface was obtained by casting a PS solution on this negative template. The resulted PS film exhibited micro/nano roughness similar to taro leaf which was responsible for the observed superhydrophobicity.

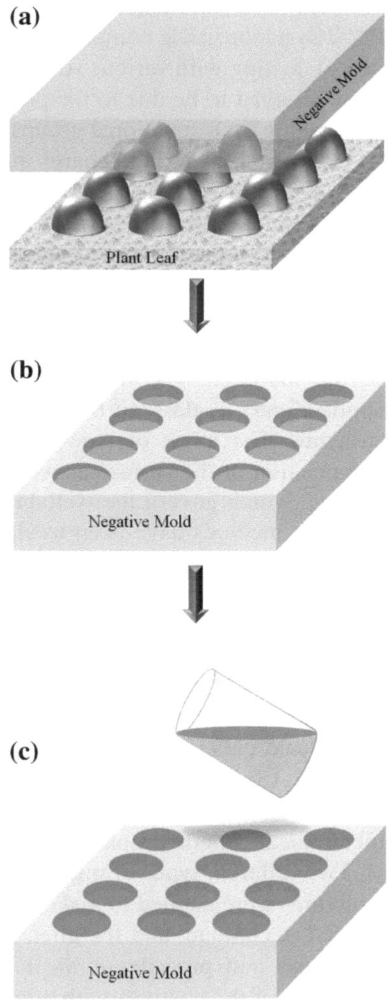

Fig. 8.14 Nanoimprint lithography. **a** Mold making. **b** Negative mold. **c** Molding

8.3.4 Roughening Through Introduction of Nanoparticles

8.3.4.1 Silica Nanoparticles

Silica nanoparticles can be used to introduce nanoscale roughness into the superhydrophobic surfaces. Since silica nanoparticles are hydrophobic, a surface modification is needed in order to show water repellency property. Hou and Wang [32] fabricated transparent coatings ranging from superhydrophilic to superhydrophobic through the casting of polystyrene (PS)–SiO_2 nanoparticle mixtures at different concentrations and different drying temperatures. The change in the PS/SiO_2 nanoparticle compositions and drying temperatures resulted in the formation of coating with various surface roughness and surface chemistry. The changes were believed to be due to the presence of either more hydrophobic PS groups or hydrophilic SiO_2 functional groups at the most outer surface of the coating.

Different low surface energy materials can be used to modify the surface of silica nanoparticles. The combination of the chemistry and the micro and nanoscale roughness of the surface-modified silica nanoparticles provided superhydrophobicity. García et al. [21] fabricated superhydrophobic surfaces from surface-modified silica nanoparticles. Silica nanoparticles were modified with four sets of organosilanes. The prepared organo-modified silica nanoparticles formed agglomerates with different sizes and polydispersities according to the chain length of the modifier. The obtained modified nanoparticle agglomerates were then applied onto a glass surface to form a rough surface possessing hierarchical structure. The effect of type and chain length of the modifier on the chemical composition of the surface as well as the micro and nano scale topography of the surface, which govern the wettability of surface was discussed.

Other methods have been used to assemble nanoparticles on a surface. Layer-by-layer (LBL) assembly, a comparatively new method, is an easy to use method and has been employed for fabrication of polymer/nanoparticle multilayer films with superhydrophobic properties. Different types of nanoparticles with different surface chemistry and different surface morphology can be applied on a surface using the LBL assembly [9, 39, 101]. Using this technique Zhao et al. [104] fabricated a hierarchical superhydrophobic surface on a polyimide substrate. Microscale patterns were generated on the polyimide surface by a micro-replica molding (Sect. 8.3.2). Then, the micro-patterned substrate was sequentially immersed in silica nanoparticles and a mixture of polyelectrolytes resulting in a hierarchical superhydrophobic surface with WCA up to 160°.

Two different types of mesoporous silica nanoparticles with different mesopore structures were applied to make superhydrophobic coatings on a glass substrate by an LBL technique, and the effect of the morphology of the nanoparticles on the roughness and porosity of the resultant coatings were investigated [56]. After treatment of the coatings with a low surface energy layer, the coating obtained by the more porous nanoparticles resulted in a better superhydrophobicity and water repellency property.

Silica nanoparticles can be combined with other micro and nanosized particles to prepare structures with efficient surface roughness. A mulberry-like composite was obtained by addition of nanosized silica particles and microsized calcium carbonate particles followed by a surface modification. The combination of micro and nanosized particles provided a hierarchical rough surface topography and the surface coated by this composite showed superhydrophobicity with high water contact angle (WCA) and very low sliding angle (SA) [92].

Other attempts have been made to fabricate hierarchical dual-size nanotextures by applying a combination of micro and nanosized particles. Using a self-assembly method [81] prepared a sphere-array of micron-sized silica on a glass substrate.

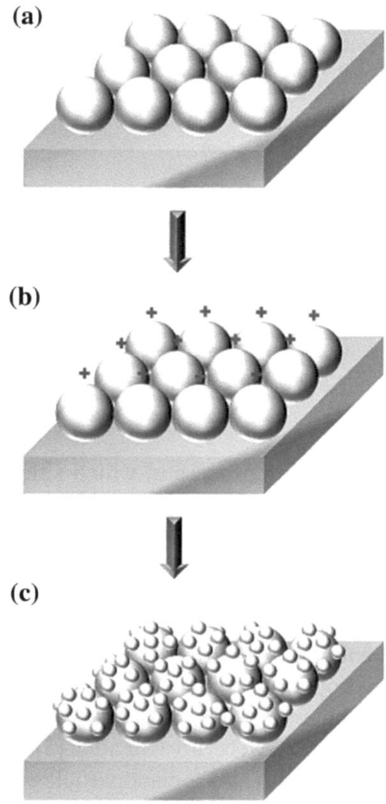

Fig. 8.15 Layer-by Layer assembly of silica nanoparticles. **a** Closed-pack array of silica microparticles. **b** Posetively charged polyelectrolyte-modified surface. **c** Dual-sized rough structure

Then, silica nanospheres with various sizes were adsorbed onto this silica micro array resulted in a set of dual micro and nanoscale rough surfaces. The effect of the size of the silica nanoparticles, assembled in the second step, on the porosity of the final coating, and the obtained WCA and SA were investigated (Fig. 8.15).

Silica nanoparticles can also be applied for the surface treatment of polymer fibers and fabrics. Hsieh et al. [34] employed silica nanoparticles through a sol–gel technique to modify the surface of micro-textured carbon fabrics (CF). A dual-scale rough surface was obtained via the combination of microscale roughness of the carbon fibers and the nanoscale roughness of the silica nanoparticles. It was shown that the concentration of the coated silica nanoparticles on the CFs has an important impact on the surface roughness which governs the superhydrophobicity.

Silica naoparticles were also incorporated into natural fibers, like cotton. Xue et al. [90] prepared superhydrophobic cotton surfaces through the introduction of functionalized silica nanoparticles. The surface of both the cotton fibers and the silica nanoparticles were first modified by different chemicals to facilitate the reaction between them. The nanoparticles were then added to the fibers followed

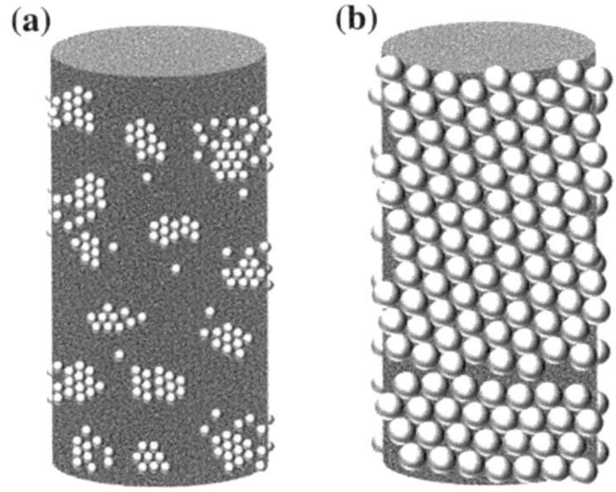

Fig. 8.16 Silica nanoparticle treated fiber **a** Non-uniform distribution of nanoparticles (*smaller* particle size) **b** Closely packed array of nanoparticles (*larger* particles size)

by a drying process in which silica nanoparticles covalently bonded to the cotton fibers. After the final treatment of the surface by a low surface energy material, a hierarchical superhydrophobic surface was obtained.

Heat treatment has been also applied to chemically bond the silica nanoparticles to the cotton fabrics [89]. Using this method, cotton fibers were treated by two sets of silica nanoparticles with different sizes, followed by a surface treatment with a cost-effective low surface energy material resulted in a superhydrophobic cotton fabric. The effect of the size of the silica naoparticles as well as the concentration of the low surface energy coating on the wetting property and WCA were investigated [3]. Figure 8.16 presents a schematic of two fibers treated with silica nanoparticles with different particle sizes and different concentrations of nanoparticles. A similar procedure was used to make superhydrophobic fabrics from the inherently hydrophobic poly (ethylene terephthalate) PET fibers. Silica nanoparticles were attached to the PET fibers through a specific reaction/interaction occured between the surface chemical groups of the PET fibers and the silica nanoparticles. Further treatment of the fabrics with a water replant agent resulted in a superhydrophobic surface with a WCA up to 158° [4].

8.3.4.2 Polymer Particles

Polymers in the form of micro and nanosized particles can be used in fabrication of superhydrophobic surfaces. Polymer nanoparticles provide the required roughness on the surface as well as the lower surface energy compared to other nanoparticles.

Bormashenko et al. [8] prepaered a triple-scale rough surface on a polyethylene (PE) film by introduction of polyvinylidene fluoride (PVDF) nanoparticles. PVDF nanoparticles were first spread on a thin polyethylene (PE) film followed by a hot embossing process in which individual PVDF nanoparticles and microscale aggregates were immobilized on the PE film resulting in a multiscale rough surface. Superhydrophobic surface with a WCA as high as 160° was prepared and a mathematical modeling was conducted based on the Cassie–Baxter wetting model.

A well-ordered structure of polymer nanoparticles provides the nanoscale roughness required for preparing superhydrophobic surfaces. Single and double-layer close-packed arrays of polystyrene (PS) nanoparticles were obtained via a spin coating process. The prepared two-dimensional nanostructure was then treated with oxygen plasma etching, which resulted in the reduction in the diameter of the PS nanoparticles up to 60%, and finally treated with a low surface energy material. Cassie's model was successfully applied to correlate the effect of size reduction of the PS nanoparticles with the solid–liquid contact and the observed WCAs [78]. Yan et al. [91] investigated rough surfaces obtained by closely packed arrays of silica and PS nano- and microspheres in a wide range of particle sizes. Three different methods of coating were employed to fabricate these surfaces. However, all of the applied methods resulted in the same well-ordered closed pack topography. After surface treatment with a fluoroalkylsilane, a low surface energy material, an increase in WCA was obtained for all surfaces due to the increased surface roughness. Cassie model was introduced to calculate the WCAs of the surfaces and good agreement was found between the theory and the experiments. The WCA values calculated for PS and silica particles are constant and do not change with particle sizes indicating that the change in particle size in closely packed array surfaces is not an effective way to tailor the wetting property. The roughness factor, r (using Eq. 8.10) for a closely packed structure is equal to 1.9. The Wenzel and Cassie–Baxter equations (Eqs. 8.9 and 8.12) can be applied to predict the water contact angles. The final form for the Wenzel equation is

$$\cos\theta = 1.9\cos\theta_0 \tag{8.13}$$

For the Cassie–Baxter equation, the f_{SL} can be calculated using the geometry of the wetted surface (Fig. 8.17). On substituting this value into Eq. 8.12 the final form can be expressed as

$$\cos\theta = \frac{2(1+\cos\theta_0)^2}{3.32 - \sin^2\theta_0 + \cos\theta_0} - 1 \tag{8.14}$$

The water contact angles, θ, expressed in Eqs. 8.13 and 8.14 are only function of θ_0, and are not function of particle size. The values obtained from Eq. 8.14 are shown to be in better agreement with the experimental results rather than the values calculated using Eq. 8.13, indicating that the water droplet on these surfaces is in Cassie–Baxter regime [91].

Ge et al. [22] proposed a simple coating method to prepare well-ordered closely packed PS nanoparticle arrays. The substrate was positioned in a colloidal suspension of PS nanoparticles, and then placed in a specific temperature and humidity to be dried. PS nanoparticles with different particle sizes were used and the effect of the particle size on the induced roughness and water repellency

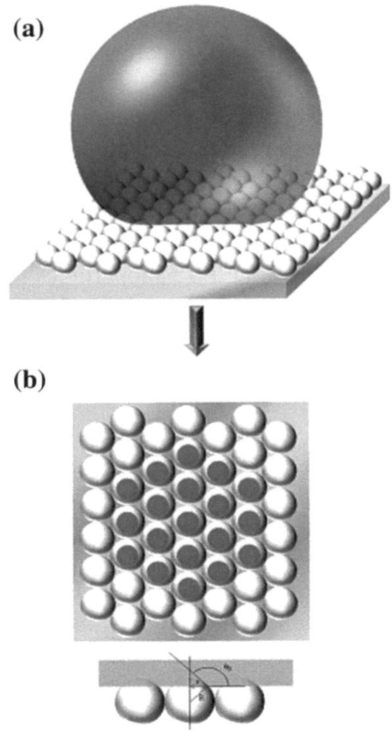

Fig. 8.17 Wetting behavior of closely packed array of polymer or silica particles. **a** Water droplet sitting on the surface. **b** *Top view* of the wetted particles and the *side view* of the angle between water and the particles (θ_0)

property was investigated. Highly hydrophobic property was obtained for all the particle size ranges.

Zhang et al. [100] used PS nanoparticles to modify both the water and oil wettability property of a polymer membrane simultaneously, which had the potential application as separation membranes. Surface of porous polyurethane (PU) foam was treated with PS nanoparticles. The applied olyurethane (PU) and PS had almost the same water and oil wettability. After modification, the dual-sized roughness obtained from the combination of microporous PU and the PS nanoparticles, resulted in a film with superhydrophobic and superoleophilic properties. The prepared composite film was successfully used to separate a water–oil mixture.

Hong and Pan [30] employed Polytetrafluoroethylene (PTFE) nanoparticles to make transparent superhydophobic surface on a UV-curable polymer. PTFE nanoparticles provided both the required hydrophobicity and nanoscale roughness, while the UV-curable polymer provided a transparent matrix for the nanoparticles. Two different methods were used to prepare micro- and nanopatterns on the substrate. In the first method, PFTE nanoparticles and the liquid polymer were mixed and then applied to a transparent substrate by a spin coating process. The micropatterns were made by a photolithography process in which, the spin coated substrate was exposure to a UV light through a photomask and then, the unexposed composite was dissolved resulted in a micro-patterned structure. In the second method, UV-curable polymer without nanoparticles was coated on the substrate and then, the PTFE nanoparticles were spray coated on the polymer surface followed by further curing to immobilize nanoparticles. Both methods obtained surfaces comprising micro- and nanoscale roughness which was responsible for the observed superhydrophobicity with WCAs higher than 150°. The surface obtained by the second method exhibited a higher WCA due to its higher surface roughness.

8.3.4.3 Carbon Nanotubes

Carbon nanotubes are allotropes of carbon in which graphite sheets have been rolled up to form hollow cylinders with diameter of a few nanometers and length up to a few centimeters. Nanotubes with their special structure have attracted a lot of interest in many application fields, such as nanotechnology, electronics, and optics in recent years. Nanotubes have been added to various polymer matrices for different physical and mechanical property enhancements. They can also be used to make nanosized roughness very similar to those of lotus leaves due to their nanosized diameter and high aspect ratio (surface/area ratio). The combination of micro- and nanotexture of the nanotubes, and their hydrophobic nature can be applied to fabricate nature-inspired superhydrophobic surfaces.

Liu et al. [57] used carbon nanotubes to modify the surface roughness and wettability of hydrophilic cotton fabrics using a dip-coating method. Nanotube clusters were first dispersed in water using an ultrasonic treatment. Cotton fabrics were then dip-coated into the nanotubes suspension resulted in the deposition and the immobilization of the nanotubes onto the fiber surfaces. The modified fabrics had hierarchical structure which mimicked the surface structure of lotus leaves. The water contact measurement, however, indicated that water was absorbed gradually by the fabrics because of the hydrophilic nature of cotton fabrics as well as the weak affinity between the nanotubes and the cotton fibers. In order to further improve the water repellency property of the fabrics, the surface of nanotubes were treated with a comparatively more hydrophobic polymer and used in the same dip-coating method to decorate the cotton fabrics. The polymer treated nanotubes resulted in a more uniform hierarchical structure with a durable WCA larger than 150°.

The surface modification of fabrics using carbon naotubes through a chemical reaction leads to more durable superhydrophobicity. Using a catalytic chemical vapor deposition, Hsieh et al. [35] fabricated superhydrophobic fabrics from intrinsically hydrophilic carbon fabrics. Carbon nanotubes were grown on the surface of the carbon fabrics by a catalytic chemical vapor deposition (CCVD) technique at 900°C. The naotube decorated fabrics further coated with a layer of fluoro-containing copolymer by a spin coating method to reduce the surface energy of the fabric surface. A dual-scale rough surface obtained from the combination of the microscale roughness of carbon fibers and the nanoscale roughness of

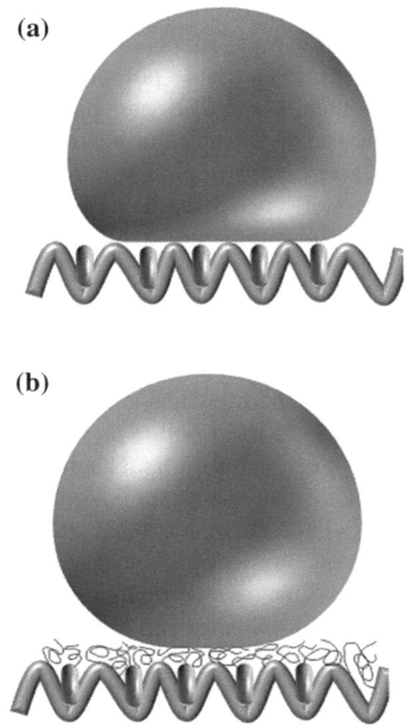

Fig. 8.18 CNT treatment of fabrics. **a** Water droplet on an unmodified fabrics. **b** Water droplet on a CNT treated fabric

nanotubes. The effect of nanoscale roughness on the wettability was investigated. A hydrophobic fabric with a WCA equal to 145° was obtained without CNT decoration, while a superhydrophobic fabric with a WCA up to 160° was reached after CNT treatment. Finally, the Cassie–Baxter model was used to investigate the effect of CNT of the surface topography and consequently on the wetting property of the fabrics. It was proposed that in case of CNT decorated fabrics, the water droplet lies on top of the CNT coils and tips with air trapped beneath them. The reduced solid–liquid contact surface exhibits superhydrophobiciy with low sliding angle (Fig. 8.18).

Beside fabrics, carbon nanotubes can also be used to fabricate superhydrophobic surfaces. Men et al. [60] investigated the effect of the introduction of various modified MWCNs into two hydrophobic polymers on the wetting properties of the obtained composites. Pristine, hydroxylic-modified and fluorocarbon modified MWCN's were combined with poly (furfuryl alcohol) (PFA) and Polytetrafluoroethylene (PTFE) at different ratios and the resultant nanocomposites were spray-coated on flat stainless steel substrates. The same dual micro- and nanoscale structures were formed for all the nanocompoosite coatings. However, a hydrophilic surface was obtained in the case of using hydroxylic-modified MWCNs and a hydrophobic substrate was obtained in the case of using pristine MWCNs, indicating the importance of the surface chemistry on the wettability.

Superhydrophobic surface was obtained only for the nanocomposites made of fluorocarbon modified MWCNs. The effect of the concentration of the MWCNs on the nanoscaled roughness was also studied. It was shown that at low concentration, the excess resin destroyed the nanoscale roughness and resulted in lower WCAs.

Other hydrophobic molecules have also been used for surface modification of CNT's [79, 102]. Yang et al. [93] used polystyrene (PS) for surface modification of MWCNs. The modified MWCNs were coated on a transparent substrate using a simple spray-coating technique resulting in a micro- and nanoscale roughness on the surface. The porous surface topography, combined with the hydrophobic nature of PS resulted in a superhydrophobic surface with an WCA of 160° and a SA as low as 3°.

8.3.5 Surface Modification by Low Surface Energy Materials

Self-assembly is a process in which disordered components of a system associate with each other and form an organized structure. Self-assembly is a versatile approach for the fabrication of nanostructures. Bhushan et al. [5] fabricated nanostructured and hierarchical surfaces by self-assembly of a thin layer of an alkane wax on smooth and micro-structured epoxy resins. Microstructured surface was prepared using a micro-molding process (presented in Sect. 8.3.2.2), and micropillar structured silicon as the master. Thin layer of the wax was then deposited on the smooth and nanostructured surfaces by evaporation and sublimation of the wax in a vacuum chamber. After coating, the specimens were heated and then immediately cooled down to interrupt the re-crystallization process, or placed at room temperature for 3 days for crystallization to be completed. Nanostractures comparable to those found on superhydrophobic natural leaves were created on the crystallized specimens. Higher WCA, lower hysteresis angle, and lower adhesive force were measured for the nanostractured surfaces compared to the smooth wax coated surfaces. Theoretical values of WCA were obtained for the surfaces based on Wenzel and Cassie–Baxter equations using the parameters obtained from the AFM maps and the SEM images. Good agreement between experimental results and the results obtained by the Cassie–Baxter model indicated that the Cassie–Baxter state is stable for the nanostructured surfaces. The same method was used to investigate the wetting property and the adhesion force of the hierarchical structures with various nanostructures. Higher WCA, lower hysteresis angle and lower adhesion force were observed for the hierarchical structures compared to nanostructures due to the farther decrease in the solid–liquid contact and the greater ability of hierarchical structures to trap more air pockets in both their levels of roughness.

Koch et al. [45] used the similar method to fabricate hierarchical surfaces with superhydrophobic and low-adhesion properties. Two different micro-patterned surfaces were prepared using both Lotus leaf and silicon micropillar array as master templates. A flat surface was prepared using a flat silicon master template for comparison. The wax used in this study was extracted from natural leaves. The process of coating of the wax on the surfaces was conducted by evaporation and sublimation of the wax in a vacuum chamber followed by allowing the crystallization of the wax in seven days. A tubular hollow structure with random orientation was formed on the surfaces after seven days. The wettability property of the prepared surfaces was investigated by measuring the static contact angle, contact angle, hysteresis and tilting angle of the surfaces. A flat surface, two micro-patterned surface prepared by lotus leaf replica and silicon replica, two hierarchical surface made by decoration of wax tubules on top of the two micro-patterned surfaces as well as a natural lotus leaf was investigated. WCA and contact angle hysteresis higher than 170° and lower than 2° was observed for the hierarchical surfaces respectively, which are comparable to those of lotus leaf. For the only micro-patterned surfaces, the WCA was higher than 150° but the contact angle hysteresis was high (27° and 29°), which prevents them from exhibiting self-cleaning ability. The values for roughness factor (R_f) and fractional liquid–air interface (f_{LA}) for different surfaces were measured using the SEM and the AFM data. The R_f and f_{LA} were higher for hierarchical surfaces than that for micro-patterned and flat surfaces, which was in agreement with the experimental results.

For many surfaces in nature such as lotus leaf, the superhydrophobicity (WCA higher than 150°) and low contact angle hysteresis are observed at the same time. However, there are superhydrophobic surfaces that exhibit high contact angle hysteresis and high adhesion. Different wetting behavior of the low adhesion and the high adhesion surfaces can be explained by having a closer look at the micro- and nanostructure as well as the chemistry of these surfaces, and provide the opportunity to fabricate artificial surfaces with tunable properties. Bhushan and Her [6] conducted a systematic investigation on the wetting behavior of two superhydrophobic rose petals with high and low adhesions. The relevant parameters responsible for the different behavior were identified, and artificial superhydrophobic surfaces with high and low adhesions were fabricated. Scanning electron microscopy (SEM) and AFM analysis indicated that the two surfaces had different micro- and nanostructure. Higher bump density and larger bump height was observed for the rose petal with low adhesion property. A Cassie–Baxter regime is favorable in this case in which, the water droplet sits on top of the asperities, resulting in low solid–liquid contact and low adhesion. For the rose petal with high adhesion property, on the other hand, lower bump density and smaller bump height was observed (Fig. 8.19). This enhanced pitch allows the water droplet to diffuse to the microstructure, resulting in high contact angle hysteresis and high adhesion. Inspired by the investigated rose petals, hirerachical surfaces with different wettability property were fabricated by self-assembly of a wax on the surface of micro-patterned epoxy substrates with the method described earlier [5]. Different micro-patterned epoxy surfaces with different pitch values, and also different amount of wax were used to fabricate hierarchical surfaces with various wetting and adhesion properties. It was observed that the combination of micro- and nanostructure of the surface governs the stable wetting state (Cassie–Baxter, Wenzel or Cassie). At high pitch values and low density of nanostructures,

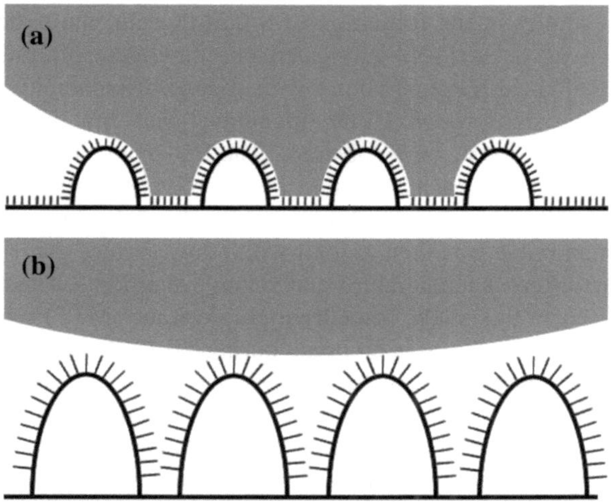

Fig. 8.19 Water droplet on two hierarchical structures **a** smaller bump height (high adhesion), **b** larger bump height (low adhesion)

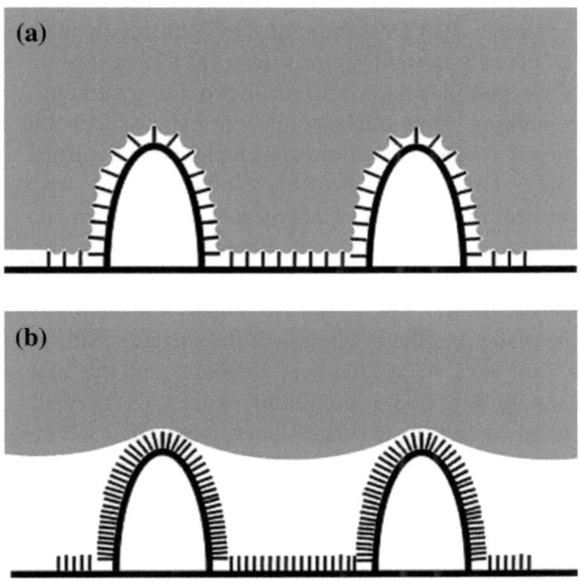

Fig. 8.20 Water droplet on two hierarchical structures, **a** low density of nanostructure (high adhesion), **b** High density of nanostructure (low adhesion)

the water droplet could diffuse to the microstructures, while the diffusion of water was not possible as the density of nanostructures increased. Two different surfaces with low adhesion and high adhesion obtained with the same microstructure but different nanostructures (Fig. 8.20).

Chemical vapor deposition (CVD) is a widely used technique for depositing materials on substrates. A chemical reaction in gaseous phase occurs at the surface, resulting in the formation of a thin layer of the material. Using a CVD process, Zheng et al. [105] fabricated polyvinylidene fluoride (PVDF) membranes with dual-sized rough structure with superhydrophobicity and self-cleaning properties. Microsized rough PVDF membrane was first prepared by solution coating of PVDF/solvent on a substrate. The PVDF membrane was then treated with high-energy oxygen ion emitted by microwave plasma which resulted in changes in both the chemistry and the topography of the surface. The oxygen plasma treated membrane was then grafted with a low surface energy material in a CVD process at different temperature and relative humidity. The low surface energy material used in this study is methyltrichlorosilane (MTS) which forms nanostructures on the surface of the membrane during the CVD process. A dual-sized rough surface was obtained by the combination of the microsized PVDF membrane and the nanosized structures obtained by CVD process. The effect of oxygen plasma treating time and oxygen plasma power in the plasma treatment process, grafting time, relative humidity, and the concentration of MTS in the CVD process on the measured WCA were investigated. A superhydrophobic membrane with self-cleaning property was obtained in the optimum condition.

In another study, Zheng et al. [106] investigated the fabrication of superhydrophobic PVFD membranes by grafting a mixture of dimethyldichlorosilane (DDS) and methyltrichlorosilane (MTS) as low surface energy materials. Instead of oxygen plasma, a sodium hydroxide treatment was used to introduce oxygen-containing functional groups on the surface, and instead of CVD process, a chemical bath deposition was employed for grafting the DDS/MTS solution in this study. The SEM and AFM results revealed that the grafted materials create a uniform distribution of nanosized asperities on top of the micro-patterned PVDF membrane. The resulted membrane had a WCA of 157° and a sliding angle of 1°.

Weibel et al. [87] investigated the preparation of superhydrophobic surfaces with high and low hysteresis angles on aluminum substrates by modifying the chemistry and the roughness of the surfaces. Microscale roughness prepared on the Al surfaces by a chemical etching, and the nanoscale roughness obtained by nanoporous anodic aluminum oxide (AAO), which are prepared in an electrochemical cell. Low surface energy treatment was conducted with an alkoxysilane functionalization and a thin layer of PTFE respectively. Different substrates were made with the combination of different chemistry and roughness treatments. It was shown that the nanostructure is necessary for the transition of the wetting state from Wenzel to Cassie–Baxter, in which the reduced solid–water contact results in low hysteresis.

8.3.6 Solution Method

Adjusting the crystallization time and nucleation rate for crystalline polymers, and the aggregation and precipitation rate in case of amorphous polymers have been shown to have a significant effect on the surface roughness of the films fabricated by solution casting method. Temperature and addition of non-solvent are two of the most important factors controlling the precipitation and the final morphology of the cast films. Lu et al. [58] investigated the preparation of low-density polyethylene (LDPE) superhydrophobic surfaces by controlling the crystallization behavior of LDPE during the film formation in a solution casting process. The LDPE was first dissolved in xylene (the solvent). A few drops of the prepared solution was then dropped onto a silicon wafer and placed in a vacuum oven at different temperatures to remove the solvent. Microporous surfaces with enhanced WCA were formed at different film formation temperatures. The microscale roughnessess of the surfaces were increased by decreasing the temperature, which subsequently increased the WCA. The increase in the roughness of the surfaces was believed to be due to the increase in crystallization time and nucleation rate [18]. To further increase the crystallization time and nucleation rate an LDPE nonsolvent (cyclohexanone) was added to the polymer solution, and the film formation was conducted at room temperature. A hierarchical micro- and nano-scale rough surface with a high WCA of 173° and low sliding angle of 1.9° was obtained.

Yuan et al. [98] used a similar method and prepared a superhydrophobic linear low-density polyethylene (LLDPE) surface with enhanced self-cleaning property compared to that for a smooth LLDPE surface. In another study [99], they prepared Superhydrophobic high-density polyethylene (HDPE) with low contact hysteresis by adding ethanol, as a precipitator, to the polymeric solution in a humid atmosphere. The effect of drying temperature on the surface roughness of the prepared films confirmed that the phase separation occurred at lower temperatures and resulted in rough surfaces [18, 58]. At high relative humidity, the water in the moisture can penetrate to the solution and act as a non-solvent for HDPE. The combination of two non-solvents resulted in the formation of more HDPE aggregates and more pore structures. After evaporation of the solvent at humid atmosphere and low temperature, a superhydrophobic surface with WCA of 160° and sliding angle of 2° was obtained.

Hou et al. [33] investigated the biomedical application of a superhydrophobic polypropylene (PP) surface prepared by a phase-inversion method introduced by Erbil et al. [18]. The anticoagulation property of the rough and smooth surfaces was studied by blood contacting experiment and SEM analysis. It was shown that under the same condition, the adhesion of platelets, blood cells, and fibrin to the superhydrophobic surface is much lower than that for the smooth surface, indicating the good blood compatibility of the PP superhydrophobic surface.

Peng et al. [69] fabricated highly hydrophobic poly-(vinylidene fluoride) (PVDF) membranes with modifying the conventional solution casting process. The polymer solution was first prepared by dissolving PVDF pellets in *N*,*N*dimethyl-acetamide (DMAc) solvent and then was spread uniformly on substrate. Two different methods were used to prepare porous membranes. In the first method, the film was immersed into a precipitation bath contains mixture of water and the solvent (water/DMAc) at different concentrations. Since water is the non-solvent for PVDF, precipitation of PVDF from the solution took place in the precipitation bath, resulting in a translucent membrane. Membranes prepared through the precipitation in water/DMAc bath was observed to have porous surfaces while it was smooth using only water as precipitation bath. The WCA was increased by 70% in this method. In the second method, the porous membrane was prepared by placing the solution coated film in the open air instead of using precipitation bath. Micro- and nanosized PVDF particles were prepared by crystallization from the solution during the gelation process, and hierarchical structure with WCA as high as 150° was obtained.

Making surface roughness by solution method can also be used to fabricate roughness on rigid polymeric surfaces rather than casting a polymer solution. Zhao et al. [104] fabricated lotus-like surfaces on smooth plates of amorphous polycarbonate (PC) by a facile solvent treatment method. Polycarbonate is a crystalline thermoplastic polymer; however, the crystallization from the melt is extremely slow due to the stiffness of its polymer chains. It has been shown that the addition of solvent increase the crystallization rate by increasing the polymer chain mobility [27]. The surface of plate was covered by a thin layer of acetone (solvent) and was allowed to evaporate at room temperature. Nanosized needle-like asperities appeared on the surface after evaporation of the solvent due to the crystallization of the plasticized chains on the surface. The WCA on the prepared nanosized rough surface was 130°, which was far from superhydrophobicity. The swollen PC surface was then treated with a non-solvent (methanol or water). Introduction of nonsolvent to the surface resulted in the precipitation of some polymer chains, which can act as nuclei for further nucleation and growth. Microsized protrusions were formed consequently beside the nanosized asperities. The final surface had a hierarchical structure with superhydrophobic characteristics.

Zhang et al. [103] fabricated PC superhydrophobic coatings on various substrates, using a "dipping and drying" method. The substrate was first dipped into the polymer solution followed by fast drying at room temperature. The PC coated substrate was then dipped into the non-solvent, where the phase separation and precipitation took place. Five different non-solvents of PC were used to study the effect of non-solvent on the phase separation and the resultant morphology. It was shown that the polymer solubility and the boiling point of non-solvents are the most important factors which control the phase separation process. The surface roughness increased as the solubility of the non-solvent decreased, and the boiling point increased.

The solution-precipitation method can also be used to fabricate rough surfaces on amorphous polymers, which normally form smooth film upon casting. Li et al. [54] fabricated superhydrophobic poly (vinyl chloride) (PVC) surface with micro- and nanoscale roughness similar to those of lotus leaf. The combination of three non-solvents was used and superhydrophobic surfaces with WCA higher than 150° were obtained in all cases. The non-solvent was first dropped onto a surface. Then, the polymer solution was coated on the non-solvent. Relative diffusion of the solvent and the non-solvent at the interphone resulted in the micro- and nanophase separation of the polymer solution, which finally resulted in multiscale surface roughness. Instead of coating the solution on top of the non-solvent layer, Chen et al. [11] added the non-solvent, ethanol, to the PVC solution. Nanosized rough surfaces were obtained and the roughness increased with the increasing of the concentration of non-solvent in solution. At 50% ethanol content a WCA of more than 150° and sliding angle of 7° resulted. Superhydrophobic surfaces were also prepared on polystyrene (PS), again an amorphous polymer, with a solution-precipitation method by using ethanol as non-solvent [97].

Shi et al. [77] fabricated poly (L-lactic acid) (PLLA) superhydrophobic surfaces by a phase-inversion method. Water and ethanol were employed as nonsolvent and three different coagulation baths; water, water/solvent, and ethanol were used to study the phase-inversion and the resulting surface morphologies. It was shown that the presence of solvent in the precipitation bath has a strong effect on the surface topography. Sponge-like porous film obtained from the precipitation in water/solvent bath, but no significant roughness was observed from the precipitation in pure water. Superhydrophobic surfaces with flower-like structures were obtained by precipitating of the surfaces in the ethanol bath.

8.3.7 Plasma, Electron, and Laser Treatment

Plasma treatment is the process of using an ionized gas to modify the surface properties of various materials. Plasma treatment of polymers changes the chemical and physical properties of the polymeric surface, and can be used to modify the wetting and adhesion properties of polymeric surfaces. Depending on the applied plasma treatment, the resultant surface can be more hydrophobic or more hydrophilic.

Kim et al. [43] used an atmospheric methane plasma treatment to fabricate hydrophobic coating on various surfaces. The chemical composition of the deposited coating indicated that, the methane plasma treatment created a surface composed of low surface energy groups. Combined with the microsized roughness of cotton fibers, a superhydrophobic surface with WCA up to 150° was obtained from cotton fabrics.

Teshima et al. [82] fabricated transparent superhydrophobic polyethylene terephthalate (PET) surfaces using an oxygen plasma treatment followed by a hydrophobic organosilane coating. Nanotextures were formed on the surface by oxygen plasma at different power levels. As the power level increased, the surface roughness increased, which subsequently degraded the optical transparency. An optimum power level was obtained in which the maximum roughness was obtained while the transparency was still high. The nanotextured films then treated with low surface energy organosilane precursors by using either a low-temperature chemical vapor deposition (CVD) or plasma enhanced CVD (PECVD). Transparent superhydrophobic surfaces with WCAs higher than 150° were obtained. The same approach was used to prepare transparent superhydrophobic Poly(methyl methacrylate) (PMMA) surface [85]. The oxygen plasma treatment was first conducted at various pressure, time, and substrate voltage to form nanotextured surfaces. Surface roughness increased with increase in substrate voltage and decrease in pressure. The surface roughness also increased with treatment time. After depositon of a thin layer of a Teflon-like coating, superhydrophobic surfaces with WCA of 152° and contact angle hysteresis less than 5° were obtained.

Sputter deposition is a widely used process for applying a thin layer of a material on a substrate. Radio frequency (RF) sputter deposition is the preferred method for the forming of thin films from insulating materials. The rf-sputter deposition has also been considered as an effective approach for the preparation of fluoropolymer thin films, which have found applications in highly hydrophobic and low friction coatings [7, 62, 83]. Sarkar et al. [74] fabricated superhydrophobic aluminum surfaces by rf-sputter deposition of a thin layer of PTFE on chemically etched aluminum surfaces. The aluminum sheets were first chemically etched in a hydrochloric acid solution for different durations. Rough surface with micron-sized pits was observed on the surfaces after etching. The roughened aluminum surfaces were then placed in a plasma reactor in which ultra thin PTFE film were deposited on the surfaces by using argon plasma. It was shown that the etching time controls the wetting properties of the resulted surfaces. As the etching time increased, the WCA increased and the contact angle hysteresis decreased until they reached to constant values. Further etching did not change the WCA and contact angle hysteresis values, indicating the formation of the maximum roughness at that value. A WCA as high as 164° and a contact angle hysteresis as low as 2° was obtained for the optimum etching time.

Gupta et al. [24] investigated a pulsed electron deposition (PED) to fabricate dual-size hierarchical PTFE coatings. The PTFE thin films were deposited on a substrate using a pulse electron gun at various electron energies. It was shown that the applied electron energy can alter the micro- and nanostructure of the prepared coatings. Single-scale rough surfaces were obtained at low electron energies, while dual-scale rough surfaces were obtained at higher electron energies. The transition between single- and double-roughness at mean electron energies was also confirmed by a sudden increase in the root mean square (RMS) roughness and WCA at the transition region. Hierarchical superhydrophobic surfaces with WCA as high as 151° were resulted at high electron energies.

Holographic lithography is a promising technique for fabrication of periodic nanostructures. In this technique, a photosensitive material is exposed to interfering laser beams, to create periodic micro- and nanostructures. The conventional holographic lithography can be modified for fabrication of nanostructured polymer surfaces. The period of patterns can be controlled by controlling the wavelength of light and incident angle of the laser beams. By using this technique, Park et al. [68] fabricated nano-patterned superhydrophobic surfaces with high and low adhesion. The holographic lithography was conducted at two different incident angles; perpendicular to the top surface of the used prism, and tilted from the perpendicular incidence of the laser beam. A face-centered cubic (FCC) and a tilted FCC lattice structure was formed in the perpendicular and the tilted incidence of the laser beam respectively. The patterned surfaces were then modified by a reactive-ion etching using CF_4 gas. Needle-like structure with a WCA of 160° and sliding angle of 3° was formed for the FCC lattice, which showed the formation of Cassie–Baxter regime. The tilted FCC surface, on the other hand, exhibited superhydrophobicity with high adhesion force, indicating the formation of Wenzel state.

8.4 Surface Characterization

An appropriate surface analysis technique to evaluate the surface properties of the fabricated polymeric surface is of great importance. Surface analysis methods can be used to verify the fabrication of surface with the desired chemical structure, wetting property, morphology and topography, and modified performance. A combination of different surface characterization techniques is frequently required to gain a comprehensive understanding of the performance of the fabricated surfaces.

Adhesion and wetting property of polymeric surfaces is governed by both the surface roughness and the surface chemistry of the outermost surface layer. One would have to study the composition and topography of this layer at a sub-nanometre scale in order to fully determine the surface properties. Various information including chemical structure, the hydrophilicity or hydrophobicity, and the topography of the polymeric surfaces can be obtained using different analysis methods [44, 72].

8.4.1 Surface Chemistry

The chemistry of surfaces has been widely studied using spectroscopic methods which reveal valuable information about the constituent elements and the chemical structure of the surface [50]. Table 8.3 presents the most common spectroscopic techniques used for surface characterization.

8.4.2 Wetting Property

Surface wettability of materials is generally characterized by measuring the contact angle of a liquid droplet on the surface (Fig. 8.6). Contact angle goniometer is used to measure the static contact angles, advancing contact angle and contact angle hysteresis. Goniometry is one of the oldest surface characterization techniques and is still commonly used. A drop of a liquid is placed on a solid surface. The surface of the drop will make a certain angle (the tangent to the point of contact) with the surface of the solid . A camera captures the shape of the water

Table 8.3 Most common spectroscopic techniques for surface characterization

Technique	Probe	Information	Comments
Auger electron spectroscopy (AES)	Electrons	Surface composition	Surface conductivity needed
X-ray photon spectroscopy (XPS)	X-rays	Chemical composition	Quantitative analysis
Secondary ion mass spectrometry (SIMS)	Ions	Surface composition, contaminations	
Infrared attenuated total reflection ATR-FTIR	Infrared light	Surface composition	

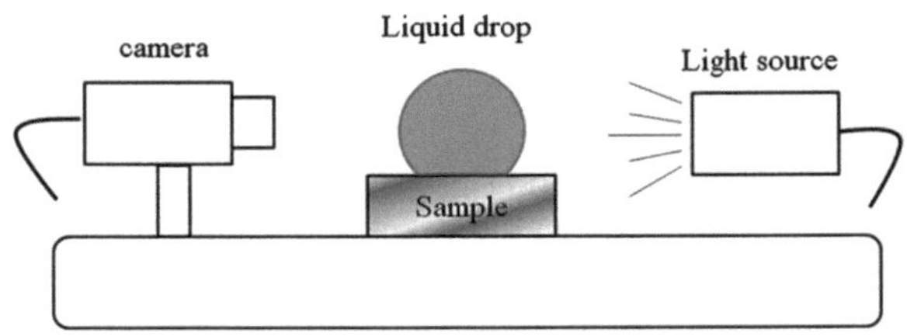

Fig. 8.21 Schematic drawing of a contact angle goniometer

droplet on the surface, and software is used to analyze the drop shape. Figure 8.21 shows the schematic of a contact angle goniometer.

8.4.3 Surface Topography and Surface Forces

8.4.3.1 Scanning Electron Microscopy

Microscopy is one of the most powerful techniques for characterizing surfaces at micro- and nanoscale. When the domain size is less than one micron, the magnification of a light microscope is no longer sufficient and electron microscopy is necessary. While optical microscopes are generally limited by magnification below 1,000×, the electron microscope can be used for magnifications up to about 10,000,000×. An electron microscope is a type of microscope that uses a beam of highly energetic electrons to illuminate the specimen and provides different information about topography, morphology, and chemical composition of a wide variety of materials. Two types of electron microscopies are distinguished: scanning electron microscopy (SEM) and transmission electron microscopy (TEM).

Scanning electron microscope (SEM) is a powerful tool not only for surface observation but also for analysis of the subsurface structure of materials. As an electron beam strikes a surface, several different signals may be generated [23]. Backscattered electrons, characteristic X-rays, secondary electrons and auger

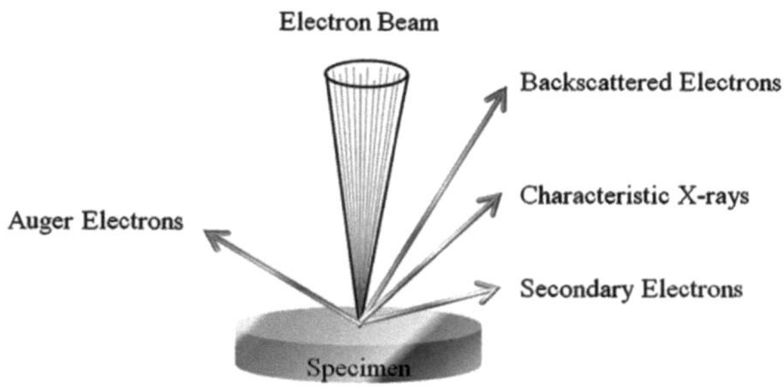

Fig. 8.22 Different signals produced by emitting of an electron beam to the surface of a specimen

electrons are the most important signals used for characterization of polymeric surfaces (Fig. 8.22).

Secondary electrons are low energy electrons which are used to form the 3D images of the surface with a resolution of less than 1 nm. Backscatter electrons are high energy (>50 eV) electrons that are reflected directly from the specimen by elastic scattering. Backscatter electrons can be used in analytical SEM along with the characteristic X-rays analysis, and provide information about the distribution of different elements on the surface.

When a high-energy electron beam strikes a surface, each element on the surface emits its own characteristic wavelength and energy spectra. So X-ray spectra [Energy dispersive spectrometer (EDS)] can be used to perform elemental analysis or chemical characterization of a surface [23]. This technique has been used for characterization of polymer coatings and polymer nanocomposites [17, 80].

8.4.3.2 Atomic Force Microscopy

Atomic Force Microscopy (AFM) is one of the most promising tools for characterization of surfaces at nanoscale. In the imaging mode, a sharp tip mounted on the end of a flexible cantilever probes the surface to obtain an image of sample surface. A laser beam is shown on the back of the cantilever and reflected onto a photodiode or a laser detector. The deflection of the cantilever is monitored by the photodetector by converting the voltage signal into distance (Fig. 8.23). As the tip approaches the surface, the interaction forces between the tip and the sample surface result in a deflection of the cantilever. Different types of adhesion forces (Sect. 8.2.1) can be responsible for the observed deflection.

The AFM can operate in three different modes: contact, non-contact, and tapping modes. In the contact mode, the tip stays in contact with the sample during scanning. For soft materials, it is likely that the AFM tip would damage the sample surface, so the non-contact modes are more desirable for these materials [51]. In the non-contact modes, on the other hand, the AFM tip is not in contact with the sample, and the change of resonance frequency of the cantilever is measured. The

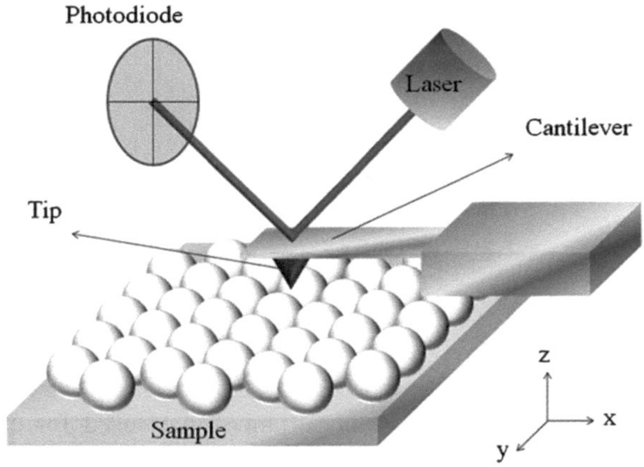

Fig. 8.23 Schematic illustration of AFM

non-contact mode is suitable for studying many soft surfaces such as biological surfaces and polymers [73]. In the tapping mode, the cantilever is oscillated above the sample with high amplitude, and the tip contacts the surface at each oscillation cycle. The change of oscillation amplitude of the cantilever is used to measure the interaction force.

AFM is a powerful technique for surface characterization of polymeric surfaces. Surface topography can be visualized, and surface roughness can be measured. Compositional imaging for heterogeneous polymers can also be done with AFM. Individual components of a heterogeneous system can be detected as the AFM tip scans over the surface. This ability is due to the senility of tip to the variations in local properties, and has a potential application in characterization of semi-crystalline polymers, block copolymers, polymer blends, and polymers composites [13].

8.5 Conclusions

Adhesive properties of polymeric materials and modern techniques of surface modification make polymers appropriate for Green Tribology applications, which require functional surfaces and the ability to control, and modify and surface properties, such as adhesion and wetting. Polymers, along with polymer composites, are appropriate materials for coating and various biomimetic applications, such as those utilizing the Lotus and gecko effects.

Acknowledgments The authors acknowledge the support of the University of Wisconsin-Milwaukee (UWM) RGI, NSF I/UCRC for Water Equipment and Policy, and UWM Research Foundation Bradley Catalyst grants.

References

1. K. Autumn, M. Sitti, Y.A. Liang, A.M. Peattie, W.R. Hansen, S. Sponberg, T.W. Kenny, R. Fearing, J.N. Israelachvili, R.J. Full, Evidence for van der Waals adhesion in gecko setae. PNAS **99**, 12252–12256 (2002)
2. R. Asmatulu, M. Ceylan, N. Nuraje, Study of superhydrophobic electrospun nanocomposite fibers for energy systems. Langmuir **27**, 504–507 (2011)
3. G.Y. Bae, B.G. Min, Y.G. Jeong, S.C. Lee, J.H. Jang, G.H. Koo, Superhydrophobicity of cotton fabrics treated with silica nanoparticles and water-repellent agent. J. Colloid Interface Sci. **337**, 170–175 (2009)
4. G.Y. Bae, Y.G. Jeong, B.G. Min, Superhydrophobic PET fabrics achieved by silica nanoparticles and water-repellent agent. Fibers Polymers **11**, 976–981 (2010)
5. B. Bhushan, K. Koch, Y.C. Jung, Nanostructures for superhydrophobicity and low adhesion. Soft Matter **4**, 1799–1804 (2008)
6. B. Bhushan, E.K. Her, Fabrication of superhydrophobic surfaces with high and low adhesion inspired from rose petal. Langmuir **26**, 8207–8217 (2010)
7. H. Biederman, S.M. Ojha, L. Holland, The properties of fluorocarbon films prepared by r.f. sputtering and plasma polymerization in inert and active gas. Thin Solid Films **41**, 329–339 (1977)
8. E. Bormashenko, T. Stein, G. Whyman, Y. Bormashenko, R. Pogreb, Wetting properties of the multiscaled nanostructured polymer and metallic superhydrophobic surfaces. Langmuir **22**, 9982–9998 (2006)
9. J. Bravo, L. Zhai, Z. Wu, R.E. Cohen, M.F. Rubner, Transparent superhydrophobic films based on silica nanoparticles. Langmuir **23**, 7293–7298 (2007)
10. A.B.D. Cassie, Contact angles. Discuss. Faraday Soc. **3**, 11–16 (1948)
11. H. Chen, Z. Yuan, J. Zhang, Y. Liu, K. Li, D. Zhao, S. Li, P. Shi, J. Tang, Preparation, characterization and wettability of porous superhydrophobic poly (vinyl chloride) surface. J. Porous. Mater. **16**, 447–451 (2009)
12. Z. Cheng, J. Gao, L. Jiang, Tip geometry controls adhesive states of superhydrophobic surfaces. Langmuir **26**, 8233–8238 (2010)
13. D. Chernoff, S. Magonov, Chapter 19: Atomic Force Microscopy, in *Comprehensive Desk Reference of Polymer Characterization and Analysis*, ed. by R. Brady (Oxford University Press, New York, 2003)
14. S.J. Choi, K.Y. Suh, H.H. Lee, A geometry controllable approach for the fabrication of biomimetic hierarchical structure and its superhydrophobicity with near-zero sliding angle. Nanotechnology **19**, 1–5 (2008)
15. B.V. Derjaguin, L. Landau, Theory of the stability of strongly charged lyophobic sols and of the adhesion of strongly charged particles in solutions of electrolytes. Acta Physicochim URSS **14**, 633–662 (1941)
16. B.V. Derjaguin, V.M. Muller, Y.P. Toporov, Effect of contact deformations on the adhesion of particles. J. Colloid Interface Sci. **53**, 314–326 (1975)
17. N. Dilsiz, R. Partch, E. Matijevic, E. Sancaktar, Silver coating of spindle- and filament-type magnetic particles for conductive adhesive applications. J. Adhes. Sci. Technol. **11**, 1105–1118 (1997)
18. H.Y. Erbil, A.L. Demirel, Y. Avci, O. Mert, Transformation of a simple plastic into a superhydrophobic surface. Science **299**, 1377–1380 (2003)
19. J. Feng, B. Huang, M. Zhong, Fabrication of superhydrophobic and heat-insulating antimony doped tin oxide/polyurethane films by cast replica micromolding. J. Colloid Interface Sci. **336**, 268–272 (2009)
20. R.A. Fisher, On the capillary forces in an ideal soil; correction of formulae given by W. B. Haines. J. Agric. Sci. **16**, 492–505 (1926)
21. N. García, E. Benito, P. Tiemblo, M.M.B. Hasan, A. Synytska, M. Stamm, Chemically guided topography in alkylsilane- and oligosiloxane-modified silica nanoparticle coatings:

from very hydrophobic surfaces to "pearl" bouncing droplets. Soft Matter **6**, 4768–4776 (2010)

22. H. Ge, Y. Song, L. Jiang, D. Zhu, One-step preparation of polystyrene colloidal crystal films with structural colors and high hydrophobicity. Thin Solid Films **515**, 1539–1543 (2006)
23. J. Goldstein, *Scanning Electron Microscopy and X-ray Microanalysis,* 3rd edn. (Kluwer Academic, New York, 2003)
24. S. Gupta, A.C. Arjunan, S. Deshpande, S. Seal, D. Singh, R.K. Singh, Superhydrophobic polytetrafluoroethylene thin films with hierarchical roughness deposited using a single step vapor phase technique. Thin Solid Films **517**, 4555–4559 (2009)
25. W.B. Haines, Studies in the physical properties of soils: IV a further contribution to the theory of capillary phenomena in soil. J. Agric. Sci. **17**, 264–290 (1927)
26. D. Han, A.J. Steckl, Superhydrophobic and oleophobic fibers by coaxial electrospinning. Langmuir **25**, 9454–9462 (2009)
27. H.R. Harron, R.G. Pritchard, B.C. Cope, D.T. Goddard, An atomic force microscope (AFM) and tapping mode AFM study of the solvent-induced crystallization of polycarbonate thin films. J. Polym. Sci. B: Polym. Phys. **34**, 173–180 (1996)
28. H. Hertz, Über die berührung fester elastischer Körper. J. Reine Angew. Math. **92**, 156–171 (1882)
29. S.H. Hong, J. Hwang, H. Lee, Replication of cicada wing's nano-patterns by hot embossing and UV nanoimprinting. Nanotechnology **20**, 1–5 (2009)
30. L. Hong, T. Pan, Photopatternable superhydrophobic nanocomposites for microfabrication. J. Microelectromech. Syst. **19**, 246–253 (2010)
31. W. Hou, Q. Wang, Stable polytetrafluoroethylene superhydrophobic surface with lotus-leaf structure. J. Colloid Interface Sci. **333**, 400–403 (2009)
32. W. Hou, Q. Wang, Wetting behavior of a SiO_2–polystyrene nanocomposite surface. J. Colloid Interface Sci. **316**, 206–209 (2007)
33. X. Hou, X. Wanga, Q. Zhua, J. Baoa, C. Mao, L. Jianga, J. Shena, Preparation of polypropylene superhydrophobic surface and its blood compatibility. Colloids Surf. B **80**, 247–250 (2010)
34. C.T. Hsieh, F.L. Wu, S.Y. Yang, Superhydrophobicity from composite nano/microstructures: Carbon fabrics coated with silica nanoparticles. Surf. Coat. Technol. **202**, 6103–6108 (2008)
35. C.T. Hsieh, W.Y. Chen, F.L. Wu, Fabrication and superhydrophobicity of fluorinated carbon fabrics with micro/nanoscaled two-tier roughness. Carbon **46**, 1218–1224 (2008)
36. J.N. Israelachvili, *Intermolecular and Surface Forces*, 2nd edn. (Academic, London, 1992)
37. H.E. Jeong, M.K. Kwak, C.I. Park, K.Y. Suh, Wettability of nanoengineered dual-roughness surfaces fabricated by UV-assisted capillary force lithography. J. Colloid Interface Sci. **339**, 202–207 (2009)
38. H.E. Jeong, R. Kwak, J.K. Kim, K.Y. Suh, Generation and self-replication of monolithic, dual-scale polymer structures by two-step capillary-force lithography. Small **4**, 1913–1918 (2008)
39. R.M. Jisr, H.H. Rmaile, J.B. Schlenoff, Hydrophobic and ultrahydrophobic multilayer thin films from perfluorinated polyelectrolytes. Angew. Chem. Int. Ed. **44**, 782–785 (2005)
40. K.L. Johnson, K. Kendall, A.D. Roberts, surface energy and the contact of elastic solids. proc.R. Soc. Lond. A **324**, 301–313 (1971)
41. L. Jiang, Y. Zhao, J. Zhai, A lotus-leaf-like superhydrophobic surface: a porous microsphere/nanofiber composite film prepared by electrohydrodynamics. Angew. Chem. Int. Ed. **43**, 4338–4341 (2004)
42. M. Kanga, R. Junga, H.S. Kima, H.J. Jinb, Preparation of superhydrophobic polystyrene membranes by electrospinning. Colloids Surf. A **313–314**, 411–414 (2008)
43. J.H. Kim, G. Liu, S.H. Kim, Deposition of stable hydrophobic coatings with in-line CH4 atmospheric rf plasma. J. Mater. Chem. **16**, 977–981 (2006)
44. D. Klee, H. Hocker, Polymers for biomedical applications, improvement of the interface compatibility. Adv. Polym. Sci. **149**, 1–57 (2000)

45. K. Koch, B. Bhushan, Y.C. Jung, W. Barthlott, Fabrication of artificial Lotus leaves and significance of hierarchical structure for superhydrophobicity and low adhesion. Soft Matter **5**, 1386–1393 (2009)
46. W. Lee, M.K. Jin, W.C. Yoo, J.K. Lee, Nanostructuring of a polymeric substrate with well-defined nanometer-scale topography and tailored surface wettability. Langmuir **20**, 7665–7669 (2004)
47. S.M. Lee, H.S. Lee, D.S. Kim, T.H. Kwon, Fabrication of hydrophobic films replicated from plant leaves in nature. Surf. Coat. Technol. **201**, 553–559 (2006)
48. S.M. Lee, T.H. Kwon, Mass-producible replication of highly hydrophobic surfaces from plant leaves. Nanotechnology **17**, 3189–3196 (2006)
49. S.M. Lee, T.H. Kwon, Effects of intrinsic hydrophobicity on wettability of polymer replicas of a superhydrophobic lotus leaf. J. Micromech. Microeng. **17**, 687–692 (2007)
50. D. Leonard, Y. Chevolot, O. Bucher, H. Sigrist, H.J. Mathieu, Part 2. N-[m-(3-(trifluoromethyl) diazirine-3-yl)phenyl]-4-(-3-thio(-1-D-galactopyrannosyl)-maleimidyl) butyramide (MAD-Gal) on diamond. Surf. Interface Anal **26**, 793–799 (1998)
51. O.M. Leung, M.C. Goh, Orientational ordering of polymers by atomic force microscope tipsurface interaction. Science **255**, 64–66 (1992)
52. A.Y.M. Lin, R. Brunner, P.Y. Chen, F.E. Talke, M.A. Meyers, Underwater adhesion of abalone. The role of van der Waals and capillary forces. Acta Mater. **57**, 4178–4185 (2009)
53. J.C. Melrose, Model calculations for capillary condensation. Am. Inst. Chem. Eng. J. **12**, 986–994 (1966)
54. X. Li, G. Chen, Y. Ma, L. Feng, H. Zhao, L. Jiang, F. Wang, Preparation of a superhydrophobic poly(vinyl chloride) surface via solvent–nonsolvent coating. Polymer **47**, 506–509 (2006)
55. X. Li, B. Ding, J. Lin, J. Yu, G. Sun, Enhanced mechanical properties of superhydrophobic microfibrous polystyrene mats via polyamide 6 nanofibers. J. Phys. Chem. C **113**, 20452–20457 (2009)
56. X. Li, X. Du, J. He, Self-Cleaning antireflective coatings assembled from peculiar mesoporous silica nanoparticles. Langmuir **26**, 13528–13534 (2010)
57. Y. Liu, J. Tang, R. Wang, H. Lu, L. Li, Y. Kong, K. Qi, J.H. Xin, Artificial lotus leaf structures from assembling carbon nanotubes and their applications in hydrophobic textiles. J. Mater. Chem. **17**, 1071–1078 (2007)
58. X. Lu, C. Zhang, Y. Han, Low-density polyethylene superhydrophobic surface by control of its crystallization behavior. Macromol. Rapid Commun. **25**, 1606–1610 (2004)
59. M. Ma, R.M. Hill, J.L. Lowery, S.V. Fridrikh, G.C. Rutledge, Electrospun poly (styrene-block-dimethylsiloxane) block copolymer fibers exhibiting superhydrophobicity. Langmuir **21**, 5549–5554 (2005)
60. X.H. Men, Z.Z. Zhang, H.J. Song, K. Wang, W. Jiang, Fabrication of superhydrophobic surfaces with poly(furfuryl alcohol)/multi-walled carbon nanotubes composites. Appl. Surf. Sci. **254**, 2563–2568 (2008)
61. R. Menini, M. Farzaneh, Production of superhydrophobic polymer fibers with embedded particles using the electrospinning technique. Polym. Int. **57**, 77–84 (2008)
62. D.J. Morrison, T. Robertson, R.F. sputtering of plastics. Thin Solid Films **15**, 87–101 (1973)
63. V.M. Muller, V.S. Yushchenko, B.V. Derjaguin, On the influence of molecular forces on the deformation of an elastic sphere and its sticking to a rigid plane. J. Colloid Interface Sci. **77**, 91–101 (1980)
64. V.M. Muller, B.V. derjaguin, Y.P. Toporov, On 2 methods of calculation of the force of sticking of an elastic sphere to a rigid plane. Colloids Surf. **7**, 251–259 (1983)
65. P. Nagaraja, D. Yao, Rapid pattern transfer of biomimetic surface structures onto thermoplastic polymers. Mater. Sci. Eng. C **27**, 794–797 (2007)
66. M. Nosonovsky, E. Bormashenko, "Lotus Effect: Superhydrophobicity and Self-Cleaning," Functional Properties of Biological Surfaces: Characterization and Technological Applications, ed. by E. Favret, N. Fuentes, (World Scientific, Singapore, 2009), pp. 43–78

67. C. Neinhuis, W. Barthlott, Characterization and distribution of water-repellent, self-cleaning plant surfaces. Ann. Bot. **79**, 667–677 (1997)
68. S.G. Park, J.H. Moon, S.K. Lee, J. Shim, S.M. Yang, Bioinspired holographically featured superhydrophobic and supersticky nanostructured materials. Langmuir **26**, 1468–1472 (2010)
69. M. Peng, H. Li, L. Wu, Q. Zheng, Y. Chen, W. Gu, Porous poly(vinylidene fluoride) membrane with highly hydrophobic surface. J. Appl. Polym. Sci. **98**, 1358–1363 (2005)
70. E. Puukilainen, H.K. Koponen, Z. Xiao, S. Suvanto, T.A. Pakkanen, Nanostructured and chemically modified hydrophobic polyolefin surfaces. Colloids Surf. A **287**, 175–181 (2006)
71. Y. Rahmawan, M.W. Moon, K.S. Kim, K.R. Lee, K.Y. Suh, Wrinkled, dual-scale structures of Diamond-Like Carbon (DLC) for superhydrophobicity. Langmuir **26**, 484–491 (2010)
72. B.D. Ratner, B.J. Tyler, A. Chilkoti, Analysis of biomedical polymer surfaces: polyurethanes and plasma-deposited thin films. Clin. Mater. **13**, 71–84 (1993)
73. T. Rhodin, Scanning probe microscopies, nanoscience and nanotechnology. Appl. Phys. A **72**, 141–143 (2001)
74. D.K. Sarkar, M. Farzaneh, R.W. Paynter, Superhydrophobic properties of ultrathin rf-sputtered Teflon films coated etched aluminum surfaces. Mater. Lett. **62**, 1226–1229 (2008)
75. M. Schoen, T. Gruhn, D.J. Diestler, Solvation forces in thin films confined between macroscopically curved substrates. J. Chem. Phys. **109**, 301–311 (1998)
76. X. Sheng, J. Zhang, Superhydrophobic behaviors of polymeric surfaces with aligned nanofibers. Langmuir **25**, 6916–6922 (2009)
77. J. Shi, N.M. Alves, J.F. Mano, Towards bioinspired superhydrophobic poly (L-lactic acid) surfaces using phase inversion-based methods. Bioinsp. Biomim. **3**, 1–6 (2008)
78. J.Y. Shiu, C.W. Kuo, P. Chen, C.Y. Mou, Superhydrophobic PET fabrics achieved by silica nanoparticles and water-repellent agent. Chem. Mater. **16**, 561–564 (2004)
79. S. Srinivasan, V.K. Praveen, R. Philip, A. Ajayaghosh, Bioinspired superhydrophobic coatings of carbon nanotubes and linear π systems based on the "bottom-up" self-assembly approach. Angew. Chem. Int. Ed. **47**, 5750–5754 (2008)
80. P.R. Start, K.A. Mauritz, Surlyn/silicate nanocomposite materials via a polymer in situ sol–gel process: morphology. J Polym Sci B: Polym Phys **41**, 1563–1571 (2003)
81. C. Sun, L.Q. Ge, Z.Z. Gu, Fabrication of super-hydrophobic film with dual-size roughness by silica sphere assembly. Thin Solid Films **515**, 4686–4690 (2007)
82. K. Teshima, H. Sugimura, Y. Inoue, O. Takai, A. Takano, Transparent ultra water-repellent poly (ethylene terephthalate) substrates fabricated by oxygen plasma treatment and subsequent hydrophobic coating. Appl. Surf. Sci. **244**, 619–622 (2005)
83. J.M. Tibbitt, M. Shen, A.T. Bell, A comparison of r.f. sputtered and plasma polymerized thin films of tetrafluoroethylene. Thin Solid Films **29**, L43–L45 (1975)
84. E.J.W. Verwey, J.Th.G. Overbeek, *Theory of the Stability of Lyophobic Colloids* (Elsevier, New York, 1948)
85. N. Vourdas, A. Tserepi, E. Gogolides, Nanotextured super-hydrophobic transparent poly(methyl methacrylate) surfaces using high-density plasma processing. Nanotechnology **18**, 1–7 (2007)
86. T. Wagner, C. Neinhuis, W. Barthlott, Wettability and contaminability of insect wings as a function of their surface sculptures. Acta Zool. **77**, 213–225 (1996)
87. D.E. Weibel, A.F. Michels, A.F. Feil, L. Amaral, S.R. Teixeira, F. Horowitz, Adjustable hydrophobicity of al substrates by chemical surface functionalization of nano/microstructures. J. Phys. Chem. C **114**, 13219–13225 (2010)
88. W. Wu, Q. Zhu, F. Qing, C.C. Han, Water repellency on a fluorine-containing polyurethane surface: toward understanding the surface self-cleaning effect. Langmuir **25**, 17–20 (2009)
89. B. Xu, Z. Cai, Fabrication of a superhydrophobic ZnO nanorod array film on cotton fabrics via a wet chemical route and hydrophobic modification. Appl. Surf. Sci. **254**, 5899–5904 (2008)
90. C.H. Xue, S.T. Jia, J. Zhang, L.Q. Tian, Superhydrophobic surfaces on cotton textiles by complex coating of silica nanoparticles and hydrophobization. Thin Solid Films **517**, 4593–4598 (2009)

91. L. Yan, K. Wang, J. Wu, L. Ye, Hydrophobicity of model surfaces with closely packed nano- and micro-spheres. Colloids Surf. A **296**, 123–131 (2007)
92. J. Yang, P. Pi, X. Wen, D. Zheng, M. Xu, J. Cheng, Z. Yang, A novel method to fabricate superhydrophobic surfaces based on well-defined mulberry-like particles and self-assembly of polydimethylsiloxane. Appl. Surf. Sci. **255**, 3507–3512 (2009)
93. J. Yang, Z. Zhang, X. Men, X. Xu, Fabrication of stable, transparent and superhydrophobic nanocomposite films with polystyrene functionalized carbon nanotubes. Appl. Surf. Sci. **255**, 9244–9247 (2009)
94. Y.E. Yoo, T.H. Kim, D.S. Choi, S.M. Hyun, H.J. Lee, K.H. Lee, S.K. Kim, B.H. Kim, Y.H. Seo, H.G. Lee, J.S. Lee, Injection molding of a nanostructured plate and measurement of its surface properties. Curr. Appl Phys. **9**, e12–e18 (2009)
95. H. Yoon, J.H. Park, G.H. Kim, A superhydrophobic surface fabricated by an electrostatic process. Macromol. Rapid Commun. **31**, 1435–1439 (2010)
96. Z. Yuan, H. Chen, J. Tang, H. Gong, Y. Liu, Z. Wang, P. Shi, J. Zhang, X. Chen, A novel preparation of polystyrene film with a superhydrophobic surface using a template method. J. Phys. D Appl. Phys. **40**, 3485–3489 (2007)
97. Z. Yuan, H. Chen, J. Tang, X. Chen, D. Zhao, Z. Wang, Facile method to fabricate stable superhydrophobic polystyrene surface by adding ethanol. Surf. Coat. Technol. **201**, 7138–7142 (2007)
98. Z. Yuan, H. Chen, J. Zhang, D. Zhao, Y. Liu, X. Zhou, S. Li, P. Shi, J. Tang, X. Chen, Preparation and characterization of self-cleaning stable superhydrophobic linear low-density polyethylene. Sci. Technol. Adv. Mater. **9**, 1–5 (2008)
99. Z. Yuan, H. Chen, J. Tang, D. Zhao, A stable porous superhydrophobic high-density polyethylene surface prepared by adding ethanol in humid atmosphere. J. Appl. Polym. Sci. **113**, 1626–1632 (2009)
100. J. Zhang, W. Huang, Y. Han, A composite polymer film with both superhydrophobicity and superoleophilicity. Macromol. Rapid Commun. **27**, 804–808 (2006)
101. L. Zhang, H. Chen, J. Sun, J. Shen, Layer-by-layer deposition of poly(diallyldimethylammonium chloride) and sodium silicate multilayers on silica-sphere-coated substrate; facile method to prepare a superhydrophobic surface. Chem. Mater. **19**, 948–953 (2007)
102. X.W. Zhang, Hydroentangling: a novel approach to high-speed fabrication of carbon nanotube membranes. Adv. Mater. **20**, 4140–4144 (2008)
103. Y. Zhang, H. Wang, B. Yan, Y. Zhang, P. Yin, G. Shen, R. Yu, A rapid and efficient strategy for creating super-hydrophobic coatings on various material substrates. J. Mater. Chem. **18**, 4442–4449 (2008)
104. N. Zhao, L. Weng, X. Zhang, Q. Xie, X. Zhang, J. Xu, Lotus-leaf-like superhydrophobic surface prepared by solvent-induced crystallization. Chem. Phys. Chem. **7**, 824–827 (2006)
105. Z. Zheng, Z. Gu, R. Huo, Y. Ye, Superhydrophobicity of polyvinylidene fluoride membrane fabricated by chemical vapor deposition from solution. Appl. Surf. Sci. **255**, 7263–7267 (2009)
106. Z. Zheng, Z. Gu, R. Huo, Z. Luo, Superhydrophobic poly(vinylidene fluoride) film fabricated by alkali treatment enhancing chemical bath deposition. Appl. Surf. Sci. **256**, 2061–2065 (2010)

Part III
Green and Sustainable Lubricants and Materials

Chapter 9
A Biomimetic Approach to Ice Friction

Anne-Marie Kietzig

Abstract This chapter addresses ice friction from a biomimicry perspective. Ice and its liquid form water are integral parts of the natural life cycle and therewith stand at the center of our existence, which makes them interesting targets for biomimetic engineering. A historic overview of friction and ice friction introduces the matter. The relevant tribological background is established and ice as a matter is discussed. On this basis the different parameters that influence ice friction are discussed. Biomimetic approaches to adapt the influence of material-related parameters to the desired amount are outlined. Furthermore, experimental methods to measure ice friction are addressed and finally the most important ice friction models are briefly introduced.

Keywords Biomimicry · Ice friction · Winter sport · Liquid-like layer · Temperature · Velocity · Relative humidity · Normal force · Contact area · Thermal conductivity roughness · Surface structure · Wetting

9.1 Introduction

Biomimicry means applying concepts found in nature to our everyday problems and thereby designing advanced products and processes, while saving energy in production and application. Lately the "green" movement has gained impact in society, politics, and economy. "Green" in this context includes concepts from

A.-M. Kietzig (✉)
Department of Chemical Engineering,
McGill University, 3610 University Street,
Montreal, QC H3A 2B2, Canada
e-mail: anne.kietzig@mcgill.ca

M. Nosonovsky and B. Bhushan (eds.), *Green Tribology*,
Green Energy and Technology, DOI: 10.1007/978-3-642-23681-5_9,

ecological sustainability, environmental consciousness, and conservation of nature in order to change our current state of living FROM nature to living in-sync WITH nature. To conform to this requirement biomimicry gives us the key to simultaneously advance our current state of living, technological progress and prosperity, while reducing our ecological footprint. Accordingly, biomimicry is particularly relevant to engineering.

The science of *tribology*, which studies the interaction of surfaces in relative motion, is highly relevant to all fields of engineering. Surfaces that are subject to relative motion are ubiquitous. Examples range from microswitches with relevance for electrical engineering to bridge dilatation compensators, which are integral construction units for civil engineers. *Ice friction*, as a sub-discipline of tribology, is a particularly interesting candidate for biomimetic engineering, since ice is abundantly found in nature. As a natural material, ice and its liquid form water are integral components of the closed loop, which stands at the center of our existence.

But how does ice friction affect us? Many people associate ice friction with slippery roads, difficult driving conditions, a greater risk of injury and getting into road accidents. From this perspective the slipperiness of ice is feared and treaded tires and shoe soles with good traction are preferred [1–3]. On the other hand, sports and outdoor activity enthusiasts highly appreciate the low friction experienced when moving over ice and/or snow, since the slipperiness of the ground is what makes so many winter sports as thrilling and enjoyable as they are [4].

Beyond these first associations, the study of ice friction is also relevant for the design and maintenance of moored floating offshore structures [5], ice breakers and cargo ships for transport through northern sea way [6], as well as sleds and vehicles for transport over snow and ice [7]. In the latter examples low friction is desired to save energy, e.g. 70% of the power of an ice breaker ship is consumed to overcome ice friction [8].

Another important application of ice friction investigations are refrigeration and heat exchanger systems. Ice slurries for thermal energy storage are environmentally more sustainable than traditional ozone depleting refrigerants. However, the flow behavior of such slurries still requires further investigations and reductions in fluid friction and pressure drop are sought [9, 10].

Furthermore, natural phenomena such as glacial movements or river ice jams are essentially ice friction processes. River ice jams can cause disruptions of hydro power generation, transportation and navigation. Waves from a glacier break-off or jam release can result in flooding, which causes major damage to property and wildlife habitat [11]. Finally, principles of ice friction also play a role in icing and its prevention on power lines, air craft, and machinery used at subzero temperatures [12].

As outlined above there are different requirements for surfaces in contact with ice depending on whether high or low friction is desired. High friction requires high adhesion between surfaces and interlocking asperities, whereas low friction results from very low adhesion forces and only marginal solid–solid contact.

The basic physical concepts underlying both directions are however the same, as will be outlined in this chapter.

Nature gives plenty of striking examples for sophisticated solutions to such typical tribological problems, which might be less energy and material consuming and more sustainable than the solutions developed by the homo sapiens over the past centuries. In the following, three intriguing examples from nature with relevance to ice friction are given [13]:

1. The tree frog's toes are composed of disk-like pads that are separated by channel-like spaces. Through these channels water gets squeezed out of the contact surfaces to reduce lubrication and enhance van der Waals forces between the contacting solid surfaces.
2. The surface of the feet of the gecko is covered with small flexible hairs that conform to the shape of any surface features for maximum contact.
3. Considering the adhesion of fluids to solids the lotus leaf stands out with its high hydrophobicity supported by a surface structure on two length scales, which allows water drops to simply roll of its surface.

All these examples are readily found outside our offices in nature along with many more clever approaches to tribological problems. The engineer just needs to open his eyes and mind, get inspired, and transfer the ideas to solve the respective adhesion- and friction-related engineering design questions.

In the following the basic understanding of ice friction will be established and wherever possible, examples for biomimetic solutions are highlighted.

9.2 Historic Overview

It has only been about 150 years since friction on ice has become of scientific interest. The study of friction, however, has a long history as shown in Fig. 9.1.

- Neanderthal Age (~200,000 B.C.): use of frictional heat to make fire by rubbing wood on wood or striking flint stones.
- ~7,000 B.C.: Scandinavian rock carving depicts the use of a sledge to transport heavy goods over frozen ground.
- ~2,400 B.C.: Egyptian carving shows facilitated sliding by the use of a lubricant, possibly water [14, 15].
- ~350 B.C.: Aristoteles (384–322 B.C.) recognizes the force of friction in his *Questiones Mechanicae* [14].
- ~1500: Leonardo da Vinci (1452–1519) studies friction quantitatively. He investigated the influence of the apparent area of contact upon frictional resistance, distinguished between rolling and sliding friction, studied the benefits of lubricants, and made the first observations of wear. In fact, he postulated the two laws of friction:

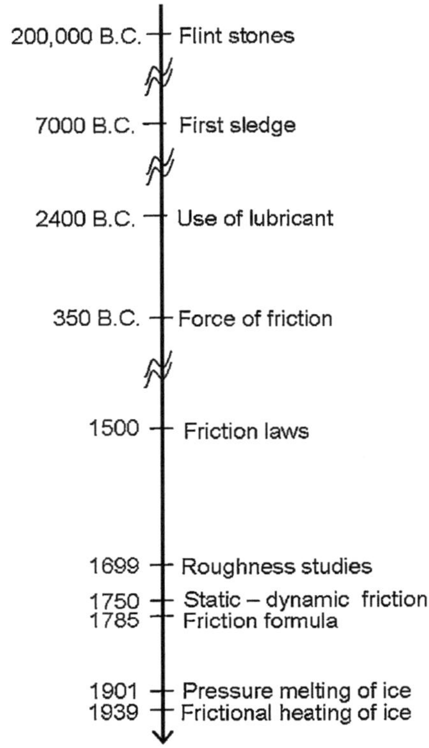

Fig. 9.1 Timeline of friction studies. (Reprinted with permission from [77]. Copyright 2010, American Institute of Physics)

- *Friction produces double the amount of effort if the weight be doubled*—The force of friction is directly proportional to the applied load.
- *The friction made by the same weight will be of equal resistance at the beginning of its movement although the contact may be of different breadth and length.*—The force of friction is independent of the apparent area of contact for a given load.

  However, since da Vinci's records were only published at the end of the nineteenth century, these laws are often referred to as Amonton's laws [14, 15].

- 1699: Guillaume Amonton (1663–1705) confirms da Vinci's findings. He identifies roughness as the primary cause of friction and defines friction as the force required to lift interlocking asperities over each other during the sliding motion [14, 15].
- 1750: Leonard Euler (1707–1783) classifies friction into static and dynamic friction and introduces the symbol μ for the coefficient of friction [14].
- 1785: Charles Augustin Coulomb (1736–1806) investigates friction as a function of various factors, such as the nature of materials in contact and surface coatings, the extent of the surface area, the normal pressure, the time of loading, and the

frictional behavior under vacuum as well as under varying ambient conditions (temperature, humidity). He also formulated friction force as an equation:

$$F_T = \mu F_N \tag{9.1}$$

where F_T is the frictional force, F_N the normal force, and μ the coefficient of friction, which was assumed to be independent of the sliding velocity (v). Sometimes the above formula is referred to as the third law of friction. However, this relationship only holds for moderate sliding velocities [14].

- 1859: Ice became a matter of scientific investigations: Faraday brought two ice cubes into contact, which instantly froze together [16]. He concluded that the ice surface is covered with a liquid-like layer. This famous experiment started research efforts to understand ice, and the role its surface plays in ice friction. Thomson explained Faraday's observations by attributing the existence of the liquid-like layer to pressure melting [17].
- 1900: Reynolds was the first to investigate systematically the matter of sliding on ice [18]. Following Thomson's ideas he wrongly attributed the low friction on ice to pressure melting; this explanation for the ease of skating was widely accepted among scientists for almost 40 years.
- 1939: Bowden and Hughes [19] find frictional heating to be the main contributor to the low friction coefficient on ice; this is today the generally accepted theory to explain ice friction.

9.3 Relevant Friction Regimes

In the following, the basic physical concepts of dry, boundary, mixed, and hydrodynamic friction are introduced with respect to the thickness of the lubricating liquid-like layer on ice. The latter greatly influences the amount of friction on ice.

9.3.1 Dry Friction

Dry friction describes the sliding contact of two surfaces in the absence of any kind of lubricating layer. Any real solid surface shows a distinct profile of surface asperities and valleys, no matter for how long it has been polished. When two asperities of different surfaces come into contact, adhesive bonds of chemical or physical nature are formed between these mating asperities. If the surfaces are moved relative to one another, the adhesive bonds are sheared. The force necessary to break the adhesive bonds between contacting asperities is the tangential friction force F_T given by:

$$F_T = \tau_c A_c \tag{9.2}$$

where τ_c is the shear strength, necessary to shear the asperity contact, and A_c is the area of real contact between the asperities of the mating surfaces.

Bowden [20] proposed that the real area of contact (A_c) between two surfaces is directly proportional to the applied load (F_N) and the softer material's hardness (H).

$$A_c = \frac{F_N}{H} \tag{9.3}$$

Hence, the friction coefficient can be written as:

$$\mu = \frac{\tau_c}{H} \tag{9.4}$$

Thus, dry friction is characterized by the work necessary to break solid surface adhesive bonds. It depends on the applied load and the hardness of the surfaces but is independent of the sliding speed [21, 22].

Real dry friction on ice under atmospheric conditions cannot exist. Even at very low temperatures a very thin liquid-like film lubricates the sliding interface, as will be further explained in a later section. This film has a thickness of a few molecular layers [23].

9.3.2 Boundary Friction

Boundary lubrication is characterized by a lubricating layer with the thickness of only a few molecular layers between the sliding surfaces [21]. Boundary lubrication on ice is characterized by the temperature (T) in the contact zone being everywhere below the melting temperature (T_m), and the thickness of the lubricating liquid-like layer (h) being far smaller than surface roughness (R) [24].

$$everywhere\,in\,contact\,zone: T < T_m, \quad h \ll R$$

The lubricating liquid-like layer reduces solid–solid contact between the surfaces. In total, the friction coefficient of boundary lubrication is typically lower than that of dry friction [21].

9.3.3 Mixed Friction

Mixed friction occurs when the surface temperature rises above the melting temperature (T_m) of ice at some points within the contact zone and the thickness of the liquid-like layer (h) is still less than the characteristic roughness of the surfaces (R) [24].

$$at\,some\,points\,in\,contact\,zone: T > T_m, \quad h < R$$

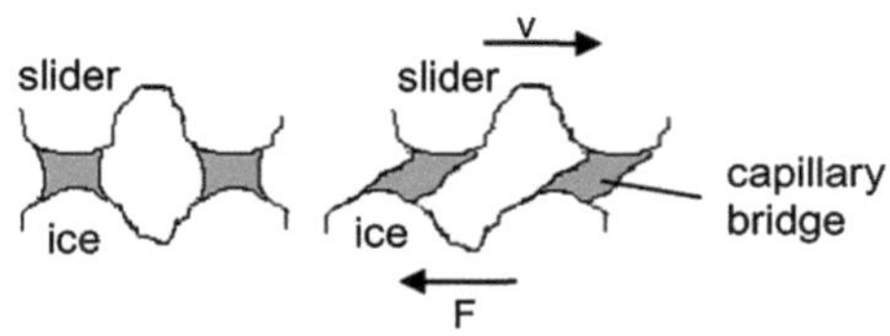

Fig. 9.2 Capillary bridges between asperities of contacting surfaces during sliding. (Reprinted with permission from [77]. Copyright 2010, American Institute of Physics)

In this regime the load of the slider is partly supported by the surface asperities and partly by the lubricating layer. It is obvious that the increased thickness of the lubricating layer reduces solid–solid adhesion and enhances the lubrication.

As a result, the decrease in friction force compared to boundary lubrication can be demonstrated in:

$$F_T = A_c \times \left[\alpha\tau_c + (1-\alpha)\eta\frac{\upsilon}{h}\right] \tag{9.5}$$

where α is the fraction of unlubricated area, τ_c the shear strength of the solid contact, υ the velocity of the slider, η the viscosity, and h the thickness of the lubricating layer [21].

However, at the same time a wetting lubricant enforces the build-up of capillary water bridges between the asperities, as illustrated in Fig. 9.2.

Capillary bridges act as bonds between the slider and ice surfaces and exercise a drag force on the slider [24]. It is important to note that they do not support the applied load. These capillary bridges act like liquid bonds and result in additional frictional resistance. Hence, capillary bridges should be taken into account to complement Eq. 9.5. The problem herein is, however, that no physical or experimental model exists to describe the contribution of capillary bridges to the friction force.

9.3.4 Hydrodynamic Friction

If everywhere in the contact zone the temperature is above the melting temperature (T_m), and the thickness of the lubricating layer between the two surfaces is greater than the height of the asperities, friction is called hydrodynamic [24].

$$\textit{everywhere in contact zone}: T > T_m, \quad h > R$$

In this friction regime the lubricating layer, and not the surface asperities, carries the applied load. If the load is very high, a part of the lubricating layer might be squeezed out between the surfaces. However, for hydrodynamic friction, it is assumed that the thickness of the lubricating layer remains greater than the height of the asperities. Following the area of real contact is identical to the surface area (A) of the slider [21]. No solid–solid contact occurs during the sliding movement.

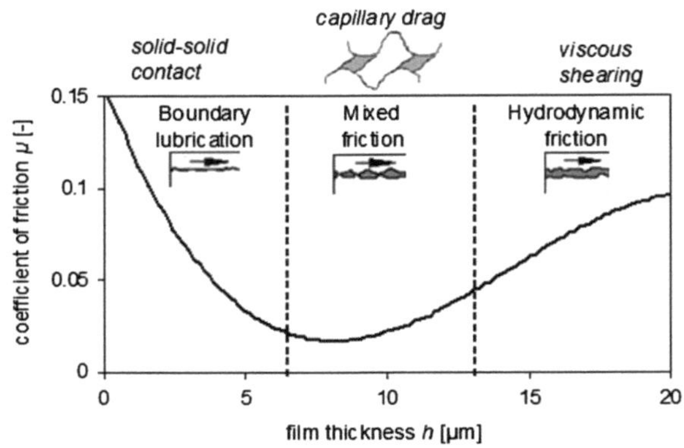

Fig. 9.3 Friction regimes relevant to ice friction depending on the thickness of lubricating layer (adapted from Bhushan [21] and Colbeck [26]). (Reprinted with permission from [77]. Copyright 2010, American Institute of Physics)

Consequently, shearing of solid–solid adhesive bonds no longer contributes to the friction force. The frictional force can be described as

$$F_T = \tau_l A \tag{9.6}$$

where τ_l is the shear strength of the lubricating liquid-like layer or an "effective" shear stress developed from shearing of the liquid-like layer. This can be simply expressed as

$$\tau_l = \eta \frac{v}{h} \tag{9.7}$$

As in the case for mixed friction capillary drag forces should be included in the case of water lubrication.

Fowler and Bejan [25] pointed out that the lubricating film under a slider on ice becomes thicker towards the trailing end. Consequently, friction mechanisms on ice can include all types of friction except for pure dry friction.

Figure 9.3 summarizes qualitatively the various regimes of ice friction. Note the drop of the coefficient of friction with the film thickness in the boundary friction regime due to reduced solid–solid contact. On the other hand, the coefficient of friction increases with film thickness as it becomes fully hydrodynamic, as discussed above. Accordingly, there is an optimal film thickness associated with minimum friction for each slider system. It is also noted that there is a smooth transition between the different regimes indicated by the dashed lines.

9.4 The Peculiar Surface of Ice

Today we know that ice is as slippery as it is because of the liquid-like layer, which is present on the ice surface even at ambient temperatures well below 0°C. There are three different mechanisms that contribute to the thickness of the liquid-like layer. These are surface melting, pressure melting, and frictional heating.

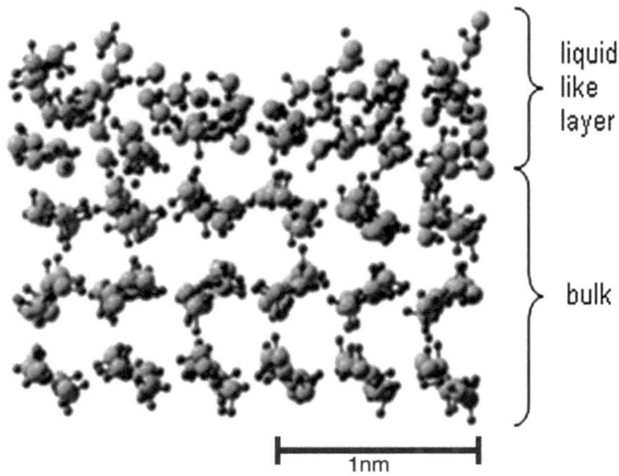

Fig. 9.4 Surface structure of ice. (Reprinted with permission from [77]. Copyright 2010, American Institute of Physics)

9.4.1 Surface Melting

Faraday [16] already suggested the existence of a liquid-like layer as an inherent part of the ice surface. The existence of this layer does not require frictional contact with another body. As it is typical for any surface, the regular bulk structure breaks down at the periphery. In the case of ice, the regular hexagonal crystal structure gives way to a more random amorphous allocation of the water molecules (Fig. 9.4).

Experimental proof for the existence of the liquid-like layer has been given with different experimental techniques, such as those presented in Fig. 9.5.

T_m denotes the melting temperature, while T stands for the temperature, at which the measurement was taken with the respective technique. As seen in Fig. 9.5 the temperature range, in which the liquid-like layer was observed in different experiments, shows a wide variability depending on the technique applied. Therefore, it is not surprising that scientists came to different conclusions regarding the general nature of the liquid-like layer and more precisely the onset temperature of its formation. Different theories attempt to shed light onto the underlying physics. However, in spite of the vast experimental evidence, the scientific explanation for the presence of the liquid-like layer is still under debate. For a more detailed review the reader is referred to Petrenko and Whitworth [23]. Here, some of the prevailing theories are summarized briefly.

Theory 1: Electrostatic Interactions
Fletcher [27, 28] has attributed the formation of the liquid-like layer to dipole/quadrupole interactions. Water molecules at the surface orient themselves in a way that their protons point outwards as a result of a favorable low energy configuration. Consequently, the hexagonal ice structure breaks down, so that a liquid-like layer covers the ice surface. The structured surface layer has to be screened by a second layer with the opposite charge to account for the uncharged body of bulk ice.

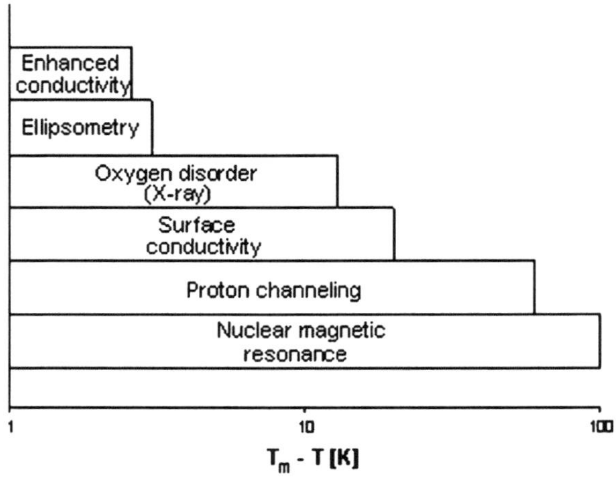

Fig. 9.5 Experimental techniques to investigate the liquid-like layer on ice (adapted from Petrenko and Whitworth [23]). (Reprinted with permission from [77]. Copyright 2010, American Institute of Physics)

Fletcher [28] has concluded that the orientation of the surface molecules produces an electric field at the surface, which is responsible for the high electric conductivity of the ice surface. However, these assumptions concerning the driving force of the surface orientation do not hold. The liquid-like layer formation cannot be attributed to a lower energy configuration, which then results in a charged double layer; but the structural breakdown is rather a result of the charged layers [23].

Theory 2: Free Surface Energy Minimization
Lacmann and Stanski [29] have attributed the existence of the liquid-like layer on nature's tendency to minimize the energy of a system. This theory of free surface energy minimization has further been advanced by Dash et al. [30]. The presence of a liquid-like layer on the solid–vapor interface is explained with the minimization of the free surface energy at the solid's surface. They have claimed that a system, whose free surface energy of a solid–vapor interface γ_{sv} is higher than the sum of the energies of the solid–liquid γ_{sl} and the liquid–vapor γ_{lv} interface, will "lower its free energy by converting a layer of the solid to liquid".

$$\gamma_{sv} > \gamma_{sl} + \gamma_{lv} \tag{9.8}$$

The thickness of the liquid-like layer would be defined by the energetic minimum of the whole system, which takes into account the reduced surface energy by wetting and the consumed energy for melting the solid. While this theory holds for many solids the problem with ice is, however, to determine the surface free energy γ_{sv} of 'dry' ice. Experimental evidence shows that the free surface energy of 'dry' ice is actually not larger than the combined surface energies of the wetted system [31–33].

Theory 3: Subsurface Pressure Melting
Fukuta [34] has suggested subsurface pressure melting as a cause of the liquid-like layer on ice. Since water molecules at the ice surface only have neighboring water

molecules from one side, they experience an inward pull; this is expressed by the surface free energy. This pull exerts a pressure on the layers below. Makkonen [33] has shown that this pressure is large enough to reduce the melting point of the ice surface layer by about 13 K. This theory is in good agreement with thermodynamics and experimental results. It should be noted that even though subsurface pressure melting might contribute to the thickness of the liquid-like layer, it fails to explain its presence at lower temperatures.

Theory 4: Molecular Dynamics
Low-energy electron diffraction (LEED) experiments coupled with molecular dynamics simulation have given further insight into the forces leading to a liquid-like layer on ice. Surface molecules tend to vibrate and rotate constantly in order to minimize dangling bonds. Kroes [35] has pointed out that through the movement of the molecules the outer ice layers become partially charged. With computer simulations and spectroscopy experiments Devlin and Buch [36] have confirmed that the breakdown of the solid structure is indeed energy minimizing, since thereby the number of dangling bonds can be reduced. Furthermore, they have found that the surface is dynamically disordered. This has been further explained by the results of Furukawa and Nada [37], which have indicated that molecules diffuse within the two outermost layers, which leads to this highly disordered surface. Finally, Marterer et al. [38] have revealed that at a temperature as low as 90 K molecular vibration is so high that the outermost atoms can no longer be detected by LEED.

Overall, the above discussion indicates that the nature of the ice surface is still not fully understood and remains a vibrant field of research.

9.4.2 Pressure Melting

Pressure melting has been considered to be the explanation of choice for the low friction coefficient on ice for many years. While it contributes to the formation of a lubricating layer close to the melting point, pressure melting does not explain the low friction on ice at lower temperatures. The pressure p exerted on the ice can be calculated by $p = F_N/A_c$. A major problem in friction studies in general is the calculation of the exact area of contact A_c. Since surfaces are never perfectly smooth, contact takes actually place between a certain number of asperities. With ice being a comparatively soft material the contact area of a slider on ice depends on the applied load. Furthermore, it also depends on the ambient temperature, because the ice softens with increasing temperatures reaching the melting point.

Consider the example of an ice-skater, who weighs about 70 kg ($F_N = 700$ N) and whose skate is in contact with the ice over an area of 1×10^{-4} m^2 (assuming a contact length of the skate with the ice of 0.1 m and width of 10^{-3} m). The pressure exerted by the skate blade on the ice amounts to 7 MPa. The phase diagram of water gives a dp/dT value of approximately −9.9 MPa/K at 0°C and 0.1 MPa.

The pressure applied by the skater of $\Delta p = 7$ MPa will only result in a change of the melting temperature of $\Delta T = -0.7$ K ($\Delta T = \Delta p/9.9$ Pa/K). Therefore, pressure melting cannot contribute much to a liquid layer on ice, as the melting temperature of ice can only be lowered by about 1 K with the herein assumed contact area. Colbeck [39] illustrated that only 0.005% of an ice skate is in actual contact with the ice for pressure melting and frictional heating to contribute equally to the heat production. The resulting very high contact pressure, however, would result in high wear rates and great losses in the film thickness by squeezing out of the lubricating layer, so that the friction coefficient would be higher than observed in experiments. Overall, these sample calculations show clearly that pressure melting, while being a valid concept, does not contribute to the high slipperiness of ice and snow as observed especially on icy roads and ice sports.

9.4.3 Frictional Heating

Bowden and Hughes [19] published on a series of ice and snow friction experiments and concluded that frictional heating plays a fundamental role in the low friction coefficient on ice and snow. Heat generated by the frictional motion raises the temperature at the contacting points to the melting temperature of ice. Hence, the ice surface melts locally at the contacting asperities, whereby a non-continuous melt water film is formed. This film contributes to the lubrication of the slider on ice. However, energy from frictional heating is also consumed through material deformation and energy losses by heat conduction into the ice and the slider; therefore, only the remainder, which can be found from thermodynamic calculations, is available for melting the ice. Furthermore, the authors point out that this lubricating layer can become continuous, if the ambient temperature is close to the melting point of ice. They also note that the observed friction at 0°C increases if a small quantity of water is added to the ice; a phenomenon well known to winter sport enthusiasts, who ski in spring conditions and have experienced the sudden breaking effect from sun exposed snow patches. The authors, however, have not drawn any further conclusions regarding the drag effect of capillary bridges in this study and as discussed previously their contribution to the coefficient of friction can be significant. In conclusion, it can be said that frictional heating is the most important contributor to the low friction on ice.

9.5 The Influence of Different Parameters on Ice Friction

Before addressing the key parameters that influence the friction and adhesion between a slider and ice, a more general example will be given to illustrate the complex interdependence of many factors, which finally defines the overall frictional resistance experienced by a body moving over ice. Most of the published

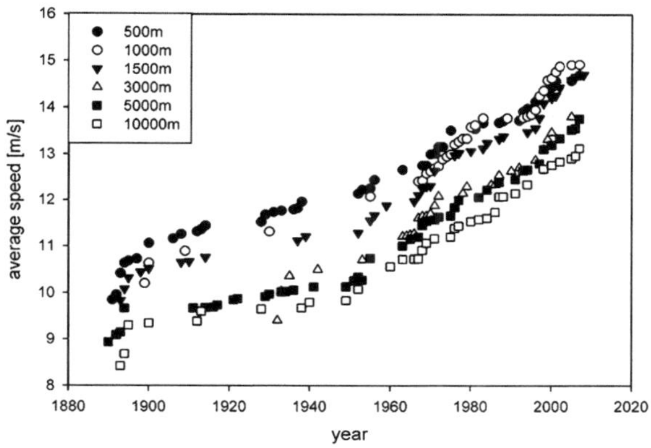

Fig. 9.6 Average speed of long track speed skating world records (men) (data from ISU [40]). (Reprinted with permission from [77]. Copyright 2010, American Institute of Physics)

data on ice friction refers to applications in winter sports. Therefore, the progress in the understanding of ice friction can be nicely illustrated with a historic overview of the different significant developments in the fields of speed skating and the race to support faster speeds and ever-more interesting athletic competitions [40]. The development of world records in long track speed-skating over time is illustrated in Fig. 9.6.

Men started racing at the end of the nineteenth century, while the first official race for women was only held in 1929. The initial sharp increase in average speed can be attributed to the process of setting a standard after the first few competitions. The second steeper increase can be observed after World War II when post war prosperity started to have an impact on athletes' performance. Until 1958 competitions were carried out on lakes or canals with natural ice. Accordingly, the ice quality depended largely on the weather. With the invention of the first ice resurfacing machine by Frank Zamboni in 1949 and the introduction of the first artificial 400 m ice rink in 1958, the athletes no longer had to cope with bumpy ice surfaces (the influence of surface roughness will be explained in detail later). The further increase in average speed in the 1970s to mid 1980s is attributed to the professionalization of the sport. Athletes started training and competing in teams with coaches and an organized support structure. Furthermore, a better understanding of the biomechanic and aerodynamic factors influencing athlete performance led to changes in the skating technique and the introduction of the first skin suit at the 1976 Olympic Games. During speed-skating the contribution of air drag is about 80%, while ice friction only contributes to about 20% to the overall frictional resistance [41]. Tightly fitting suits with certain nature-imitating patterns proved successful in suppressing air friction. In 1988, a zigzag pattern was introduced, followed by a smooth/rough texture biomimicking the shark skin in the 1999/2000 season. Further significant events, which changed the sport forever, were the introduction of indoor ice rinks. The first indoor rink dates back to 1985 in East Berlin, which apart from better sheltering for the spectators, also allowed athletes to train more on the ice and to be less dependent on weather conditions.

The introduction of the biomechanically favorable Klapskate during the 1996/1997 season led to the latest push in world records. The race for the faster speed remains unbroken; it is to be seen what the next innovations will bring along.

Ice friction experiments carried out by different researchers shed light on the influence that different parameters have on the friction coefficient. Following, results from various sources are compared to explain the influence single process-inherent and material-related parameters have on ice friction. In some cases such comparisons were proven difficult as experimental findings were performed under different operating conditions.

9.5.1 Process-Inherent Parameters

As process-inherent parameters we consider those influence factors, which are defined by ambient conditions (temperature, humidity) or the nature of the process (velocity, normal force, area of contact) and which, therefore, often cannot be influenced in real-life applications. However, careful experiments have been carried out to gain a better understanding of their individual effects and interactions with other factors.

9.5.1.1 Temperature

From the above discussion on the influence of the lubricating layer thickness on the friction coefficient and our general experience with ice, it is evident that temperature has a strong influence on the friction coefficient. Some researchers merely report a decrease in the friction coefficient with increasing temperature [2, 42–47], which is the case when friction is dominated by boundary friction conditions [42, 48]. However, with Fig. 9.3 in mind one can anticipate that across the whole range of friction regimes the friction coefficient on ice will eventually increase with temperature after passing through a minimum; this has been confirmed by several researchers as shown in Fig. 9.7 [2, 42–52]. The coefficient of friction decreases first with increasing temperature and rises again when the temperature approaches 0°C. A minimum coefficient of friction is seen between −2 and −7°C. Obviously, at lower temperatures the friction is dominated by solid–solid contact, typical for the ice friction curve to the left of the minimum, as illustrated in Fig. 9.3. At temperatures close to the melting point the thickness of the lubricating liquid-like layer becomes large enough to not only facilitate sliding but also add to the resistance through the build-up of capillary bridges, as is typical for the mixed friction regime. However, the location of the friction minimum and therewith the onset of this increase in friction depends largely on the slider material, the normal load, and the linear sliding velocity. Table 9.1 summarizes the differences in the operating parameters between the experiments and therewith provides explanations for the differences between the four sets of data. In the

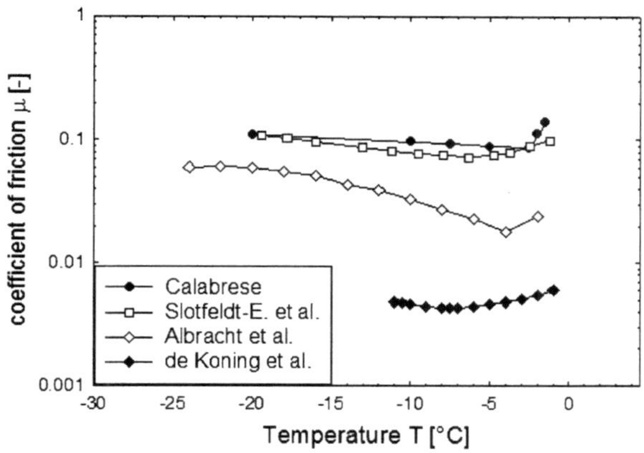

Fig. 9.7 Temperature dependence of the friction coefficient; for experimental parameters refer to Table 9.1. (Reprinted with permission from [77]. Copyright 2010, American Institute of Physics)

Table 9.1 Experimental parameters for the data plotted in Fig. 9.7

Reference	Slider material	F_N (N)	υ (m/s)	A_c (mm^2)
Calabrese [51]	Steel ring (AISI 1018)	889.6	<1	1,235
Slotfeldt-E et al. [50]	HDPE slider block	100	0.3	15,000
Albracht et al. [52]	Cr-steel pin	1	0.13	≈2
de Koning et al. [41]	Steel skate blade ("Viking special")	700	8	≈400

following sections it is discussed how sliding velocity, normal force, area of contact, and slider material influence the coefficient of friction.

9.5.1.2 Sliding Velocity

With increasing sliding velocity the coefficient of friction decreases. This was first reported in 1953 by Bowden [53]. Evans et al. [44], who were the pioneers in modeling ice friction mathematically, confirmed these findings both experimentally and theoretically. Many other researchers found the same dependency using different experimental setups and materials [46, 48, 51, 54–56]. Figure 9.8 illustrates some of these results.

The higher the sliding velocity the more the frictional heat produced. Accordingly, more heat is available to melt the ice surface, which results in a thicker lubricating melt-water layer that facilitates the sliding motion; this is the case in the boundary friction regime and also the mixed friction regime before drag forces outweigh the benefits from a thicker lubricating layer (compared to Fig. 9.3). Overall, the five different sets of data shown here all agree with the predictions of theoretical models, which describe the velocity dependence in the boundary regime with $\mu \propto 1/\sqrt{\upsilon}$. Evans et al. [44] (for high velocities), [46, 57]

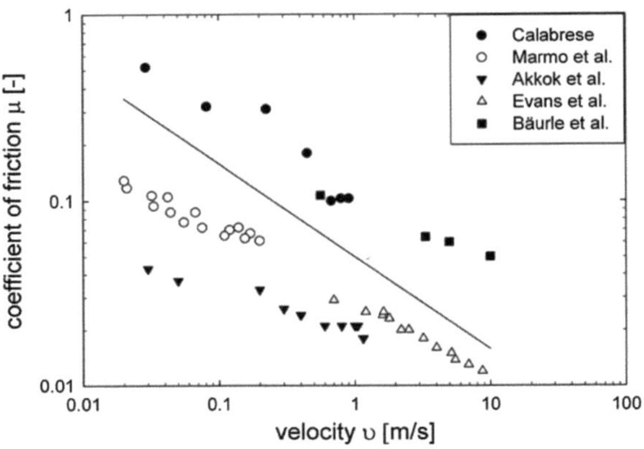

Fig. 9.8 Velocity dependence of the friction coefficient (enhanced lubrication); for experimental parameters refer to Table 9.2. (Reprinted with permission from [77]. Copyright 2010, American Institute of Physics)

Table 9.2 Experimental parameters for the data sets plotted in Fig. 9.8

Reference	Slider material	F_N (N)	T (°C)	A_c (mm^2)
Evans et al. [44]	Mild steel rod	45.4	−11.5	≈300
Marmo et al. [66]	Ice hemisphere over steel	2.1/2.4	−11.5	≈2
Akkok et al. [46]	Steel cylinder	75	−20	≈50
Calabrese [51]	Steel ring (AISI 1018)	889.6	−18	1,235
Bäurle et al. [48]	PE block	84	−10	≈200

(for friction dominated by thermal conductivity), [46, 58]. Other experimental results, which are not plotted here, show a similarly decreasing trend in the coefficient of friction with increasing velocity. The differences among the various sets can again be attributed to different experimental setups and operating conditions, i.e. slider material, normal force, temperature, and apparent contact area (see Table 9.2 for details).

Once drag forces considerably contribute to the overall friction force in the mixed friction and especially in the hydrodynamic friction regime (to the right of the minimum in Fig. 9.3) the scaling of the friction coefficient with linear velocity changes significantly. With ice on ice experiments at temperatures above −5°C Okasanen and Keinonen [57] showed for the first time that the coefficient of friction increases with velocity. They extended the mathematical model developed by Evans et al. [44] and found that the thickness of the melt water layer and therewith the coefficient of friction is proportional to $\upsilon^{1/2}$ for temperatures close to the melting point. Other researchers have confirmed these findings with different slider materials on ice as illustrated in Fig. 9.9 [48, 49, 52, 59]. The increase in friction with velocity can be explained through the increase in drag forces from

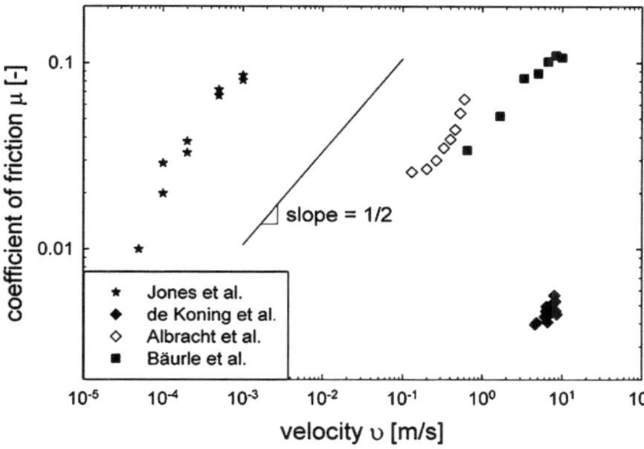

Fig. 9.9 Velocity dependence of the friction coefficient (added drag by melt water); for experimental parameters refer to Table 9.3. (Reprinted with permission from [77]. Copyright 2010, American Institute of Physics)

Table 9.3 Experimental parameters for the data plotted in Fig. 9.9

Reference	Slider material	F_N (N)	T (°C)	A_c (mm^2)
Jones et al. [59]	Formica block	196.2	≈0	15,000
Albracht et al. [52]	High alloy steel pin	1	−7	≈2
Bäurle et al. [48]	PE block	52	≈0	1,000
de Koning et al. [49]	Steel skate blade ("Viking special")	706	−4.6	≈400

shearing the lubricating layer in the hydrodynamic friction regime. The onset of this increase is obviously largely dependent on the size, weight, and material of the construction of the slider, as well as the ambient temperature of the experiment. Again, differences in the experimental setup and operating conditions of the different sets of data are summarized in Table 9.3.

In conclusion, it can be stated that by varying temperature or velocity widely all friction regimes from boundary to hydrodynamic can be identified in ice friction. Figure 9.10 shows the combined effect of temperature and velocity on ice friction. These results were found with a rotational experimental device as detailed elsewhere [60]. The friction coefficient initially decreases sharply with increasing velocity for all temperatures; however, for this particular experimental setup a slight increase in the friction coefficient could be noticed for velocities above 1 m/s, which can be attributed to added drag through capillary bridges, as explained above. For small velocities it can be clearly seen that friction is dominated by different mechanisms with increasing temperature. Furthermore, at around −4°C a minimum in the friction coefficient can be observed. For temperatures below this minimum friction decreases with increasing temperature due to enhanced lubrication and reduced solid–solid contact. Close to the melting

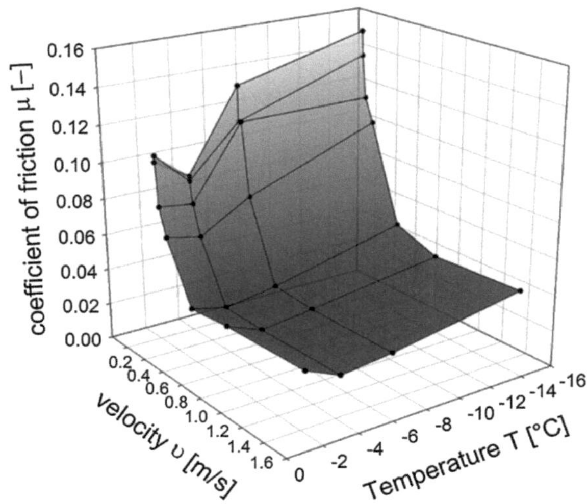

Fig. 9.10 3D plot of temperature and velocity dependence of the friction coefficient (AISI 304L, $F_N = 3$ N, $A = 147$ mm^2) [60]. (Reprinted with permission from [60]. Copyright 2009, American Institute of Physics)

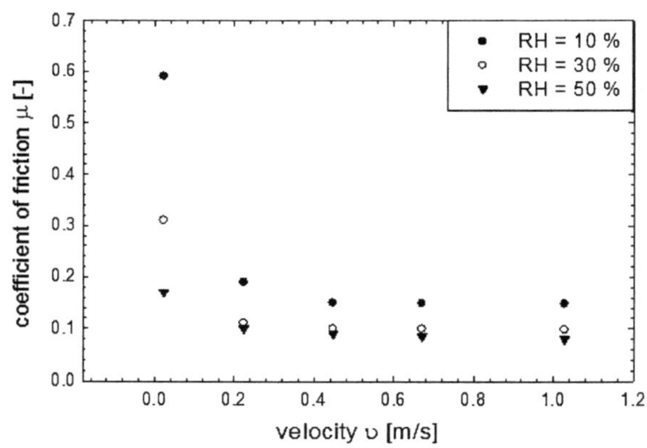

Fig. 9.11 The influence of relative humidity on the friction coefficient with $F_N = 177.9$ N, $T = -29$°C (data from Calabrese [51]). (Reprinted with permission from [77]. Copyright 2010, American Institute of Physics)

point additional resistance from capillary bridges and viscous shearing of the melt film increase friction again as discussed before.

9.5.1.3 Relative Humidity

From the above discussion on the influence of the thickness of the melt-water layer it is clear that relative humidity (RH) must play a significant role in ice friction. However, most likely due to the experimental difficulties of controlling the latter, relative humidity remains a factor that has hardly been investigated with respect to ice friction. In 1980 Calabrese [51] performed friction experiments with steel sliding against ice at −29°C and different relative humidity conditions, as seen in Fig. 9.11.

Relative humidity has a strong influence on the onset of the sliding movement. The higher the humidity the higher the number of capillary bridges, which build up between the surfaces. For the onset of motion this is synonymous with a more lubricated sliding interface and a lower frictional resistance. AFM experiments on silicon substrates have shown [61] that while an increase in relative humidity aids lubrication, it simultaneously also adds to the viscous resistance of the overall friction force, as discussed before and shown in Fig. 9.3 past the friction minimum. Unfortunately, however, no further experimental data exist on the influence of humidity on ice friction at higher temperatures than the one used by Calabrese. One can hypothesize that humidity will have less influence on the coefficient of ice friction at temperatures close to the melting point, where the existing lubricating layer is already considerably thicker than at low temperatures. However, this remains to be seen experimentally.

9.5.1.4 Normal Force

The above discussion illustrated that the applied normal force (F_N) and the apparent area of contact between the slider body and the ice surface are important factors for the resulting coefficient of friction.

It is generally accepted in the literature that the friction coefficient of a slider against ice decreases with increasing normal force at a given temperature and velocity. Bowden and Hughes [19](ice–ice), [62](PE-snow), [46–48, 52]. While Oksanen and Keinonen [57] have shown the same trend with their ice against ice experiments at low temperatures (−15°C) and slow sliding velocities ($\upsilon = 0.5$ m/s), they have found that at temperatures close to the melting point (−1°C), the decrease in the friction coefficient with increasing normal force is less pronounced (Fig. 9.12a and Table 9.4 for experimental conditions). Also, Derjaguin [47], who has carried out experiments with a steel slider, has shown that for increasing loads at temperatures close to the melting point the friction coefficient becomes independent of the normal force. Similarly, Calabrese [51] found a coefficient of friction for a steel slider, which showed complete independence of normal force for applied loads above 400 N. With the investigation of different slider materials, Akkok et al. [46] have shown that the coefficient of friction of a glass slider indeed decreases with normal force, while the corresponding results with a steel slider show no dependence. Likewise, Albracht et al. [52] confirmed the decreasing trend of the coefficient of friction with normal force for a PTFE slider, while the friction coefficients of their aluminum alloy and chrome-nickel-steel sliders show independence of the applied normal force (Fig. 9.12b and Table 9.4 for experimental conditions).

Summarizing, the different findings from different researchers indicate that the friction coefficient of high load sliders made of high surface energy materials is independent of the applied normal force in experiments carried out at higher

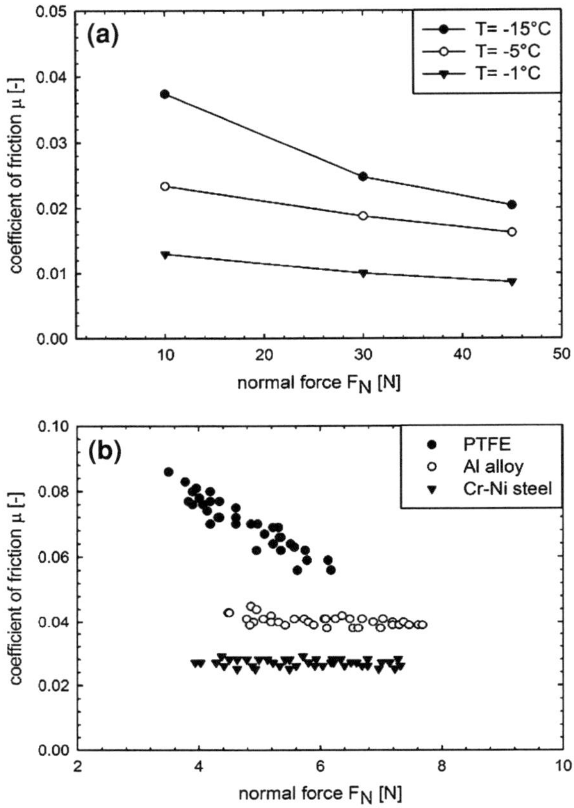

Fig. 9.12 Normal force dependence of the friction coefficient. **a** data from Oksanen and Keinonen [57]. **b** data from Albracht et al. [52]; for experimental parameters refer to Table 9.4. (Reprinted with permission from [77]. Copyright 2010, American Institute of Physics)

Table 9.4 Experimental parameters for the data plotted in Fig. 9.12a and b

Reference	Slider material	υ (m/s)	T (°C)	A_c (mm^2)
Oksanen and Keinonen [57]	Ice on ice	0.5	−15	11,475
			−5	
			−1	
Albracht et al. [52]	PTFE	1	−7	≈2
	Al alloy			
	Cr-Ni steel			

speeds and/or temperatures closer to the ice melting point. These results point towards hydrodynamic friction, where complete wetting of the slider dictates the results.

All parameters, which have been discussed so far, can be classified as process inherent. In the following, a group of parameters are examined, which are more easily controllable and where lessons can be learned from nature to achieve the desired friction on ice.

9.5.1.5 Apparent Area of Contact

The concept of the apparent area of contact (A) is closely related with roughness and surface structure. In the following, we consider the apparent area of contact as a process inherent factor, which is defined by the geometry of the slider. However, it is important to keep in mind that the apparent contact area can be significantly influenced by changing the roughness and structure of a surface. The latter will be discussed in detail in the following section.

The first investigations on the influence of the contact area on the resulting ice friction coefficient go back to the experiments of Bowden and Hughes [19] from 1939. While their experimental results have shown little dependence on the geometric contact area, Bäurle et al. [63] have recently studied the influence of the geometric size of the slider under more controlled conditions and have shown a clear trend. Figure 9.13 illustrates Bäurle's experimental data with exponential curves fitted through the data points. The friction coefficient increases with increasing geometric contact area of the slider. The exponential growth curves show that the coefficient of friction tends to become independent of the applied force with larger contact areas, which explains Bowden and Hughes' observations. However, Bäurle's results for the strong dependence of the friction coefficient for small contact areas disagree considerably with Leonardo da Vinci's second law of friction, which implies that the friction coefficient is independent of the apparent area of contact, and also with Bowden's [20] discussion of the influence of the area of contact on the coefficient of friction. The explanation for Bäurle's results [63] lies in the nature of ice. For a very small geometric contact area the actual contacting asperities are located closer together, so that a larger amount of frictional energy is produced per unit area. This results into a thicker lubricating layer per unit area and an actual contact of close to 100%. In contrast, the contacting points for a larger geometric area of the slider are more spread out. Thus, for the same experimental conditions a slider with a small contact area will already have entered the hydrodynamic friction regime, while another slider with larger contact area will still experience friction phenomena defined by asperity interactions. Figure 9.13 illustrates once more that a slider with larger load shows lower friction, due to a greater amount of contacting points, which contribute to frictional heating and therewith to a thicker water layer. In conclusion, this discussion illustrates the importance of considering the apparent contact area of different slider samples when comparing results of different researchers, as is done here.

Nature gives us many examples for an optimized number of contacting points to achieve the desired amount of friction, as first introduced in the introduction and further elaborated in the following sections with regard to the influenceable parameter's surface roughness and surface structure.

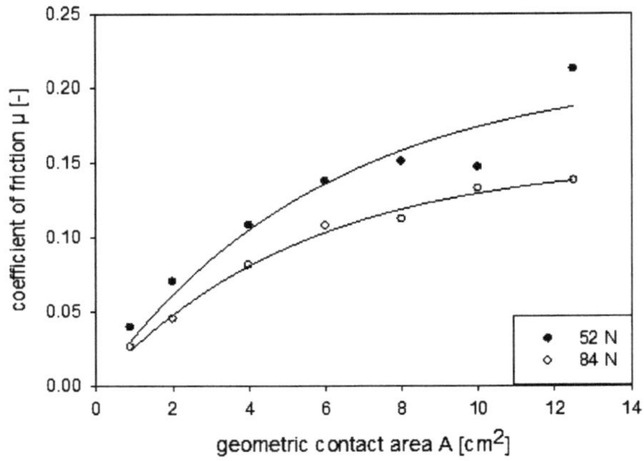

Fig. 9.13 Area and normal force dependence of the friction coefficient with T = −5°C, υ = 3.3–5 m/s (data from Bäurle et al. [63]). (Reprinted with permission from [77]. Copyright 2010, American Institute of Physics)

9.5.2 Material-Related Parameters

In contrast to process-inherent parameters material-related parameters are externally controllable by the choice of material or the preparation method. Their respective effects will be outlined below and practical approaches as well as biomimetic attempts to influence them are shown.

9.5.2.1 Thermal Conductivity

To support the frictional heating theory, Bowden and Hughes [19] investigated the influence of the slider's thermal conductivity (λ) on the friction coefficient on ice. They compared the friction arising from a hollow ski with a copper surface to the one of the same ski construction filled with mercury. The thermal conductivity of mercury (8 W/mK) is about 320 times higher than the one of air (0.025 W/mK). The experiments have shown that the mercury filled ski resulted in higher friction compared to the hollow air-filled ski. Although no further details are given about the exact experimental conditions, this result implies that the friction of a good thermal conductor is higher because less heat is available at the interface to melt the ice. Experiments carried out by Itagaki et al. [43] with steels of different thermal conductivity have supported this relationship between thermal conductivity and the friction coefficient. Albracht et al. [52], however, were not able to find a significant influence of thermal conductivity on ice friction in their experiments with sliders made from different materials.

Generally, as is the case for wettability studies, the isolated effect of thermal conductivity cannot be assessed by experiments with different materials, as the change in material always brings along a change in other material parameters such as wettability and material hardness. Nevertheless, all analytical models on ice friction take account of thermal conductivity as a major factor to describe

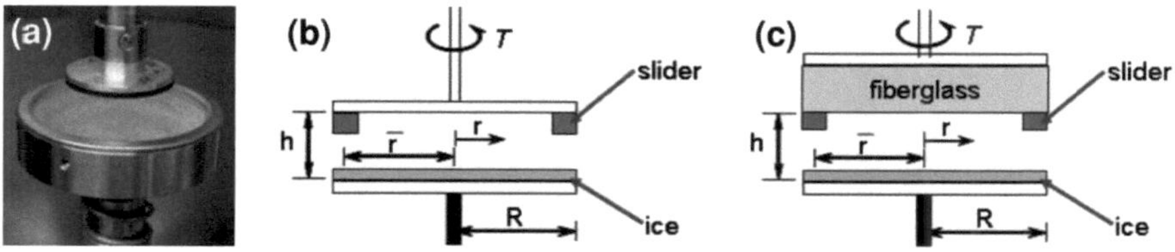

Fig. 9.14 Rheometer **a** and **b** with newly designed friction fixture for ice friction experiments; **c** with fiberglass insulation [64]. (Reprinted with permission from the Journal of Glaciology with permission of the International Glaciological Society.)

the thickness of the melt-water film and therewith the resulting coefficient of friction.

Inspired by Bowden and Hughes's early experiments, we have recently quantified the influence of thermal conductivity on the coefficient of ice friction across the relevant friction regimes without changing but instead thermally insulating the slider material [64]. Ice friction tests have been run across a wide range of temperatures and velocities with and without fiberglass insulation (thermal conductivity of 0.04 W/mK) between the stainless steel slider and the aluminum fixture, as depicted in Fig. 9.14.

Figure 9.15 shows the results in 3-D bar charts for a Stellite 6B slider. Figure 9.15a shows comparable trends of the coefficient of friction with temperature and velocity as discussed in Fig. 9.10 for AISI 304L. Figure 9.15b illustrates the ice friction behavior of the Stellite 6B slider insulated with a fiberglass disk. At low velocities the friction coefficient on ice is considerably reduced by the insulation in comparison to Fig. 9.15a. The fiberglass prevents the frictional heat from being conducted away from the interface through the slider and slider fixture, but instead traps the heat at the interface, which quickly raises the interface temperature to the melting point, and therefore ensures a thicker lubricating layer, which again results in lower friction in the boundary friction regime. Furthermore, comparing the two graphs, it is noticeable that the friction minimum is shifted to lower speeds and that the minimum with the fiberglass insulation is characterized by a higher friction coefficient than is the case for the noninsulated Stellite 6B slider, which can be explained by added capillary drag from the thicker melt-water layer. Clearly, the effect of thermal conductivity decreases with increasing temperature and velocity and, therefore, a thicker melt-water layer.

Overall, it is clear that the careful choice of the slider material and possibly the selection of a thermal insulator can easily influence the resulting friction.

9.5.2.2 Roughness

In 1699, Amonton related friction and roughness [14] and concluded that smooth surfaces show less friction. This relationship generally holds true and has been confirmed by several researchers with various materials and under largely different

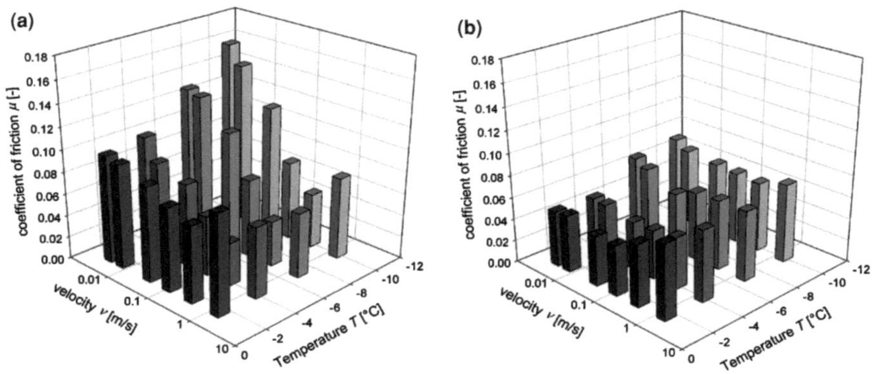

Fig. 9.15 3D bar chart of temperature and velocity dependence of the ice friction coefficient for **a** Stellite 6B and **b** Stellite 6B insulated with a fiberglass disk [64]. (Reprinted with permission from the Journal of Glaciology with permission of the International Glaciological Society.)

experimental conditions [43, 51, 60, 65, 66]. Marmo et al. [66] explain that roughness leads to a decreased actual thickness of the lubricating film. Melt water gets trapped in valleys between asperities, and a less uniform temperature profile exists due to hot spots at asperity tips. Furthermore, it is intuitive that an increased surface roughness leads to an increased surface area and more interlocking asperities during the sliding movement, which increases the wear rate and overall friction, which is also illustrated in Fig. 9.21. This basic relationship is used in the design of tread tires and foot wear; while the desire for the contrary—very low friction—supported the invention of the first ice resurfacing machine (Zamboni) for skating competitions, as outlined above.

9.5.2.3 Surface Structure

While the concepts of roughness and surface structure are closely related, we distinguish between the two here, reasoning that roughness is of a more random nature than a specific structure or pattern. Itagaki et al. [43] found in their experiments that their samples with polished grooves in sliding direction showed similar low friction as highly polished sliders. Kietzig et al. [60] recently examined the influence of surface structure more closely and found that for the same roughness value R_a of 600 nm the orientation of polishing marks plays a significant role on ice friction. Figure 9.16 illustrates the difference in the two ice slider samples: one shows random polishing marks in all directions, while the other shows almost concentric grooves, which correspond with the sliding direction.

The results from ice friction tests with both slider types are shown in Fig. 9.17. Sliders with random marks show the typical dependence of the coefficient of friction on velocity with $\mu \propto \upsilon^{-1/2}$, as already discussed in Sect. 9.5.1.2.

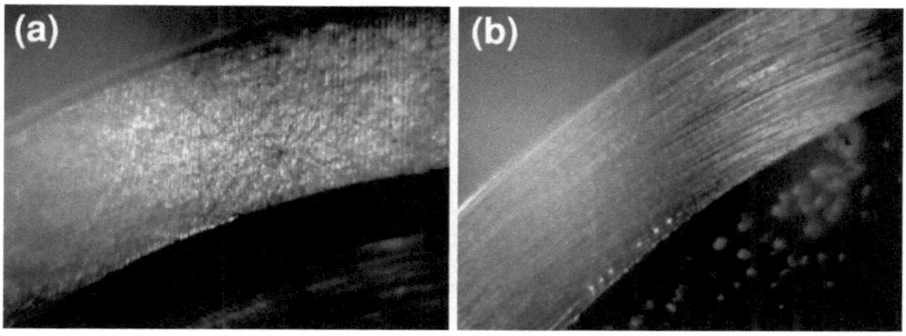

Fig. 9.16 Optical microscopy images of AISI 304L slider with **a** random polishing marks, **b** concentric grooves. (Reprinted with permission from [60] Copyright 2009, American Institute of Physics)

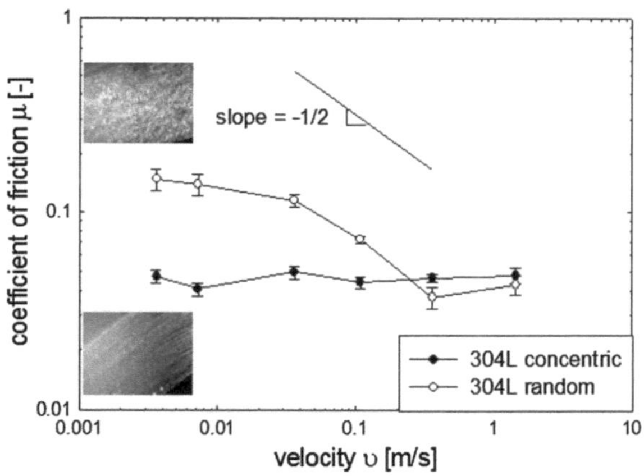

Fig. 9.17 AISI 304L slider with different surface structure at −7°C [60]. (Reprinted with permission from [60] Copyright 2009, American Institute of Physics)

A minimum in the coefficient of friction can be observed. The slider with the concentric grooves, however, shows an almost constant coefficient of friction, which appears to be independent of sliding velocity. The far lower coefficient of friction at low velocities of the slider with grooves in sliding direction is explained by ice surface asperities running smoothly within these concentric grooves and thereby causing less interlocking than the random polishing marks of the other slider type.

Increasing speed, however, does not significantly change the measured friction coefficient for the slider rings with concentric grooves. It seems like the decrease in friction from enhanced lubrication is immediately balanced off by added capillary drag of melt water being trapped in the grooves. These experiments have been carried out with a ring slider; therefore, the melt water has no easy way to escape the concentric grooves. When the amount of melt water increases and can no longer be retained within the size of the grooves at the applied normal force, it is squeezed out. Accordingly, reduced asperity interaction through melt

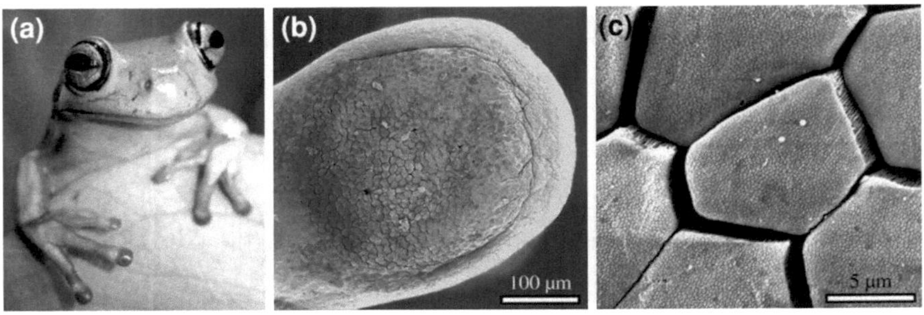

Fig. 9.18 Morphology of tree frog toe pads. **a** Tree frog (Litoria caerulea), **b** SEM image of toe pad, **c** SEM image of channels between hexagonal epithelial cells [67]. (Reprinted with permission from the Journal of The Royal Society Interface)

water and increased capillary drag seem to compensate one another with increasing velocity.

An interesting example for the above introduced strategies of contact area, roughness, and surface structure optimization is the pattern of a snow tire, whose design has been inspired by the natural example of the toe pads of tree frogs and geckos. The tree frog's toes are composed of disk-like pads, which are separated by channel-like spaces. Through these channels water is squeezed out of the contact interface to reduce lubrication and enhance van der Waals forces between the contacting solid surfaces. This particular surface design enables the frog to adhere to vertical surfaces in moisture-rich environments [13, 67]. Gecko feet are covered with microscaled bristles, which are flexible and allow for adaptation to any surface profile thereby maximizing the actual contact between the foot and the surface [68]. Snow tires combine lessons learnt from both natural examples. They are characterized by a comparatively coarse pattern with about 0.5–1 cm wide channels, which primarily serve the purpose to direct away vast amounts of water and therewith prevent aquaplaning. In addition, many finer channels are found on top of the coarse profile, which are designed to act like the gaps between the epithelial cells of a frog's toe pad, as illustrated in Fig. 9.18.

These small channels are especially important for traction on snowy or icy surfaces, where a small amount of melt water underneath the tire can already causes significant slippage and result in fatal accidents. In addition, bristles on top of the tire profile are intended to enhance adhesion and interlocking and thereby increase traction. Clearly, in these examples the desired friction regime is to the left of the minimum shown in Fig. 9.3.

9.5.2.4 Wettability

Again it was the pioneer of ice friction studies, Bowden [53], who conducted the first experiments with different materials on the influence of wettability on the friction coefficient of ice. He came to the conclusion that friction was highest for surfaces that wet easily, especially close to the melting point. This can be explained by the enhanced build-up of capillary bridges between the sliding surfaces, which becomes especially important in the mixed and hydrodynamic friction regimes. However, the different degrees of hydrophobicity were achieved by changing the slider material. Therefore, the impact of hydrophobicity cannot be investigated independently of other material inherent parameters, such as thermal conductivity and material hardness. In addition, it is well understood in the wetting research community that surface roughness and regular surface patterns have a significant influence on the resulting surface wettability. However, Bowden did not report on the surface roughness and structure of the different materials used.

Investigating friction on snow in particular, Colbeck [26] recognized the importance of capillary forces on snow friction. He has found that snow grains that do not support the slider's load directly are bonded to the slider surface by capillary bridges. As outlined above increased sliding speed induces a larger melt water production, which again enhances the build-up of capillary bridges. Further investigations of the adhesion between water and static rough, hydrophilic, and hydrophobic surfaces have shown that hydrophobicity clearly reduces the capillary bonding [69, 70].

The fundamental investigations on the influence of roughness and wettability on ice friction are also supported by common practices used by winter sports enthusiasts, which are approaches copied from natural examples as outlined in the following.

The water repellency of the superhydrophobic lotus leaf is attributed to its unique surface structure: microscale surface bumps are covered with a nanoscale hairy wax. This surface structure on two length scales results in air being trapped between the surface asperities, so that water drops are supported by a solid-air composite surface, which results in high hydrophobicity [71].

In ski sports, a certain surface structure in combination with a fluoro-based ski wax is typically applied to the ski base to reduce friction. First, the bases of downhill skis commonly show V-shaped, diagonal, or linear grinding patterns, which are designed to channel melt water away from the interface and thereby decrease friction by reducing capillary drag [26]. Second, a hydrophobic wax is applied to reduce water adhesion to the surface. Furthermore, this wax treated surface is roughened by brush strokes to expose little hairs, which support higher hydrophobicity following the example of the lotus leaf [71]. The disadvantage in the application of waxes or soft coatings to a sliding surface is their susceptibility to wear and abrasion. Compared to snow sports, ice friction applications are characterized by higher ratios of load per contact area. In addition, the hardness of the ice surface is typically higher than that of snow. Therefore, the application of waxes to ice slider surfaces is even more problematic.

Kietzig et al. [60, 72] have recently shown that metal sliders with roughness induced hydrophobicity significantly reduce ice friction, without the need for an additional coating. The particular structure is copied from the superhydrophobic lotus leaf and created on a steel slider by femtosecond laser irradiation. The laser

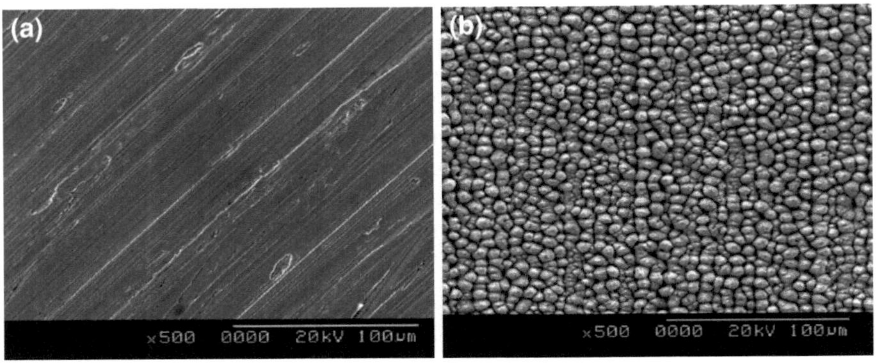

Fig. 9.19 SEM image of SS304L ice slider surfaces **a** polished, **b** irradiated by a femtosecond laser with a fluence of 3.16 J/cm^2 (magnification × 500). (Reprinted with permission from [60] Copyright 2009, American Institute of Physics)

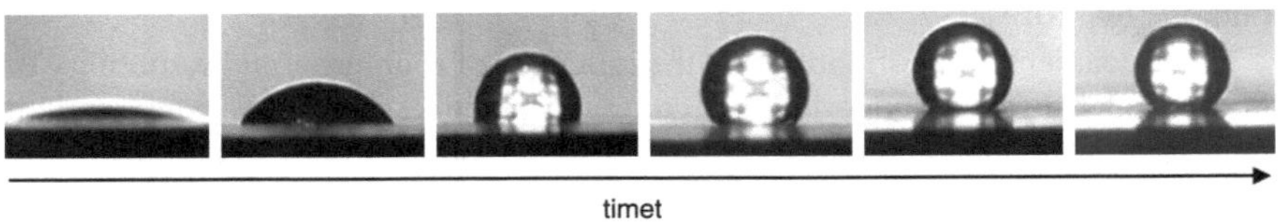

Fig. 9.20 Evolution of surface wettability with time on SS304L after having been irradiated by a femtosecond laser with a fluence of 3.16 J/cm^2

irradiated slider surfaces exhibit a controlled pattern with roughness on two length scales. Microscaled bumps are covered by nanoscaled ripples. The roughness of the irradiated slider surface is about twice as rough (R_a value) as the polished sliders' (Fig. 9.19).

While the laser irradiated surface is initially superhydrophilic it becomes superhydrophobic with time. After the laser process the irradiated surface is highly reactive. A surface reaction with CO_2 from ambient air causes the change in wetting behavior, which can be explained by the accumulation of a nonpolar carbonaceous screening layer, which successively covers the inherent hydrophilicity of the polar metal oxide surface. The timescale of this process is dependent on the underlying material and the environment and is in the range of hours to days as explained elsewhere [73]. The evolution of the wettability is illustrated in Fig. 9.20.

The laser irradiation only influences the surface chemistry of the topmost 10 nm of the metal surface. Accordingly, with this process it has become possible to investigate the influence of wettability on ice friction without modifying other material parameters, such as the thermal conductivity and hardness. It therefore, presents a unique opportunity to investigate the influence of capillary drag on ice friction more closely.

Figure 9.21 shows results from ice friction experiments with the polished and laser patterned slider, as introduced in Fig. 9.19. While the polished slider is hydrophilic with a contact angle of 81°, the laser patterned slider showed an average contact angle of 134° on its surface.

In the low velocity range the irradiated slider exhibits higher friction coefficients for all temperatures. This is easily understood considering the comparatively higher surface roughness, which leads to increased asperity interactions in the boundary friction regime. In agreement with Colbeck's findings outlined above the influence of surface wettability becomes most obvious for the experiments carried out at −1.5°C (Fig. 9.21c). While the coefficient of friction of the polished slider follows the typical trend with increasing speed (initial decrease, and after passing through a minimum the subsequent increase), the irradiated slider's coefficient of friction continuously decreases with increasing sliding speed.

A significant question, however, poses the durability of the delicate roughness structure under load and frictional motion. Kietzig and al. have carried out ice friction experiments with a ring slider on a rotational tribometer [60]. Figure 9.22 illustrates the wear tracks on the surface after 1.500 km of ice friction tests (equivalent to the results shown in Fig. 9.21a–c). First, our images confirm that the actual contact is less than the geometric contact area, due to the real nature of slider and ice surface, as discussed in Sect. 9.5.1.5. The width of the wear track is on average 50–80 μm, while the slider width is 2 mm. Furthermore, it can be seen that while the tip of the microscaled laser-created bumps are flattened, their sides maintain the hydrophobicity supporting structure. The average sessile contact angle from a water droplet of 1 μl has been measured to be about 121° along the wear track. Overall, the results indicate that the surface maintains its overall hydrophobicity, while it is, however, subject to wear.

Hydrophobic surfaces in frictional contact with ice are used to reduce friction in the regimes past the friction minimum of Fig. 9.3, where capillary drag outweighs the friction reduction by lubrication. These regimes are characterized by significant melt water production, as outlined before. It is therefore advisable to design surfaces with hydrophobic water channels (Sect. 9.5.2.3) to reduce the arising drag. The hydrophobic surfaces, however, should ideally not bear load but simply be in contact with the produced, channeled, and squeezed out melt water. This is due to the fact that hydrophobic coatings or as shown here microstructures are likely to become subject to wear under frictional contact. As an example, Fig. 9.23 shows several ice sports blades and indicates in red surface areas, which are not load-bearing but affected by capillary drag from melt water.

In conclusion, wettability being a material property and also a function of surface roughness is a significant factor to consider when conducting ice friction experiments. In addition, as shown here, a biomimetic approach to render surface wettability by influencing roughness and surface structure is an interesting sustainable alternative to fluoro waxes and lubricants.

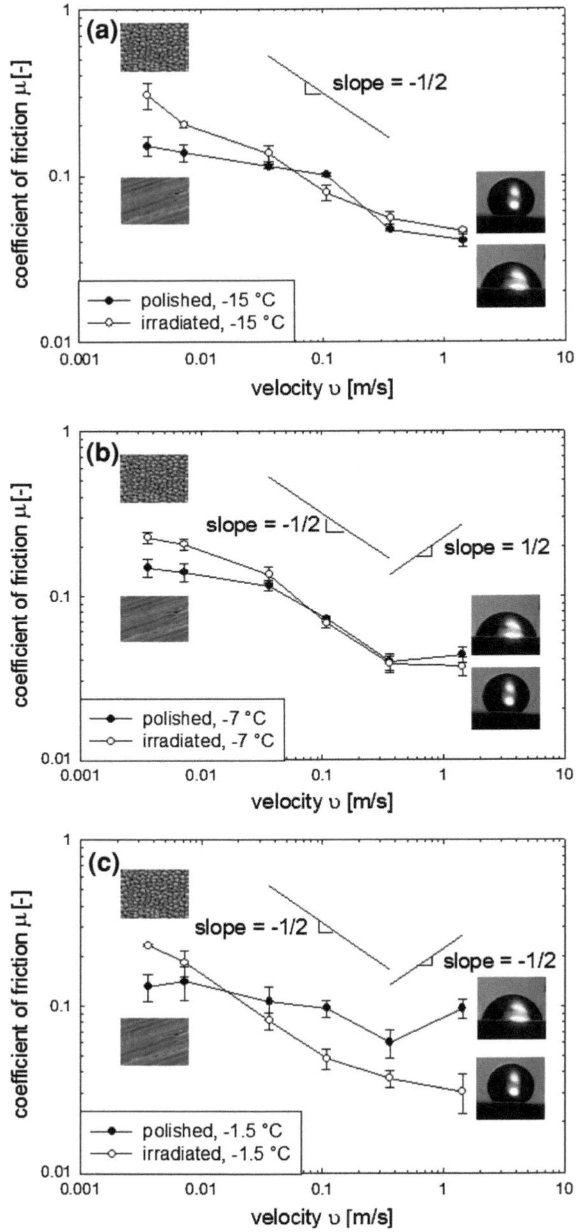

Fig. 9.21 Friction of hydrophilic, polished and hydrophobic, laser irradiated ASIS 304L sliders at **a** −15°C, **b** −7°C, and **c** −1.5°C [60]. (Reprinted with permission from [60] Copyright 2009, American Institute of Physics)

Fig. 9.22 Slider surface at 500 × and 4 K magnification **a** before ice friction test; **b** after ~1.5 km ice friction test

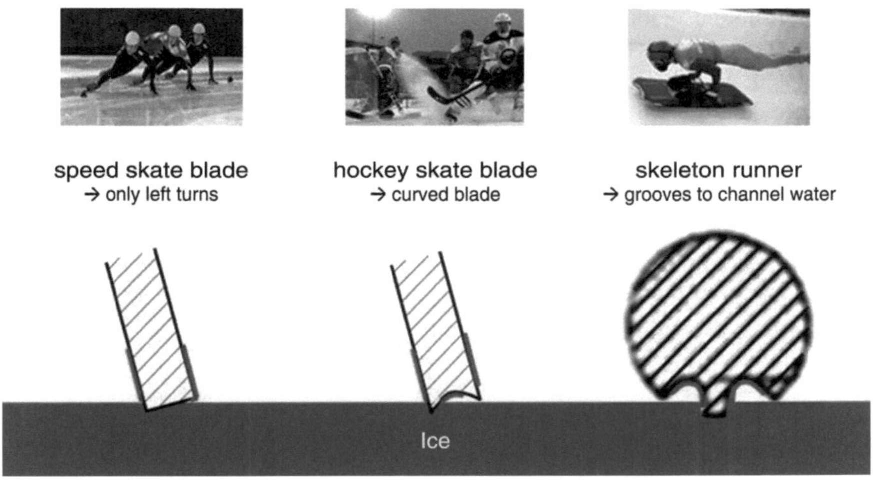

Fig. 9.23 Examples for non-load-bearing surfaces of ice sports blades, which rendered hydrophobic will assist to reduce friction on ice

9.6 Ice Friction Experiments

The previous section has shown that experiments carried out with different instrumental setups can result in considerably different conclusions depending on the experimental conditions and the exact settings for the many different parameters discussed before. Various experimental configurations, their advantages and disadvantages are summarized and discussed below.

9.6.1 Real-Life Experiments

As real-life experiments we consider tests with the actual slider object, on which the ice friction research focuses. Colbeck et al. [70] investigated ice skating with an instrumented skate. They used thermocouples to assess the temperature at the skate surface and therewith gave experimental proof for the frictional heating theory. However, it was impossible to assess the friction coefficient with this skate.

De Koning et al. [49] were the first to attempt real-life analysis of ice skating with an instrumented skate. They developed ice skates, which are instrumented with strain gages and transducers. These skates enabled the measurement of the push-off and friction force during skating. An experienced ice skater skated on them on different indoor and outdoor ice skating rinks.

The advantage of real-life experiments is that real skating conditions are investigated. However, these experiments provide limited control over various variables. At different ice rinks different ice-making procedures are used, and different locations imply different air temperatures, relative humidities, and water qualities, which will all have an effect on the measured friction coefficient, as outlined in the last section. In addition, experiments involving humans require a big sample size to ensure statistical significance, as the results are greatly dependent on the skaters' daily performance, i.e. the skating technique might not be the same on different days at different locations, which results in different loads and pressures as well as different velocities.

9.6.2 Slider Models

Using slider models in ice friction measurements eliminates the human factor in the experiment. Bowden [53] used small sledges with rounded front edges cut from different materials to measure friction on snow and ice. He applied a certain procedure to prepare the ice surface. It is unclear how the sledges were accelerated.

Kuroiwa [54] reported ice friction measurements with real skate blades mounted to a frame. This slider model was ejected from a catapult to reach its sliding speed; thereby the velocity can be more closely monitored.

Slotfeldt-Ellingsen and Torgersen [50] and Itagaki et al. [43] also used automatically accelerated slider models of different size and weight to measure friction on ice. In contrast to Bowden [53] and Kuroiwa [54] they also applied an ice making procedure to further limit variability.

Experiments with slider models increase the controllability of parameters such as the slider material used, velocity, and load. However, other parameters, such as ice and air temperature, still contribute to error in the measurements. One particular problem of slider models is that their sliding track cannot be completely predicted. Therefore, the slider model is likely to take different routes over the ice in each experiment. If this unpredictability is to be eliminated by using prepared tracks, the friction on the sidewalls of the track will contribute to the overall measured friction.

9.6.3 Linear Experimental Devices

Linear experimental devices to measure ice friction share the characteristic that the movement of the slider on the ice surface is guided by a control mechanism during the experiment. In comparison to slider models this solves the problem of the unpredictable sliding track. Different linear devices were used by Jones et al. [59]. Furthermore, the ice making procedure, the load and velocity settings are all well controlled [56, 59]. Montagnat and Schulson [55] and Ducret et al. [65] also ensured a certain temperature setting by conducting the experiment in a cold room or freezer unit. Compared to other laboratory equipment, as will be explained later, linear experimental devices have the advantage that the ice slider is continuously in contact with a fresh ice surface along the length of the ice track, which typically defines the length of one experimental run.

9.6.4 Rotational Experimental Devices

Different types of rotational devices were used to measure friction on ice. As in the case for linear experimental devices, load and velocity can easily be adjusted by the experimenter. One advantage of these setups is the compact size of the equipment. This facilitates the use of cold boxes and temperature chambers, which allows for close surveillance of the ice and air temperature and even relative humidity. Furthermore, artificial ice surfaces can be created following an ice making procedure to limit variability.

For their experiments Kozlov and Shugai [24] have used a rotary viscometer, with a flat metal ring sliding against a hollow cylinder. This setup necessitates measurements of the vertical displacement of the ring, since the rotation melts the ice at the cylinder wall.

A ring or round disk, which rotates against a stationary ice sample, has been used in many other experiments. Strausky et al. [74], Buhl et al. [62], Liang et al. [45] and Kietzig et al. [60] successfully applied this setup for their analysis of ice friction. While these setups enable good controllable settings and limited variability, the sample continuously slides over the same ice. Accordingly, edge effects in front of the slider sliding over a fresh ice surface cannot be assessed as seen under real ice friction conditions.

Evans et al. [44] and Petrenko [8] used a lathe setup with an ice cylinder rotating against a slider sample. Accordingly, not the whole circumference of the ice cylinder is in contact with the slider. This setup allows a particular area on the ice cylinder to refreeze before the next contact with the slider occurs.

Another way of maintaining a relatively small setup and also investigating contact with fresh ice was chosen by Bäurle et al. [48]. They have used a similar setup as described above with an ice turntable and a stationary sample mounted to an arm above it. The ice track with a diameter of 1.60 m is large enough to ensure that the ice surface refreezes before the next pass under the sample. Bowden and Hughes [19], Oksanen and Keinonen [57], and Lehtovaara [75] had previously worked with similar versions of this setup with a sample mounted to an arm over a rotating ice surface, which permits horizontal displacement of the slider, so as to ensure that the samples get into contact with fresh ice after each revolution.

Akkok et al. [46], Calabrese [51], and Albracht et al. [52] used rotational experimental devices, which consisted of a stationary ice surface with a rotating sample. Calabrese [51] used a ring slider, which again means continuous contact between the slider and the ice and therefore does not enable analysis of edge effects. Akkok et al. [46] used a ball on disk setup, which allows the ice to refreeze after the slider passes. Albracht et al. [52] even avoided sliding over the same track by using a spiral track for their pin slider.

In conclusion, laboratory instruments are better suited for the investigation of the different parameters that influence ice friction. They ensure good control over the settings such as ice and air temperature, humidity, velocity, and load. In addition, the use of a certain ice making procedure further limits undesired variability in the results. However, a drawback of compact laboratory devices is often seen in the continuous ice-slider contact or the fact that the slider slides in the same track. While this might not have an impact on the overall analysis and understanding of ice friction, it is rather different to what is seen in actual ice friction applications. Accordingly, field experiments are best suited to upscale and verify laboratory tests. Typically, a larger number of replicates have to be run to ensure reproducibility of the experiment and to ensure statistical significance and eliminate variation resulting from the human factor. Studies comparing results obtained from real-life experiments involving humans with those from a lab setup are most welcome.

9.7 Ice Friction Models

Figure 9.3 and the above discussion on the isolated effect of the various parameters illustrate that the matter of ice friction includes different competing processes with many interactions. Therefore, it is not surprising that no single conclusive model exists, which describes ice friction across all regimes. Nonetheless, there has been a major effort to model and thereby better understand and predict ice friction. The most interesting models are briefly introduced here.

Evans et al. [44] derived the first theoretical explanation for the dependence of the friction coefficient on frictional heating. They defined the total frictional force F_T in terms of heat generated per unit displacement. Accordingly, the total frictional heat is described as the sum of three heat components.

$$\mathrm{F}_T = \mathrm{F}_S + \mathrm{F}_I + \mathrm{F}_M \tag{9.9}$$

These are the heat conducted away from the interface through the slider material F_S, the amount of heat, which diffuses into the ice F_I, and the heat that remains for melting the surface F_M. By assuming that the surface is at the melting temperature of ice and taking the length of contact from their experimental observations, this results into the following equation for the friction coefficient μ

$$\mu = \frac{A\lambda_S(T_m - T_o)}{F_N \upsilon} + \frac{B(T_m - T_o)}{F_N\sqrt{\upsilon}} + \mu_M \tag{9.10}$$

with T_m and T_o being the melting and the ambient temperature respectively, λ_S being the thermal conductivity of the slider, υ being the sliding velocity, A being a constant depending on the actual contact area, and B being a constant depending on the actual contact area, thermal conductivity and diffusivity of ice. Since, it is impossible to find the contribution of the melt friction coefficient μ_M directly, the authors point out that an upper limit can be derived from experiments. Furthermore, from experimental and theoretical considerations they find the thickness of the lubricating layer to be smaller than the combined surface roughness, which indicates an overall mixed friction regime. Since the slider constantly moves over fresh ice, the temperature gradient $\Delta T_{\mathrm{ice}}(= |T_{\mathrm{contact}} - T_{\mathrm{ice}}|$; absolute value of the temperature difference between the contacting interface and the bulk ice) is generally greater than the one for the slider $\Delta T_{\mathrm{slider}}(= |T_{\mathrm{slider}} - T_{\mathrm{contact}}|$; absolute value of the temperature difference between the bulk slider material and the contacting interface), except for the case of a very conductive slider material. Their experiments with copper, Perspex™, and steel indicate that 40–60% of the frictional heat is conducted away from the surface through the slider independent of the ambient temperature. In addition, the experimental results from the authors have shown that the measured friction coefficient depends on the applied normal force according to $\mu \propto F_N^{-\frac{1}{3}}$. These findings disagree, however, with their own model (Eq. 9.10), which is explained by the model's deficiencies in fully describing the dependence of the actual contact area on load.

Moreover, Evans' et al. experimental results at $-11.5°C$ showed a dependence of the friction coefficient on velocity according to $\mu \propto 1/\sqrt{\upsilon}$. The authors found that the dependency of the friction coefficient on velocity differs at higher and lower speeds. For velocities lower than 3.16 m/s the friction coefficient corresponds to $\mu \propto 1/\upsilon$ indicating that heat conduction into the slider F_S dominates, whereas for higher velocities F_I takes over and $\mu \propto 1/\sqrt{\upsilon}$ explains the experimental results more closely. The results indicate again that with increasing slider speed other mechanisms gain importance for friction on ice. However, other mechanisms, such as fluid mechanics of the squeezed lubricating film, were not included in the theory. Another interesting observation is that the authors observed increased wear of the ice track above $-2°C$ with significant softening of the ice. As mentioned in the previous section, wear of the slider surface will generally be higher on new fresh ice than observed in their studies, where a rod formed a track on an ice cylinder. However, Evan et al.'s model does not take energy losses by wear processes into account.

Oksanen and Keinonen [57] further elaborated this first model. With the assumption that the lubricating layer is the main origin of frictional resistance, the authors combine the theory of Evans et al. [44] with hydrodynamic friction. They assume that frictional motion results in non uniform heat transfer and derive a model for the friction coefficient. The frictional heat Q_f generated by the motion during a certain time interval b/υ is

$$Q_f = \mu F_N \upsilon \frac{b}{\upsilon} \tag{9.11}$$

with b being the length of a contacting point. Equating this with the heat consumption equations from Evans et al. [44] results into:

$$\mu = \frac{n^{1/4} A_c^{3/4}}{F_N} \left\{ \begin{array}{l} \frac{1}{2}\frac{1}{(2\upsilon)^{1/2}} \left[\Delta T_I (\lambda_I c_I \rho_I)^{1/2} + \Delta T_S (\lambda_S c_S \rho_S)^{1/2} \right] + \\ \quad + \left(\frac{1}{8\upsilon} \left[\Delta T_I (\lambda_I c_I \rho_I)^{1/2} + \Delta T_S (\lambda_S c_S \rho_S)^{1/2} \right]^2 + \eta_0 \upsilon h \rho_0 \right)^{1/2} \end{array} \right\} \tag{9.12}$$

where n stands for the number of contacting points, A_c for the actual area of contact, λ_I and λ_S for the thermal conductivity, c_I and c_S for the specific heat capacity, and ρ_I and ρ_S for the density of ice or the slider material respectively, ρ_0 for the density of, and η_0 for the viscosity of water.

The authors identify two regions with different relationships between the friction coefficient and the velocity. In the case of a great temperature gradient for the ice ΔT_{ice} thermal conductivity dominates over viscous shearing (the first part of Eq. 9.12). The same is true when the thickness of the lubricating layer is very small, suggesting that the produced heat is mainly conducted away and not available for melting the ice. The velocity dependency of the friction coefficient is then described by $\mu \propto 1/\sqrt{\upsilon}$ in the model. If the temperature gradients for the ice

ΔT_{ice} and the slider ΔT_{slider} are both small, meaning that the ambient temperature is close to 0°C, friction is governed by viscous shearing and melting of the ice (the second part of Eq. 9.12). The model gives a velocity dependency of $\mu \propto \sqrt{\upsilon}$ in this case. Furthermore, the authors point out that in the mixed region between the two cases discussed above the strength of the effects is determined by the velocity. Thermal conductivity plays the greater role at low velocities. At high velocities the time for heat conduction is reduced, which results in a greater amount of heat available for melting the ice. Although the model indicates increased frictional resistance close to the melting temperature, it does not mention a potential contribution of capillary drag forces. Furthermore, energy losses due to wear mechanisms and squeeze out of the lubricating layer are ignored in the model. Nevertheless, the proposed relationships between the sliding velocity and the resulting friction coefficient correspond reasonably well to experimental results, as shown in Sect. 9.5.1.2.

The theoretical approach of Akkok et al. [46] does not consider the melting temperature as an upper bound for the surface temperature in contrast to the analytical models introduced above. They see the upper bound as determined by the softening temperature of ice, since once the softening temperature has been reached, frictional motion wears the surface so much that the originally touching materials are no longer in contact. This is important, since no energy is consumed in the phase change through this kind of wear process, but all heat is conducted away. The heat is assumed to be only conducted into the ice and not through the slider. Based on these assumptions Akkok et al. derived the following model

$$\mu = C\frac{A_c(T_c - T_I)}{F_N}\left(\frac{\lambda_S \rho_S c_S}{\upsilon b}\right)^{\frac{1}{2}} \tag{9.13}$$

where C is a constant, A_c the actual area of contact, T_c and T_I are the temperature at the contact and the ice respectively. This model and further experiments confirm the results of Evans et al. [44] and Oksanen and Keinonen [57] with $\mu \propto 1/\sqrt{\upsilon}$. Furthermore, they emphasize that the velocity dependency changes at high temperatures, when hydrodynamic friction dominates. Their model describes the dependency of the friction coefficient on load by $\mu \propto 1/\sqrt[4]{F_N}$ for partial contact. However, the regression analysis on their experimental data gives $\mu \propto 1/\sqrt{F_N}$. The results of Evans et al. [44] lie between these two. This difference indicates again that effects like the squeeze of the lubricating layer might play an important role here.

The analytical models introduced so far recognize heat conduction on the one hand and friction dominated by viscous shearing on the other hand as two extreme cases in the determination of the overall frictional resistance. However, other important mechanisms are left on the side. Squeeze flow, for example, is not considered, which decreases the thickness of the lubricating layer at high loads and fast speeds.

Stiffler [58, 76] included squeeze as a type of wear in his model. He derived the following friction model for a conducting surface:

$$\mu = \frac{2\lambda_S A_c(T_m - T_0)}{F_N(\pi\alpha_S l_c \upsilon)^{1/2}} \tag{9.14}$$

with α_S being the thermal diffusivity of the slider material and l_c the characteristic length of the contact. This approach assumes hydrodynamic friction with a thickness of the lubricating layer that is larger than the combined roughness of both surfaces. Stiffler mentions that this assumption is unrealistic for an ice skater and concludes that the model is not suitable to describe ice sport applications.

While we are mainly concerned with ice friction, it is important to include a model, which attempts to describe snow friction in this context. Colbeck's [26] comprehensive model for snow friction recognizes the different mechanisms of friction acting at the same time:

$$\mu = \mu_S + \frac{\mu_D \mu_W}{\mu_D + \mu_W} \tag{9.15}$$

with μ_S being friction due to capillary drag, μ_D being dry friction due to asperity interactions, and μ_W being wet friction due to shearing of the water film. The calculation of μ_W is based on the shear stress calculation for a Newtonian fluid with $\mu_W = c\mu_o \upsilon/h$, with μ_o being the water viscosity at 0°C, and c a constant that considers the area of contact but is based on the assumption that snow friction is independent of the applied load. The latter was found to not hold true for small sliders on ice as explained in Sects. 9.5.1.4 and 9.5.1.5. The dry friction coefficient μ_D is based on the heat flow calculations, first introduced by Evans et al. [44]. It is summarized in $\mu_D = \varepsilon e^{-\beta h}$, where β and ε are coefficients. While ε is not further defined, Colbeck [26] discusses four different cases of heat flow to derive a value for β. These four cases are (1) heat flow into the ice only, (2) heat flow at contacts only, (3) entire lower surface at 0°C, and (4) heat flow assuming an average temperature gradient. For an exact calculation for a particular slider information about the actual area of contact and the division of heat flow into slider and ice would be needed. μ_S is approximated by $\mu_S = \gamma h^3$, with γ being a constant. Although this model is based on several assumptions regarding the constants, it gives a good example of the interaction of the different mechanisms controlling friction on snow and also ice.

Bäurle et al. [63] recently developed a numerical model specifically for their tribometer experiment (explained in Sect. 9.5.1.5) but with observations valid for general ice friction problems. The analysis is based on equating heat generation with possible energy dissipation mechanisms. Considering a non uniform frictional contact with some areas being almost dry and some areas covered by lubricating water, the total friction force F_T is calculated by

$$F_T = \mu_{\text{dry}} F_N \frac{A_{\text{dry}}}{A_{\text{app}}} + \frac{\eta \upsilon \Delta_{\text{rel}}}{h_{\text{wf}}}(A_{\text{app}} - A_{\text{dry}}) \tag{9.16}$$

with h_{wf} being the thickness of the lubricating water film; A_{dry} and A_{app} being the dry and apparent contact areas respectively, and Δ_{rel} the relative real contact area. A constant temperature is assumed within the water film. However, the heat diffusion process depends on the change of the water film thickness h_{wf} along the ice-slider interface which is described as

$$\frac{\partial h_{\mathrm{wf}}}{\partial t} = \frac{1}{L}\left(\frac{\eta v^2}{h_{\mathrm{wf}}} - k \times \partial_z T|_{z=0}\right) - \frac{8h_{\mathrm{wf}}^3 \sigma_0}{3\eta D^2} \tag{9.17}$$

with L being the volumetric latent heat of fusion and $\partial_z T|_{z=0}$ the average temperature gradient at the interface. The last term describes the loss of lubricating water by squeeze, with σ_0 being the perpendicular pressure and D the contact spot diameter, which changes during the contact according to $D = D_0\sqrt{\frac{\Delta_{\mathrm{rel}}}{\Delta_{\mathrm{rel},0}}}$, with D_0 being the initial static contact sport diameter, and $\Delta_{\mathrm{rel},0}$ being the initial relative real contact area. The model is based on one-dimensional calculations of heat generation and conduction perpendicular to the interface into slider and ice. Changing conditions along the ice-slider interface are taken into account by conducting several calculations along the contact length. While describing the general trend of decreasing friction with increasing temperature reasonably well, the model fit to experimental data was improved by including a correction factor to allow for three-dimensional heat flow into the slider. Through Δ_{rel} melting and roughness are considered in this numerical model. h_{wf} is related to Δ_{rel} through the bearing ratio curve of the ice surface. Furthermore, a rough slider moving over rough ice results in intermittent contact of the surfaces allowing sections of the surfaces to cool before the next contact, which reduces h_{wf}. The authors determine Δ_{rel} by fitting their model to their experimental results. This model allows for implications of slider parameters such as thermal conductivity and roughness on the resulting ice friction.

Overall, there are only few analytical models for approximating the friction coefficient on ice. A common problem with all these models is the estimation of the real contact area. An error can easily be introduced to the calculation, since the size and density of the surface asperities are unknown and have to be approximated. At the same time, adjusting the value for this variable is an easy way to adjust model results to fit experimental data. However, the different models provide good insight into the main effects interacting in frictional heating. Other important mechanisms are not yet fully included in the theoretical analysis. The contribution of capillary drag forces to frictional resistance is not separated from overall mixed and hydrodynamic friction due to the lack of a sound physical understanding. Also, the influence of material parameters, such as surface wettability and roughness, on drag force, squeeze flow, and the overall friction coefficient still need to be investigated in greater detail and included into a model.

9.8 Challenges for the Future

Biomimicry means, first of all, getting inspired by nature. This can simply represent the engineering of one part of a product or process after a natural example. However, taking this idea to the next step, biomimicry also suggests a holistic view considering the entire life cycle of a process or product and conforming to the general requirement to develop a "green" process or product. The examples, which we find in nature, are exactly those that have succeeded in an overall process optimization and satisfy entirely our requirement of being "green". Those species and processes that did not conform to this standard have been sorted out by evolution.

To illustrate what this evolution process should mean to engineering we consider exemplarily the design of ice breaker ship hulls and the related fuel consumption of these vessels. Primarily, the objective is to reduce the friction of the ship hull surface against ice and therewith to decrease the imminent fuel consumption of the vessel through intelligent surface design and the use of adequate materials. However, it is important to also consider the environmental footprint of the construction and disposal processes related to the new nature-inspired vessel hull. If the related energy requirements exceed the reductions achieved during operation with the biomimetic surface, the engineers and researchers need to go back to the draft table and redesign the whole product and process until net benefits become obvious.

As an example for successful biomimicry with relation to ice friction let us briefly reflect on non-wax cross-country skis. The ski base pattern combines grooves and channels, as exhibited by the toe pads of geckos and tree frogs to eliminate superficial melt water from the interface for better adhesion when climbing or during the kick phase. Furthermore, a hairy structure on the ski base enhances hydrophobicity following the example of the lotus leaf to reduce capillary suction. Overall, these skis do not require the application of waxes, which typically wear off during motion and need to be frequently reapplied. The environmental benefits of the no-wax technology are evident, since ski waxes are typically based on fluoropolymers. When waxes wear off on the snow surface, these fluoropolymers will remain in nature—initially on the snow surface and once the snow melts on the ground. From there they migrate into our water sources. Accordingly, non-wax cross-country skis are the green alternative combining biomimicry and tribology.

On the laboratory scale we have found ways to solve ice friction and/or adhesion problems by following the biomimetic approach. We have, for example, succeeded to increase melt water lubrication without externally heating the surface and to overcome frictional resistance without using high pressure. The next challenge is the scale up of these fundamental ideas and proof-of-concept studies to the industrial scale and integrating them into a sustainable life cycle.

References

1. W.R. Chang et al., Ergonomics **44**, 1217 (2001)
2. A.D. Roberts, J.C. Richardson, Wear **67**, 55 (1981)
3. A. Klein-Paste, N.K. Sinha, Tribol. Int. **43**, 1145 (2010)
4. S.C. Colbeck et al., Rev. Geophys. **13**, 435 (1975)
5. R.C. Woolgar, D. Bruce Colbourne, Ocean Eng. **37**, 296 (2010)
6. N. Nakazawa, T. Terashima, H. Saeki, T. Ono, in *Proceedings of the* 12th *International Conference on Port and Ocean Engineering under Arctic Conditions*, vol. 1 (1993), p. 97
7. J. Weale, J.H. Lever, Cold Reg. Sci. Technol. **52**, 166 (2008)
8. V.F. Petrenko, J. Appl. Phys. **76**, 1216 (1994)
9. H. Kumano, T. Hirata, R. Shouji, M. Shirakawa, Int. J. Refrig. **33**, 1540 (2010)
10. J.P. Bedecarrats, F. Strub, C. Peuvrel, Int. J. Refrig. **32**, 1791 (2009)
11. S. Beltaos, Cold Reg. Sci. Technol. **62**, 83 (2010)
12. L. Cao, A.K. Jones, V.K. Sikka, J. Wu, D. Gao, Langmuir **25**, 12444 (2009)
13. M. Scherge, S.N. Gorb, Phy. Today **55**, 53 (2002)
14. D. Dowson, *History of Tribology*, 2nd edn. (Professional Engineering, London, 1998)
15. B.N.J. Persson, *Sliding Friction: Physical Principles and Applications*, 2nd edn. (Springer, Berlin, 2000)
16. M. Faraday, Philos. Mag. **17**, 162 (1859)
17. J. Thomson, Proc. R. Soc. Lond. Ser. A, Math Phys. Sci. **10**, 151 (1859)
18. O. Reynolds, *Papers on Mechanical and Physical Subjects* (University Press, Cambridge, 1900)
19. F.P. Bowden, T.P. Hughes, Proc. R. Soc. Lond. Ser. A. Math Phys. Sci. **172**, 280 (1939)
20. F.P. Bowden, Proc. R. Soc. Lond. Ser. Math. Phys. Sci. **212**, 440 (1952)
21. B. Bhushan, *Introduction to Tribology* (Wiley, New York, 2002)
22. F.P. Bowden, D. Tabor, *The Friction and Lubrication of Solids*, 3rd edn. (Oxford University Press, New York, 2001)
23. V.F. Petrenko, R.W. Whitworth, *Physics of Ice* (Oxford University Press, New York, 1999)
24. I.I. Kozlov, A.A. Shugai, Fluid Dyn. **26**, 145 (1991)
25. A.J. Fowler, A. Bejan, Int. J. Heat Mass Transfer **36**, 1171 (1993)
26. S.C. Colbeck, J. Glaciol. **34**, 78 (1988)
27. N.H. Fletcher, Philos. Mag. **7**, 255 (1962)
28. N.H. Fletcher, Philos. Mag. **18**, 1287 (1968)
29. R. Lacmann, I.N. Stranski, J. Cryst. Growth **13–14**, 236 (1972)
30. J.G. Dash, F. Haiying, J.S. Wettlaufer, Rep. Prog. Phys. **58**, 115 (1995)
31. C.A. Knight, Philos. Mag. **23**, 153 (1971)
32. M. Elbaum, S.G. Lipson, J.G. Dash, J. Cryst. Growth **129**, 491 (1993)
33. L. Makkonen, J. Phys. Chem. B **101**, 6196 (1997)
34. N. Fukuta, J. Phys. vol. 48 (Paris, 1987), pp. 503–509
35. G.J. Kroes, Surf. Sci. **275**, 365 (1992)
36. J.P. Devlin, V. Buch, J. Phys. Chem. **99**, 16534 (1995)
37. Y. Furukawa, H. Nada, J Phys. Chem. B **101**, 6167 (1997)
38. N. Materer et al., Surf. Sci. **381**, 190 (1997)
39. S.C. Colbeck, Am. J. Phys. **63**, 888 (1995)
40. ISU, International Skating Union ISU, http://www.isu.org(2009)
41. J.J. De Koning, H. Houdijk, G. De Groot, M.F. Bobbert, J. Biomech. **33**, 1225 (2000)
42. D.D. Higgins, B.A. Marmo, C.E. Jeffree, V. Koutsos, J.R. Blackford, Wear **265**, 634 (2008)
43. K. Itagaki, G.E. Lemieux, N.P. Huber , J. Phys. vol. 48 (Paris, 1987), pp. 297–301
44. D.C.B. Evans, J.F. Nye, K.J. Cheeseman, Proc. R. Soc. London Ser. A, Math. Phys. Sci. **347**, 493 (1976)
45. H. Liang, J.M. Martin, T.L. Mogne, Acta Mater. **51**, 2639 (2003)
46. M. Akkok, C.M.M. Ettles, S.J. Calabrese, J. Tribol. **109**, 552 (1987)

47. B.V. Derjaguin, Wear **128**, 19 (1988)
48. L. Bäurle, D. Szabó, M. Fauve, H. Rhyner, N.D. Spencer, Tribol. Lett. **24**, 77 (2006)
49. J.J. De Koning, G. De Groot, G.J. Van Ingen Schenau, J. Biomech. **25**, 565 (1992)
50. D. Slotfeldt-Ellingsen, L. Torgersen, J. Phys. D Appl. Phys. **16**, 1715 (1983)
51. S.J. Calabrese, R. Buxton, G. Marsh, Lubr. Eng. **36**, 283 (1980)
52. F. Albracht, S. Reichel, V. Wlnkler, H. Kern, Materialwissenschaft und Werkstofftechnik **35**, 620 (2004)
53. F.P. Bowden, Proc. R. Soc. London, Ser. A, Math. Phys. Sci. **217**, 462 (1953)
54. D. Kuroiwa, J. Glaciol. **19**, 141 (1977)
55. M. Montagnat, E.M. Schulson, J. Glaciol. **49**, 391 (2003)
56. B.A. Marmo, J.R. Blackford, C.E. Jeffree, J. Glaciol. **51**, 391 (2005)
57. P. Oksanen, J. Keinonen, Wear **78**, 315 (1982)
58. A.K. Stiffler, J. Tribol. **108**, 105 (1986)
59. S.J. Jones, H. Kitagawa, K. Izumiyama, H. Shimoda, Ann. Glaciol. **19**, 7 (1994)
60. A.-M. Kietzig, S.G. Hatzikiriakos, P. Englezos, J. Appl. Phys. **106**, 024303 (2009)
61. Y. Ando, in *Scanning Probe Microscopy/Optical Memories*, vol. 43 (Japan Society of Applied Physics, Japan, 2004) , pp. 4506–4510
62. D. Buhl, M. Fauve, H. Rhyner, Cold Reg. Sci. Technol. **33**, 133 (2001)
63. L. Bäurle, T.U. Kaempfer, D. Szabó, N.D. Spencer, Cold Reg. Sci. Technol. **47**, 276 (2007)
64. A.M. Kietzig, S.G. Hatzikiriakos, P. Englezos, J. Glaciol. **56**, 473 (2010)
65. S. Ducret, H. Zahouani, A. Midol, P. Lanteri, T.G. Mathia, Wear **258**, 26 (2005)
66. B.A. Marmo, I.S. Farrow, M.P. Buckingham, J.R. Blackford, Proc. IME. J. Mater. Des. Appl **220**, 189 (2006)
67. W. Federle, W.J.P. Barnes, W. Baumgartner, P. Drechsler, J.M. Smith, J. R. Soc. Interface **3**, 689 (2006)
68. H. Zeng et al., Langmuir **25**, 7486 (2009)
69. S.C. Colbeck, Surf. Coat. Technol. **81**, 209 (1996)
70. S.C. Colbeck, L. Najarian, H.B. Smith, Am. J. Phys. **65**, 488 (1997)
71. W. Barthlott, C. Neinhuis, Planta **202**, 1 (1997)
72. A.M. Kietzig, S.G. Hatzikiriakos, P. Englezos, Langmuir **25**, 4821 (2009)
73. A-M. Kietzig, M.N. Mirvakili, S. Kamal, S.G. Hatzikiriakos, P. Englezos, J. Adhesion Sci. Technol. **25**, 1293 (2011)
74. H. Strausky, J.R. Krenn, A. Leitner, F.R. Aussenegg, Appl. Phys. B: Lasers Opt. **66**, 599 (1998)
75. A. Lehtovaara, Wear **115**, 131 (1987)
76. A.K. Stiffler, J. Tribol. Trans. ASME **106**, 416 (1984)
77. A.-M. Kietzig, S.G. Hatzikiriakos, P. Englezos, J. Appl. Phys. **107**, 081101 (2010)

Chapter 10
Green Lubricants: Role of Additive Size

Pradeep L. Menezes, Michael R. Lovell, M. A. Kabir, C. Fred Higgs III and Pradeep K. Rohatgi

Abstract The depletion of the world's crude oil reserve, increased oil prices, and the demand to protect the environment against pollution exerted by hydraulic and gear oils have brought about renewed interest in the development and use of green lubricants. In this light, many automotive and manufacturing industries are actively seeking out new green lubricants. Although, many green lubricants have exhibited excellent properties, more improvement in their friction and wear performance are still needed for them to become mainstream. Consequently, environmental friendly additive materials are being included in green lubricant formulations to improve their friction and wear properties. In this study, a review of the tribological behavior of the green lubricants presently available has been performed. Overall, the review indicates that green lubricants can significantly outperform conventional lubricants with respect to frictional and wear performance. In addition, the review shows that the size and composition of the lubricant additives play an important role in determining a green lubricant's overall performance.

Keywords Green lubricants · Powder additives · Friction · Wear

P. L. Menezes · M. R. Lovell (✉)
Department of Industrial Engineering, University of Wisconsin-Milwaukee, Milwaukee, WI 53211, USA
e-mail: mlovell@uwm.edu

M. A. Kabir · C. F. Higgs III
Department of Mechanical Engineering, Carnegie Mellon University, Pittsburgh, PA 15213, USA

P. K. Rohatgi
Department of Materials Engineering, University of Wisconsin-Milwaukee, Milwaukee, WI 53211, USA

M. Nosonovsky and B. Bhushan (eds.), *Green Tribology*,
Green Energy and Technology, DOI: 10.1007/978-3-642-23681-5_10,

10.1 Introduction

The United States consumes more oil than any other country in the world. According to the Geographic Information Systems Laboratory [1], the United States oil consumption exceeds 20 million barrels per day, which represents more than 25% of the world's total consumption. Despite predictions that the United States will exhaust its supply of oil in as little as 40 years, the demand for oil is predicted to continue to increase over the next several decades. In the United States, 54% of the oil is consumed as automotive lubricants (engine oils and transmission fluids) and 44% of the oil is consumed as industrial lubricants (hydraulic fluids and gear oils) [2, 3]. Nearly all lubricants used in the automotive and manufacturing sectors are oil or grease based. These products are not typically environmentally friendly or biodegradable and can introduce significant quantities of pollutants into the waste stream. The long-term impacts that these lubricants have are cumulative and are ultimately detrimental to humans, plants, fish, and wildlife. Due to their potential negative impacts, the Environmental Protection Agency and other government agencies have imposed increasingly stringent regulations on the use, containment, and disposal of oil and grease-based lubricants. These regulations have changed the landscape of the lubrication marketplace over the last decade as new lubricants need to be developed to address a combination of environmental, health, economic, and performance challenges. Since the disposal costs of traditional lubricants are also increasing exponentially, it has become essential to develop and implement lubricants that come from natural resources.

10.2 Background

A lubricant is a substance introduced between two moving surfaces to primarily reduce friction, improve efficiency, and reduce wear. A lubricant may also serve the function of dissolving or transporting foreign particles, carrying away contaminations and debris, preventing corrosion or rust, sealing clearances, and dissipating heat [4–6]. Typically lubricants contain 90% base oil and less than 10% additives. In general, there are three categories of lubricants–liquid, solid, and gaseous. Liquid lubricants may be characterized in a variety of methods. One of the most common method is by the type of base oil used. The most common types of base oils are lanolin (wool grease, natural water repellant), water, mineral oils (derived from crude oil), vegetable oils (derived from plants and animals), and synthetic oils (consisting of chemical compounds which are synthesized from compounds other than crude oil).

Solid (dry) lubricants include grease and powders. Some of the commonly known powder lubricant materials are graphite, molybdenum disulphide (MoS_2), tungsten disulphide (WS_2), and titanium dioxide (TiO_2). Solid lubricants offer lubrication at temperatures (up to 350°C) that are higher than many liquid and oil-based lubricants can attain. Solid lubricants such as Teflon or PTFE (poly tetra fluoro ethylene) are typically used as a coating layer to provide a non-stick surface. Gaseous lubricants have a much lower viscosity than liquid or solid lubricants. They also exhibit lower heat capacity and higher compressibility than liquid or solid lubricants. Some examples for gaseous lubricants are air, technical gases, steam or liquid–metal vapors.

In the automotive sector, the most commonly used fluids are engine oils (petrol and diesel engine oils), automatic transmission fluid, gearbox fluids, brake fluids, and hydraulic fluids. In the manufacturing sector, a wide range of oils are utilized including hydraulic oils, air compressor oils, gas compressor oils, gear oils, bearing and circulating system oils, refrigerator compressor oils, and steam and gas turbine oils. In aviation, gas turbine and piston engine oils are the most common.

Additives are substances used to improve the performance of lubricants. Additives are selected based on their ability to reduce friction and wear, increase viscosity, improve viscosity index, resist corrosion and oxidation, increase lifetime, and minimize contamination. The main families of additives are antioxidants, anti-wear formulations, anti-foaming agents, corrosion inhibitors, detergents, demulsifying/emulsifying, extreme pressure, friction modifiers, metal deactivators, and viscosity index improvers [5].

10.3 The Green Lubricants

The term "green lubricants" is generally used for lubricants manufactured from vegetable or other natural sources. The other common names for green lubricants are "environmentally acceptable", "environmentally benign", and "environmentally friendly". The development of green lubricants to displace petroleum-based products is presently an emerging area of research in the engineering community. Among other factors, this development has been driven by dramatic increases in the price of crude oil and the movement towards using biodegradable oils from renewable resources. It is generally assumed that green lubricants must not be toxic and must biodegrade in a relatively short time. With better biodegradable and toxicity properties than conventional petroleum-based products, green lubricants have tremendous potential for use in the industrial sector [7–17]. Although, industry standards do not yet exist, engineers using green lubricants must carefully match applications with specific green lubricant products.

10.3.1 The Green Liquid Lubricants

Green liquid lubricants are lipid materials that are liquid at room temperature. Most of the green liquid lubricants are produced from vegetables (e.g. rapeseed/canola, corn, or soybean oil), synthetic esters, polyalkylene glycols, or severely hydrotreated petroleum-based oils. Vegetable oils have a long history of use as lubricants—together with animal fats—as they dominated the lubrication industry before the petroleum age [11]. Problems of limited oxidative stability and poor low temperature performance gradually led to their replacement by fossil oils and later by synthetic oils. When compared to mineral and synthetic oils, vegetable oils have a number of distinct advantages including higher lubricity, lower volatility, higher shear stability, detergency and dispersancy, and a substantially higher viscosity index. In fact, vegetable oils offer a large range of kinematics viscosity, from 2.37 to 8.53 mm^2/s at 100°C and from 7.85 to 35.01 mm^2/s at 40°C [17]. Today, vegetable oils represent a nearly inexhaustible source of raw materials for lubricants. In 2004, the world production of vegetable oil exceeded 100 million tons [9] and most of this oil was used by the food industry, with only a small percentage entering the lubrication market, making it an under-utilized commodity. Presently, the highest volume of green liquid lubricant production is for hydraulic applications.

The physical structure of 'green' vegetable oils is very different from that of traditional lubricants. Plant oils are composed of triacylglycerol, which consists of three fatty acids esterified to a glycerol molecule. Most oilseed crops produce triacylglycerol containing a mixture of fatty acids that have varying chain lengths; the most common lengths of carbon chains in these triacylglycerol are 16, 18, and 20. Figure 10.1 shows the structure of triglycerides with varying chain lengths of fatty acids. It is the fatty acid composition of the oil that determines its characteristics. In most lubricant applications, vegetable oils are used either as the intact triacylglycerol molecule or the oil is reacted with methanol to produce methyl esters and glycerol [11]. A triacylglycerol molecule containing saturated 18 carbon fatty acids is solid at room temperature. The presence of the double bonds makes the oil liquid, but also increases oxidative instability. The recent introduction of ultra-high oleic (>80%) varieties of oilseed crops such as sunflower and canola has made a new type of oil available with characteristics far superior to the oil from existing temperate crops. Oils highly enriched in this monounsaturated fatty acid have improved lubricity and an oxidative stability that surpasses Group 1 petroleum-based oils [11]. Despite their significant promise, these improved vegetable oils still have lower oxidation stability and poorer low-temperature performance than many conventional oils [18, 19]. These and other limitations must be further researched and addressed before a new class of green lubricants can be used to replace traditional grease and oil lubricants.

10.3.2 The Green Solid (Powder) Lubricants

The term "powder lubricants" is usually given to lamellar solids that have low interlayer friction [20]. Over the past 30 years, dry powder lubricants have been employed in a variety of sliding contacts as an alternative to conventional liquid lubricants. Some of the interesting attributes of powder lubricants are that they

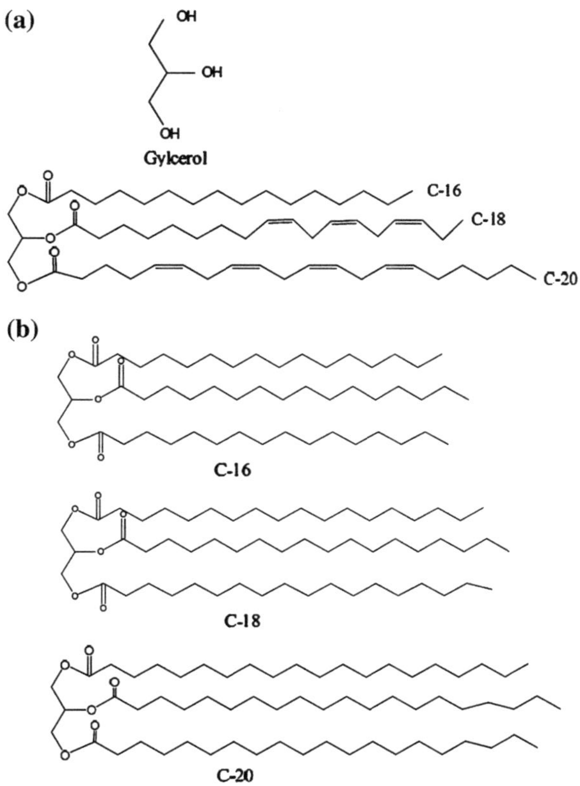

Fig. 10.1 **a** Triglycerides present with various mixtures (i.e. 16, 18, and 20) of fatty acids. They are either saturated or unsaturated esters of fatty acids. **b** some chemical structures with different chain lengths (i.e. 16, 18, and 20) of fatty acids

(1) adhere to surfaces to form a protective boundary layer that prevents wear [21–25], (2) act as a lubricant in sliding contacts by accommodating relative surface velocities [26–30], and (3) lubricate at high temperatures [31–35]. Several of these lubricants—molybdenum disulfide, tungsten disulfide, titanium oxide, and boron nitride—were evaluated for their lubrication behavior at a variety of sizes and found to exhibit excellent tribological performance [36–41].

10.3.2.1 Green Boric Acid Powder Lubricant

Efforts are being made to develop and implement powder lubricants that come from natural resources. As reported in the literature, solid lubricants such as boric acid have excellent lubrication properties without requiring expensive disposal [42–45]. Thus, boric acid is another often overlooked lamellar solid that has been found to be an effective green powder lubricant. The effectiveness of boric acid can be attributed to its low friction and shear strength values. As reported in the literature [46], the shear strength of boric acid has been experimentally determined to be 23 MPa, and its coefficient of friction has been measured to be less than 0.02 in ambient environments. These values are very similar to the more commonly used and synthetic-based molybdenum disulphide, which has a measured shear strength value of 24 MPa. Two other important characteristics of boric acid for use as a lubricant are that it is readily available and environmentally safe. In its solid form, boric acid is a weakly acidic white powder that is soluble in water (about 27 wt% in boiling water and about 6% at room temperature), soft, ductile, stable, free flowing, and easily handled. It is very inexpensive, as finely ground technical grade boric acid powder (>99% pure) is commercially available for less than US$ 4.5 per kilogram. The Environmental Protection Agency has established that boric acid is benign [10, 47] and the Clean Water Act does not classify it as a pollutant. In fact, a dilute water solution of boric acid is also commonly used as a mild antiseptic and eyewash. In addition, the industrial marketplace has already accepted boric acid as an engineering material. The United States is the world's largest producer of boron compounds, as large domestic reserves of boron materials reside in lake sediments and brines. The present markets for boric acid and boric oxide in the United States include glass making (78%), fire retardants (9%), agricultural fertilizers (4%), and industrial applications such as metal plating and finishing, paints and pigments, electroplating and cosmetics (9%).

Examining its physical characteristics, boric acid is the common term for orthoboric acid (H_3BO_3), which is a hydrate of boric oxide (B_2O_3). When in contact with water, boric oxide will readily hydrate, converting into the lamellar solid boric acid. Under atmospheric pressure, boric acid dehydrates above 170°C and reverts back to a substantially less lubricious form of boric oxide. Boric oxide is known to soften around 400°C at atmospheric pressure. As shown in Fig. 10.2, the fundamental molecular structure of boric acid (below 170°C) allows it to act as an effective solid lubricant film. When crystallized, boric acid forms weak Van der Waals bonds between individual layers and strong hydrogen bonds within a layer. Such a bonding structure makes the structural properties of boric acid highly anisotropic. Therefore, when tangentially loaded, the individual lamellae slide relatively easy over one another. This is in contrast to the normal direction where the boric acid has a relatively high load carrying capacity. Hence, when properly aligned with a substrate, boric acid will exhibit minimal friction and provide effective separation between surfaces. In fact, it has been shown that the friction coefficient of sliding interfaces with a boric acid film decreased with increasing pressure [48].

Considerable research related to the solid lubrication properties of boric acid have been carried out over the past several decades [42, 43, 47, 49, 50]. These works have primarily focused on the performance of boric acid in high-temperature applications. Research specifically aimed at the use of boric acid in engineering systems has also been undertaken [44, 45, 51]. These studies have indicated that boric acid's unique layered inter-crystalline structure makes it a very promising solid lubricant material because of its relatively high load carrying capacity and low steady-state friction coefficient. Other important research has investigated the use of boric acid as a lubricant in manufacturing processes such as forming, drilling,

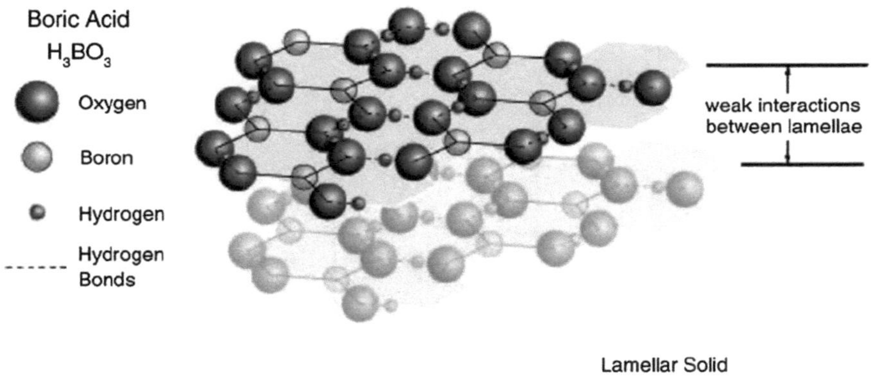

Fig. 10.2 Lamellar structure of boric acid

and machining. In metal forming applications [52–55], it was shown that boric acid powder provided very low friction (0.04) between an aluminum workpiece and a steel forming tool. During these processes, the post-fabrication cleaning of the boric acid lubricant was environmentally safe, nontoxic, and water soluble. In drilling experiments with sapphire tools [53], results indicated that the addition of boric acid to distilled water increased the rate of drilling of polycrystalline alumina by a factor of two. In addition, boric acid was found to help reduce friction and corrosion when mixed with cutting and grinding fluids [46, 56]. Using high speed pin-on-disk experiments [46], the performance of boric acid during the sliding of an alumina pin on an AISI-M50 bearing steel disk at a maximum central Hertzian contact pressure of 1.9 GPa was examined. For sliding speeds of approximately 1 m/s, the resulting friction revealed a reduction in friction from 0.4 for the dry contact to approximately 0.15 for the boric acid lubricated contact (see Fig. 10.3). As illustrated in the figure, continuous powder delivery of the boric acid successfully produced a lubricous surface film and low friction until the delivery of powder was halted. It is very important to note that the success of boric acid as a lubricant in the above studies was dependent on the continuous replenishment of the boric acid into the contact region.

Despite its known benefits, it is demonstrated that there are difficulties associated with utilizing boric acid alone as a lubricant in repeated or extended duration contacts [10]. These drawbacks are due to the fact that boric acid powder can be dehydrated or forced out of the contact zone during sliding contact. Thus, the widespread use of boric acid has been limited by the fact that it must be continuously replenished, which requires the development of spray and recovery systems that would significantly change, in for example, the present manufacturing process lines. Therefore, a logical alternative to replace oil/grease lubricants is to incorporate an environmentally friendly liquid lubricant that carries green solid additives such as boric acid.

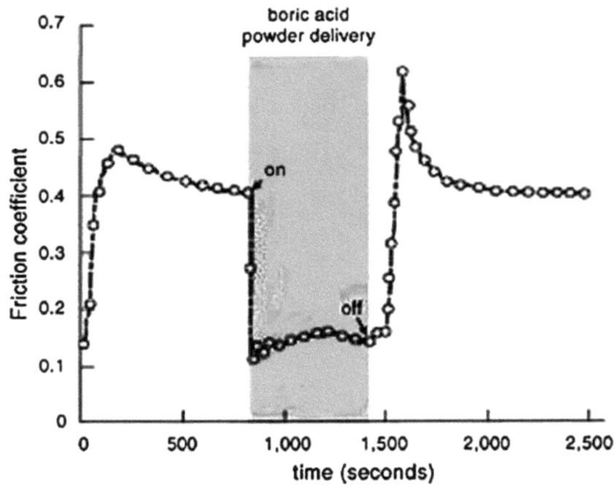

Fig. 10.3 Friction performance under boric acid lubricated conditions [46]

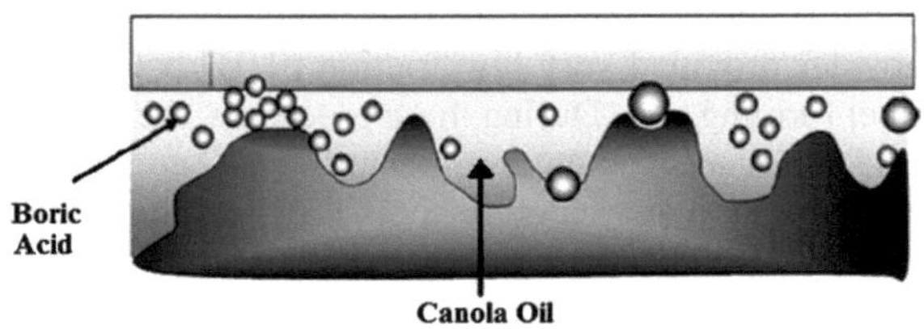

Fig. 10.4 Green liquid lubricant (i.e. canola oil) carries green solid lubricant (i.e. boric acid) providing a mixed lubrication system

10.3.3 Green Powder-Based Green Lubricants

As conceptually depicted in Fig. 10.4, an environmentally friendly liquid lubricant (i.e. canola oil) can carry the green solid lubricant (third body boric acid particles) to provide a mixed lubrication system between contacting surfaces. Hence, efforts have been made to analyze a green, petroleum-free lubricant that was produced by mixing two environmentally benign components—canola oil and boric acid powder [14]. The boric acid powder and canola oil lubricant mixture demonstrates multifunctional lubrication performance, where surfaces are separated by a liquid lubricant film and protected by a solid powder. In the previous work [14], micron-scale (100–700 μm) boric acid particles were added to canola oil. To determine the optimum composition of the boric acid–canola oil lubricant mixture, a series of experiments were conducted at different volume fractions of canola oil with a constant particle size distribution. Based on these experiments, a mixture of 7% volume was found to outperform the other lubricant mixtures. Experiments were conducted to study the tribological properties for this lubricant mixture using a pin-on-disk apparatus with spherical copper pins and aluminum disks. Figure 10.5 shows the variation of friction with sliding distance. As depicted in Fig. 10.5,

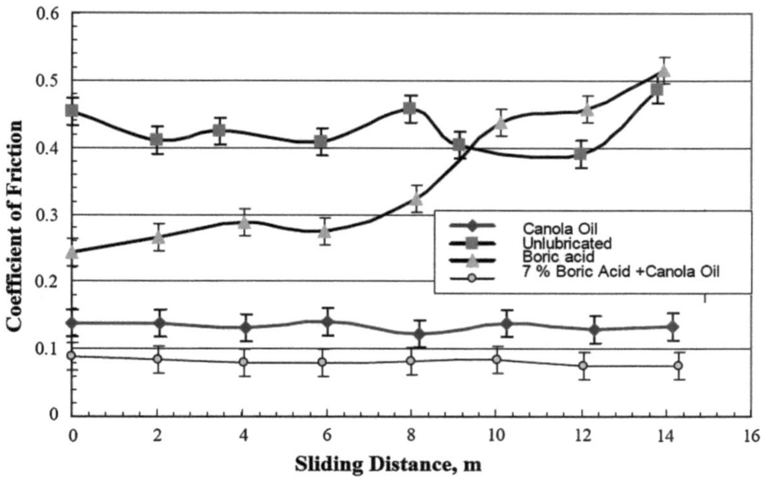

Fig. 10.5 Variation of coefficient of friction with sliding distance for various lubricants when copper pins slide against aluminum disks [14]

the unlubricated and boric acid cases yielded the highest friction coefficients over the sliding distances tested. In the unlubricated case, the friction is large because the surface interaction is dominated by metal to metal asperity contact. In the boric acid condition, the powder lubricant's performance is poor because the solid lubricant particles are forced out of the contact interface during each sliding cycle; hence, the boric acid only provides separation of asperities at the early stages of sliding but ultimately generates the same frictional value as the unlubricated case. In fact, Fig. 10.5 shows that the friction coefficient steadily increases for the boric acid case from an initial value of 0.24 to more than 0.50 at the end of the experiments. The most effective lubricants in the experiments were the canola oil and canola oil–boric acid mixture. For these cases, the surface tension properties of the canola oil allowed the lubricant to remain in the contact interface and partially separate the contacting asperities for the duration of the experiments. The decreased asperity interaction is signified by the fact that the friction coefficient is relatively low for these cases and remains constant or slightly decreases over the sliding distances tested. Furthermore, as illustrated in Fig. 10.5, the lubricant mixture of 7% boric acid and canola oil clearly showed the lowest friction coefficient. Thus, it was found that the canola oil and boric acid mixture exhibited multifunctional lubrication behavior. Essentially, the viscosity and surface tension properties of the canola oil minimized surface interaction and the friction coefficient while the powder particles formed a protective boundary layer that coated the tribo-surfaces.

10.4 Green Lubricants for Automotive and Manufacturing Industries

In the automotive industry, one of the best ways to increase fuel efficiency is to reduce vehicle weight by using lightweight aluminum or magnesium alloyed parts. Forming these alloyed parts can be very difficult due to high friction, extreme pressure requirements, and the inability of conventional lubricants to prevent wear during metalworking. Also, most conventional lubricants are flammable and contain chlorine, phosphorus, and sulfur-bearing additives that are potentially hazardous: removing these lubricants from finished products and treating them for disposal is difficult and costly. Thus, it has long been recognized that metal forming operations have significant environmental impact. Despite several decades of research on the subject, lubricants used in forming operations continue to pose substantial technological and economic challenges to the tribological community. The present challenges primarily stem from various factors such as (1) Nearly all lubricants used in sheet metal forming are made from petroleum-based resources which are not environmentally friendly and are becoming substantially more expensive, (2) Forming lubricants often carry a health risk to humans both during primary use and in their disposal, (3) Over the past several decades, there has not been a novel lubricant introduced in the manufacturing community to significantly increase the formability. Thus, use of boric acid-based green lubricant has the potential to overcome these problems. Its exceptionally low friction coefficient prevents the metals from sticking or transferring to the die or roll surfaces. Boric acid greatly reduces the friction and wear of dies and molds and at the same time provides an ultra-smooth surface finish on final products. After metal-forming operations, parts can be rinsed in water to remove the excess lubricants—no toxic or flammable solvents are necessary. The use of boric acid can decrease the unit cost for automotive parts because the near-perfect finished products do not require secondary machining or grinding. Other potential automotive applications of boric acid include its use as a lubricant for gears and bearings. Boric acid may also be mixed with existing liquid and solid lubricants.

Efforts have been made to evaluate the interfacial friction characteristics of canola oil and boric acid lubricants in metal forming operations. To simulate their performance, a strip tensile friction simulator [16] was utilized. This apparatus has been shown to effectively model the strain behavior encountered in a deep drawing process and accurately replicates the frictional effects experienced during sheet bending. By means of the simulator, the friction coefficient between a die (carbon steel) and a steel sheet was measured for four different lubricant conditions: unlubricated, canola oil (viscosity of 33 cSt at 20°C), transmission fluid (viscosity of 0.028 Pa s at 40°C), and a combination of boric acid and canola oil (viscosity of 0.276 Pa s at 20°C). As shown in Fig. 10.6, a lubricant consisting of canola oil mixed with boric acid crystals (5 wt% and 100 μm average particle size) significantly outperformed several other lubricants with respect to the measured friction coefficient in the specialized experiments. This lubricant had a steady-state friction

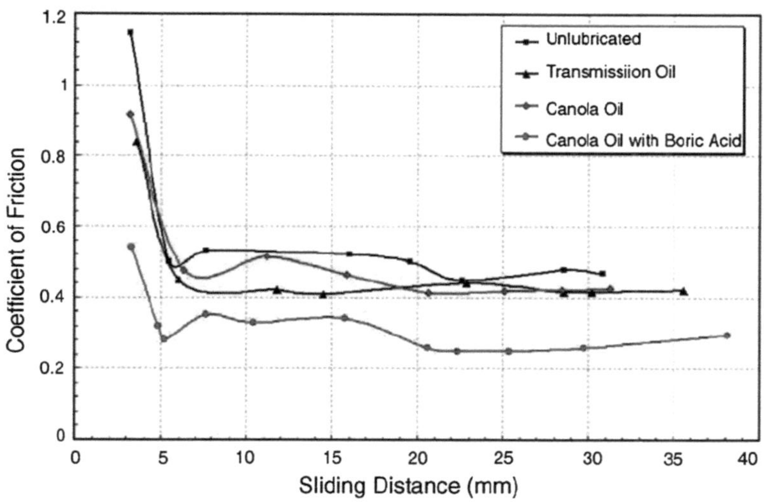

Fig. 10.6 Variation of coefficient of friction with sliding distance for various lubricants obtained in the metal forming experiment [16]

value that was 54% lower than the unlubricated condition and 44% less than the transmission fluid case. Such a finding indicated that the novel lubricant may be used in manufacturing processes where replenishment is not feasible and disposal of non-environmentally friendly lubricants is expensive.

Studies were also conducted to determine the relative performance of boric acid-based lubricants and other oil combinations in a commercial brake valve assembly [10]. Pin-on-disk experiments were performed under the same operating conditions experienced by the brake valve assembly. To represent the slide valve, in the experiments, the disk samples were prepared from copper alloy C84400. Likewise, to represent the bushing material, a pin was constructed from copper alloy C93200. Figure 10.7 shows the variation of coefficient of friction with sliding distance obtained in the pin-on-disk experiment. As depicted in Fig. 10.7, the base materials in the absence of a lubricant yielded the highest friction coefficients over the sliding distance tested. Figure 10.7 also demonstrates that boric acid alone was an ineffective lubricant for the operating conditions examined. The increase in friction in boric acid case is caused by the fact that the amount of boric acid in the contact interface gradually diminishes with time. It is important to note, however, that in terms of the final friction coefficient values, the boric acid was significantly lower than the unlubricated case. Although the boric acid did not maintain an optimal separation level within the contact interface, a better frictional performance was attained because a portion of the lubricious boric acid remained in the contact region. For the transmission fluid case, the coefficient of friction was found to increase over the sliding range tested. This increase was attributed to the degradation of the lubricant film over time. Further, a lubricant combination of

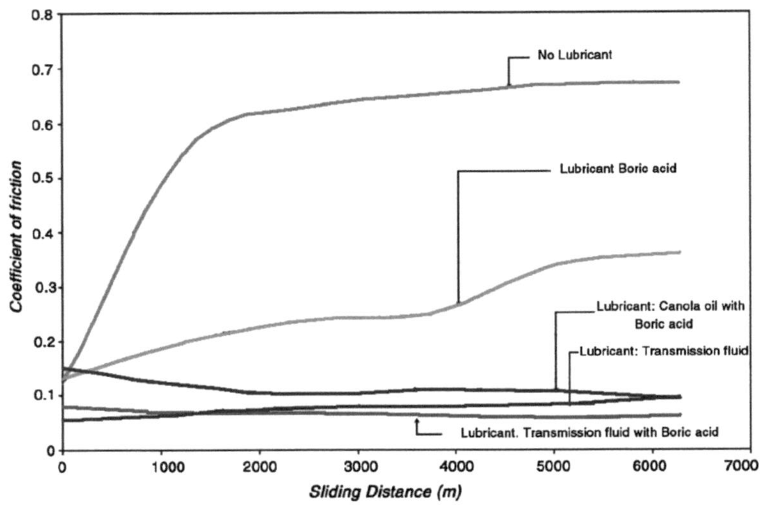

Fig. 10.7 Variation of coefficient of friction with sliding distance for various lubricants when copper alloy pins slide against copper alloy disks [10]

transmission fluid and boric acid (5% weight) was found to provide optimum friction properties as the friction coefficient actually decreased over a significant portion of the testing period. The friction coefficients in a combined boric acid and canola oil lubricant were comparable but slightly larger than the transmission oil and the combined boric acid and transmission fluid cases. The boric acid and canola oil case showed no degradation over the entire range of the sliding tests and demonstrated the lowest wear rate of any of the lubricants tested. In fact, the wear rate was nearly 50% lower than the combined transmission fluid and boric acid lubricant mixture. The better wear performance was attributed to the fact that the boric acid and canola oil did not degrade over the entire range of sliding experiments. Such a finding is extremely important as it indicates that the lubricant can be used in extended duration applications where replenishment is not feasible.

10.5 The Size of the Additive and the Scale Effect on the Lubrication

In the literature, considerable effort has been made to study the effect of powder lubricant size (micron-scale, submicron-scale, and nano-scale) on system performance [36–38, 41]. Numerous studies have indicated that sub-micron-particulate additives of graphite, MoS_2, and other lamellar solids have improved the performance of base oils [36, 57–59]. Sub-micron forms of these powder lubricants have been synthesized using chemical vapor deposition, liquid precipitation, electronic

Fig. 10.8 Variation of coefficient of friction with wear time for lubricants containing common microns sized MoS_2 (c-MoS_2), nano sized MoS_2 (n-MoS_2), and mixture of c-MoS_2 and n-MoS_2 particles [36]

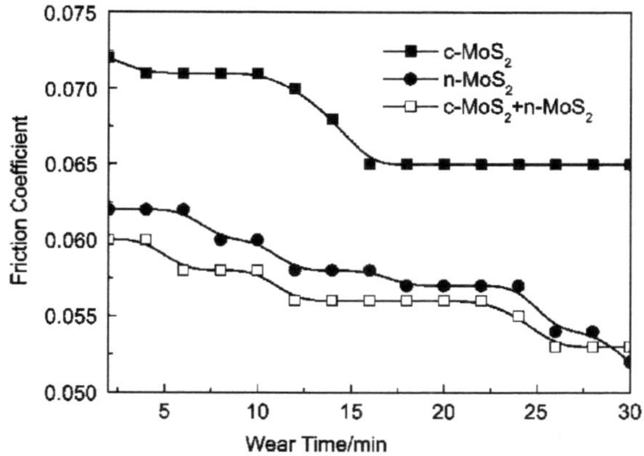

chemistry deposition, thermal desulphurization, and mechanical grinding or milling. In many cases, sub-micron scale powder lubricants offer an enhanced tribological performance when compared to micron-scale powder lubricants because their size is on the order of the asperities for surfaces [60–64]. Recently, it has been shown that the lubricant containing a mixture of micro- and nano-sized particles demonstrated better tribological performance than lubricants containing either micro- or nano- sized particles alone. Hu [36] studied the size effect of MoS_2 particles on tribological performance. In the experiments, nano-MoS_2 (n-MoS_2) particles and commercial common-MoS_2 (c-MoS_2) particles (approximately 1.5 μm in diameter) were dispersed in liquid paraffin with different concentrations and ratios in order to study their lubrication capacity, friction reduction, and wear resistance. The tribological experiments were carried out by a four-ball tribometer in which extreme pressure, wear scan diameter, and friction coefficient were measured. Figure 10.8 shows the variation of coefficient of friction with wear time for lubricants containing c-MoS_2, n-MoS_2, and mixture of c-MoS_2 and n-MoS_2 particles. The experimental results showed that the loading capacity of liquid paraffin was increased with MoS_2 particle content. The liquid paraffin containing the mixture of n-MoS_2 and c-MoS_2 particles had a better wear resistance, friction-reducing performance, and extreme pressure property than the liquid paraffin containing either c-MoS_2 or n-MoS_2 particles. When the optimal mixing ratio of n-MoS_2 and c-MoS_2 is 20 wt%, was studied, the loading capacity reached the highest value. The difference in the tribological performance between n-MoS_2 and c-MoS_2 was attributed to the surface and interfacial size effect of nano-particles and the formation of molybdenum trioxide thin film on the contacting surface. The effect of lubricant mixture containing c-MoS_2 and n-MoS_2 on friction and wear reduction was not fully explained in the study.

In other work, Narayan [65] studied the effect of lubricants having micro-particles and nano-particles and their combinations on fuel efficiency. The nano-particles were chosen from a class of hard materials, such as alumina, silica, ceria, titania, diamond, cubic boron nitride, and molybdenum oxide. The micro-particles were chosen from a class of materials of layered structures, such as graphite, hexagonal boron nitride, magnesium silicates (talc), and molybdenum disulphide. The nano-micro combinations were chosen from the same materials. This group of materials included zinc oxide, copper oxide, molybdenum oxide, graphite, talc, and hexagonal boron nitride. The nano- and micro-particle combinations improved coefficients of friction, fuel efficiency (up to 35%) and exhaust emissions (up to 90%).

10.6 The Role of Boric Acid Additive Size on Green Lubricants' Performance

Based on the success of using micron-scale boric acid particles as an additive, efforts were made to determine the relative performance of sub-micron and nano scale boric acid particles in canola oil for metal forming applications. Despite several patents [66, 67] on the subject, there are no studies in the archival literature that quantitatively investigate the performance of boric acid particle additives in the colloidal size range. Recently, Lovell et al. [68] studied the tribological performance of nano-scale (20 nm), sub-micron (average size 600 nm), micron (average size 4 μm), and combined sub-micron and micron boric acid additives in base canola oil. As a basis for comparison, 5 μ size MoS_2 powder additives mixed with canola oil was also studied. Figure 10.9 shows the scanning electro micrographs of micron, sub-micron and nano sized boric acid particles. Once prepared, friction and wear experiments were carried out on the lubricants using the pin-on-disk apparatus at ambient conditions. The pin-on-disk investigation utilized spherical Cu pins (6.5 mm diameter) and Al disks (66 mm diameter). In the experiments, a 100 g normal load was applied and a constant sliding velocity of 120 mm/s was utilized. The friction force was continuously recorded for a sliding distance of 7,500 m (65,000 disk revolutions). Although this sliding distance adequately captured the variation and performance of the lubricants over time, it does not represent the maximum distance for which the lubricants should ultimately be investigated; automotive and industrial applications often operate for several 100 h and over extremely long sliding distances. Figure 10.10 shows the variation of coefficient of friction with sliding distance for the five lubricants studied. Although the exact sliding distance location of the minimum friction varies for each lubricant tested, Fig. 10.10 shows that—with the exception of 20 nm boric acid—each of the lubricants followed a similar trend where the coefficient of friction significantly decreases to a minimum value before slightly increasing over the remainder of the sliding distance. Such a trend suggests that the powder additives in the canola oil form a thin 'protective layer' between the pin and disk surfaces as the micron and sub-micron scale particles are squeezed into the contact interface. This powder layer, which helps carry the load and provides

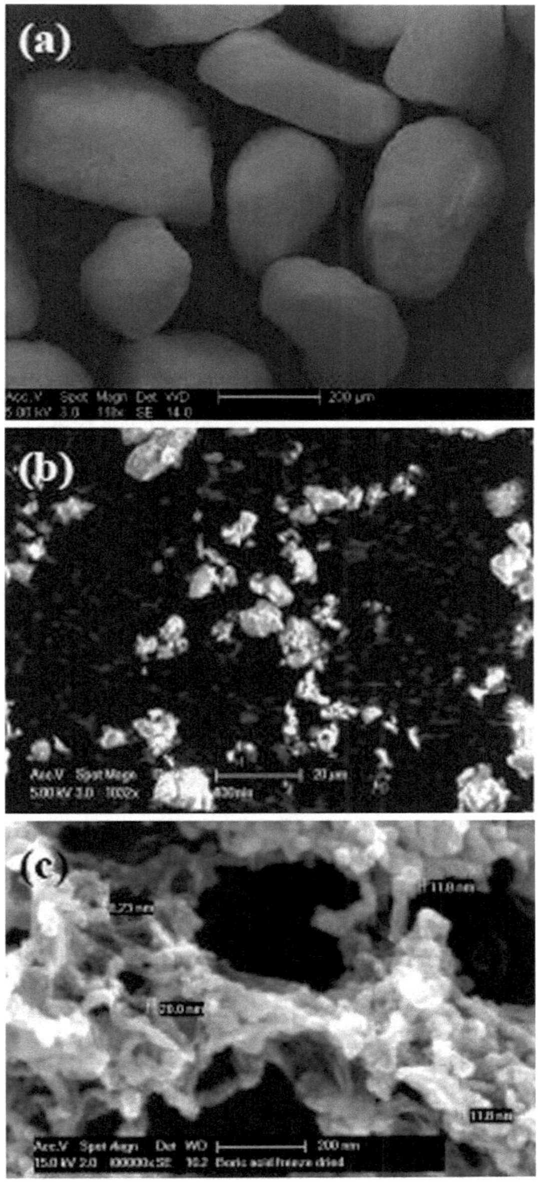

Fig. 10.9 Scanning electron micrograph of **a** 250 μ, **b** 4 μ, and **c** nano (20 nm) boric acid particles [68]

low shear resistance, increases over time as more particles are forced between the contacting asperities which naturally round-off over time. The fact that the friction increases after a minimum value is attained demonstrates that the boric acid and MoS_2 particles 'run-out' and are no longer available to enter the contact interface;

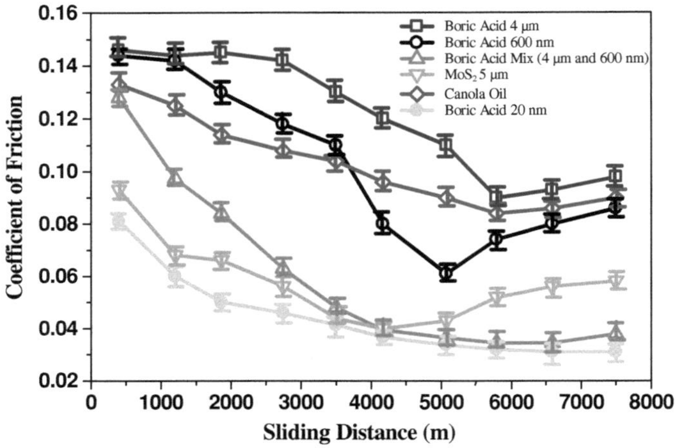

Fig. 10.10 Variation of coefficient of friction with sliding distance for various lubricants [68]

this causes the protective layer to wear away over time which leads to the increase in friction at the latter stages of the experiments. For the case of the 20 nm boric acid particles, the shape of the curve in Fig. 10.10 is unique because the friction continuously decreases to an asymptotic value at the end of the sliding experiments. It is possible that if the experiments had been run to a greater sliding distance, the 20 nm particles would have exhibited a behavior similar to the other additives. Unlike the other additive cases, however, the 20 nm particles go to colloidal suspension within the canola oil; this allows the lubricant particulates to easily enter between the asperities of the pin and disk throughout the sliding experiments. Thus, the 20 nm solution is able to prevent asperity interaction over the entire sliding distance so that no degradation occurs over time. The 20 nm particle lubricant also opens the possibility of additional friction reduction mechanisms that are unique to nano-particles [69].

In terms of the magnitude, the coefficient of friction substantially varied with the type of additives used with the canola oil. As shown in Fig. 10.10, the best frictional performance was exhibited by the 20 nm boric acid particles, followed closely by the MoS_2 powder additives (0.5–10 μm) and a combination of sub-micron (0.6 μm) and micron-scale (4 μm) boric acid additives. The largest measured coefficient of friction over the sliding distance tested was found for the micron (4 μm) boric acid particles and the sub-micron (0.6 μm) boric acid powder. As described previously, the 20 nm boric acid lubricant combination exhibits the lowest friction because it consists of a colloidal solution that is able to enter into the pin/disk contact interface and provide optimum lubrication without degrading over time. Considering the mixed sub-micron and micron-size powder additives, the reason for the stark differences in the frictional performances of the additives was explained by examining the physical phenomena at play between the different

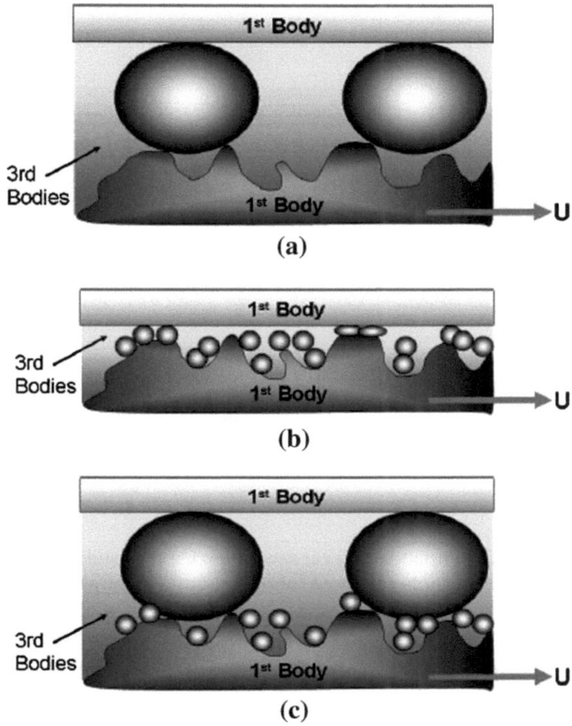

Fig. 10.11 Schematic diagram of contacting interface in the presence of oil with **a** micro-scale, **b** sub-micro, and **c** mixed sub-micro and micro-scale powder particles [68]

sized particles. Figure 10.11a–c shows the schematic diagram of contacting interface in the presence of oil with micro-scale, sub-micro, and mixed sub-micro and micro-scale powder particles, respectively. Boric acid additives that are larger than the asperities size decrease friction by carrying some of the load between contacting asperities (see Fig. 10.11a). The smaller sub-micron-scale particles, in contrast, will 'fill' the asperity valleys and provide a thin, smooth, solid lamellar film between the contacting surfaces (see Fig. 10.11b). Since the protective sub-micron-particulate additives are a better lubricant than individual micron-scale particles, it exhibited a lower coefficient of friction. As shown in Fig. 10.11c, the combined sub-micron and micron-scale powder additives exhibited strong frictional performance because both physical phenomena were at play; the small particles formed a protective boundary film while the larger particles helped support the load while being sheared with their well-known low interlayer friction. It is important to note that an identical trend was observed by Hu [36] while studying the frictional behavior of 1.5 μm, 30 nm, and mixed 1.5 μm and 30 nm MoS_2 additives in paraffin oil.

Figure 10.12 shows the wear volume results for the different lubricant mixtures. As shown in Fig. 10.12, the 20 nm boric acid particle lubricant clearly demonstrated the best wear resistance. In fact, the wear rate of the 20 nm lubricant was at

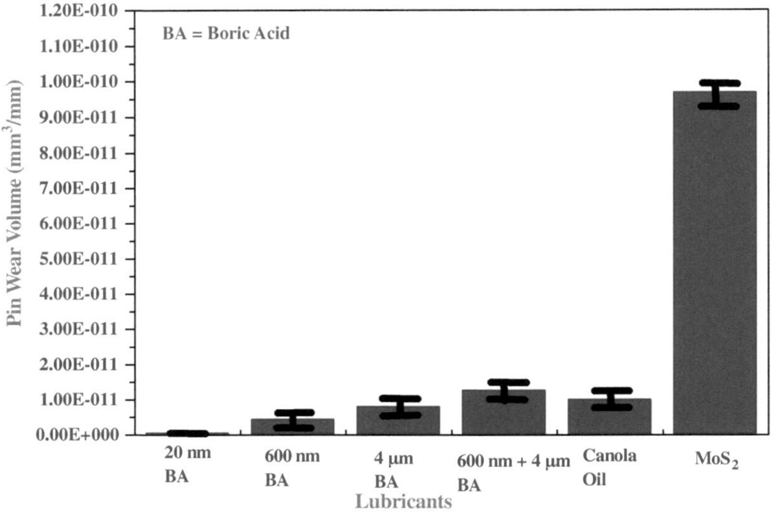

Fig. 10.12 Variation of wear rate of the Cu pins slid against Al disks in the presence of various additives in canola oil-based lubricants at the interface [68]

least an order of magnitude lower than the other lubricants. This again highlights the benefits of placing the nano-meter scale particles into a colloidal solution that is able to readily enter the pin-disk contact interface throughout the experiments. In Fig. 10.12, it is interesting to note that unlike the coefficient of friction, the lubricant mixtures with the combined sub-micron and micron-size powder additives exhibited higher wear than the individual micron-boric acid and sub-micron size boric acid powder. The sub-micron size boric acid powder exhibited the lowest wear rate. While this trend might not seem intuitive, similar phenomena have been observed in reports on dry (i.e. no carrier fluid or oil) powder lubrication. For example, a tribo-system consisting of two surfaces separated by dry third-body powder has been known to exhibit multiple tribological regimes as a function of the particle size (relative to the surface roughness) [22, 70]. If the third-body powder particles are smaller than a critical size P_{d1}, the surfaces may begin to experience adhesive wear, and if they are greater than the higher critical size P_{d2}, they will likely experience abrasive wear. This is because below P_{d1}, the particles are smaller than the asperities and can compact and form solid 'rigid-like' bodies that can promote adhesive wear. Above P_{d2}, in contrast, the particles are greater than or equal to the surface roughness which leads to more abrasive wear. When the particle size range is $P_{d1} < P_d < P_{d2}$, a regime known as "quasi-hydrodynamic" powder lubrication [70] exists, where the particles undergo shearing to accommodate surface velocity differences similar to that of hydrodynamic fluids and oils. Thus, the combination of sub-micron and micron mixtures in our experiments will promote wear due to the larger micron-particles being squeezed between the pin and disk while sitting on top of the smaller sub-micron-particles (see Fig. 10.11c). The fact that the micron-boric acid powder exhibits lower wear than the combined sub-micron and micron boric acid is likely due to the contact between the micron-boric acid powder additive and the disk surface being more disposed to induce the layered-lattice shearing behavior of boric acid powder. It is noteworthy that the MoS_2 additives, which consisted of particle sizes between from 0.5 to 10 μ, exhibited significantly greater wear than any of the boric acid additive results. In principal, the MoS_2-based lubricants followed a similar trend and can be explained by the combined sub-micron and micron sized boric acid additive discussion above. The wear rate for MoS_2, however, increases to a much greater level than the combined boric acid particles and will be the focus of future research by the authors.

10.7 Conclusion

Millions of gallons of petroleum products are improperly disposed off or accidentally lost into the environment each year. Such losses have resulted in increased regulations on oil producers and users at the federal, state, and local levels. These regulations, increased costs, and other environmental concerns are the driving demand for the use of green lubricants. Green lubricants are bio-based, biodegradable, renewable, sustainable, energy conserving, and non-toxic and have undergone significant improvements in performance and reduction in cost. In this study, a review on the tribological behavior of green lubricants was made. It was reported that the tribological performance of green lubricants was improved by the addition of additives. It was shown that green powder lubricant additives in liquid lubricants improved tribological performance. In addition, the size of the additive played an important role in determining the lubricant performance. The lubricant containing the mixture of micron- and submicron-sized particles had a better friction performance than the lubricant containing either micron- or submicron-sized particles. It is hypothesized that the lubricant mixture containing sub-micron- and micron-sized powder additives may suggest a novel lubrication technology with favorable friction characteristics. Further, the green lubricant containing nano-sized additive showed the best tribological performance when compared to the lubricant having either micron- or submicron-sized additives.

References

1. http://maps.unomaha.edu/peterson/funda/sidebar/oilconsumption.html
2. http://www.carbohydrateeconomy.org/library/admin/uploadedfiles/Harvesting_Lubricants.htm
3. http://researchwikis.com/Lubricants_Marketing_Research

4. B. Bhushan, *Principles and Applications of Tribology* (Wiley, New York, 1999)
5. B. Bhushan, *Introduction to Tribology* (Wiley, New York, 2002)
6. K.C. Ludema, *Friction, Wear, Lubrication: A Text Book in Tribology* (CRC Press, New York, 1996)
7. J.W. Bartz, Lubricants and the environment. Tribol. Int. **31**(1–3), 35–47 (1998)
8. A. Barriga, A. Igartua, A. Aranzabe, Sunflower based grease for heavy duty applications. *In Proceedings World Tribology Congress III*, Washington, D.C., USA, 2005, pp. 481–482
9. N. Battersby, Environmentally acceptable lubricants: current status and future opportunities. *In Proceedings World Tribology Congress III*, Washington, D.C., USA, 2005, pp. 483–484
10. P. Deshmukh, M.R. Lovell, W.G. Sawyer, A. Mobley, On the friction and wear performance of boric acid lubricant combinations in extended duration operations. Wear **260**(11–12), 1295–1304 (2006)
11. J. Grushcow, High oleic plant oils with hydroxy fatty acids for emission reduction. *In Proc. World Tribology Congress III*, Washington, D.C., USA, 2005, pp. 485–486
12. J. Grushcow, M.A. Smith, Next generation feedstocks from new frontiers in oilseed engineering. *In Proceedings World Tribology Congress III*, Washington, D.C., USA, 2005, pp. 487–488
13. N. Jayadas, N. Prabhakaran, G. Ajithkumar, Vegetable oils as base oil for industrial lubricants- evaluation oxidative and low temperature properties using TGA, DTA and DSC. *In Proceedings World Tribology Congress III*, Washington, D.C., USA, 2005, pp. 539–540
14. M.A. Kabir, C.F. Higgs III, M.R. Lovell, A pin-on-disk experimental study on a green particulate-fluid lubricant. J. Tribol. **130**(4), 041801–041806 (2008)
15. M. Lovell, C.F. Higgs III, A.J. Mobley, A novel particulate-fluid lubricant for environmentally benign forming processes. *In Proceedings World Tribology Congress III* Washington, D.C., USA, 2005 pp. 829–830
16. M.R. Lovell, C.F. Higgs III, P. Deshmukh, A. Mobley, Increasing formability in sheet metal stamping operations using environmentally friendly lubricants. J. Mater. Process. Technol. **177**(1–3), 87–90 (2006)
17. C. Puscas, G. Bandur, D. Modra, R. Nutiu, Considerations about using vegetable oils in lubricants. *In Proceedings World Tribology Congress III*, Washington, D.C., USA, 2005 pp. 915–916
18. A.M. Petlyuk, R.J. Adams, Oxidation stability and tribological behavior of vegetable oil hydraulic fluids. Tribol. Trans. **47**(2), 182–187 (2004)
19. W.E. Neff, T.L. Mounts, W.M. Rinsch, H. Konishi, M.A. El-Agaimy, Oxidative stability of purified canola oil triacylglycerols with altered fatty acid compositions as affected by triacylglycerol composition and structure. J. Am. Oil Chem. Soc. **71**(10), 1101–1109 (1994)
20. E.Y.A. Wornyoh, V.K. Jasti, C.F. Higgs III, A review of dry particulate lubrication: powder and granular materials. J. Tribol. **129**(2), 438–449 (2007)
21. D.W. Dareing, S. Atluri, Traction behavior and physical properties of powder graphite lubricants compacted to hertzian pressure levels. Tribol. Trans. **40**(3), 413–420 (1997)
22. H. Heshmat, Quasi-hydrodynamic mechanism of powder lubrication: part III: on theory and rheology of triboparticulates. Tribol. Trans. **38**(2), 269–276 (1995)
23. V. Johnson, G. Vaughn, Investigation of the mechanism of MoS_2 lubrication in vacuum. J. Appl. Phys. **27**(10), 1173–1179 (1956)
24. J.K. Lancaster, Lubrication by transferred films of solid lubricants. ASLE Trans. **8**, 146–155 (1965)
25. J.K. Lancaster, Anisotropy in the mechanical properties of lamellar solids and its effect on wear and transfer. Wear **9**, 169–188 (1966)
26. M. Brendle, P. Turgis, S. Lamouri, A general approach to discontinuous transfer films: the respective role of mechanical and physicochemical interactions. Tribol. Trans. **39**(1), 157–165 (1996)
27. M. Godet, The third-body approach: a mechanical view of wear. Wear **100**, 437–452 (1984)
28. M. Godet, D. Play, D. Berthe, An attempt to provide a unified treatment of tribology through load carrying capacity, transport and continuum mechanics. J. Lubr. Technol. **102**, 153–164 (1980)

29. H. Heshmat, J.F. Dill, Traction characteristics of high-temperature, powder-lubricated ceramics (Si_3N_4/SiC). Tribol. Trans. **35**(2), 360–366 (1992)
30. C.F. Higgs III, C. Heshmat, H. Heshmat, Comparative evaluation of MoS_2 and WS_2 as powder lubricants in high speed, multi-pad journal bearings. J. Tribol. **121**, 625–630 (1999)
31. P.W. Centers, The role of oxide and sulfide additions in solid lubricant compacts. Tribol. Trans. **31**(2), 149–156 (1987)
32. H. Heshmat, High-temperature solid-lubricated bearing development-dry powder lubricated traction testing. J. Propuls. Power **7**(5), 814–820 (1991)
33. H. Heshmat, W. Shapiro, High temperature, unbalanced, dry, contact face seal interfacial phenomenon and design considerations. Tribol. Trans. **45**(4), 235–242 (1989)
34. H. Heshmat, J.F. Walton, The basics of powder lubrication in high-temperature powder-lubricated dampers. J. Eng. Gas Turbine Power **115**(2), 372–382 (1993)
35. C.F. Higgs III, J. Tichy, Granular flow lubrication: continuum modeling of shear behavior. J. Tribol. **124**, 499–510 (2004)
36. X. Hu, On the size effect of molybdenum disulfide particles on tribological performance. Ind. Lubr. Tribol. **57**(6), 255–259 (2005)
37. X.G. Hu, S.L. Hu, Y.S. Zhao, Synthesis of nanometric molybdenum disulphide particles and evaluation of friction and wear properties. Lubr. Sci. **17**(3), 295–308 (2005)
38. F. Ilie, C. Tita, Tribological properties of solid lubricant nanocomposite coatings on base of tungsten disulphide nanoparticles. Tribologia **27**(4), 5–11 (2008)
39. S. Prasad, J.S. Zabinski, Tribology of tungsten disulphide (WS_2): characterization of wear-induced transfer films. J. Mater. Sci. Lett. **12**(18), 1413–1415 (1993)
40. A. Shankara, P.L. Menezes, K.R.Y. Simha, S.V. Kailas, Study of solid lubrication with MoS_2 coating in the presence of additives using reciprocating ball-on-flat scratch tester. Sadhana **33**(3), 207–220 (2008)
41. X. Shao, W. Liu, Q. Xue, The tribological behavior of micrometer and nanometer TiO_2 particle-filled poly (phthalazine ether sulfone ketone) composites. J. Appl. Polym. Sci. **92**(2), 906–914 (2004)
42. A. Erdemir, Tribological properties of boric acid and boric-acid-forming surfaces: Part II: mechanisms of formation and self-lubrication films on boron- and boric oxide-containing surfaces. Lubr. Eng. **47**(3), 179–184 (1991)
43. A. Erdemir, G.R. Fenske, R.A. Erck, F.A. Nicholas, D.E. Busch, Tribological properties of boric acid and boric-acid-forming surfaces: Part 1: crystal chemistry and mechanism of self - lubrication of boric acid. Lubr. Eng. **47**(3), 168–178 (1991)
44. A. Erdemir, M. Halter, G.R. Fenske, Preparation of ultralow-friction surface films on vanadium diboride. Wear **205**, 236–239 (1997)
45. J. Wei, A. Erdemir, G. Fenske, Dry lubricant films for aluminum forming. Tribol. Trans. **43**(3), 535–541 (2000)
46. T. Barton, J. Steffens, W.G. Sawyer, T.L. Schmitz, J.C. Ziegert, M.R. Lovell, In situ solid lubricant deposition for environmentally benign forming. *In Proceedings STLE Annual Meeting*, Toronto, Canada, 2004
47. W.G. Sawyer, J.C. Ziegert, T.L. Schmitz, T. Barton, In situ lubrication with boric acid: powder delivery of an environmentally benign solid lubricant. Tribol. Trans. **49**(2), 284–290 (2006)
48. A. Erdemir, R.A. Erck, J. Robles, Relationship of hertzian contact pressure to friction behavior of self-lubricating boric acid films. Surf. Coat. Technol. **49**, 435–438 (1991)
49. R.L. Johnson, H.B. Sliney, Ceramic surface films for lubrication at temperatures of 2000 F. Ceram. Bull. **41**, 504–508 (1962)
50. M.B. Peterson, S.L. Murray, J.J. Florek, Consideration of lubricants for temperatures above 1000 F. ASLE Trans. **2**, 225–234 (1960)
51. A. Erdemir, O.L. Eryilmaz, G.R. Fenske, Self-replenishing solid lubricant films on boron carbide. Surf. Eng. **15**(4), 291–295 (1999)
52. W.T. Branneen, G.D. Burt, R.A. McDonald, *Phosphite amine lubricant for metal working and machining*. US Patent No 4965002, 1990

53. H. Liang, S. Jahanmir, Boric acid as an additive for core-drilling of alumina. J. Tribol. **117**(1), 65–73 (1995)
54. K.P. Rao, J.J. Wei, Performance of a new dry lubricant in the forming of aluminum alloy sheets. Wear **249**(1–2), 85–92 (2000)
55. K.P. Rao, C.L. Xie, A comparative study on the performance of boric acid with several conventional lubricants in metal forming processes. Tribol. Int. **39**(7), 663–668 (2006)
56. T.E. Fischer, H. Tomizawa, Interaction of tribochemistry and micro fracture in the friction and wear of silicon nitride. Wear **105**, 29–45 (1985)
57. H.D. Huang, J.P. Tu, L.P. Gan, C.Z. Li, An investigation on tribological properties of graphite nanosheets as oil additive. Wear **261**(2), 140–144 (2006)
58. Q. Sunqing, D. Junxiu, C. Guoxu, A review of ultrafine particles as antiwear additives and friction modifiers in lubricating oils. Lubr. Sci. **11**(3), 217–226 (1999)
59. Z. Xiaodong, F. Xun, S. Huaqiang, H. Zhengshui, Lubricating properties of cyanex 302-modified MoS_2 microspheres in base oil 500SN. Lubr. Sci. **19**(1), 71–79 (2006)
60. L. Cizaire, B. Vacher, T. Le Mogne, J.M. Martin, L. Rapoport, A. Margolin, R. Tenne, Mechanisms of ultra-low friction by hollow inorganic fullerene-like MoS_2 nanoparticles. Surf. Coat. Technol. **160**(2–3), 282–287 (2002)
61. V. Leshchinsky, E. Alyoshina, M. Lvovsky, Y. Volovik, I. Lapsker, R. Tenne, L. Rapoport, Inorganic nanoparticle impregnation of self lubricated materials. Int. J. Powder Metall. **38**(5), 50–57 (2002)
62. V. Leshchinsky, R. Popovitz-Biro, K. Gartsman, R. Rosentsveig, Y. Rosenberg, R. Tenne, L. Rapoport, Behavior of solid lubricant nanoparticles under compression. J. Mater. Sci. **39**(13), 4119–4129 (2004)
63. L. Rapoport, N. Fleischer, R. Tenne, Fullerene-like WS_2 nanoparticles: superior lubricants for harsh conditions. Adv. Mater. **15**(7–8), 651–655 (2003)
64. L. Rapoport, V. Leshchinsky, M. Lvovsky, I. Lapsker, Y. Volovik, Y. Feldman, R. Popovitz-Biro, R. Tenne, Superior tribological properties of powder materials with solid lubricant nanoparticles. Wear **255**(7–12), 794–800 (2003)
65. J. Narayan, *Lubricant having nanoparticles and microparticles to enhance fuel efficiency and a laser synthesis method to create dispersed nanoparticles*. U.S. Patent 0042751Al, 2009
66. A. Erdemir, *Lubrication from mixture of boric acid with oils and greases*. U.S Patent 5431830, 1995
67. A. Erdemir, *Method to improve lubricity of low-sulfur diesel and gasoline fuels*. U.S. Patent 6783561, (2004)
68. M.R. Lovell, M.A. Kabir, P.L. Menezes, C.F. Higgs III, Influence of boric acid additive size on green lubricant performance. Phil. Trans. R. Soc. A **368**, 4851–4868 (2010)
69. L. Rapoport, Y. Bilik, Y. Feldman, M. Homyonfer, S.R. Cohen, R. Tenne, Hollow nanoparticles of WS_2 as potential solid-state lubricants. Nature **387**, 791–793 (1997)
70. H. Heshmat, The rheology and hydrodynamics of dry powder lubrication. Tribol. Trans. **34**(3), 433–439 (1991)

Chapter 11
Natural Oil-Based Lubricants

K. R. Sathwik Chatra, N. H. Jayadas and Satish V. Kailas

Lubricants and lubrication were inherent in a machine ever since man invented machines. It was water and natural esters like vegetable oils and animal fats that were used during the early era of machines. During the late 1800s, the development of the petrochemical industry put aside the application of natural lubricants for reasons including its stability and economics. The growing awareness of the lower biodegradability and higher toxicity of petrochemical-based lubricants created the requirements of the best possible protection of nature. The recent research on the adverse effect of mineral oil-based lubricants on the environment has reconfirmed its role in polluting groundwater for up to 100 years and its effects on reducing the growth of trees and the life span of aquatic life [1]. This awareness, of the use of ecofriendly processes and materials, increases interest in Tribology for the use of natural esters in lubrication processes [2]. The development of the retro parade attitude in the lubricant industry and its customers with more environmental awareness, keen to prefer products which do not diminish the world's finite resource of mineral hydrocarbons and which have a minimal adverse effect on the environment, created an opportunity to use naturally available ecofriendly lubricants [3]. The potential candidates for ecofriendly lubricants include vegetable oils, animal fats and synthetic esters. Although animal fats are also considered biodegradable the most common mineral oil substitutes consist of vegetable oils and synthetic esters [4]. The economical concerns and price stability edge the potential use of vegetable oils as lubricants over synthetic esters. With the

K. R. Sathwik Chatra · S. V. Kailas (✉)
Department of Mechanical Engineering, Indian Institute of Science,
Bangalore 560012, India
e-mail: satish.kailas@gmail.com; satvk@mecheng.iisc.ernet.in

N. H. Jayadas
Department of Mechanical Engineering, Cochin University of Science
and Technology, Ernakulam, Kerala, India

M. Nosonovsky and B. Bhushan (eds.), *Green Tribology*,
Green Energy and Technology, DOI: 10.1007/978-3-642-23681-5_11,

notion that we live on a planet with finite resources, we have to think about the coming generations and work for sustainable development in the field of Tribology. This chapter has key concepts like the advantageous and inherent limitations of vegetable oils over mineral oils, possible application of vegetable oil in the field of Tribology, composition and structure of vegetable oils and use of different vegetable oils as bioderived lubricants with their properties and functions.

11.1 Advantages and Inherent Limitations of Vegetable Oils

Vegetable oils are the semisolid or liquid plant products composed of glycerides of fatty acids. The advantages and disadvantages of vegetable oils over mineral-based oils are given below.

11.1.1 Advantages

- *Good Solvents*:
  The polar nature of vegetable oils makes them fairly good solvents, which flushes dirt and wears particles of metal surfaces.
- *Good lubricity properties*:
  The polar nature of vegetable oils gives them a greater affinity for metal surfaces than non-polar mineral oils [4]. Vegetable oils show good boundary lubrication over mineral oils.
- *Less CO_2 production*:
  Vegetable oils are positive on CO_2 level, considering that the released CO_2 in the atmosphere from their combustion is only a part (ranging from 30 to 50%) of that absorbed from plants during their growth [5].
- *Load Carrying capacity*:
  Stronger adsorption of long chain fatty acids present in vegetable oils to metallic surface assures higher load carrying capacity of vegetable oils over mineral-based oils [6].
- *Thermal Properties*:
  Thermal properties of vegetable oils are improved due to the presence of a large number of carboxylate groups in the ester molecules and thereby a thicker lubricating film can be maintained in highly loaded and high slip contacts. Higher thermal conductivity increases the amount of energy transferred from the elasto hydrodynamic contact to the surrounding material and fluid, thereby lowering the temperature in the contact zone and increasing viscosity [7].
- *Low volatility*:
  The high molecular weight of the triacylglycerol molecule and the narrow range of viscosity change with temperature ensure the low volatile nature of vegetable oils [8].

- *High Viscosity Index*:
  High linearity of the oils permits the triglycerides to maintain stronger intermolecular interactions with increasing temperature than branched hydrocarbons or esters. This ensures the high viscosity index of vegetable oils [9].
- *Low emission of CO and Hydrocarbons*:
  Vegetable oil-based lubricants are more effective in reducing the emission levels of carbon monoxide and hydrocarbons [10].
- *Better fire resistance*:
  The higher flash and fire points of vegetable oils than mineral-based oils and animal fats indicate their better fire resistance.
- *Biodegradability*:
  Vegetable oils are readily and completely biodegradable into carbon dioxide and water molecules. Biodegradability of vegetable oil and mineral oil were tested in Finland by pouring 10 L of the oils into the ground. Vegetable oils completely disappeared after 6–7 weeks but 20% of mineral oils were left even after 4 months of experiments [11].
- *Toxicity*:
  Vegetable oils composed of fatty acids are generally non toxic to aquatic and terrestrial environments.
- *Renewability*:
  The sources of vegetable oil are crops and they promote self-reliance as ample production capacity exists but the source of mineral oil is a finite mineral deposit [11].

11.1.2 Disadvantages

- *Poor low temperature fluidity and higher pour point*:
  The presence of long chain and saturated fatty acids of vegetable oils makes them crystallize at higher temperatures than mineral oils. Higher melting point of vegetable oils is also for the same reason. Highly saturated fatty acids containing vegetable oils such as coconut and palm oils show higher pour point. The presence of double bonds in the monounsaturated and polyunsaturated fatty acids forms a kink structure due to which they crystallize at a lower temperature than saturated fatty acids [8].
- *Hydrolytic Stability*:
  Vegetable oils are highly susceptible to hydrolysis in the presence of water to produce corrosive acidic breakdown products as compared to mineral-based oils. The carboxylic acids formed by hydrolysis may corrode metal bearings, affect seals and shorten life [2]
- *Poor resistance to foaming*:
  Presence of foam reduces the lubrication properties of a lubricant. Vegetable oils produce higher foam than mineral-based oils.

- *Limited resistance to oxidation in storage*:
  The presence of mono-unsaturation and poly-unsaturation promotes the auto-oxidation of vegetable oils during storage. Another cause for auto-oxidation is the presence of beta proton in the triglyceride structure. The electron shift in the beta carbons weakens the C–O bonds and breaks off the ester bond to form carboxylic acids. Refined vegetable oils are devoid of natural antioxidants that are present in unrefined oils which make the former more susceptible to oxidation [12, 13].
- *Tendency to clog the filter*:
  The high viscosity of vegetable oils and oxidized vegetable oils due to the presence of polymerized products cause the clogging of filters.
- *Poor thermal stability compared to mineral and synthetic oils*:
  Vegetable oils undergo oxidation at higher temperatures. In general, vegetable oils can be used up to a maximum temperature of 120°C.
- *Narrow range of viscosities*:
  One of the biggest limitations of vegetable oils as potential lubricants is their narrow range of available viscosities.

11.1.3 Application of Vegetable Oils

The use of vegetable oils is encouraged in applications where recycling is not easy or the usage is a complete loss one. The examples include the following:

- Chain bar and saw frame oils
- Concrete mold release oils
- Outboard two-stroke engine oils
- Chain and conveyor lubricants
- Greases
- Corrosion preventive oils
- Wire rope lubricants
- Hydraulic fluids
- Compressor oils
- Milking machine lubes
- Engine oils
- Gear and transmission lubes
- Metal working Fluids
- Machine tool lubricants
- Truck centralized system lubricants
- Lubricants for agricultural equipment
- Lubricants for sewage treatment plants, water weir plants and lock gate mechanisms
- Lubricants for food machinery

- Lubricants for snow mobiles and sky run maintenance equipment
- Railway wheel tread lubricants

The current major areas of applications where biodegradable lubricants have been put to real use:

- *Boating two-stroke lubricants*: Due to stringent rules in Western Europe and in the US, mineral-based boating two-stroke lubricants have been replaced with vegetable ester-based lubricants. Ethylhexyl esters of partially hydrogenated rice bran oil and karanja oil are good candidates for such applications [14].
- *Forestry and Chain saw lubricants*: Chain bar and saw frame oils have been completely replaced by vegetable-based oils such as rapeseed and sunflower oils containing formulation primarily in Europe, US, Canada and Australia. It provided improved lubrication at much lower application rates and also increased the life of saw blades. The use of this fluid has eliminated respiratory and skin problems in operators attributed to the petroleum products [14].
- *Concrete mould release fluids*: The use of biodegradable fluids for concrete mould release fluids has started. About 15% of the total lubricant application for this purpose has been already replaced by biodegradable lubricants [14].
- *Hydraulic system oils*: Hydraulic systems involved in agriculture, forestry, off highway construction, waterways, open cast mining, quarrying etc., can cause significant pollution to the environment during leakage. At present the highly visible areas where their use is well advanced in certain countries are fluids for hydraulic system on canals and waterways, agriculture tractors, other farm equipments and off highway construction equipment, particularly in the European Unions, US and Japan. The share of rapidly biodegradable hydraulic oils increased from 5% in 1998 to more than 50% by 2003 [14].
- *Manufacturing industries*: Since 1996 Germany has taken the initiative to use biodegradable lubricants in manufacturing industries. Another major industry built a diesel engine which uses only ester-based cutting fluids, hydraulic fluids, gear oils etc.

The areas for possible applications of biodegradable lubricants in the near future are turbine oils for hydro electric plants metalworking fluids, engine oils and Rail Curve Grease.

The possible potential applications for various vegetable oils are given in Table 11.1 [15].

11.1.4 Composition and Structure of Vegetable Oils

Vegetable oils are the semisolid or liquid plant products composed of glycerides of fatty acids. They are termed triglycerides and are esters of glycerol (Fig. 11.1) [16]. In this figure R_1, R_2 and R_3 refer to the hydrocarbon chains of fatty acids. Generally, all alkali refined vegetable oils are more than 99% of triglycerides, with

Table 11.1 Possible applications of vegetable oils

Type of oil	Applications
Canola oil	Hydraulic oils, tractor transmission fluids, metal working fluids, food grade lubes, penetrating oils, chain bar lubes
Castor oil	Gear lubricants, greases
Coconut oil	Gas engine oils
Olive oil	Automotive lubricants
Palm oil	Rolling lubricant- steel industry, grease
Rapeseed oil	Chain saw bar lubricants, air compressor- farm equipment, biodegradable greases
Safflower oil	Light colored paints, diesel fuel, resins, enamels
Linseed oil	Coating, paint, lacquers, varnishes, stains
Soybean oil	Lubricants, bio-diesel fuels, metal castings/workings, detergents, pesticides, disinfectants, plasticizers, hydraulic oil
Jojoba oil	Grease, cosmetic industry, lubricant applications
Sunflower oil	Grease, diesel fuel substitutes
Tallow oil	Steam cylinder oils, soaps, cosmetics, lubricants, plastics

Fig. 11.1 Structure of a triglyceride

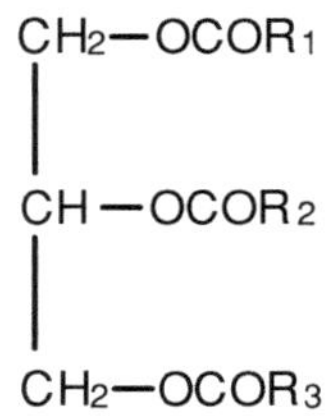

Where R_1, R_2 and R_3 refers to the hydrocarbon chains of fatty acids.

the exception of jojoba oil and a few other exotic oils. The chemical structure of vegetable oils is relatively uniform as opposed to mineral oils. The alcohol component (Glycerin) is the same in all vegetable oils. The fatty acid components are plant specific and therefore the chemical nature of vegetable oils is variable [17]. The variability in the fatty acid proportion not only depends upon the type of plant but also on the region, the climate and on the availability of nutrients for growth.

Triglycerides or triacylgylcerols are fatty acid esters. Fatty acids are composed of straight chain carbon molecules with a range of 8–24 molecules. They are classified according to the presence of double bonds and special groups. Fatty acids with no double bonds are classified as saturated fatty acids. If it contains one double bond like oleic acid it is called mono-unsaturated. Poly-unsaturated fatty acids contain more than one double bond in their molecular structure. Few types of fatty acids contain special functional groups in their structure such as hydroxyl group in ricinoleic acid and epoxy groups in vernolic acid. The general structure of triglyceride may be represented as below.

Mannekote et al. [18], performed tribological experiments on fresh vegetable oil and aged vegetable oil samples. The results indicated that the more saturated fatty acid containing vegetable oils like coconut and palm oil showed less wear scar diameter than less saturated fatty acids containing vegetable oils like castor and sunflower oils, due to their boundary lubrication capabilities. Whereas, for coefficient of friction more unsaturated fatty acid containing vegetable oils resulted in less coefficient of friction than less unsaturated fatty acids. This result is attributed to the formation of oxidation products like free fatty acids upon ageing which facilitate the formation of metallic soaps, thus reducing the coefficient of friction [18].

11.1.5 Vegetable Oils as Bioderived Lubricants; Its Properties and Applications

Vegetable oils are mainly classified into two categories:

1. Edible vegetable oils: These oils are produced in large amounts worldwide and are mainly used for edible purposes. Examples of edible oils include coconut oil, palm oil, sunflower oil, corn oil, cottonseed oil, mustard oil, peanut/groundnut oil, olive oil, sesame oil, safflower oil, rice bran oil, babassu oil, soybean oil etc.
2. Non-Edible oils: These oils are mainly used for commercial purposes than for edible purpose. The commonly available non-edible oils include castor oil, honge oil, jojoba oil, jatropa oil, etc.

11.1.6 Edible Oils as Bioderived Lubricants

11.1.6.1 Coconut Oil

Coconut oil is extracted from the kernel or meat of matured coconuts harvested from the coconut palm (cocos nucifera). The primary sources of coconut oil are the tropical regions of the world. The major coconut oil producing countries in 2009 include Philippines, Indonesia, India, Brazil, Srilanka, Thailand, Mexico, Vietnam, Papua New Guinea, Malaysia and Tanzania. Coconut oil belongs to a unique group of vegetable oils called lauric oils. More than 90% of fatty acids of coconut oil are saturated. The saturated character of the oil imparts a strong oxidative stability. The high resistance to oxidation ensures its weak tendency to form deposits, sludge and corrosive by-products in engine oil and industrial applications. Also, it is more resistant to undesirable viscosity increases during use. The inherent limitation of coconut oil is its higher pour point. The fatty acid chains in triglycerides of coconut oil are nearly parallel to one another. This enables close packing of the molecule during cooling leading to a gel-like structure which entraps the low melting constituents. The pour point of coconut oil can be brought down by inducing branched aromatic hydrocarbon molecules of high molecular weight so as to prevent close packing of molecules during cooling [19]. Mannekote et al. [20, 21], compared coconut oil and palm oil as lubricants in four-stroke engines with commercially available engine oils. The results showed that anti-wear properties of fresh vegetable oil and engine oils were comparable at ambient conditions. The deviation in tribological properties of vegetable oils to commercially available engine oil with usage was linked to its oxidation instabilities and absence of additives. The results of exhaust gas analysis showed a lesser concentration of carbon and nitrogen oxides in vegetable oil-based lubricants. They also reported improvement in fuel efficiency with vegetable oil-based lubricants. Due to oxidation of vegetable oil samples in crankcase, the kinematic viscosity and coefficient of friction increase with the usage of oil and with time mixing of the fuel to lubricants decreases the flash point [22]. Fatty acid composition and physical, chemical and tribological properties of coconut oil [23, 24] are given in Tables 11.2 and 11.3.

11.1.6.2 Sunflower Oil

Sunflower oil often referred to as sunflowerseed oil or sunoil is obtained from the seed of the plant helianthus annus, which is native to North America. Sunflower oil is currently the fourth largest edible oil commodity traded on the world market following palm oil, soybean oil and coconut oil [25]. Recently, high oleic acid crops have been developed whose oil has higher oxidative stability than the standard genotypes [26]. The major sunflower growing regions in the world include Russia, Ukraine, France, Romania, Hungary, Bulgaria, Spain, Italy and others. The Asian sunflower growing regions include China, India and Pakistan [27]. Sunflower oil contains the highest level of alpha tocopherols, the most active form of vitamin E, which will act as an antioxidant during storage to overcome the auto oxidation of fatty acids [28]. Fatty acid composition of common sunflower oil, high oleic sunflower oil and high oleic and high palmitoleic sunflower oil [29] are given in Table 11.4.

Experimental results from Krzan et al. showed that high oleic sunflower oil recorded lesser coefficient of friction than mineral-based oil but promoted higher wear scar diameter. Gear protection properties of sunflower oil are better or equivalent to mineral-based oils. Sunflower oil was stable during spur gear test system test for about 700 h of working and no excessive wear was observed which indicates effective lubrication in the gear box maintained during operation. The presence of tocopherols improves the oxidation stability of unrefined sunflower oil as compared to refined sunflower oils. The addition of ascorbic acid to refined sunflower oil improves its stability by about 2% [30]. Cultivation of high oleic

Table 11.2 Fatty acid profiles of different vegetable oils [38, 129–131]

Fatty acid name	Fatty acid composition of vegetable oil (in %)														
	Coconut oil	Soybean oil	Corn oil	Cotton seed oil	Mustard oil	Peanut oil	Rice Bran oil	Safflower oil	Sesame oil	Babasso oil	Palm oil	Castor oil	Honge oil	Jatropa oil	Rubber seed oil
Caprylic acid	9.5	–	–	–	–	0.1	–	–	–	4.8–6.5	0.1	–	–	–	–
Capric acid	4.5	–	–	–	0.1	0.1	0.2	–	–	2.7–6.6	0.1	–	–	–	–
Lauric acid	51.0	–	–	–	0.2	0.6	0.2	–	0.4	44.1–45.8	0.1	–	–	5.9–9.8	–
Myristic acid	18.5	0.3	0.1–1.7	0.8	0.1	0.3	0.7	0.16	0.2	15.5–19.9	1.2	0.01	–	2.7–4.3	–
Palmitic acid	7.5	10.8	7.8–11.0	22.3–24	2.6	13.3	27.6	7.59	11.7	6.9–8.5	42.1	1.64	3.7–7.9	13.5–26.8	8.56
Palmitoleic acid	–	0.3		0.4	0.2	0.1	0.5	0.28	0.2	–	0.6		–	–	–
Stearic acid	3.0	3.2	2.5–3.6	2–2.2	1.1	2.1	4.1	2.42	5.2	2.7–5.5	4.3	1.85	2.4–8.9	6.1–10.4	10.56
Oleic acid	5.0	24.0	25–31	16.7–17	12.9	47.8	47.5	11.04	41.4	11.9–18.1	43.1	5.06	44.5–71.3	21.8–22.4	22.95
Recinoleic acid	–	–	–	–	–	–	–	–	–	–	–	82.88	–	–	–
Linoleic acid	1.0	54.4	54–60	55–57	13.8	29.2	16.1	70.46	39.4	1.4–2.8	8.6	6.42	10.8–18.3	31.4–47.4	37.28
Linolenic acid	–	6.8		0.3	8.8	–	0.2	0.1	0.4	–	–	0.30		–	19.22
Arachidic acid	–	0.1	0.4	–	0.7	1.2	0.8	0.36	0.4	–	–	–	–	–	–
Behenic acid	–	–		–	–	2.9	–	–	0.6						
Erucic acid					48.7										

Table 11.3 Properties of vegetable oils [62, 132, 133, 134–135]

Properties	Coconut oil	Sunflower oil	Corn oil	Cotton seed oil	Mustard oil	Peanut oil	Rice bran oil	Safflower oil	Soybean oil
Iodine value (Wijs method)	6–8	110–143	122–131	103–116	109–113	90.1–94.8	99–108	145	144.8
Saponification value (mg KOH/g)	248–265	188–194	189–195	190–198	168–177	188.2–187.8	180–195	191	180–200
Unsaponifiable matter (in %)	1.5	1.5	1.5	–	1.20	0.22–0.27	4.0	0.5	1.0
Free fatty acid (in %)	2.0–2.05	0.10	0.05–2.5	0.05	1.07–1.11	1.44–1.47	0.05	0.03	0.05
Percent of saturation	90–94	9–11	10–16	25–27	5–6	17–24	15-20	8–12	16
Percent of unsaturation	6–10	89–91	84–90	73–75	94–95	73–78	80–85	88–92	84
Flash point (°C)	240	274	280	320	316	336	327	322	254
Melting point (°C)	21	−17.0	–	–	–	–	–	–	−12
Cloud point (°C)	27	7.20	−11	–	–	–	1.1	–	−3.9
Pour point (°C)	20	−15.0	−26.0	−4	–	–	−6.6	–	−12.2
Smoke point (°C)	232	227	–	221	–	–	254		160
Specific gravity at 25/25°C (g/cm^3)	0.915–0.917	0.918–0.923	0.916	0.920	0.918	0.913–0.914	0.921	0.920	0.916–0.922
Refractive index at 25°C	1.448–1.450	1.473	1.475	–	–	–	–	1.4744	1.465–1.475
Peroxide value (mg/kg)	1.50	4.7	–	1.00–0.920	–	–	1.0	–	–
Viscosity index	140–150	172–176	237	–	205	–	–	–	170
Viscosity in cst									
40°C	36.2	37.1	32.4	32.7	44.1	–	–	–	32.93
100°C	6.76	7.68	8.06		9.4	–	–	–	8.08
Wear scar diameter (μm)	541.7	–	–	–	–	–	–	–	–
Avg CoF	0.060	–		–	–	–	–	–	–

(continued)

Table 11.3 (continued)

Properties	Babasso oil	Palm oil	Castor oil	Honge oil	Jatropa oil	Rubber seed oil	Jojoba oil	Sesame oil
Iodine value (Wijs method)	16.1	64–65	82–88	88.5	105.2	80–135	81.7–82.98	82.45
Saponification value (mg KOH/g)	225.6	225–227	180–182	185	198	210	92–94	195
Unsaponifiable matter (in %)	–	–	0.7	0.8	0.8	1.7	48.3	–
Free fatty acid (in %)	0.06	6.99	0.90	5.06	1.76	2.6	0.32–1.10	2.0
Percent of saturation	79–87	53	3.5	14–16	22–26	20–21	–	11–18
Percent of unsaturation	13–21	47	96.5	82–84	74–78	79–80	–	82–89
Flash point (°C)	112	304	260	230	117–133	198	294–300	328
Melting point (°C)	24	–	–18	–	–	–	–	–
Cloud point (°C)	–	32.55	–	15	–	–	–	1.0
Pour point (°C)	–	–	−31.7	−2 to −5	−8	–	1–9	−10
Specific gravity at 25/25°C (g/cm^3)	0.8795	0.910	0.961–0.963	0.924–0.927	0.919	0.910	0.8642–0.8662	0.916–0.921
Refractive index at 25°C	–	–	1.4792	–	–	–	1.4648	–
Peroxide value (mg/kg)	–	–	3.87	8.3	–	–	4.41–4.76	–
Viscosity index	–	–	130–140	–	–	–	233	–
Viscosity (cst)								
40°C	–	42.66	203.92	38.3	24.5–30.68	66.2–76.4	24.92	25.78
100°C	–	8.65	34.92	–	–	–	6.43	–
Wear scar diameter (μm)	–	580	–	–	–	–	–	–

Table 11.4 Fatty acid composition of common sunflower oil, high oleic sunflower oil and high oleic and high palmitoleic sunflower oil

Fatty acid name	Fatty acid composition of common sunflower oil (in %)	Fatty acid composition of high oleic sunflower oil (in %)	Fatty acid composition of high oleic high palmitoleic sunflower oil (in %)
Palmitic acid	6.0–8.0	4.7	25.50
Palmitoleic acid	–		7.3
Stearic acid	1.5–3.0	3.7	1.9
Oleic acid	23–28	72.2	59.7
Linoleic acid	60–68	17.0	2.2
Linolenic acid	–	–	–
Others	1	2	–

sunflower oil cultivar, a variation in genotype can improve the thermal and oxidation stability of sunflower oils [31].

11.1.6.3 Corn Oil

Corn oil or maize oil is oil extracted from the germ of corn or Zea mays L. It is semidrying oil noted for its high content of linoleic acid. The major corn growing countries include, Argentina, Brazil, China, Egypt, India, Indonesia, Mexico, Nigeria, Philippines, Romania, Serbia, Thailand, South Africa, United states and other European countries. The fatty acid composition of corn oil [32] is given in Table 11.2 and the physical, chemical and tribological properties of corn oil [33] are given in Table 11.3.

11.1.6.4 Cottonseed Oil

Cottonseed oil is a vegetable oil derived from the seeds of cotton plants of various species, mainly Gossypium hirsutum and Gossypium herbaceum. It contains gossypol, a phenolic compound, which acts as an antimicrobial and insect resistance compound. The fatty acid profile of cottonseed oil generally consists of 70% unsaturated fatty acids including 18% monounsaturated (oleic acid), 52% polyunsaturated (Linoleic acid) and 26% saturated fatty acids. The four largest cottonseed oil producing countries in 2008/2009 crop year accounted for about two-thirds of the world production, which are China-32%, India-24%, United States-8% and Turkey-3%. The fatty acid composition of cottonseed oil is given in Table 11.2 and the physical, chemical and tribological properties of cottonseed oil are given in Table 11.3.

11.1.6.5 Mustard Oil

Mustard oil or Indian mustard oil is canola quality oil extracted from Brassica junctia L. It is being developed as a complementary oilseed crop to canola (Brassica napus) for cultivation in hot and low rainfall areas, where canola does not grow well. The major mustard oil producing countries are India and Bangladesh. The fatty acid composition of mustard oil is given in Table 11.2 and the physical, chemical and tribological properties of mustard oil are given in Table 11.3.

11.1.6.6 Peanut or Groundnut Oil

Peanut oil, also called groundnut oil, is highly desirable oil for deep frying. It is made from legumes called peanuts, and is known for its high smoke point. The cultivated peanut was likely first domesticated in the valleys of the Paraguay and Prarana rivers in the Chaco region of South America. Today, peanut is widely distributed and has been adopted in various countries of the world. The most important countries which produce peanut oil are India, China, USA, West and South Africa and Brazil. The fatty acid composition of peanut oil is given in Table 11.2 and the physical, chemical and tribological properties of peanut oil are given in Table 11.3.

11.1.6.7 Rice Bran Oil

Rice bran oil is the oil extracted from the germ and inner husk of rice. The typical composition of crude rice bran oil is 81.3–84.3% triglycerides, 2–3% diglycerides, 5–6% monoglycerides, 2–3% free fatty acids, 0.3% waxes, 0.8% glycolipids, 1.6% phospholipids, 4% unsaponifiables. In comparison with other vegetable oils, crude rice bran oil tends to contain higher levels of non-triglyceride components, most of which are removed during further refining processes. Oleic and linoleic fatty acids constitute more than 80% of the fatty acids of glycerides. Rice bran oil is similar to peanut oil in fatty acid composition with a saturation level slightly higher than the conventional soybean oil. The much lower linolenic acid content of rice bran oil makes it more stable to oxidation than soy oil. Fatty acid composition and the physical, chemical and tribological properties of rice bran oil [34] are given in Tables 11.2 and 11.3.

11.1.6.8 Safflower Oil

Safflower oil is produced from a highly branched, herbaceous, thistle-like annual plant called *Carthamus tinctorius*. Safflower crop is mostly produced in areas or regions possessing a warm and dry climate with suitable moisture content in the atmosphere. Safflower producing countries include India, China, Mexico, Canada, Australia and USA. The fatty acid composition of safflower oil is given in Table 11.2 and the physical, chemical and tribological properties of safflower oil are given in Table 11.3.

11.1.6.9 Sesame Oil

Sesame oil is also called gingelly oil or til oil produced from the plant Sesamum indicum L. It is an oilseed herbaceous crop of the Pedaliaceae family. Sesame also contains a class of unusual compounds known as lignans, comprising sesamin, sesamolin and a small amount of sesamol. It shows high stability to oxidation compared with other vegetable oils. It can also be used to prevent lipid oxidation in soybean, rapeseed and sesame oil [35]. Sesame is recognized as the oldest crop in the world. India and China are the world's largest producers of sesame, followed by Burma, Sudan, Mexico, Ethiopia, Nigeria, Venezuela, Turkey and Uganda. The fatty acid composition of sesame oil is given in Table 11.2 and the physical, chemical and tribological properties of sesame oil are given in Table 11.3.

11.1.6.10 Babassu Oil

Babassu oil is a semidrying, clear golden yellow vegetable oil, which is extracted from the seeds of the babassu palm (Attalea speciosa). The babassu palm mainly grows in the Amazon regions of South America. The fatty acid composition and properties are very similar to coconut oil. Hence it is increasingly being used as a substitute for coconut oil.

More than 70% of these esters are saturated. Thus it has a higher stability against oxidation than a blend of unsaturated fatty acid esters, since unsaturation of the vegetable oils and their corresponding esters increase their susceptibility to oxidation [36]. The fatty acid composition of babassu oil is given in Table 11.2 and the physical, chemical and tribological properties of babassu oil are given in Table 11.3.

11.1.6.11 Palm Oil

Palm oil is produced by a tropical perennial palm tree which belongs to the family Aracaceae. Palm oil contains a high content of saturated fatty acids. It is semisolid at room temperature. The major palm oil producing countries include Indonesia, Malaysia, Columbia, Benin, Kenya and Ghana. As of 2009, Indonesia was the largest producer of palm oil, surpassing Malaysia in 2006, producing more than 20.9 million tons. The fatty acid composition of palm oil is given in Table 11.2 and the physical, chemical and tribological properties of palm oil are given in Table 11.3.

11.1.6.12 Soybean Oil

Soybean oil is the largest grown crop in the world and it is the world's most important oilseed. The United states produces soybean oil in large quantities and other countries which produce soybean oil are Brazil, Canada, India, China, Indonesia, Paraguay, South Korea, Thailand, Italy, Romania and Argentina. The chemical composition and the physical, chemical and tribological properties are given in Tables 11.2 and 11.3.

11.1.7 Non Edible Oils as Bioderived Lubricants

11.1.7.1 Castor Oil

Castor oil is the oil extracted from the plant Ricinus communis L.; popularly known as castor bean oil, higuerilla, mamona etc., has been feedstock, which has a great demand in the pharmaceutical and chemical industry. The principal producers of castor oil are India and Brazil; their combined production makes up about 90% of the annual world production of castor oil. Other producers of this nonvolatile oil from the castor plant include the following countries, China, Russian Federation, Ukraine, Costa Rica, Romania, Ecuador, Pakistan, Thailand, Philippines, Paraguay, Ethiopia, Sudan, Tanzania and Mexico. In 2000, the United Nations Food and Agriculture Organization (FAO) reported that the world's production of castor oil totaled to 1,227,669 metric tons. The fatty acid composition and the physical, chemical and tribological properties of castor oil [37] are given in Tables 11.2 and 11.3.

11.1.7.2 Honge Oil

Honge oil or Karanja oil is derived from the Pongamia pinnata, which is native to India. Pongamia is widely distributed in tropical Asia. The major producing countries include East Indies, Philippines and India. The fatty acid composition of Honge oil is given in Table 11.2 and the physical, chemical and tribological properties of honge oil are given in Table 11.3.

11.1.7.3 Jatropa Oil

Jatropa oil is extracted from the plant species Jatropa, which is a multipurpose bush or small tree belonging to the family of Euphorbiacea. It is believed to have originated in Central America and now it is widespread all over the tropical and subtropical world. The plant is perennial, drought resistant and can grow in

Table 11.5 Fatty acid composition of Jojoba oil

Fatty acid name	Fatty acid composition of jojoba oil (in %)
Saturated acids	1.64
Palmitoleic acid	0.24
Oleic acid	0.66
Eicosanoic acid	30.30
Dicosanoic acid	14.20
Eicosanol	14.60
Dicosanol	33.70

marginal lands and even in saline soils. The physical and chemical properties of jatropa oil [38, 39] are given in Table 11.3.

11.1.7.4 Jojoba Oil

Jojoba oil is obtained from the jojoba plant (Simmondsia chinesis L.) It belongs to the family buxaceae, which produces seeds that contain 50–60 weight percent of inedible long chain wax esters that are suitable as a potential base stock [40]. Jojoba originated from the Sonoran desert (southwest USA and Northern Mexico) and grows today in several countries as a commercial crop. Jajoba oil is like whale sperm oil which is a liquid wax. It consists primarily of wax esters of eicosenoic acid (C20) and eicosenol (C20) and docosenol (C22). It contains a very small amount of saturated material and the main constituents are mostly mono unsaturated [41, 42]. The physical and chemical properties of jojoba oil are given in Table 11.3. The fatty acid composition of jojoba oil is given in Table 11.5.

The use of jojoba oil, with its high polarity and straight and longer chain length of 40–42 carbon atoms, as a component of lubricant reduces wear because of strong adsorption of the polar molecules on the metal surfaces. It also shows a good coefficient of friction of about 0.0445 [43]. The viscosity of jojoba oils falls within SAE 20 range and the oil can be used as a base stock without any dilution. The viscosity index is quite high compared with HVI paraffinic base stocks which can improve lubrication especially at higher temperatures [44]. Saponification and iodine values of jojoba oil are relatively low compared to other vegetable oils, such as castor, soybean and rapeseed oil, suggesting better stability. The low and constant foam tendency of jojoba oil assures its stableness in mechanical systems, such as high speed gearing, high volume pumping and splash lubrication [45].

11.2 Chemical Modifications for Natural Oils

A steady increase in the demand for environmentally friendly lubricants by consumers and strict government regulations to use such lubricants opens an opportunity to use vegetable or natural oils as biodegradable lubricants [46].

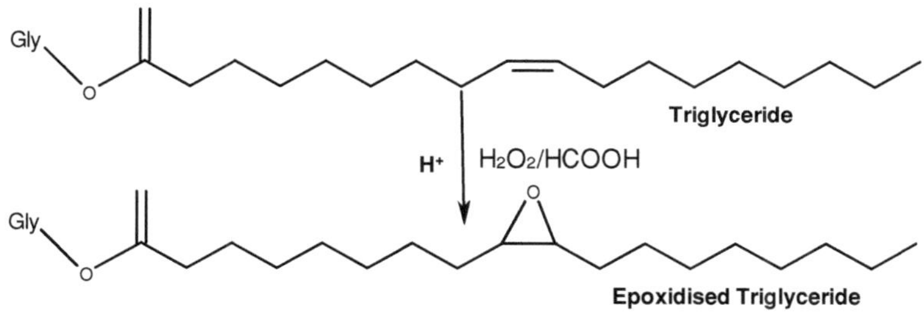

Fig. 11.2 Epoxidation of soybean oil

Natural oil-based lubricants generally have many advantages over mineral-based oils, including low volatility, better inherent lubricity property, higher viscosity index, better bulk modulus, better flash and fire point and better solvency for additives and oxidation products [47]. The widespread use of vegetable oils for lubricant formulation is restricted with its inherent limitations, including insufficient thermal and oxidation stability, poor low temperature fluidity and hydrolytic instability [48]. Poor oxidative stability of vegetable oils is primarily because of the presence of bisallylic protons. These protons are highly susceptible to radical attack and subsequently undergo oxidation degradation to form polar-oxy compounds. This oxy polymerization process ultimately results in insoluble deposit formation and an increase in oil acidity and viscosity. Vegetable oil-based lubricants show poor corrosion protection [49] and the presence of ester linkage make these oils susceptible to hydrolytic breakdown [50]. Low temperature studies have also shown that most vegetable oils undergo cloudiness, precipitation, poor flow and solidification at cold temperature [51]. Chemical and genetic modification of vegetable oils can overcome these shortcomings by reducing or complete removal of unsaturation in vegetable oils. The triglyceride structure of vegetable oils has double bonds, allylic carbons and ester groups, which are potential sites for chemical modification. This part of the chapter will focus on the different chemical modifications of vegetable oils for suitable lubricant applications.

11.2.1 Epoxidation

Epoxidation is a chemical reaction in which an oxygen atom is joined to an olefinically unsaturated molecule to form cyclic, three-membered ether (Fig. 11.2). The products of epoxidation are known as oxirane compounds or epoxides. In the case of vegetable oils the double bonds present in the unsaturated fatty acids of triglyceride structure are potential sites for chemical modifications and can be epoxidized with hydrogen peroxide, either in acetic acid or formic acid to give epoxidized vegetable oil. Epoxidized vegetable oil is a promising intermediate for ring opening reactions [52]. These epoxy sites of epoxidized vegetable oil can convert into the diester derivatives and thioether derivatives in either one- or two-step reactions [53, 54].

The only commercial source of epoxidized oils is the Prilezhaev reaction which uses short chain percarboxylic acids. Other methods for the epoxidation of unsaturated fatty compounds include use of dioxiranes, the generation of per-acids from aldehydes and molecular oxygen, chemoenzymatic self-epoxidation catalysed by immobilized Novozyme 435 and transition metal catalyzed C=C epoxidation using alkyl hydro peroxides. In recent years many catalytic systems were developed for the epoxidation of unsaturated fatty acids by using hydrogen peroxide as oxidant. Catalysts based on titanium, manganese, tungsten and rhenium have been developed for epoxidation [55].

The fatty acid profile of vegetable oils also affects the kinetics of epoxidation. In the triglyceride structure, the double bonds of oleic acid and linoleic acid were equally reactive and the double bonds of linolenic acid were approximately three times more reactive than oleic and linoleic acids. The steric and electronic effects caused fatty acids with different levels of unsaturation to have different reactivities. The effect of steric factors is such that double bonds further from the glycerol center are more reactive than double bonds near the glycerol center. The fatty acids with higher degree of unsaturation have a higher rate constant of epoxidation due to higher number of double bonds. As the number of double bonds increases, the electron density increases, this causes an increase in the rate constant. The double bonds on the fatty acid methyl esters were more reactive than the double bonds of the respective fatty acids of triglycerides [56].

Gan et al. [57] prepared epoxy esters of palmolein perooxyformic acid generated in situ. Analysis of thermal and oxidation properties of synthesized epoxy esters of palmolein were done in an oven at 120°C and the loss of mass, acid, iodine value, percent of oxirane, hydroxyl and peroxide values were monitored periodically. Epoxidised esters had higher densities and lower volatilities, and were more resistant towards oxidation than their unepoxidised counterpart. The rate of mass loss for epoxy ester of palmolein decreased with increasing chain length of the alkyl group. Epoxidation increased thermal oxidative stabilities and reduced the rate of increase in acid value. Epoxides produced less hydroperoxides and were thermally more stable.

Mungroo et al. [58] synthesized epoxides from canola oil. The reaction occurred optimally at a temperature of 65°C, an acetic acid to ethylenic unsaturation molar ratio of 0.5:1, a hydrogen peroxide to ethylenic unsaturation molar ratio of 1.5:1, and a catalyst acidic ion exchange resin loading of 22 weight percent of total canola oil used. The successful epoxidation reaction produced about 90% of oxirane conversion. The epoxides of Canola oil can be used as high temperature lubricants and the products obtained from ring opening of Canola oil epoxides can be employed as low temperature lubricants.

Dinda et al. [59] studied epoxidation of cottonseed oil using in situ generated peroxy acetic acid from hydrogen peroxide and glacial acetic acid in the presence of liquid inorganic acid catalysts. The optimum condition for epoxidation of cottonseed oil includes a temperature range of 50–60°C, hydrogen peroxide to unsaturation molar ratio range of 1.5–2.0, acetic acid to unsaturation mole ratio of about 0.5 and sulphuric acid loading of about 2% by weight of the aqueous phase. It was possible for them to obtain up to 78% relative conversion to oxirane with very less oxirane cleavage by in situ technique.

Goud et al. [60] carried out epoxidation of karanja oil with peroxyacetic acid that was generated in situ from aqueous hydrogen peroxide and glacial acetic acid. The optimal epoxidation level was obtained in a shorter time at a moderate reaction temperature range of 55–65°C, at which the hydrolysis rate was insignificant. The optimal concentration of hydrogen peroxide appeared to be 1.5 mol of hydrogen peroxide per mole of ethylenic unsaturation and about 0.5 of acetic acid to ethylenic unsaturation ratio. The results showed that the oxidation rate was not substantially affected by stirring speeds beyond 1,500 rpm, which indicates the reaction is free from mass transfer resistance beyond 1,500 rpm under the given conditions of temperature and catalyst loading.

The epoxidation kinetics of rubber seed oil by peroxy acetic acid generated in situ were studies at various temperatures by Okieimen et al. [61]. It was found that epoxidation with almost complete conversion of unsaturated carbon and negligible oxirane cleavage can be attained by the in situ technique. The rate constant for epoxidation for epoxidation of rubber seed oil was found to be of the order of 10^{-6} L/mol/sec and activation energy of epoxidation of 15.7 kcal/mol was determined. The results from this study showed show that the epoxidation of rubber seed oil by peroxyacetic acid generated in situ can be carried out at moderate temperatures with minimum epoxides degradation. The kinetic and thermodynamic parameters of epoxidation obtained from this study indicate that an increase in the process temperature would increase the rate of epoxides formation.

Adhvaryu et al. [62] demonstrated the improved performance of epoxidized soybean oil over soybean oil and genetically modified high oleic soybean oil in certain high temperature lubricant applications. Epoxidized soybean oil showed very high viscosity at 40°C than conventional and genetically modified soybean oil. The presence of more polar structure and higher molecular weight compared with conventional and genetically modified soybean oil results in stronger intermolecular interaction, which enhance the lubricity in a dynamic system. Studied oxidation and thermal properties of epoxidized soybean oil show that induction time for deposit formation is roughly twice that of high oleic soybean oil under similar oxidation temperature of 175°C. The percent insoluble deposit becomes stable after 2 h of oxidation with epoxidized oil remaining low compared with other oils. The peak temperature (T_p) and onset temperature(T_o) combination of these oils are soybean oil is 180.3 and 167.8°C, high oleic soybean oil is 201.1 and 185.8°C and for epoxidized soybean oil is 214.9 and 188.1°C. The oxidation induction time (I_t) measurement under isothermal condition for epoxidized soybean oil at 195°C showed greater stability over soybean oil at 150°C and high oleic soybean oil at 160°C. Epoxidized soybean oil showed excellent coefficient of friction reducing properties due to the formation of stable polymeric film on the metal surface during boundary lubrication. The film formation is achieved through

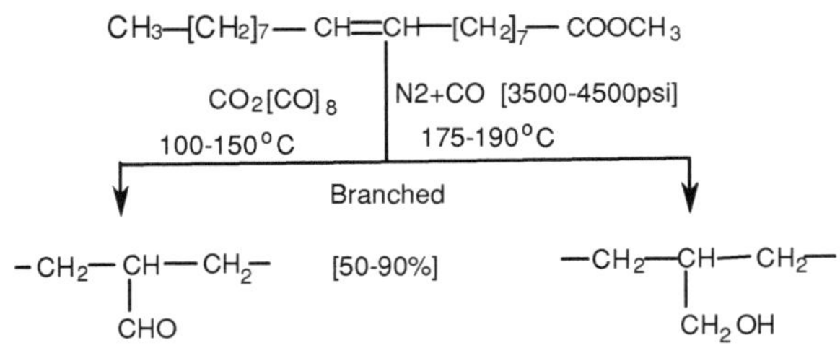

Fig. 11.3 Conventional hydroformylation with cobalt-based catalysts

the –O– cross linking on the metal surface, and thus reducing friction between the moving parts.

Erhan et al. [63] converted epoxy ring opened sites of soybean oil into ester derivatives either in one-step reaction. Results show the pour point decreased with an increasing chain length of ester branching. The branching groups at optimum length at the epoxy carbon removed the unsaturation sites and also imposed spacing from other triacylglycerol molecules interfering with the formation of the macro crystalline structure. Experimental results found the branching group with a chain length of at least six carbons to be the most effective for imposing the derived molecular spacing and thus imparting the most desired pour point properties. The wear scar diameter decreased with an increasing chain length of branching in chemically modified soybean oil products. The coefficient of friction of all chemically modified products was in the range of 0.11–0.12 and is a considerable improvement over base oil used.

Sharma et al. [64] incorporated sulfur using a chemical reaction of epoxidized vegetable oil and common thiols, resulting in formation of a hydroxy thioether derivative of vegetable oils. The antioxidants activity of these sulfides increased with the number of sulfur atoms. Sulfides with long hydrocarbon chains were found to be good antioxidants over cyclic sulfides. Hydroxy Thioether derivatives of vegetable oils showed very less coefficient of friction than commercially available anti-wear additives. Although sulfur could enhance the tribological performances it is not an environmentally acceptable element.

11.2.2 Hydroformylation

Hydroformylation involves the addition of an aldehyde to olefinic double bond. Commercially, homogeneous metal complexes based on cobalt and rhodium are used as catalysts (Fig. 11.3). Pryde [65] hydroformylated the unsaturated fatty materials with Rhodium-triphenyl phosphine as catalysts, which had a number of advantages over cobalt-carbonyl catalyzed reactions, such as, reaction can be carried out at pressure as low as 200 psi and at each double bond location in a polyunsaturated fatty acid along with high yield of conversion.

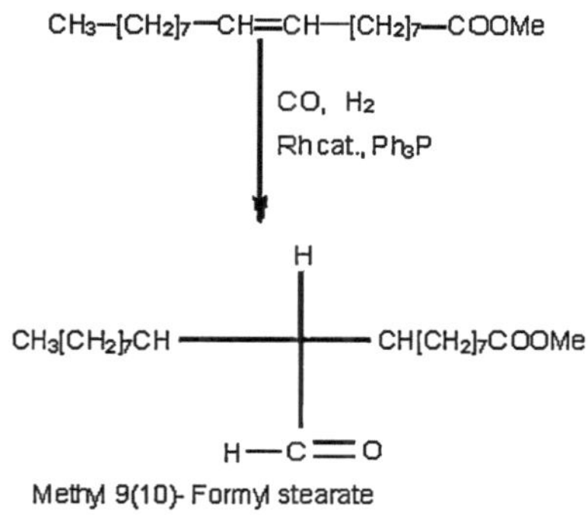

Fig. 11.4 Selective hydroformylation of methyl oleate with rhodiumtriphenyl phosphine catalyst

Hydroformylation also allows the branching of unsaturated fatty acid chains. Friedrich et al. [66] studied Hydroformylation of methyl oleate with a recycled Rhodium catalyst. Methyl oleate was hydroformylated to methyl formylstearate at 120°C and 850–900 psi with a 1:1 mol mixture of hydrogen and carbon monoxide. Catalysts were able to reactivate and could be used till ten runs. The reaction time required for conversion increased from 40 to 56 min as the catalyst level decreased from 250 to 181 mg due to recycling.

Frankel [67] studied a highly selective system for the hydroxyformylation of methyl oleate into methyl 9 (10)-formyl stearate in high yields (Fig. 11.4). A rhodium catalyst in the presence of tripenyl phosphine is used with oleic esters, acids or triglycerides. The reaction was carried out at 95–110°C with a 1:1 mixture of hydrogen and carbon at 500–2000 psi with or without a solvent such as toluene. The formyl stearate obtained in 90–99% conversion from oleate can be either hydrogenated or reduced to hydroxyl methyl stearate or oxidized to carboxy stearate.

11.2.3 Hydrogenation

Hydrogenation changes liquid oil made of triglycerides of polyunsaturated fatty acids to one containing mostly monounsaturated fatty acids. This procedure improves the functionality and flavor stability of the oil. Hydrogenation of vegetable oils increases the stability and improves the shelf life of oils [68]. In the industry the predominantly used catalysts are heterogeneous catalysts such as carrier catalysts (palladium on activated carbon), skeleton catalysts (Raney nickel) or metal oxide catalysts (copper chromium) [69]. Johansson et al. [70] studied hydrogenation of soybean oil and rapeseed oils with fresh copper chromite catalyst, the rate of reaction −d (IV)/dt varies extensively with time. These variations are ascribable to changes in phase composition of the catalyst during its reduction.

Mondal et al. [71] studied the effect of fatty acid profile on the rate of sodium formate catalyzed hydrogenation of soybean, peanut, corn and olive oils. The rate

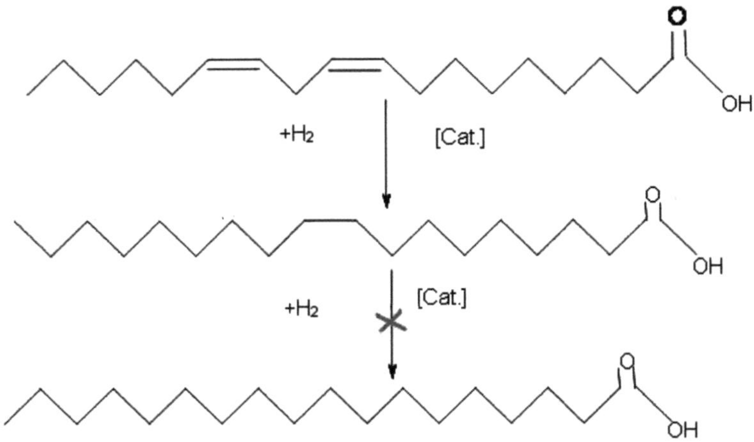

Fig. 11.5 Selective hydrogenation of linoleic acid

constant for linolenic acid hydrogenation is the highest, while the rate constant for oleic acid is the lowest that predicts the higher degree of unsaturation in the fatty acids resulting in more reactive species. The initial rate of hydrogenation was highest in the case of soybean oil followed by peanut and corn oils. Olive oil has the lowest initial rate of hydrogenation, while corn oil has the second lowest overall hydrogenation rate. This observation attributed to the belief that the presence and amount of linolenic acid affect the rate of hydrogenation.

Martinelli et al. [72] performed the hydrogenation of castor oil using a catalytic transfer hydrogenation system in which palladium on carbon was the catalyst and limonene was the solvent and hydrogen donor. The optimum conditions used for highest percentage of castor oil modification was a temperature of 178°C with 1% Pd/C and an oil/limonene ratio of 1:3. The optimized system presented very good reproducibility and showed 100% conversion of the ricinoleate. Spectrometer studies of hydrogenated castor oil indicated that hydrogenation was accompanied by dehydrogenation leading to a mixture of 12-hydroxy and 12-keto stearic derivatives.

Kitayama et al. [73] studied catalytic activity and selectivity of Linoleic acid hydrogenation with catalysts like, Pt, Pd, Ru and Ir supported on Al_2O_3 (Fig. 11.5). The products of hydrogenation reaction include stearic acid (18:0) and ten different octadecenoic isomers (18:1). The hydrogenation on platinum-group metal catalysts showed that 12-position double bond in linoleic acid has higher reactivity than the 9-position double bond. Ru/Al_2O_3 exhibited the highest activity for isomerization of linoleic acid with the noble metal catalysts. Conjugated octadecadienoic acid isomers have been observed in products for the reaction on Pt/Al_2O_3, Ru/Al_2O_3 and Ir/Al_2O_3.

Ravasio et al. [74] studied selective hydrogenation of vegetable oils and their methylesters with different supported copper catalysts. All ex situ reduced catalysts prepared by the chemisorption–hydrolysis (CH) method show better performances than those prepared by incipient wetness technique. Using an 8% Cu/SiO_2 catalyst the trienic component C18:3 can be eliminated and the dienic one lowered from 22 to 3–5% without increasing the stearic C18:0 content and limiting the cis/trans-isomerisation process. These oils, with a C18:1 content up to 88% show remarkable oxidation stability and keep fluid down to −15°C.

Behr et al. [75] studied the selective hydrogenation of using homogeneous transition metal catalysts. They showed that selectivity can be controlled by metal carbonyls, platinum-tin systems or catalysts of the Ziegler-type (iron, cobalt or nickel salts activated by triethylaluminum). Neutral precious metal-phosphine complexes such as the Wilkinson catalysts, $RhCl(PPh_3)_3$, or cationic complexes (e.g. $[Rh(cod)(PPh_3)_2]BPh_4$) were tested, but only moderate success appeared in selective hydrogenetaion. The best results were achieved with a palladium nano-catalyst, which is easily accessible from the pre-catalyst palladium dichloride.

11.2.4 Metathesis of Unsaturated Fats and Oils

The metathesis of alkenes is a metal-catalyzed exchange of alkylidenic groups (=CR2). The Greek word "metathesis" means "rearrangement" or "displacement". This reaction is in principle a reversible reaction. That means that at most a 50% yield is achievable if no procedural intervention occurs. Oleochemical metathesis can be distinguished and divided into different types: In self-metathesis (Fig. 11.6), the same fatty substrate reacts with itself, in the cross-metathesis a fatty substrate reacts with e.g., a petrochemical Alkene. Example for the self metathesis is the formation of 9-octadecene and 9-octadecendioic dimethyl ester from oleic acid methyl ester.

The co-metathesis of erucic acid or oleic acid methyl ester with short-chain olefins such as ethane (ethenolysis) or 2-butene produces unsaturated fatty acid methyl esters of chain lengths C10–C15 and the corresponding olefins. Warwel et al. [76] performed the co-metathesis of oleic acid methyl ester and ethene to obtain methyl-9-decenoate (Fig. 11.7) with a yield of >80% over $Re_2O_7{\cdot}B_2O_3/AL_2O_3{\cdot}SiO_2{\cdot}SnBu_4$ as catalyst. The resulting fatty acid methylesters such as: Methyl-9-decenoate and methyl-13-tetradecenoate are of interest for the manufacture of polyamides, polyesters, lubricants and for flavours and fragrances [77].

11.2.5 Dimerisation or Oligomerisation

Dimerisation of unsaturated fatty acid is a liquid phase batch process, which was studied by Jonson et al. [77] with montomorillonite clay as heteregeneous catalyst in the reaction. The major product groups are dimmers, trimers and isostearic acid. These fractions of the product are applied in lubricants, adhesives, epoxy coatings, printing inks and cosmetics. Homogeneous catalysts have been used in

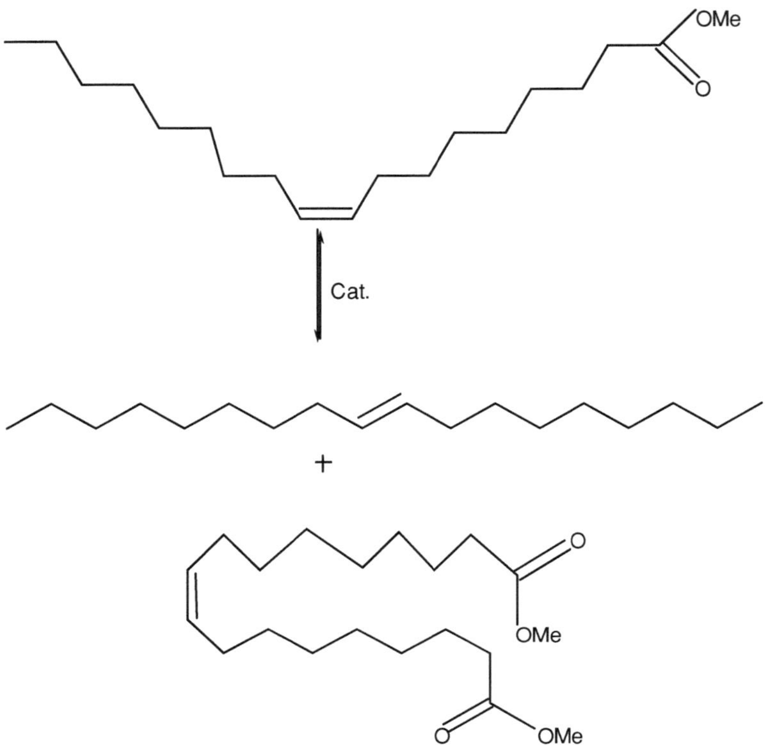

Fig. 11.6 An example for self-metathesis of oleic acid methyl ester

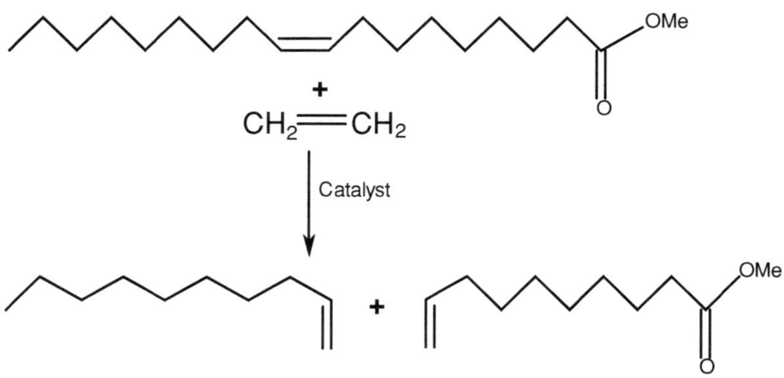

Fig. 11.7 Metathesis of oleic acid methyl ester and ethane

dimerisation, but are only active with the methyl esters of the fatty acids as reactants. Dimers of fatty acids can be made using radical initiation w5x, but this leads to an unfavorably high amount of trimers and higher oligomers, compared

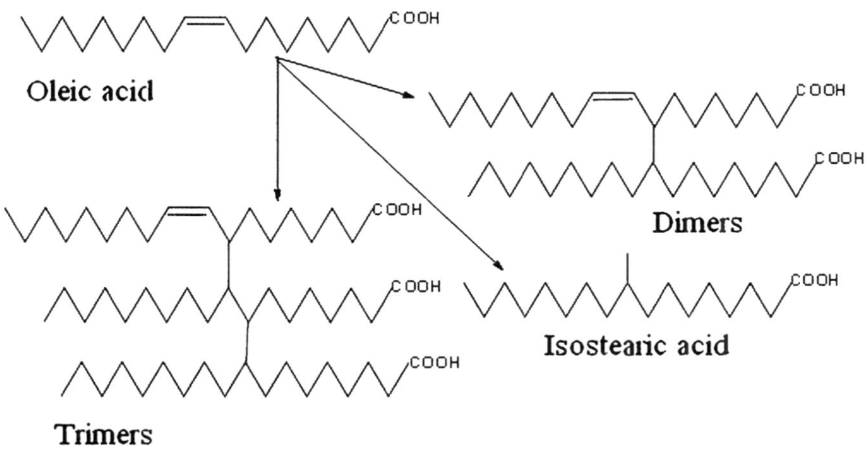

Fig. 11.8 Dimerisation reaction scheme

with the montmorillonite clay catalysed process. Montmorillonite clay is by far the most active heterogeneous catalyst found in the scientific and patent literature. Reaction scheme for Dimerisation process is given in Fig. 11.8.

Isostearic acid and its derivatives showed good pour point and thermo-oxidative properties like oleic acid, therefore, are ideal raw materials for lubricants, cosmetics and plastic additives [78].

11.2.6 Acyloxylation

This is another method for modifying vegetable oils by adding carboxylic acids to the double bonds of unsaturation fatty acids. Use of sterically hindered carboxylic acids for the modification of vegetable oils improves their hydrolytic and oxidative stability. Black et al. [79] performed acetoxylation of methyl oleate with a resin catalyst. They reacted in methyl oleate with acetic acid in the presence of a reticulated cation exchange resin to produce methyl aeetoxystearate. For an optimum yield of 42% reaction was carried out for 42 h with an acid: ester ratio of 50:1 and reaction temperature of 118°C. They also tried acyloxylation of other carboxylic acids but yields were in comparison with acetic acid acyloxylation.

11.2.7 Friedel–Crafts Alkylation of Alkenes

The Friedel–Crafts alkylation is another possible method for synthesis of branched fatty acids and their derivatives. Biermann et al. [80] performed the reaction of oleic acid with isopropyl chloroformate in the presence of

Fig. 11.9 Friedel–Crafts Alkylation of oleic acid

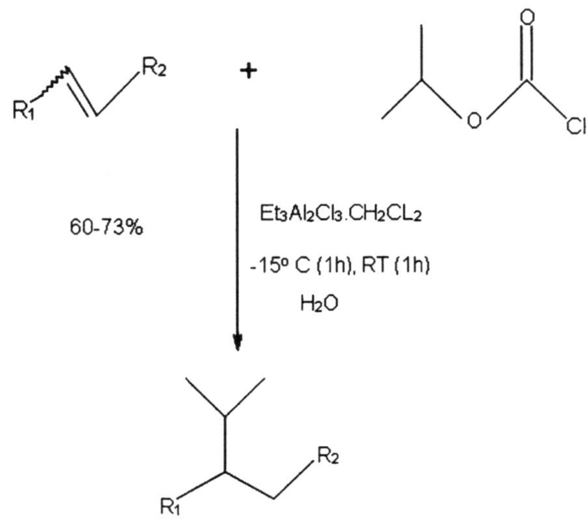

ethylaluminum sesquichloride ($Et_3Al_2Cl_3$) which gave after 2 h a 1:1 regioisomeric mixture of 9- and 10-isopropyloctadecanoic acid in 73% yield (Fig. 11.9).

11.2.8 Friedel–Crafts Acylation of Unsaturated Fatty Acids

Friedel–Crafts acylation is another method for functionalization of unsaturated fatty acids. Friedel–Crafts acylation of alkenes are induced by Lewis acids like $AlCl_3$, $SnCl_4$ or $ZnCl_2$; the acylation often leads to a mixture of β,γ–unsaturated ketones, α,β-unsaturated ketones and β–cholro ketones. The $EtAlCl_2$-induced acylation of oleic acid with acetyl chloride produces a mixture with approximately similar quantities of 9- and 10-regioisomers as adducts with the yield of about 55% (Fig. 11.10) [81].

11.2.9 Transesterification

Transesterification is an alcoholysis in which one alcohol is capable of displacing another alcohol from an ester in a process similar to hydrolysis, except that alcohol is used instead of water. This process has been widely used to reduce the high viscosity of triglycerides. Transesterification is one of the reversible reactions and proceeds essentially by mixing the reactants. However, the presence of a catalyst (a strong acid or base) accelerates the conversion. The transesterification reaction is represented by the general equation,

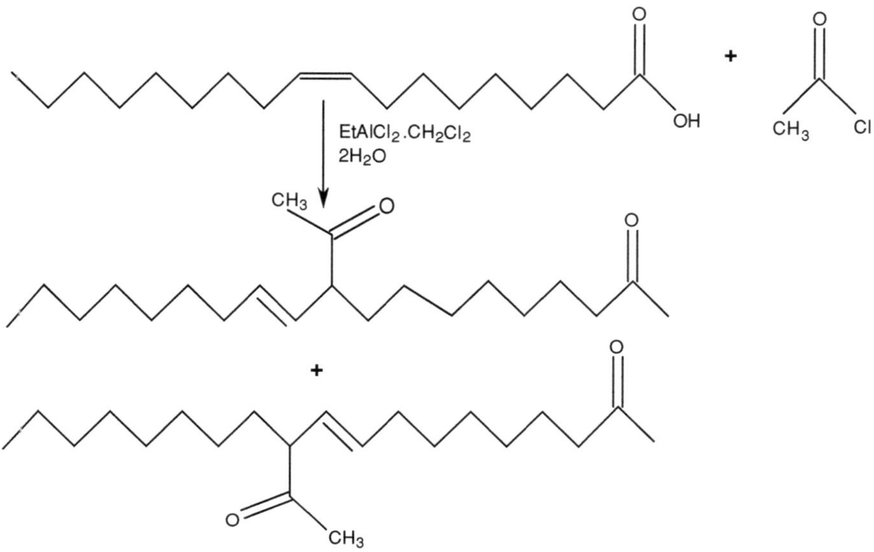

Fig. 11.10 Freidel craft acylation of oleic acid

$$\underset{\text{Ester}}{RCOOR^1} + \underset{\text{Alcohol}}{R^2OH} \longleftrightarrow \underset{\text{Ester}}{RCOOR^2} + \underset{\text{Alcohol}}{R^1OH}$$

Transesterification of triglycerides produce fatty acid alkyl esters and glycerol. The glycerol layer settles down at the bottom of the reaction vessel. Diglycerides and monoglycerides are the intermediates in this process. The mechanism of transesterification is described as follows:

$$\text{Triglyceride} + R^1OH \leftrightarrow \text{Diglyceride} + RCOOR^1$$
$$\text{Diglyceride} + R^1OH \leftrightarrow \text{Monoglyceride} + RCOOR^1$$
$$\text{Monoglyceride} + R^1OH \leftrightarrow \text{Glycerol} + RCOOR^1$$

The stepwise reactions are reversible and an excess of alcohol is used to shift the equilibrium towards the formation of esters. In the presence of excess alcohol, the forward reaction is pseudo-first order and the reverse reaction is found to be second order. Transesterification reaction can be catalyzed by base catalyst and also by acid catalyst. It was found that base catalysed transesterification reactions are faster than acid catalysed reactions [82].

11.2.9.1 Mechanism of Acid-Catalyzed Transesterification

Transesterification can be catalyzed by Brownsted acids, preferably by sulfonic and sulfuric acids. These catalysts give very high yields in alkyl esters but the reactions are slow, requiring typically temperatures above 100°C and more than

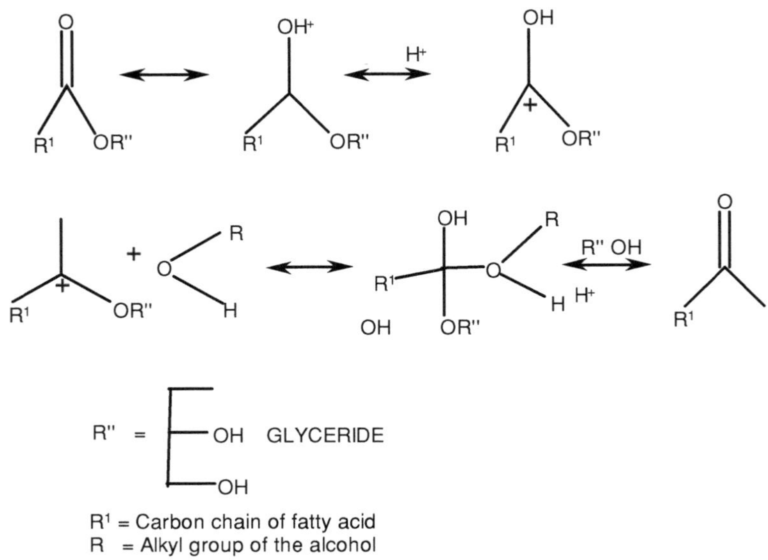

Fig. 11.11 The general equation for acid catalysed transesterification reaction

3 h to complete the conversion. The mechanism of acid catalyzed transesterification of vegetable oil (for a monoglycerides) is shown in Fig. 11.11. However, it can be extended to di- and tri-glycerides. The protonation of carbonyl group of the ester leads to the carbocation, which after a nucleophilic attack of the alcohol produces a tetrahedral intermediate. This intermediate eliminates glycerol to form a new ester and to regenerate the catalyst [83].

11.2.9.2 Mechanism of Alkali-Catalyzed Transesterification

The first step involves the attack of the alkoxide ion to the carbonyl carbon of the triglyceride molecule, which results in the formation of a tetrahedral intermediate. The reaction of this intermediate with an alcohol produces the alkoxide ion in the second step. In the last step the rearrangement of the tetrahedral intermediate gives rise to an ester and a diglyceride. The mechanism of alkali catalysed transesterification reaction is given in Fig. 11.12.

The transesterification process can make use of other alcohols such as ethyl, butyl and isopropyl alcohol. Among these alcohols, iso propanol is the only one with a branched structure. Transesterification is the process of separating the fatty acids from their glycerol backbone to form fatty acid esters and free glycerol. The reaction of a triglyceride with methyl alcohol using potassium hydroxide (KOH) as a catalyst to produce methyl ester is as follows (Fig. 11.13) [84].

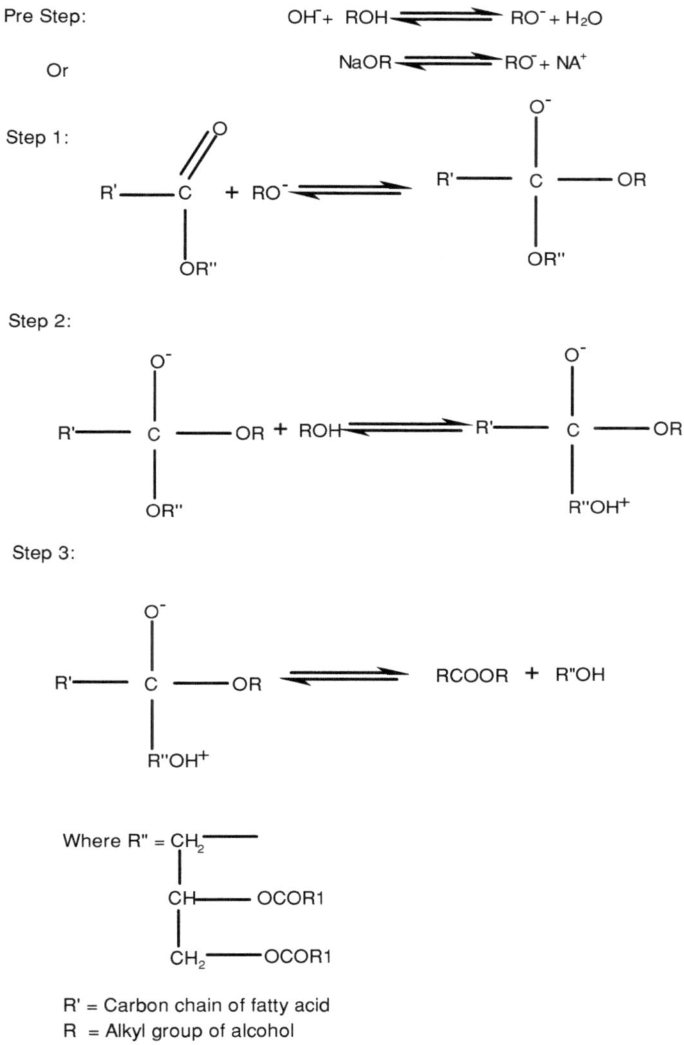

Fig. 11.12 The general equation for alkali catalysed transesterification reaction

Transesterification reaction for soybean oil using frying oil and tallow with methanol was carried out in 2L glass reactor equipped with a glass anchor-shaped stirrer, a type-K thermocouple, a water condenser and a funnel, and surrounded by a heating mantle controlled by a proportional integral derivative (PID) temperature controller device; molar ratio of oil to alcohol used was 0.8:6 with sodium methoxide (1% w/w of oil) as catalyst, the complete conversion of oil into ester took less than 2 h at 600 rpm of mixing rotation [85]. Transesterification reaction can be carried out with different catalysts like acids, alkaline metal hydroxides,

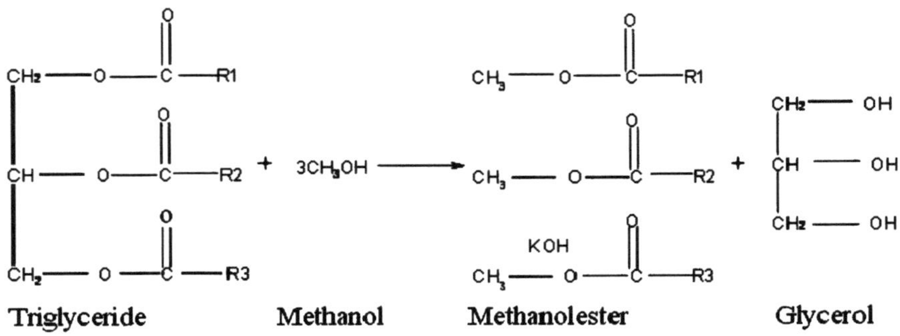

Fig. 11.13 The reaction of a triglyceride with methyl alcohol to produce methyl ester

and alkoxides, carbonates, and enzymes, non-ionic bases such as amines, amide guanidine and zeolite as heterogeneous catalysts [86]. Pyrone complexes of different metals such as tin, lead and zinc were also used as catalyst for transesterification of vegetable oils; catalytic activity was higher when oil with short chain fatty acids or with a high unsaturation degree and short and linear chain alcohols were used [87]. Ninety-nine percent of the conversion occurred with 0.125 mol/kg concentration of base catalyst like sodium hydroxide and sodium methoxide for palm oil transesterification at 60.5°C with 1:10 oil to alcohol molar ratio. Sodium methoxide was performed best with the same ester conversion at 1:6 molar ratios [88]. Transesterification of palm oil with sodium metal yields 99% of product, while activated silica gel was not suitable for reaction. Decrease in the stirring speed decreases the reaction rate [89]. At 75°C temperature, nearly 1% catalyst to weight of triglyceride and 1:9 molar ratios between *Cynara cardunculus* L oil to ethanol gives the satisfactory conversion of oil into ester with improved separation of ester layer from glycerol layer during separation which increases the final yield of ester product [90].

Acid catalyst was most efficient for transesterification of palm oil with methanol than base catalyst, even though it was required in less quantity for the reaction but separation of ester layer from glycerol was difficult upon completion of reaction [91]. The acid catalysed transesterification reaction was suitable for low grade high free fatty acid containing oils, such as crude rice bran oil [92]. Ester product for castor oil and cotton seed oil with ethanol was constant with variation in concentration of acid catalyst; variation in concentration of base catalyst increases the ester yield. A rise in temperature increases the rate of reaction for acid catalysed reactions [93].

The effect of mixing was studied with the motionless mixture and high shear mixture; both are able to convert 97% of vegetable oil into methyl ester. The increase in catalyst concentration increases the ester conversion but it increases the solubility of ester in glycerol layer [94]. The conversion of triglyceride, diglyceride and monoglyceride appear to be second order up to 30 min of reaction time and reaction rate constants were 0.018–0.191 and were higher at higher temperature in a batch reactor for transesterification of palm oil. The 1 and 1.2% of KOH as catalyst showed no significant difference for the conversion of palm oil triglyceride into methyl ester [95]. Alcoholysis for sunflower oil with methyl alcohol, ethyl alcohol and butyl alcohol was carried out near the boiling point of each alcohol with 0.5 wt% of sodium methoxide as catalyst. After 1 m of reaction, greater ester yield was obtained with butyl alcohol (88%), because of higher reaction temperature and at the end of 60 min ester yield of all the three alcohols are almost the same [96]. The addition of 5% of palm oil methyl ester to lubricant oil reduces both wear and friction tested with four ball tribometer and addition of above 5% ester derivative of palm oil causes some oxidation and chemical corrosion [97]. Analysis of fuel lubricity and base fuel properties of transesterified esters with sodium methoxide and sodium ethoxide showed that ethyl ester had more lubricity than methyl ester of the same vegetable oils and all esters showed increased kinematic viscosity for increasingly longer carbon chains [98]. Mixtures of methanol and ethanol increased the rate of transesterification reaction and produced methyl as well as ethyl esters. The lubricity of ethyl ester was better than that of methyl when adding 1 vol% to the base fuel. Similarly, lubricity of mixed esters of methanol and ethanol (3:3 ratios) was better than that of pure methyl ester [99].

11.3 Different Laboratory Scale Test Methods for Natural Oil-Based Lubricants

Laboratory procedures have been developed to aid in the selection of the right lubricants for the right application. A laboratory test method must be meaningful. It must simulate the most important conditions of the lubricant operation. The results must be measurable and must be compared against a standard or a reference fluid. Variables that may affect the test results must be controlled. This part of the chapter provides an overview of the methods available to conduct laboratory scale tests for natural oil-based lubricants. Complete, step-by step procedures for many of these tests are provided in the references.

11.3.1 Chemical and Physical Tests

11.3.1.1 Specific Gravity

The ASTM D5355 [100] method is used to measure the specific gravity of all oils and liquid fats that are used in lubricants. The specific gravity is one parameter that can be used to monitor the quality of these lubricant products. The method to determine the specific gravity is by calculating the ratio of the weight of a unit volume of the sample to the weight of a unit volume of water at 25°C.

11.3.1.2 Viscosity

Viscosity is an important property of a lubricant, as it affects the film thickness and, thus, the wear rate of sliding surfaces. It is used for the identification of individual grades of oil and for monitoring the changes occurring in oil while in service. Viscosity increase usually indicates that oil used has deteriorated by oxidation or by contamination, while a decrease usually indicates dilution by lower viscosity oil or by fuel. Kinematic viscosities of the fluids can be measured with ASTM D445 [101] method. The kinematic viscosity is the product of the measured flow time and the calibration constant of the viscometer. The flow time can be measured by measuring the flow of fixed volume of liquid under gravity through a capillary of calibrated viscometer at known temperature. The dynamic viscosity, h, can be obtained by multiplying the kinematic viscosity, n, by the density, p, of the liquid.

11.3.1.3 Acid and Base Number

Both new and used lubricant products contain few compounds, maybe as additives or results of degradation during service. The relative amount of these compounds can be calculated by titrating with acids or bases. The acid number is the measure of the amount of acid substance present in the oils and base number is the measure of the base constituents in the oil. Acid and base number gives the quality of newly formulated lubricants and also provides information about thermo-oxidative and bio degradation of lubricants. ASTM method D664 measures the acid number of lubricating oils by potentiometric titration method. The method uses the mixture of toluene and isopropyl alcohol with a small amount of water to dissolve the sample and titrated potentiometrically with alcoholic KOH using a glass indicating electrode and a reference electrode or a combination electrode. End points are taken only at well-defined inflections in the resulting curve obtained during meter reading. ASTM D974 [102] method determines both acid and base number by the color indicator titration method. The method uses the similar chemical mixture as D664 method to dissolve the sample and to determine the acid and base number; the resulting single-phase solution is titrated at room temperature with standard alcoholic base or alcoholic acid solution, respectively, to the end point indicated by the color change of the added *p*-naphtholbenzein solution. After the addition of indicator, if the sample assumes a yellow–orange color, proceed for acid number test, if it becomes green or green–black, proceed for base number test. ASTM D3339 [103] method is especially intended for cases in which the amount of sample available to be analyzed is too small to allow for accurate analysis by Test Methods D974 or D664. This method is a semi micro color indicator titration method, which uses a similar solvent as the previous methods to dissolve the oils and the resulting single-phase solution is titrated at room temperature under a nitrogen atmosphere with standardized 0.01 M KOH in isopropyl alcohol to the stable green color of the added *p*-naphtholbenzein indicator.

11.3.1.4 Viscosity Index

The viscosity index is the measure of the variation in kinematic viscosity due to changes in the temperature of a lubricant product between 40 and 100°C. A higher viscosity index indicates a smaller decrease in kinematic viscosity with increasing temperature of the lubricant. ASTM D2270 [104] is the widely used method to calculate viscosity indices of different lubricants by using the respective kinematic viscosities at two different temperatures.

11.3.1.5 Saponification Value

ASTM D5558 [105] method covers the procedure to measure Saponification values of fats and oils used in the natural oil-based lubricants. In this method fats and oils are saponified with alcoholic KOH by boiling the solution gently and steadily. The resulting solution with indicator will be titrated against hydrochloric acid.

11.3.1.6 Iodine Value

The iodine value is a measure of the unsaturation of fats and oils and is expressed in terms of the number of centigrams of iodine absorbed per gram of sample. The ASTM D5554 [106] method provides the procedure to determine the iodine value of oils used in natural lubricants, which do not contain conjugated double bonds.

11.3.1.7 Free Fatty Acid Value

Free fatty acids present in the oils used for development of natural oils-based lubricants can be measured with ASTM D5555 [107] method. In this method the desired amount of sample is mixed with ethyl alcohol and indicator. The solution is titrated against alkali to obtain the amount of free fatty acids in the sample.

11.3.1.8 Refractive Index

ASTM D1747 [108] method provides the method to measure the refractive indexes of transparent and light-colored viscous hydrocarbon liquids and melted solids that have refractive indexes in the range between 1.33 and 1.60, and at temperatures from 80 to 100°C. This method measures the refractive index by the critical angle method using monochromatic light from sodium lamp.

11.3.1.9 Water Content in the Lubricating Oil

Presence of water in the range from 0.02 to 2% in lubricating oil can be determined with ASTM D4377 [109] method. In this method homogenized oil with solvent will be titrated to an electrometric end point using Karl Fischer reagent. ASTM D4928 [110] uses the similar method to determine the water content in the lubricating oils in the 0.005–0.02 mass percent ranges.

11.3.1.10 Flash and Fire Points

The flash point is the lowest liquid temperature at which application of the test flame causes the vapors of the test specimen of the sample to ignite. Fire point, is a temperature above the flash point, at which the test specimen will support combustion for a minimum of 5 s. ASTM D92 [111] method finds out the flash and fire point in cleveland open cup tester by heating the 70 ml of specimen rapidly at the beginning and at a slower constant rate as the flash point is approached. At specified intervals a test flame is passed across the cup to determine the flash point. ASTM D7094 [112] method covers the determination of the flash point by a continuously closed cup tester utilizing a specimen size of 2 ml, cup size of 7 ml, with a heating rate of 2.5°C per minute.

11.3.1.11 Low Temperature Properties

1. *Pour Point*:
   ASTM D97 [113] method provides the procedure to find the pour point of lubricating oils. The method follows the cooling of a preheated sample at the specified rate and examined at intervals of 3°C for flow characteristics. The lowest temperature at which movement of the specimen is observed is recorded as the pour point.
2. *Cloud Point*:
   Cloud point is the temperature of a liquid specimen when the smallest observable cluster of hydrocarbon crystals first occurs upon cooling under prescribed conditions. ASTM D2500 [114] method measures the cloud point cooling the specimen at a specified rate and observing for cloudy appearance at the bottom of the glass jar.

11.3.1.12 Dilution Stability Determination

ASTM D1401 [115] method provides a guide for determining the water separation characteristics of oils subject to water contamination and turbulence. The test methods include stirring of 40 ml of specimen and 40 ml of water for about 5 min at

54°C. The time required for water separation is recorded for every 5 min or at some specification time limit. If complete separation or emulsion reduction to 3 mL or less does not occur after standing for 30 min or some other specification time limit, the volumes of oil (or fluid), water and emulsion remaining at the time are reported.

11.3.1.13 Foam Tests

The tendency of oils to foam can be a serious problem in systems such as high-speed gearing, high-volume pumping and splash lubrication. Inadequate lubrication, cavitation and overflow loss of lubricant can lead to mechanical failure ASTM D3601 [116] (Bottle Test) method provides the procedure to find out the foaming tendency of low viscous oils. The increased volume of sample is measured by determining the increase in the total height of the test fluid, including foam, after vigorous shaking of the emulsion at 25 ± 1°C.

ASTM D3519 [117] (Blender Test), measures the increase in height of the sample in a glass jar including foam after the completion of blending for about 30 s at 25 ± 1°C.

Another ASTM method D892 [118] intends to determine the foaming tendency of oils by blowing air to the sample at a constant rate for 5 min and then allowed to settle for 10 min at 24°C. The volume is measured at the end of the period.

11.3.1.14 Hydrolytic Stability

ASTM D2619 [119] (Beverage Bottle Method) provides the information about relative stability of Hydraulic fluids in the presence of test conditions. Hydrolytically unstable hydraulic fluids form acidic and insoluble contaminants which can cause hydraulic system malfunctions due to corrosion, valve sticking or change in viscosity of the fluid. The method uses 75 g of sample and 25 g of water and a copper test specimen sealed in a pressure-type beverage bottle. The bottle is rotated, end for end, for 48 h in an oven at 93°C. Hydrolytic stability will be determined by measuring viscosity, weighting insoluble and weight o copper strips.

11.3.1.15 Thermal Stability

Thermal stability characterizes physical and chemical property changes which may adversely affect the oil's lubricating performance. ASTM D2070 [120] method characterizes the thermal stability of hydraulic oils. The method uses a beaker containing test oil, with copper and iron rods placed in an aluminum block in an electric gravity convection oven for 168 h at a test temperature of 135°C. The result is obtained by determining the discoloration of copper and iron rods and by measuring the quantity of sludge formed. Another ASTM method D6375 [121] uses thermo gravimetric analyses to determine the evaporation loss of lubricating oils.

The sample is placed in a TGA pan holding to TGA pan holder and quickly heating the sample to between 247 and 249°C under a stream of air, and then held isothermal for an appropriate time. Throughout this process, the TGA monitors and records the mass loss experienced by the specimen due to evaporation.

11.3.1.16 Oxidation Stability

ASTM D2272 [122] provides a method to determine the oxidation stability of the oils by rotating the pressure vessel. The test oil, water and copper catalyst coil, contained in a covered glass container, are placed in a vessel equipped with a pressure gauge. The vessel is charged with oxygen to a gauge pressure of 620 kPa, placed in a constant-temperature oil bath set at 150°C or dry block taken at 150° and rotated axially at 100 rpm at an angle of 30° from the horizontal. The number of minutes required to reach a specific drop in gauge pressure is the oxidation stability of the test sample.

Adhvaryu et al. [123], studied oxidation stability of alkoxylated triacylglycerol by using Thin Film Micro Oxidation test and by pressure differential scanning calorimetry method.

11.3.1.17 Lubricity Tests

ASTM method D3233 [124] (Pin and Vee-Block test) provides the procedure to perform measurement of extreme pressure properties of lubricating fluids. The test method consists of running a rotating steel journal at 290 ± 10 rpm against two stationary V-blocks immersed in the lubricant sample. Load is applied to the V-blocks by a ratchet mechanism and the load-fail value obtained is the criteria for the level of load-carrying properties.

Extreme pressure properties of lubricating fluids can also be measured with ASTM D2783 [125] method. The tests are operated in a four-ball tester machine with a rotating speed of 1760 ± 40 rpm. The machine and test lubricant are brought to 18–35°C and then a series of tests of 10 s duration are made at increasing loads until welding occurs.

ASTM D4172 [126] method determines the wear preventive characteristics of lubricating oil. The test is performed in a four-ball tester machine with the force of 392 N. The temperature of the test lubricant is regulated at 75°C and the top ball is rotated at 1200 rpm for 60 min. Lubricants are compared by using the average size of the scar diameters worn on the three lower clamped balls.

11.3.1.18 Biodegradability and Eco-Toxicity

ASTM D6081 [127] method covers the procedures to be used in the preparation of lubricants or their components for toxicity testing in aquatic systems. This method discusses the techniques for the preparation of test materials for toxicity testing of poorly water soluble lubricants or their components under acute or chronic exposure conditions with fish, large invertebrates or algae. Other methods for aquatic toxicity are OECD fish embryo toxicity tests and OECD acute fish toxicity tests. In Fish Embryo Toxicity tests, Zebra fish embryos are individually exposed in 24-well micro titer plates to a range of concentrations of the test substance. The test is initiated immediately after fertilization and is continued for 48 h. Lethal effects, as described by four apical end points, are determined by comparison with controls to identify the LC50 values. In Fish Acute Toxicity test, the fish are exposed to the test substance preferably for a period of 96 h. Mortalities are recorded at 24, 48, 72 and 96 h and the concentrations which kill 50% of the fish (LC50) are determined.

ASTM D5864 [128] method measures the biodegradation by quantifying the percentage of carbon dioxide produced by lubricant sample, which is kept under aerobic aquatic condition with microorganisms. This value is then compared to the theoretical amount of CO_2 which could be generated if all of the carbon in the test material were converted into CO_2. Biodegradability is expressed as a percentage of theoretical CO_2 production.

References

1. W.J. Bartz, Lubricants and the environment. Tribol. Int. **31**(1–3), 35–47 (1998)
2. P.V. Joseph, S. Deepak, D.K. Sharma, Study of some non-edible vegetable oils of Indian origin for lubricant application. J. Synth. Lubr. **24**, 181–197 (2007)
3. B. Wilson, Lubricants and functional fluids from renewable sources. Ind. Lubr. Technol. **50**(1), 6–15 (1998). (January/February)
4. R.L. Goyan, R.E. Melley, P.A. Wissner, W.C. Ong, Biodegradable lubricants. Lubr. Eng. **54**(7), 10–17 (1998)
5. L. Lazzeri, M. Mazzoncini, A. Rossi, E. Balducci, G. Bartolini, L. Giovannelli, R. Pedriali, R. Petroselli, G. Patalano, G. Agnoletti, A. Borgioli, B. Croce, L. D'Avino, Biolubricants for the textile and tannery industries as an alternative to conventional mineral oils: an application experience in the Tuscany province. Ind. Crops Prod. **24**, 280–291 (2006)
6. C.W. Lea, European development of lubricants from renewable sources. Ind. Lubr. Tribol. **54**(6), 268–274 (2002)
7. A. Pettersson, High-performance base fluids for environmentally adapted lubricants. Tribol. Int. **40**, 638–645 (2007)
8. S.Z. Erhan, B.K. Sharma, Z. Liu, A. Adhvaryu, Lubricant base stock potential of chemically modified vegetable oils. J. Agric. Food Chem. **56**(19), 8919–8925 (2008)
9. S. Asadauskas, J.H. Perez, J.L. Duda, Lubrication properties of castor oil–potential base stock for biodegradable. Lubr. Eng. **53**(12), 35–40 (1997)
10. H.H. Masjuki, M.A. Maleque, A. Kubo, T. Nonaka, Palm oil and mineral oil based lubricants—their tribological and emission performance. Tribol. Int. **32**, 305–314 (1999)
11. M.A. Maleque, H.H. Masjuki, S.M. Sapuan, Vegetable based biodegradable lubricating oil additives. Ind. lubr. technol. **55**(3), 137–143 (2003)
12. S.Z. Erhan, S. Asadauskas, Lubricant basestocks from vegetable oils. Ind. Crops Prod. **11**, 277–282 (2000)
13. V.K. Bhatia, A. Chaudhry, G.A. Sivasankaran, R.P.S. Bisht, M. Kashyap, Modification of jojoba oil for lubricant formulations. JAOCS **67**(1), 1–7 (1990)

14. O.N. Anand, V.K. Chhibber, Vegetable oil derivatives: environment-friendly lubricants and fuels. J. Synth. Lubr. **23**, 91–107 (2006)
15. Y.M. Shashidhara, S.R. Jayaram, Vegetable oils as a potential cutting fluid—an evolution. Tribol. Int. **43**, 1073–1081 (2010)
16. I.I. Ştefanescu, C. Calomir, G. Chirita, On the future of biodegradable vegetable lubricants used for industrial trybosystems. The annals of university "dunărea de jos" of galaţi fascicle VIII (2002)
17. B. Krzan, J. Vizintin, Tribological properties of an environmentally adopted universal tractor transmission oil based on vegetable oil. Tribol. Int. **36**, 827–833 (2003)
18. J.K. Mannekote, S.V. Kailas, Influence of chemical structure on the boundary lubrication properties of vegetable oils. ASME 2010 10th biennial conference on engineering systems design and analysis (ESDA2010), (Istanbul, Turkey, ESDA2010-25070), pp. 633–637, 12–14 July 2010
19. N.H. Jayadas, K. Prabhakaran Nair, Coconut oil as base oil for industrial lubricants—evaluation and modification of thermal, oxidative and low temperature properties. Tribol. Int. **39**, 873–878 (2006)
20. J.K. Mannekote, S.V. Kailas, Performance evaluation of vegetable oils as lubricant in a four stroke engine. *World Tribology Conference 2009*, Kyoto, Japan, D-215, p. 331, 12–15 Sept 2009
21. J.K. Mannekote, S.V. Kailas, Experimental investigation of coconut and palm oils as lubricants in four stroke engines. Tribol. Online **6**(1), 76–82 (2011)
22. J.K. Mannekote, S.V. Kailas, R.T. Naik, Condition monitoring of vegetable oils used in a 4 stroke engine as lubricants. *Tribo-India Conference on Tribology of Automotive Systems*, 11, 12 Dec 2009
23. Jan Cloin, Coconut oil as a biofuel in Pacific Islands. Refocus July/August 2005
24. P.J. Singh, J. Khurma, A. Singh, Preparation, characterisation, engine performance and emission characteristics of coconut oil based hybrid fuels. Renew. Energy **35**, 2065–2070 (2010)
25. W.K. Trotter, W.D. Givan, Economics of sunflower oil use m the United States production and use in the United States. JAOCS **48**, 442–449 (1971)
26. Z. Flagella, T. Rotunno, E. Tarantino, R. Di Caterina, A. De Caro, Changes in seed yield and oil fatty acid composition of high oleic sunflower (Helianthus annuus L.) hybrids in relation to the sowing date and the water regime. Eur. J. Agron. **17**, 221–230 (2002)
27. G.A. Pereyra-Irujo, N.G. Izquierdo, M. Covi, S.M. Nolasco, F. Quiroz, L.A.N. Aguirrezábal, Variability in sunflower oil quality for biodiesel production: a simulation study. Biomass Bioenergy **33**, 459–468 (2009)
28. S.A. Smith, R.E. King, D.B. Min, Oxidative and thermal stabilities of genetically modified high oleic sunflower oil. Food Chem. **102**, 1208–1213 (2007)
29. S. Marmesat, M. Mancha, M.V. Ruiz-Méndez, M.C. Dobarganes, Performance of sunflower oil with high levels of oleic and palmitic acids during industrial frying of almonds, peanuts, and sunflower seeds. JAOCS **82**(7), 505–510 (2005)
30. I. Stefanescu, C. Calomir, C. Gheorghies, C. Spanu, Study on tribological properties of vegetable sunflower oil used as possible ecological lubricant, The Annals Of University "Dunărea De Jos" Of Galaţi Fascicle Viii, 2005, Issn 1221-4590, Tribology
31. G.A. Pereyra-Irujo, L.A.N. Aguirrezabal, Sunflower yield and oil quality interactions and variability: Analysis through a simple simulation model. Agric. For. Meteorol. **143**, 252–265 (2007)
32. J. Fredric, R.J. Baur, J.B. Brown, The fatty acids of corn oil, vol. 67, pp. 1899–1900 (1945)
33. L. Pop, C. Puscas, G. Bandur, G. Vlase, R. Nut iu, Basestock oils for lubricants from mixtures of corn oil and synthetic diesters. JAOCS **85**, 71–76 (2008)
34. F. Marini, F. Balestrieri, R. Bucci, A.L. Magr, D. Marini, Supervised pattern recognition to discriminate the geographical origin of rice bran oils: a first study. Microchem. J. **74**, 239–248 (2003)

35. A. Saydu, M.Z. Duz, C. Kaya, A.B. Kafadar, C. Hamamci, Transesterified sesame (Sesamum indicum L.) seed oil as a biodiesel fuel. Bioresou. Technol. **99**, 6656–6660 (2008)
36. N.A. Santos, M.L.A. Tavares, R. Rosenhaim, F.C. Silva, V.J. Fernandes Jr, A.G. Souza, Thermogravimetric and calorimetric evaluation of babassu biodiesel obtained by the methanol route. J. Therm. Anal. Calorim. **87**(3), 649–652 (2007)
37. L. Canoira, J.G. Galea, R. Alcantara, M. Lapuerta, R. Garcıa-Contreras, Fatty acid methyl esters (FAMEs) from castor oil: production process assessment and synergistic effects in its properties. Renew. Energy **35**, 208–217 (2010)
38. J.S. de Oliveira, P.M. Leite, L.B. de Souza, V.M. Mello, E.C. Silva, J.C. Rubim, S.M.P. Meneghetti, P.A.Z. Suarez, Characteristics and composition of jatropha gossypiifolia and Jatropha curcas L. oils and application for biodiesel production. Biomass Bioenergy **33**, 449–453 (2009)
39. E.T. Akintayo, Characteristics and composition of Parkia biglobbossa and Jatropha curcas oils and cakes. Bioresour. Technol. **92**, 307–310 (2004)
40. S.N. Shah, B.K. Sharma, B.R. Moser, S.Z. Erhan, Preparation and evaluation of jojoba oil methyl esters as biodiesel and as a blend component in ultra-low sulfur diesel fuel. Bioenerg. Res. **3**, 214–223 (2010)
41. H. Gisser, J. Messina, D. Chasan, Jojoba oil as a sperm oil substitute. Wear **34**, 53–63 (1975)
42. A. Apelblat, J. Wisniak, G. Shapiro, Physical properties of (jojoba oil + n-hexane) compared with other (vegetable oil + n-hexane) mixtures. J. Chem. Thermodyn. **40**, 1477–1484 (2008)
43. R.P.S. Bisht, G.A. Sivasankaran, V.K. Bhatia, Additive properties of jojoba oil for lubricating oil formulations. Wear **161**, 193–197 (1993)
44. G.A. Sivasankaran, R.P.S. Bisht, V.K. Jain, M. Gupta, A. Sethuramiah, V.K. Bhatia, Jojoba-oil-based two-stroke gasoline engine lubricant. Tribol. Int. **21**(6), 327–333 (1988)
45. M. Allawzi, M.K. Abu-Arabi, H.S. Al-zoubi, A. Tamimi, Physicochemical characteristics and thermal stability of Jordanian jojoba oil. JAOCS **75**(1), 57–62 (1998)
46. S.L. Pearson, J.E. Spagnoli, Environment lubricants–an overview of onsite applications and experience. Lubr. Eng. **56**(4), 40 (2000)
47. B. Wilson, Lubricants and functional fluids from renewable sources. Ind. Lubr. Technol. **50**(1), 6–15 (1998)
48. B.K. Sharma, A. Adhvaryu, Z. Liu, S.Z. Erhan, Chemical modification of vegetable oils for lubricant applications. JAOCS **83**(2), 129–136 (2006)
49. S. Onkawa, A. Konishi, H. Hatano, K. Ishihama, K. Tanaka, M. Awamura, Oxidation and corrosion characteristics of vegetable base biodegradable hydraulic oils. SAE Tech. Pap. 951038, 55–63 (1995)
50. B.N. Rhodes, W. Mammel, P. Landis, F.L. Erickson, Water rejection of vegetable oil base stocks for tractor/hydraulic fluids. SAE Tech. Pap. 952073, 1–4 (1995)
51. S. Asadauskas, S.Z. Erhan, Depression of pour points of vegetable oils by blending with diluents used for biodegradable lubricants. JAOCS **76**(3), 313–316 (1999)
52. H.-S. Hwang, S.Z. Erhan, Modification of epoxidized soybean oil for lubricant formulations with improved oxidative stability and low pour point. JAOCS **78**(12), 1179–1184 (2001)
53. S.Z. Erhan, B.K. Sharma, Z. Liu, A. Adhvaryu, Lubricant base stock potential of chemically modified vegetable oils. J. Agric. Food Chem. **56**, 8919–8925 (2008)
54. B.K. Sharma, A. Adhvaryu, S.Z. Erhan, Friction and wear behavior of thioether hydroxy vegetable oil. Tribol. Int. **42**, 353–358 (2009)
55. J. Sepulveda, S. Teixeira, U. Schuchardt, Alumina-catalyzed epoxidation of unsaturated fatty esters with hydrogen peroxide. Appl. Catal. A **318**, 213–217 (2007)
56. J.L. Scala, R.P. Wool, Effect of FA composition on epoxidation kinetics of TAG. JAOCS **79**(4), 373–378 (2002)
57. L.H. Gan, K.S. Ooi, L.M. Gan, S.H. Goh, Effects of epoxidation on the thermal oxidative stabilities of fatty acid esters derived from palm olein. JAOCS **72**(4), 439–442 (1995)

58. R. Mungroo, N.C. Pradhan, V.V. Goud, A.K. Dalai, Epoxidation of canola oil with hydrogen peroxide catalyzed by acidic ion exchange resin. JAOCS **85**, 887–896 (2008)
59. S. Dinda, A.V. Patwardhan, V.V. Goud, N.C. Pradhan, Epoxidation of cottonseed oil by aqueous hydrogen peroxide catalysed by liquid inorganic acids. Bioresour. Technol. **99**, 3737–3744 (2008)
60. V.V. Goud, N.C. Pradhan, A.V. Patwardhan, Epoxidation of karanja (*pongamia glabra*) oil by H_2O_2. JAOCS **83**(7), 635–640 (2006)
61. F.E. Okieimen, O.I. Bakare, C.O. Okieimen, Studies on the epoxidation of rubber seed oil. Ind. Crops Prod. **15**, 139–144 (2002)
62. A. Adhvaryu, S.Z. Erhan, Epoxidized soybean oil as a potential source of high-temperature lubricants. Ind. Crop Prod. **15**, 247–254 (2002)
63. S.Z. Erhan, B.K. Sharma, Z. Liu, A. Adhvaryu, Lubricant base stock potential of chemically modified vegetable oils. J. Agric. Food Chem. **56**, 8919–8925 (2008)
64. B.K. Sharma, A. Adhvaryu, S.Z. Erhan, Friction and wear behavior of thioether hydroxy vegetable oil. Tribol. Int. **42**, 353–358 (2009)
65. E.H. Pryde, JAOCS **61**(2), 419–425 (1984)
66. J.P. Friedrich, G.R. List, V.E. Sohns, Hydroformylation of methyl oleate with a recycled rhodium catalyst and estimated costs for a batch process. J. Am. Oil Chem. Soc. **50**, 455–458 (1973)
67. E.N. Frankel, Methyl 9(lO)-formylstearate by selective hydroformylation of oleic oils. JAOCS **48**, 248–253 (1971)
68. M.W. Balakos, E.E. Hernanolez, Catalyst characteristics and performance in edible oil hydrogenation. Catal. Today **35**, 415–425 (1997)
69. A. Behr, H. Schmidke, Selektive hydrierung ungesattigter modellverfindungen mit solvesstabilisierten palladium-kolloidkatalysatoren. Chem.-1ng.-Tech. **65**, 568–569 (1993)
70. L.E. Johansson, S.T. Lundin, Copper catalysts in the selective hydrogenation of soybean and rapeseed oils: I. The activity of the copper chromite catalyst. JAOCS **56**, 974–980 (1979)
71. K. Mondal, S.B. Lalvani, A second-order model for catalytic-transfer hydrogenation of edible oils. JAOCS **77**(1), 1–8 (2000)
72. M. Martinelli, R. de Cassia de Souza Schneider, V.Z. Baldissarelli, M.L. von Holleben, E.B. Caramao, Castor oil hydrogenation by a catalytic hydrogen transfer system using limonene as hydrogen donor. JAOCS **82**(4), 279–283 (2005)
73. Y. Kitayamaa, M. Muraokaa, M. Takahashia, T. Kodamaa, E. Takahashib, M. Okamurac, Catalytic hydrogenation of linoleic acid over platinum-group metals supported on alumina. JAOCS **74**(5), 525–529 (1997)
74. N. Ravasio, F. Zaccheria, M. Gargano, S. Recchia, A. Fusi, N. Poli, R. Psaro, Environmental friendly lubricants through selective hydrogenation of rapeseed oil over supported copper catalysts. Appl. Catal. A **233**, 1–6 (2002)
75. A. Behr, A. Westfechtel, J.P. Gomes, Catalytic processes for the technical use of natural fats and oils. Chem. Eng. Technol. **31**(5), 700–714 (2008)
76. S. Warwel, P. Bavaj, M.R. Klass, B. Wolff, H. Eierdanz, *Perspektiven nachwachsender Rohstoffe in der Chemie* (VCH, Weinheim, 1996), p. 119
77. R.W. Johnson, E. Fritz, *Fatty Acids in Industry* (Marel Dekker, New York, 1988), p. 667
78. H. Wagner, R. Luther, T. Mang, Lubricant base fluids based on renewable raw materials their catalytic manufacture and modification. Appl. Catal. A **221**, 429–442 (2001)
79. L.T. Black, R.E. Beal, Acetoxylation of methyl oleate with a resin catalyst. JAOCS **44**, 310–312 (1967)
80. U. Biermann, J.O. Metzger, Friedel–crafts alkylation of alkenes: ethylaluminum sesquichloride induced alkylations with alkyl chloroformates. Angew Chem. Int. Ed. **38**(24), 3675–3677 (1999)
81. J.O. Metzger, U. Biermann, Alkylaluminium dichloride induced friedel-crafts acylation of unsaturated carboxylic acids and alcohols. Liebigs. Ann. Chem. 645–650 (1993)

82. S. Asadauskas, H.P. Joseph, J.D. Larry, Lubrication properties of castor oil potential basestock for biodegradable lubricants. Lubr. Eng. **53**(12), 35–40 (1997)
83. L.C. Meher, S.D. Vidya, S.N. Naik, Technical aspects of biodiesel production by transesterification—a review. Renew. Sustain. Energy Rev. **10**, 248–268 (2006)
84. P.S. Wang, The production of isopropyl esters and their effects on a diesel engine. Master of Science thesis to Iowa State University (2003)
85. R. Alcantara, J. Amores, L. Canoira, E. Fidalgo, M.J. Franco, A. Navarro, Catalytic production of biodiesel from soy-bean oil, used frying oil and tallow. Biomass and Bioenergy **18**, 515–527 (2000)
86. U. Schuchardta, S. Ricardo, M.V. Rogerio, Transesterification of vegetable oils: a review. J. Braz. Chem. Soc. **9**(1), 199–210 (1998)
87. F.R. Abreu, G.L. Daniella, H.H. Elias, W. Carlos, A.Z.S. Paulo, Utilization of metal complexes as catalysts in the transesterification of Brazilian vegetable oils with different alcohols. J. Mol. Catal. A: Chem. **209**, 29–33 (2004)
88. M.W. Formo, Ester reactions of fatty materials. JAOCS **3**(11), 548–559 (1954)
89. C.Y. May, Transesterification of palm oil: effect of reaction parameters. J. Oil Palm Res. **16**(2), 1–11 (2004)
90. J.M. Encinar, J.F. Gonzalez, J.J. Rodrıguez, A. Tejedor, Biodiesel fuels from vegetable oils: transesterification of cynara cardunculus l oils with ethanol. Energy Fuels **162**, 443–450 (2002)
91. E. Crabbe, C. Nolasco-Hipolito, G. Kobayashi, K. Sonomoto, A. Ishizaki, Biodiesel production from crude palm oil and evaluation of butanol extraction and fuel properties. Process Biochem. **37**, 65–71 (2001)
92. O. Rachmaniah, J. Yi-Hsu, R.V. Shaik, T. Ismojowati, A.S. Musfil, A study on acid-catalyzed transesterification of crude rice bran oil for biodiesel production (2001)
93. M.P. Simoni, R.M. Mario, R. Carlos, E.C. Wolf, G.E.S. Silva, M.A. Lima, J.I.S. Coimbra, H.V.C. Sandra, Ethanolysis of castor and cottonseed oil:a systematic study using classical catalysts. JAOCS **83**(9), 819–822 (2006)
94. H. Noureddin, D. Zhu, Kinetics of transesterification of soyabean oil. Biocatal. Art. **74**(11), 1457–1563 (1997)
95. D. Darnoko, M. Cheryan, Kinetics of palm oil transesterification in a batch reactor. JAOCS **77**(12), 1263–1268 (2000)
96. A.W. Schwab, M.O. Bagby, B. Freedman, Preparation and properties of diesel fuels from vegetable oils. Fuel **66**, 1372–1378 (1987)
97. M.A. Maleque, H.H. Masjuki, S.M. Sapuan, Vegetable-based biodegradable lubricating oil additives. Ind. Lubr. Technol. **55**(3), 137–143 (2003)
98. D.C. Drown, K. Harper, E. Frame, Screening vegetable oil alcohol esters as fuel lubricity enhancers. JAOCS **78**(6), 579–585 (2001)
99. M.G. Kulkarni, A.K. Dalai, N.N. Bakhshi, Transesterification of canola oil in mixed methanol/ethanol system and use of esters as lubricity additive. Bioresour. Technol. **98**, 2027–2033 (2007)
100. ASTM D5355
101. ASTM D445
102. ASTM D974
103. ASTM D3339
104. ASTM D2270
105. ASTM D5558
106. ASTM D5554
107. ASTM D5555
108. ASTM D1747
109. ASTM D4377
110. ASTM D4928
111. ASTM D92
112. ASTM D7094

113. ASTM D97
114. ASTM D2500
115. ASTM D1401
116. ASTM D3601
117. ASTM D3519
118. ASTM D 892
119. ASTM D2619
120. ASTM D2070
121. ASTM D6375
122. ASTM D2272
123. A. Adhvaryu, Z. Liu, S.Z. Erhan, Synthesis of novel alkoxylated triacylglycerols and their lubricant base oil properties. Ind. Crops Prod. **21**, 113–119 (2005)
124. ASTM D3233
125. ASTM D2783
126. ASTM D4172
127. ASTM D6081
128. ASTM D5864
129. http://www.cottonseed.com/publications/csobro.asp, as on March 26, 2011
130. B. Sreenivasan, Component fatty acids of some oils and fats and composition. JAOCS **45**, 259–265 (1968)
131. V.S. Yaliwal, S.R. Daboji, N.R. Banapurmath, P.G. Tewari, Production and utilization of renewable liquid fuel in a single cylinder four stroke direct injection compression ignition engine. Int. J. Eng. Sci. Tech. **2**(10), 5938–5948 (2010)
132. G. Fontaras, T. Tzamkiozis, E. Hatziemmanouil, Z. Samaras, Experimental study on the potential application of cottonseed oil—diesel blends as fuels for automotive diesel engines. Trans. I Chem. E **85**(B5), 396–403 (2007)
133. E.J. CAMPBELL, Sunflower oil. JAOCS **60**(2), 387–392 (1983)
134. S. Mia, N. Ohno, Prospect of mustard and coconut oil as environment friendly lubricant for Bangladesh. in *Proceedings of International Conference on Environmental Aspects of Bangladesh* (ICEAB10), Japan, Sept. 2010, pp. 120–121
135. F.L. Jackson, H.E. Longenecker, The fatty acids and glycerides of babassu oil. Oil Soap **21**, 73–75 (1944)

Chapter 12
Studies on the Tribological Behavior of Natural Fiber Reinforced Polymer Composite

Pradeep L. Menezes, Pradeep K. Rohatgi and Michael R. Lovell

Abstract The gradual development of asbestos in automotive friction materials in many parts of the world has sparked the onset of extensive research and development into safer alternatives. The development of green friction products for automotive application is important to minimize the environmental impacts caused by asbestos-based products. Natural fibers have been used to reinforce materials for over thousand years. More recently, they have been employed in combination with plastics. Natural fibers are environmentally friendly, fully biodegradable, abundantly available, renewable and cheap and have low density. Natural fiber reinforced polymer composites have emerged as a potential environmentally friendly and cost-effective option to synthetic fiber reinforced composites. The availability of natural fibers and ease of manufacturing have tempted researchers to study their feasibility of reinforcement and to what extent they satisfy the required specifications for good reinforced polymer composite for tribological applications. In this study, a review on the tribological behavior of natural fiber reinforced composites was made to understand their usability for various automotive applications.

Keywords Friction · Wear · Natural fiber · Polymer composite

P. L. Menezes · M. R. Lovell
Department of Industrial Engineering, University of Wisconsin-Milwaukee, Milwaukee, WI 53201, USA

P. K. Rohatgi (✉)
Department of Materials Engineering, University of Wisconsin-Milwaukee, Milwaukee, WI 53201, USA
e-mail: prohatgi@uwm.edu

M. Nosonovsky and B. Bhushan (eds.), *Green Tribology*,
Green Energy and Technology, DOI: 10.1007/978-3-642-23681-5_12,

12.1 Introduction

Commercially available automotive brake pads are considered to be organic friction materials since the matrix of these complex composites is made by one or more polymers [1]. The friction materials usually contain four classes of ingredients: binders, reinforcements, friction modifiers and fillers. Developing a successful friction material is to find the best balance among many factors yielding acceptable performance, costs and environmental friendliness. Asbestos fiber, used to be the dominant friction fiber enhanced polymeric composites in brake pads, brake linings, brake couplings, etc., has been forbidden due to harmful environment and human health issues. According to the regulations against hazardous ingredients in the United States and Europe, several raw materials usually used in commercial friction materials could have a potential negative environmental impact. Friction components made of chemicals such as antimony trisulfide, copper, lead, tin, potassium titanate whisker silicon carbide whisker, and others have been extensively used in the past [2, 3]. Increased environmental awareness and consciousness throughout the world has developed an increasing interest in natural fibers and its application in various fields. Natural fibers are now considered as serious alternative to synthetic fibers for use in various automotive fields. The use of natural fibers as reinforcing materials in both thermoplastic and thermoset matrix composite provides positive environmental benefits with respect to ultimate disposability and best utilization of raw materials. During the last few years, a series of works have been done to replace the conventional synthetic fibers with natural fiber composites [4–10]. For instance, hemp, sisal, jute, cotton, flax and broom are the most common fibers used to reinforce polymers such as polyolefins, polystyrene, and epoxy resins. In addition, fibers such as sisal, jute, coir, oil palm, bamboo, wheat and flax straw, waste silk and banana have proved to be good and effective for reinforcement in the thermoset and thermoplastic matrices. The advantages of natural fibers over traditional reinforcing materials such as glass fiber and carbon fiber are their specific strength properties, enhanced energy recovery, high toughness, non-corrosive nature, good thermal properties, reduced tool wear, reduced dermal and respiratory irritation. Table 12.1 shows the mechanical properties of natural fibers as compared to conventional reinforcing fibers. It has been reported that natural fiber reinforced composite have properties similar to traditional synthetic fiber reinforced composite [11, 12].

Natural fibers, as reinforcement, have attracted the attention of researchers due to their advantages over other established materials. They are environmentally friendly, fully biodegradable, abundantly available, renewable and cheap and have low density. Plant fibers are light compared to glass, carbon and aramid fibers. The biodegradability of plant fibers can contributes to a healthy ecosystem while their low cost and high performance fulfills the economic interest of industry. Natural fiber reinforced plastics, by using biodegradable polymers as matrices, can be the most environmental friendly materials. A number of automotive components previously made with glass fiber composites are now being manufactured using environmentally friendly composites [13]. In the United States and Europe,

Table 12.1 Mechanical properties of natural fibers as compared to conventional reinforcing fibers [10]

Fiber	Density (g/cm^3)	Elongation (%)	Tensile Strength (MPa)	Young's Modulus (Gpa)
Cotton	1.5–1.6	7.0–8.0	287–597	5.5–12.6
Jute	1.3	1.5–1.8	393–773	26.5
Flax	1.5	2.7–3.2	345–1,035	27.6
Hemp	–	1.6	690	–
Ramie	–	3.6–3.8	400–938	614–128
Sisal	1.5	–	1,000	40.0
Coir	1.2	30.0	175	4.0–6.0
Viscose (cord)	–	11.4	593	11.0
Soft Wood Kraft	1.5	–	1,000	40.0
E-glass	2.5	2.5	2,000–3,500	70.0
S-glass	2.5	2.8	4,570	86.0
Aramid (normal)	1.4	3.3–3.7	3,000–3,150	63.0–67.0
Carbon (standard)	1.4	1.4–1.8	4,000	230–240

car manufactures are aiming to make every component of their vehicles either recyclable or biodegradable. Although natural fibers and their composites are environmental friendly and renewable these composites have several drawbacks. They have poor wettability, incompatibility with some polymeric matrices and high moisture absorption [14]. Thus, composite materials made with the use of unmodified plant fibers frequently exhibit unsatisfactory mechanical properties. To overcome this, in many cases, a surface treatment or compatibilizing agents need to be used prior to composite fabrication. The properties can be improved both by physical and chemical treatments [15].

During the past decade, a number of significant industries such as the automotive, construction and packaging industries have shown massive interest in the progress of new natural fiber reinforced composite materials. However, little information concerning the tribological performance of natural fiber reinforced composite material has been reported. Hence, in this study, a review on the tribological performance of natural fiber reinforced composite is presented.

12.2 Tribological Behavior of Green Composites

12.2.1 Tribological Behavior of Sisal Fiber Reinforced Resin Brake Composites

In automotive industry, non-asbestos fiber-based components such as brake pads and lining couplings, etc. have been in demand due to environmental and human health concerns. The alternatives to asbestos fibers include mineral fiber, metal fiber and artificial polymer fibers such as glass fiber, alumina fiber, carbon fiber, steel fiber,

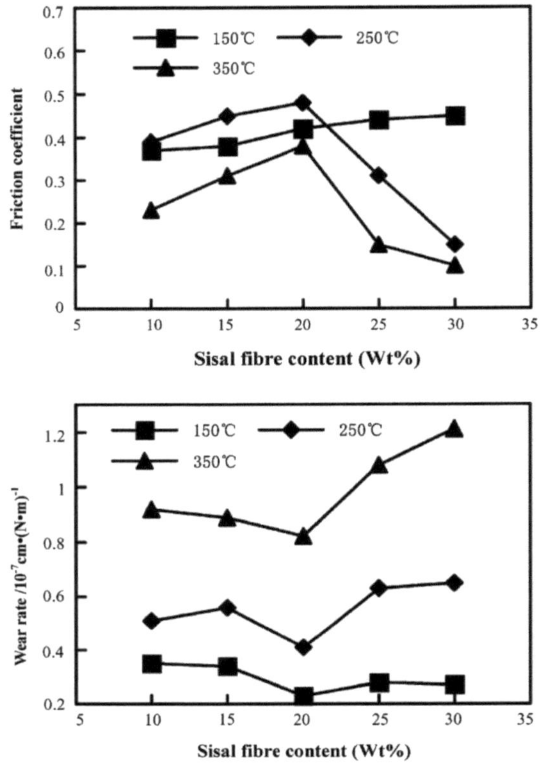

Fig. 12.1 Variation in friction coefficient and wear rate with sisal fiber content [16]

aramid fiber and their combinations. Recently, sisal fibers are found to be a good alternative because they are biodegradable, inexpensive, and lightweight and exhibit high specific mechanical properties. As the temperature of friction surface during braking could go up to 750°C, the sisal fibers can easily decompose at this temperature. Furthermore, its poor wettability, incompatibility with resins and high moisture absorption restrict its use. However, incorporating certain treatments to improve its chemical and structural properties, sisal fiber reinforced friction composite exhibits properties equivalent to the commercial friction composite.

Extensive research has been focused on the potential use of sisal fibers reinforced composites for tribological applications. Xin et al. [16] studied the friction and wear properties of sisal fiber reinforced with phenolic resin composites at various temperatures. The chemical and structural properties of sisal fiber were improved with modification. The variation of coefficient of friction and wear rate with sisal fiber content is shown in Fig. 12.1. Compared with asbestos and mineral/steel fiber, sisal fiber reinforced friction composites shows better performance with respect to coefficient of friction and wear at different friction temperatures. Thus, the sisal is an ideal substitute of asbestos for brake pads.

The orientation of natural fiber in the composite also influences the tribological behavior as seen in the traditional composites. Chand and Dwivedi [17] studied the

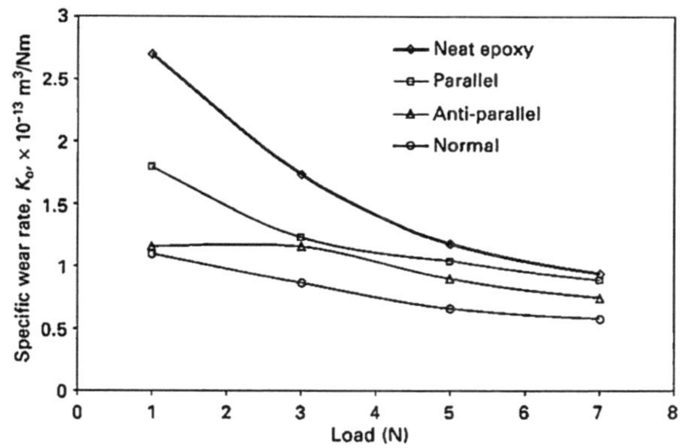

Fig. 12.2 Specific wear rate of sisal-epoxy composite [17]

influence of fiber orientation on wear behavior of sisal fiber reinforced epoxy composites. Figure 12.2 shows the variation in specific wear rate with load for the case of sisal-epoxy composite. The specific wear rate was found to decrease with applied load for all materials. The specific wear rate was highest for the neat epoxy and decreases for parallel then anti-parallel and least for the fibers aligned normal to the sliding direction.

Efforts were also made to study the influence of fiber modification using certain treatments on friction and wear. Chand and Dwivedi [18] studied the sliding wear and friction characteristics of sisal fiber reinforced polyester composites with and without silane modification. They found that silane treatment plays a significant role for reducing the wear and increasing the friction coefficient as shown in the Fig. 12.3.

12.2.2 Tribological Behavior of Sugarcane Fiber Reinforced Composites

El-Tayeb [19] studied the wear rate and friction coefficient of chopped sugarcane fiber reinforced polyester (C-SCRP) and chopped glass fiber reinforced polyester (C-GRP) composites. The variation in wear rate and friction coefficient for different fiber length is shown in Fig. 12.4. Results of wear rate and friction coefficient proved that SCRP composite is a promising composite which can be used as a competitive to GRP composite.

12.2.3 Tribological Behavior of Jute Fiber Reinforced Composites

El-Sayed et al. [20] studied the tribological behavior of jute fiber reinforced polyester composite. The variation in friction coefficient and wear rates with the fiber orientation is shown in the Fig. 12.5. An increase in fiber volume fraction

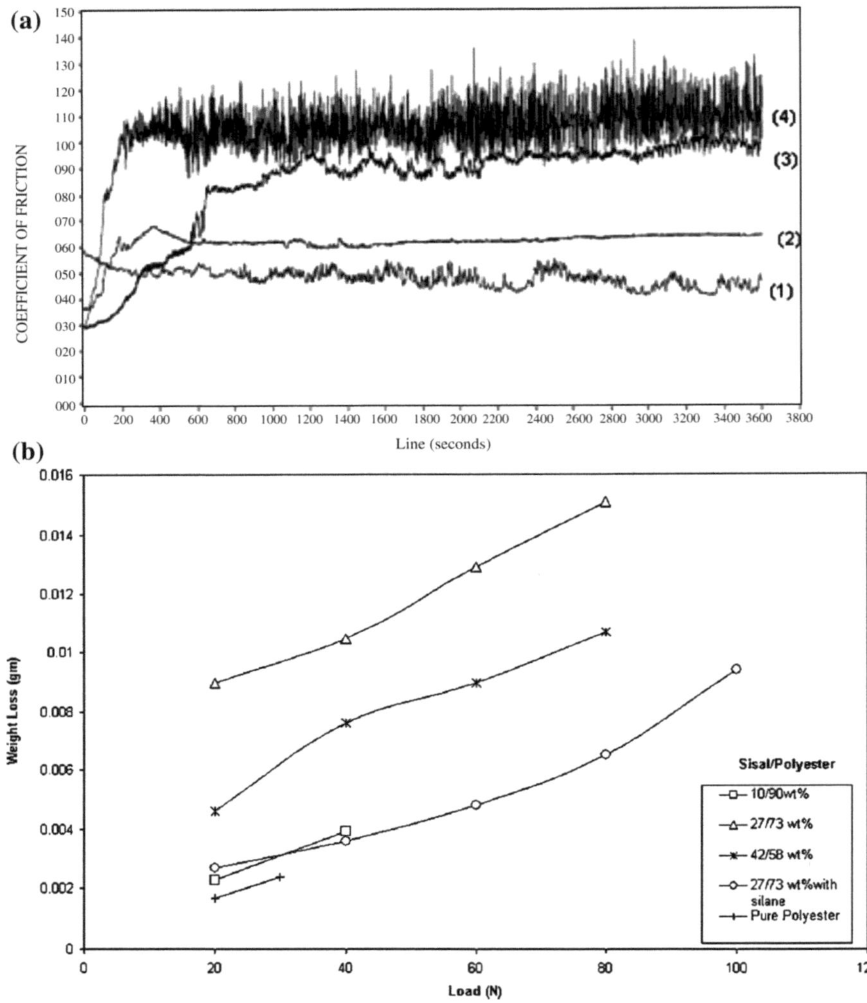

Fig. 12.3 **a** Variation of coefficients of friction with time for (*1*) polyester (*2*) 10 wt% sisal fiber composites (*3*) 27 wt% sisal fiber composites (*4*) silane-treated sisal fiber composites. **b** Plots between weight loss and applied load for polyester resin, sisal polyester and silane-treated sisal polyester (27 wt%) composites under sliding motion [18]

increases the friction coefficient of the composite and decreases its wear rate. This composite is now extensively used in bearing applications.

Chand and Dwivedi [21] investigated the effect of coupling agent on abrasive wear behavior of chopped jute fiber reinforced polypropylene (PP) composites.

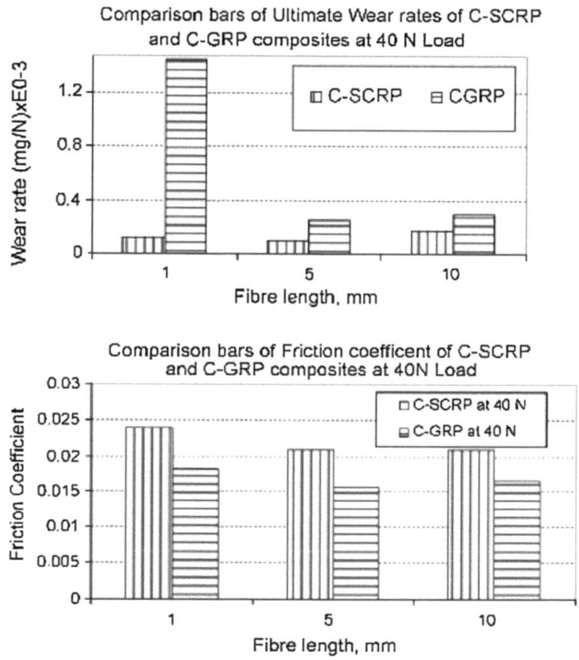

Fig. 12.4 Wear rate and friction coefficient of C-SCRP and C-GRP composites [19]

Figure 12.6 shows the variation in wear rate with load for untreated jute fiber reinforced composites (UT), maleic anhydride-grafted polypropylene (MA-g-PP) solution-treated jute fiber reinforced PP composites (CT) and MA-g-PP melt-mixed jute fiber reinforced PP composites (MT). Use of coupling agent gives better wear resistance as compared to without the use of coupling agent. It has also been found that addition of MA-g-PP coupling agent during melt mixing gives better wear resistance as compared to the jute PP composites having MA-g-PP solution-treated jute fibers.

12.2.4 Tribological Behavior of Cotton Fiber Reinforced Composites

Eleiche et al. [22] studied the effect of unidirectional cotton fiber reinforcement on friction and wear characteristics of polyester. It was found that friction coefficient increases and wear rate decreases with increasing fiber volume fraction as shown in the Fig. 12.7. This type of composites is also used in bearing applications.

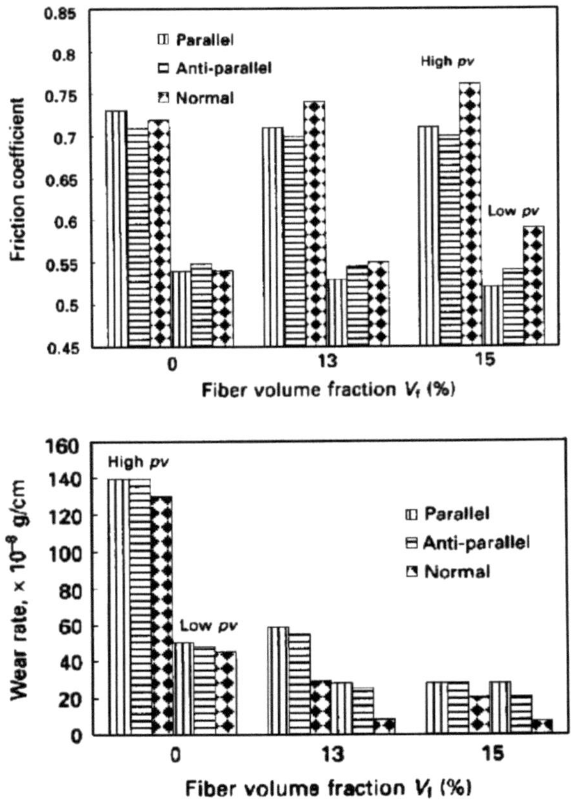

Fig. 12.5 Variation of friction coefficient and wear rate with fiber volume fraction [20]

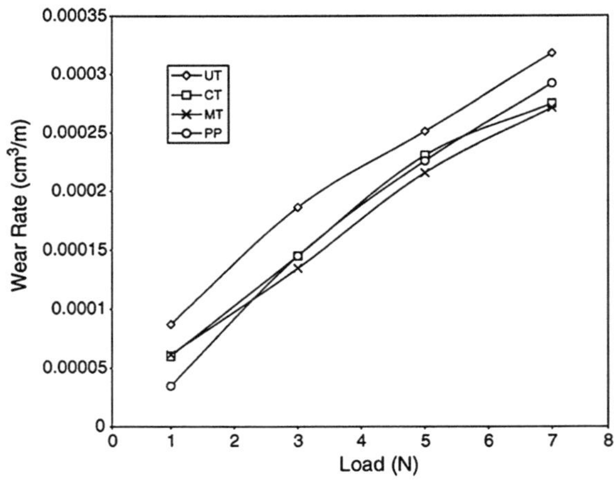

Fig. 12.6 Variation in wear rate with load for untreated jute fiber reinforced composites (UT), MA-g-PP solution-treated jute fiber-reinforced PP composites (CT) and MA-g-PP melt-mixed jute fiber-reinforced PP composites (MT) [21]

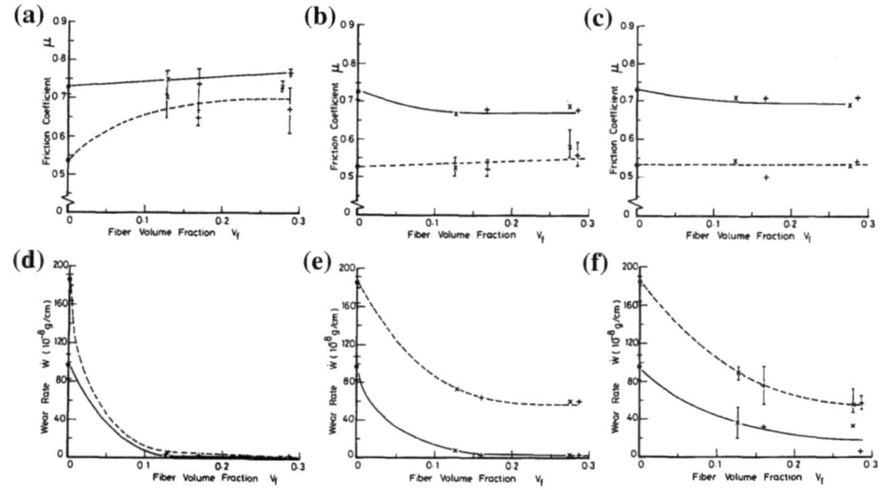

Fig. 12.7 Friction coefficient (**a**–**c**) and wear rate (**d**–**f**) of cotton polyester composite as a function of fiber volume friction for **a**, **d** normal, **b**, **e** longitudinal, **c**, **f** transverse orientation of fibers. *Dashed lines* are for speed of 32 m/s while *solid lines* shows data at a sliding speed of 10 m/s [22]

12.2.5 Tribological Behavior of Bio-Waste Reinforced Polymer Composites

Sabat and Rajak [23] studied the tribological behavior of biowaste reinforced polymer composite. In their study, natural fibers such as chicken feather fibers are used for reinforcement in epoxy and polyester matrix composite. They found that addition of 20% short feather fiber improved the wear resistance than that of pure matrix (epoxy or polyester). The coefficient of friction was also increased with increasing fiber content.

12.2.6 Tribological Behavior of Seed Oil Palm Fibers Reinforced Polyester Composite

Yousif [24] studied the tribological performance of two different polyester composites based on woven glass reinforced polyester (WGRP) and seed oil palm reinforced polyester (S-OPRP). Figure 12.8 shows the variation of friction coefficient and wear rate with sliding distance. The results revealed that S-OPRP composites showed very high friction coefficient compared to WGRP. The S-OPRP composite with 35 vol.% exhibited a promising wear result. Thus,

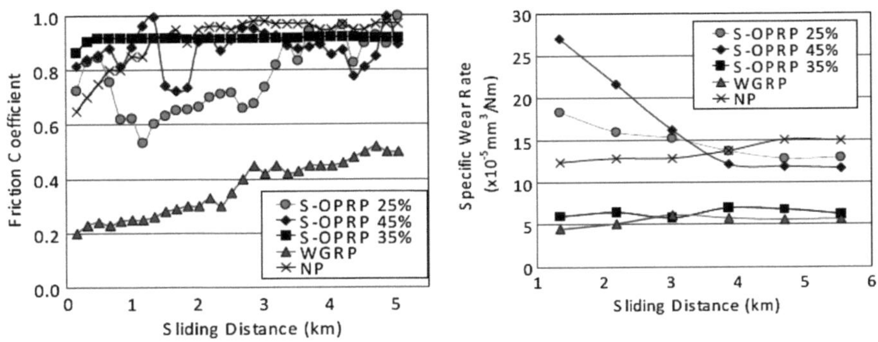

Fig. 12.8 Variation of friction coefficient and specific wear rate with sliding distance [24]

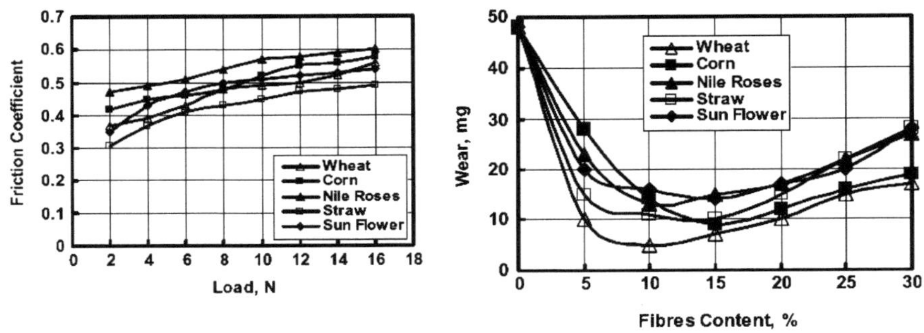

Fig. 12.9 Friction coefficient and wear of the natural composite reinforced by different fibers [25]

the woven glass fibers can be replaced by the seed oil palm fibers in polymeric composites reinforcements. The wear resistance of the synthetic and natural composite was found to be almost the same. The wear mechanisms for S-OPRP composites were predominated by micro-cracks, deformation and pulled-out of fibers while in the WGRP composite was abrasive in nature.

12.2.7 Tribological Performance of Natural Composites Reinforced by Different Fibers

Mohmoud et al. [25] studied the tribological behavior of asbestos-free friction composites reinforced by natural fibers such as wheat, corn, nile roses, straw and sun flower. Figure 12.9 shows the variation of friction and wear of friction composite reinforced by different fibers. The authors found that the friction and wear performance of natural fiber reinforced composites were much better than glass fiber reinforced composites and thus the asbestos can be replaced by natural fibers in reinforcing the friction composites. They also found that reinforcing the tested composites by the natural fibers caused significant increase in friction coefficient compared to that reinforced by glass fibers. Fibers of nile roses displayed the highest friction values compared to the other tested natural fibers. The highest values of friction coefficient were displayed by composites reinforced by the fibers of 15 wt% corn, 15 wt% nile roses, 10 wt% straw, 5 wt% wheat and 20 wt% sun flower. In addition, minimum wear values were observed for composites reinforced by 10 wt% wheat fibers.

12.2.8 Influence of Ingredients on Friction and Wear of Composites

Cho et al. [26] studied the effects of ingredients on tribological characteristics of a brake lining. More specifically, the composite was developed using a non-asbestos organic-based friction material for an automotive brake system. With increasing ingredients such as phenolic resin, Mgo and cashew the friction coefficient increases as shown in the Fig. 12.10. The wear rate is slightly increased for cashew; however, there is a substantial reduction in noise occurrence with increasing cashew content as shown in the Fig. 12.10. Thus, cashew can be used as prospective ingredients for future friction materials.

Mutlu [27] investigated the tribological properties of brake pads by using rice straw (RS) and rise husk (RH) dust as ingredients. Table 12.2 shows the friction coefficient and wear rate data for different percentage of RS and RH brake pads. It was found that the tribological properties were significantly improved by the addition of RS and RH dust in the composites and concluded that these composites can be effectively used in brake pad formulations.

Attempts were also made to use the boric acid particle in the phenolic matrix. Boric acid is environmentally safe and inexpensive. Mutlu et al. [28] studied boric acid effect in phenolic composites on tribological properties in brake linings. They found that boric acid phenolic composites showed higher and stable friction coefficient. However, at higher temperatures (i.e., over 300°C), coefficient of friction was decreased due to poor thermal stability of boric acid. Thus, boric acid additive was positively affected the braking performance over 300°C.

12.2.9 Use of Green Composites as Anti-Friction Materials

Natural fiber composites are also used as anti-friction (low friction and low wear rate) materials. Efforts have been made to study tribological behavior of graphite modified polyester–cotton composites [29]. Figure 12.11 shows the variation in

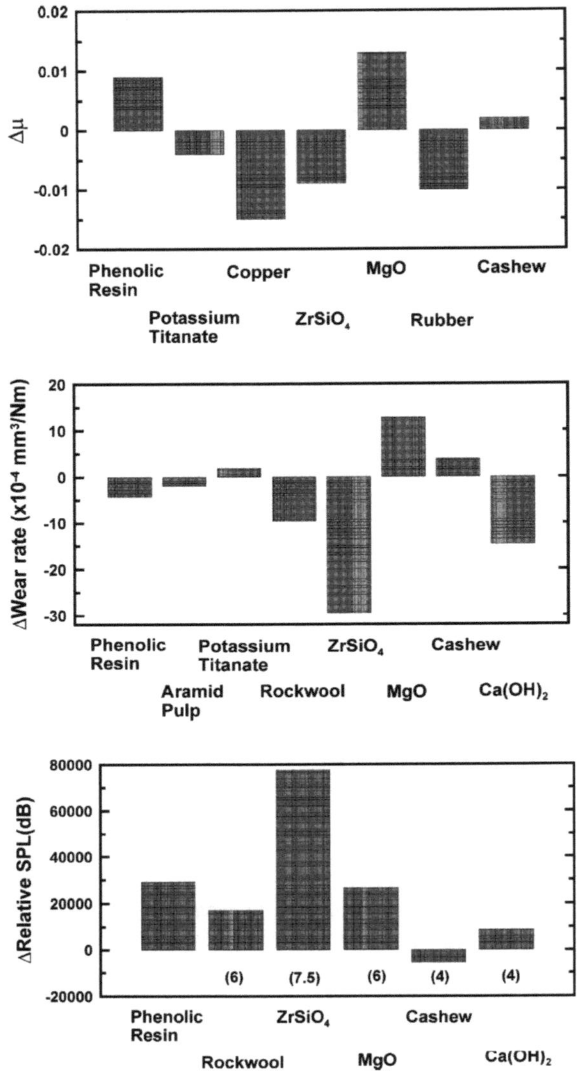

Fig. 12.10 Relative influence of ingredients on friction, wear rate and noise propensity [26]

Table 12.2 Tribological properties of RS- and RH-based composite materials [25]

Sample code	Mean of friction coefficients	SD	Wear rate (g mm^{-2})
RS4	0.315	0.0106	0.853×10^{-3}
RS20	0.347	0.0168	1.214×10^{-3}
RH4	0.341	0.0099	0.964×10^{-3}
RH20	0.381	0.0120	1.041×10^{-3}

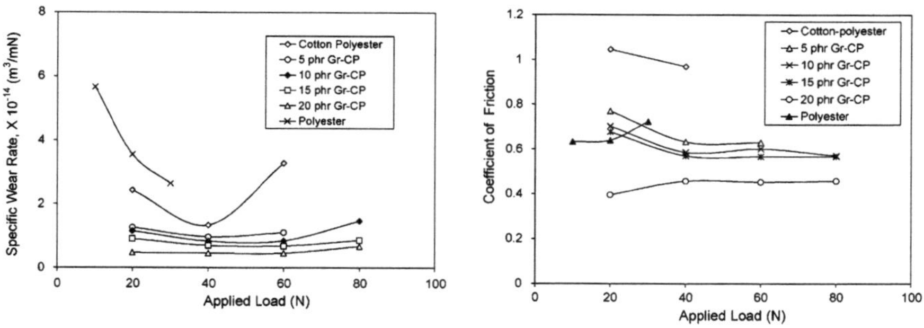

Fig. 12.11 Variation in specific wear rate and friction coefficient with applied load for various amounts of graphite filled cotton reinforced polyester composites [29]

Fig. 12.12 Specific wear rate versus sliding distance at different applied loads for **a** OPRP **b** NP [30]

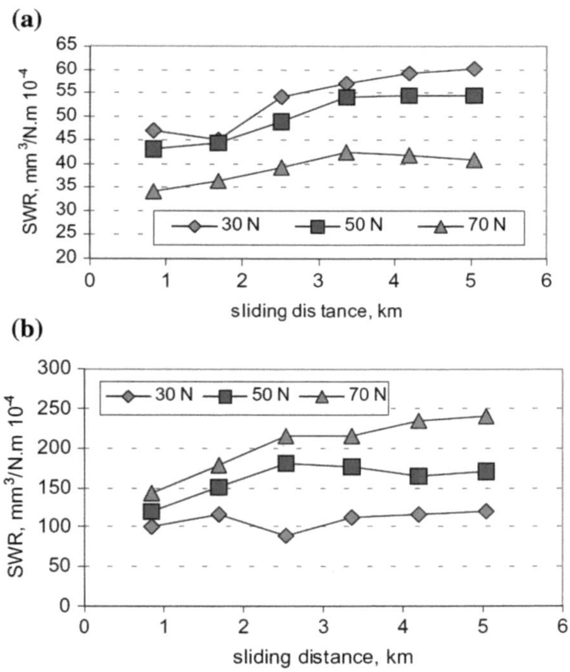

specific wear rate and friction coefficient with applied load for various amounts of graphite filled cotton reinforced polyester composites. The specific wear rate of polyester reduced on reinforcement of cotton and on addition of graphite. The coefficient of friction of polyester resin increased on cotton reinforcement and reduced significantly on addition of graphite in cotton–polyester composites.

Yousif and El-Tayeb [30] studied wear and friction characteristics of oil palm fiber reinforced polyester (OPRP) composite and neat polyester (NP). Figures 12.12 and 12.13, respectively, show the variation in specific wear rate and

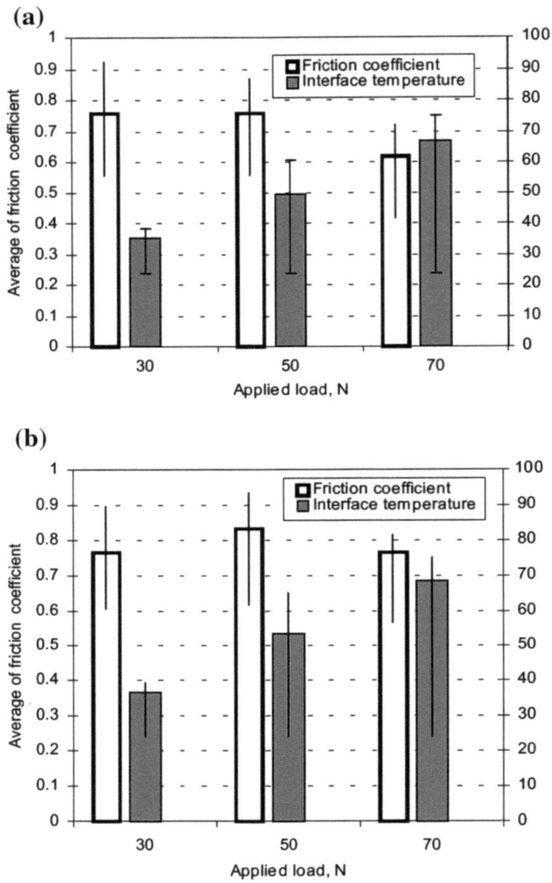

Fig. 12.13 Averages of friction coefficient and interface temperature versus applied load for **a** OPRP **b** NP [30]

friction coefficient with sliding distance for OPRP and NP. It was found that the presence of oil palm fiber in the polyester enhanced the wear property by about three to four times compared to neat polyester. In addition, the friction coefficient of OPRP composite was less by about 23% than that of the neat polyester.

Yousif et al. [31] studied the effect of chemical treatment on tribo-performance of coir fiber reinforced polyester (CFRP) composite. The fibers were treated by alkaline and bleaching methods. Figure 12.14 shows the variation of friction coefficient and wear rates with sliding distance for CFRP composites with different chemical treatments. It was found that untreated CFRP composites showed higher friction and wear due to the debonding of the coir fiber embedded in the matrix. Surface modification (either bleached or alkaline treated) on coir fibers exhibited good interfacial adhesion when used as a reinforcement in polyester composites. Alkaline-treated CFRP composites have low friction and better wear resistance compared to the bleached treated and untreated CFRP composites. The high

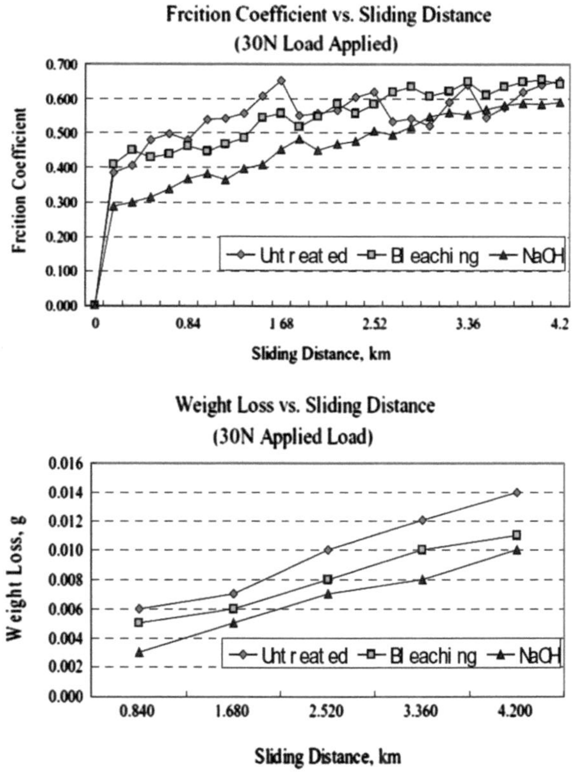

Fig. 12.14 Variation of friction coefficient and wear rates with sliding distance for coir fiber reinforced polyester composite with different chemical treatments [31]

adhesion bonding between the matrix and the coir fiber assists in protecting the rubbing surface from worn.

12.3 Conclusions

Environmental concerns are driving demand for biodegradable materials such as plant-based natural fiber reinforced polymer composites. These composites are fast replacing conventional materials in many applications, especially in automobiles, where friction and wear are important. In this study, a review on the tribological behavior of natural fiber reinforced polymer composite was made. More specifically, tribological behavior of sisal, sugarcane, jute, cotton, coir, palm, wheat, corn, nile roses, straw and sun flower fibers reinforced polymer composites was studied. Influence of fiber orientation, fiber volume fraction and fiber length for various loads, sliding distance and temperature on friction and wear are reported. Influence of other natural ingredients in the matrix materials and the uses of green composite as anti-friction material are also reported. The following conclusion can be drawn from this study.

- Natural fiber reinforced composites can be effectively used as friction materials in the brake applications.
- Natural ingredients improve tribological performance and hence can be used for friction materials.
- Natural composites can also be used as anti-friction materials.

References

1. D. Chan, G.W. Stachowiak, Review of automotive brake friction materials. Proc. Inst. Mech. Eng. D: J. Automob. Eng. **218**, 953–966 (2004)
2. V. Roubicek, H. Raclavska, D. Juchelkova, P. Filip, Wear and environmental aspects of composite materials for automotive braking industry. Wear **265**, 167–175 (2008)
3. K. Malachová, Z. Pavlíčková, R. Holuša, J. Kubačková, V. Mička, D. MacCrimmon, P. Filip, Wear mechanism in automotive brake materials, wear debris and its potential environmental impact. Wear **267**, 807–817 (2009)
4. A.K. Bledzki, J. Gassan, Composites reinforced with cellulose based fibre. Prog. Polym. Sci. **24**, 274 (1999)
5. A.K. Mohanty, M. Misra, L.T. Drzal, Sustainable bio-composites from renewable resources: opportunities and challenges in the green materials world. J. Polym. Environ. **10**, 19–26 (2002)
6. S. Joseph, M.S. Sreekalab, Z. Oommena, P. Koshyc, S. Thomas, A comparison of the mechanical properties of phenol formaldehyde composites reinforced with banana fibres and glass fibres. Compos. Sci. Technol. **62**, 1857–1868 (2002)
7. P.J. Roe, M.P. Ansel, Jute reinforced polyester composites. J. Mater. Sci **20**, 4015–4020 (1985)
8. M. Baiardo, E. Zini, M. Scandola, Flax fibre-polyester composites. J. Compos. A **35**, 703–710 (2004)
9. J. George, M.S. Sreekala, S. Thomas, A review on interface modification and characterization of natural fibre reinforced plastic composites. Ploym. Eng. Sci. **41**, 1471–1485 (2002)
10. S. Taj, M.A. Munawar, S. Khan, Natural fiber-reinforced polymer composites. Proc. Pak. Acad. Sci. **44**, 129–144 (2007)
11. P. Wambua, U. Ivens, I. Verpoest, Natural fibers: can they replace glass in fiber-reinforced Plastics? Compos. Sci. Technol. **63**, 1259–1264 (2003)
12. N. Chand, M. Fahim, Tribology of natural fiber polymer composites, Woodhead publishing Limited, UK (2008)
13. H. Larbig, H. Scherzer, B. Dahlke, R. Pol-trock, Natural fiber reinforced foams based on renewable resources for automotive interior applications. J. Cell. Plast. **34**, 361–379 (1998)
14. A. Vazguez, J. Riccieri, L. Carvalho, Interfacial properties and initial step of the water sorption in unidirectional unsaturated polyester/vegetable fiber composites. Polym. Compos. **20**, 29–37 (1999)
15. S. Luo, A. Netravali, Mechanical, thermal properties of environment-friendly "green" composites made from pineapple leaf fibers and poly (hydroxybutyrate-co-valerate) resin. Polym. Compos. **20**, 367–378 (1999)
16. X. Xin, C.G. Xu, L.F. Qing, Friction properties of sisal fibre reinforced resin brake composites. Wear **262**, 736–741 (2007)
17. N. Chand, U.K. Dwivedi, Influence of fiber orientation on high stress wear behavior of sisal fiber-reinforced epoxy composites. Polym. Compos. **28**, 437–441 (2008)
18. N. Chand, U.K. Dwivedi, Sliding wear and friction characteristics of sisal fibre reinforced polyester composites: effect of silane coupling agent and applied load. Polym. Compos. **29**, 280–284 (2008)
19. N.S.M. El-Tayeb, A study on the potential of sugarcane fibers/polyester composite for tribological applications. Wear **265**, 223–235 (2008)

20. A.A. El-Sayed, M.G. El-Sherbiny, A.S. Abo-El-Ezz, G.A. Aggag, Friction and wear properties of polymeric composite materials for bearing Applications. Wear **184**, 45–53 (1995)
21. N. Chand, U.K. Dwivedi, Effect of coupling agent on abrasive wear behaviour of chopped jute fibre-reinforced polypropylene composites. Wear **261**, 1057–1063 (2006)
22. A.M. Eleiche, G.M. Amin, The effect of unidirectional cotton fibre reinforcement on the friction and wear characteristics of polyester. Wear **112**, 67–78 (1986)
23. R.K. Sabat, R.K.Rajak, Sliding wear behavior of bio waste reinforced polymer composite, graudate (B.Tech) Thesis, National Institute of Technology Rourkela, India, 2009
24. B.F. Yousif, Replacing of glass fibres with seed oil palm fibres for tribo-polymeric composites. Tribol. Mater. Surf. Interfaces **2**, 99–103 (2008)
25. M. Mahmoud, M.O. Mousa, M. Khashaba, W. Ali, Effect of natural fiber reinforcing friction composite, Internal report, Taif University, KSA, 2009
26. M.H. Cho, S.J. Kim, D. Kim, H. Jang, Effects of ingredients on tribological characteristics of a brake lining: an experimental case study. Wear **258**, 1682–1687 (2005)
27. I. Multu, Investigation of tribological properties of brake pads by using rice straw and rice husk dust. J. Appl. Sci. **9**, 377–381 (2009)
28. I. Mutlu, C. Oner, F. Findik, Boric acid effect in phenolic composites on tribological properties in brake linings. Mater. Des. **28**, 480–487 (2007)
29. S.A.R. Hashmi, U.K. Dwivedi, N. Chand, Graphite modified cotton fibre reinforced polyester composites under sliding wear conditions. Wear **262**, 1426–1432 (2007)
30. B.F. Yousif, N.S.M. El-Tayeb, The effect of oil palm fibers as reinforcement on tribological performance of polyester composite. Surf. Rev. Lett. **14**, 1095–1102 (2007)
31. B.F. Yousif, O.B. Leong, L.K. Ong, W.K. Jye, The effect of treatment on tribo-performance of CFRP composites. Recent Patents Mater. Sci. **2**, 67–74 (2009)

Chapter 13
Water Lubricated Bearings

R. Pai and D. J. Hargreaves

Abstract The increasing ecological awareness and stringent requirements for environmental protection have led to the development of water lubricated bearings in many applications where oil was used as the lubricant. The chapter details the theoretical analysis to determine both the static and dynamic characteristics, including the stability (using both the linearised perturbation method and the nonlinear transient analysis) of multiple axial groove water lubricated bearings. Experimental measurements and computational fluid dynamics (CFD) simulations by the Tribology research group at Queensland University of Technology, Australia and Manipal Institute of Technology, India, have highlighted a significant gap in the understanding of the flow phenomena and pressure conditions within the lubricating fluid.

An attempt has been made to present a CFD approach to model fluid flow in the bearing with three equi-spaced axial grooves and supplied with water from one end of the bearing. Details of the experimental method used to measure the film pressure in the bearing are outlined. The lubricant is subjected to a velocity induced flow (as the shaft rotates) and a pressure-induced flow (as the water is forced from one end of the bearing to the other). Results are presented for the circumferential and axial pressure distribution in the bearing clearance for different loads, speeds and supply pressures. The axial pressure profile along the axial groove located in the loaded part of the bearing is measured. The theoretical analysis shows that smaller the groove angle better will be the load-carrying

---

R. Pai (✉)
Department of Mechanical and Manufacturing Engineering,
Manipal Institute of Technology,
Manipal University, Manipal 576104, India
e-mail: raghuvir.pai@manipal.edu

D. J. Hargreaves
School of Engineering Systems, Queensland University of Technology,
G.P.O. Box 2434, Brisbane, 4001, Australia

M. Nosonovsky and B. Bhushan (eds.), *Green Tribology*,
Green Energy and Technology, DOI: 10.1007/978-3-642-23681-5_13,

capacity and stability of these bearings. Results are compared with experimentally measured pressure distributions.

Keywords Water lubricated bearings · Stability · Computational fluid dynamics · Environmental protection · Marine applications · Piston pumps · Experimental details

Nomenclature

C	Radial clearance (m)
D	Diameter of the bearing (m)
$D_{\text{rr}}\ D_{\Phi\Phi}\ D_{\text{r}\Phi}\ D_{\Phi\text{r}}$	Damping coefficients
$\bar{D}_{rr}\ \bar{D}_{\phi\phi}\ \bar{D}_{r\phi}\ \bar{D}_{\phi r}$	Non-dimensional damping coefficient, $\bar{D}_{ij} = \frac{D_{ij}C\omega}{LDp_s}$
e	Eccentricity (m)
F	Friction force (N); Non-dimensional friction $\bar{F} = \frac{F}{2LCp_s}$
F_r, $\bar{F}_r$	Dynamic force along radial direction, $\bar{F}_r = \frac{F_r}{2LCp_s}$
F_{ϕ}, $\bar{F}_{\phi}$	Dynamic force along ϕ direction, $\bar{F}_{\phi} = \frac{F_{\phi}C^2}{\eta\omega R}$
h, $\bar{h}$	Local film thickness (m), $\bar{h} = \frac{h}{c} = 1 + \varepsilon\cos\theta$
$K_{\text{rr}}\ K_{\Phi\Phi}\ K_{\text{r}\Phi}\ K_{\Phi\text{r}}$	Stiffness coefficients
$\bar{K}_{rr}\bar{K}_{\phi\phi}\bar{K}_{r\phi}\bar{K}_{\phi r}$	Non-dimensional stiffness coefficient, $\bar{K}_{ij} = \frac{K_{ij}C\omega}{LDp_s}$
L	Length of the bearing (m)
L_{cav}	Dimensionless width of analytical bubble
M, $\bar{M}$	Mass parameter (kg), $\bar{M} = \frac{MC\omega^2}{LDp_s}$
p, $\bar{p}$	Film pressure (N/m^2), $\bar{p} = p/p_s$
p_s	Supply pressure (N/m^2)
$\bar{p}_1, \bar{p}_2$	Perturbed pressures
q_p	Flow on pressure side
q_{cav}	Flow in cavitation zone
q_t	Flow due to boundary movement
Q	Volume Flow rate (m^3/sec); Non-dimensional volume flow rate $\bar{Q} = \frac{2Q\eta L}{C^3 p_s D}$
R	Radius of the bearing (m)
t	Time (s)
U	Journal peripheral speed, ωR (m/s)
W	Steady state load capacity (N)
$\bar{W}$	Non-dimensional load capacity, $\bar{W} = W/LDp_s$
$W_{\text{r,t}}$	Components of load along and perpendicular to the line of centres (N)
$\bar{W}_r$, $\bar{W}_t$	Non-dimensional load capacity, $\bar{W}_{r,t} = W_{r,t}/LDp_s$
ε	Eccentricity ratio, e/C
$\varepsilon_1\ \Phi_1$	Perturbation parameters

η	Coefficient of absolute viscosity of lubricant (Ns/m^2)
θ, $\bar{z}$	Non-dimensional co-ordinates, $\theta = \frac{x}{R}, \bar{z} = z/L$, θ measured from line of centre
θ^*	Co-ordinate in the circumferential direction measured from centre of the groove
$\theta_1\theta_2$	Angular coordinates at which the film cavitates and reforms, respectively
λ	Whirl ratio, $\lambda = \omega_p/\omega$
τ	Non-dimensional time, $\tau = \omega_p t$
θ	Filling gap coefficient
Λ	Bearing number $\Lambda = 6\eta\omega/[p_s(C/R)^2]$
Φ	Attitude angle (rad)
Ψ	Assumed attitude angle
ω	Journal rotational speed (rad/s)
ω_p	Frequency of journal vibration (rad/sec)
$( )_0$	Steady state value

13.1 Introduction

All rotating machines require a shaft supported on a bearing. The bearings are one of the most important machine elements in a machine. They not only support the rotor weight and operating loads, but also influence the dynamic behaviour of the rotor. Lubricated journal bearings have been widely used since the Egyptian Empire (in chariots lubricated by animal fat) and formed a critical element in the Industrial Revolution. The number of journal bearings currently in use worldwide, well exceeds the population of the earth.

The design of journal bearings is considered important to the development of rotating machinery. A journal bearing consists of a shaft completely or partially surrounded by a sleeve. Differences in their respective diameters result in a small clearance between the shaft and the sleeve—see Fig. 13.1. Bearings use a lubricant between the bearing surfaces to drastically reduce the friction force and resultant wear which in turn prevents large temperature gradients. Sliding surface bearings have the common aim of ensuring a good hydrodynamic oil film lubrication when operating, thereby avoiding contact with the shaft. Commonly oil is used, but other liquids such as water can also act as a lubricant. Although the viscosity of water is about 30–40 times less than mineral oils, it can be used with appropriate design considerations. To facilitate the hydrodynamic process, the constant supply of lubricant is fed into the bearing through a hole or groove to maintain the thin fluid film between the shaft and bearing. As a load is applied to the shaft it is carried on the thin layer of lubricant and not on the shaft or bearing/journal surfaces in contact.

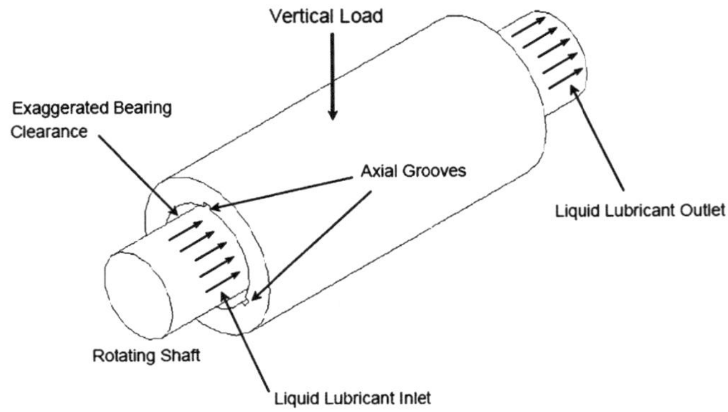

Fig. 13.1 A schematic diagram of the clearance space of the bearing showing two representative axial groove

A water lubricated bearing consists of a cylindrical shell and liner, made from hard plastics from a casting process and rubber. Rubber and water make the perfect relationship for a bearing. The natural resilience of rubber gives the bearing its shock, vibration and noise absorption properties. These bearings have a synthetic rubber liner with internal grooves often called flutes which increase the performance of the bearing in the water as the lubricant. During the rotation of the shaft, a layer of water is created between the shaft and bearing, providing lubrication. The water enters the bearing through the longitudinal grooves and moves radially between the shaft and the bearing face in a thin film. This film, or wedge, that has developed, prevents the shaft from coming into contact with the bearing. The flow of water reduces the wear on the shaft due to the low friction and takes away the heat generated in the bearing. The internal grooves of the bearing are shaped in such a way that a hydrodynamic water wedge is formed between the shafts and bearing, even at very low shaft speed. The unique shape of these grooves in the bearing gives it excellent resistance to abrasion. This helps to reduce wear on the bearing surfaces in environments where sand and other abrasive particulate are held in suspension, as found in shallow water. The bearing is designed to let these abrasive particles across its surface and into the grooves where it is flushed out by the water feed pressure. This system prevents the particles from getting embedded into the bearing surface and causing severe wear on the shaft.

The increasing ecological awareness, stringent requirements for environmental protection, fines for emission of noxious substances to sea and economical calculations have resulted in ship designers and manufacturers searching for inexpensive, simple and reliable design solutions for building new ships and modernising existing ones. The significance and size of the contribution of water lubricated journal bearings to the development of ecologically sustainable transport technologies and power generation systems is surprisingly large and widespread. When the number of boats and ships using international waterways is taken into account, along with the typical location of power stations near the fragile ecosystems of inland waterways, storage dams or estuaries, this "hidden infrastructure" technology takes on major importance.

In marine applications, shaft journal bearings are expected to last the life of the ship—in operational terms it is usually regarded as a minimum of 100,000 h of running time. Thus reliability is of paramount importance and is greatly emphasised in the design feature and in carrying out a regular maintenance programme—which is labour intensive and costly. The lubricating medium (usually oil) has an operating life and needs to be frequently replenished or changed. Most lubricating seals are prone to minor leaks particularly when operating in harsh conditions and this has a severe effect on the environment. Most operators follow a general rule that if the performance of the bearing is not compensated in any way, minor oil leaks are often ignored due to the high costs and downtime involved in repair.

In a report published by the National Research Council (*NRC*) of the U.S., National Academy of Sciences, the average total worldwide annual release of petroleum (oils) from all known sources to the sea has been estimated at 1.3 million tonnes. However, the range is wide, from a possible 470,000 tonnes to a possible 8.4 million tonnes per year. According to the report, the main categories of sources contribute to the total input as follows:

- natural seeps: 46%
- discharges from consumption of oils (operational discharges from ships): 37%
- accidental spills from ships: 12%
- extraction of oil: 3%

As pointed out in comments to these figures, "oil pollution from ships probably reached its peak in 1979. Despite the publicity that oil spills always attract, even in 1979 only a small fraction of the oil entering the sea came from tanker accidents. Most came from routine operations, and discarded lubricants—such as engine oil poured into drains—accounted for a much higher percentage of the total. Since 1979, the amount of oil getting into the sea as a result of shipping operations has declined dramatically". Although from a wide variety of sources, a significant amount is discharged from the oil-lubricated bearings that help run the engine and associated systems. Pollutants such as these have a direct effect on India's or for that matter the coastline and waterways of many countries the world over.

For example, in Queensland boats and ships release about 4,500,000 litres of oil into the marine environment each year [1, 2]. This figure is approximately 1.0 billion tonnes world wide [3]. Although from a wide variety of sources, a significant amount is discharged from the oil-lubricated bearings that help run the engine and associated systems. Pollutants such as these have been shown to have a direct effect on the coastline, mangroves and waterways. Furthermore, the global interest in water-based lubrication technologies is very strong.

13.2 Why Water Lubrication?

The use of water as a bearing lubricant dates back to the earliest days of shaft-driven propellers on ships. However, as supply of good quality lignum vitae bearings diminished, ship owners pushed marine engineers for new systems with more predictable wear rates and extended periods between dockings. The result was the development of systems in the 1940s using mineral-oil-lubricated white metal bearings in a sealed stern tube. Bearing wear rates decreased, especially when operating in dirty abrasive water; and, mineral oil was a good fluid to develop a hydrodynamic bearing film. As an added benefit, mineral oil also protected the tail shaft from corrosion. These improvements resulted in extended docking cycles and, while stern-tube seals were not always reliable, oil leakage or "consumption" was not considered to be a major concern. Water being the perfect lubrication medium is preferred particularly for marine craft because of its natural abundance, non-compressibility, cooling properties and low coefficient of friction when between bearing strip and shaft. These types of bearings find application in water power plants and pumps in which a pumped liquid, usually water, serves as a lubricating medium. They are also required in mining industry, water conditioning stations, Navy ships, boats and submarines.

Water lubrication has many advantages over oil lubrication, both in terms of environmental sustainability and design and operating efficiency:

- Oil-based bearings are hazardous to the environment, not only due to potential oil release into waterways and the land as a result of mechanical malfunction, spillage or leakage but also because of the problems associated with safe disposal following use.
- Cooling is required in many systems that contain bearings, e.g. a turbine compartment, and unlike oil, water can function as both a coolant and a lubricant. This is because of the low frictional drag coefficient due to the low viscosity of water. Furthermore, by appropriate design of the lubrication system, water can control the temperature of a mechanical assembly within a very narrow temperature range, if required.
- Water lubrication is simple in design and the cost of equipment can be kept low due to the elimination of any special design features to accommodate the lubricant. More importantly, water lubrication removes the need for complex oil sealing arrangements. This allows bearings shafts to be shorter and stiffer, eliminating problems of misalignment.
- Water lubricated bearings are uniquely suited to many common applications such as marine, pumps and hydroelectric turbines because lubricant (water or process fluid) is supplied at one end and flows through the bearing along the multiple axial grooves. When marine vessels travel through the water the speed of the vessel forces water through the axial grooves keeping the bearing lubricated.

Public awareness of the preservation of the environment continues to increase in the twenty-first century, particularly with ship owners, operators and their customers. This has led to increased legislation defining the requirements for the marine industry, and to both corporate and personal liability for breaking these laws. For example, in Europe, the UK government has increased fines for oil discharge by 400%. In the U.S., The Oil Pollution Act has established a framework for financial liability for costs and damages resulting from even a minor spill. Many other countries are also enacting their own legislation to protect the marine environment. Compared to a mishap involving an oil tanker spill, it may appear that oil leaking from a stern tube would be of minor consequence. However, incidences of vessels being detained and significant fines being assessed resulting directly from stern tube oil leakage are now becoming a thing of the past.

13.3 Review of Literature

The first designs of these bearings had a plain bore, but this was modified several times in order to improve the operation of the bearing. The most popular design used today is the straight fluted bearing. This design of bearing consists of a number of load-carrying lands or staves, separated by flutes orientated axially in the bearing (See Fig. 13.2). The flutes supply the bearing with lubricant, which enters at one end of the bearing and leaves at the other. The flexible rubber bearing liner is bonded onto the outer rigid shell. Environmental concerns with oil and grease lubricated bearings have brought about an increase in the use of water lubricated bearings.

Journal bearing design and performance has been a major research area for the Tribology Research Group at Manipal Institute of Technology [4–10], and Queensland University of Technology (QUT) for many years [11–17].

Water-based bearing lubrication has been an important theme in this endeavour and has been investigated through undergraduate student projects and national and international industrial consulting. In 1990, a test rig [11] was developed at QUT to assess the wear performance of non-metallic, water lubricated journal bearings. The rig was subsequently enhanced to determine the pressure distribution in the water film and the optimum groove size and number. Conventional practice is to assume a linear pressure distribution along the groove [12, 13]. Subsequent experimental work by Hargreaves et al. [16] has cast this assumption into doubt (Fig. 13.3). Indeed, preliminary Computational Fluid Dynamics (CFD) analysis by Pai et al. [17] suggests that the liquid lubricant flows into the clearance space in the unloaded section of the bearing periphery and out of the clearance in the loaded section. This represents a highly complicated flow phenomenon.

Pai et al. [17] modelled the clearance volume of the water bearing for single and three grooves using CFD to study the fluid flow phenomenon. Their objective of study was to predict the pressure in the lubricant (water), measure the pressure in the lubricant (water) film in the journal bearing operating with various conditions.

Fig. 13.2 Typical straight fluted bearing

The inlet pressure was set to 50 kPa, normal to the boundary and the outlet pressure was set to 42 kPa. The bearing shell was modelled as a moving wall with absolute motion of 0 rad/s. The water in the clearance volume was modelled as type 'fluid' with rotation axis origin and direction same as that of the journal. The segregated solver was used for the solution and the flow was assumed to be laminar and steady. There was a difference in the predicted pressure contours and the experimentally measured contours. Axial pressure variations along the groove agreed with experimental trends from a qualitative perspective.

The experimental measurements of pressure in Fig. 13.3 indicate that the pressure remains constant along the groove and drops sharply at the inlet and exit rather than following a linear distribution along the groove. The groove flow in the unloaded part of the bearing is less significant to the bearing performance because the pressures there are much less than in the loaded part, and therefore have little contribution to load-carrying capacity.

Some research on the groove arrangement in journal bearings has been conducted [18, 19]. They analysed the performance of journal bearings with oil grooves that were positioned at the maximum pressure location. They discovered that positioning the grooves at the maximum pressure location will cause 30–70 percent reduction in the load capacity of the bearing. However these results are not applicable to journal bearings with multiple axial grooves supplied with water from one end of the bearing only (Fig. 13.1). Theoretical calculations have also been undertaken to verify and augment the experimental data generated [16, 17], see Fig. 13.3.

The flow in a journal bearing supply groove has an important role in determining the performance of the journal bearing, represented by its load-carrying capacity and energy consumption. The flow in the grooves and land area determines the pressure field along the grooves and over the land area. Estimation of this pressure will give the load capacity (that can be supported by the lubricant film), coefficient of fluid friction and volume rate of flow. The data produced from measuring the lubricant flow is also used to determine the pressures at the groove

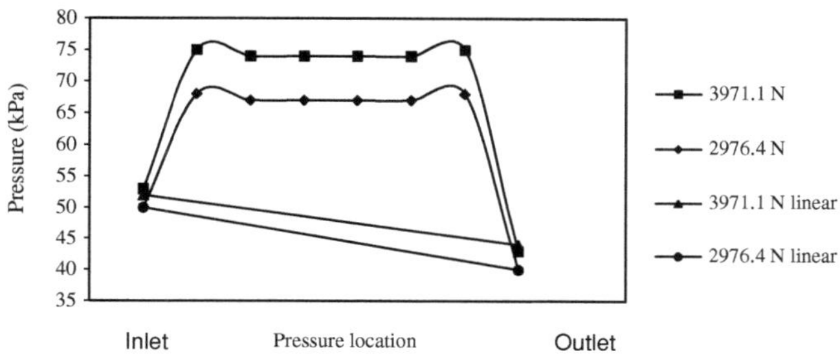

Fig. 13.3 Measured axial pressure distribution shown together with the linear variation in the groove (Hargreaves et al. [16])

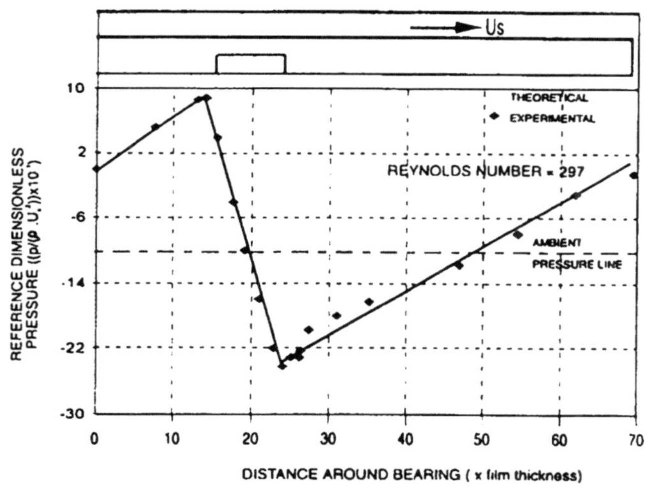

Fig. 13.4 Laminar flow pressure distribution for flow across a bearing with a step (Salimonu et al. [20])

edges, which is an important consideration in the detail design of these bearings. It is important to note that there is no published literature on the type of flow phenomenon illustrated in Fig. 13.4, even though journal bearings have been used as stern-tube bearings in ships for many years. A much simpler bearing configuration has been studied experimentally by Salimonu et al. [20] and Burton [21]. In that work, the pressure profile across a blockage (rather than a groove) in a conventional bearing with shaft rotation and no axial pressure-induced flow along the groove was measured. It was evident that pressure jumps occurred near to and across the blockage (Fig. 13.3). This implies that such jumps will occur for the bearing being studied in the proposed project. This in turn emphasises the importance of the boundary conditions at the groove in any theoretical modelling used to predict bearing performance.

Some design procedures for axial groove journal bearings have been reported [22, 23] but once again, these do not cater for multiple axial grooves supplied with lubricant from one end only.

The configuration of the water lubricated bearing is similar to the submerged bearing studied by Pai and Majumdar [24]. The lubricant (water) is fed from one end through multiple axial grooves. Groove depth is kept at 3 mm but the number and size of grooves may vary. The lubricant film is generated in the non-grooved (land) region (see Fig. 13.1). The flow in this region may be both circumferential and axial. The lubricant flows out from the bearing ends axially [25]. They found that the maximum pressure in the clearance space of the bearing does not occur at the central plane but shifts closer to the outlet side of the bearing. This is because the lubricant is supplied axially. At the inlet, flow into the bearing takes place only in the unloaded region. At the outlet, flow takes place out of the bearing in the loaded region. The flow into and out of the bearing is maintained very much similar to that in a submerged bearing The pressure generated due to wedge action in the clearance space supports the applied load without metal-to-metal (or solid-to-solid) contact. Although the pressure in the grooved region has been shown in previous experiments to be constant, the pressure in the land region will be generated due to axial and circumferential flow components.

Majumdar et al. [26] studied the steady state and dynamic characteristics including whirl stability of water lubricated journal bearings having three axial grooves theoretically. The governing equation is the Reynolds equation in two dimensions which is solved for pressure numerically for the already defined boundary conditions. The Swift—Stieber boundary conditions are adopted in case cavitation occurs in the divergent portion. The pressure distribution in the clearance space gives the steady state characteristics in terms of load-carrying capacity, attitude angle, volume rate of flow and coefficient of fluid friction. These data help in design of the bearing. The fluid film bearings are prone to whirl instability to some extent; a further study was undertaken to ascertain the dynamic characteristics in terms of stiffness and damping of the fluid film and stability of the rotor bearing system. A first order perturbation method was used to find the dynamic pressures in the clearance space. The stiffness and the damping characteristics were calculated from the real and imaginary parts of the integrated pressures. These coefficients are then used in the equation of motion of a rigid rotor for estimating the mass parameter, a measure of stability. The already existing experimental data was verified by the theoretical solution. The pressure along the groove was found to be constant. It was seen that both load-carrying capacity and stability improve for the 18° angle grooved bearing, whereas friction variable is less. As expected, load capacity increases with the increase in eccentricity. Higher journal speeds leads to higher pressure development and results in a sharp fall in pressure at the bearing ends. High eccentricity will also give rise to high pressure development and will lead to higher flow rate. The coefficient of friction will increase with higher speed. This is due to the shearing action at high speeds. The mass parameter and whirl ratio are used as the measure of stability. The stability also increases sharply at a very high eccentricity ratio. It was concluded that bearing having smaller groove angle gives higher load capacity due to high pressure in the land region. The stability also improves when small groove angles are used.

Pai et al. [7, 8, 27] discussed theoretically the stability characteristics of water lubricated journal bearings having multiple axial grooves. A nonlinear analysis of a rigid rotor supported on journal bearings under unidirectional constant load, unidirectional periodic load and variable load was carried out. The time-dependent Reynolds equation in two dimension for incompressible fluid was solved numerically by the finite difference method, with Jakobsson-Floberg-Olsson boundary conditions, to obtain the hydrodynamic forces. Using these forces the equation of motion are solved by the fourth order Runge–Kutta method to predict the transient behaviour of the rotor. The analysis gives the orbital trajectory within the clearance circle. It was found that better pressure value distributions were obtained when JFO model was used as boundary conditions than using Reynolds boundary conditions. This helps in better prediction of bearing stability. The locus of the journal with a constant load has smaller excursions of the journal centre, when compared with a periodic load superimposed on the system. The trajectory for the variable rotating load was complex, but the journal locus lies within the clearance circle, indicating the bearing system can operate without any problem. Pai et al. [9] also studied the stability of four and six axial groove bearings using a nonlinear transient approach using the method described above.

Cabrera et al. [28] studied the film pressure distribution in water lubricated rubber journal bearings. The measurements of pressure indicated that the film pressure profiles are very different from those of the conventional rigid bearings. The relatively low pressures in the film caused significant rubber deflections but too low to produce viscosity changes. Integration of the pressure in the bearings showed that they operate in the regime of mixed lubrication. The behaviour of the bearings was theoretically investigated using CFD and compared with the experimental values conducted on the test rig designed (See Fig. 13.5).

The bearing test rig was used for 50 mm diameter bearings and length-to-diameter ratio of 2. The test shaft was made of chrome-plated stainless steel and supported by two hybrid bearings. The test loads were applied using a motorised unit, a de-coupling spring and a load cell. The loading unit allowed loads of up to 1,000 N to be applied. The tests were carried out with a shaft surface speed of 2 m/s and loads in the range 0–500 N. Production bearings manufactured by Silvertown UK Ltd. were used for this investigation.

The commercial CFD software, FIDAP, was selected as it allowed the solution of coupled structural and fluid problems. It has been found from experimental measurements and confirmed using CFD that complex film pressure distributions exist over the loaded staves. These distributions are a result of the interaction of the elastic bearing surface deflections with the film pressures. Peak pressures in the bearings are greatly reduced compared with conventional bearings.

Wojciech Litwin [29] has extensively studied the performance of water lubricated bearing through his experimental studies. The experimental setup is as shown in the Fig. 13.6. The diameter of the sea water resistant steel shaft was

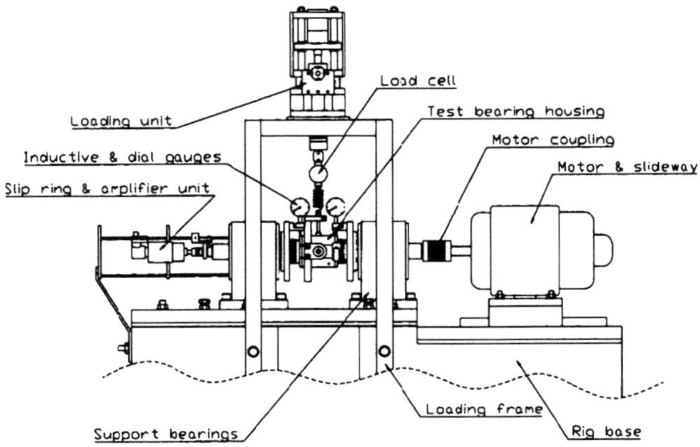

Fig. 13.5 Dynamic test rig layout (Cabrera et al. [28])

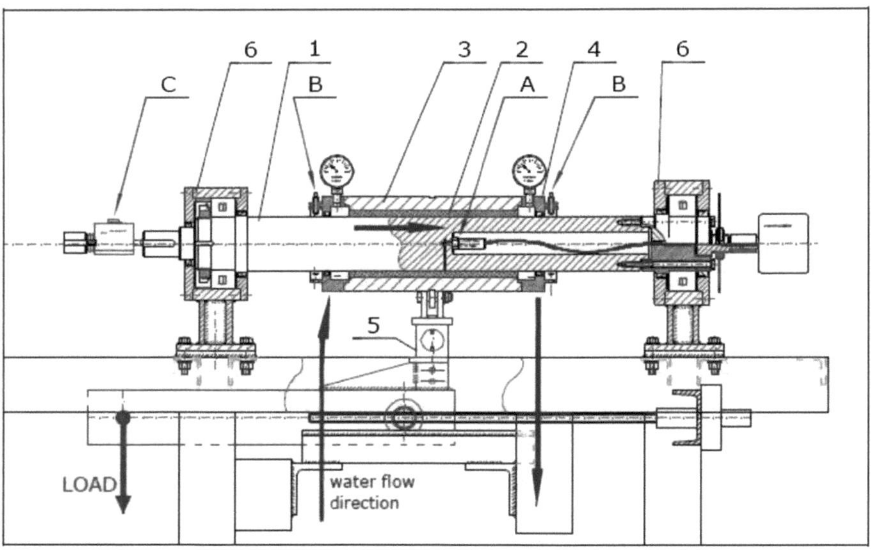

Fig. 13.6 Stern-tube test rig (Litwin [29]). *1* Main shaft ø100 mm; *2* tested bearing bush; *3* baring sleeve; *4* covers with sealing; *5* static load leaver; *6* roller bearings; *A* pressure pickup sensor; *B* touch less distance sensors; *C* torqmeter

100 mm and the length being 700 mm. The shaft was connected with a clutch and torque meter to a electric engine so as to vary the rotating speed. The examined bearing is set in the steel sleeve and processed along with it. The sleeve is closed on both ends with the sealed covers which make it possible for the lubricating and cooling water flow inside it. This enables the bearing to work in different pressures

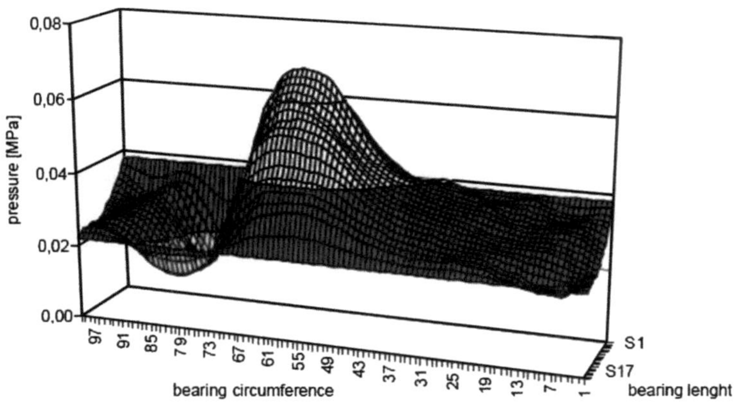

Fig. 13.7 Measured pressures in water film in water lubricated bearing (Litwin [29])

up to the pressure of 0.6 MPa (Fig. 13.7). The load is exerted on the bearing in the static or dynamic way. The parameters which could be varied are as follows:

- Shaft rotation speed from 0 to 11 rev/s.
- The radial load from 1 to 5 kN.
- The pressure of the water supplying the bearing, varying from 0 to 0.6 MPa.

These conditions are similar to the real small ship main shaft bearing.

The pressure measuring (Fig. 13.6) system was a pressure sensor mounted inside the shaft and the signals were transmitted using a wireless system. The trajectory of the shaft was determined by having two pairs of non-contact transducers with an accuracy of 1 μm on either side of the shaft (Fig. 13.8).

The second test rig called the Propeller Shaft bearing test stand is similar to the previous test rig except for the number of bearings to be tested are two. This test rig has the capability of measuring the fluid film pressure in the mid plane of the bearing.

Apart from the experimental tests the theoretical research was also conducted. The calculation results were obtained by using the author's original software based on hydrodynamic lubrication theory applicable to elastic bush material (EHL). The software is composed of two modules: the programme based on hydrodynamic lubrication theory, written in FORTRAN language, and the module based on the finite element method (FEM) in the form of macro to a commercial software. The calculations are performed in an iterative mode. A satisfactory result was obtained after twenty iterations (Fig. 13.9). Results of the calculations were verified experimentally. The author confirms that the applied calculation method is correct. In the author's opinion the calculation error does not exceed 20%.

Andersson et al. [30] have studied the load-carrying capability of water lubricated ceramic journal bearing. The possibilities of using advanced ceramics in water lubricated bearings were studied by performing tests on journal bearing lubricated with water. The materials studied were two aluminas (Al_2O_3) a zirconia toughened

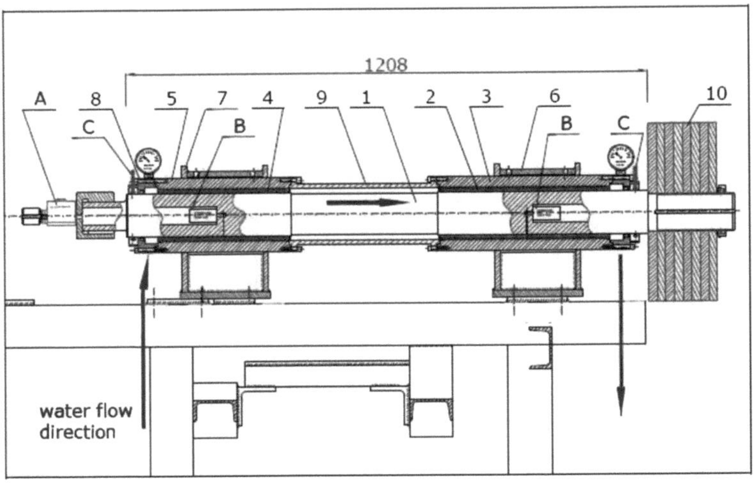

Fig. 13.8 Stern-Tube test stand with two bearings [29]. *1* Main shaft ø100 mm. *2, 3* Stern-tube bearing. *4,5* Front bearing. *6, 7* Supports. *8* Covers with sealing. *9* Connecting pipe. *10* Radial load—discs. *A* Torqmeter. *B* Pressure pickup sensors. *C* Touch less distance sensors

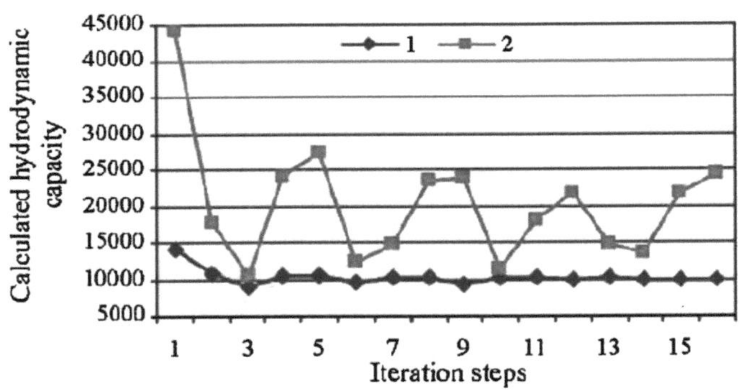

Fig. 13.9 Calculated values of load-carrying capacity of a bearing in function of number of successive iteration steps [29]. *1st series*—Stable solution was achieved. *2nd series*—A convergent result was not achieved

alumina (ZTA), a partially stabilised zirconia (PSZ), a sintered silicon carbide (SiC). The tribological tests were performed using a journal bearing test rig built for this particular purpose. The ceramic test bearing was fixed into a sealed floating housing connected to a closed lubrication circuit. A shaft sleeve made of the same ceramic was attached to a steel shaft to be rotated. The test bearing housing was loaded by dead weights. During the test, the friction force and the temperature on the outside of the ceramic bearing were recorded. The tests were carried out with φ 40 mm $\times$ φ 30 mm $\times$ 40 mm shaft sleeves made of two aluminas (Al_2O_3), a ZTA, a partially stabilized zirconia (PSZ), a SiC, a reaction bonded silicon carbide (SiSiC) and $\beta^{\prime}$-sialon rotating in φ 56 mm × φ 40 mm × 20 mm journal bearing made out of same materials. The test was performed with a pair of new specimen and with de-ionised water as the lubricant. The water pressure was in the range of 3.5–6 bar and the feed rate 0.3–7 × 10^{-6} m^3/s depending on the wear degree of the ceramic specimens. The temperature of the water before entering the bearings was 28 ± 5°C. Prior to testing all the specimens were ultrasonically cleaned and dried. The entire test was run at 0.4 m/s sliding velocity, which is equivalent to a rotational speed of 192 revolutions per minute. The load applied on the bearing was between 0.5 and 4.8 kN. It was found that tribological behaviour of the different ceramics in the present tests differed very strongly. It was found that silicon carbides performed well even at the highest loads applied. Both the SiC and the SiSiC materials experienced very low wear; their stable coefficient of friction was below 0.01 and their sliding surfaces became polished. Hence their load-carrying capability is at least 4.8 and 3.5 kN, respectively. The load-carrying capability of Al_2O_3 and ZTA ceramics was limited to 0.8–1.3 kN. The PSZ and sialon materials failed at the lowest load applied. To conclude we can summarise that evaluation of ceramics for possible use in sliding bearing assemblies lubricated with water and the best results were obtained with SiC and SiSiC, due to their ability to become tribochemically polished combined with their good thermal conductivity. These materials can be recommended for water lubricated bearing systems provided that the mechanical strength of the ceramics is sufficient for the application. In recent years, Yoshimoto et al. [31, 32] considered axial load capacity and stability of water lubricated hydrostatic conical bearings with spiral grooves.

13.4 Theory

13.4.1 Reynolds Equation

The equation governing the lubricant flow in the bearing is the Reynolds equation in two dimensions for an incompressible fluid. The equation can be written as

$$\frac{\partial}{\partial x}\left(h^3\frac{\partial p}{\partial x}\right)+h^3\frac{\partial^2 p}{\partial z^2}=6\eta U\frac{\partial h}{\partial x}+12\eta\frac{\partial h}{\partial t} \tag{13.1}$$

By substituting the following,

$\theta = x/R$, $\bar{z} = z/L$, $\bar{p} = p/p_s$, $\tau = \omega_p t$ and $\bar{h} = h/C = 1 + \varepsilon\cos(\theta^* - \psi)$

Then the non-dimensionalised Eq. 13.1 can be written as:

$$\frac{\partial}{\partial\theta}\left(\bar{h}^3\frac{\partial\bar{p}}{\partial\theta}\right)+\frac{1}{4}\left(\frac{D}{L}\right)^2\bar{h}^3\left(\frac{\partial^2\bar{p}}{\partial\bar{z}^2}\right)=\Lambda\left(\frac{\partial\bar{h}}{\partial\theta}\right)+2\Lambda\lambda\left(\frac{\partial\bar{h}}{\partial\tau}\right) \tag{13.2}$$

where Bearing number, $\Lambda = 6\eta\omega\Big/\left[p_s(C/R)^2\right]$ and Whirl ratio, $\lambda = \omega_p/\omega$.

13.4.2 Steady State Characteristics

The time-dependent term is dropped under steady state condition. Then the Reynolds equation for incompressible fluid will be

$$\frac{\partial}{\partial\theta}\left(\bar{h}_0^3\frac{\partial\bar{p}_0}{\partial\theta}\right)+\frac{1}{4}\left(\frac{D}{L}\right)^2\bar{h}_0^3\left(\frac{\partial^2\bar{p}_0}{\partial\bar{z}^2}\right)=\Lambda\left(\frac{d\bar{h}_0}{d\theta}\right) \tag{13.3}$$

Further reducing the Eq. 13.3, we get

$$\bar{h}_0^3\frac{\partial^2\bar{p}_0}{\partial\theta^2}+3\bar{h}_0^2\frac{\partial\bar{h}_0}{\partial\theta}\frac{\partial\bar{p}_0}{\partial\theta}+\frac{1}{4}\left(\frac{D}{L}\right)^2\bar{h}_0^3\frac{\partial^2\bar{p}_0}{\partial\bar{z}^2}-\Lambda\frac{\partial\bar{h}_0}{\partial\theta}=0 \tag{13.4}$$

13.4.2.1 Boundary Conditions

The domain boundary conditions for the three axial grooved journal bearing are $\bar{p}_0=1$ at $\bar{z}=0$ at the grooves and elsewhere $\bar{p}_0=0$ and $\bar{p}_0=0$ at $\bar{z}=1$ for all θ.

As the hydraulic resistance is low, the lubricant will predominantly escape into the axial grooves since the inlet chamber is filled in with lubricant having a supply pressure p_s. It is therefore assumed that the boundary conditions at the inlet end ($\bar{z}=0$) are $\bar{p}_0=1$ at the grooves and $\bar{p}_0=0$ elsewhere (Fig. 13.10). Figure 13.11 shows the model of the bearing showing three axial grooves.

13.4.2.2 Cavitation Boundaries

Reynolds boundary condition

$$\bar{p}=0,\quad \partial\bar{p}/\partial\theta=0 \tag{13.5}$$

Jakobsson-Floberg-Olsson boundary condition modified by Kicinski [33]:

The JFO boundaries were applied only in land (lobe) regions, the cavitation zone is applied for the full length of the bearing as indicated in Fig. 13.12. The pressure is allowed to vary linearly in the groove from inlet to outlet.

With regard to the cavitation zone it is assumed that

(a) The flow in the real cavitated zone, which includes several more or less regular 'finger like' bubbles correspond to the flow in the theoretical zone with only one analytical 'bubble'.
(b) All water flows around the analytical bubble, which has a flat surface in the direction of film thickness and the shape of the analytical bubble resulting from the continuity equation will be defined by its width.
(c) Within the cavitation zone the pressure is equal to the atmospheric pressure.

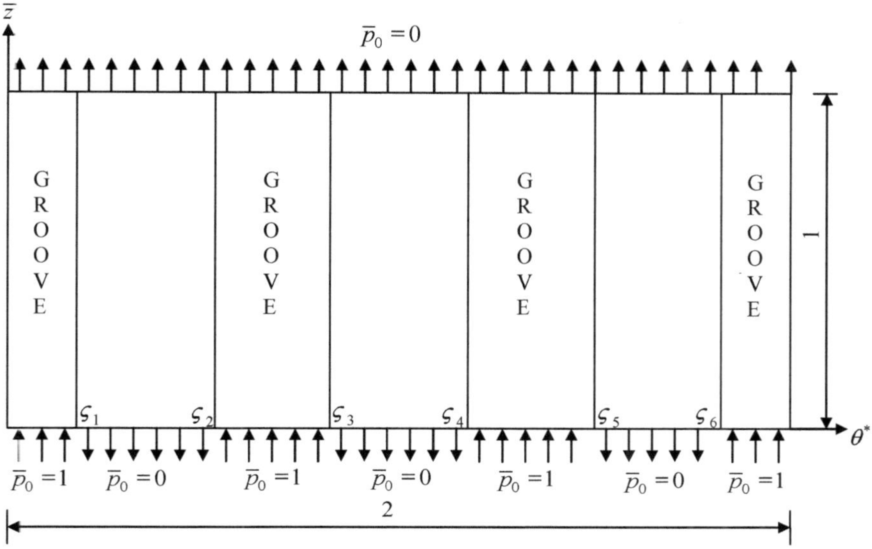

Fig. 13.10 Developed view of the bearing showing the boundary conditions (Majumdar et al. [26])

Fig. 13.11 Model of the clearance volume showing 3-axial groove

The model of the cavitation zone based on the above assumption is shown in Fig. 13.12 and is as given in Ref. [24].

The water gap filling coefficient is defined as

$$\beta = 1 - L_{\text{cav}} \tag{13.6}$$

where L_{cav} is the dimensionless width of the analytical bubble

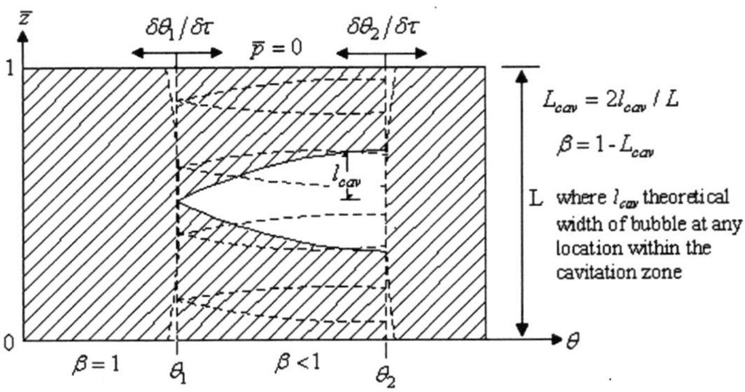

Fig. 13.12 Theoretical model of cavitation zone (Pai et al. [24])

$$L_{cav} = 2L_{\text{cav}}/L \tag{13.7}$$

If the water fills the gap completely then $L_{\text{cav}} = 0$ and the condition of flow continuity means that in the element $L\Delta x$ during the time interval at $\Delta\tau$ the following must be satisfied.

$$\frac{1}{2}UL\frac{\partial(\beta h)}{\partial x}\Delta x\,\Delta\tau = L\frac{\partial(\beta h)}{\partial \tau}\Delta x\,\Delta\tau \tag{13.8}$$

After rearrangement

$$U\frac{\partial(\beta h)}{\partial x} + 2\frac{\partial(\beta h)}{\partial \tau} = 0 \tag{13.9}$$

Non-dimensionlising

$$\frac{\partial(\beta\bar{h})}{\partial\theta} + 2\frac{\partial(\beta\bar{h})}{\partial\tau} = 0 \tag{13.10}$$

The analytical solution of Eq. 13.10 gives the value of the filling coefficient at a point of time $\tau = \tau_0 + \Delta\tau$, when θ_1 and θ_2 define the limits of the cavitation zone. For the given initial condition at the time τ_0.

$$\bar{h} = \bar{h}_0(\theta, \tau_0)$$

$$\beta = \beta_0(\theta, \tau_0)$$

Hence

$$\beta\left(\theta + \frac{1}{2}\Delta\tau,\ \tau_0 + \Delta\tau\right) = \frac{\beta_0(\theta,\tau)\bar{h}_0(\theta,\tau)}{\bar{h}\left(\theta + \frac{1}{2}\Delta\tau, \tau_0 + \Delta\tau\right)} \tag{13.11}$$

Equation 13.11 indicates that the flow $\beta\bar{h}$ in the cavitation zone at the moment τ depends on the flow $\beta_0\bar{h}_0$ at the moment τ_0 directly preceding the moment τ.

The boundaries of the cavitation zone are determined by satisfying the flow continuity at the boundaries, neglecting the curvature of the boundary in the z direction (i.e. $\partial\bar{p}/\partial\bar{z} = 0$).

The flow on the pressure side,

$$q_p = L\left(\frac{Uh_{1/2}}{2} - \frac{h_{1/2}^3(\partial p/\partial x)_{1/2}}{12\eta}\right), \tag{13.12}$$

is balanced by the flow in the cavitation zone and an additional flow because of boundary movement with time. (The subscript 1/2 indicates that the quantities determined concern the cavitation and reformation boundaries).

The flow in the cavitation zone is given by

$$q_{\text{cav}} = \frac{L\beta Uh_{1/2}}{2} \tag{13.13}$$

The flow due to boundary movement is

$$q_t = L(1-\beta)h_{1/2}\frac{\partial x}{\partial t} \tag{13.14}$$

Therefore

$$q_p = q_{\text{cav}} + q_t \tag{13.15}$$

Substituting Eqs. 13.12, 13.13 and 13.14 in Eq. 13.15, we get

$$\left(\frac{U}{2} - \frac{\partial x_{1/2}}{\partial t}\right)(1-\beta) = \frac{h_{1/2}^2(\partial p/\partial x)_{1/2}}{12\eta} \tag{13.16}$$

Non-dimensionalizing Eq. 13.16 gives

$$\left(\frac{1}{2} - \frac{\partial\theta_{1/2}}{\partial\tau}\right)(1-\beta) = \frac{\bar{h}_{1/2}^2(\partial\bar{p}/\partial\theta)_{1/2}}{12} \tag{13.17}$$

The filling coefficient values for a given time step are determined from the flow continuity Eq. 13.10 in the cavitation zone for that time step.

From the above reference the solution of Eq. 13.10 indicates that the flow $\beta\bar{h}$ in the cavitation zone at a given time step depends on the flow $\beta_0\bar{h}_0$ at the preceding time step.

Moreover, it is assumed that the pressure drop is linear from the supply end to the outlet end. Equation 13.4 is solved numerically by the finite difference method using the SOR scheme satisfying both the boundary conditions of Eqs. 13.5 and 13.10.

In a conventional cylindrical bearing the coordinate θ in the circumferential direction is taken from the position of maximum film thickness. Here in the grooved bearing this position needs to be found beforehand. This is done by assuming the attitude angle ψ by an arbitrary and the coordinate θ^* is measured from the vertical position, as shown in Fig. 13.13. Using this ψ the film thickness equation is written as $\bar{h} = 1 + \varepsilon\cos(\theta^* - \psi)$ and Eq. 13.4 is solved for pressure using the boundary conditions. The load components along the line of centres and its perpendicular directions are found from:

$$\bar{W}_r = \left(\frac{W_r}{LDp_s}\right) = -\left[\int_0^1\int_{\varsigma_1}^{\varsigma_2} \bar{p}_0\cos(\theta^* - \psi)\,\mathrm{d}\theta\,\mathrm{d}\bar{z} + \int_0^1\int_{\varsigma_3}^{\varsigma_4} \bar{p}_0\cos(\theta^* - \psi)\,\mathrm{d}\theta\,\mathrm{d}\bar{z} + \int_0^1\int_{\varsigma_5}^{\varsigma_6} \bar{p}_0\cos(\theta^* - \psi)\,\mathrm{d}\theta\,\mathrm{d}\bar{z}\right] \tag{13.18}$$

$$\bar{W}_t = \left(\frac{W_t}{LDp_s}\right) = \left[\int_0^1\int_{\varsigma_1}^{\varsigma_2} \bar{p}_0\sin(\theta^* - \psi)\,\mathrm{d}\theta\,\mathrm{d}\bar{z} + \int_0^1\int_{\varsigma_3}^{\varsigma_4} \bar{p}_0\sin(\theta^* - \psi)\,\mathrm{d}\theta\,\mathrm{d}\bar{z} + \int_0^1\int_{\varsigma_5}^{\varsigma_6} \bar{p}_0\sin(\theta^* - \psi)\,\mathrm{d}\theta\,\mathrm{d}\bar{z}\right] \tag{13.19}$$

The load capacity $\bar{W}$ and attitude angle ϕ_0 are calculated by the following equations:

$$\bar{W} = \left[\bar{W}_r^2 + \bar{W}_t^2\right]^{1/2} \tag{13.20}$$

$$\phi_0 = \tan^{-1}\left(\frac{\bar{W}_t}{\bar{W}_r}\right) \tag{13.21}$$

The attitude angle calculated from Eq. 13.21 is compared with the assumed value of attitude angle (ψ) after each calculation. The value of ψ is modified with a small increment of ψ and the Reynolds equation is solved using this modified value until ψ is equal to ϕ_0. The volume rate of flow and coefficient of friction are calculated from the pressure distribution found earlier. A flowchart indicating the above calculation procedure is given in Fig. 13.14.

The flow balance will take place in the following way:

Flow into the bearing ($\bar{z} = 0$) across the grooves is due to inlet supply pressure $\bar{p}_s = 1$. Flow out comprises flow (1) from the exit end ($\bar{z} = 1$) for all θ, and (2) from the inlet end ($\bar{z} = 1$) excluding the groove regions, where the pressures are kept at a constant value of $\bar{p}_s = 1$ shown in Fig. 13.12.

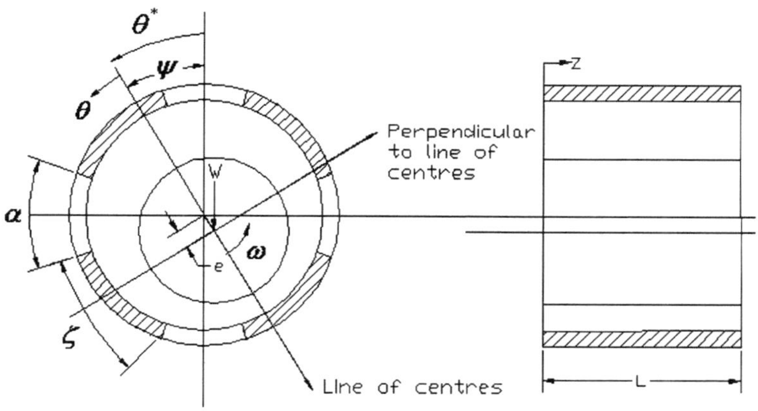

Fig. 13.13 Four axial groove journal bearing and the coordinate system (Pai et al. [7]).

Then the Q volume rate of flow is

$$Q = \left[-\int_{\varsigma_1}^{\varsigma_2} \frac{h_0^3}{12\eta} \frac{\partial p_0}{\partial z}\bigg|_{z=0} R\mathrm{d}\theta - \int_{\varsigma_1}^{\varsigma_2} \frac{h_0^3}{12\eta} \frac{\partial p_0}{\partial z}\bigg|_{z=1} R\mathrm{d}\theta \right] + \left[-\int_{\varsigma_3}^{\varsigma_4} \frac{h_0^3}{12\eta} \frac{\partial p_0}{\partial z}\bigg|_{z=0} R\mathrm{d}\theta - \int_{\varsigma_3}^{\varsigma_4} \frac{h_0^3}{12\eta} \frac{\partial p_0}{\partial z}\bigg|_{z=1} R\mathrm{d}\theta \right] + \left[-\int_{\varsigma_5}^{\varsigma_6} \frac{h_0^3}{12\eta} \frac{\partial p_0}{\partial z}\bigg|_{z=0} R\mathrm{d}\theta - \int_{\varsigma_5}^{\varsigma_6} \frac{h_0^3}{12\eta} \frac{\partial p_0}{\partial z}\bigg|_{z=1} R\mathrm{d}\theta \right] \tag{13.22}$$

The dimensionless form of flow rate can be written as:

$$\bar{Q} = \left(\frac{2Q\eta L}{C^3 p_s D} \right) = \left[-\frac{1}{12} \int_{\varsigma_1}^{\varsigma_2} \bar{h}_0^3 \frac{\partial \bar{p}_0}{\partial \bar{z}}\bigg|_{\bar{z}=0} \mathrm{d}\theta - \frac{1}{12} \int_{\varsigma_1}^{\varsigma_2} \bar{h}_0^3 \frac{\partial \bar{p}_0}{\partial \bar{z}}\bigg|_{\bar{z}=1} \mathrm{d}\theta \right] + \left[-\frac{1}{12} \int_{\varsigma_3}^{\varsigma_4} \bar{h}_0^3 \frac{\partial \bar{p}_0}{\partial \bar{z}}\bigg|_{\bar{z}=0} \mathrm{d}\theta - \frac{1}{12} \int_{\varsigma_3}^{\varsigma_4} \bar{h}_0^3 \frac{\partial \bar{p}_0}{\partial \bar{z}}\bigg|_{\bar{z}=1} \mathrm{d}\theta \right] + \left[-\frac{1}{12} \int_{\varsigma_5}^{\varsigma_6} \bar{h}_0^3 \frac{\partial \bar{p}_0}{\partial \bar{z}}\bigg|_{\bar{z}=0} \mathrm{d}\theta - \frac{1}{12} \int_{\varsigma_5}^{\varsigma_6} \bar{h}_0^3 \frac{\partial \bar{p}_0}{\partial \bar{z}}\bigg|_{\bar{z}=1} \mathrm{d}\theta \right] \tag{13.23}$$

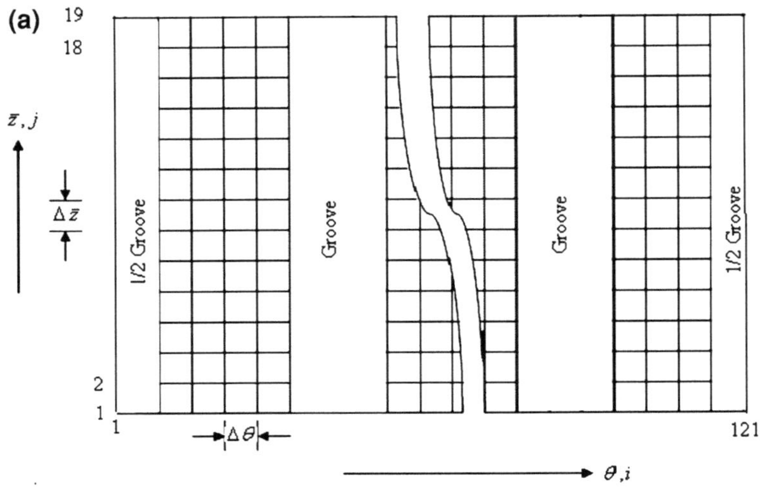

Fig. 13.14 **a** A grid used in the analysis (Pai et al. [8]). **b** Flow chart (Majumdar et al. [26])

while calculating the first integral of $\bar{Q}$ the portions of the grooves at the entrance ($z = 0$) region are excluded.

The Friction force F can be found from

$$F = \int_0^1 \int_{\varsigma_1}^{\varsigma_2} \gamma R \, \mathrm{d}\theta \, \mathrm{d}z + \int_0^1 \int_{\varsigma_3}^{\varsigma_4} \gamma R \, \mathrm{d}\theta \, \mathrm{d}z + \int_0^1 \int_{\varsigma_5}^{\varsigma_6} \gamma R \, \mathrm{d}\theta \, \mathrm{d}z \tag{13.24}$$

The dimensionless form of friction force can be written as

$$\bar{F} = \left(\frac{F}{2LCp_s}\right) = \left[\int_0^1 \int_{\varsigma_1}^{\varsigma_2} \left(\bar{h}_0 \frac{\partial \bar{p}_0}{\partial \theta} \mathrm{d}\theta \mathrm{d}\bar{z} + \frac{\Lambda}{12} \mathrm{d}\theta \mathrm{d}\bar{z}\right) + \int_0^1 \int_{\varsigma_3}^{\varsigma_4} \left(\bar{h}_0 \frac{\partial \bar{p}_0}{\partial \theta} \mathrm{d}\theta \mathrm{d}\bar{z} + \frac{\Lambda}{12} \mathrm{d}\theta \mathrm{d}\bar{z}\right) + \int_0^1 \int_{\varsigma_5}^{\varsigma_6} \left(\bar{h}_0 \frac{\partial \bar{p}_0}{\partial \theta} \mathrm{d}\theta \mathrm{d}\bar{z} + \frac{\Lambda}{12} \mathrm{d}\theta \mathrm{d}\bar{z}\right)\right]$$

For one axial grooved journal bearings, only the first term will be used in Eqs. 13.18, 13.19 and 13.22–13.25, whose integral limits are ς_1 to ς_2. Similarly, for two axial grooved bearings, the first two terms will be used in Eqs. 13.18–13.25, whose integral limits will be ς_1 to ς_2 for the first term and ς_3 to ς_4 for the second term.

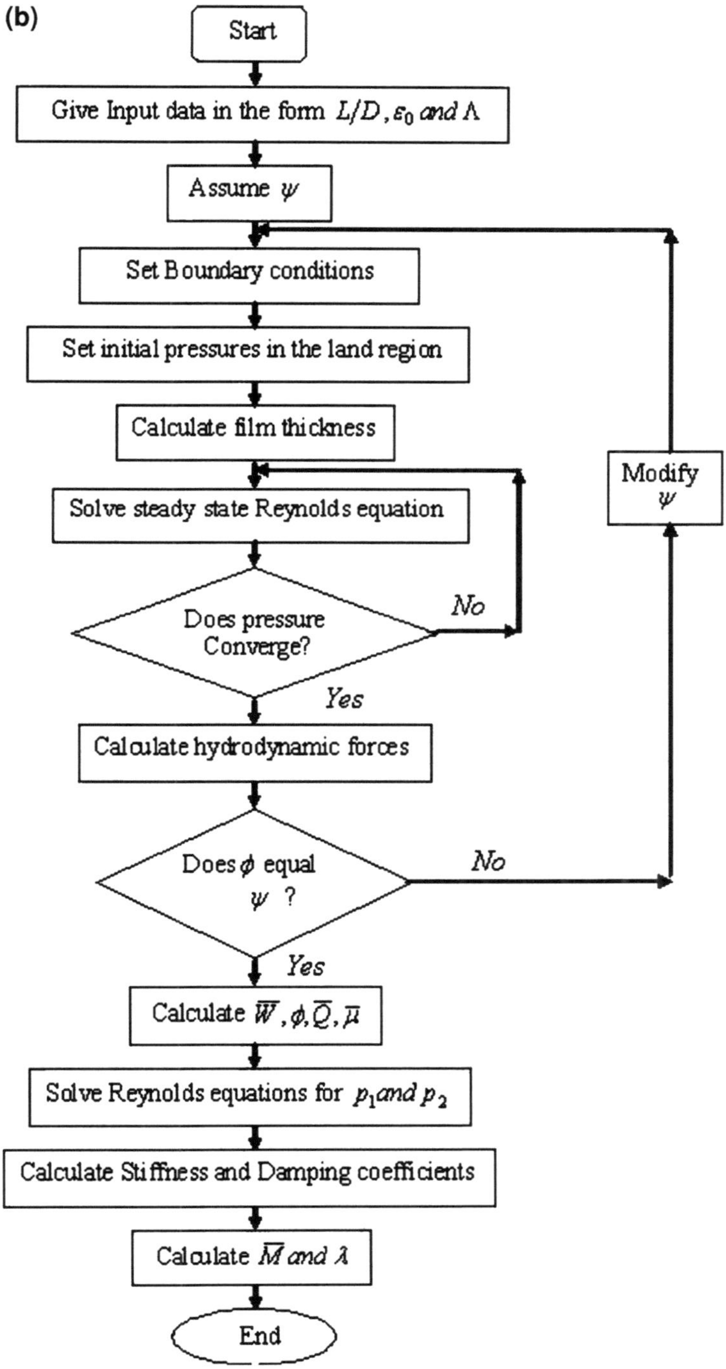

Fig. 13.14 (continued)

Likewise, the number of terms used in these equations would increase as the number of grooves in the bearing increases with the appropriate limits of integrals.

13.4.3 Dynamic Characteristics

For dynamic condition, the Reynolds equation can be written as

$$\frac{\partial}{\partial\theta}\left(\bar{h}^3\frac{\partial\bar{p}}{\partial\theta}\right)+\frac{1}{4}\left(\frac{D}{L}\right)^2\bar{h}^3\left(\frac{\partial^2\bar{p}}{\partial\bar{z}^2}\right)=\Lambda\left(\frac{\partial\bar{h}}{\partial\theta}\right)+2\Lambda\lambda\left(\frac{\partial\bar{h}}{\partial\tau}\right) \tag{13.26}$$

The journal performs a periodic motion of small amplitude $\mathrm{Re}(C\varepsilon_1 e^{i\tau})$ along the line of centres and $\mathrm{Re}(C\varepsilon_0\phi_1 e^{i\tau})$ perpendicular to the line of centres around its steady state position ε_0 and ϕ_0.

It is assumed that at the beginning of instability, the position of the journal centre can be defined as a steady state value (ε_0, ϕ_0) together with a harmonic vibration of frequency ω_p; thus

$$\varepsilon=\varepsilon_0+\varepsilon_1 e^{i\tau}$$

$$\phi=\phi_0+\phi_1 e^{i\tau}$$

where

$$|\varepsilon_1|\langle\langle\varepsilon_0$$

$$|\phi_1|\langle\langle\phi_0$$

Under this condition (i.e. for small amplitude of vibration) first-order perturbation will be valid. The pressure and film thickness can be written as:

$$\bar{p}=\bar{p}_0+\varepsilon_1 e^{i\tau}\bar{p}_1+\varepsilon_0\phi_1 e^{i\tau}\bar{p}_2 \tag{13.27}$$

$$\bar{h}=\bar{h}_0+\varepsilon_1 e^{i\tau}\cos\theta+\varepsilon_0\phi_1 e^{i\tau}\sin\theta \tag{13.28}$$

By the substitution of Eqs. 13.27 and 13.28 into Eq. 13.26 and retaining up to first linear terms, we get the following three equations:

$$\varepsilon_0\ :\bar{h}_0^3\frac{\partial^2\bar{p}_0}{\partial\theta^2}+3.0\bar{h}_0^2\frac{\partial\bar{h}_0}{\partial\theta}\frac{\partial\bar{p}_0}{\partial\theta}+\frac{1}{4}\left(\frac{D}{L}\right)^2\bar{h}_0^3\frac{\partial^2\bar{p}_0}{\partial\bar{z}^2}-\Lambda\frac{d\bar{h}_0}{d\theta}=0 \tag{13.29}$$

$$\begin{aligned}\varepsilon_1 e^{i\tau}:\ &\bar{h}_0^3\frac{\partial^2\bar{p}_1}{\partial\theta^2}+3.0\bar{h}_0^2\cos\theta\frac{\partial^2\bar{p}_0}{\partial\theta^2}-3.0\bar{h}_0^2\sin\theta\frac{\partial\bar{p}_0}{\partial\theta}+6.0\bar{h}_0\cos\theta\frac{\partial\bar{h}_0}{\partial\theta}\frac{\partial\bar{p}_0}{\partial\theta}\\&+3.0\bar{h}_0^2\frac{\partial\bar{h}_0}{\partial\theta}\frac{\partial\bar{p}_1}{\partial\theta}+\frac{1}{4}\left(\frac{D}{L}\right)^2\bar{h}_0^3\frac{\partial^2\bar{p}_1}{\partial\bar{z}^2}+\frac{3}{4}\left(\frac{D}{L}\right)^2\bar{h}_0^2\cos\theta\frac{\partial^2\bar{p}_0}{\partial\bar{z}^2}\\&+\Lambda\sin\theta-2i\Lambda\lambda\cos\theta=0\end{aligned} \tag{13.30}$$

$$\varepsilon_0\phi_1 e^{i\tau} : \bar{h}_0^3 \frac{\partial^2 \bar{p}_2}{\partial\theta^2} + 3.0\bar{h}_0^2 \sin\theta \frac{\partial^2 \bar{p}_0}{\partial\theta^2} + 3.0\bar{h}_0^2 \cos\theta \frac{\partial \bar{p}_0}{\partial\theta} + 6.0\bar{h}_0 \sin\theta \frac{\partial \bar{h}_0}{\partial\theta}\frac{\partial \bar{p}_0}{\partial\theta}$$
$$+ 3.0\bar{h}_0^2 \frac{\partial \bar{h}_0}{\partial\theta}\frac{\partial \bar{p}_2}{\partial\theta} + \frac{1}{4}\left(\frac{D}{L}\right)^2 \bar{h}_0^3 \frac{\partial^2 \bar{p}_2}{\partial\bar{z}^2} + \frac{3}{4}\left(\frac{D}{L}\right)^2 \bar{h}_0^2 \sin\theta \frac{\partial^2 \bar{p}_0}{\partial\bar{z}^2}$$
$$- \Lambda\cos\theta - 2i\Lambda\lambda.\sin\theta = 0 \tag{13.31}$$

Equations 13.30 and 13.31 are solved numerically satisfying the modified boundary conditions of Eqs. 13.5 and 13.10, by knowing the steady state pressure distribution.

13.4.4 Stiffness and Damping Coefficients

The dynamic pressures p_1 and p_2 are produced due to the dynamic displacements of the journal centre $\mathrm{Re}(C\varepsilon_1 e^{i\tau})$ parallel to and $\mathrm{Re}(C\varepsilon_0\phi_1 e^{i\tau})$ perpendicular to the line of centres. The components of the dynamic load due to the dynamic pressure p_1 along and perpendicular to the line of centres can be written as

$$\begin{aligned}(W_1)_r &= \int_0^L\int_0^{2\pi} p_1 R\cos\theta \, \mathrm{d}\theta \, \mathrm{d}z \\ (W_1)_\phi &= \int_0^L\int_0^{2\pi} p_1 R\sin\theta \, \mathrm{d}\theta \, \mathrm{d}z\end{aligned} \tag{13.32}$$

It is found that the fluid film, which supports the rotor, is equivalent to a spring and a dashpot system. Since the journal executes small harmonic oscillation about its steady state position, the dynamic load-carrying capacity can be expressed as a spring and a viscous damping force as given below:

$$\begin{aligned}-(W_1)_r\varepsilon_1 e^{i\tau} &= K_{rr}Y + D_{rr}\frac{\mathrm{d}Y}{\mathrm{d}t} \\ -(W_1)_\phi\varepsilon_1 e^{i\tau} &= K_{\phi r}Y + D_{\phi r}\frac{\mathrm{d}Y}{\mathrm{d}t},\end{aligned} \tag{13.33}$$

where the position of the journal centre is given by $Y = C\varepsilon_1 e^{i\tau}$. Then

$$\begin{aligned}(\bar{W}_1)_r &= \frac{(W_1)_r}{LDp_s} = \int_0^1\int_0^{2\pi} \bar{p}_1 \cos\theta \, \mathrm{d}\theta \, \mathrm{d}\bar{z} \\ (\bar{W}_1)_\phi &= \frac{(W_1)_\phi}{LDp_s} = \int_0^1\int_0^{2\pi} \bar{p}_1 \sin\theta \, \mathrm{d}\theta \, \mathrm{d}\bar{z}\end{aligned} \tag{13.34}$$

Using Eqs. 13.33 and 13.34, we can write

$$
\begin{aligned}
-(\bar{W}_1)_r &= -\int_0^1\int_0^{2\pi} \bar{p}_1 \cos\theta \,\mathrm{d}\theta\,\mathrm{d}\bar{z} = \frac{K_{rr}C}{LDp_s} + i\frac{D_{rr}C\omega_p}{LDp_s} \\
-(\bar{W}_1)_\phi &= -\int_0^1\int_0^{2\pi} \bar{p}_1 \sin\theta \,\mathrm{d}\theta\,\mathrm{d}\bar{z} = \frac{K_{\phi r}C}{LDp_s} + i\frac{D_{\phi r}C\omega_p}{LDp_s}
\end{aligned}
\tag{13.35}
$$

Since p_1 is complex, the dynamic load $\bar{W}_1$ can be expressed in terms of real and imaginary parts as

$$
\bar{W}_1 = \mathrm{Re}(\bar{W}_1) + i\mathrm{Im}(\bar{W}_1) \tag{13.36}
$$

Hence we get,

$$
\begin{aligned}
\bar{K}_{rr} &= -\mathrm{Re}\left[\int_0^1\int_0^{2\pi} \bar{p}_1 \cos\theta\,\mathrm{d}\theta\,\mathrm{d}\bar{z}\right] \\
\bar{K}_{\phi r} &= -\mathrm{Re}\left[\int_0^1\int_0^{2\pi} \bar{p}_1 \sin\theta\,\mathrm{d}\theta\,\mathrm{d}\bar{z}\right] \\
\bar{D}_{rr} &= -\mathrm{Im}\left[\int_0^1\int_0^{2\pi} \bar{p}_1 \cos\theta\,\mathrm{d}\theta\,\mathrm{d}\bar{z}\right]\Big/\lambda \\
\bar{D}_{\phi r} &= -\mathrm{Im}\left[\int_0^1\int_0^{2\pi} \bar{p}_1 \sin\theta\,\mathrm{d}\theta\,\mathrm{d}\bar{z}\right]\Big/\lambda,
\end{aligned}
\tag{13.37}
$$

where, $\bar{K}_{ij} = \frac{K_{ij}C}{LDp_s}$ and $\bar{D}_{ij} = \frac{D_{ij}C\omega}{LDp_s}$.

In the same way, considering the dynamic displacement of the journal centre along ϕ direction, the following expressions result.

$$
\begin{aligned}
\bar{K}_{\phi\phi} &= -\mathrm{Re}\left[\int_0^1\int_0^{2\pi} \bar{p}_2 \sin\theta\,\mathrm{d}\theta\,\mathrm{d}\bar{z}\right] \\
\bar{K}_{r\phi} &= -\mathrm{Re}\left[\int_0^1\int_0^{2\pi} \bar{p}_2 \cos\theta\,\mathrm{d}\theta\,\mathrm{d}\bar{z}\right] \\
\bar{D}_{\phi\phi} &= -\mathrm{Im}\left[\int_0^1\int_0^{2\pi} \bar{p}_2 \sin\theta\,\mathrm{d}\theta\,\mathrm{d}\bar{z}\right]\Big/\lambda \\
\bar{D}_{r\phi} &= -\mathrm{Im}\left[\int_0^1\int_0^{2\pi} \bar{p}_2 \cos\theta\,\mathrm{d}\theta\,\mathrm{d}\bar{z}\right]\Big/\lambda
\end{aligned}
\tag{13.38}
$$

To study the stability of a rigid rotor supported on this bearing the stiffness and damping coefficients thus obtained can be used.

13.4.5 Stability Criteria

The shaft is supposed to be rigid, the mass of the rotor to be equal to 2 M, and the rotor to be supported by 2 identical journal bearings. Under steady load, the journal centre is displaced from the bearing centre to the steady operating position. When the supporting bearings are rigid, shaft vibration amplitude varies with shaft speed. Rigidly supported rotors cannot be operated at a critical speed and can become hung on the critical when attempting to drive through.

The stability of the journal is analysed by combining the equations of motion and the film forces F_r and F_ϕ, assuming the rotor to be rigid and having mass M.

$$F_r + W\cos\phi - MC\left[\frac{\mathrm{d}^2\varepsilon}{\mathrm{d}t^2} - \varepsilon\left(\frac{\mathrm{d}\phi}{\mathrm{d}t}\right)^2\right] = 0 \tag{13.39}$$

$$F_\phi - W\sin\phi - MC\left[\varepsilon\frac{\mathrm{d}^2\phi}{\mathrm{d}t^2} + 2\frac{\mathrm{d}\varepsilon}{\mathrm{d}t}\frac{\mathrm{d}\phi}{\mathrm{d}t}\right] = 0 \tag{13.40}$$

After substituting for ε, ϕ and nondimensionalising, we get

$$\bar{F}_r + \bar{W}\left(\cos\phi_0 - \phi_1 e^{i\tau}\sin\phi_0\right) + \bar{M}\lambda^2\varepsilon_1 e^{i\tau} = 0 \tag{13.41}$$

$$\bar{F}_\phi - \bar{W}\left(\sin\phi_0 + \phi_1 e^{i\tau}\cos\phi_0\right) + \bar{M}\lambda^2\varepsilon_0\phi_1 e^{i\tau} = 0 \tag{13.42}$$

$$\left(-\bar{M}\lambda^2 + \bar{K}_{rr} + i\lambda\bar{D}_{rr}\right)\varepsilon_1 + \left(\bar{W}\sin\phi_0 + \bar{K}_{r\phi}\varepsilon_0 + i\lambda\,\varepsilon_0\bar{D}_{r\phi}\right)\phi_1 = 0 \tag{13.43}$$

$$\left(\bar{K}_{\phi r} + i\lambda\,\bar{D}_{\phi r}\right)\varepsilon_1 + \left(-\bar{M}\lambda^2\varepsilon_0 + \bar{W}\cos\phi_0 + \bar{K}_{\phi\phi}\varepsilon_0 + i\lambda\,\varepsilon_0\bar{D}_{\phi\phi}\right)\phi_1 = 0 \tag{13.44}$$

For non-trivial solution we get,

$$\begin{vmatrix} -\bar{M}\lambda^2 + \bar{K}_{rr} + i\lambda\,\bar{D}_{rr} & \bar{W}\sin\phi_0 + \bar{K}_{r\phi}\varepsilon_0 + i\lambda\varepsilon_0\bar{D}_{r\phi} \\ \bar{K}_{\phi r} + i\lambda\,\bar{D}_{\phi r} & -\bar{M}\lambda^2\varepsilon_0 + \bar{W}\cos\phi_0 + \bar{K}_{\phi\phi}\varepsilon_0 + i\lambda\varepsilon_0\bar{D}_{\phi\phi} \end{vmatrix} = 0 \tag{13.45}$$

By simplifying we get,

$$\bar{M} = \frac{1}{\lambda^2\left(\bar{D}_{\phi\phi} + \bar{D}_{rr}\right)}\left[\left(\bar{D}_{rr}\bar{K}_{\phi\phi} + \bar{K}_{rr}\bar{D}_{\phi\phi}\right) - \left(\bar{K}_{\phi r}\bar{D}_{r\phi} + \bar{D}_{\phi r}\bar{K}_{r\phi}\right)\right.$$
$$\left. + \frac{\bar{W}}{\varepsilon_0}\left(\bar{D}_{rr}\cos\phi_0 - \bar{D}_{\phi r}\sin\phi_0\right)\right] \tag{13.46}$$

$$\bar{M}^2\lambda^4 - \lambda^2\left[\bar{M}\left(\frac{\bar{W}\cos\phi_0}{\varepsilon_0} + \bar{K}_{\phi\phi} + \bar{K}_{rr}\right) + \left(\bar{D}_{rr}\bar{D}_{\phi\phi} - \bar{D}_{\phi r}\bar{D}_{r\phi}\right)\right] + \left(\bar{K}_{rr}\bar{K}_{\phi\phi} - \bar{K}_{\phi r}\bar{K}_{r\phi}\right) + \frac{\bar{W}}{\varepsilon_0}\left(\bar{K}_{rr}\cos\phi_0 - \bar{K}_{\phi r}\sin\phi_0\right) = 0 \quad (13.47)$$

Equations 13.46 and 13.47 are linear algebraic equations in $\bar{M}$ and λ and solution of these will give $\bar{M}$ and λ. The speed of journal calculated from this value of $\bar{M}$ is the threshold speed, above which the bearing system will be unstable. The whirl speed (speed of the journal centre) can be found from the whirl ratio for the above journal speed. The journal centre rotates in the same direction as that of the journal.

13.4.6 Solution Technique

By solving Eq. 13.4, satisfying the boundary conditions, the film pressure distribution can be found. For this purpose, the equation is written in finite difference form, using a standard notation. A numerical solution of the finite difference equation satisfying the boundary conditions is obtained, using a successive over-relaxation scheme in the rectangular grid and adopting the five-point formula.

The bearing film is represented by an orthogonal grid because it is assumed that there is no variation of pressure across the film. A typical grid used in the analysis is shown in Fig. 13.14a. In this representation a grid point is designated as (i, j) where i and j represent the coordinates along the x (θ) and z ($\bar{z}$) directions, respectively.

The second-order derivatives of pressure are approximated by central difference as

$$\frac{\partial^2\bar{p}}{\partial\theta^2} = \frac{\bar{p}_{i+1,j} - 2\bar{p}_{i,j} + \bar{p}_{i-1,j}}{(\Delta\theta)^2}$$

$$\frac{\partial^2\bar{p}}{\partial\bar{z}^2} = \frac{\bar{p}_{i,j+1} - 2\bar{p}_{i,j} + \bar{p}_{i,j-1}}{(\Delta\bar{z})^2}$$

$$\frac{\partial\bar{p}}{\partial\theta} = \frac{\bar{p}_{i+1,j} - \bar{p}_{i-1,j}}{2(\Delta\theta)}$$

The first-order derivatives of pressure at the rupture and reformation boundaries are replaced by the three-point backward and forward difference quotients, respectively. The pressures are integrated using Simpson's 1/3rd rule to get the non-dimensional hydrodynamic force components (Fig. 13.14b).

13.5 Nonlinear Transient Analysis

The basic differential equation for pressure distribution in the bearing clearance under dynamic conditions for incompressible fluid may be written as

$$\frac{\partial}{\partial x}\left(h^3\frac{\partial p}{\partial x}\right)+h^3\frac{\partial^2 p}{\partial z^2}=6\eta U\frac{\partial h}{\partial x}+12\eta\frac{\partial h}{\partial t} \tag{13.48}$$

Using the following substitutions: $\theta = x/R$, $\bar{z} = z/L$, $\bar{p} = p/p_s$, $\tau = \omega_p t$ and $\bar{h} = h/C = 1 + \varepsilon\cos(\theta^* - it\psi)$

Equation 13.48 can be non-dimensionalised as:

$$\frac{\partial}{\partial\theta}\left(\bar{h}^3\frac{\partial\bar{p}}{\partial\theta}\right)+\frac{1}{4}\left(\frac{D}{L}\right)^2\bar{h}^3\frac{\partial^2\bar{p}}{\partial\bar{z}^2}=\Lambda\frac{d\bar{h}}{d\theta}+2\Lambda\lambda\frac{\partial\bar{h}}{\partial\tau} \tag{13.49}$$

Equation 13.49 can be rewritten as

$$\frac{\partial}{\partial\theta}\left(\bar{h}^3\frac{\partial\bar{p}}{\partial\theta}\right)+\frac{1}{4}\left(\frac{D}{L}\right)^2\bar{h}^3\frac{\partial^2\bar{p}}{\partial\bar{z}^2}=\Lambda(-\varepsilon\sin(\theta^*-it\psi))+2\Lambda\lambda(\dot{\varepsilon}\cos(\theta^*-it\psi)) \tag{13.50}$$

where $\dot{\varepsilon} = \partial\varepsilon/\partial\tau$.

13.5.1 Solution Technique

Equation 13.50 is solved, satisfying the boundary conditions, to find the film pressure distribution. For this purpose the equation is written in finite difference form, using a standard notation. A numerical solution of the finite difference equation satisfying the boundary conditions is obtained, using a successive over-relaxation scheme in the rectangular grid and adopting the five-point formula.

The first order derivatives of pressure at the rupture and reformation boundaries are replaced by the three-point backward and forward difference quotients respectively. At each grid point the pressure gradient is evaluated and then used in Eq. 13.16. The boundaries are located at those points at which the filling coefficient 'β' takes a value of one. A solution to the two-dimensional Reynolds equation satisfying the boundary conditions is obtained by Gauss–Seidel iteration.

We start by assuming a pressure distribution at the internal grid points. The film thickness at each grid point is evaluated and then Reynolds equation is integrated between the boundaries determined by Eq. 13.16. The pressure distribution is obtained by successive over-relaxation scheme.

To compute the pressure numerically, the initial pressure distribution is taken to be zero at all the grid points and iteration is started. After each iteration the computed grid pressures are modified by the over-relaxation factor. The solution is supposed to have converged if it satisfies the convergence criterion.

$$\left|\frac{\sum\bar{p}_{i,j}^{n+1}-\sum\bar{p}_{i,j}^{n}}{\sum\bar{p}_{i,j}^{n+1}}\right|\leq 0.001$$

where ‘n’ is the number of iterations.

The value of the over-relaxation factor chosen determines the speed of convergence. The over-relaxation factor is selected by trial and error and a value of 1.57 gives a good rate of convergence for the present analysis.

At first, the differential coefficients of eccentricity ratio $[\dot{\varepsilon}]$ and attitude angle $\left[\dot{\phi}\right]$ with respect to time are set equal to zero. This gives us the pressure distribution under steady state conditions. The pressures are integrated using Simpson's 1/3rd rule to get the non-dimensional hydrodynamic force components.

The forces are computed from

$$\bar{F}_r = \int_0^1 \int_{\theta_2}^{\theta_1} \bar{p} \cos(\theta^* - \psi) \mathrm{d}\theta \, \mathrm{d}\bar{z} \tag{13.51}$$

$$\bar{F}_\phi = \int_0^1 \int_{\theta_2}^{\theta_1} \bar{p} \sin(\theta^* - \psi) \mathrm{d}\theta \, \mathrm{d}\bar{z} \tag{13.52}$$

13.5.2 Stability Analysis

Solving the equation of motion, one can compute ε, $\dot{\varepsilon}$, ϕ and $\dot{\phi}$ for the next time step, starting from the steady state position. The equations of motion are written as

$$MC\left\{\frac{\mathrm{d}^2\varepsilon}{\mathrm{d}t^2} - \varepsilon\left(\frac{\mathrm{d}\phi}{\mathrm{d}t}\right)^2\right\} = F_r + W\cos\phi \tag{13.53}$$

$$MC\left(\varepsilon\frac{\mathrm{d}^2\phi}{\mathrm{d}t^2} + 2\frac{\mathrm{d}\phi}{\mathrm{d}t}\frac{\mathrm{d}\varepsilon}{\mathrm{d}t}\right) = F_\phi - W\sin\phi \tag{13.54}$$

The dimensionless forms of Eqs. 13.53 and 13.54 are

$$\bar{M}\ddot{\varepsilon} - \bar{M}\varepsilon\dot{\phi}^2 - \frac{\bar{F}_r}{\bar{W}_a} - \cos\phi = 0 \tag{13.55}$$

$$\bar{M}\varepsilon\ddot{\phi} + 2\bar{M}\dot{\varepsilon}\dot{\phi} - \frac{\bar{F}_\phi}{\bar{W}_a} + \sin\phi = 0 \tag{13.56}$$

where $\bar{M} = MC\omega^2/LDp_s$ and $\bar{W}_a = W_aC^2/\eta\omega R^3L$. Equations 13.55 and 13.56 are second-order differential equations in ε and ϕ. These are solved by using a fourth–order Runge–Kutta method for constant values of $\bar{M}, \bar{F}_r$ and $\bar{F}_\phi$. The integration scheme gives the eccentricity ratio, attitude angle and their derivatives. These are used in the solution of the Reynolds equation for the next time step to obtain the pressure distribution and hence the force components. The force components are then used in the solution of the equations of motion. The procedure of solving the Reynolds equation and the equations of motion is repeated until a sufficient number of ε and ϕ values are obtained so that a trajectory can be drawn. The nature of the trajectory helps in determining the status of the system.

13.5.3 Unidirectional Constant Load

The hydrodynamic forces under steady state conditions are found using a set of initial conditions. The equations of motion are then solved to get the values of ε, $\dot{\varepsilon}$, ϕ and $\dot{\phi}$ for the next time step. The equations of motion are written as

$$\ddot{\varepsilon} = \varepsilon\dot{\phi}^2 + \frac{\bar{F}_r}{\bar{M}\bar{W}_a} + \frac{\cos\phi}{\bar{M}} \tag{13.57}$$

$$\ddot{\phi} = -\frac{2\dot{\varepsilon}\dot{\phi}}{\varepsilon} + \frac{\bar{F}_\phi}{\bar{M}\varepsilon\bar{W}_a} - \frac{\sin\phi}{\bar{M}\varepsilon} \tag{13.58}$$

For each time step the hydrodynamic forces are determined and are used in the solution of equations of motion for that time step. While the mass parameter and steady state load are held constant over the entire analysis period.

The procedure of integration gives us the values of ε and ϕ values at each time step, from which a trajectory of the journal centre can be obtained. This helps in indicating the status of the system.

13.5.4 Unidirectional Periodic Load

The external load applied was assumed to be of the form

$$\bar{W} = \bar{W}_a\left\{1 + \sin\left(\frac{\tau}{2}\right)\right\} \tag{13.59}$$

where $\bar{W}_a$ is the steady state load corresponding to the eccentricity ratio $\varepsilon_0 = 0.8$. Equation 13.59 is used to calculate the external load value at each time step.

The equations of motion are written as

$$\ddot{\varepsilon} = \varepsilon\dot{\phi}^2 + \frac{\bar{F}_r}{\bar{M}\bar{W}_a} + \frac{[1 + \sin(\tau/2)]\cos\phi}{\bar{M}} \tag{13.60}$$

$$\ddot{\phi} = -\frac{2\dot{\varepsilon}\dot{\phi}}{\varepsilon} + \frac{\bar{F}_\phi}{\bar{M}\varepsilon\bar{W}_a} - \frac{[1 + \sin(\tau/2)]\sin\phi}{\bar{M}\varepsilon} \tag{13.61}$$

The hydrodynamic forces F_r and F_ϕ are computed for each time step. Integration of the equations of motion using the values of $\bar{M}$ gives us the $\varepsilon, \dot{\varepsilon}, \phi$ and $\dot{\phi}$ for the next time step. The journal centre locus is obtained by repeating the above procedure.

13.5.5 Variable Rotating Load

This analysis is for Radial piston pumps. The pump has a rated output of 10.0 L/min at 300 bar and the shaft is vertical with a speed of 1,450 rpm. Peak load on each piston is 3,980 N, attained at the maximum lift of the cam (8 mm).

The cam lift at 10° crank interval was determined graphically considering L/D = 1.0, eccentricity as 8 mm and that the cam follows a simple harmonic motion. At 180° crank angle, it was assumed that the load on the bearing is lowest. For the next stroke, the lift of the cam and the non-dimensionlised load values repeat in a reverse order. The results were non-dimensionalised using $\bar{W} = WC^2/\eta\omega R^3 L$ and were used in the equations of motion along with the mass parameter and steady state load to obtain the journal centre trajectory. The hydrodynamic forces were computed for each time step. It should be remembered that the time step has been chosen to match the applied load. A solution of equations of motion written in appropriate form gives the $\varepsilon, \dot{\varepsilon}, \phi$ and $\dot{\phi}$ values for each time step.

$$\ddot{\varepsilon} = \varepsilon\dot{\phi}^2 + \frac{\bar{F}_r}{\bar{M}\bar{W}_a} + \frac{\bar{W}\cos\phi}{\bar{M}\bar{W}_a} \tag{13.62}$$

$$\ddot{\phi} = -\frac{2\dot{\varepsilon}\dot{\phi}}{\varepsilon} + \frac{\bar{F}_\phi}{\bar{M}\varepsilon\bar{W}_a} - \frac{\bar{W}\sin\phi}{\bar{M}\varepsilon\bar{W}_a} \tag{13.63}$$

This helps in getting a trajectory of the journal centre and hence in determining the status of the system.

13.6 Modelling and Analysis Using Computational Fluid Dynamics

Extensive research has been conducted in the area of journal bearings over many years for various operating conditions and geometry, effects of different types of lubricants (oil and water), different numbers (zero, one and three) and positions of the grooves and flow of lubricants between shaft and bearing. To minimize the experimental time and cost of predicting the performance of the journal bearing operating under wide variety of conditions has led to the development of

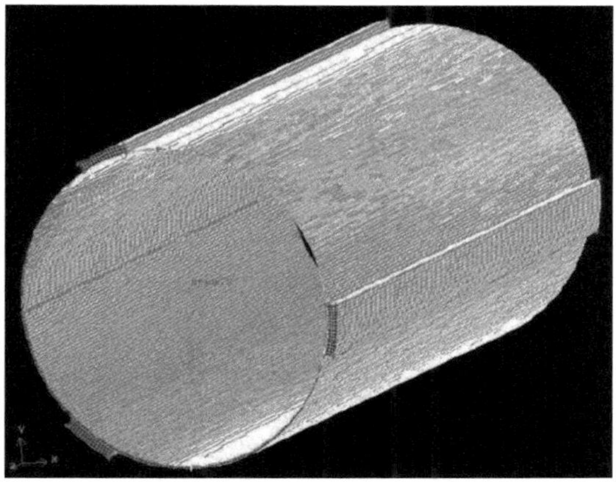

Fig. 13.15 Meshed clearance volume of a 3-axial groove water lubricated bearing (Pai et al. [25])

numerical methods such as CFD. This method has proved to be very useful tool in this research area (Fig. 13.15).

A software package FLUENT 5.0 was used to predict the velocity distributions in the bearing. The CFD software is based on the four fundamental equations of fluid mechanics including turbulent and compressible flow. The computational domain is discretised into control volumes and the equations solved for pressure and velocity. Some difficulty was encountered on setting the pressure boundary conditions, in defining the mesh (because of the aspect ratio in lubrication problems that is long thin volumes over which the equations need to be solved) and the solution of the system of equations.

The segregated solver was used for the solution and the flow was assumed to be laminar and steady. Under-relaxation factors were required for both pressure and velocity equations. An interesting feature of the solution technique is that the angular velocity of the journal surface is varied in steps. The converged solution of the first step is used as the start point for the next step. The step size of the angular velocity is determined by halving the full angular velocity, which is again halved and so on. The halving process is stopped if it is determined that there are sufficient number of steps. If the step size is too small it may lead to unnecessary computations. It is necessary to arrive at a solution by varying the angular velocity in steps because of the curvature of the bearing surfaces. The solution is converged if the normalised residuals are of the order of 1×10^{-3}.

13.7 Experimental Details

A bearing test rig, shown schematically in Fig. 13.16 and described in [15, 16], was developed to characterise the non-metallic journal bearing performance when being lubricated by water. The rig is configured to test bearings with 100 mm bore diameters and length of 200 mm at speeds up to 1,500 rpm and loads up to

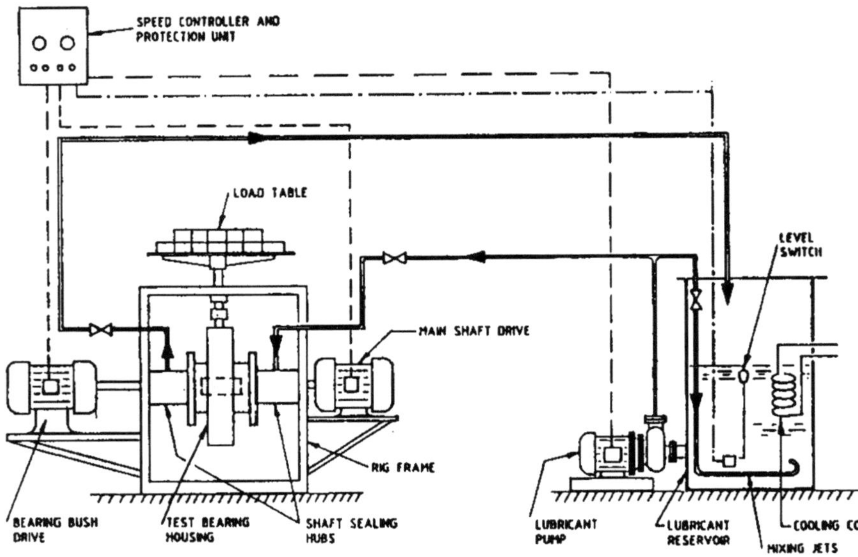

Fig. 13.16 Schematic layout of the bearing test rig (Tay [15], Hargreaves et al. [16])

9,810 N. The bearing test rig was designed and built to provide reliable measurements for journal bearings with any number of grooves which extend from one end to the other end. For all tests reported here, the bearing material was an elastomeric material made from thermosetting resins which are three-dimensional cross-linked condensation polymers.

The non-metallic bearing was mounted in a steel sleeve with an interference fit and the sleeve-bearing assembly was fixed into the housing. Pressures were measured along the groove located in the loaded region of the bearing and around the circumference of the bearing half-way along its length for various operating conditions—see Fig. 13.17. There were 16 tapping points circumferentially and 7 axially. Special note was made of pressures near to the groove edges. Flexible tubes were fitted to each tapping point and taken to a manifold where the pressure was measured by a pressure transducer, one at a time. An air release valve was located near to the manifold.

The load was applied by a dead weight to the top of the bearing housing thereby ensuring that the minimum film thickness was located near to the top of the bearing. Only bearings with three equi-spaced axial grooves along the complete length of the bearing were tested. The angular extent of each groove was 36 or 18°, with the groove depth being maintained at 0.003 m.

The bearing material has some unique properties and special care was required in establishing the actual bearing clearance. Using the manufacturer's recommendations for changes in geometry due to the interference fit (with the steel sleeve),

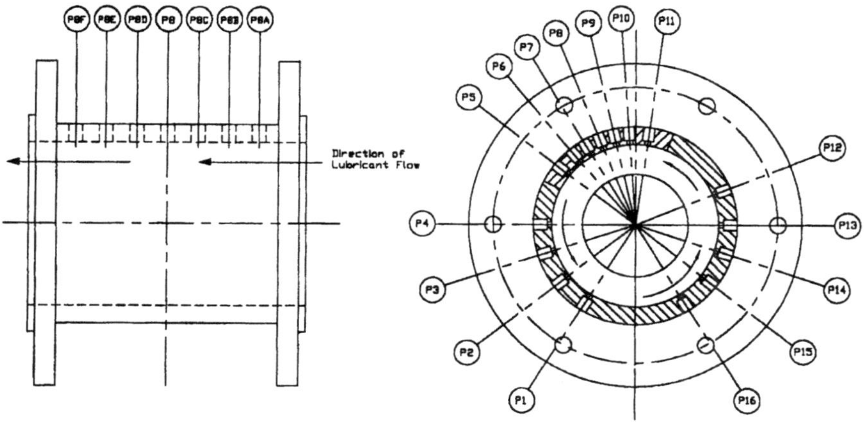

Fig. 13.17 Location of pressure measuring points (Hargreaves et al. [16])

thermal expansion and swelling due to contact with water, the actual diametral clearance was calculated to be 0.86×10^{-3} m.

The lubricant water was fed from one end of the bearing through each of the three axial grooves and around the space between the shaft and the bearing bush. This pressure-fed situation, together with the rotation of the shaft, means that the flow conditions inside the bearing consisted of a combination of Couette flow (shaft velocity induced) and the axially fed Poiseuille flow. Published data on this arrangement was not available. Static measurement capabilities include the flow rate, the inlet pressure and the outlet pressure to and from the bearings, rise in water temperature as it passes through the bearing, the shaft speed, the applied load and the power consumed by the three-phase electric motor.

The supply of lubricating water was maintained at around 7.4 L/min for different loads at different lubricant supply pressures. The flow rate was measured by transferring the water which flowed to the tank to a pail where it could be measured within one minute. Two different shaft speeds were established for each set of bearings. For shaft speeds of 1,500 and 1,000 rpm, a constant lubricating supply pressure of 100 kPa, 75 kPa or 50 kPa was applied. Different loads were added to the load table for recording of the internal bearing pressures. Lead weights were used to apply the load. The mass of the load table was taken as 8 kg and was accounted for in the load increment.

The driving power of the three-phase electric motor was determined for each load using the 'Clamp On Power Factor Tester' by recording the input voltage, the current from each phase and the power factor. The formula for calculating the input power is $\sqrt{3}$ VIcos θ. The exact shaft speed was measured with a handheld photosensitive tachometer for each load added to the load table. Also the inlet and outlet temperature of the water were recorded by connecting the thermocouples to an electronic analogue thermometer. Bearing pressures at various measuring points were recorded and its tapping points were denoted as P1 and P16 (Figs. 13.16 and 13.17).

13.8 Results and Discussion

13.8.1 Results From Finite Difference Analysis

13.8.1.1 Nonlinear Transient Analysis

The analysis gives the orbital trajectory within the clearance circle, which is not obtainable using a simplified theory (linearised perturbation theory). The stability of the system was studied by simulating the journal centre trajectory and thereby estimating a stability parameter which is the function of speed.

Figures 13.18 and 13.19 show the journal locus for constant load for different groove angles (36 and 12°). For constant load it is found that the journal locus attains a stable position. For a periodic load the journal locus does not end in a limit cycle indicating the system is neither stable nor unstable. Any disturbance can make the system unstable. For the variable rotating load, the trajectory of the journal centre is complex, but lies within the clearance circle.

The three-dimensional plot of the pressure variation in the bearing clearance for Reynolds boundary condition for groove angle 36° is shown in Fig. 13.20. A stability diagram is prepared with the variation of mass parameter and eccentricity ratio for groove angle 36 and 12° is shown in Fig. 13.21. This diagram shows the stable and unstable region for the water lubricated journal bearing with a unidirectional constant load at various eccentricity ratios. To obtain a stability diagram, the journal is released at different eccentricity ratios and the mass parameter varied. Each time the mass parameter is changed, a trajectory is obtained using computer software and the mass parameter at which the change from stable to unstable takes place is noted. The procedure is repeated to get the critical mass parameters at various eccentricity ratios. It is seen that smaller groove angles show more stability.

13.8.1.2 Static Characteristics and Linear Perturbation Analysis

This section presents the various aspects of boundary conditions and the effect of groove angles on the stability for a range of bearing numbers and eccentricity ratios.

Figure 13.22 shows the load-carrying capacity and attitude angle of a bearing having 36° groove angle with Reynolds boundary condition. As the journal speed increases the bearing number and load-carrying capacity both increase. The load-carrying capacity increases with increase in eccentricity ratios. As the eccentricity ratio increases the attitude angle decreases. For smaller groove angles there is an

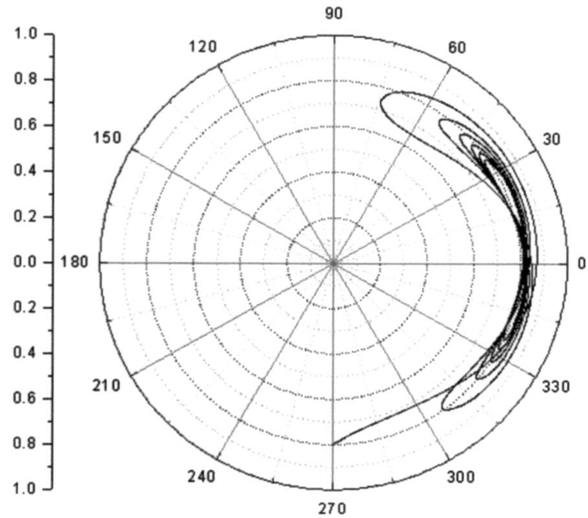

Fig. 13.18 Journal centre locus for unidirectional constant load (Pai et al. [8])

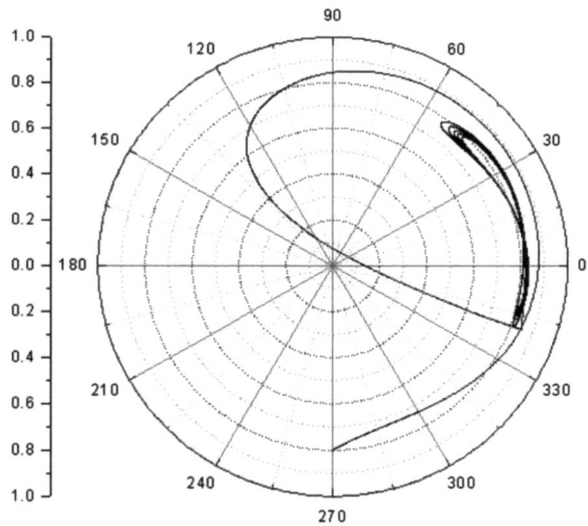

Fig. 13.19 Journal centre locus for unidirectional constant load (Pai et al. [8])

increase in the load-carrying capacity. Therefore, as the groove angle decrease there is an increase in the load-carrying capacity. The improved load is due to higher pressure development in the larger land area due to the smaller grooves. The volume flow rate increases with bearing number and eccentricity ratio. As the journal speed increases the coefficient of friction also increases. The rise in friction is particularly high at low eccentricity ratio, as shown in Fig. 13.23. For smaller groove angles there is a decrease in friction variable. It is seen that smaller groove angle shows a decrease in volume flow rate and friction.

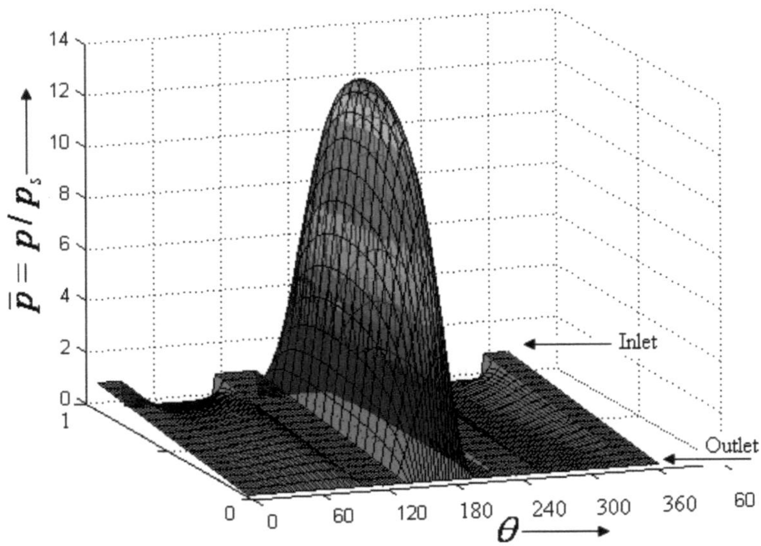

Fig. 13.20 Pressure distribution in bearing clearance for groove angle of 360, L/D = 1.0, $\varepsilon_0 = 0.8$, Bearing number = 10 (Pai et al. [8])

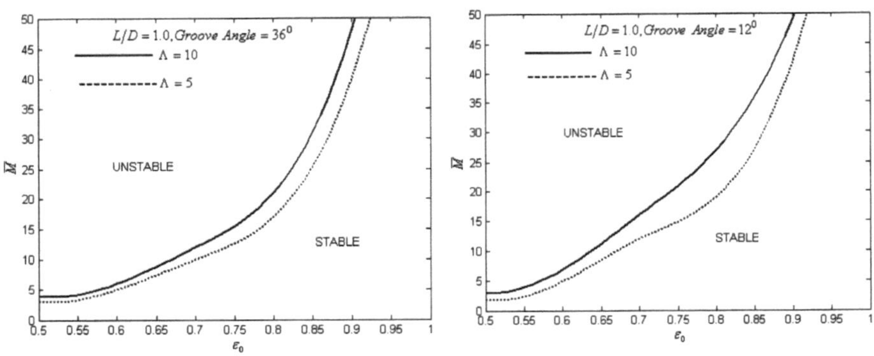

Fig. 13.21 Variation of mass parameter with eccentricity ratio

Figure 13.24 gives the variation of load capacity with bearing number for various length-to-diameter ratios and eccentricity ratios. It is seen that as the speed increases load-carrying capacity also increases with increase in bearing number and L/D ratio. For smaller groove angle, as the groove angle decreases there is an increase in load-carrying capacity.

The Eqs. 13.46 and 13.47 are linear algebraic equations in $\bar{M}$ and λ solutions of these will give $\bar{M}$ and λ. The speed of journal calculated from this value of $\bar{M}$ is the threshold speed, above which the bearing system will be unstable. The whirl speed

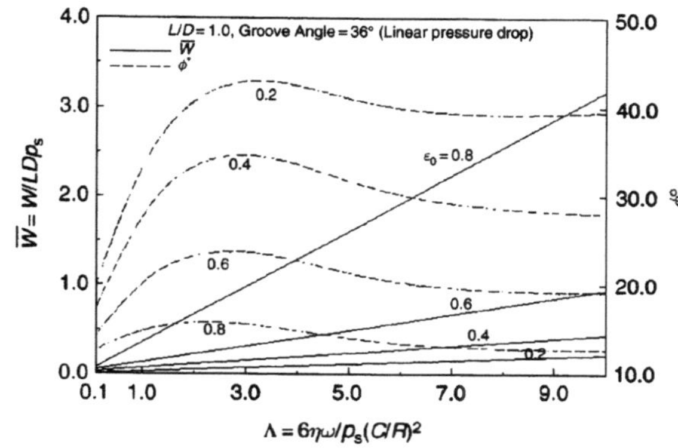

Fig. 13.22 Variation of load capacity and attitude angle with Bearing number (Λ) for various eccentricity ratios ε_0 (Majumdar et al. [26])

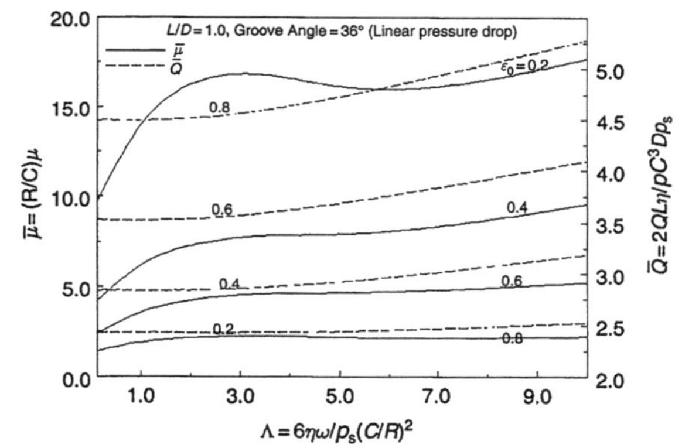

Fig. 13.23 Variation of friction variable and volume flow rate with bearing number (Λ) for various ε_0 (Majumdar et al. [26])

(speed of journal centre) can be found from the whirl ratio for the above journal speed. The journal centre rotates in the same direction as that of the journal. Figure 13.25 shows the variation in mass parameter $\bar{M}$ and whirl ratio λ (the measure of stability).

13.8.2 Experimental Results

The test rig was used to measure the circumferential and axial pressure profiles (Table 13.1) around and along the bearing for various conditions of load, speed, geometry of groove and Poiseuille pressure supply. Theoretical calculations using ESDU 85028 [22] in conjunction with ESDU 84031 [23] were performed to locate the position of the minimum film thickness through the attitude angle and eccentricity ratio. It should be noted that these procedures are based on two axial grooves of finite length positioned at right-angles to the load line and the conventional method of supplying the lubricant that is via the two axial grooves. Table 13.2 shows the results for the 36° groove bearing.

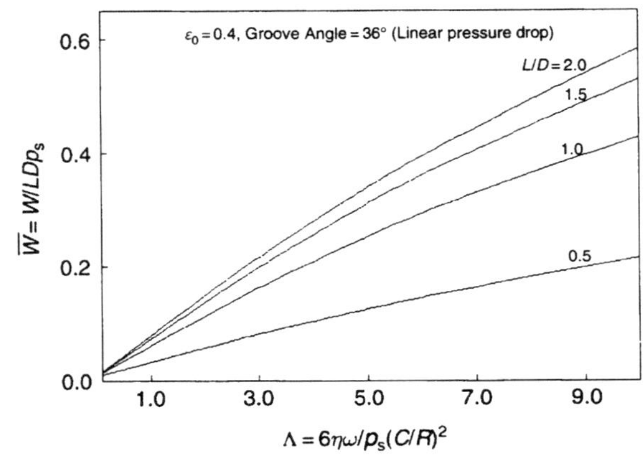

Fig. 13.24 Variation of load capacity with bearing number (Λ) for various L/D (Majumdar et al. [26])

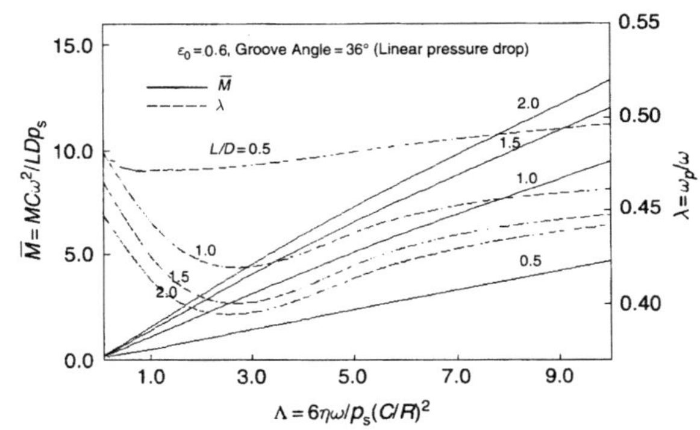

Fig. 13.25 Variation of mass parameter M and whirl ratio λ with bearing number (Λ) for various L/D (Majumdar et al. [26])

Inspection of these figures and tables leads to the following observations.

- Pressure tapping points P14 and P13 are located in a groove at 90° to the load line in the convergent section of the bearing. Pressures for each load are higher than the lubricant supply pressure and are equal in magnitude.
- P12 and P11 are in the converging section of the bearing for all load conditions. Pressure always increases as the film thickness reduces, that is from P12 to P11.
- P10 and P6 are located inside the groove located in the loaded section of the bearing and the pressure always remains constant, again at a value higher than the supply pressure.
- P5 and P3 are located in the diverging section of the bearing and apart for the "no load" case, exhibit a constant pressure at a value higher than the supply pressure. The increased
- Groove Location: P1–P2, P6–P10 and P13–P14, P1–P16 are not equally spaced around the bearing (refer Fig. 13.26)
- pressures for 0 N load probably indicates that the bearing and shaft arrangement wasn't set with zero eccentricity.

Table 13.1 Pressure and readings for 1,000 rpm and 100 kPa-see Fig. 13.17 for locations of pressure tapping points (All pressures are in kPa) [16]

Actual load, N	P1	P2	P3	P4	P5	P6	P7	P8	P9	P10	P11	P12	P13	P14	P15	P16	P8A	P8B	P8C	P8D	P8E	P8F
0	119	119	171	197	123	123	122	121	121	121	121	118	118	119	142	278	121	121	121	121	121	121
995.7	127	126	124	121	124	128	126	126	127	127	129	133	128	128	138	138	132	132	132	132	132	132
1984.6	133	133	130	124	133	135	132	133	133	133	208	207	136	135	139	137	134	134	134	134	134	134
2976.4	134	134	130	129	133	136	133	134	134	134	247	240	137	136	142	137	135	135	135	135	135	135
3971.1	134	134	130	136	135	136	133	134	134	135	300	172	137	137	144	137	135	135	135	135	135	135

Actual load (N)	Flow rate (L/min)	Outlet pressure (kPa)	Inlet pressure (kPa)	Inlet temp. (°C)	Outlet temp. (°C)	RPM
0	7.2	99	109	26	28.5	1,034
995.7	7.2	108	114	26.5	29	1,033
1984.6	7.2	112	120	28	31	1,034
2976.4	7.2	112	121	29	32	1,034
3971.1	7.2	112	121	30.5	34	1,034

Table 13.2 Theoretical calculations for the 36° groove bearing [16]

Load (N)	Speed (RPM)	Eccentricity ratio	Minimum film thickness (mm)	Attitude angle (degrees)
995.7	1,040	0.925	0.0365	28
1984.6	1,040	0.980	0.0097	18
2976.4	1,040	0.990	0.0049	15
3971.1	1,040	0.995	0.0024	13
995.7	1,490	0.920	0.0389	31
1984.6	1,490	0.950	0.0243	22
2976.4	1,490	0.965	0.0170	20
3971.1	1,490	0.970	0.0146	18

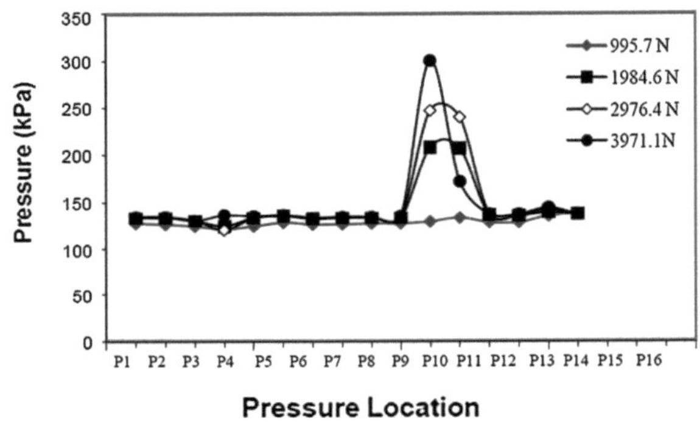

Fig. 13.26 Circumferential pressure distribution of 3 mm, 36° groove (1,000 rpm, 100 kPa) (Hargreaves et al. [16])

- P2 and P1 are located in the third axial groove and pressures were again constant and higher than the supply pressure.
- P16 and P15 are located either in the diverging or the converging section of the bearing depending on the load applied, that is, the resulting attitude angle and this is reflected in the measured pressures.
- P8A to P8F measure the pressure along the groove, and exhibit almost constant pressures at values higher than the supply pressure. There is a sharp drop from the pressure in the groove to the inlet and outlet values.
- Power drawn by the electric motor generally increased as the load and speed increased.
- The temperature rise in the water as it passed through the bearing increased as the load increased and was higher for 1,500 rpm than for 1,000 rpm.

13.8.3 Results From CFD Analysis

The result of the CFD analysis is presented in Fig. 13.27.

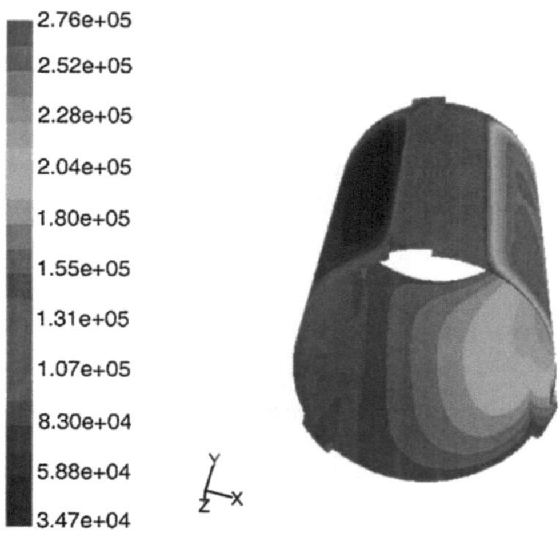

Fig. 13.27 Pressure contours from FLUENT 5.0 for 1,000 rpm, 50 kPa 2976.35 N, 3 mm, 36° groove (Pressure in Pascals) (Pai et al. [25])

13.9 Summary and Conclusions

The advantages of axial groove journal bearings with water as a lubricant over less environmentally friendly oil-lubricated bearings are becoming increasingly recognised and understood. In terms of regular use, preventive maintenance and operating costs can be potentially reduced. For example, the oil in lubricating systems has to be replaced at specified intervals. The usage of oil results in sludge deposits and in accordance to preventive maintenance schedule the machine parts have to be overhauled.

However, recent research has indicated a major difference between the theoretical assumptions regarding the pressure characteristics of the lubricant flow in the grooves, and experimental observations. Once the fundamental reasons for this difference are understood, it will be possible to develop techniques and models to optimise bearing designs for maximum robustness and efficiency.

Hence the primary outcome from the studies presented in this section will be a solid foundation for theoretical modelling of these bearings, as well as interpretation of experimental measurements of journal performance. These results will enable better design of water lubricated journal bearings and thrust bearings.

The outcome will have a large impact on marine usage of motors and rotating equipment world. In addition the spin off will have a great national benefit and a strong potential for attracting industrial interest to exploit the technology further particularly in the designing aspects.

In addition to environmental benefits, the economic potential benefits of optimally designed water lubricated bearings are enormous when the power consumption of pumps, turbines and vessels is taken into account. This is primarily due to the lower viscosity of water, which results in lower running temperature and decreases frictional drag on the mechanical components. This in turn reduces the power required to drive these systems.

This research will be important to many industries and communities in reducing the environmental impact due to mineral-oil- based lubricants. The results of the research can be applied in the design and implementation of water bearings in power stations using water turbines and/or circulating water pumps, which are in locations where oil lubricant spillage or leakage could cause environmental damage. Oil can be an environmental hazard not only in the case of mechanical malfunction, but also in its safe disposal.

References

1. Environment Australia, *Emissions estimation technique manual for aggregate emissions from commercial ships/boats and recreational boats* (Environment Australia, 1999)
2. Queensland Department of Transport, Recreational vessels registered in queensland: by local authority and length. Qld. Dept. Transpt. (2001)
3. Queensland Transport: Marine Pollution Section, Marine oil pollution: its potential impact and control. Qld. Transpt. (1989)
4. R. Pai, Measurement of Clearance of a fluid film bearing, in *Proceedings of the 24th National Conference on Fluid Mechanics*, B. E. College, Howrah, India, 1997, pp. 26–28
5. R. Pai, D.W. Parkins, Measurement of eccentricity and attitude angle of a journal. Presented at the 3rd national conference on fluid machinery, 1998, pp. 12–13
6. R. Pai, D.W. Parkins, Measurement of power absorbed in fluid film bearings. Presented at the 26th national conference on fluid power and fluid mechanics, IIT Kharagpur, 1999, pp. 15–16
7. R.S. Pai, R. Pai, Stability of four-axial and six-axial grooves with water lubricated journal bearings under dynamic load. Proc. IMechE J. Eng. Tribol. **222** (2008). (Part J)
8. R.S. Pai, R. Pai, Non-linear transient analysis of multiple axial groove water lubricated journal bearing. Proc. IMechE J. Eng. Tribol. **222** (2008). (Part J)
9. R.S. Pai, R. Pai, B.C. Majumdar, D.J. Hargreaves, Comparison of static *performance* characteristics of 3-axial groove water lubricated journal bearings: reynolds and JFO models. in *Proceedings of the International Tribology Conference AUSTRIB*, Brisbane, 2006
10. V. Ranjan, R. Pai, D.J. Hargreaves, Stiffness and damping coefficients of 3-axial groove water lubricated bearing using perturbation technique. in *Proceedings of the 5th EDF and LMS Poitiers Workshop, "Bearing Behavior Under Unusual Operating Conditions"*, 2006, pp. Q1–Q6
11. J.F. Sprengel, D.J. Hargreaves, A test rig for the performance assessment of non-metallic water lubricated bearings. International Tribology Conference, Brisbane, Institutions of Engineers, Australia National Conference Publication NCP 90/14, 1990
12. D.J. Hargreaves, A.S. Elgezawy, Optimisation of grooving arrangements in water lubricated non-metallic journal bearings. Trans. IEAust ME18 3 (1993), pp. 245–251
13. A.S. Elgezawy, A theoretical and experimental study of water lubricated non-metallic journal bearings, Queensland University of Technology, Australia, 1996
14. P. Gleeson, Measurement of pressure distribution in a water lubricated journal bearing, Queensland University of Technology, Australia, 1996
15. Y.S. Tay, Performance of grooved water lubricated journal bearing, undergraduate thesis, Queensland University of Technology, Australia, 1997
16. D.J. Hargreaves, G. Chadwick, Experimental and theoretical analyses of combined couette and poiseuille flows in journal bearings with multiple axial grooves. in *Proceedings of the International Tribology Conference*, Nagasaki, 2000
17. R. Pai, D.J. Hargreaves, R.J. Brown, Modelling of fluid flow in a 3-axial groove water bearing using computational fluid dynamics. in *Proceedings of 14th Australasian Fluid Mechanics Conference*, Adelaide University, Adelaide, Australia, 2001

18. D. Vijayaraghavan, T.G. Keith, Effect of type and location of oil groove on the performance of journal bearings. Tribol. Trans. **35**, 98 (1992)
19. P. Shelly, C. Ettles, Solutions for the load capacity of journal bearings with oil grooves, holes, reliefs or chamfers in non-optimum positions, in *Proceeding of the Institution of Mechanical Engineers*, vol. C56/71 (1971), pp. 38–46
20. D.O. Salimonu, A.L. Kistler, R.A. Burton, Turbulent flows near steps in a thin channel with one moving wall. Trans. ASME **99**, 224–229 (1977)
21. R.A. Burton, The straight grooved bearing with turbulence and inertia. Private communication, 2002
22. ESDU International, Calculation methods for steadily loaded axial groove hydrodynamic journal bearings—super-laminar operation, Engineering sciences data item, 85028, London, 1984
23. ESDU International, Calculation methods for steadily loaded axial groove hydrodynamic journal bearings. Engineering sciences data item, 84031, London, 1985
24. R. Pai, B.C. Majumdar, Stability submerged oil film journal bearings under dynamic load. Wear **146**, 125–135 (1991)
25. R. Pai, D.J. Hargreaves, Modelling of fluid flow in the grooves of a 3-axial groove water bearing using computational fluid dynamics. in *Proceedings of 6th International Tribology Conference AUSTRIB*, University of Western Australia, Perth, Australia, 2002, pp. 407–411
26. B.C. Majumdar, R. Pai, D.J. Hargreaves, Analysis of water lubricated journal bearings with multiple axial grooves. Proc. I MECH E. J: J. Eng. Tribol. **218**(2), 135–146 (2004)
27. R.S. Pai, R. Pai, D. J. Hargreaves, in *Static and Dynamic Characteristics of a 3-axial Groove Water Lubricated Journal Bearing*, 4th International Conference on Tribology for Equipment Reliability ICIT, Mumbai, vol. 15–18, 2004, pp. 355–361
28. D. L. Cabrera, N. H. Woolley, D. R. Allanson, Y. D. Tridimas, Film pressure distribution in water lubricated rubber journal bearings. Proc. IMechE, J. Eng. Tribol. **219** (2005)
29. W. Litwin, Water lubricated bearings of ship propeller shafts—problems, experimental tests and theoretical investigations. Pol. Maritime Res. 4(62) **16**, 42–50 (2009)
30. P. Andersson, P. Lintula, Load carrying capability of water lubricated ceramic journal bearings. Tribol. Int. **27**, 315–321 (1994)
31. S. Yoshimoto, Y. Anno, M. Tamura, Y. Kakiuchi, K. Kimura, Axial load capacity of water lubricated hydrostatic conical bearings with spiral grooves. Trans. ASME J. Tribol. **118**, 893–899 (1996)
32. S. Yoshimoto, S. Oshima, S. Danbara, S. Shitara, Stability of water lubricated hydrostatic conical bearings with spiral grooves for high speed spindles. Trans. ASME J. Tribol. **124**, 398–405 (2002)
33. J. Kicinski, Influence of the flow pre-history in the cavitation zone on the dynamic characteristics of slide bearings. Wear **111**, 289–311 (1986)

Chapter 14
Green Tribological Behavior of Waxes, Adhesives and Lubricants

W. Li, X. H. Kong, M. Ruan, F. M. Ma, X. H. Zuo and Y. Chen

Abstract General characteristics of waxes, adhesives and lubricants as well as the recent fundamental investigations on their physical and mechanical behavior are introduced. The current R & D status for new type/generation of waxes, adhesives, and lubricants from natural products is reviewed, with an emphasis on their tribological applications. In particular, some crucial issues and challenges on the technological improvement and the materials development are discussed. Based on the current predicted shortage of energy resources and environmental concerns, prospective research on the development of green waxes, adhesives, and lubricants is suggested.

Keywords Waxes · Adhesives · Lubricants · Green materials

W. Li (✉) · X. H. Kong · M. Ruan · F. M. Ma · X. H. Zuo · Y. Chen
Hubei Key Laboratory of Environmental Pollution Control and Resources Reuse and School of Chemical and Materials Engineering,
Huangshi Institute of Technology, Huangshi 435003,
People's Republic of China
e-mail: wenl@ualberta.ca

X. H. Kong
e-mail: xkong@ualberta.ca

M. Ruan
e-mail: rm@126.com

F. M. Ma
e-mail: fmma1@yahoo.cn

X. H. Zuo
e-mail: xhzuo@126.com

Y. Chen
e-mail: ychen21@163.com

M. Nosonovsky and B. Bhushan (eds.), *Green Tribology*,
Green Energy and Technology, DOI: 10.1007/978-3-642-23681-5_14,

14.1 Waxes

14.1.1 Introduction

The term wax is used to denote linear or branched-chain of hydrocarbon (up to about C40H82) mixed with organic and inorganic material [1]. From the applied perspective, waxes are defined as mixtures of lipophilic compounds that are solid at room temperature, range from transparent to opaque, and are ductile and easy to polish [2].

Waxes are used as additives into lubricants, adhesives, coatings, sealants, impregnation materials and adjuvants, etc. A wide range of commercial products rely on waxes, including automobiles, textiles, papers and specialty inks, pesticides, candles, plastics and wood–plastic composites, furniture and shoe polish, household cleaners, cosmetics, dental treatment products, drugs and food (chewing gum, cheese packaging, confectionery coating) [2]. Waxes are currently generated by (1) chemical syntheses, (2) obtained from geological deposits originating from fossils, or (3) obtained from living organisms (recent waxes). The vast majority of waxes are based either on alkane or on ester structures. Synthetic waxes are mainly generated by the Fischer–Tropsch process ($CO + H_2$) and olefin (ethylene, propylene) polymerization, producing mixtures of normal and branched alkanes [3]. Fossil waxes are extracted from crude oil and coal deposits, and consist of alkanes and alkyl ester mixtures (together with the corresponding free acids and alcohols), respectively.

Beeswax and wool wax are the prime commodities of natural waxes from animal sources [4]. The most important plant sources for commercial production of waxes are carnauba, candelilla, ouricouri, sugar cane and the desert shrub jojoba. All the green waxes are relatively rich in aliphatic esters, with varying overall chain lengths of both acyl and alkyl groups, and contains characteristic admixtures of cinnamates, hydroxyesters and lactones, steryl esters, estolides and alkanes [5–7]. Waxes found in the cuticle of plants are mainly composed of alkenes, fatty alcohols, secondary alcohols, ketones and fatty acids. They consist of carbon chains predominately of a length between C20 and C35 and are derived from very long chain fatty acids [8].

Physical parameters used to characterize waxes include hardness, cure speed, melting point or range, melting point, pour point, viscosity, surface tension, adhesive strength, optical transparency and durability, and thermal expansion coefficient [9–12]. Molecular weight largely controls these fundamental properties for most wax products. For example, Alpha olefin (AO) waxes with low molecular weight are relatively soft but have desirable high melt flow, whereas, high molecular weight waxes such as polyethylene waxes are extremely hard, have desirable high physical strength, but their viscosity is usually undesirably very high. The degree of crystallinity of waxes is also a significant parameter in defining their properties [13]. Scarce studies are found in the literature relating composition and chemical structure to crystallinity [14–16].

Waxes tend to contain wax esters as major components, and other lipid constituents which can vary greatly with the source of the waxy material, including hydrocarbons, sterol esters, aliphatic aldehydes, primary and secondary alcohols, diols, ketones, β-diketones, triacylglycerols (TAGs) and many more. Wax esters are esters of long chain fatty alcohols and fatty acids. Wax esters have lubrication properties that are superior to triglycerides, due to their high oxidation stability and resistance to hydrolysis. The main natural sources of wax esters have been the seeds of jojoba [17] and spermaceti oil (sperm whale oil) [18, 19]. The sulfurized form of spermaceti oil was used as an additive in many lubricant applications [20]. Wax esters constitute around 30% of the sebum and meibum that is produced in the sebaceous and meibomian glands and are secreted onto the surface of the skin and eye of some mammals [21, 22]. Sebum has important functions in lubricating the skin and contributes to the development of a water barrier. [23–25].

14.1.2 Present R & D Progress

Vegetable oil is widely studied as a desirable and important candidate as raw material for wax production [26, 27]. Vegetable oil-based waxes are mainly achieved by partially or fully hydrogenating vegetable oil (PHVO or FHVO). For example, castor wax is produced by the hydrogenation of pure castor oil with nickel catalyst [28]; partially hydrogenated soy bean oil gives soy wax [29]; alcoholized palm oil gives palm wax [30] and jojoba wax is prepared by the hydrogenation or interesterification of Jojoba oil [31]. Some vegetable oils, such as sunflower oil [32], canola oil [33] and Rice bran oil [34] contain waxes which are undesirable in the edible oil and are therefore removed during the purification process.

TAGs are the major component in most vegetable oils and waxes. Many commercially utilized TAGs and free fatty acids are obtained preferably from plant sources, including soybean, cotton seed, corn, sunflower, canola and palm oils [35]. The TAG component is a complex mixture of TAG compounds, which very often are derivatives of C16 and/or C18 fatty acids [36]. Whether one refers to TAGs as wax, fat, or oil depends upon the ambient temperature at which the characterization is made. The physical properties of a TAG are primarily determined by the chain length of the fatty acyl chains, the amount and type (cis or trans) of unsaturation present in the fatty acyl chains, and the distribution of different fatty acyl chains among the TAGs that make up the fat or oil [37]. The highly saturated TAGS are typically solid at room temperature, while those with high proportion of unsaturated fatty acyl chains tend to be liquid.

Hydrogenation of TAGs can change the phase statue by varying the degree of unsaturation, and the melting, crystallization and fluidity characteristics of a TAG can be varied by changing the distribution of fatty acyl chains in the TAG, e.g., by blending together materials with different fatty acid profiles. Chemical modification of vegetable oil or partially hydrogenated vegetable oil also is used to attach functional groups on the acyl chain to achieve practical applications, such as in candles [26, 27, 37–39] emulsion coatings [40] and hot melt adhesives (HMAs) [35]. Despite such efforts at modification, many of the physical characteristics of waxes still prevent them from being used successfully commercially. For example, the major problems for soy wax used as candle materials are its greasy surface texture and brittle structure. Also poor melting and recrystallization properties of soy wax result in delayed resolidification and therefore excessive dripping of wax. Fully hydrogenated soybean oil (SBO) exhibits improved melting properties but owing to its fragile texture, it is not yet acceptable in most candle applications. Rezaei et al. [26] improved the solidification properties of KLX (a wax composed of fractionated hydrogenated soy and cottonseed oils) candles by adding free fatty acid (C16, C18, C18:1), but another additive is needed to improve its compressibility. Wang and Wang [27] introduced functional groups into the fatty acyl chain of PHVO by epoxidation, ring-opening reaction and esterification to improve its cohesiveness and thermal properties. The melting range of modified HPVO was increased; however, its hardness was lower compared to those of commercial paraffin wax.

Investigation of structure–property relationship of vegetable oil-based waxes would be advantageous to direct the modification of available vegetable oil-based waxes based on given purposes. There is great potential in exploring the possibility to mimic the properties of natural waxes. Research can be directed to provide methods and develop alternatives to complex lipids from a low cost source, such as fats and oils. It is hypothesized that waxes with integrated properties can be achieved by changing the composition of the waxes. A recent study, for example, reported semi-synthetic lipid mixtures that mimic the complex lipid composition, organization and thermotropic behavior of vernix caseosa lipids, a waxy white substance found coating the skin of newborn humans [41].

It has been proposed recently to use agricultural products for the development of green HMAs. However, to date ideal green raw materials for the replacement of the petroleum-based ingredients such as paraffin wax in a HMA formulation have never been identified. Very recently, we propose to use soybean wax to replace the paraffin wax and further conduct an experimental investigation on its crystalline structure, thermal and mechanical behavior and crystallization morphology by XRD, DSC, NMR, Texture Analyzer and PLM, respectively, with an emphasis on the comparison between soybean wax and paraffin wax. We demonstrate that compared to the paraffin wax, the soybean wax shows a natural difference in molecular and crystalline structures. Such an essential structural difference may result in different performances between a formulated HMA containing the soybean wax and a conventional HMA containing paraffin waxes, but this may also offer a promising advantage to modify the functionalities of conventional HMAs, such as hardness. We further show that the soybean wax exhibits some similar or even unique characteristics in melt/crystallization temperature and onset time of crystallization, hardness and microstructures. The above results indicate that a green HMA containing soybean waxes may keep an original or exhibit, at least, an acceptable adhesion performance such as strength and hardness, since its microstructure is similar to that of the paraffin waxes and can be adjusted by heat treatment parameters such as cooling rate. In particular, from the viewpoint of hardness, the soybean wax may be a practical substitute of paraffin waxes since it exhibits a good mechanical performance (At least, its toughness is better than that of the paraffin 158 waxes). We therefore suggest that the soybean wax could be a suitable substitute of the paraffin wax, although it is still an open issue whether or not a green HMA containing the soybean wax would show a better practical adhesion performance. Furthermore, we argue that aiming at a specific purpose, such as improvement of green compositions, biodegradability or low-temperature application, the soybean wax is promising for the formulation of a green HMA with an acceptable functionality.

14.1.3 Prospective Research

The cosmetic industry is the largest user of wax today. Jojoba is a main source for wax in this industry. The other major industry using waxes is the pharmaceutical sector. Cost and availability suggest a need to develop alternative waxes to petroleum oil the dominating raw material for lubricants. Jojoba-type of wax is a suitable choice. Unlike most seed-bearing plants, the oil accumulating in the jojoba seed contains almost all wax esters (97%) and less than 3% TAGs [42–44]. The main wax esters in jojoba oil are C38–C44, which are in turn composed of mono unsaturated fatty acids and fatty alcohols of 20 and 22 carbon chain length (with one double bond on each side of the ester bond). Jojoba oil contains no trans isomers and is highly resistant to oxidation. In particular, it requires little or no purification and can absorb larger amounts of sulfur. It also imparts unique and beneficial properties when used for a variety of industrial purposes. This oil has outstanding qualities for lubrication applications but is a too expensive commodity for use in other products than specialized and high value product areas such as cosmetics and/or specialty lubricants. Furthermore, the current price and availability of this wax make them unsuitable for widespread use for the lubricant market. As a result, lubricant applications of jojoba are very limited. These are special applications such as high pressure, high temperature lubricants for use in gearboxes, differentials, crankcase lubricant and cutting oils. With the synthetic ester segment of the lubricants the market is quite substantial with polyalfaolefins at around 0.5 million tonnes per year and organic esters 0.05 million tonnes per year.

To change the current situation, it is therefore necessary to develop economic, versatile, green waxes for lubricant applications. C22–30 branched-chain fatty acids are very interesting in formulating lubricants for lowering the melting temperature of the fatty acids and thereby the pour point of base oils without adding oxidative instability. The presence of a branched-chain fatty acid as well as a double bond on the hydrocarbon chain broadens the liquid–solid phase transition by reducing the melting temperature. The branching is much more effective if it is positioned internally rather than terminally [45]. Branched structure can lower melting point, increase hardness and viscosity, but linear polymers have higher melting point. Also, the introduction of elements (such as S, P, Si and N) can have effects on the properties of waxes. Phosphorous or sulfurized derivatives can improve friction, wear and extreme-pressure properties; silicone waxes may provide the properties of wide range thermal stability, resistance to oxidation, low surface tension, gas permeability, excellent dielectric properties, physiological inertness and moisture resistance [46, 47]. And the end functional groups (OH, COOH) may optimize the properties of waxes similar to addition of additives as reported by Rezaei et al. [26]. Hydroxy fatty acids are presently used in lubrication formulas as additives in order to improve low-temperature, improve viscosity index and oxidative stability properties in way very similar to branched fatty acids. Bio-based hydroxy fatty acids are synthetically produced by introducing hydroxy groups in the hydrocarbon chains of common plant oils by a chemical catalyst [48].

Therefore, to develop ideal green waxes, it is expected that changes in the composition of a natural vegetable oil by blending different oils, or adding acid or alcohol then applying reactions, such as interesterification and hydrogenation can vary the resultant properties of waxes. To this end, a series of triglycerides can be synthesized by the reaction of fatty acids containing different length and/or different degree of saturation in fatty chains (C12–C22:x, x = 0,1,2) with 1,3-dihydroxylacetone with or without catalysis and their wax-like physical properties could be obtained. This would be very helpful for waxes blends, modification of waxes based on the given purposes. The resulting compounds can also be used as additives to optimize certain property of waxes.

Particularly, the fatty acids (C12–C22:x, x = 0, 1, 2), 1,3-dihydroxylacetone can be employed as starting materials. Two or three steps reaction of fatty acids with 1,3-dihydroxylacetone and one-step reduction of ketone by $NaBH_4$ may be carried out based on the targeted compound. Ultimately, one can develop a database of triglyceride stocks linking molecular ensembles, processing protocols and physical functionality. To produce jojoba-type wax esters one can use long chain fatty alcohols that can be transesterified with long chain fatty acids. To synthesize jojoba-type compounds, unsaturated fatty acids with a carbon–carbon bond at position 9 (omega-9), such as Oleic acid (*cis*-9-octadecenoic acid, 18:1), Eicosenoic acid (*cis*-11-eicosenoic acid, 20:1) and Erucic acid (*cis*-13-docosenoic acid, 22:1) and Nervonic acid (*cis*-15-tetracosenoic acid, 24:1) can be employed as starting materials. The fatty acids could be reduced by LiAlH4 to give relative fatty alcohols. Then the resulting fatty alcohols are reacted with the original fatty acids to give targeted wax esters. A maximum of 16 wax (esters) can therefore be synthesized and should be tested for lubricating properties and compared to jojoba wax ester. These wax esters could be considered for much larger applications such as automotive, transmission and hydraulic fluids.

14.2 Adhesives

14.2.1 Introduction

Adhesive or glue is a compound in a liquid or semi-liquid state that adheres or bonds items together. The types of materials that can be bonded using adhesives are virtually limitless, but they are especially useful for bonding thin materials, such as metals, plastics and wood [49–56]. Since adhesives are used to adhere to the same or two different substrates, the most important concern should be paid to the adhesive or bonding property at interfacial regions where adhesive coatings or films locate. In most applications, acceptable bonds are those that produce failure of one or both substrates when pulled apart. Adhesion depends on a number of factors. The most important factors are the wet-out of the substrate, particularly by the polymer component of the adhesive system, and the specific adhesion with the substrate. To study the open- and set-time, which are crucial technological or process parameters and can determine various properties and functionalities of an adhesive, the effect of waxes has been intensively investigated. It was found that both open- and set-time are primarily dependent on the choice of wax and its compatibility in an adhesive system. In general, waxes with a higher recrystallization temperature and with a marginal compatibility in a HMA system will set more quickly. In particular, a recent patent argued that N-α olefin waxes instead of paraffin waxes can substantially improve open-time [50, 51]. However, since the effect of wax and its interactions with other additives are extremely complicated, in practice, both open- and set-time also depend strongly on material properties of other components of the adhesive (e.g., Slark et al. [52–55] used a high molar weight acrylic polymer to achieve long open-time). In addition, waxes are also critical to HMAs performance in terms of thermal stability and mechanical properties.

Most adhesives are synthesized from petroleum base stocks, which have many advantages such as high bond strength and high water resistance. But there are significant environmental problems associated with the use of various adhesive since most of them contain urea formaldehyde and/or phenol formaldehyde and a chemical agent that causes health hazards [50–53]. For instance, in a conventional solvent-based adhesive formulation, solvents such as toluene and trichloroethane are volatile organic compounds (VOCs), which makes them ideal as a carrier fluid, and also causes environmental and safety concerns [54, 55]. Another environmental issue that we are facing is a serious global supply shortage for petrochemical products because petroleum resources are finite and are becoming limited. Meanwhile, such petrochemical products are not biodegradable and hence very harmful to our environments. It is therefore necessary to explore a strategy to replace petroleum-derived ingredients in adhesive formulations.

Fortunately, nature has provided us an opportunity. Natural products are abundant and always available from renewable resources. Utilization of these products as adhesive will help to overcome these problems and add value to agricultural by-products. Natural adhesives such as starch, casein and protein were used as wood adhesives as early as 1900s [56, 57]. The exploratory use of these adhesives mainly for plywood and fiberboard production has attracted increasing interest from both academic and industrial aspect because of their specific advantages, including ease of handling and processability by both hot and cold press [58]. However, their utilization is currently limited because of low durability, especially low water resistance. To overcome such problems, scientists are trying to develop new polymeric adhesives from renewable resources [59, 60].

14.2.2 Present R & D Progress

Adhesives are traditionally produced by reacting petroleum-based polyols with isocyanates. Since the beginning of the twenty first century, the increasing concern over environmental pollution has forced the industry to develop environmentally friendly adhesives [61]. For such adhesives, the green requirements mainly include: renewability; biodegradability and repulpability (i.e., water dispersibility and releasability). Although for many years researchers have been trying to develop polyurethane coatings [62, 63] and interpenetrating networks from natural oils [64–66], there are very few studies dealing with the use of vegetable oil-based polyurethanes as adhesives. Studies on new natural polymer-based adhesives are necessary for the development of adhesives in general. To the best of our knowledge, until fairly recently very few examples could be found in the literature where seed oil derivatives were being employed as adhesives. It is also worth mentioning that these products are not used industrially/commercially as yet, but some use of them can be envisaged in the future. One instance of adhesive systems produced from potato starch/natural oil-based polyester polyol and toluene 2,4-di isocyanate was reported by Desai and co-workers [51, 52]. They found that the shear strength of the vegetable oil-based adhesive performed better than the commercially available adhesives for wood joints. They also discovered that as the hydroxyl value of the polyol increased, and bond strength increased due to formation of a more highly cross-linked structure. This factor was also confirmed by Somani et al. [53] who produced adhesives by reacting a castor oil derivative and glycols (several types) with isocyanate adducts. They reported that adhesives made from castor oil-based polyol with shorter glycols segments having a NCO/OH ratio 1:3 gave excellent shear strength compared to other homologous adhesives and was 10 times higher than that found for the commercially available wood adhesives. Again, these are due to the fact that the cross-linking density of this adhesive is the highest, which indicates that cross-linking density of the polyurethane materials has significant effect on the performance or the behavior of a wood adhesive system.

In both of the cases mentioned above, the location of the hydroxyl groups of the vegetable oils is in the middle of the fatty acid chains. The pendant chains act as dangling chains when these polyols are cross-linked, resulting in significant steric hindrance to further cross-linking, and hence decreasing the cross-linking density. Recently, with the advent of new types of polyol with terminal functional groups from canola oil using ozonolysis and hydrogenation technology, the effect of pendant chains has been significantly reduced [62–65, 67, 68]. This type of polyol has been used successfully to produce polyurethane which had better thermo-mechanical and mechanical properties than the corresponding polyurethane made from commercially available bio-based polyol. Furthermore, it has been reported that the cross-linking density of canola oil polyol-based polyurethane is much higher than that of castor oil-based polyurethane with the same NCO/OH molar ratio [66].

14.2.3 Prospective Research

Based on the above literature review, we therefore suggest to develop green adhesives from natural oil derivatives. In particular, it is very important to synthesize canola oil-based adhesives, which should be friendly to human being and can be applied to various wood substrates. For this purpose, it is necessary to investigate the physical and chemical properties of the adhesives, especially adhesive strength, green strength, chemical resistance and thermal stability. Furthermore, it is also needed to use various methodologies (such as dynamic mechanical analysis and swelling), to study the effect of various OH/NCO ratios on the physical and chemical properties of the developed adhesives, to determine the role of chemical structure of different polyols (e.g., dangling chains) in mechanical and thermal mechanical properties of the resulted adhesives, and to evaluate the feasibility of cross-linking density of the adhesives. Based on these fundamental investigations, one can optimize the formulation and design new type green adhesives.

Specifically, to investigate mechanical properties and the comprehensive performance of the obtained canola oil-based polyurethane adhesives, various wood substrates (hardwood and softwood) can be chosen as typical adherends. To further investigate bonding or adhesion behavior of the developed polyurethane adhesives and their compatibility with different substrates, other materials such as rubber, ceramic and even metals and alloys can also be used as adherends. Based on different adherends materials, an interesting comparative study on various chemical, physical and mechanical behaviors of adhesive joints or their interfacical regions should be conducted both experimentally and theoretically. Physical functionalities and green requirements of the new oil-based adhesives can be addressed experimentally although available approaches are insufficient. In particular, the viscoelastic properties and thermal stability should be measured (e.g., using TA Instruments DMA or Rheometer). Based on the experimental measurements, physical mechanisms responsible for the viscosity and thermal behavior can be suggested, and predictive models may be further proposed to link joint/bonding structure to these specific physical properties.

14.3 Lubricants

14.3.1 Introduction

A lubricant is a substance (often a liquid) introduced between two moving surfaces to reduce the friction between them, improving efficiency and reducing wear. It may also have the function of dissolving or transporting foreign particles and of distributing heat. Most lubricants currently originate from petroleum base stocks that are toxic to the environment, making it increasingly difficult for safe and easy disposal. It is claimed that 50% of all lubricants sold worldwide end up in the environment via spill, volatility or total loss applications, which contaminate soil, air and drinking water [69]. Environmental protection from pollution can be performed by either preventing undesirable losses, reclaiming and recycling mineral oil lubricants, or using environmentally friendly lubricants. Because of the concerns about pollution, as well as the depletion of petroleum reserves, the uncertainly in petroleum supply and increasingly strict government regulations, there has been increasing demand for environmentally friendly lubricants in the past decade [70]. The non-toxic lubricant composition is especially useful for application to sports equipment which comes into contact with the human body. Another application concerns the use of the non-toxic lubricant composition in connection with food processing equipment which comes into direct contact with the food substrate [71].

Vegetable oils can be considered environmentally friendly because these materials are non-toxic, renewable resources and lower cost alternatives to synthetic fluids. They have low volatility, excellent lubricity, favorable viscosity temperature characteristics and higher solubilizing capacity for contaminants and additives than mineral base oils [70]. The ester linkages deliver inherent lubricity and enable the oils to adhere to metal surfaces. In all of these technical properties, the vegetable oils are comparable or better than mineral oils [69]. Vegetable oils have demonstrated to be potential as use for biodegradable lubricants in applications that include engine oils, hydraulic fluids, and transmission oils [72–76]. Biodegradability provides an indication of the persistence of any particular substance in the environment and is the yardstick for assessing the eco friendliness of substances [77]. However, extensive vegetable oils use is restricted due to their poor fluid flow behavior, solidification at low-temperatures [78, 79], low oxidative [80–83] and hydrolytic stability. The most serious problem is their poor oxidative stability, primarily due to the presence of bis allylic protons. These protons are highly susceptible to radical attack and subsequently undergo oxidative degradation to form polar oxy compounds, which ultimately result in insoluble deposit formation and an increase in oil acidity and viscosity [70]. Vegetable oils also show poor corrosion protection and the presence of ester functionality render these oils susceptible to hydrolytic breakdown.[1] Some of these problems can be resolved by avoiding or modifying polyunsaturation in triacylglycerol structures of vegetable oils. Genetic modification [84] by reducing or eliminating unsaturation [76, 85], and use of additives (antioxidants and pour point depressants) [86] can overcome these shortcomings to some extent. Another way to improve these properties of vegetable oils is chemical modification of fatty acid chains of triglycerides at sites of double bond and carboxyl groups. One group of the polyesters based on fatty acids are so-called estolides. Estolides result from ester bond formation between a hydroxyl or olefinic group of one fatty acid and the terminal carboxyl group of a second fatty acid. These and other estolides are biodegradable and frequently discussed to be applicable as lubricants [87, 88]. The other examples are hydrogenation [89], esterification [75, 90], epoxidation [78, 87, 91], metathesis [92–94], acylation [84] and transesterification [95, 96] that all have been used in order to synthesize an improved product.

14.3.2 Present R & D Progress

Some vegetable oils such as SBO, linseed oil and rapeseed oil have a high degree of unsaturation, depending on the amount of linoleic and linolenic acid derivatives. As a result, the thermo-oxidative stability of these oils is poor and leads to polymerization, resulting in gummy and resinous products at elevated temperatures. Their use as lubricants without reduction of unsaturation can cause deposit formation, corrosive action, and damage with relatively short useful service life. One of the several ways to reduce unsaturation is to partially hydrogenate [89, 97] these vegetable oils to improve their service lives without affecting the freezing points to a large extent. Epoxidation [98–101] is another convenient method to improve the poor thermo-oxidative stability caused by the presence of unsaturated double bonds in vegetable oils.

Additionally, neat vegetable oils also pose some problems when subjected to extended use in internal combustion engine. These problems are attributed to high viscosity, low volatility and polyunsaturated character of neat vegetable oils, which could be substantially reduced, by subjecting chemical modifications at the carboxyl group in triacylglycerol molecule of vegetable oils. Transesterification [95, 102, 103] is one of the common ways to chemically break the molecule of the raw vegetable oils into their methyl or ethyl esters with glycerol as a by-product. Transesterification reactions of the partially hydrogenated and cyclized ester vegetable oils are important because these reactions yield monoesters of vegetable oils with better thermal stability and lower freezing points than the vegetable oils as such [104–106]. The glycerol obtained as a by-product, commands as much market value or more than the monohydric alcohols used for transesterification. Therefore, transesterification leads to value addition without much cost escalation.

The most efficient method to improve the lubricating properties of vegetable oils is to combine both reactions at double bond and carboxyl position of to modify the triacylglycerol molecules in vegetable oils. Erhan and co-workers [70] have chemically modified SBO to produce SBO-based lubricants with good oxidative stability and low pour point. They [78, 91, 99] performed the reaction of epoxidized SBO with various alcohols in the presence of an acid as a catalyst to give a ring-opened intermediate product with the structure of–CH(OR1)CH(OH)—where the R group is different alkyl group. Then they esterified the hydroxyl groups by acid anhydride. Based on different alkyl group, they found that pour point decreased and these changes significantly increased the performance of lubricant material. Later on, the same group [70] also reported a simple one-step synthesis of potentially useful acyl derivatives of SBO. These derivatives were prepared using epoxidized SBO and various common acid anhydrides. The process retained the vegetable oil structure while it allowed for removal of polyunsaturation in the fatty acid chain. The lubricants formulated with these chemically modified SBO derivatives exhibited superior low-temperature flow properties, improved thermo-oxidative stability, and better tribological properties [69].

Also Kodali at Cargill central research [107] has introduced the family of asymmetric diester compounds from renewable raw materials. He claimed that these lubricants are non-toxic, biodegradable, cost-effective and environmentally friendly that exhibit excellent lubrication properties such as low viscosity, high flashpoint and high oxidative stability.

Recently, Erhan and co-workers [108] reported the synthesis and evaluation of a series of diesters derived from oleic acid in three-step reactions which include esterification of oleic acid, epoxidation of double bond and ring-opening esterification. The achieved diesters have one ester group at the end of the chain and another one in the middle of the chain. It was discovered that low-temperature properties of these diesters were improved with the increasing chain length of both esters. Decreasing chain length of the mid-chain ester had a positive influence on the oxidation stability. The other effort of this group was to prepare oleochemicals from commercially available methyl oleate and common organic acids; and determine their lubricant properties [69]. These branched oleochemicals are characterized as 9(10)-hydroxy-10(9)-ester derivatives of methyl oleate. These derivatives showed improved low-temperature properties, over olefinic oleochemicals, as determined by pour point and cloud point measurements.

In the similar process, Salimon and Salih [109] have reported several diesters compounds, synthesized from commercially available oleic acid and common fatty acids. The key step in the three-step synthesis of oleochemical diesters entails a ring-opening of epoxidized oleic acid with different fatty acids (octanoic, nonanoic, lauric, myristic, palmatic, stearic and behenic acids) using *p*-toluenesulfonic acid (PTSA) as catalyst. All of these produced compounds are solid except 9(10)-hydroxy-10(9)-(octanoyloxy) stearic acid and 9(10)-hydroxy-10(9)-(nonanoyloxy) stearic acid. The esterification reaction of these compounds with octanol was further carried out in the presence of 10 mol-% H_2SO_4 giving diester compounds. This study described a systematic approach to improve the low-temperature pour property of oleic acid derivative. The presence of long chains alkyl group in the middle of the molecule has strong influence on the pour point property by disrupting the stacking process and results in improved low-temperature behavior. Therefore it is plausible to use the prepared products as biolubricant base stock oil. Also this group [110] has published another similar paper that synthesized the same procedure but using 2-ethyl hexanol in final esterification. They found out that one of the products, the behenic ester of 2-ethylhexyl hydroxy stearate showed pour point, flash point and viscosity indices of −53, 161°C and 215 cp, respectively, which was favorable properties in the synthesis of a bio-based lubrication base fluid. Overall, the data indicated that most of these synthesized derivative compounds had significant potential as lubricant base oil.

Erhan and co-workers [111–113] also reported the preparation of branched-chain ethers from oleic acid in three-step synthesis. First esterification of Oleic acid followed by epoxidation of double bond and then ring-opening step by an alcohol in acidic medium. Low-temperature performance has been studied for different derivatives [111, 112]. Further studies such as oxidation stability and lubricate behavior have been performed [113]. The same group also studied the oxidative stability of synthesized epoxidized methyl oleate (EMO), epoxidized methyl linoleate (EMLO) and epoxidized methyl linolenate (EMLEN), as well as that of a commercial epoxidized SBO, and epoxidized 2-ethylhexyl soyate [114]. They found that under some conditions, epoxy fatty methyl esters epoxides showed increased stability and friction reducing properties which enable them to outperform their methyl ester analogs, and therefore have significant potential to be used as a fuel additive or lubricating fluid.

Because of the development of a suitable molecule of lubricant via chemical synthetic pathway alone will be expensive and a lengthy process, Erhan and co-workers [115] applied molecular modeling of desired compounds and subsequent computation of their minimum energy profile, steric environment and electron charge density distribution, prior to actual synthesis, can shed valuable information on their physicochemical properties. They claimed that based on such information, chemical synthesis can be focused only on the promising molecules. Calculations based on equilibrium geometries were optimized using AM1 semiempirical molecular orbital models. They observed that ring-opening of the triacylglycerol epoxy group and subsequent derivatization of the epoxy carbons can improve the oxidation and low-temperature stability of these synthetic lubricant base oils.

14.3.3 Prospective Research

Based on the above review on the literatures, it can be seen that the efficient way to improve the physical properties of oil-based lubricants is to alter the chemical structure of the substances. Particularly, high degree of branching can lower viscosity index, whereas high linearity can lead to high viscosity index, and relatively poor low-temperature characteristics. In contrast, low saturation can limit oxidation stability, whereas high saturation can result in outstanding oxidation stability. Therefore, there are two main methods to develop ideal green lubricants. One is the modification of the carboxyl group, where the carboxyl group (COOH) of the fatty acid chain is the etherified to glycerol esters and the fatty acids can be obtained by cleavage of such esters, leading to the improved oxidation stability and viscosity. The process is performed on a large scale and is normally catalyzed with acidic or basic catalysts. Another is dodifications of the fatty acid chain. Due to high reactivity, the double bonds of unsaturated fatty acids and their derivatives can form excellent starting points for further modification, such as elective hydrogenation, dimerization/oligomerization, formation of C–C and C–O bonds (branching).

Based on the two general principles, it is expected that both of SBO and oleic acid may be ideal candidates for synthesis of the vegetable oil-based lubricants because for the two oils, epoxidation of double bond, and then conversion of the epoxide ring to ether or ester groups by ring-opening reaction can be easily realized. In the case that alkyl oleate is used as the starting material, OR and OH group should be placed on the double bond site and the physical properties of resulting lubricant should be the best one. Based on this point, it is therefore suggested to synthesize the diester compounds from oleic acid and different diols as the precursor for design of green lubricants with comprehensively excellent tribological properties.

Specially, the first step is the esterification of oleic acid by diol reactant. In this step two ester groups in the middle of chain can be achieved. The second step is the epoxidation of two double bonds to produce a middle compound. Last step should be the ring-opening of epoxide group. To optimize the final compound, it could be helpful to use different diols two groups in the last step. It is expected that two OH groups in the chain may cause the lower cloud point in these compounds. If the cloud points are not satisfied, the other possible way could be the conversion of OH group to ether or ester. Optimization of the reaction condition can be justified based on the evaluation of the physical properties of resulted compounds, including density, viscosity, viscosity index, flash point, pour point, cloud point and oxidative stability as well as tribological performance.

14.4 Summary

Over the past years substantial efforts have been devoted to make environment friendly green waxes, adhesives and lubricants in order to meet multiple technological, economical and social requirements. Such objective is becoming more and more imperative and urgent in the future. It is of considerable interest in both academic research and practical applications to explore the use of renewable resources or various natural products such as crops and vegetables to replace petroleum-derived ingredients in the current products. However, to use the natural resources in a feasible and economical manner we are facing a great challenge for the development of suitable raw materials and/or ingredients for new generation of waxes, adhesives and lubricants. To achieve the proposed aim and to overcome the challenges, first of all is to develop various new chemical methods to synthesize natural ingredients based on different natural products, e.g., natural waxes from various plant oils such as canola, soybean, cottonseed and palm. Furthermore, it is also necessary to employ various approaches to completely understand the correlations between composition, microstructure and property of a wax, adhesive or lubricant and the interactions between various ingredients in a blending system, with an emphasis on the various relationships between functionality, structural parameter and processing condition, such as crystal size, shape, volume fraction, modulus, strength, toughness and setting speed. Based on the review of the current R & D, it has been recognized that thorough theoretical research is evidently lacking in all the aspects of this field, particularly, synthesis, composition/ingredient design, structure–property relationships, and engineering performance.

Acknowledgments The authors would like to acknowledge the financial support from the Natural Science Foundation of Hubei Province of China (2010CDA026), and the Science Foundation of Hubei Provincial Department of Education of China (Z20104401).

References

1. A. Fasano, L. Fusi, S. Correra, Mathematical models for waxy crude oils. Meccanica **39**, 441–482 (2004)
2. R. Jetter, L. Kunst, Plant surface lipid biosynthetic pathways, their utility for metabolic engineering of waxes, hydrocarbon biofuels. Plant J. **54**, 670–683 (2008)
3. H. Schulz, Short history, present trends of Fischer–Tropsch synthesis. Appl. Catal. A **186**, 3–12 (1999)
4. A.P. Tulloch, Beeswax—structure of esters, their component hydroxy acids diols. Chem. Phys. Lipids **6**, 235–265 (1971)
5. P.J. Holloway, Surface lipids of plants and animals, in *Handbook of Chromatography, Lipids*, ed. by H.K. Mangold, G. Zweig, J. Sherma (CRC Press, Boca Raton, FL, 1984), pp. 347–380
6. I. Basson, E.C. Reynhardt, An investigation of the structures, molecular-dynamics of natural waxes. 2. Carnauba wax. J. Phys. D: Appl. Phys. **21**, 1429–1433 (1988)
7. M. Regert, J. Langlois, S. Colinart, Characterisation of wax works of art by gas chromatographic procedures. J. Chromatogr. A **1091**, 124–136 (2005)
8. L. Kunst, A.L. Samuels, Biosynthesis, secretion of plant cuticular wax. Prog. Lipid Res. **42**, 51–80 (2003)
9. E.S. Kim, P.G. Mahlberg, Glandular cuticle formation in cannabis (cannabaceae). Am. J. Bot. **82**, 1207–1214 (1995)
10. M. Anwar, H.U. Khan, S.P. Nautiyal, K.M. Agrawal, B.S. Rawat, Solubilised waxes, their influence on the flow properties of lube oil base stocks. Pet. Sci. Technol. **17**, 491–501 (1999)
11. T. Imai, K. Nakamura, M. Shibata, Relationship between the hardness of an oil-wax gel, the surface structure of the wax crystals. Colloid surf. A: Physicochem. Eng. Asp. **194**, 233–237 (2001)
12. M. Kobayashi, M. Saitoh, K. Ishida, H. Yachi, Viscosity properties, molecular structure of lube base oil prepared from Fischer–Tropsch waxes. J. Jpn. Pet. Inst. **48**, 365–372 (2005)
13. D. Petitjean, J.F. Schmitt, V. Laine, M. Bouroukba, C. Cunat, M. Dirand, Presence of isoalkanes in waxes and their influence on their physical properties. Am. Chem. Soc. 697–701 (2008)
14. I. Basson, E.C. Reynhardt, The structure, melting of paraffinic Fischer–Tropsch waxes. Chem. Phys. Lett. **198**, 367–372 (1992)
15. D.L. Dorset, Crystallography of waxes—An electron diffraction study of refined, natural products. J. Phys. D: Appl. Phys. **30**, 451–457 (1997)
16. M. Dirand, V. Chevallier, E. Provost, M. Bouroukba, D. Petitjean, Multicomponent paraffin waxes, petroleum solid deposits: structural and thermodynamic state. Fuel **77**, 1253–1260 (1998)
17. T.K. Miwa, Structural determination, uses of jojoba oil. J. Am. Oil Chem. Soc. **61**, 407–410 (1984)
18. C.J. Challinor, R.J. Hamilton, K. Simpson, Sperm whale head oil. I. The ester fraction. Chem. Phys. Lipids **3**, 145–151 (1969)
19. G.F. Spencer, W.H. Tallent, Sperm whale oil analysis by gas chromatography, mass spectrometry. J. Am. Oil Chem. Soc. **50**, 202–206 (1973)
20. H.J. Nieschlag, G.F. Spencer, R.V. Madrigal, J.A. Rothfus, Synthetic wax esters, diesters from crambe, limnanthes seed oils. Ind. Eng. Chem. Prod. Res. Dev. **16**, 202–207 (1977)
21. D.T. Downing, J.S. Strauss, Synthesis, composition of surface lipids of human skin. J. Invest. Dermatol. **62**, 228–244 (1974)
22. T. Nikkari, Comparative chemistry of sebum. J. Invest. Dermatol. **62**, 257–267 (1974)
23. N. Nicolaides, J.K. Kaitaranta, T.N. Rawdah, J.I. Macy, F.M. Boswell, R.E. Smith, Meibomian gland studies - comparison of steer, human lipids. Invest. Ophthalmol. Vis. Sci. **20**, 522–536 (1981)

24. P.J. Driver, M.A. Lemp, Meibomian gland dysfunction. Surv. Ophthalmol. **40**, 343–367 (1996)
25. J.P. McCulley, W.E. Shine, Meibomian gland and tear film lipids: Structure, function and control. In: D.A. Sullivan, M.E. Stern, K. Tsubota, D.A. Dartt, R.M. Sullivan, B.B. Bromberg (eds.), (Kluwer Academic/Plenum Publ, 2002), pp. 373–378
26. K. Rezaei, T. Wang, L.A. Johnson, Hydrogenated vegetable oils as candle wax. J. Am. Oil Chem. Soc. **79**, 1241–1247 (2002)
27. L.P. Wang, T. Wang, Chemical modification of partially hydrogenated vegetable oil to improve its functional properties for candles. J. Am. Oil Chem. Soc. **84**, 1149–1159 (2007)
28. M. Zajcew, The hydrogenation of fatty oils with palladium catalyst. 1. Hydrogenation of castor oil. J. Am. Oil Chem. Soc. **35**, 475–477 (1958)
29. E.N. Frankel, E.A. Emken, H. Itatani, J.C. Bailar, Homogeneous hydrogenation of methyl linolenate catalyzed by Platinum-Tin complexes. J. Org. Chem. **32**, 1447 (1967)
30. A. Hassan, A. Hassan, G.G. Borsinger, U.S. 6811824, (2003)
31. T.K. Miwa, Gas chromatograms of synthetic liquid waxes prepared from seed triglycerides of limnanthes, crambe and lunaria. J. Am. Oil Chem. Soc. **49**, 673–674 (1972)
32. A.A. Carelli, L.M. Frizzera, P.R. Forbito, G.H. Crapiste, Wax composition of sunflower seed oils. J. Am. Oil Chem. Soc. **79**, 763–768 (2002)
33. X.J. Hu, J.K. Daun, R. Scarth, Characterization of wax sediments in refined canola oils. J. Am. Oil Chem. Soc. **70**, 535–537 (1993)
34. S. Mezouari, S.P. Kochhar, K. Schwarz, K. Eichner, Effect of dewaxing pretreatment on composition, stability of rice bran oil: Potential antioxidant activity of wax fraction. Eur. J. Lipid Sci. Technol. **108**, 679–686 (2006)
35. G.G. Borsinger, A. Hassan, Wax for hot melt adhesive application. 6890982, (2005)
36. T.A. Murphy, K.S. Derby, Vegetable Oil Based Wax Composition. (2004)
37. T.A. Murphy, K.S. Derby, Vegetable Oil Based Wax Composition. Application#: US 6,824,572 B2, (2004)
38. J.E. Nash, D.G. Evans, D.M. Hoganson, Connector system for vessels ducts, lumens or hoolw organs and methods of use. Application#: 6063114, (2000)
39. B.Y. Tao, Vegetable Lipid- Based Composition and Candle. Application: 6284007, (2001)
40. G.G. Borsinger, A. Hassan. Wax Emulsion Coating. Application: 7267743, (2007)
41. R. Rissmann, M.H.M. Oudshoorn, E. Kocks, W.E. Hennink, M. Ponec, J.A. Bouwstra, Lanolin-derived lipid mixtures mimic closely the lipid composition, organization of vernix caseosa lipids. Biochim. Biophys. Acta, Biomembr. **1778**, 2350–2360 (2008)
42. J. Wisniak, *The Chemistry and Technology of Jojoba Oil* (AOCS Press, Boulder, Il, 1987), p. 272
43. J.A. Retamar, G. Di Giacomo et al., Plants and natural products as ecological mediators. Jojoba oil: a liquid wax. Essenze, Derivati Agrumari **65**, 523–535 (1995)
44. P. Cappillino, R. Kleiman, C. Botti, Composition of chilean jojoba seeds. Ind. Crops Prod. **17**, 177–182 (2003)
45. A. Gibbs, J.G. Pomonis, Physical-properties of insect cuticular hydrocarbons -the effects of chain-length, methyl-branching and unsaturation. Comp. Biochem. Physiol B: Biochem. Mol. Biol. **112**, 243–249 (1995)
46. H. Maciejewski, A. Wawrzynczak, M. Dutkiewicz, R. Fiedorow, Silicone waxes -synthesis via hydrosilylation in homo- and heterogeneous systems. J. Mol. Catal. A: Chem. **257**, 141–148 (2006)
47. G. Gu, Z. Wu, Z. Zhang, F. Qing, Tribological properties of fluorine-containing additives of silicone oil. Trib. Int. **42**, 397–402 (2009)
48. D.S. Ogunniyi, Castor oil: vital industrial raw material. Bioresour. Technol. **97**, 1086–1091 (2006)
49. K.L. Mittal, A. Pizzi, *Handbook of Adhesive Technology*, 2nd edn. (M. Dekker, New York, 2003)
50. K.H. Badri, A.H. Ujar, Z. Othman, F.H. Sahaldin, Shear strength of wood to wood adhesive based on palm kernel oil. J. Appl. Polym. Sci. **100**(3), 1759–1764 (2006)

51. S.D. Desai, A.L. Emanuel, V.K. Sinha, Biomaterial based polyurethane adhesive for bonding rubber, wood joints. J. Polym. Res.-Taiwan **10**, 275–281 (2003)
52. S.D. Desai, J.V. Patel, V.K. Sinha, Polyurethane adhesive system from biomaterial-based polyol for bonding wood. Int. J. Adhes. Adhes. **23**, 393–399 (2003)
53. K.P. Somani, S.S. Kansara, N.K. Patel, A.K. Rakshit, Castor oil based polyurethane adhesives for wood-to-wood bonding. Int. J. Adhes. Adhes. **23**, 269–275 (2003)
54. R.D. Mitchell, Raw materials for hot melt adhesives in the twenty first century. Adhes. Age **42**, 24–27 (1999)
55. D. De Gray, PUR adhesives offer solutions for assembly challenges. Adhes. Age **41**, 23–24 (1998)
56. A. Pizzi, *Advanced Wood Adhesives Technology* (Marcel Dekker, New York, 1994)
57. A. Pizzi, *Wood Adhesives: Chemistry and Technology* (Marcel Dekker, New York, 1983)
58. A. Pizzi, Recent developments in eco-efficient bio-based adhesives for wood bonding: opportunities and issues. J. Adhes. Sci. Tech. **20**, 829–846 (2006)
59. N. John, R. Joseph, Rubber solution adhesives for wood-to-wood bonding. J. Appl. Polym. Sci. **68**, 1185–1189 (1998)
60. M.F. Vallat, N. Bessaha, J. Schultz, J. Maucourt, C. Combette, Adhesive behavior of polyurethane-based materials. J. Appl. Polym. Sci. **76**, 665–671 (2000)
61. C.W. Paul, M.L. Sharak, M. Blumenthal, Green adhesives. Adhes. Age **42**, 34–39 (1999)
62. X. Kong, S.S. Narine, Physical properties of polyurethanes plastic sheets produced from polyols from canola oil. Biomacromolecules **8**, 2203–2209 (2007)
63. X. Kong, J. Yue, S.S. Narine, Physical properties of canola oil based polyurethane networks. Biomacromolecules **8**, 3584–3589 (2007)
64. X.H. Kong, S.S. Narine, Physical properties of sequential interpenetrating polymer networks produced from canola oil-based polyurethane and poly(methyl methacrylate). Biomacromolecules **9**, 1424–1433 (2008)
65. X.H. Kong, S.S. Narine, Sequential interpenetrating polymer networks produced from vegetable oil based polyurethane and poly (methyl methacrylate). Biomacromolecules **9**, 2221–2229 (2008)
66. L.H. Sperling, *Interpenetrating Polymer Networks and Related Materials* (Plenum Press, New York, 1981)
67. S.S. Narine, X. Kong, L. Bouzidi, P. Sporns, Physical properties of polyurethanes produced from polyols from seed oils. II-foams. J. Am. Oil Chem. Soc. **84**, 65–72 (2007)
68. S.S. Narine, X. Kong, L. Bouzidi, P. Sporns, Physical properties of polyurethanes produced from polyols from seed oils. I-elastomers. J. Am. Oil Chem. Soc. **84**, 55–63 (2007)
69. S.Z. Ethan, B.K. Sharma, Z.S. Liu, A. Adhvaryu, Lubricant base stock potential of chemically modified vegetable oils. J. Agric. Food Chem. **56**(19), 8919–8925 (2008)
70. B.K. Sharma, Z.S. Liu, A. Adhvaryu, S.Z. Erhan, One-pot synthesis of chemically modified vegetable oils. J. Agric. Food Chem. **56**(9), 3049–3056 (2008)
71. G.L. Coffey, H.B. Coffey, Nontoxic lubricant composition. U.S. Patent 5,691,285, (1997)
72. A. Arnsek, J. Vizintin, Scuffing load capacity of rapeseed-based oils. Lubr. Eng. **55**, 11–18 (1999)
73. A. Arnsek, J. Vizintin, Lubricating properties of rapeseed-based oils. Synth Lubr **16**, 281–296 (1999)
74. A. Arnsek, J. Viztntin, Fitting resistance of rapeseed-based oils. Lubr. Eng. **57**(3), 17–21 (2001)
75. S.Z. Erhan J. Am. Oil Chem. Soc., A. Adhvaryu, B.K. Sharma, Chemically functionalized vegetable oils. In synthetics, mineral oils, and bio-based lubricants: chemistry and technology, Rudnick, L. R., Ed. CRC Press: Boca Raton, FL, 361-387 (2006)
76. L.A.T. Honary, An investigation of the use of soybean oil in hydraulic systems. Bioresour. Technol. **56**(1), 41–47 (1996)
77. E.O. Aluyor, K.O. Obahiagbon, M. Ori-jesu, Biodegradation of vegetable oils: A review. Sci. Res. Essays **4**(6), 543–548 (2009)

78. H.S. Hwang, S.Z. Erhan, Modification of epoxidized soybean oil for lubricant formulations with improved oxidative stability and low pour point. J. Am. Oil Chem. Soc. **78**(12), 1179–1184 (2001)
79. S. Li, J. Blackmon, A. Demange, T.C. Jao, Linear sulphonate detergents as pour point depressants. Lubr. Sci. **16**, 127–137 (2004)
80. A.R. Coscione, W.E. Artz, Vegetable oil stability at elevated temperatures in the presence of ferric stearate and ferrous octanoate. J. Agric. Food Chem. **53**(6), 2088–2094 (2005)
81. R.O. Dunn, Effect of antioxidants on the oxidative stability of methyl soyate (biodiesel). Fuel Process. Technol. **86**(10), 1071–1085 (2005)
82. S.Z. Erhan, B.K. Sharma, J.M. Perez, Oxidation and low temperature stability of vegetable oil-based lubricants. Ind. Crops Prod. **24**(3), 292–299 (2006)
83. J.Z. Adamczewska, D. Wilson, Development of Ecologically Responsive Lubricants. J. Synth. Lubr. **14**, 129–142 (1998)
84. M.A. Schmidt, C.R. Dietrich, E.B. Cahoon, Biotechnological Enhancement of Soybean Oil for Lubricant Applications, in *Synthetics, mineral oils, and bio-based lubricants: chemistry and technology*, ed. by L.R. Rudnick (CRC Press, Boca Raton FL, 2006), pp. 389–398
85. S. Asadauskas, J.M. Perez, J.L. Duda, Oxidative stability and antiwear properties of high oleic vegetable oils. Lubr. Eng. **52**(12), 877–882 (1996)
86. B.K. Sharma, J.M. Perez, S.Z. Erhan, Soybean oil-based lubricants: a search for synergistic antioxidants. Energy Fuels **21**(4), 2408–2414 (2007)
87. A. Birova, A. Pavlovicova, J.C. Lubricating Oils Based on Chemically Modified Vegetable Oils. **18**, 291–299 (2002)
88. M.A.R. Meier, J.O. Metzger, U.S. Schubert, Plant oil renewable resources as green alternatives in polymer science. Chem. Soc. Rev. **36**(11), 1788–1802 (2007)
89. J.W. King, R.L. Holliday, G.R. List, J.M. Snyder, Hydrogenation of vegetable oils using mixtures of supercritical carbon dioxide and hydrogen. J. Am. Oil Chem. Soc. **78**(2), 107–113 (2001)
90. K.M. Doll, B.K. Sharma, S.Z. Erhan, Synthesis of branched methyl hydroxy stearates including an ester from bio-based levulinic acid. Ind. Eng. Chem. Res. **46**(11), 3513–3519 (2007)
91. H.S. Hwang, A. Adhvaryu, S.Z. Erhan, Preparation and properties of lubricant basestocks from epoxidized soybean oil and 2-ethylhexanol. J. Am. Oil Chem. Soc. **80**(8), 811–815 (2003)
92. E. Verkuijlen, F. Kapteijn, J.C. Mol, C. Boelhouwer, Heterogeneous metathesis of unsaturated fatty-acid esters. J. Chem. Soc. Chem. Commun. **1**(7), 198–199 (1977)
93. S.Z. Erhan, M.O. Bagby, T.C. Nelsen, Drying properties of metathesized soybean oil. J. Am. Oil Chem. Soc. **74**(6), 703–706 (1997)
94. R.A. Holser, K.M. Doll, S.Z. Erhan, Metathesis of methyl soyate with ruthenium catalysts. Fuel **85**(3), 393–395 (2006)
95. A. Birova, A. Pavloviova, J. Cvenro, Lubricating oils based on chemically modified vegetable oils. J. Synth. Lubr. **18**, 291–299 (2002)
96. R. Yunus, A. Fakhru'l-Razi, T.L. Ooi, S.E. Iyuke, J.M. Perez, Lubrication properties of trimethylolpropane esters based on palm oil and palm kernel oils. Eur. J. Lipid Sci. Technol. **106**(1), 52–60 (2004)
97. E.D. Dehault, A. Demoulin, Partial hydrogenation of poly-unsaturated fatty materials. J. Am. Oil Chem. Soc. **61**(2), 195–200 (1984)
98. A. Adhvaryu, S.Z. Erhan, J.M. Perez, Tribological studies of thermally and chemically modified vegetable oils for use as environmentally friendly lubricants. Wear **257**(3–4), 359–367 (2004)
99. W. Castro, J.M. Perez, S.Z. Erhan, F. Caputo, A study of the oxidation and wear properties of vegetable oils: Soybean oil without additives. J. Am. Oil Chem. Soc. **83**(1), 47–52 (2006)
100. N.J. Fox, G.W. Stachowiak, Vegetable oil-based lubricants—a review of oxidation. Tribol. Int. **40**(7), 1035–1046 (2007)

101. P.S. Lathi, B. Mattiasson, Green approach for the preparation of biodegradable lubricant base stock from epoxidized vegetable oil. Appl. Catal. B **69**(3–4), 207–212 (2007)
102. D. Hörner, Recent trends in environmentally friendly lubricants. J. Synth. Lubr. **18**, 327–347 (2006)
103. O.N. Anand, J. Mehta, T.S.R. Prasada Rao, Lubricant components from vegetable oils of indian origin. J. Synth. Lubr. **15**, 97–106 (1998)
104. V.V. Bokade, G.D. Yadav, Synthesis of Bio-diesel and Bio-lubricant by transesterification of vegetable oil with lower and higher alcohols over heteropolyacids supported by clay (K-10). Process Saf. Environ. Prot. **85**(B5), 372–377 (2007)
105. H.S. Hwang, S.Z. Erhan, Synthetic lubricant basestocks from epoxidized soybean oil and Guerbet alcohols. Ind. Crop Prod. **23**(3), 311–317 (2006)
106. E. Uosukainen, Y.Y. Linko, M. Lamsa, T. Tervakangas, P. Linko, Transesterification of trimethylolpropane and rapeseed oil methyl ester to environmentally acceptable lubricants. J. Am. Oil Chem. Soc. **75**(11), 1557–1563 (1998)
107. D.R. Kodali, High performance ester lubricants from natural oils. Ind. Lubr. Tribol. **54**(4), 165–170 (2002)
108. B.R. Moser, B.K. Sharma, K.M. Doll, S.Z. Erhan, Diesters from oleic acid: Synthesis, low temperature properties, and oxidation stability. J. Am. Oil Chem. Soc. **84**(7), 675–680 (2007)
109. J. Salimon, N. Salih Oleic, Acid diesters: synthesis, characterization and low temperature properties. Eur. J. Sci. Res. **32**, 216–222 (2009)
110. J. Salimon, N. Salih, Improved low temperature properties of 2-Ethylhexyl 9(10)-Hydroxy-10(9)-Acyloxystearate derivatives. Eur. J. Sci. Res. **31**, 583–591 (2009)
111. B.R. Moser, S.Z. Erhan, Synthesis and evaluation of a series of alpha-hydroxy ethers derived from isopropyl oleate. J. Am. Oil Chem. Soc. **83**(11), 959–963 (2006)
112. B.R. Moser, S.Z. Erhan, Preparation and evaluation of a series of alpha-hydroxy ethers from 9, 10-epoxystearates. Eur. J. Lipid Sci. Technol. **109**(3), 206–213 (2007)
113. B.R. Moser, S.Z. Erhan, Branched chain derivatives of alkyl oleates: Tribological, rheological, oxidation, and low temperature properties. Fuel **87**(10–11), 2253–2257 (2008)
114. B.K. Sharma, K.M. Doll, S.Z. Erhan, Oxidation, friction reducing, and low temperature properties of epoxy fatty acid methyl esters. Green Chem. **9**(5), 469–474 (2007)
115. A. Adhvaryu, B.K. Sharma, H.S. Hwang, S.Z. Erhan, J.M. Perez, Development of biobased synthetic fluids: Application of molecular modeling to structure-physical property relationship. Ind. Eng. Chem. Res. **45**(3), 928–933 (2006)

Chapter 15
Tribo-Chemical Activation of Green Eco-Cements

Konstantin Sobolev

Abstract Due to growing global demand for cement, the production of cement worldwide has significantly increased in the past 15 years, and this trend is the most significant factor affecting the technological and manufacturing advancements in the cement industry. While the increase in demand for cement reflects the growth of national economies, the production of cement clinker is ecologically harmful because it consumes considerable energy and natural resources, and it emits many pollutants into the atmosphere. Therefore, new ways to produce high volumes of cement clinker with less energy and less impact on the environment is greatly needed. One such approach is the production of tribo-chemically activated, high-volume mineral additive (HVMA) cement, which helps to improve the ecological compatibility of cements. This "green" technology is based on the intergrinding of portland cement clinker, gypsum, mineral additives, and a special complex admixture. Tribo-chemical activation increases the compressive strength of ordinary portland cements, improves the durability of cement-based materials, can be processed with a high volume of inexpensive indigenous mineral additives or industrial by-products, and which reduces energy consumption per unit of the cement produced. Additional ecological advantages for green HVMA cements include higher strength, better durability, less pollution at the clinker production stage, and fewer industrial by-products placed in landfills.

K. Sobolev (✉)
Department of Civil Engineering and Mechanics,
University of Wisconsin-Milwaukee,
EMS 939 3200 North Cramer Street,
Milwaukee, WI 53211, USA
e-mail: sobolev@uwm.edu

M. Nosonovsky and B. Bhushan (eds.), *Green Tribology*,
Green Energy and Technology, DOI: 10.1007/978-3-642-23681-5_15,

15.1 Introduction

The most recent developments in cement and concrete technology can be summarized as follows [1–5]:

- Following a historical cycle of cement and concrete development, there is a strong need to increase strength. Continuous breakthroughs in knowledge allow the design and application of super high-strength concrete with a compressive strength of up to 250 MPa, high flexural strength, and remarkable ductility [2–9]. As a result, the design and application of effective structures built with high-strength concrete has been phenomenal in the past years, demonstrating strong growth potential.
- The use of industrial by-products and waste (IBPW) as mineral additives has become a valuable segment of cement and concrete technology. According to wide-scale investigations [2–6, 10–16], the performance of concrete with controlled volumes of IBPW can be significantly improved. Well-investigated mineral additives include granulated blast furnace slag, fly ash, and silica fume, which not only result in concrete with improved properties and economical effectiveness, but they also improve the eco- balance of cement and concrete.
- The use of chemical admixtures is essential in producing modern concrete. Added to the concrete mixture, relatively small amounts of chemical admixtures radically alter the behavior of fresh or hardened concrete. Modern admixtures can help realize almost any desired property of concrete [7–9, 17–19].
- Application of nanotechnology and nano materials is a relatively new direction focused on better understanding of the cement-based materials at the nanolevel and the application of nanomaterials (such as nano-SiO_2 particles) in cement and concrete [20–23].

As a result of these developments, the concept of high-performance concrete (HPC) has been put forward and successfully applied. It is generally considered that concrete with improved properties (i.e., workability, strength, permeability, and durability) over the conventional levels can be classified as HPC. According to Forster [9], HPC is "a concrete made with appropriate materials combined according to a selected mix design and properly mixed, transported, placed, consolidated, and cured so that the resulting concrete will give excellent performance in the structure in which it will be exposed, and with the loads to which it will be subjected for its design life."

To realize the HPC concept, a variety of chemical admixtures and mineral additives are necessary, and modern concrete batching plant technology requires adequate equipment for precise control, dispatching, dosing, and batch processing [8, 9, 17–19]. The technology of high-performance cement was developed in order to simplify HPC production, thereby further extending the frames of its application [5, 6, 18, 24].

A novel approach for improving cement performance includes the application of chemical admixtures (modifiers) at the stage of cement grinding. Air-entraining,

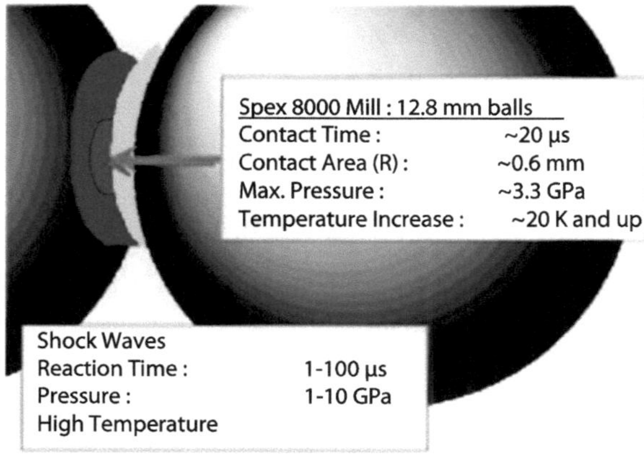

Fig. 15.1 The mechanism of tribo-mechanically induced solid state reactions

hydrophobic, plasticized cements, and the family of cements manufactured with grinding aids are early examples of tribo-chemical and mechanochemical activation of cement surface layers resulting in the desired performance. The theory of tribo-chemical and mechano-chemical activation (TMCHA) has been applied to process nanopowders, pigments, fillers, binders, ceramic, and ferromagnetic materials, but has a limited use in cement modification.

TMCHA activation is used to describe the chemical conversions in solids induced by a mechanical process such as milling or grinding. The mechanical processing usually results in the formation of dislocations and other defects in the structure of the material. In the case of TMCHA, the mechanical impacts cause the development of elastic, plastic, and shear deformations leading to fracture, amorphization, and even chemical reactions in the solid state. Characteristic features of TMCHA are summarized in Fig. 15.1. Ball milling, and especially high-energy ball milling, breaks down the crystallinity of solid reactants and provides a transfer of mass required for chemical reactions. In addition, high pressure and shear stress facilitate both the phase transitions and the chemical transformations of solids. The energy in the form of various lattice defects, accumulated by the solid during the mechanical processing, can support/trigger various chemical transformations.

The application of advanced cements with high strength and superior durability is an attractive method of controlling concrete properties and the design of high-performance concrete. Such cements require the application of special clinkers, or they can be manufactured using blended cement technology (e.g., shrinkage compensating/expansive cements, high-early-strength cements, regulated-set cements). Advanced cements such as Ultimax cement, Pyrament, energetically modified cement, and silica fume cements are developed using this principle [6, 20, 24].

Tribo-chemistry can be used to improve the strength of conventional cement. For example, low water demand binder is produced by intergrinding cement and a dry modifier at a high energy. The application of a specially selected admixture-modifier at a relatively high dosage (about 4%) resulted in a cement with both reduced water demand and high strength [20, 24].

Fig. 15.2 Scanning electron microphotograph of TCA HP cement

Supersilica, a complex admixture for application in cement technology has been developed. It is based on a reactive silica-based sorbent, an effective surfactant, and some minor corrective components. Supersilica, a reactive silica-based complex admixture (RSA), was produced using this principle. It was proposed that, when added during the cement grinding process, RSA modifies the surface of cement particles and promotes the formation of highly reactive layers on the surface of cement grains. Scanning electron microscopy helps to reveal some details of interaction between RSA and cement, such as micro-or nano-indentations on the surface of cement particles (Fig. 15.2). Furthermore, the silica component of RSA also acts as a micro-filler and participates in a pozzolanic reaction. The tribo-chemical activation of cement with RSA results in a new product—high-performance cement.

High-performance cement can be defined as a material manufactured by the tribo-chemical activation of certain proportions of clinker, gypsum, a complex admixture (RSA) and, optionally, a mineral additive of industrial (IBPW) or natural origin. The application of HP cement imparts high strength and extreme durability to the concrete or mortar; and its high strength can be used for engineering cement with a high volume of mineral additives. The flowchart for the grinding unit for manufacturing HP/HVMA cement is presented in Fig. 15.3.

15.2 Tribo-Chemical Activation of Cement

The effect of tribo-chemical activation (TCA) on the properties of cement systems was investigated [20]. Three different modifiers were used in the experimental program: sulfonated naphthalene formaldehyde condensate (SNF, used in a dry form), sulfonated melamine formaldehyde condensate (SMF, used in a dry form), and polyacrilate/polycarboxylate superplasticizer (PAE, 31% concentration).

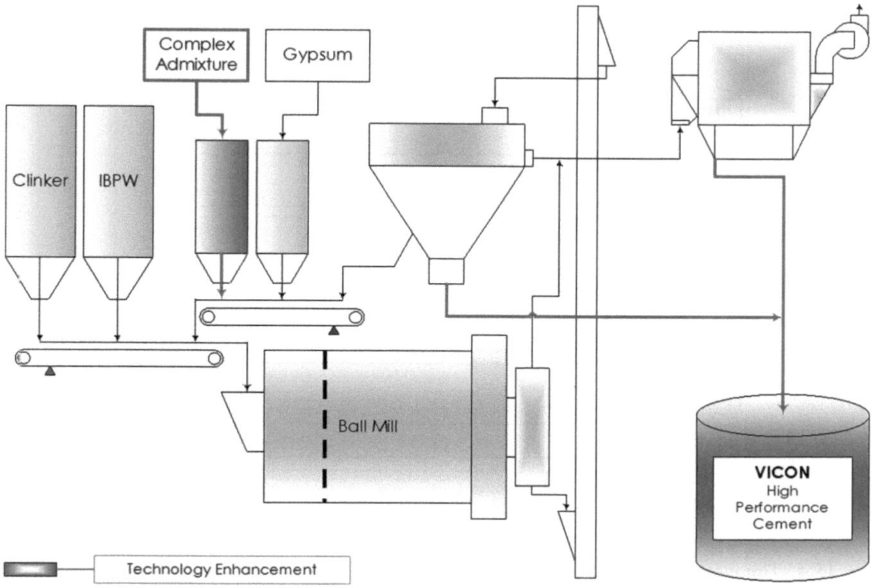

Fig. 15.3 High-performance cement technology

Table 15.1 Composition and compressive strength of tribo-chemically activated cements

Cement type	Composition (%)		Flow (mm)	Compressive strength (MPa)			
	Silica fume	Supersilica		1 day	7 days	28 days	90 days
NPC	–	–	195	52.7	69.3	84.7	72.7
SFC	8	–	208	44.4	71.3	89.0	88.8
HPC-SNF	–	8	211	54.2	76.6	93.8	98.6
HPC-SMF	–	8	230	40.3	91.5	115.4	124.4
HPC-PAE	–	8	370	57.4	88.1	105.5	110.8

The modifiers were processed into complex admixture (Supersilica) by intermixing and granulating with silica fume (SF) at a ratio of 1:10, by weight [20]. Portland cement clinker conforming to ASTM Type I requirements and natural gypsum were used as the components of tribo-chemically activated (TCA) HP cements (Table 15.1). These cements were obtained by intergrinding clinker, gypsum, and complex admixture (or SF) in a standard ball mill for 3 h.

The monitoring of the crystalline structure (determined by X-ray diffraction) and particle size distribution (determined by laser diffraction method) was performed for the products obtained at different stages of intergrinding at the time intervals of 1, 2, and 3 h. Mortars were prepared at the sand-to-cement ratio (S/C) of 1 and water-to-cement ratio (W/C) of 0.3 (Table 15.1). The investigation of compressive strength of obtained cements was realized using 51 × 51 × 51 mm cubes following the ASTM C109 procedure. In addition to compressive strength, the flow of mortars was tested

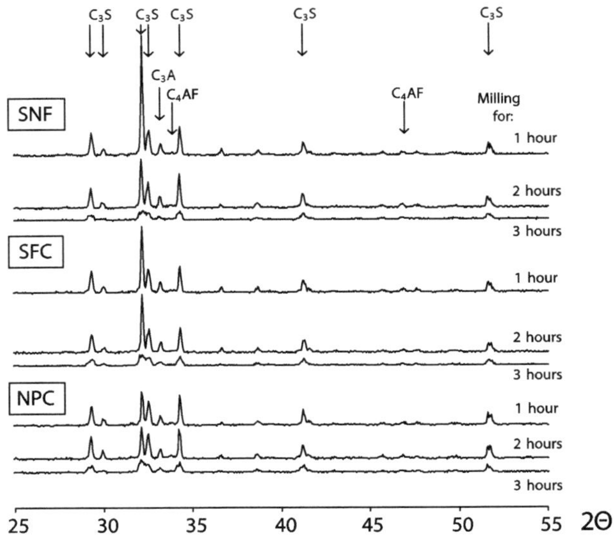

Fig. 15.4 The effect of surface amorphization due to TCA [20]

using the flow table (according to ASTM C109); a 400 × 400 mm glass plate was used on the tabletop to accommodate the large flows.

The experiment demonstrates that the TCA method can be effectively used to modify cement properties and to design novel groups of cement-based materials such as HP cements, low water demand binders, and high-performance cements. Continuous grinding in a ball mill for 3 h results in significant amorphization of the surface of investigated cements (Fig. 15.4). Even after milling for 1 h, the main diffraction peaks of C_3S were still sharp, with higher intensity levels obtained by the cement with SNF modifier; after milling for 2 and 3 h, the intensity of C_3S peaks diminishes considerably (Fig. 15.4). These results correspond with the particle size reduction process (Table 15.2).

It can be observed that SNF was the most effective modifier to reduce the particle size and increase the surface area of investigated cements (Table 15.2). After milling for 3 h, the median particle size (D_{50}) and specific surface area of SNF cement were 12.5 μm and 415 m^2/kg (vs. 13.6 μm and 377 m^2/kg for reference SF cement), respectively. Still, the efficiency of grinding in a ball mill was not very impressive—in spite of the particle size reduction by up to 40% (achieved with the additional 2 h milling when SNF/SMF modifiers were used), the surface area of investigated cements was not significantly affected (Table 15.2). The cement milled with PAE-based modifier was coarser than the reference cements.

The majority of reported data on TCA HP cements describes the behavior of such cements at a reduced W/C of less than 0.3, usually at the levels near to normal consistency [20]. It was demonstrated that the strength of wide-range compositions obtained at a different S/C and W/C follows the law of W/C (Table 15.1; Fig. 15.5).

Table 15.2 The effect of milling on particle size and surface area of investigated cements

Cement type	Duration of milling (min)	Superficial area (m^2/kg)	Average diameter D_{50} (μm)
NPC	60	227	22.1
	120	302	16.5
	180	340	12.8
SFC	60	256	21.6
	120	311	16.9
	180	377	13.6
HPC-SNF	60	285	20.8
	120	388	13.5
	180	415	12.5
HPC-SMF	60	285	20.8
	120	388	13.5
	180	415	12.5
HPC-PAE	60	247	24.6
	120	274	22.2
	180	300	18.8

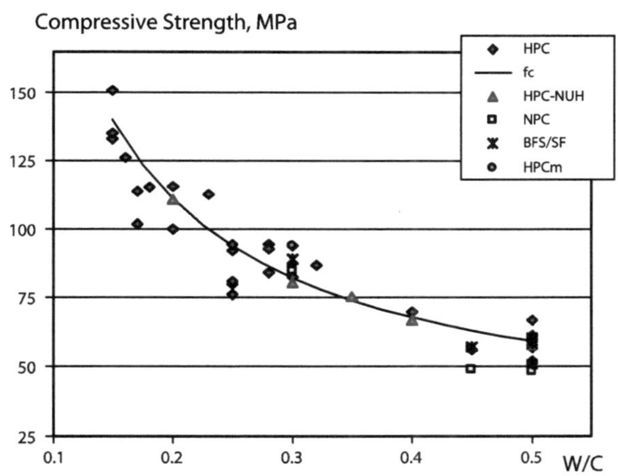

Fig. 15.5 Strength of TCA HP cements as a function of W/C [20]

The necessity of using mortars with low S/C (<2, usually 1) for investigating superplasticized and high-strength compositions was already discussed [24]. This investigation was conducted at W/C = 0.3 in order to compare the effects of different modifiers on strength and workability (determined as a flow). It is assumed that mortars with higher flow will follow the W/C law (Fig. 15.5) and increase the strength at lower W/C when tested for constant flow (for example, as it is required by the GOST 310/ASTM C928). This adopted test procedure also accounts for the reduction of the superplasticizing efficiency of the modifier due to the change of oligomer's chain size and molecular weight distribution that occurs during TCA.

The flow of all investigated mortars was higher than that of reference portland- and SF-based cements (Table 15.1). While the SNF-based modifier almost lost its

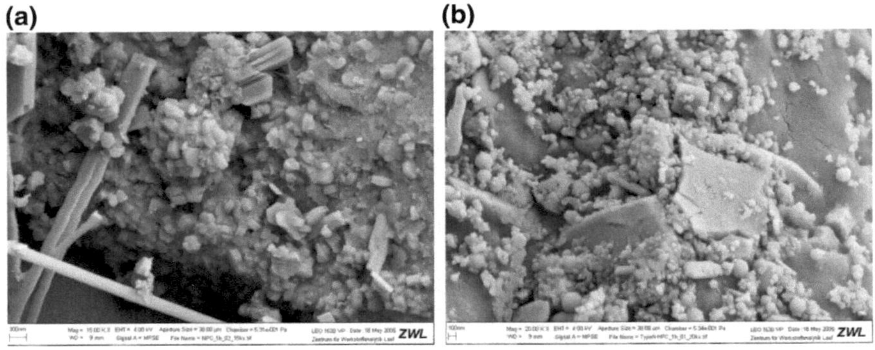

Fig. 15.6 The nanostructure of a NPC and b HP cement at early-stages of hydration [20]

plasticizing effect (vs. SF cement), the SMF and, especially, PAE modifiers preserved their supreme plasticizing efficiency.

A significant difference between NPC and TCA HP cements at the early stage of hydration (1 h) can be revealed with cryogenic SEM (Fig. 15.6). It can be observed that during this early period, NPC is hydrating rapidly and generating significant quantities of portlandite; in the same period, HP cement with SNF admixture still maintains the dormant period without visible formation of portlandite and C–S–H (Fig. 15.6). Such behavior can help to tailor the performance of cement systems to induce the controlled workability patterns and, at the same time, preserve the intensive rates of strength development.

The 28-day compressive strength of all investigated cements was very high, at the range of 93–115 MPa vs. the strength of reference cements of 72–89 MPa. The highest strength of 115.4 MPa was demonstrated by SMF-modified cement. It was reported that the strength of HP cements increases significantly from the very early ages of hardening (apparently, from 5 to 12 h) [20]; however, when tested at the same W/C, the 1-day strength of SNF and PAE modified cements was improved only by 5 and 12% vs. reference portland cement. On the other hand, the strength of these mortars was significantly higher than that of SF cement. The application of SMF additive reduced the 1-day compressive strength to 40.3 MPa. Meanwhile, the additives based on SMF and PAE were most effective in improving the 7-day strength, up to 91.5 and 88.1 MPa, respectively.

The 90-day strength of SMF-based mortar was quite remarkable, i.e., 124.4 MPa; this cement demonstrated steady growth during the investigated period throughout. Reduction of 90-day strength vs. 28-day level was observed for NPC; this is a common observation for finely ground cements, which can be related to the accelerated hydration and consequent stress accumulation.

In summary, tribo-chemical activation is a very effective method to manipulate the structure of portland cement-based systems at the nanolevel. The developed modifiers form the organo-mineral nanolayers (as mono/bi-molecular layers) or nano-grids (when the functional groups are attached only to the active centers) on the surface of cement particles. Using this method, the performance of the resulting cement can be tailored as desired—from high-strength materials to the products with controlled workability.

15.3 Technology of HP/HVMA Cement: High Strength and Durability

The high-strength phenomenon was used to engineer cement with HVMA. Because of its high strength, large amounts (up to 70%) of portland cement clinker can be replaced with indigenous mineral additives such as natural pozzolanic materials, sand, limestone, granulated blast furnace slag, fly ash, broken glass, and ceramic [6, 18, 20]. As a result two types of HP cement can be distinguished:

- Basic HP cement (Type A)
- Blended HP/HVMA cement (Type B)

Here, blended Type B cement covers a wide range of HP cements with a mineral additive. The content of a mineral additive in blended HP cement varies with the specified level of properties and the type of additive used. Wide-scale research [6, 18] has demonstrated that HP cements with a mineral additive content within the standard limits (25–50%) can be produced; furthermore, the contents of mineral additives in HVMA cement can be increased (Fig. 15.7).

The application of granulated blast furnace slag (GBFS) in blended HP cement provides a very high resistance to chemical attack [6]. Test results have demonstrated the possibility of producing blended HP cement with a compressive strength of more than 80 MPa at a BFS content of up to 50% (Fig. 15.8). As a result, TCA HP cement technology helps to boost the performance of blended cements.

Two possible approaches to HVMA cement production are optimal and economical, which are presented in Fig. 15.9. New types of mineral admixtures (such as sand, glass cullet, and waste ceramics) can be used as an ingredient in blended HP/HVMA cement, Fig. 15.10.

The performance of tribo-chemically activated HVMA eco-cements was reported [18]. The composition of HVMA cements included a ternary blend of NPC, Supersilica, and finely ground mineral additive (FGMA). The following FGMA materials were used in the experimental program: limestone, natural sand, perlite, fly ash, blast furnace slag, waste glass, spent catalyzer, and a waste of alum process. To realize the tribo- chemical activation, the HVMA cements were manufactured by intergrinding all components in a high-energy vibrating mill for 20 min.

The FGMA content in HVMA cements was varied from 15 to 60%. Two major groups of HVMA cement were investigated (Fig. 15.9):

- With optimum content of Supersilica at 15% of Supersilica in HVMA cement;
- With economical content of Supersilica at Supersilica-to-NPC ratio of 15:85.

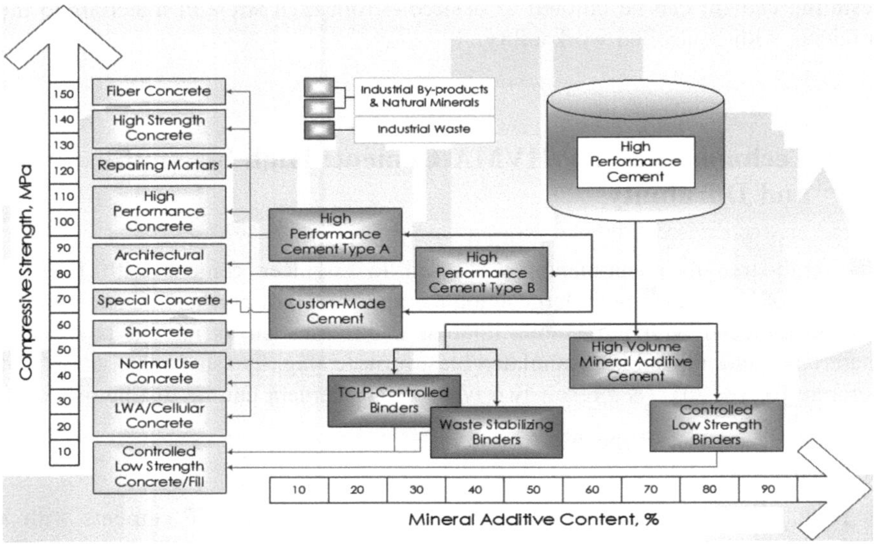

Fig. 15.7 Performance and application of HP/HVMA cements

Fig. 15.8 Strength development of TCA HP/HVMA cement (vs. reference NPC)

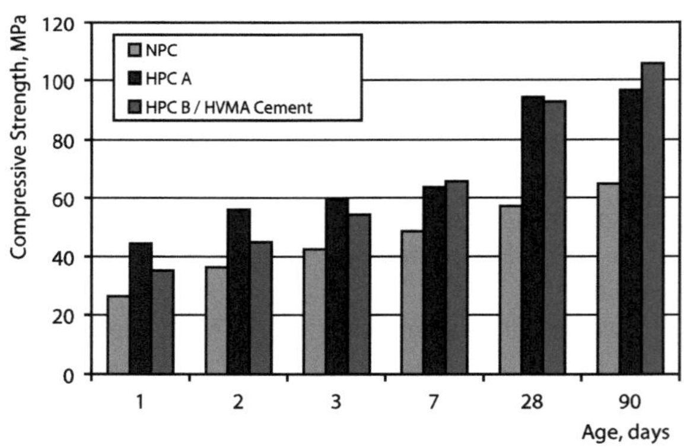

The strength characteristics of HVMA cements were determined in accordance with an accelerated test procedure: the strength of mortar consisting of cement, standard sand, and water (in quantity, required to obtain flow of 106–115 mm) at a cement-to-sand ratio of 1:1 using 4 × 4 × 16 cm prismatic samples, cured for 8 h at 80°C in a steam chamber.

It was found that at 15% Supersilica TCA HP cement has optimal performance with a compressive strength of 135.4 MPa and flexural strength of 18.1 MPa (at W/C of 0.14) [18, 20]. Replacing the portland cement component with FGMA at optimal Supersilica content increases the compressive strength (Fig. 15.10):

- 135.8 MPa at limestone content of 60%
- 140.5 MPa at sand content of 60%

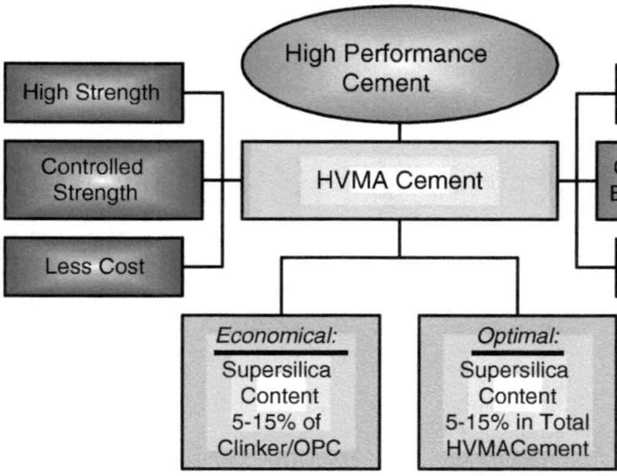

Fig. 15.9 TCA HVMA eco-cement production options: optimal and economical

- 145.2 MPa at fly ash content of 15%
- 165.6 MPa at waste glass content of 30%

A further increase in the FGMA content above these limits leads to a proportional reduction of strength (Fig. 15.10). The TCA HVMA cements with a strength level of normal cement can be manufactured with up to 45% FGMA at a reduced Supersilica content (Supersilica-to-NPC ratio of 15:85).

15.4 Eco-Efficiency of Green Eco-Cements

According to Cembureau's reports [1], the world's production of cement has increased by about 50% in the past 10 years. The growing demand for cement is the most significant factor affecting technological development and updating of the manufacturing facilities in the cement industry [2–5, 10]. However, the existing technology of portland cement is ecologically harmful because it consumes considerable energy and natural resources, and emits a number of pollutants such as CO_2, NO_x, and SO_x. With annual production of 2.6 billion tons, the cement industry contributes up to 8% to global CO_2 emissions [1–5]. Theoretically, the production of 1 ton of portland cement releases 900 kg of CO_2; which makes the cement industry an important sector for CO_2 emission mitigation strategies.

If the cement industry uses existing technologies to increase the production of bulk cement, products will have the same level of performance at the cost of increasing the consumption of raw materials and energy. The demand from growing markets cannot be met by simply expanding cement industry's current capacity [5], which will substantially degrade the environment, or by factoring the environmental damage from materials as an economic cost. Sustainable development is the future of the cement industry's success; therefore, new products must either perform better or use fewer raw materials and less energy.

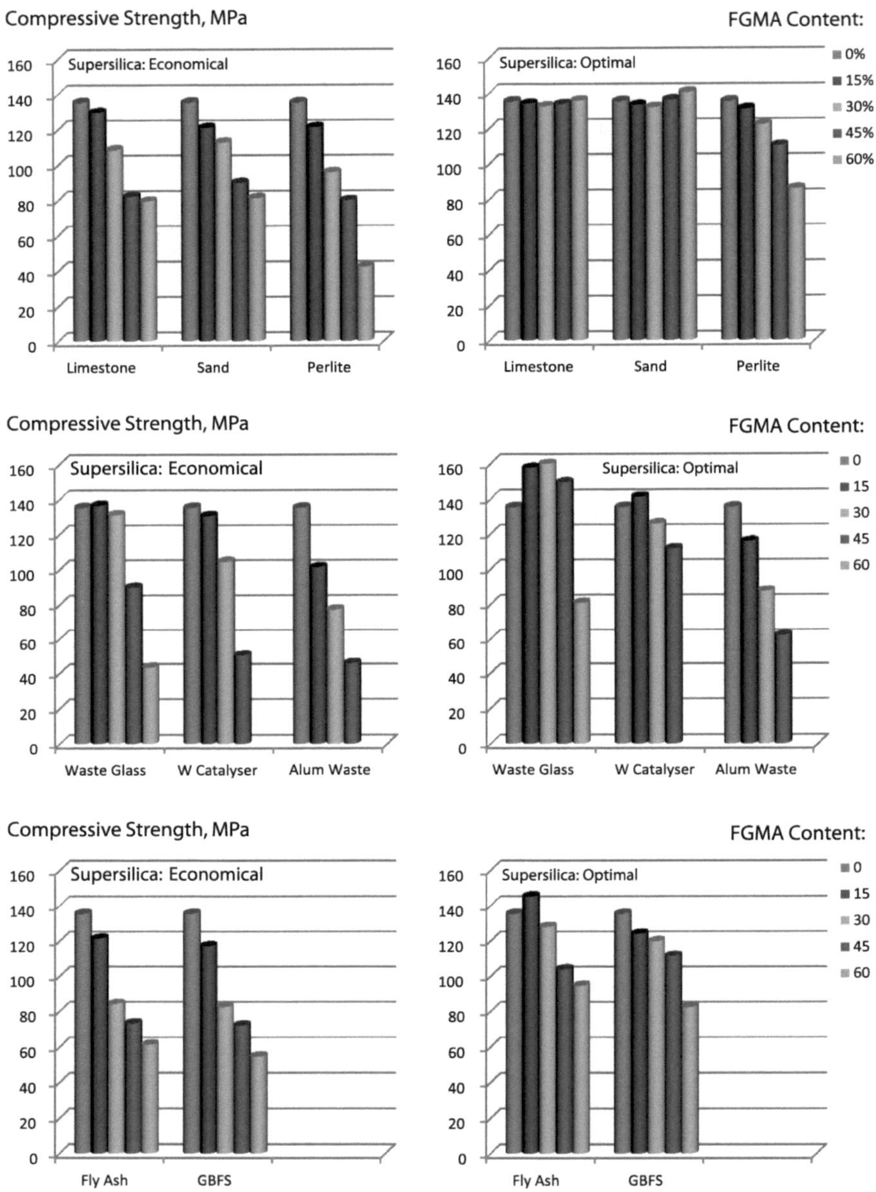

Fig. 15.10 Effect of mineral additives on the compressive strength of HVMA cement

Sustainable development can be defined as simultaneously achieving social, economic, and environmental objectives [12–16]. For the construction industry, sustainability can be presented as:

- Social progress that meets the needs of humankind
- Economic growth
- Effective protection of the environment
- Efficient use of resources

The concept of eco-efficiency was formulated at the 1992 United Nations Conference on Environment and Development, or Earth Summit. The World Business Council for Sustainable Development (WBCSD) has proposed the following definition: "Eco-efficiency is reached by the delivery of competitively priced goods and services that satisfy human needs and bring quality of life, while progressively reducing ecological impacts and resource intensity throughout the life cycle, to a level at least in line with the earth's estimated carrying capacity" [13–15]. WBCSD has identified seven elements of eco-efficiency:

- Reduce the material intensity of products
- Reduce the energy intensity of products
- Reduce toxic emissions
- Increase the engineering performance or ecological benefit
- Enhance material recyclability (recycling ratio)
- Maximize sustainable use of renewable resources
- Extend product durability
- Increase the service life of the products

For the cement industry, these elements of eco-efficiency can be presented in the following guidelines [5]:

- Transfer wet cement technology to dry/semi-dry process and implement technologies for low/reduced-temperature (mineralized/belite) clinker with consequent energy/fuel savings;
- Produce stronger cement and consequently apply stronger concrete in structural members with reduced cross-sections, thereby reducing the amount of concrete raw materials, structural steel, and energy used and reducing bulk consumption of cement and concrete;
- Produce more durable cement-based materials with a significantly increased service life and reduced maintenance costs;
- Apply mineral additives or industrial by-products in blended cement, and reduce energy consumption and the corresponding emissions per unit of the cement produced;
- Implement modern technologies for energy and materials recovery and for reducing heat, dust, and pollutant emissions.

Among the major concerns of the misuse of mineral additives is overdosing which results in reduced early strength, increased water demand of concrete mixtures, and reduced freezing and thawing resistance. The possible adverse effect of blended cements on the properties of concrete must be precisely evaluated in order to meet the specific requirements of a particular construction project.

From a historical perspective, cement made by the Romans around 100 A.D. was highly durable, evidenced by the remains of many ancient structures still standing in Rome today. The earliest reference to the remarkable durability of Roman cement and concrete was made by the architect and engineer Marcus Vitruvius Pollio (25 B.C.). According to Vitruvius, the secret to the durability of Roman cement is based on mixing slaked lime with pozzolana, a volcanic ash from Mount Vesuvius, which resulted in a binder that hardened under water. Modern blended cements also use pozzolana as a mineral additive or cement-replacing material, which is intergrinded or blended with portland cement. At the same time, using volumes of pozzolana that exceed certain limits must be avoided, because it may have a detrimental effect on concrete performance (especially strength, permeability, and durability), despite the evident savings from energy and raw materials.

Blended cements incorporating different mineral admixtures or IBPW, can partly replace the cement clinker, thereby meeting the challenges of increasing bulk production and conserving energy [2–5, 18, 20–24]. In the case of acceptable CO_2 emission levels, which are currently "frozen" by the EC and US at 2010 levels, the share of conventional portland cement in the market must be drastically reduced by the 2015 [2–5].

Furthermore, updating the existing manufacturing facilities extensively for clinker production (which comprises an essential part of the inferior cements) consumes the bulk of capital investments for many cement manufacturers and yields a very slow return; therefore, the expansion of an existing cement plant requires a proportionally high rate of investment. The major part of this investment is associated with the installation of heavy equipment and construction works; however, in the case of green HVMA cement, new investments are required only to upgrade the grinding unit. This may increase the production capacity by 40–50% without additional increase of clinker output.

Since green HVMA cement uses about 30–50% less clinker than inferior cements, it creates less ecological damage, reduces carbon dioxide and other emissions at the source; and uses IBPW materials economically that would otherwise be transported to landfill sites.

15.5 Conclusions

Tribo-chemically activated green eco-cements use less clinker than existing cements, thereby meeting today's environmental and economic challenges by increasing bulk production and conserving energy. TCA green cements also contribute to the reduction of carbon dioxide and other emissions at source.

The production and application of green blended cement helps to reduce environmental degradation connected with the construction industry. Considering the superior properties of TCA binders, the most important ecological aspects associated with eco-cement application include:

- Reduced cross-sections of structures for concrete raw materials, structural steel, and energy savings;
- Reduced construction costs due to the lighter weight of structures and foundations, reduced total energy consumption, less formwork amount, reduced maintenance costs, less equipment loading, and increased speed of erection;
- Significantly increased service life of structures and reduced cost of repairing works due to high durability;
- Replacement of generally used structural steel, polymers, and natural stone;
- Use of industrial by- products and waste in green HP/HVMA cements; reduced energy consumption per unit of the cement produced;
- Decreased land degradation from industrial by-products and waste.

References

1. Cembureau Annual Report, 1994–1995, European Cement Association, World Cement, pp. 20–21, August 1995
2. P.-C. Aitcin, Cements of yesterday and today concrete of tomorrow. *Cement and Concrete Research*, No.30 (2000), pp. 1349–1359
3. J.-C. Roumain, S.L. Sarkar, Cements of the 21st century. *Cement and Concrete Technology in the 2000-s: II International Symposium*, Istanbul (2000), pp. 32–37
4. R.N. Swamy, Holistic design of concrete technology the only route to durability and sustainability in construction. *Cement and Concrete Technology in the 2000-s: II International Symposium*, Istanbul (2000), pp. 58–71
5. K. Sobolev, T.R. Naik, Performance as a factor for sustainability of the cement industry. *CANMET/ACI International Symposium on Sustainable Development of Cement and Concrete*, Toronto, Canada (2005), pp. 295–312
6. K. Sobolev, Mechano-chemical modification of cement with high volumes of blast furnace slag. Cem. Concr. Compos. **27**(7–8), 848–853 (2005)
7. S.P. Shah, Resent trends in the science and technology of concrete; concrete technology: new trends, industrial applications. *Proceedings of the International RILEM Workshop*, E&FN Spon, London (1993), pp. 1–18
8. P. Fidjestol, K. Knudsen, High performance concrete for durability—using microsilica; ERMCO-95. *Proceedings of the XIth European Ready Mixed Concrete Congress*, Istanbul (1995), pp. 379–389
9. S.W. Forster, High-performance concrete: stretching the paradigm. Concr. Int. USA **16**(10), 33–34 (1994)
10. P. Dountran, Present situation of cement standardization in Europe, blended Cements, ed. by G. Frohnsdorff, ASTM ST 897 (1986) pp. 144–154
11. European Environmental Agency http://org.eea.eu.int/documents/Issuerep/envcom/alternative.htm
12. N. Howard, Sustainable construction—the data, Centre for Sustainable Construction/BRE Report CR258/99, March 2000
13. A. Yates, Quantifying the business benefits of sustainable buildings—summary of existing research findings, Centre for Sustainable Construction/BRE, February 2001
14. Eco-efficiency Indicators: A first set of eco-efficiency indicators for industry pilot study, Anite Systems/Eurostat (1999)
15. H. Verfaillie, R. Bidwell, Measuring Eco-Efficiency a Guide to Reporting Company Performance, WBCSD (2000)

16. G. Wischers, K. Kuhlmann, Eco-balance of Cement and Concrete; BFT, No 11 (1991), pp. 33–40
17. V.S. Ramachandran, *Concrete Admixtures Handbook*, 2nd edn. (Noyes Publications, New Jersey, 1995)
18. K. Sobolev, M. Arikan, High-volume mineral additive eco-cement. Am. Ceram. Soc. Bull. **81**(1), 39–43 (2002)
19. V.M. Malhotra, High-performance high-volume fly ash concrete, high performance high-strength concrete: material properties. *Structural Behavior and Field Application* (Perth, 1998), pp. 97–122
20. K. Sobolev, I. Flores, R. Hermosillo, L.M. Torres-Martínez, Application of nanomaterials in high-performance cement composites. *Proceedings of ACI Session on Nanotechnology of Concrete: Recent Developments and Future Perspectives—2006* (Denver, ACI SP-254, 2008), pp. 93–120
21. K. Sobolev, M. Ferrada-Gutiérrez, How nanotechnology can change the concrete world: part 2. Am. Ceram. Soc. Bull.**11**, 16–19 (2005)
22. F. Sanchez, K. Sobolev, Nanotechnology in concrete: a review. Construct. Build. Mater. **24**(11), 2060–2071 (2010)
23. I. Flores, K. Sobolev, L.M. Torres, P.L. Valdez, E. Zarazua, E.L. Cuellar, Performance of cement systems with nano-SiO_2 particles produced using Sol-gel method. *TRB First International Conference in North America on Nanotechnology in Cement and Concrete*, Irvine, 5–7 May 2010
24. K. Sobolev, The effect of complex admixtures on cement properties and development of a test procedure for the evaluation of high-strength cements. Adv. Cem. Res. **AC15**(2), 65–75 (2003)

Chapter 16
Tribological Properties of Fly Ash-Based Green Friction Products

Pradeep K. Rohatgi, Pradeep L. Menezes and Michael R. Lovell

Abstract Growing concern for the environment coupled with the increasing cost of petro-based resources and advancements in the fields of biotechnology, nanotechnology and materials science, and engineering has led to the development of green materials for various applications. Fly ash is a particulate waste by-product formed as a result of coal combustion in power plants. Worldwide, more than 65% of fly ash produced from coal power stations is disposed off in landfills and ash ponds. The recycling of fly ash has become an increasing concern in recent years due to increasing landfill costs and the current interest in sustainable development. The use of fly ash as a filler or reinforcement for composites is desirable from an environmental standpoint. Recently, fly ash was successfully used in metal matrix composites to reduce overall weight and these composites are successfully used in automotive and aerospace applications. The polymer matrix composites developed using fly ash can be used as low cost green friction materials. This study is a review on the tribological behavior of fly ash-based green friction composites to understand their usability for various automotive applications.

Keywords Friction · Wear · Fly ash · Green composite

P. K. Rohatgi (✉)
Department of Materials Engineering, University of Wisconsin-Milwaukee, Milwaukee, WI 53201, USA
e-mail: prohatgi@uwm.edu

P. L. Menezes · M. R. Lovell
Department of Industrial Engineering, University of Wisconsin-Milwaukee, Milwaukee, WI 53201, USA

M. Nosonovsky and B. Bhushan (eds.), *Green Tribology*,
Green Energy and Technology, DOI: 10.1007/978-3-642-23681-5_16,

16.1 Introduction

The automotive industry has come a long way from the first cast iron steam engine to the latest carbon fiber structure in super cars like the Ferrari 360, Chevy Camero SSX, Audi R8 GT to mention a few. The ultimate goal in vehicle design is weight reduction thereby enhancing its fuel efficiency. However, a vehicle is only as good as its braking system. Brakes are one of the most important safety and performance components in automobiles. The purpose of friction brakes is to decelerate a vehicle by transforming the kinetic energy of the vehicle into heat, via friction, and dissipating that heat to the surroundings. There are two common types of friction brakes—drum/shoe brakes and disk/pad brakes. The design of the brakes affects heat flow, reliability, noise characteristics, and ease of maintenance. The Society of Automotive Engineers (SAE) has developed a Friction Identification System for Brake Linings and Brake blocks (see http://standards.sae.org). The codes and the associated friction coefficient values are listed in Table 16.1. However, in recent years, the value of edge codes has become controversial in light of the growing recognition that brake frictional response, and the apparent friction coefficient, are dependent not only on the material composition, but also on the environment, the mechanics of the system, and the duty cycles to which the brakes are subjected. Some of the limitations of edge codes are (1) no indication of wear resistance, (2) failure of the test specimen to represent full-sized brake pads, and (3) the inability to use edge codes as the sole selection criterion for a replacement brake material have also contributed to the aforesaid issue. In order to achieve the required properties for brakes, most brake materials are not composed of single elements or compounds, but rather are composites of many materials. More than 2,000 different materials and their variants are now used in commercial brake components. Table 16.2 summarizes the historical composition of automotive friction brake materials [1].

Most of the brake materials used in American cars are based on a metal fiber reinforced phenolic resin matrix and are called semimetallic. There are, however, also other types of lining materials, categorized into metallic, organic, and carbon-based. Most of the brake materials are typically a composite of a number of different ingredients [2–7]. These ingredients are categorized into four broad classes: (1) binder materials—most commonly used is modified and unmodified phenolic resins (2) structural materials—mostly a combination of synthetic (e.g., aramid), metallic (e.g., steel wool), ceramic (e.g., chopped glass fibers), mineral fibers (e.g., rock wool), and natural fibers (e.g., cellulose) (3) fillers, and (4) frictional additives/modifiers. The binder materials bind together the rest of the ingredients, structural materials provide the structural reinforcement to the composite matrix, fillers make up the volume of the brake materials while keeping the costs down, and friction modifiers stabilize the coefficient of friction. The above constituents are expected to perform synergistically in an optimized brake material to provide a consistent friction and wear performance at varying operating conditions, and should generate less noise and vibrations, and should have low manufacturing cost. Even a slight difference of a percent or two of ingredient

Table 16.1 SAE recommended practice codes and associated friction coefficients for brakes [1]

Code	Friction coefficient
C	10.15
D	>0.15 but ≤0.25
E	>0.25 but ≤0.35
F	>0.35 but ≤ 0.45
G	>0.45 but ≤ 0.55
H	>0.55
Z	Under classified

concentration can significantly affect the performance, so composition balance is very important.

In 1986, the United States Environmental Protection Agency announced a proposed ban on asbestos due to its harmful health hazards. Before the ban, asbestos was the most preferred filler in brake materials. In vehicle brakes its content varied between 30 and 70%. Asbestos had a few engineering characteristics that made it very desirable for inclusion in brake materials. (1) Asbestos is thermally stable up to 500°C, (2) It helps regenerate the friction surface during use, (3) Silicates produced by asbestos are harder and more abrasive than asbestos, (4) It insulates thermally, (5) It is strong, flexible, and mostly cheap, (6) The fibrous character remains intact until about 1,400°C. After the ban on the usage of asbestos in materials, most friction material suppliers and vehicle manufacturers started to shift away from asbestos to non-asbestos materials. Researchers have contended to come up with an equally efficient alternative. This led to the replacement of asbestos by calcium silicate, calcium aluminum silicate, mica, barites, cashew dust, basalt fiber, rockwool, Fiberfrax® ceramic fiber, polyacrylonitrile (PAN), polyester, chopped glass fiber, and aramid fibers. On the other hand, to reduce the weight, cast iron was also replaced in brake disk and drum materials by aluminum-based metal matrix composites. As aluminum-based metal matrix composites are not resistant to high temperatures they are only used on the rear axles of automobiles because the energy dissipation requirements are not as severe compared with the front axle. Since then research is underway to determine the optimum composition of these materials. Efforts have been made to study the friction and wear of aluminum/SiC composite and these composites were also compared with the cast iron properties [8–12]. Similarly the effectiveness of hollow ceramic microspheres has been studied in six brake pad formulations [1, 13, 14]. Further, the development of fly ash-based composites for friction products is underway due to its improved properties, low cost, and environmental issues.

16.2 Fly Ash Material and Characteristics

Fly ash is one of the residues created during the combustion of coal in coal-fired power plants and comprises fine particles that rise with the flue gases. In the past, fly ash produced from coal combustion was simply entrained in flue gases and

Table 16.2 Historical compositions of automotive friction brake materials [1]

Material description	Application(s)	Approximate year
Cast iron on steel	Railroad car brake blocks and tires	prior to1870s
Hair or cotton belting (limited by charring at about 300°F)	Wagon wheels and early automobiles	ca. 1897
Woven asbestos with brass and other wires for increased strength and performance	Automobiles and trucks	ca. 1908
Molded limings with shorter chrysotile fibers, brass particles, and low-ash bituminous coal	Automobiles and trucks	ca. 1926
Dry-mix molded material to replace cast iron brake blocks that produced metallic dust that shorted electric train rails	London underground	ca. 1930
Flexible resin binders developed along with more complex formulations	Brake drum limings	1930s
Resin bonded metallic brake limings	Industrial and aircraft applications	1950s
Glass fibers, mineral fibers, metal fibers, carbon, and synthetic fibers to provide semi-metallics with higher performance than asbestos (beginning of safety issues with asbestos)	Automotive and trucks	1960s
Non-asbestos (fiberglass) materials	Brake drums on original equipment cars	1980s
Suggested use of carbon fibers	Automotive brakes	1991

dispersed into the atmosphere. This created environmental and health concerns that prompted laws which have reduced fly ash emissions to less than 1% of ash produced. Coal-fired power plants all over the world generate huge amounts of fly ash each year, 70 million tons of which are produced in the United States alone. Fly ash is a waste by-product material that must be disposed off or recycled. Worldwide, more than 65% of fly ash produced is disposed off in landfills or in ash ponds. Only 40% of all fly ashes generated in the United States find beneficial applications in concrete, embankments, road subbase, etc. [15, 16]. The recycling of fly ash has become an increasing concern in recent years due to increasing landfill costs and the current interest in sustainable development. Also, there have been few attempts at value-added application of fly ash in composite fabrications [17–19]. In recent years, research on fly ash as a filler and reinforcement in metal-matrix composites (MMCs) and polymer-matrix composites (PMCs) has been growing.

Fly ash particles are classified into two types, precipitator and cenosphere. Generally, the solid spherical particles of fly ash are called precipitator fly ash. It is composed of the crystal compound such as quartz, mullite, and hematite, and glassy compound such as silica glass, and other metal oxides. The precipitator fly

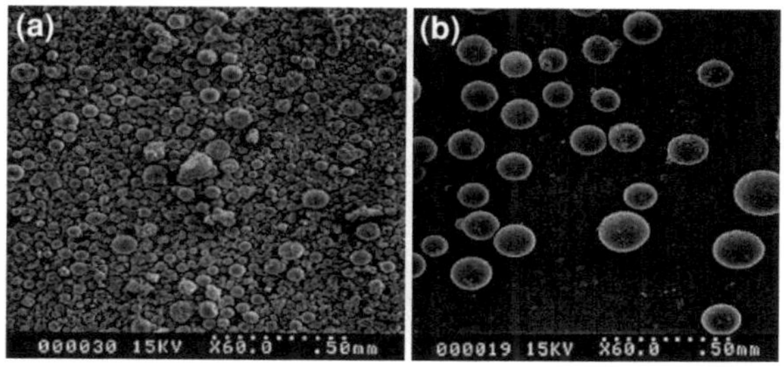

Fig. 16.1 Scanning electron micrograph of **a** Precipitatory fly ash (*left*), **b** Cenosphere fly ash (*right*) [20]

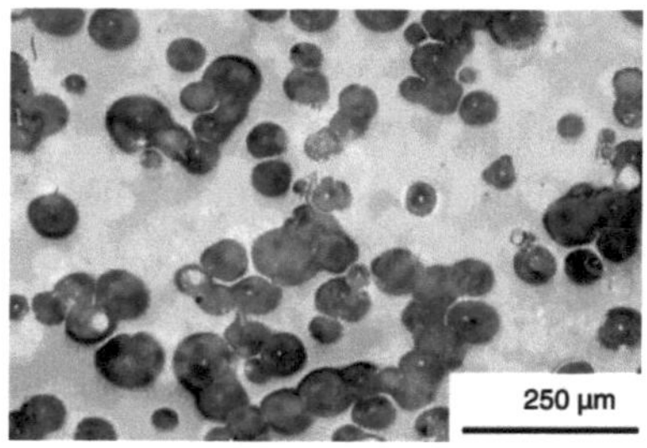

Fig. 16.2 Optical micrograph of a typical polyester/cenosphere composite containing 29.5% fly ash [21]

ash can improve various properties of matrix materials including stiffness, strength, and wear resistance. Cenosphere fly ash, which consists of hollow fly ash particles, can be used for ultra-light materials due to its significantly low density, which is approximately 0.6 g/cm^3 compared to the density of metal matrix, which is in the range of 1.6–11.0 g/cm^3. Figure 16.1 shows scanning electron micrographs of precipitatory and cenosphere fly ash composites [20]. However, due to the lower density of cenospheres, the particles rise to the surface when incorporated in the matrix as shown in Fig. 16.2 [21].

Fly ash particles possess certain characteristics that make them suitable for use in friction composites as a filler material. Compositionally fly ash is mainly a mixture of SiO_2, Al_2O_3, $CaSO_4$, and un-burnt carbon. They are composed of fine-sized particles (mean particle size of 10–30 µm) with uniform physical and engineering characteristics. Since fly ash particles are typically generated above 1,000°C, they can be expected to provide thermally stable bulk at high temperature environment which is a highly desirable characteristic for friction composite materials. The specific heat of fly ash particles is also high ($\approx$800 kJ/kg K). This attribute of the fly ash may help store the excess heat generated at the braking interface of lining and rotor. The poor conductivity of fly ash may help in preventing the heat generated to travel towards the backing plate side of the brake, which may vaporize the braking fluid. Also, fly ash is readily available at negligible cost. Fly ash has the potential to improve the abrasion resistance and porosity of the material while simultaneously enhancing the specific performance. Thus, it could be fundamentally expected that their incorporation would possibly cater to various functional roles which otherwise are expected from a set of fillers as

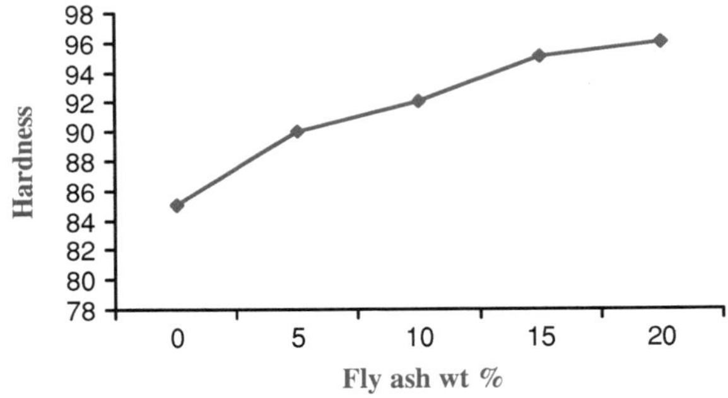

Fig. 16.3 Variation in hardness (shore durometer) with wt% of precipitator fly ash in polyester matrix composite [26]

ingredients. The above-mentioned characteristics present fly ash as a potential ingredient in a brake material. The successful incorporation of fly ash or fly ash derivatives into friction material formulations could greatly reduce the cost of the friction material. On the downside, fly ashes all over the world differ mineralogically from each other due to variability in quality of coal used and overall combustion processes adopted in the respective power plants.

Many researchers are exploring the use of fly ash as reinforcing filler in polymer matrix composites. More specifically, fly ash has been successfully incorporated into both thermoset and thermoplastic matrices [22, 23]. Studies on the mechanical properties of these composites show that fly ash improves mechanical properties like the modulus, tensile strength, hardness, abrasion resistance of the polymer matrix composites [24, 25]. A typical variation in hardness with fly ash content is shown in Fig. 16.3 [26]. In unsaturated polyester resin matrix precipitator fly ash has improved the abrasion resistance and hardness with increasing amounts of fly ash in the composite [26]. Some efforts had already been made to incorporate fly ash in brake pad composites and several materials were optimized with fly ash content up to 25 wt% [27, 28]. Hee and Filip [27] studied the performance of ceramic enhanced phenolic matrix brake lining materials for automotive application. They showed that formulated brakes' test samples, which included up to 25 wt% of fly ash, outperformed the commercially available original brake pads in full-scale automotive dynamometer tests (SAE J2430). Thus, the authors [27] concluded that the fly ash by-product can be effectively used in brake lining formulations when properly combined with other additives. Extensive intellectual property search reveal a couple of US patent documents stating that fly ash has successfully been incorporated into some friction composites up to 50% by volume in combination with phenolic resin as binder, carbon fiber, aluminium fibers, rock wool, and Kevlar fibers (15% by volume) [29, 30]. The document also claims the fly ash to binder ratio that was achieved was 2.5:1. Similarly, in another independent classified document the development of friction materials comprising $\approx$12% of fibers by volume and a major fraction of fly ash-based derivatives as friction modifiers instead of normal space fillers [31]. Thus, the successful incorporation of fly ash or its derivatives into friction material formulations could

Table 16.3 Fractions used to form friction composite [33]

Composite ID	Polymer binder (volume %)	Slag fibers (volume %)	Scrubber sludge (volume %)	By-product (volume %)
Basic matrix	45	22	33	
Fly ash	42	20	30	8 (fly ash)
Bottom ash	42	20	30	8 (bottom ash)
Carbon-1	42	20	30	8 (unburned carbon-1)
Carbon-2	42	20	30	8 (unburned carbon-2)

greatly reduce the cost of the friction material. Several patents highlighting the possible use of fly ash as a key ingredient in friction material sector are already reported though they remain obscure from the general material researchers as their outcomes are highly sensitive from the techno-commercial and ecological points of view [30–32]. However, the partial/complete realization of such an ideology of utilization of industrial waste conforming to "green" norms conceptually depends on the performance assessment and its fluctuation due to variation in the nature of the waste material which may vary widely as the grades of coals. Hence, to address this problem of performance alterations of compositional origin and precisely due to change in the nature of added fly ash into friction composites is of key concern.

Based on the aforementioned success in incorporating fly ash in friction composites, it was also proposed to incorporate more than 50 wt% of fly ash in brake lining formulations. Application of fly ash in intimately mixed hybrid composites as friction materials ideally conform to such an economy driven ideology where the amount of fillers in the composites can be up to 70% other than the binder, friction modifiers and specialty additives, and fibrous reinforcements.

16.3 Fly Ash-Based Composites for Friction Braking Application

Malhotra et al. [28] studied the effect of fly ash and bottom ash on the frictional behavior of a base composite matrix consisting of phenolic resin, slag fiber, and scrubber sludge. Table 16.3 shows the various fractions used to form friction composite. Figure 16.4 shows the variation of friction coefficient with time for basic matrix composite and commercial automotive brake pads. It can be seen that the commercial brake composite showed a stabilized friction value of around 0.3. However, this was not the case for the basic matrix composite which showed strong variations and lower friction value. Figure 16.5 shows the frictional behavior of the composite with and without fly ash. As shown in Fig. 16.5, the incorporation of fly ash particles in friction composites increases the overall coefficient of friction. Figure 16.6 shows the frictional behavior of fly ash and bottom ash-based composite. It can be seen that the frictional response was more

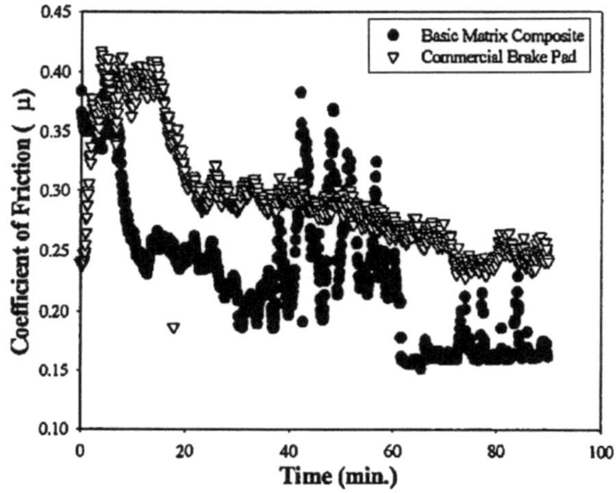

Fig. 16.4 Friction values obtained for basic matrix composite and commercial brake pad [33]

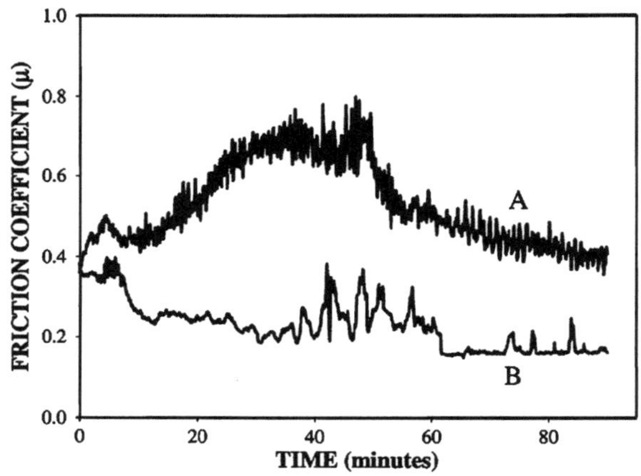

Fig. 16.5 The frictional behavior of: *A* frictional composite formulated from binder, slag fibers, scrubber sludge (30 vol%), and fly ash (8 vol%) and *B* basic matrix formulated from binder, slag fibers, scrubber sludge (33 vol%) [28]

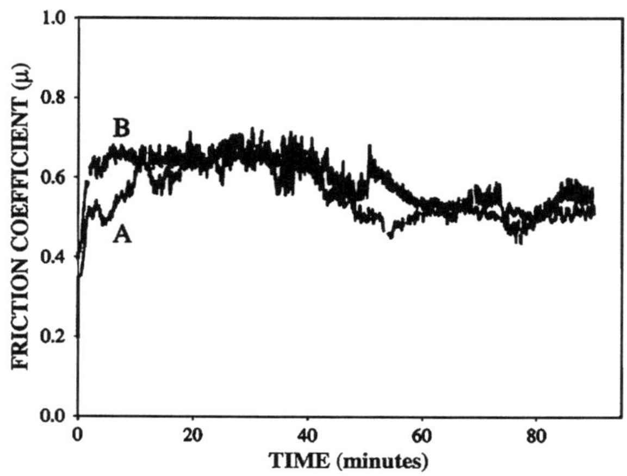

Fig. 16.6 The frictional behavior of a composite fabricated from binder, slag fibers, scrubber sludge (25 vol%) and *A* fly ash (20 vol%) and *B* bottom ash (20 vol%) [28]

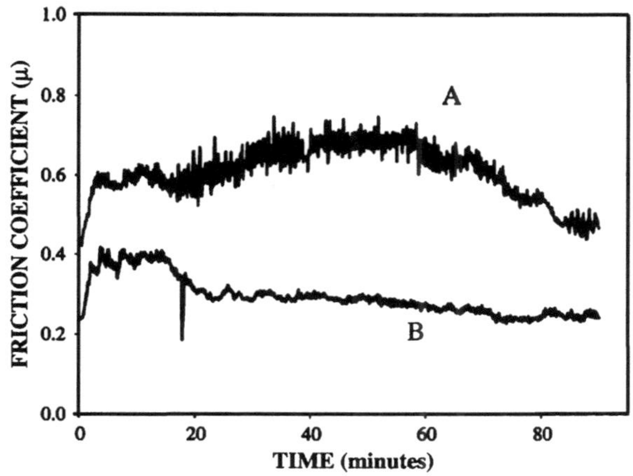

Fig. 16.7 The frictional behavior of: *A* frictional composite fabricated from binder, slag fibers, scrubber sludge (30 vol%), and bottom ash (8 vol%) and *B* a commercial automotive brake which besides other ingredients contained asbestos fiber [28]

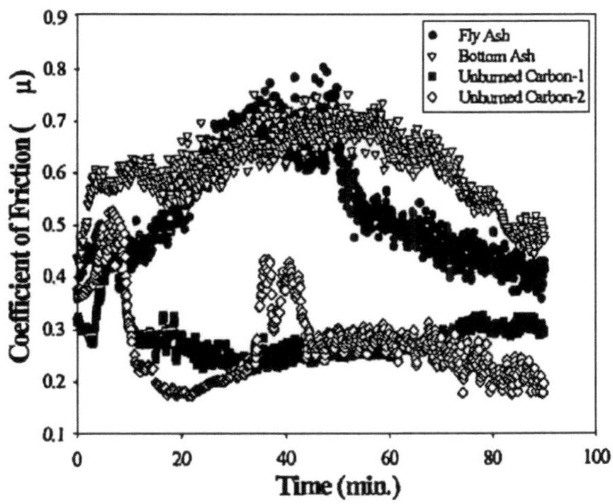

Fig. 16.8 Friction values obtained for fly ash, bottom ash, and uncarbon-based composites [33]

or less the same for both types of ash materials. Figure 16.7 shows the frictional behavior of bottom ash-based composite and asbestos fiber-based composite. It was found that the bottom ash-based composite shows better frictional performance than the asbestos fiber-based composites. It was concluded that the behavior of ash particles was similar to abrasives like Al_2O_3, Cr_2O_3, and SiC. The frictional and wear performance of this composite was compared with a commercial automotive brake material and it was reported that with 20 vol% of fly ash or bottom ash and 30 vol% scrubber sludge composites performed similarly to that of actual commercial brake pads. In addition to this study, Malhotra et al. [33] studied the influence of fly ash, bottom ash, and unburn carbon (extracted from fly ash) on the frictional property of the brake composites. The addition of 8 vol% fly ash, bottom ash, or unburn carbon affected the frictional performance of the basic composites as shown in Fig. 16.8. The unburn carbon-based composites gave

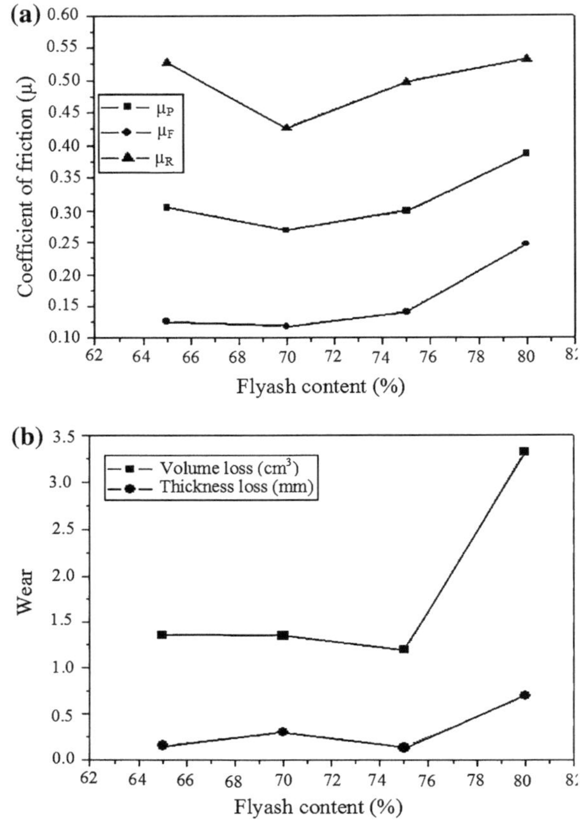

Fig. 16.9 Friction performance of the composites: **a** μ_P (*P*–Performance), μ_F (*F*–Fade) and μ_R (*R*–Recovery) versus fly ash content of the composite. **b** Wear of the composites as a function of fly ash content in the composites [34]

much lower friction values when compared to fly ash and bottom ash-based composites. However, the unburn carbon-based composites not only induced stable frictional behavior but also substantially reduced the noise and wear of brake composite.

Dadkar et al. [34] studied tribological properties of fly ash-filled and aramid fiber reinforced (5 wt%) hybrid polymer matrix composites for friction braking applications. Figure 16.9 shows the variation in friction coefficient and wear with fly ash content. The coefficient of friction was observed to be highest in the composite with the highest amount of fly ash. Wear was largely dominated by the intrinsic nature of the material followed by the temperature rise in the disc.

Mohanty and Chugh [35] studied the tribological performance of fly ash-based composites for automotive brake lining application with up to 65 wt% of fly ash content in combination with graphite, aramid, and potassium titanate whiskers as structural reinforcement along with copper powders. It was found that the brake composites based on fly ash can be $\approx$50–60% lighter than the mostly available commercial brake linings for similar friction, wear, and temperature performance

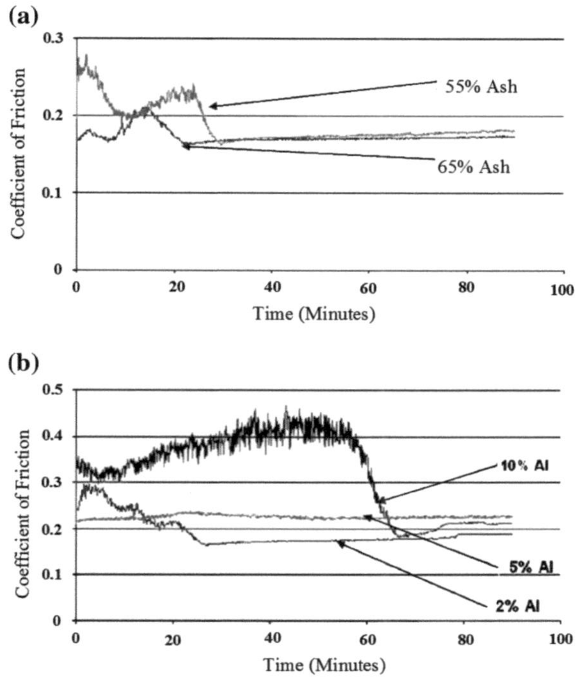

Fig. 16.10 Variation in friction coefficient with time of **a** fly ash, and **b** fly ash and Al fibers-based composite [35]

under dry and wet conditions. Figure 16.10 shows the variation in friction with time for fly ash-based composites with and without aluminum fiber content. The composite made of only fly ash and phenolic resin showed low friction coefficient and high wear rates. Incorporating the aluminum fibers in the matrix improved the friction and wear behavior of the composites. It was also stated that incorporating higher amount of fly ash to the composite significantly improves the tribological behavior. Hence, sufficient structural reinforcement to the composite matrix that contains a large percentage of fly ash particles can be obtained by including aramid pulp and potassium titanate in the composition. Hence, a green friction product was developed by eliminating aluminum fibers and including fly ash in phenolic resin, aramid pulp, and potassium titanate. It was reported that the addition of graphite to the composition helps the fly ash composite sample maintain stable friction performance and lower wear rates. Figure 16.11 shows the variation in friction in the absence and presence of graphite ingredient. The authors [35] also stated that a metallic constituent in the form of copper fiber/powder is needed to scrape off the excess lubricating layer generated by graphite. The temperature developed at the contact interface was also studied for baseline (brake lining composite) and fly ash composites. It was found that both the fly ash and baseline composites witness similar temperature rise behavior as shown in Fig. 16.12.

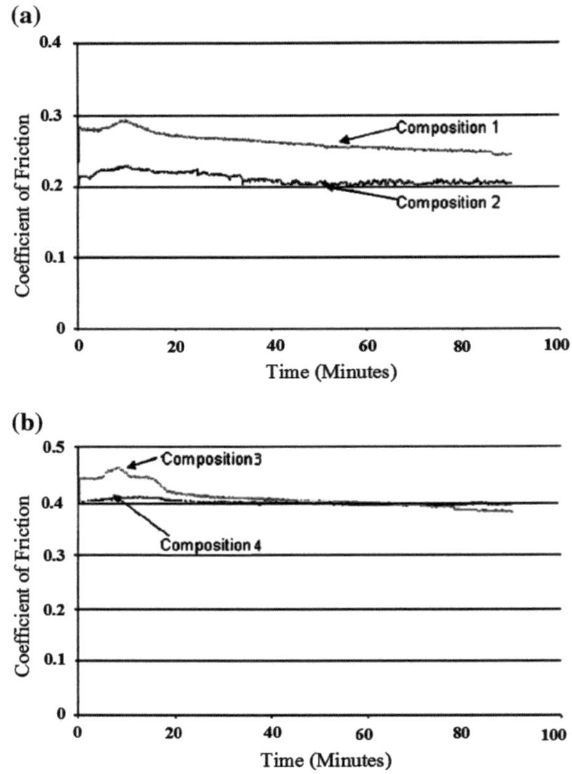

Fig. 16.11 Variation in friction with time **a** before, and **b** after adding graphite to the composite [35]

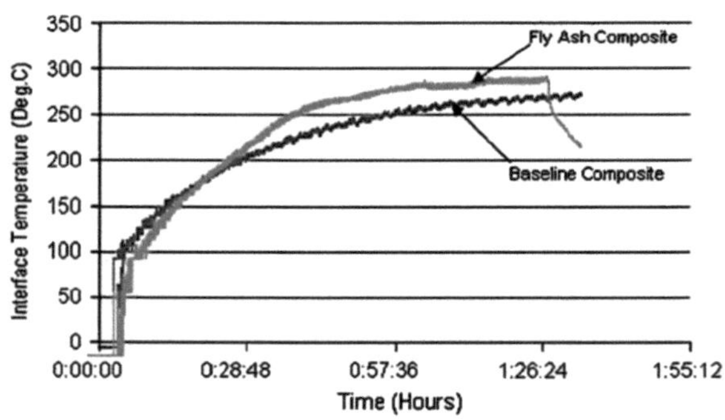

Fig. 16.12 Temperature at the friction interface as a function of time for fly ash and baseline composites [35]

16.4 Fly Ash-Based Composites for Antifriction Materials

Chauhan et al. [36] studied the tribological behavior of glass fiber reinforced vinyl ester composites filled with fly ash particulates. Figure 16.13 shows the variation in coefficient of friction and wear rates with fly ash content. The results showed

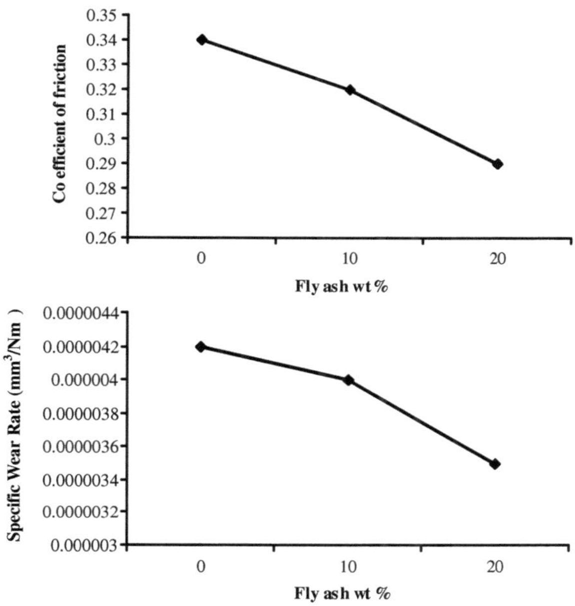

Fig. 16.13 Variation in coefficient of friction and wear rate with fly ash content in hybrid PMC [36]

that the inclusion of fly ash as filler material in glass vinyl ester composites decreases the coefficient of friction and increases the wear resistance of the glass vinyl ester composites significantly. Zhang et al. [37] studied the friction and wear behavior of glass fiber and fly ash reinforced monomer casting nylon composites. The results showed that the tribological properties of composites were significantly affected by the proportion of glass fiber and fly ash. The glass fiber and fly ash reinforced composites had lower friction coefficients and better wear resistance than monomer casting nylon. Ray and Gnanamoorthy [38] studied the friction and wear behavior of vinylester resin matrix composites filled with fly ash particles (40 and 50 wt%). They found that the friction coefficient and wear rate were much higher in neat resin compared to composites. Among the composites, a better wear resistance was shown by the 40% fly ash filled composites with a lower wear loss, lesser linear wear, and a minimum value of the coefficient of friction. Thus, these composite can be used as antifriction material.

16.5 Conclusions

Fly ash is a particulate waste material formed as a result of coal combustion in power plants. The recycling of fly ash has become an increasing concern in recent years due to environmental issues. There are many ways of fly ash utilization. One of them is to include fly ash in the composites to reduce the overall weight of the final product. Thus, the use of fly ash as a filler or reinforcement for composites is very desirable from an environmental standpoint and cost reduction. The composites developed from fly ash showed improved mechanical and tribological performance, specifically for brake application. Hence, in this study, a review on the tribological performance on fly ash-based green friction composite was made. The following conclusion can be drawn from this study.

- Fly ash-based composites showed improved friction and wear behavior when compared to that of actual commercial brake materials.
- The developed compositions can be 50–60% lighter than the current commercial friction products for similar friction, wear and temperature performance under dry and wet conditions.
- The potential for development of commercial grade friction products using fly ash as an ingredient is high.

References

1. P.J. Blau, Compositions, functions, and testing of friction brake materials and their additives, Report by Oak Ridge National Laboratory, USA, 2001
2. P.J. Blau, J.C. McLaughlin, Effects of water films and sliding speed on the frictional behavior of truck disc brake materials. Tribol. Int. **36**, 709–715 (2003)
3. P.J. Blau, H.M. Meyer III, Characteristics of wear particles produced during friction tests of conventional and unconventional disc brake materials. Wear **255**, 1261–1269 (2003)
4. H. Jang, K. Ko, S.J. Kim, R.H. Basch, J.W. Fash, The effect of metal fibers on the friction performance of automotive brake friction materials. Wear **256**, 406–414 (2004)
5. W. Krenkel, F. Berndt, C/C–SiC composites for space applications and advanced friction systems. Mater. Sci. Eng. A **412**, 177–181 (2005)
6. J. Bijwe, Composites as friction materials: recent developments in non-asbestos fiber reinforced friction materials–a review. Polym. Compos. **18**, 378–396 (1997)
7. M.G. Jacko, P.H.S. Tsang, S.K. Rhee, Automotive friction materials evolution during the past decade. Wear **100**, 503–515 (1984)
8. R.K. Uyyuru, M.K. Surappa, S. Brusethaug, Effect of reinforcement volume fraction and size distribution on the tribological behavior of Al-composite/brake pad tribo-couple. Wear **260**, 1248–1255 (2006)
9. R.K. Uyyuru, M.K. Surappa, S. Brusethaug, Tribological behavior of Al-Si-SiC_p composites/automobile brake pad system under dry sliding conditions. Tribol. Int. **40**, 365–373 (2007)
10. K. Laden, J.D. Guérin, M. Watremez, J.P. Bricout, Frictional characteristics of Al–SiC composite brake discs. Tribol. Lett. **8**, 237–247 (2000)
11. S. Zhang, F. Wang, Comparison of friction and wear performances of brake material dry sliding against two aluminum matrix composites reinforced with different SiC particles. J. Mater. Process. Technol. **182**, 122–127 (2007)
12. N. Natarajan, S. Vijayarangan, I. Rajendran, Wear behaviour of A356/25SiC_p aluminium matrix composites sliding against automobile friction material. Wear **261**, 812–822 (2006)
13. A. D'souza, Filled polymer composites, US Patent 0104943, 2007
14. S. Steinmetz, H.-D. Elison, J. Meiers, M. Kraus, O. Runge, Friction material, European Patent EP1694979B1, 2010
15. D.C. Adriano, A.L. Page, A.A. Elseewi, A.C. Chang, I. Straughan, Utilization and disposal of fly ash and other coal residues in terrestrial ecosystems: a review. J. Environ. Qual. **9**, 333–344 (1980)

16. R.S. Iyer, J.A. Scott, Power station fly ash–a review of value-added utilization outside of the construction industry. Resour. Conserv. Recycl. **31**, 217–228 (2001)
17. R.Q. Guo, P.K. Rohatgi, D. Nath, Preparation of aluminium-fly ash particulate composite by powder metallurgy technique. J. Mater. Sci. **32**, 3971–3974 (1997)
18. P.K. Rohatgi, R.Q. Guo, B.N. Keshavaram, Cast aluminum alloy–fly ash composites. Key Eng. Mater. **104–107**, 283–292 (1995)
19. R.Q. Guo, P.K. Rohatgi, D. Nath, Compacting characteristics of aluminium-fly ash powder mixtures. J. Mater. Sci. **31**, 5513–5519 (1996)
20. T. Matsunaga, J. Kim, S. Hardcastle, P. Rohatgi, Crystallinity and selected properties of fly ash particles. Mater. Sci. Eng. A **325**, 333–343 (2002)
21. P.K. Rohatgi, T. Matsunaga, N. Gupta, Compressive and ultrasonic properties of polyester/fly ash composites. J. Mater. Sci. **44**, 1485–1493 (2009)
22. K.T. Varughese, B.K. Chaturvedi, Fly ash as fine aggregate in polyester based polymer concrete. Cem. Concr. Compos. **18**, 105–108 (1996)
23. R.T. Hemmings, R.L. Hill, B.J. Cornelius, Filler comprising fly ash for use in polymer composites, US Patent 0032707Al, 2003
24. K.W.-Y. Wong, R.W. Truss, Effect of fly ash content and coupling agent on the mechanical properties of fly ash-filled polypropylene. Compos. Sci. Technol. **52**, 361–368 (1994)
25. S. Bose, P.A. Mahanwar, Effect of fly ash on the mechanical, thermal, dielectric, rheological and morphological properties of filled nylon 6. J. Miner. Mater. Charact. Eng. **3**, 65–89 (2004)
26. G. Subhashini, B.H. Renuka, H.R. Imamkhasim, Studies on fly ash characteristics and mechanical properties of fly ash polyester matrix composite, Internal Report, CPRI, India, 2004
27. K.W. Hee, P. Filip, Performance of ceramic enhanced phenolic matrix brake lining materials for automotive brake linings. Wear **259**, 1088–1096 (2005)
28. V.M. Malhotra, P.S. Valimbe, M.A. Wright, Effects of fly ash and bottom ash on the frictional behavior of composites. Fuel **81**, 235–244 (2002)
29. J. Pan, Y. Naerheim, T.T.-L. Liao, P. Min, Brake pad lining filler material, US patent 0055126, 2003
30. R.B. Pond, Metal composites with fly ash incorporated therein and a process for producing the same, US patent 4888054, 1989
31. Y.P. Chugh, Friction materials comprising coal combustion and coal gasification byproducts, US patent 0121474, 2008
32. B.K. Satapathy, A. Majumdar, B.S. Tomar, Optimal design of fly ash filled composite friction materials using combined analytical hierarchy process and technique for order preference by similarity to ideal solutions approach. Mater. Des. **31**, 1937–1944 (2010)
33. V.M. Malhotra, P.S. Valimbe, M.A. Wright, Fabrication of automotive brake composite from unburned carbon. ACS. Fuel Div. **45**(3), 504–508 (2000)
34. N. Dadkar, B.S. Tomar, B.K. Satapathy, Evaluation of fly ash-filled and aramid fibre reinforced hybrid polymer matrix composites (PMC) for friction braking applications. Mater. Des. **30**, 4369–4376 (2009)
35. S. Mohanty, Y. Chugh, Development of fly ash-based automotive brake lining. Tribol. Int. **40**, 1217–1224 (2007)
36. S.R. Chauhan, A. Kumar, I. Singh, P. Kumar, Effect of fly ash content on friction and dry sliding wear behavior of glass fiber reinforced polymer composites: a taguchi approach. J. Miner. Mater. Charact. Eng. **9**, 365–387 (2010)
37. S.H. Zhang, G. Chen, C. Cui, C. Mi, F. Tian, Study on friction and wear behavior of glass fiber and fly ash reinforced MC nylon composites. Advan. Tribol. Part 3, II 460–463 (2010)
38. D. Ray, R. Gnanamoorthy, Friction and wear behavior of vinylester resin matrix composites filled with fly ash particles. J. Reinf. Plast. Compos. **26**, 5–13 (2007)

Chapter 17
Self-Lubricating Behavior of Graphite Reinforced Metal Matrix Composites

Pradeep L. Menezes, Pradeep K. Rohatgi and Michael R. Lovell

Abstract Lubricants are extensively used between the contacting surfaces to reduce friction and wear. These lubricants are usually toxic and not readily biodegradable and thus these lubricants can cause considerable damage to the environment. The use of external lubricants can be eliminated by designing self-lubricating composite materials. These composite materials have the ability to achieve low friction and wear at the contact surfaces without any external supply of lubrication during sliding. The metal matrix composites reinforced with various self-lubricating particles such as graphite, molybdenum disulfide are being used as self-lubricating materials for various engineering applications. In this paper, the tribological behavior of metal matrix composites reinforced with graphite particles has been reviewed. More specifically, aluminum-graphite, magnesium-graphite, copper-graphite, silver-graphite and nickel-graphite composites have been studied. The influence of various variables on the friction coefficient and wear rate is discussed. It was found that the amount of graphite released on the wear surface forms a tribo film on the contact surfaces. This reduces the overall friction coefficient and wear rate. The formation and retention of this tribo layer on the sliding surface as well as its composition, area fraction, thickness and hardness are important factors in controlling the friction and wear behavior of the material and depend on the nature of the sliding surface, the test condition, environment and the graphite content in the composite. The presence of graphite tribo layer also

P. L. Menezes · M. R. Lovell
Department of Industrial Engineering, University of Wisconsin-Milwaukee,
Milwaukee, WI 53201, USA
e-mail: mlovell@uwm.edu

P. K. Rohatgi (✉)
Department of Materials Engineering, University of Wisconsin-Milwaukee,
Milwaukee, WI 53201, USA
e-mail: prohatgi@uwm.edu

M. Nosonovsky and B. Bhushan (eds.), *Green Tribology*,
Green Energy and Technology, DOI: 10.1007/978-3-642-23681-5_17,

increases the seizure resistance and enables to run under boundary lubrication without galling.

Keywords Friction · Wear · Self-lubrication · Metal matrix composites

17.1 Introduction

A lubricant is a material used to facilitate relative motion of solid bodies by minimizing friction and wear between interacting surfaces. A lubricant functions by introducing a layer of material with lower shear strength than the surfaces themselves between the sliding surfaces [1]. In some lubricated systems, the lubricant may not completely prevent asperity contact, although it will reduce it and may also reduce the strength of the junctions formed. In other cases, the lubricant completely separates the surfaces and no asperity junctions are formed at all [2]. In general, there are three categories of lubricants—liquid, solid and gaseous.

Nearly all lubricants used in the automotive and manufacturing sectors are oil or grease-based. These oils include engine oils, transmission fluids, hydraulic fluids and gear oils. These products are not typically environmentally friendly or biodegradable and can introduce significant quantities of pollutants into the waste stream [3]. Thus, disposal of this very large volume of material obviously represents an environmental burden. Hence, it is very important to develop either green lubricants or composite materials which have self-lubrication properties.

Composite materials are a special class of materials and are made from two or more constituent components with significantly different physical or chemical properties which remain separate and distinct at the macroscopic or microscopic scale within the finished structure. In the case of two constituent components, one with the highest volume fraction is considered as the matrix whereas the other is the reinforcement modifying the properties of the matrix. In metal matrix composites, the matrix is made of metal and the reinforcement may be a different metal or another material, such as a ceramic or organic compound. When the composite has more than two materials then it is called a hybrid composite. Metal matrix composites are synthesized by dispersing a reinforcing material into a metal matrix. Sometimes, the reinforcements are coated to prevent a chemical reaction with the matrix. The matrix is the monolithic material into which the reinforcement is embedded, and is completely continuous. The reinforcement material is embedded into the matrix, usually in the form of particles, fibers, whiskers or wires. The most common reinforcing materials are alumina and silicon carbide. The reinforcement can be either continuous or discontinuous. Discontinuous metal matrix composites can be isotropic. The reinforcement does not always serve a purely structural task, but used to change physical properties such as friction coefficient, wear or thermal conductivity. Metal matrix composites have been used in the aerospace, aircraft and automotive industries because they possess many potential advantages over monolithic materials, including higher specific strength and stiffness, higher wear resistance, higher thermal conductivity and lower coefficient of thermal expansion.

When solid lubricants such as graphite or molybdenum disulfide (MoS_2) are introduced at the sliding interfaces, the friction and wear are observed to decrease. Graphite and MoS_2 are important solid lubricants, and these have a layered structure [4]. The layer consists of flat sheets of atoms or molecules, and the structure is called a layer-lattice structure. The atoms are strongly bonded within the sheets and weakly bonded between the layers. The important effect is that the materials can shear more easily parallel to the layers than across them. They can therefore support relatively heavy loads at right angle to the layers while still being able to slide easily parallel to the layers. This property is being effectively used for lubrication process. It is also important that the solid lubricant should adhere strongly to the bearing surface; otherwise it would be easily rubbed away and gives very short service life. Graphite which consists of carbon atoms arranged in a layer like structure, displays a very low coefficient of friction ranging from 0.1 to 0.2 while sliding on another clean surface, thus suggesting that it can be used as a solid lubricant. It is reported that graphite is an effective lubricant additive for its anti-corrosion, high-temperature endurance and self-lubricating properties derived from its lamellar structure [5]. The slippage between lamellar endows the graphite excellent lubricating properties under heavy loads [6].

One of the major issues in using graphite as a solid lubricant lies in maintaining a continuous supply of graphite which acts as a solid lubricant between two sliding surfaces. Such a continuous supply of graphite is more easily maintained in the case of fluid lubricants when compared to solid lubricants. An interesting innovation to ensure a supply of graphite between sliding surfaces is to incorporate graphite into the metal matrix of one of the sliding components by forming a metal matrix composite. Thus, self-lubricating composites are the materials in which solid lubricants such as graphite, MoS_2 etc., are introduced as reinforcement in the matrix during preparation.

It is well known that during sliding, wear particles are produced at the contacting interface. Sometimes these wear particles act as third body abrasive and the friction and wear rates increase with time. Hence, lubricants are generally used to reduce the friction coefficient and wear rate. Self-lubricating composites have unique nature in that the wear particles formed at the interface acts as solid lubricants and decrease the friction coefficient and wear rate. Under sliding conditions, the metal-graphite composite becomes self-lubricating because of the transfer of the graphite embedded in its matrix to the tribo surfaces and formation into a thin film of graphite which prevents direct contact between the mating surfaces. Thus friction and wear are reduced and this process eliminates usage of any types of external lubricants. The success of metal matrix graphite particle composites depends on the ability of the graphite particles to come out of their embedded state in the matrix and spread evenly in the form of a solid lubricating film over the tribological surface to provide lubrication.

Metal matrix graphite particle composite combines the properties of strength, hardness, and abrasion resistance of the matrix alloy with the natural lubricity of graphite. The matrix alloy may also be chosen to impart good electrical and thermal conductivity to the composite. Much of the early work on metal matrix graphite particle composites was done using powder metallurgy which is expensive and limited in the sizes that can be produced. Rohatgi and coworkers [7–10] first introduced graphite as a solid lubricant in aluminum matrices by casting routes, involving mixing the molten alloy with graphite particles to make a uniform suspension and followed by casting. The problem of graphite rejection by liquid aluminum is always encountered in the casting, caused by density differences and poor wettability between the matrix and reinforcement. These problems have been overcome to a large extent at the laboratory scale by the use of metal coatings such as Ni and Cu on the particle of graphite, and by the addition of reactive elements (e.g., Mg and Ti) to the melt. This work was subsequently extended to other solid lubricant dispersions but aluminum-graphite by far has the most potential for commercial applications. Cast aluminum-graphite composites for antifriction applications were developed beginning in 1966. The cast aluminum-graphite alloys have a unique structure, in which graphite particles are in the interdendrite regions and the alloys have been reported to have superior tribological properties compared with similar alloys prepared by powder metallurgy techniques. These alloys have been cast into tribological components using sand casting, pressure die casting, gravity die casting and centrifugal casting. The particulate composites of graphite and a matrix alloy are characterized by (1) the composition and microstructure of the matrix alloy, (2) the size, volume fraction and distribution of particles, and (3) the nature of the interface between the matrix and the dispersed graphite. In addition, the composites may have defects such as voids and porosity which, unless accounted for, will result in an erroneous estimate of area of contact and thereby influence friction and wear. Further, the strength of the composite will reduce with increased porosity content, enhancing wear rate.

In metal matrix composites, graphite presumably imparts improved tribological properties to the composite through the formation of a graphite-rich film on the tribo-surface which provides solid lubrication [11, 12]. This lubricating film is formed as a result of shearing graphite particles located immediately below the sliding surface of the composite. This graphite-rich lubricant film helps to reduce the magnitude of shear stress transferred to the material underneath the contact area, alleviates the plastic deformation in the subsurface region, prevents a metal-to-metal contact and acts as a solid lubricant between two sliding surfaces. Therefore, it helps in reducing friction and wear and improves seizure resistance of the composites. The formation and retention of this tribo layer on the sliding surface as well as its composition, area fraction, thickness and hardness are important factors in controlling the wear behavior of the material and depend on the nature of the sliding surface, the test condition, environment and the graphite content in the composite. The formation of the lubricating layer at the sliding surface becomes thicker when addition of graphite to the base alloy increases and it is this graphite lubricating layer that is responsible for playing an effective role

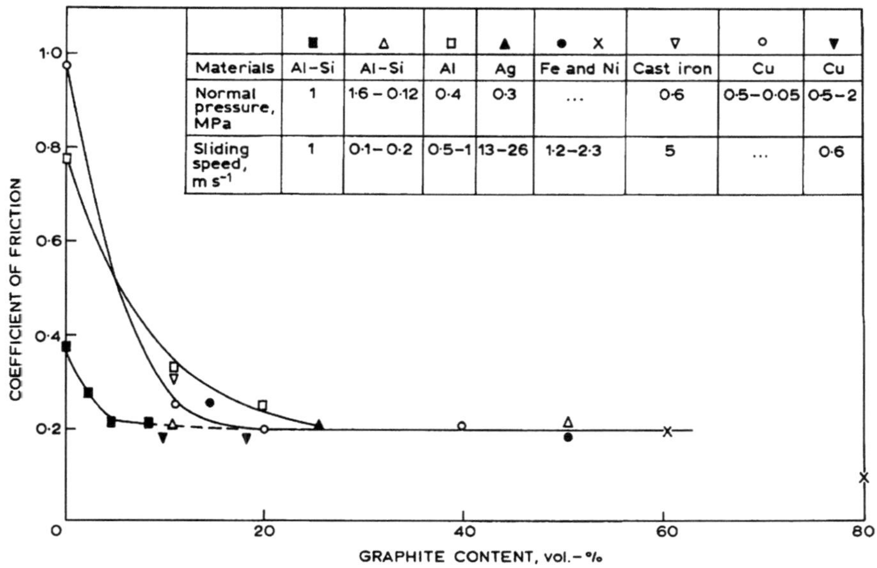

Fig. 17.1 Variation of coefficient of friction with graphite content for composites with different base alloys [12]

for keeping the friction and wear behavior of the composite low. The formation of the solid lubricating film also depends on the matrix characteristics, for example, its deformability helping the process of transfer of graphite to the tribo-surface, adhesion of graphite film to the matrix, and the presence of an environment which permits graphite to spread in the form of a film and act as a solid lubricant. Earlier works have also identified that with increasing graphite content, the lubricating film formed on the lubricating surface lowers the wear rate [11, 12]. Several researchers [13, 14] have also observed the solid lubricating film formed on the wearing surface and reported that graphite particulate in the composite helps in reducing the friction coefficient and increases the anti-seizing quality and thus improve tribological behavior of the base alloy.

In metal matrix graphite composites, graphite particles are embedded in the matrix, and the formation of a graphite film will take place by transfer of the graphite from the particles embedded in the matrix on to the tribo surfaces during initial periods of sliding. The observed friction and wear behavior will, therefore, have two distinct stages: (1) transient state, while a graphite film is being established, and (2) steady state, when a stable graphite film (in the dynamical sense of being continuously replenished to make up for the wear loss) has formed. The variation of coefficient of friction with graphite content observed in a number of particulate graphite composites with aluminum, copper, iron, nickel and silver as the matrix is shown in Fig. 17.1 [12]. The tests were carried out with different loads and sliding velocities. All the composites, with the exception of the silver matrix, were tested against steel counter surface, for the conditions indicated in the Figure. It can be seen that when the graphite content in the composites exceeds 20 vol%, the friction coefficients observed in different composites are virtually independent of the matrix and the graphite content, and appear to attain a constant value close to 0.2. The elemental graphite has a friction coefficient of 0.18 and this increases with the desorption of adsorbed vapors. Thus, it appears that both the mating surfaces, including the graphitic composite and the counterfaces like steel, become completely smeared with graphite, and a friction coefficient close to that of pure graphite against itself is observed, regardless of the matrix.

In this paper, self-lubricating behavior of different graphite reinforced metal matrix composites are reviewed. More specifically, aluminum-graphite, magnesium-graphite, copper-graphite, silver-graphite and nickel-graphite composites on tribological properties have been discussed. The influence of the matrix alloy composition, the graphite content and its shape and size and test variables such as contact pressure, sliding velocity, temperature, current density, counterpart materials and environment on the tribological behavior of metal matrix composites are discussed. The process of graphite-rich film formation on the tribo surfaces of these composites was discussed as the superior tribological properties of metal matrix graphite composites are strongly dependent on the formation of graphite-rich film on tribo surfaces. It was showed that the friction and wear rate in the metal matrix graphite particle composites are significantly reduced when compared to alloys as a result of the incorporation of graphite particles. These composites have been extensively used in antifriction applications and illustrate the mechanism of solid lubrication by graphite. Selected applications of metal matrix graphite particle composites and their industrial potential are also discussed.

17.2 Self-Lubricating Behavior of Aluminum-Graphite Composites

Aluminum alloys are promising materials in high technology automotive fields owing to their excellent specific mechanical and physical properties such as low density, good resistance to corrosion and low thermal expansion. However, their low resistance to wear under poor lubricating conditions and their severe seizure and galling under boundary lubrication conditions are the main concerns for their high performance tribological applications. With solid lubricant particle dispersion in the matrix of aluminum alloy, this material exhibits good potential for resistant to wear and consequently becomes more suitable for tribological application [15, 16].

Considerable efforts have been put into incorporating lubricating particles in aluminum alloy matrices to improve friction, wear and anti-seizing properties. Earlier researchers have identified graphite as a suitable solid lubricating material for the preferred applications. These self-lubricating aluminum alloy-graphite particulate composites have received attention because of their low friction and

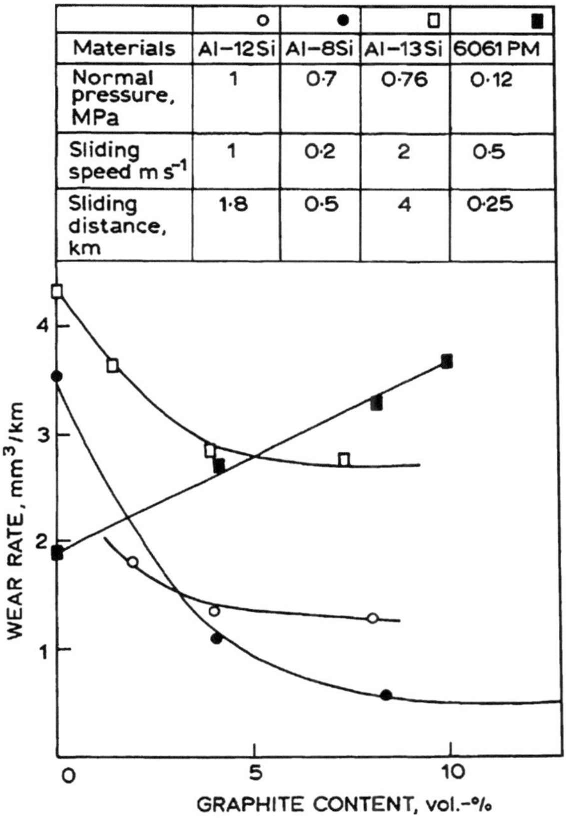

Fig. 17.2 Variation of wear rate with graphite content in aluminum alloy matrix [12]

wear, reduced temperature rise at the wearing contact surface, improved machinability, excellent anti-seizure effects, low thermal expansion and high damping capacity. The variation of coefficient of friction with graphite content for the Al–Si–graphite composite is shown in Fig. 17.1. Figure 17.2 shows the variation of wear rate with graphite content for different Al–Si alloy reinforced with graphite. It can be seen that the friction coefficient and wear rate decreases with increasing graphite content, and levels off once the film of graphite completely covers the sliding surface. It is likely that the wear rate may increase when graphite content in the matrix reaches very high values because of the decrease in strength of the matrix at very high graphite levels and the formation of thick graphite film of thickness exceeding the critical value, which itself can wear by delamination within the film in a manner similar to that of bulk graphite. It is interesting to see in Fig. 17.2 that for the case of 6061 alloy prepared by powder metallurgy method, the wear rate increases with increasing graphite content. Recently, Baradeswaran and Elayaperumal [17] studied the effect of graphite content on tribological behavior of aluminum alloy-graphite composite. The 6061

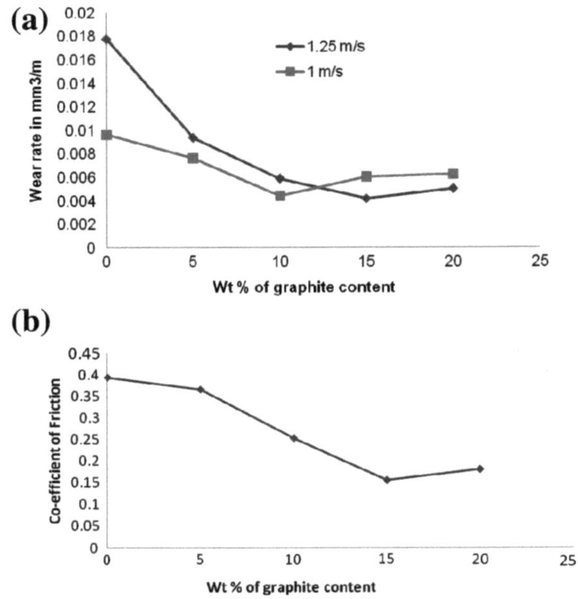

Fig. 17.3 Variation of **a** wear rate, and **b** friction coefficient with graphite content [17]

aluminum alloy-graphite composite with graphite particle dispersions up to 20% was used. Here, the composite was prepared by casting method. Interestingly, the wear rate of the composite decreased with increasing graphite content [Fig. 17.3a] which is in contrast to the result presented by the previous authors (Fig. 17.2) [12]. The coefficient of friction of 6061 aluminum alloy-graphite composite was also found to decrease with addition of graphite particles and recorded a 2.5 times lower value than the base alloy [Fig. 17.3b]. It was also reported that the wear loss was found to decrease with increasing sliding speed. Like other Al–Si alloy-graphite composite, the 6061 aluminum alloy exhibits its potential to act effectively as a self-lubricating material under dry sliding conditions.

It is reported above that the wear rate is also a function of sliding velocity in these composites. Figure 17.4 shows the variation of wear rate with sliding velocity in graphitic composites of Al–Si alloys, and compares it with that of the respective matrix alloys without any graphite. As the sliding speed increases, the interface temperature also increases resulting in (1) the formation of oxides on the sliding surface, and (2) decrease in flow stress. In addition there may be microstructural changes such as dissolution of precipitates, which will also be reflected in the wear behavior. In the figure, it is interesting to note that the composite with only 5 vol% graphite retains more or less the same trend of wear rate with sliding speed as that of the matrix. However, the composite with 15 vol% graphite shows a completely different trend to that of the matrix; this is an indication that the sliding surface might have been largely covered by graphite film so that the wear rate becomes almost unaffected by changes in sliding speed under the

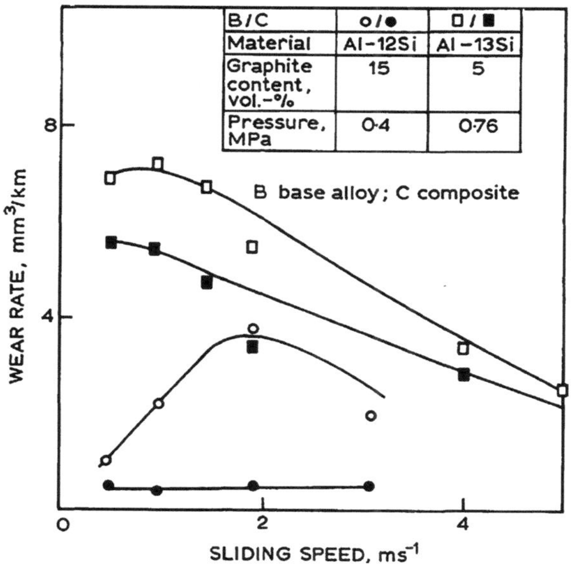

Fig. 17.4 Variation of wear rate with sliding speed for Al–Si alloy-graphite particle composites and corresponding base alloys [12]

given test conditions. When the graphite content is low, the sliding interface is, for the most part, not covered by the graphite-rich film, and the wear characteristics are similar to that of the matrix.

Normal load is another important parameter that influence wear rate of the composites. The variation of wear rate with normal load in Al–Si alloys composites is shown in Fig. 17.5. Results showed that the wear rate increases with increasing normal loads for all materials.

Efforts have been made to study various aluminum-graphite composites using different testing variables and conditions. For example, Akhlaghi and Zare-Bidaki [18] studied the influence of graphite content on the dry sliding and oil impregnated sliding wear behavior of Al 2024-graphite composites. The variation in the measured coefficient of friction and wear rate with the weight percent of graphite in the composites for both dry sliding and oil impregnated sliding are shown in Fig. 17.6. It was found that an increase in graphite content reduced the coefficient of friction for both dry and oil impregnated sliding, but this effect was more pronounced in dry sliding. Under dry sliding, the wear rate of the Al 2024-5 wt% graphite is about ten times lower than that for the base alloy. However, for composites with 10 wt% or more graphite particles addition, the wear rate increases. The change in the wear rate for a certain amount of graphite addition was caused by the following two competing factors. First, the beneficial effect of the graphite addition in reducing the wear of the composites due to formation of a thin lubricating graphite-rich film on the tribo-surface and second, the adverse effects of graphite addition in formation of porosity and cracks as well as the deterioration of mechanical properties resulting in enhanced delamination.

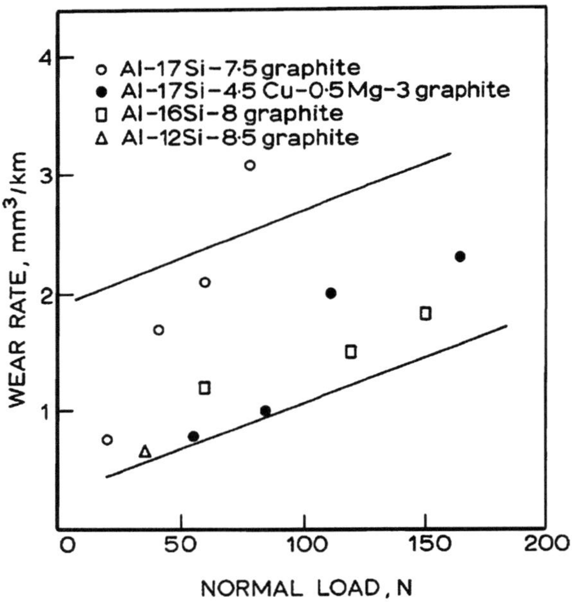

Fig. 17.5 Variation of wear rate with normal load for Al–Si alloy-graphite particle composites [12]

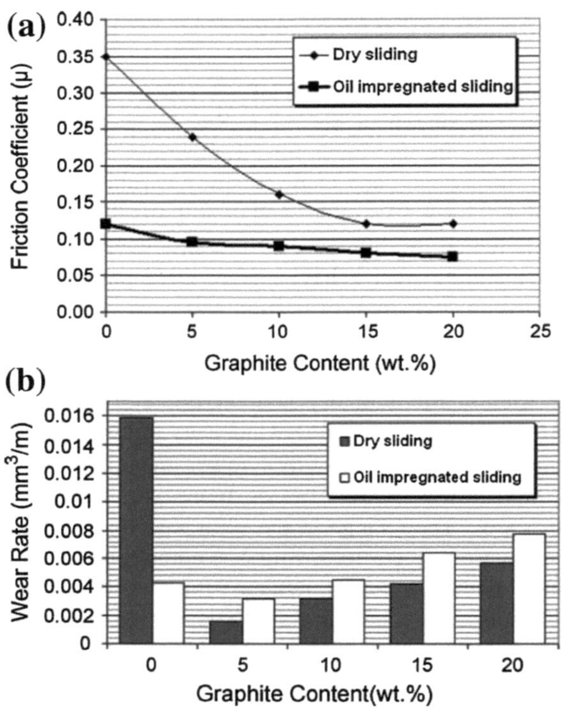

Fig. 17.6 The variation of **a** friction, and **b** wear rate with the graphite content in the composites for both dry sliding and oil impregnated sliding [18]

As shown in figure, for dry sliding, the friction coefficients starts off at 0.35 for the base alloy and decreases with increasing graphite content reaching to a final value of about 0.12 for composites containing 15 wt% graphite which is about one-third that of the base alloy. However, the figure shows that there is no significant difference in friction coefficient between composites with 15 wt% and that with 20 wt% graphite addition. The reason for this reduction in coefficient of friction is attributed to the presence of the smeared graphite layer at the sliding surface of the wear sample which acts as a solid lubricant. This lubricant film prevents direct contact of the two surfaces.

Studies were also made to investigate the tribological behavior of hybrid aluminum-graphite composites. Suresha and Sridhara [19, 20] studied wear characteristics of hybrid aluminum matrix composites reinforced with graphite and silicon carbide particulates. In tribological applications demanding similar strength requirement, the authors [19, 20] reported that Al–SiC–graphite hybrid composites are better substitutes to Al–graphite or Al–SiC composites owing to improved wear resistance and friction coefficient as a result of combined reinforcement of SiC and graphite particulates.

Attempts were also made to study the tribological behavior of the composites with higher graphite content under various environments. Goto and Uchijo [21] conducted wear tests of the Al–Si alloy impregnated graphite composite containing a high content (56 vol%) of graphite against bearing steel in various gas environments to investigate the wear mechanisms. The composite exhibited a decrease in wear at high loads in moist air. The wear rate is very high at low relative humidity (RH), decreases with increasing RH to a minimum at middle RH, and increases slightly at higher RH. The wear rate in moist argon is approximately equal to one-third of the wear rate in moist oxygen. At the RH for the minimum wear, the wide, compacted films consisting of graphite and metallic wear particles were formed on the disk sliding surface due to smearing of the particles. These films prevent the sliding surfaces from metal-to-metal contact. The entrance of wear particles onto the contact surfaces causes the pin lifting, leading to an apparent decrease in wear.

The influence of graphite particle size on the tribological performance of the composites was also investigated. Jinfeng et al. [22] studied the effect of graphite particle reinforcement on dry sliding wear of SiC/Gr/Al composites. The 40%SiC/5%Gr/Al composites with various-sizes graphite addition were fabricated by squeeze casting technology, and their friction and wear properties were investigated. Results showed that after the addition of graphite the friction coefficient of composites decreased and the wear resistance increased by 170–340 times. In addition, wear resistance was improved with increasing of graphite particle size, which is attributed to the enhancement of integrity of lubrication tribo layer composed of a complex mixture of iron oxides, graphite as well as fractured SiC particles and some fine particles containing aluminum.

It is well known that there have been a number of publications in the literature on the sliding wear and friction of aluminum alloy-graphite composites. Unfortunately, different researchers have used different experimental parameters for hardness and roughness of the counterface, sliding speed, normal load and the test environment, making it difficult to quantify the effect of graphite content on the tribological performance of the composites. In addition, comparing empirical wear data to theoretical generalizations of wear behavior is often difficult because of the widely different test conditions employed by different researchers to characterize the tribological properties of the composites. In spite of a lack of universal testing procedure, useful generalizations concerning wear behavior of different materials, including composites, have been applied by constructing wear mechanism maps. Wear maps serve as predictive tools to draw meaningful conclusions relative to wear behavior under different test conditions. Specifically, some of the variations between different studies can be overcome by utilizing normalized test parameters such as non-dimensional wear rate, load and sliding velocity. Hence, normalized wear rate (i.e., composite to the base matrix alloy) was introduced to analyze the data from different studies. Figure 17.7 is a compilation of the normalized friction coefficient and wear rate data of aluminum alloy-graphite composite showing the reduction in friction and wear volume due to graphite particle dispersion [23]. As shown in figure, the aluminum alloy-graphite composites typically exhibit much lower friction coefficient compared with that of the matrix alloy. The coefficient of friction decreases considerably, up to about 3% by weight of graphite, and thereafter remains constant at about 0.2 (0.5 on a normalized scale).

Although aluminum-graphite composites possess a number of superior tribological properties, there are some drawbacks. For instance, graphite loses its lubricity in dry environment. In many aerospace applications where the tribo-components are expected to perform in a vacuum, this could be a limitation.

Aluminum alloy composites have applications in cylinder block, journal bearings, connecting rods and pins, fan bushings and internal combustion engines. Aluminum–silicon composites are used in piston and liner in two stroke and four stroke engines in passenger and racing cars.

17.3 Self-Lubricating Behavior of Magnesium-Graphite Composites

Like aluminum, magnesium alloys have been used in automobile and aerospace industries due to their low density, high specific strength and stiffness, good damping characteristic, excellent machineability and castability. However, its corrosion and wear resistance limit it to be used as widely as aluminum alloy. The most commonly used magnesium alloy is AZ91 alloy.

Attempts have been made to study the tribological performance of magnesium-graphite composites. Qi [24] studied the influence of graphite particle content on the friction and wear characteristics of AZ91 magnesium alloy matrix composite. The variation of the friction coefficient of the composites normalized with respect to that of the base alloy as a function of applied load against the hardened steel is

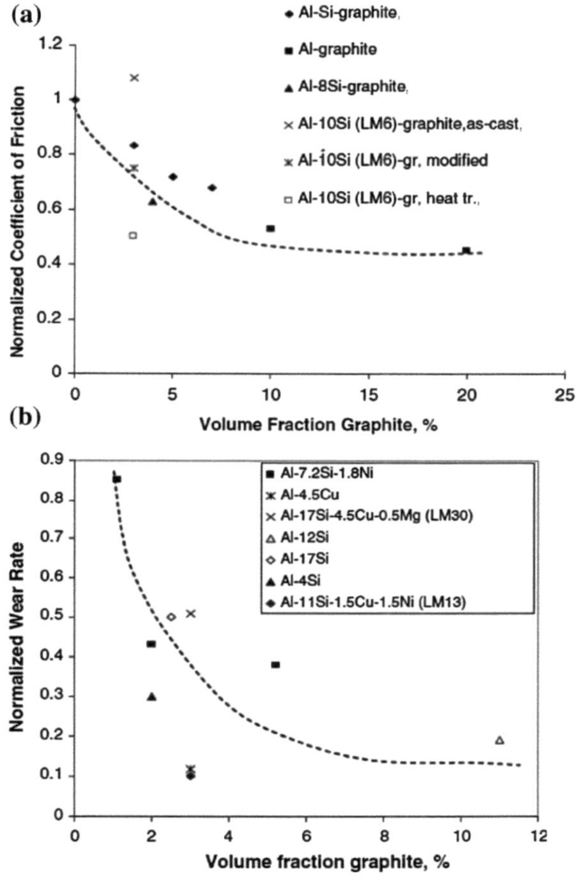

Fig. 17.7 Normalized **a** friction coefficient, and **b** wear rate of aluminum-graphite composites [23]

shown in Fig. 17.8a. It can be seen that the friction coefficients of composites are much lower than that of the matrix alloy. The graphite content presents significant effect on the friction coefficient of the specimens and the friction coefficients reduce with increasing graphite content. As regards to the wear rate, the wear behavior of composites containing different contents of graphite particles has been determined and compared with that observed in the base alloy. Figure 17.8b shows the wear mass loss of the composites as well as the base alloy specimens as a function of the applied load. The results showed that the wear resistances of graphite containing composite are much better than those of the matrix under similar testing conditions. The wear mass loss of each composite specimen reduces with the increase in graphite content. The authors reported that a continuous black lubricating film forms progressively on the worn surface during sliding, which effectively limits the direct interaction between the composite tribo-surface and the

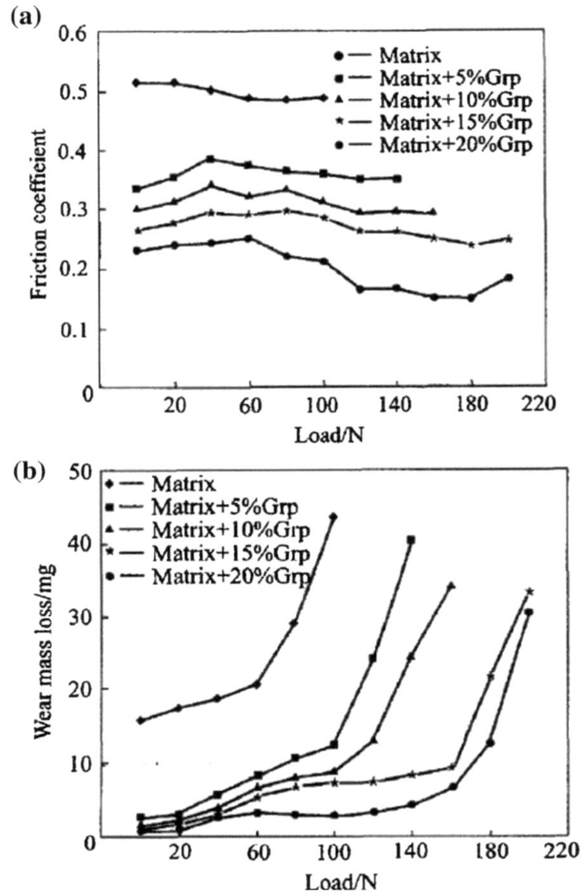

Fig. 17.8 Relation between **a** friction coefficient and, **b** wear mass loss with load [24]

counterpart, and also remarkably delays the transition from mild wear to severe wear for magnesium alloy composites.

As regards to the influence of particle size on the tribological performance, Zhang et al. [25] studied the effect of graphite particle size on wear property of graphite and Al_2O_3 reinforced AZ91D-0.8%Ce composites. Figure 17.9 shows the variation of wear loss with load. It was observed that the embedded graphite in the matrix acts as lubricant and decreases the wear loss. The wear resistance of the composites increases as the graphite particle size increases. At low load the composites have similar wear loss; at high load the composite with the largest graphite particle size has the best wear resistance. The wear mechanism of all the composites at low load is abrasive wear and oxidation wear; at high load, the wear mechanism of the composites changes to delamination wear. In another study, Zhang et al. [26] investigated the effect of graphite content on wear property of graphite/Al_2O_3/Mg–9Al–1Zn–0.8Ce composites. More specifically, the composites with the graphite

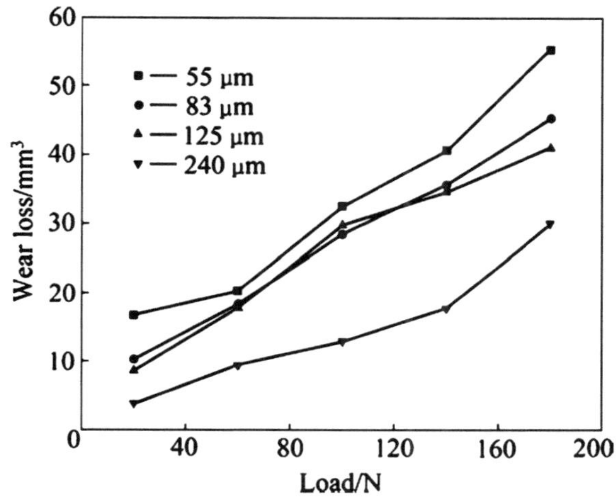

Fig. 17.9 Variations of wear loss with load of composites [25]

content of 5, 10, 15 and 20% were investigated. They found that the wear resistance of the composites increases with increasing the graphite content. Again, the wear mechanism of the composites at low load and high loads were mainly abrasive/oxidation and delamination wear, respectively. As Ce plays an important role in the magnesium-graphite composite, Zhang et al. [27] further investigated wear properties of graphite and Al_2O_3 reinforced AZ91D-Ce*x* composites. They found that the wear resistance of the composites increases with increasing Ce content as shown in the Fig. 17.10.

Magnesium alloy-graphite composites found applications in pistons and space structural applications.

17.4 Self-Lubricating Behavior of Copper-Graphite Composites

Copper metal matrix composites reinforced with particulate graphite are the potential candidate materials for electrical sliding contact application such as brushes in the electric motor and generator [28–35]. In the case of low voltage and high current densities, typically for sliding parts of welding machines, it is required to employ materials with a very high specific electrical conductivity, good thermal conductivity and low friction coefficient. Such conditions are fulfilled only by copper-graphite composite material. Copper-graphite particulate composites retain combined properties of copper, i.e., excellent thermal and electrical conductivities, and properties of graphite, i.e., solid lubricating and low thermal expansion coefficient.

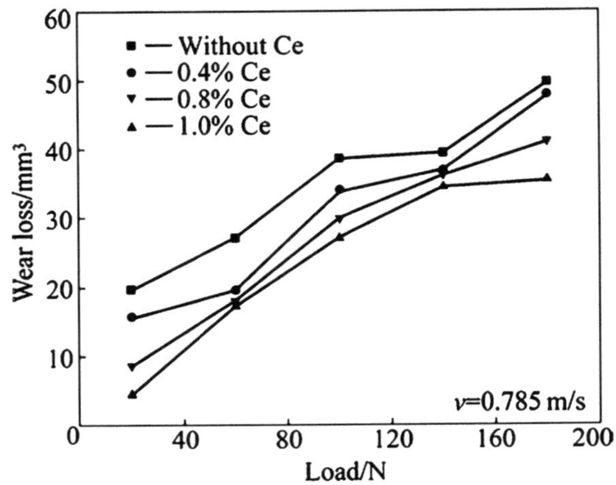

Fig. 17.10 Variations of wear loss of the composites with loads as a function of Ce content in the magnesium-graphite composite [27]

Considerable efforts have been made to study the tribological performance of copper-graphite composites. Tu et al. [36] studied the tribological behavior of zinc-coated graphite-copper matrix self-lubricating composite contact strip (ZGCCS) and common graphite-copper matrix self-lubricating composite (GCCS). Figure 17.11 shows the variation of wear rate with sliding velocity and electrical current. It was observed that the wear rate of both the ZGCCS and the GCCS increased with the increment of the current density and the sliding velocity. However, the ZGCCS exhibited wear resistance superior to the GCCS, indicating the electroplating Zn on graphite played an important role in improving the wear resistance and the wear rate of the former was only one-third of the latter at the sliding velocity of 41.7 m/s and with the current density of 285 A/cm^2. Arc erosion wear, oxidative wear and adhesive wear were the dominant mechanisms during the electrical sliding process. The arc erosion of the composite contact strip was remarkably reduced by the electroplating of Zn on graphite since the Zn-coated graphite particles were homogeneously distributed in the Cu matrix. Kováčik et al. [37] investigated the effect of composition on the friction coefficient of copper-graphite composites in the range of 0–50 vol% of graphite at constant load. The study was specifically done to determine the critical graphite content above which the coefficient of friction of the composite remains almost composition independent and constant. Hence, uncoated copper-graphite composite in the composition range of 0–50 vol% of graphite and coated composites with 30 and 50 vol% of graphite were prepared and their tribological properties were measured. Figure 17.12 shows the composition dependence of friction coefficient of (a) uncoated and (b) coated copper-graphite composites. It was found that with increasing concentration of graphite the coefficient of friction and wear rate of coated and uncoated composites at first decreases until certain critical concentration threshold of graphite is reached. Then the coefficient of friction of composites becomes independent on the composition and corresponds to the dynamic

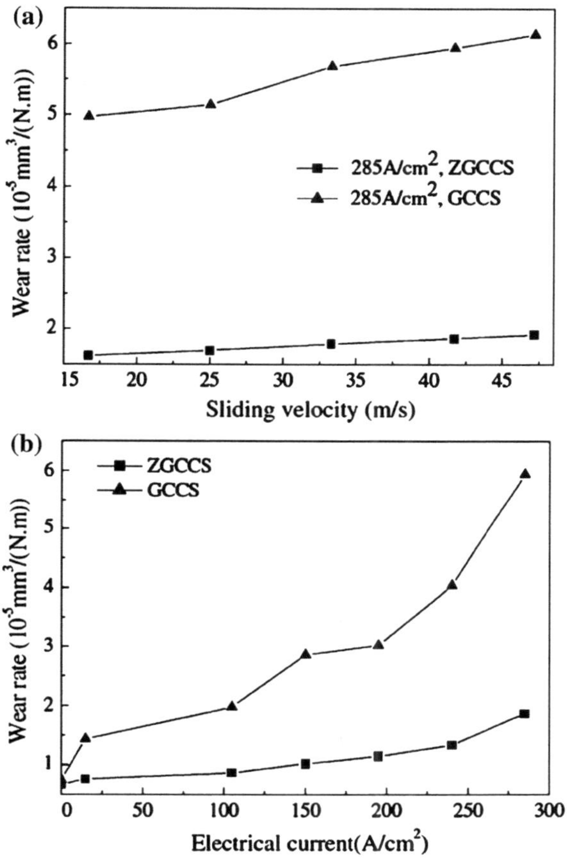

Fig. 17.11 The variation of wear rate with **a** sliding velocity, and **b** electrical current [36]

coefficient of friction of used graphite material (0.15–0.16 for used graphite) while the wear rate decreases further. The results were also compared with work done by Moustafa et al. [31] on copper-graphite composites.

Efforts have also been made to study the friction and wear performance of copper-graphite composites against different counterpart materials. Ma et al. [38] investigated sliding tribological behaviors of copper-graphite composite against different counterparts, specified as 2024 aluminum alloy, AZ91D magnesium alloy and Ti-6Al-4V titanium alloy. Figure 17.13 shows the variation of friction coefficient and wear rate with sliding speed. It was found that the tribological performance of copper-graphite composite strongly depended on its counterpart materials. Copper-graphite composite could provide friction reduction in sliding against 2024 and Ti-6Al-4V. Copper-graphite composite was a good self-lubricating material in sliding against AZ91D at low speeds. The authors described that the transfer layer of copper-graphite composite on counterface is the key to friction reduction. Wear mechanism of copper-graphite composite was related to the transfer (from copper-

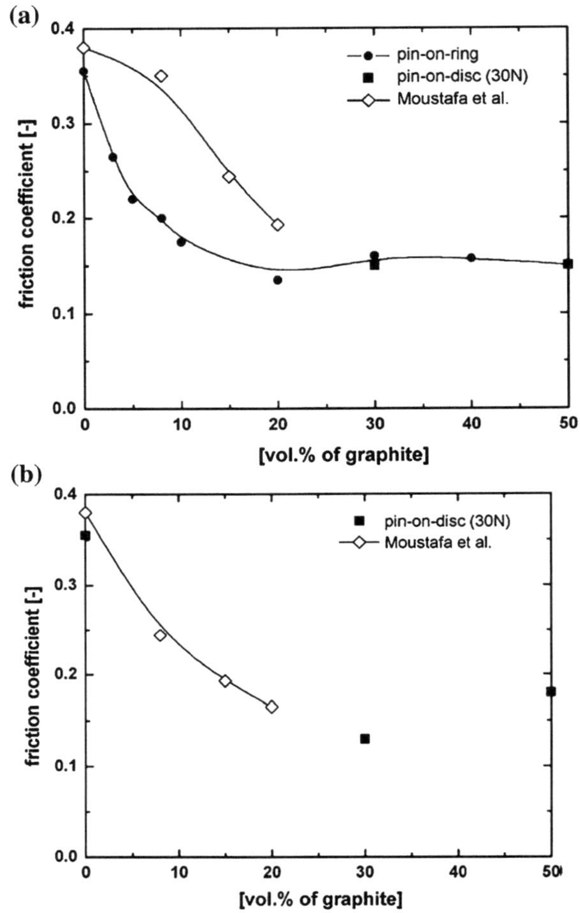

Fig. 17.12 Composition dependence of friction coefficient of **a** uncoated, and **b** coated copper-graphite composites [37]

graphite composite to light weight alloys) and counter-transfer (from light weight alloys to copper-graphite composite).

Studies were also made to investigate the influence of various self-lubricating particles on tribological behavior of copper-graphite composites. Chen et al. [39] studied the tribological properties of solid lubricants (graphite, h-BN) for copper-based friction composites. Copper-based friction composites containing graphite at weight fractions in the range of 0, 2, 5, 8 and 10%, corresponding to the hexagonal boron nitride (h-BN) at weight fractions in the range of 10, 8, 5, 2 and 0%, were fabricated by a powder metallurgy hot press method, respectively. Figure 17.14 shows the variation of coefficient of friction and wear rates for different normal loads. Results indicate that lubrication effects of graphite are superior to those of h-BN. With the increase of graphite content wear rates and friction coefficient decreased significantly.

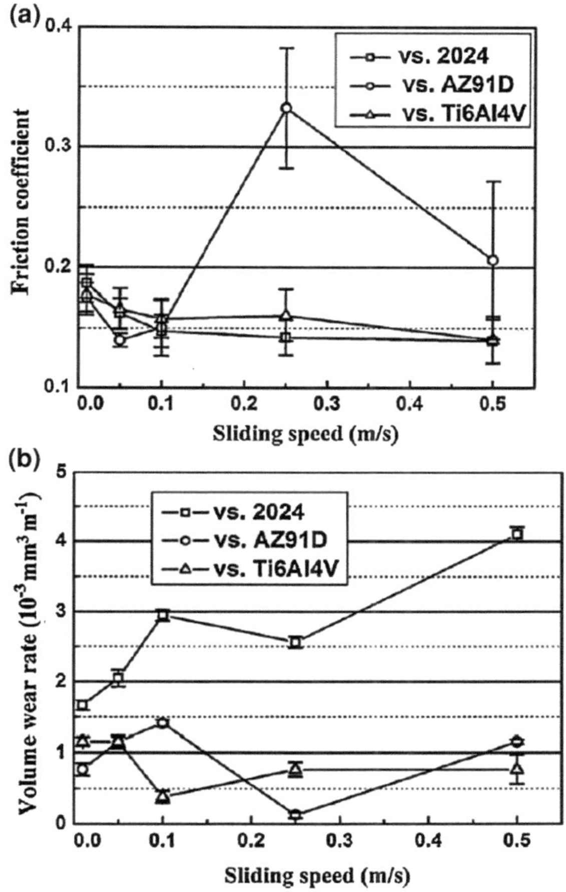

Fig. 17.13 **a** Friction coefficients. **b** Wear rates of copper-graphite composite sliding against 2024, AZ91D and Ti-6Al-4V [38]

Different heat treatments were used to improve the tribological properties of copper-graphite composites. Rajkumar and Aravindan [40] studied friction and wear properties of microwave-heat-treated copper-5 wt% graphite composites and untreated copper-graphite composites. Figure 17.15 shows the variation of friction coefficient and wear rates of untreated and microwave-treated copper-graphite composites. It was found that the microwave-heat treated composites exhibited reduced coefficient of friction and specific wear rate when compared to the untreated ones. The untreated copper-graphite composites exhibited the highest coefficient of friction and specific wear rate due to weaker interface contact. Microwave heat treating is helpful for the formation of steady graphite layer at the contact region of tribo-surface, thereby reducing coefficient of friction and specific wear rate. Further, Rajkumar and Aravindan [41] studied the tribological performance of microwave sintered copper–TiC–graphite hybrid composites. Figure 17.16 shows the variation of friction coefficient and wear rate of hybrid composite with normal load. Coefficient

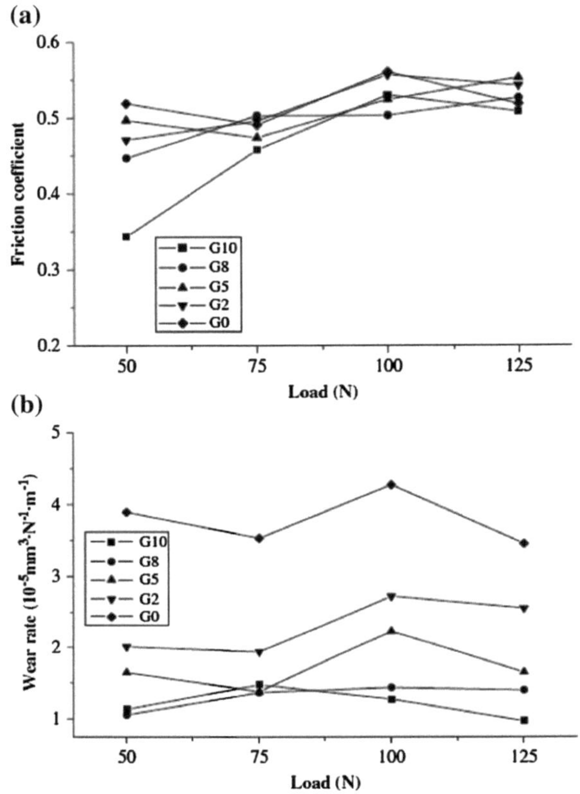

Fig. 17.14 Variation of **a** friction coefficient, and **b** wear rates with loads for various copper-based composites [39]

of friction and wear rate of hybrid composites and unreinforced copper increases with increase in normal load. Coefficient of friction and wear rates of hybrid composites are lower than those of unreinforced copper. Wear rate of hybrid composites is reduced with increasing %TiC and %graphite, due to the cooperative effect offered by both the reinforcements. Increased content of TiC reinforcement for a given volume fraction of graphite, leads to higher coefficient of friction. Coefficient of friction of hybrid composites is decreased with increase in % graphite reinforcement. The presence of mixed layer particularly in higher content graphite hybrid composites was found to have influence on the tribological properties. Increasing sliding velocity increased the coefficient of friction and wear rate of hybrid composites. Wear mechanism involved in Cu–TiC-graphite hybrid composites for given volume fraction of graphite and 5% TiC was oxidative wear with plastic deformation. The wear mechanism operating in hybrid composite with 15% TiC and 5% graphite hybrid composites was oxidative wear with high strained delamination and it was oxidative wear with delamination wear for 15% TiC and 10% graphite. Ramesh et al. [42] analyzed tribological performance of novel cast copper–SiC–Gr hybrid composites. Coefficient of friction of hybrid composites was higher than that of copper

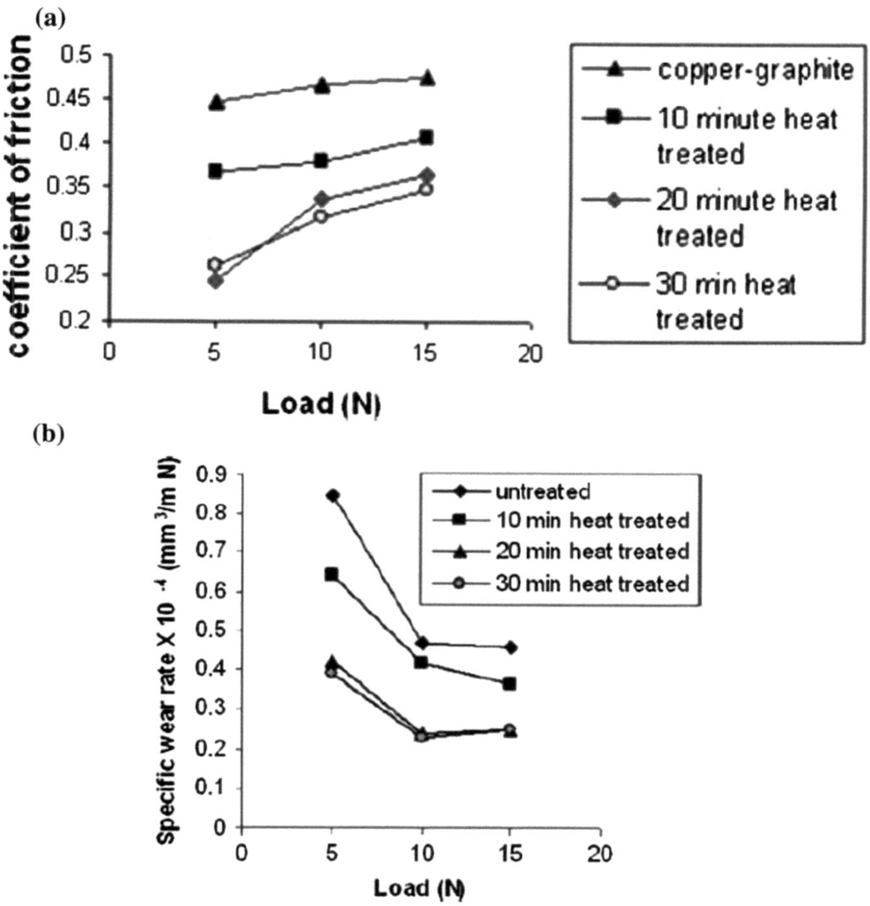

Fig. 17.15 Variation of **a** friction coefficient, and **b** wear rates of untreated and microwave treated copper-graphite composites [40]

and wear rates of hybrid composites are lower when compared to copper. Increased content of hard reinforcement for a given volume fraction of soft reinforcement leads to higher coefficient of friction and lower wear rates of hybrid composites.

As regards to the influence of sliding speed on the tribological performance of copper-graphite composite, Ma and Lu [43] studied the effect of sliding speed on surface modification and tribological behavior of copper-graphite composite. Figure 17.17 shows the variation of coefficient of friction and wear rates for different sliding speeds. The results showed that the friction coefficient and wear rate of copper-graphite composites are largely dependent on sliding speed. When the speed exceeds the critical value, a transition of the friction and wear regime occurs. The formation of a lubricant layer on the contact surface is regarded as an

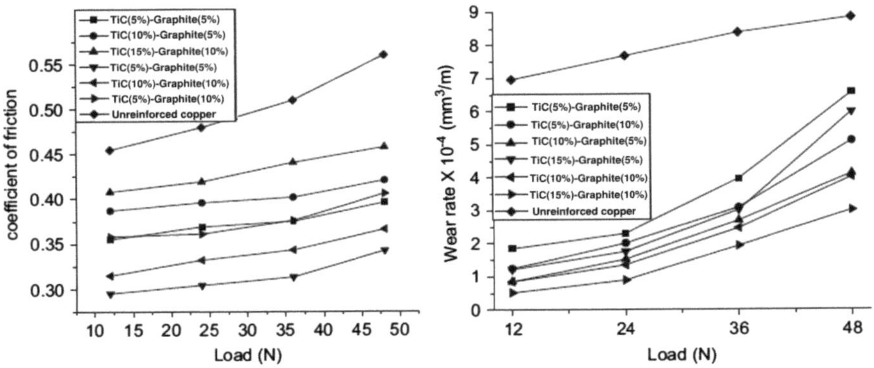

Fig. 17.16 Variation of friction coefficient and wear rate of hybrid composite with normal load [41]

important characteristic for the good tribological performance of copper-graphite composites. Due to a large strain gradient in subsurface deformation zone, the graphite-rich lubricant layer can easily form on the sliding surface when the speed is lower than the critical value. At speeds exceeding the critical value, however, formation of the lubricant layer is difficult due to the delamination wear by the high strain rate. The wear mechanism is found to be mild wear caused by ratchetting at speeds less than the critical value and severe wear induced by delamination at speeds exceeding the critical value. Ma and Lu [44] also studied the effect of surface texture on transfer layer formation and tribological behavior of copper-graphite composite. Three kinds of texture, namely Parallel Groves (PG), Random Groves (RG) and Polished Surfaces (PS), were prepared on the surface of steel discs. The influences of surface textures on the friction and wear behaviour of copper-graphite composite were investigated under both low and high load condition at a fixed speed. They found that the textures had different ratcheting effects on the contact surface of copper-graphite composite and thus influence friction and wear behaviors as shown in the Fig. 17.18.

Efforts have also been made to study the tribological performance of copper-graphite composite at high temperature. Zhan and Zhang [45] studied the role of graphite particles in the high-temperature wear of copper hybrid composites against steel. Dry sliding wear properties of copper matrix composite reinforced with SiC and graphite (Gr) particles were tested in the temperature range 373–723 K. Figure 17.19 shows the variation of friction coefficient and wear rate for the two composites. It was found that the addition of graphite particles simultaneously decreased the wear rates of the composite and effectively avoided the occurrence of severe wear up to 723 K. The friction coefficient of copper hybrid composite was more stable and lower than that of SiC/Cu. It was concluded that the graphite particle was an effective addition agent for copper matrix composite applied in high-temperature sliding wear condition.

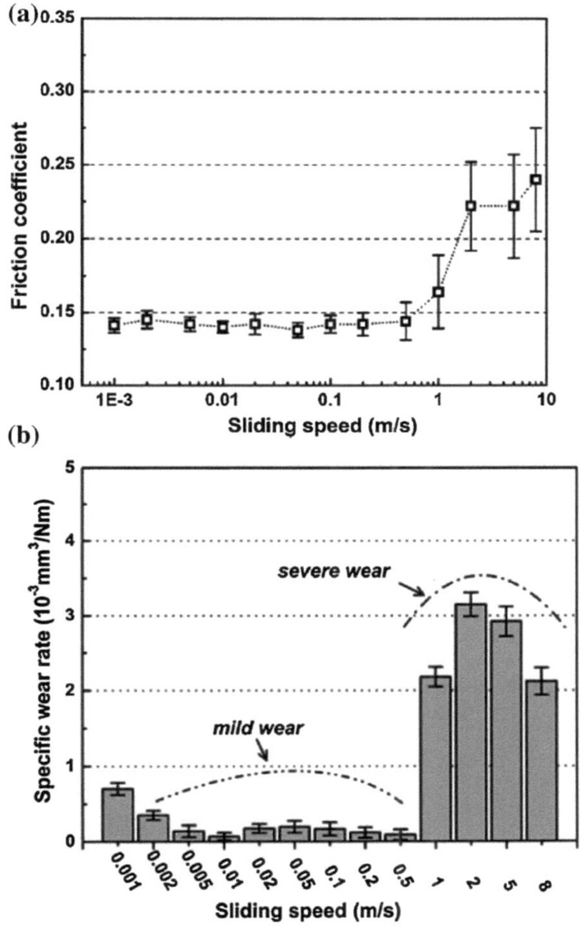

Fig. 17.17 Variation of **a** average friction coefficient, and **b** wear rate of copper-graphite composites at different sliding speeds [43]

Various novel composites were developed to enhance self-lubricating properties for specific application. Chen et al. [46] developed new composites with 1 and 2 wt% graphite for the application of frictional bearing under the high speed low-load sliding frictional conditions. More specifically, new Cu-based self-lubricating composites with 1 and 2 wt% graphite that were prepared using atomized Cu–10Ni–3Sn–3Pb (wt%) powder as the matrix alloy with 0.5 wt% Y_2O_3 as the matrix. Figure 17.20 shows the variation of friction and wear rate. They found that the composite with 1 wt% graphite possesses better mechanical and frictional properties, while the composite with 2% graphite possesses better self-lubricating properties.

For use in maglev transportation system, Ma et al. [47] studied the sliding wear behavior of copper-graphite composite material using a specially designed sliding

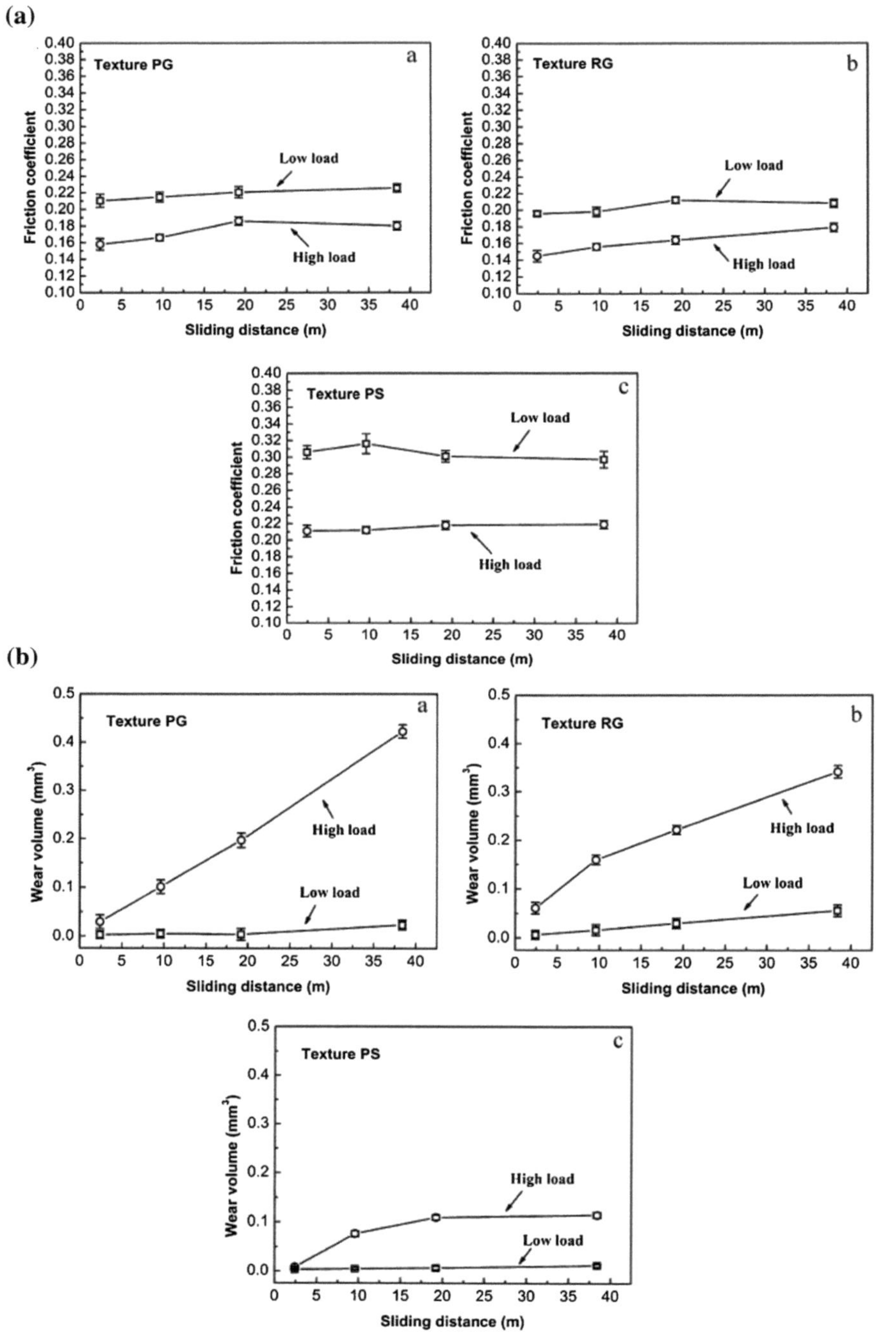

Fig. 17.18 Variation of average friction coefficient and wear rates with sliding distance for the pin sliding against various surface textures [44]

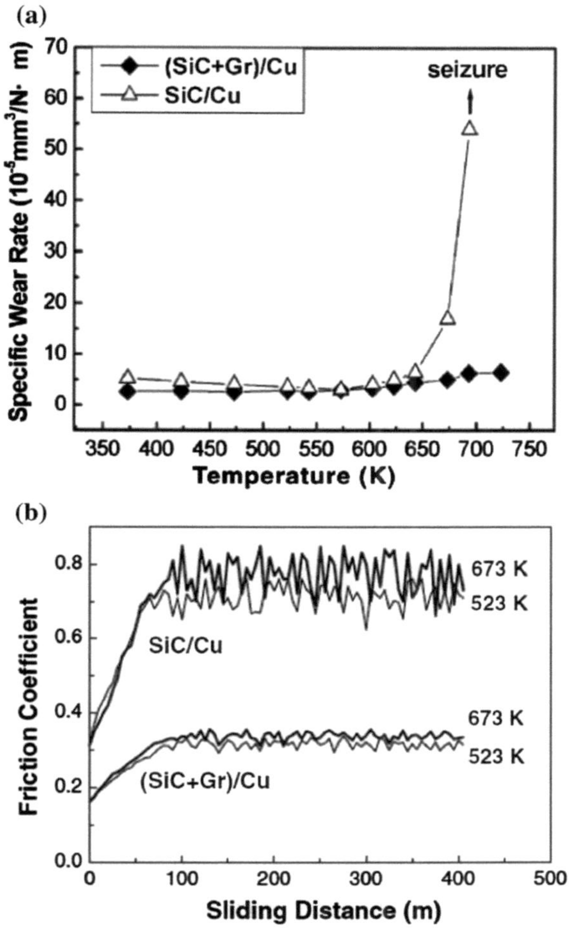

Fig. 17.19 **a** Variation of specific wear rate with temperature. **b** Variation of friction coefficient with sliding distance for various temperatures [45]

wear apparatus, which simulates the tribological conditions of sliding current collectors in a maglev system. The material was slid against a stainless steel band under unlubricated conditions. Within the studied range of normal pressure and electrical current, the wear loss increased with the increasing normal pressure and electrical current. More specifically, at a sliding speed of 50 km/h after sliding of cumulative distance 100 km without electrical current, the wear loss under a stress of 40 N is about five times to that with 20 N [Fig. 17.21a]. The wear loss almost doubled under electrical current, compared to the case without electrical current at an applied stress of 20 N at a sliding speed of 25 km/h [Fig. 17.21b]. Adhesive wear, abrasive wear and electrical erosion wear are the dominant wear mechanisms during the electrical sliding wear processes. The authors concluded that the present results provides a better understanding principle of design suitable sliding counter parts for the current collection device in maglev systems.

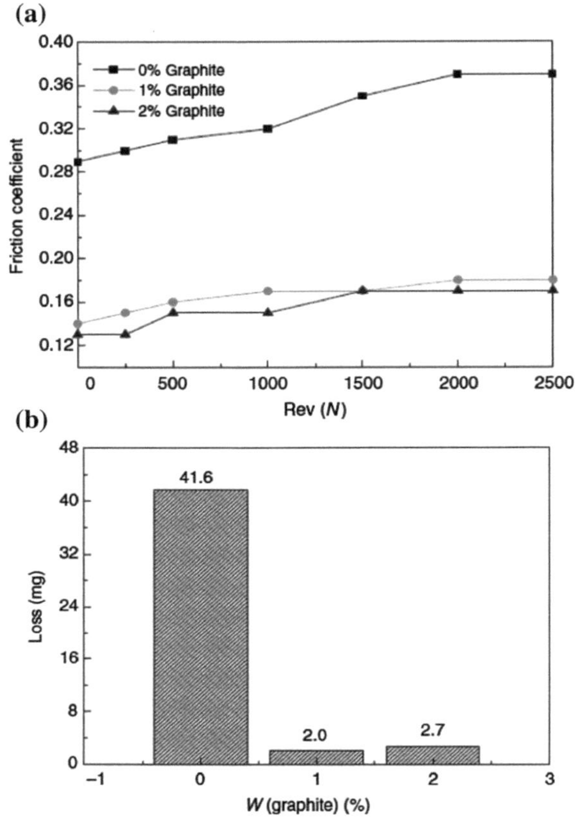

Fig. 17.20 **a** Friction coefficient. **b** Wear loss of the composites with different graphite contents [46]

Copper-graphite composites are used in plane bearings for trucks, cranes, bulldozers and automotives; drive shafts, bearings; stoker chain, screw conveyor, roller conveyor and roller bearings; spherical bushings for automotive transmission; slider for electric cars, overhead railways, dams and flood gates; slide bushing for discharge valves in hydraulic turbines; electrical contacts and brushes.

17.5 Self-Lubricating Behavior of Silver-Graphite Composites

Several metal matrix composites have been designed for use in electrical contacts by combining constituents which impart excellent wear resistance with constituents which have high electrical conductivity. Apart from copper base-graphite composites, another composite which is widely used in this field include silver base-graphite composites. The friction and wear behavior under sliding in the

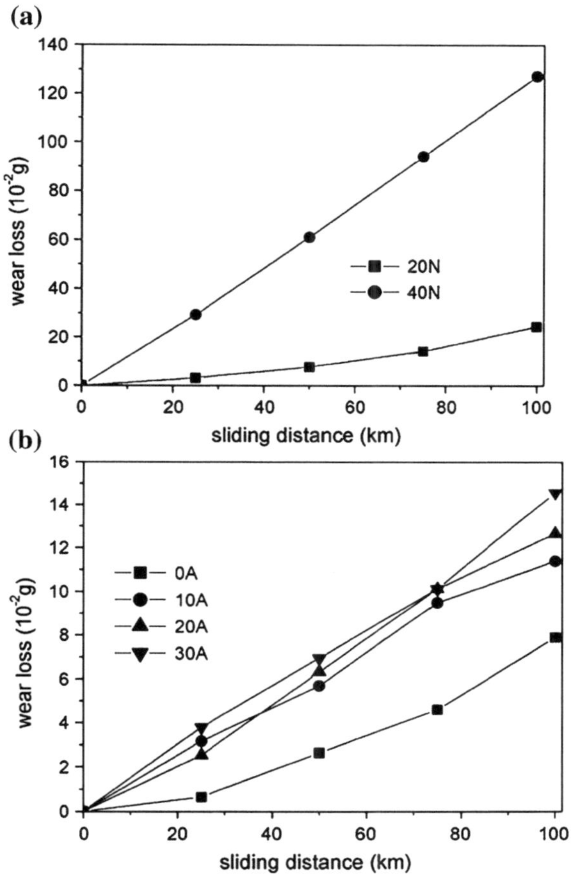

Fig. 17.21 Variation of wear loss with sliding distance as function of **a** normal load, and **b** current [47]

electrical contacts during flow of current complicates when compared to that under purely mechanical sliding conditions. The amount of materials removed under sliding electrical contact is the sum of contributions from purely mechanical wear in the absence of current arc erosion and an increment of mechanical wear resulting from softening of the matrix by local heating as a result of arcing. As reported in the literature [12, 48], in silver-graphite systems, the wear rate of composites increased with an increase of current density at both low and high temperatures, as shown in Fig. 17.22a. The figure shows that when there is a transition in the mode of sliding wear with temperature the current density for the same wear is considerably higher while sliding at higher temperatures, compared with that observed during sliding at ambient temperature. The coefficient of friction decreases with the current density, as shown in Fig. 17.22b, both in air and in a CO_2 atmosphere.

The influence of pressure on tribological behavior of the silver-graphite composites has been studied. Feng et al. [49] investigated the effect of pressure on the

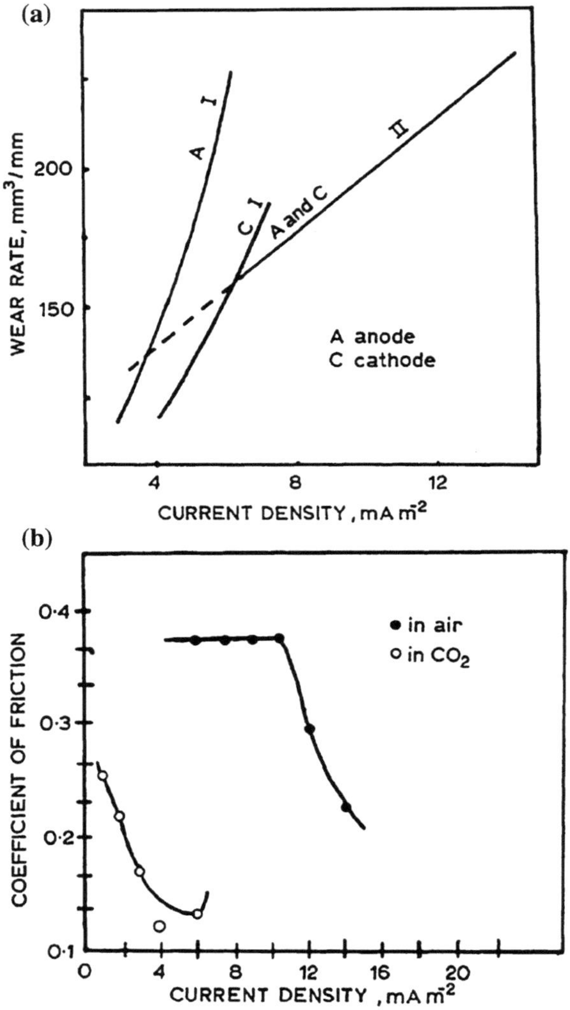

Fig. 17.22 **a** Variation of wear rate with current density at ambient and elevated temperatures; I is observed below 100°C, but II is observed at elevated temperatures, above 100°C. **b** Reduction in coefficient of friction of Ag-25 vol% graphite composite with current density in air and CO_2 atmosphere [12]

friction and wear properties of carbon nanotube-silver-graphite composite with 10 A/cm^2 and without electrical current. The results show that the wear of composite increase with the increase of pressure under mechanical wear, but the wear of composite varies with the pressure in the shape of U under electrical wear [Fig. 17.23a]. The electrical wear is higher than mechanical wear by 6–20 times. The differences between the no-current and with-current wear is due to the Joule heat released in the friction zone which leads to breakdown of the lubricating film, roughening of the surface and intensification of the adhesive interaction at the contact spots. The friction coefficient of composite with current is greater than that without current as shown in Fig. 17.23b.

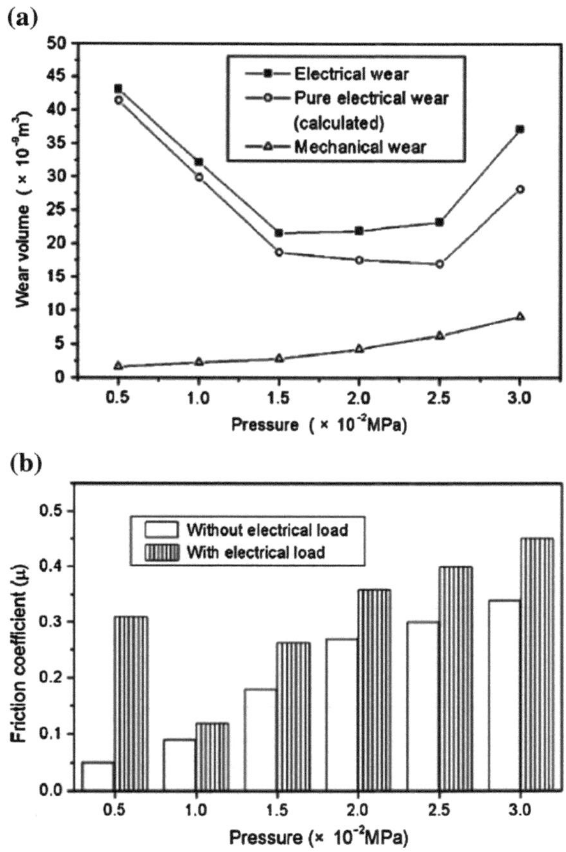

Fig. 17.23 Effect of pressure on **a** wear rate, and **b** friction coefficient of carbon nanotube reinforced silver-graphite composite under mechanical and electrical wear [49]

Silver-graphite composite brushes are marked by very low noise level, low and stable contact resistance, low friction and high conductivity. In modern inertial energy storage devices (homopolar motor generators), a large amount of energy is stored and the energy is delivered in the form of low voltage, high current pulses. Silver-graphite brushes are suitable for such systems. Silver-graphite brushes suppress radio interference noise level and are useful for slip rings, segmented rings and other applications where special requirements justify the high cost.

17.6 Self-Lubricating Behavior of Nickel-Graphite Composites

Nickel-graphite composites are widely used in the high-efficiency engine due to their excellent high-temperature performance. Considerable efforts have been made to study the effect of various operating parameters on the tribological

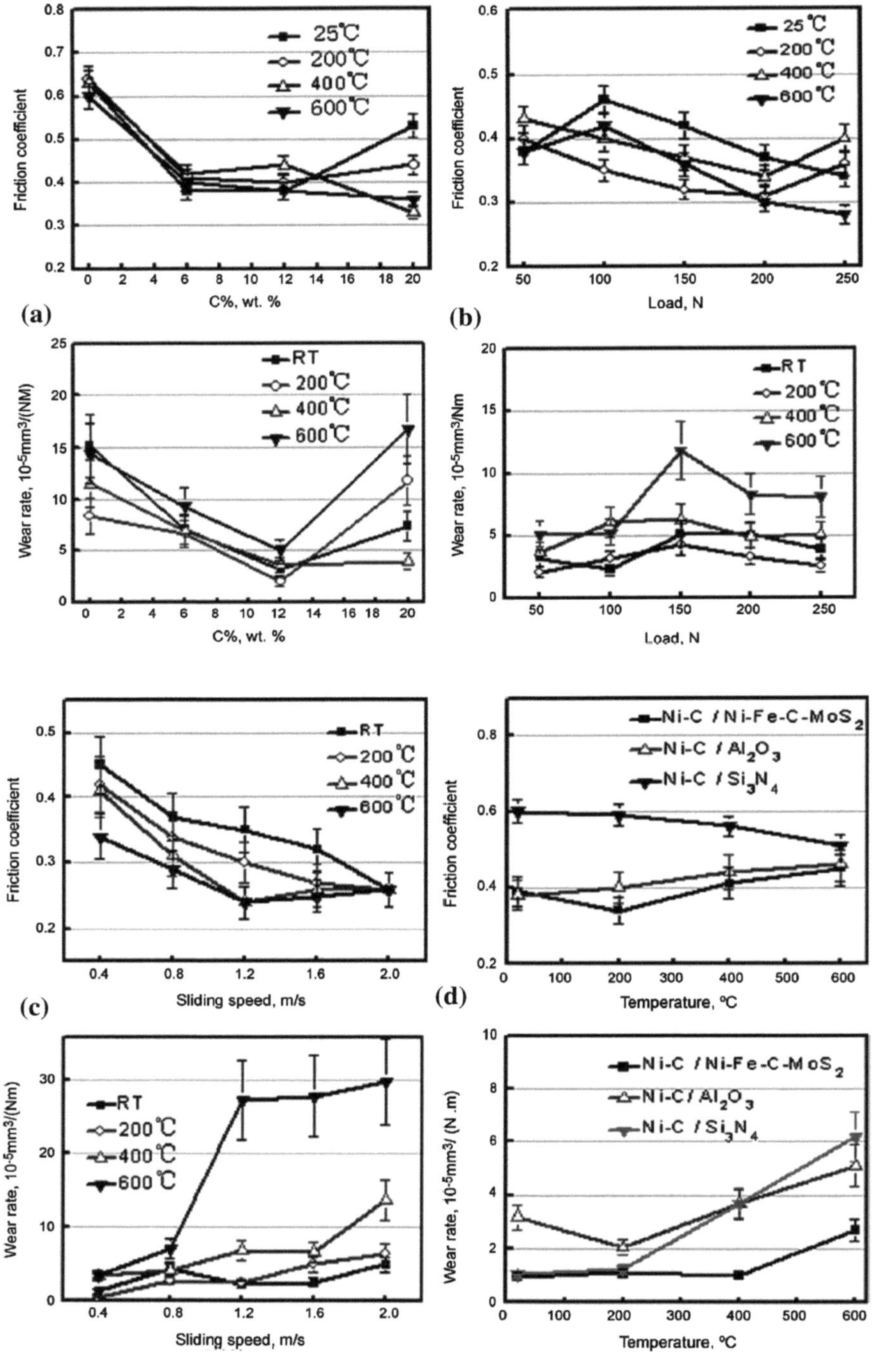

Fig. 17.24 Variation of friction coefficients and wear rates for different **a** graphite content, **b** load, **c** speed, and **d** counterpart materials [50]

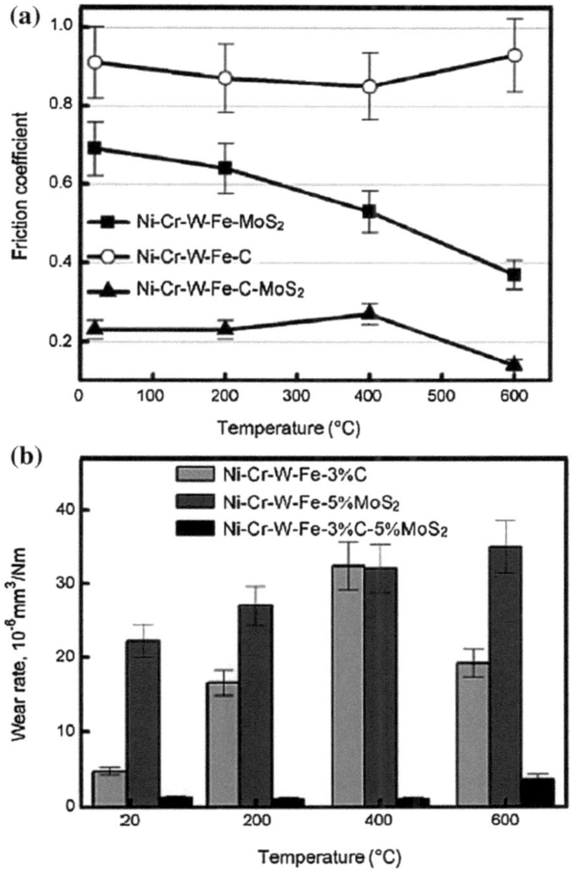

Fig. 17.25 Variation of **a** friction coefficients, and **b** wear rates in the composites containing graphite and MoS_2 as lubricants [51]

behavior of nickel-graphite composites. Li and Xiong [50] studied the tribological performance of graphite-containing nickel-based composite as function of temperature, load, speed and counterface. Figure 17.24 shows the variation of friction coefficients and wear rates for different graphite content, load, speed and counterpart materials. It was observed that the friction and wear properties are significantly improved by adding graphite in nickel-based alloy. The optimum adding amount of graphite is 6–12 wt%. The friction coefficients of graphite-containing composite decrease with the increase of load and sliding speed while the wear rates increase with the increasing temperature and sliding speed. The lower friction coefficients and wear rates are obtained when the composite rubs against nickel-based alloy containing molybdenum disulfide when compared to alumina and silicon nitride counter surfaces. Further, Li and Xiong [51] studied the tribological properties of nickel-based self-lubricating composite with different temperatures. More specifically nickel-based self-lubricating composites with graphite and

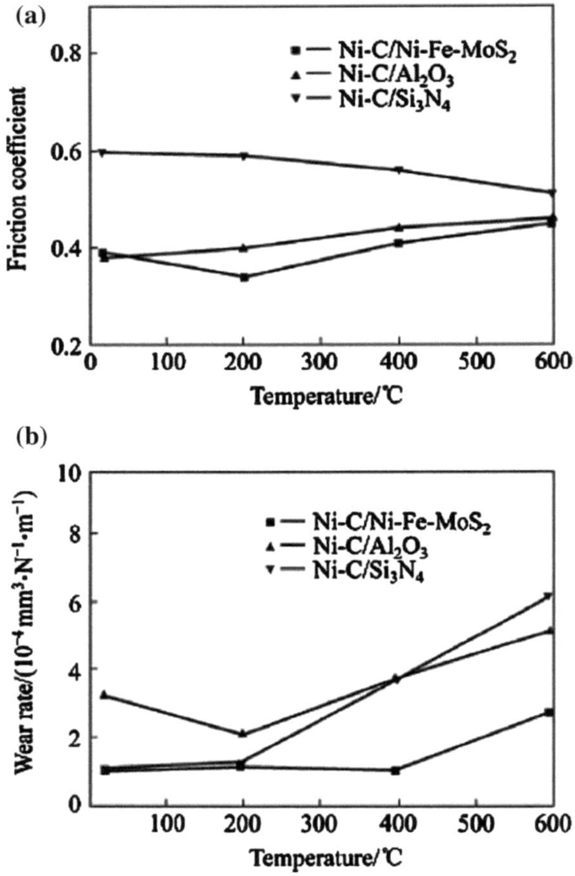

Fig. 17.26 Curves of **a** friction coefficient, and **b** wear rate with temperature against different counter faces [52]

molybdenum disulfide as lubricant were prepared. It was found that chromium sulfide and tungsten carbide were formed in the composite by adding molybdenum disulfide and graphite, which were responsible for low friction and high wear resistance at elevated temperatures, respectively. Figure 17.25 shows the variation of friction and wear rate for different composites. The Ni–Cr–W–Fe–C–MoS_2 composite possessed excellent self-lubricating properties over a wide range of temperature as a result of synergistic lubricating effect of graphite and molybdenum disulfide. The graphite played the main role of lubrication at room temperature, while sulfides were responsible for low friction at high temperature. Li et al. [52] studied tribological properties of nickel-graphite composite against different counterfaces at elevated temperatures. Figure 17.26 shows the curves of friction coefficients and wear rates of 12% graphite-containing composite varied with temperature when rubbing against Al_2O_3 ceramic, Si_3N_4 ceramic, and Ni–Fe–C–MoS_2 alloy under the load of 50 N and the velocity of 0.8 m/s. Compared with the

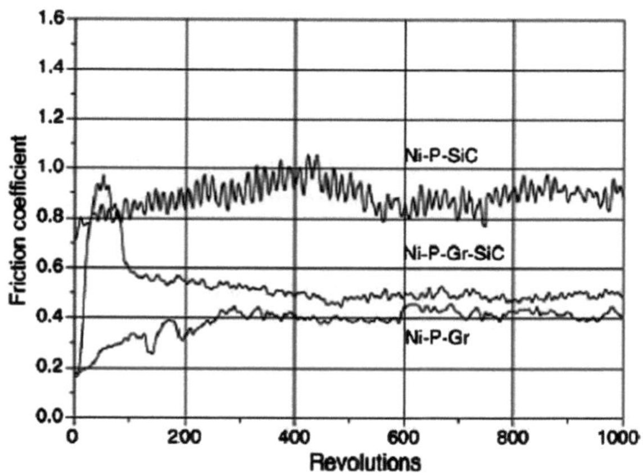

Fig. 17.27 The variation of friction coefficient for various Ni–P composites [53]

counterface of alumina and silicon nitride, the friction coefficients and wear rates are lower when the composite rubs against nickel-based alloy containing molybdenum disulfide. Wu et al. [53] studied the dry friction and wear behavior of Ni–P matrix composite with graphite and/or SiC particles. Figure 17.27 shows the variation of friction for the three composites. It was found that the Ni–P–Gr had the lowest friction coefficient. The wear rate of the composites was found to be 0.5447, 0.0015 and 0.0152 (10^{-3} mm^3 m^{-1}) for Ni–P–Gr, Ni–P–SiC and Ni–P–Gr–SiC, respectively indicating that the Ni–P–SiC had the lowest wear rate. By comparison with Ni–P–Gr and Ni–P–SiC, the overall results indicate that hybrid Ni–P–Gr–SiC composite presented well anti-friction and wear resistance which is resulted from a graphite-rich mechanical mixed layer (GRMML) formed on the contact surface. The GRMML formed on the worn surface was responsible for the good antifriction properties and SiC particles mixed with graphite played a load-bearing role in protecting GRMML from shearing easily.

17.7 Conclusions

Self-lubricating composite materials have the ability to sustain low friction and wear without any external supply of lubricants. The lubricants that are used for the tribological applications in automotive and manufacturing sectors are oil or grease-based. These lubricants can introduce significant quantities of pollutants into the environment. In addition, friction and wear often lead to heat and chemical contamination to the environment. The development of self-lubricating composite materials is very important for green or environment-friendly tribology. Thus, in this work, the tribological behavior of metal matrix composites reinforced with graphite particles has been reviewed. More specifically, aluminum-graphite, magnesium-graphite, copper-graphite, silver-graphite and nickel-graphite composites on tribological properties have been discussed. The influence of various parameters such as graphite content, particle size, normal pressure, sliding speed, sliding distance, surface texture, temperature, current density, environment and counterpart materials on the friction coefficient and wear rate is discussed. The important conclusions based on this study are as follows.

1. The friction and wear rate in the metal matrix graphite particle composites are significantly reduced when compared to alloys as a result of the incorporation of graphite particles.
2. The friction and wear rate of the composite decreases with increasing graphite particle content.
3. The friction coefficient and wear rate of the metal matrix graphite composites are largely dependent on sliding speed, current density, surface texture, environment and normal pressure.
4. The wear resistance of the composites increases with increasing graphite particle size.
5. The lubrication effects of graphite in the composite are found superior when compared to other lubricant particles.
6. Graphite particles were also found an effective addition agent for the composite operating at elevated temperature sliding condition.
7. Heat treatment like microwave can improve the tribological performance of the metal matrix graphite composites.
8. Metal matrix graphite hybrid composites showed optimum tribological performance when compared to unreinforced composites.

References

1. K.C. Ludema, *Friction, Wear, Lubrication: A Textbook in Tribology* (CRC Press, Boca Raton, 1996)
2. G.W. Stachowiak, A.W. Batchelor, *Engineering Tribology* (Butterworth-Heinemann, UK, 2005)
3. M.R. Lovell, M.A. Kabir, P.L. Menezes, C.F. Higgs III 3, Influence of boric acid additive size on green lubricant performance. Philos Trans. A. Math. Phys. Eng. Sci. **368**, 4851–4868 (1929)
4. A.I. Brudnyi, A.F. Karmadonov, Structure of molybdenum disulphide lubricant film. Wear **33**(2), 243–249 (1975)
5. C. Donnet, A. Erdemir, Historical developments and new trends in tribological and solid lubricant coatings. Surf. Coat. Technol. **180–181**, 76–84 (2004)
6. P.J. Bryant, P.L. Gutshall, L.H. Taylor, A study of mechanisms of graphite friction and wear. Wear **7**(1), 118–126 (1989)
7. B.C. Pai, P.K. Rohatgi, S. Venkatesh, Wear resistance of cast graphitic aluminium alloys. Wear **30**(1), 117–125 (1974)
8. P.K. Rohatgi, B.C. Pai, Seizure resistance of cast aluminium alloys containing dispersed graphite particles of various sizes. Wear **59**(2), 323–332 (1980)

9. B.P. Krishnan, N. Raman, K. Narayanaswamy, P.K. Rohatgi, Performance of an Al–Si-graphite particle composite piston in a diesel engine. Wear **60**(1), 205–215 (1980)
10. S. Biswas, P.K. Rohatgi, Tribological properties of cast graphitic-aluminium composites. Tribol. Int. **16**(2), 89–102 (1983)
11. F. Akhlaghi, A. Zare-Bidaki, Influence of graphite content on the dry sliding and oil impregnated sliding wear behavior of Al 2024-graphite composites produced by in situ powder metallurgy method. Wear **266**(1–2), 37–45 (2009)
12. P.K. Rohatgi, S. Ray, Y. Liu, Tribological properties of metal matrix-graphite particle composites. Inter. Mater. Rev. **37**(3), 129–152 (1992)
13. P.R. Gibson, A.J. Clegg, A.A. Das, Wear of cast Al–Si alloys containing graphite. Wear **95**(2), 193–198 (1984)
14. S. Das, S.V. Prasad, T.R. Ramachandran, Microstructure and wear of cast (Al–Si alloy)-graphite composites. Wear **133**(1), 173–187 (1989)
15. A.K. Jha, S.V. Prasad, G.S. Upadhyaya, Sintered 6061 aluminium alloy–solid lubricant particle composites: sliding wear and mechanisms of lubrication. Wear **133**(1), 163–172 (1989)
16. A.K. Jha, S.V. Prasad, G.S. Upadhyaya, Dry sliding wear of sintered 6061 aluminium alloy–graphite particle composites. Tribol. Int. **22**(5), 321–327 (1989)
17. A. Baradeswaran, A. Elayaperumal, Effect of graphite content on tribological behaviour of aluminium alloy–graphite composite. Eur. J. Sci. Res. **53**(2), 163–170 (2011)
18. F. Akhlaghi, A. Zarebidaki, Influence of graphite content on the dry sliding and oil impregnated sliding wear behavior of Al 2024–graphite composites produced by in situ powder metallurgy method. Wear **266**(1–2), 37–45 (2009)
19. S. Suresha, B.K. Sridhara, Friction characteristics of aluminium silicon carbide graphite hybrid composites. Mater. Des. (2011, in press)
20. S. Suresha, B.K. Sridhara, Wear characteristics of hybrid aluminium matrix composites reinforced with graphite and silicon carbide particulates. Compos. Sci. Technol. **70**(11), 1652–1659 (2010)
21. H. Goto, K. Uchijo, Wear mechanism of Al–Si alloy impregnated graphite composite under dry sliding. Wear **259**(1–6), 613–619 (2005)
22. L. Jinfeng, J. Longtao, W. Gaohui, T. Shoufu, C. Guoqin, Effect of graphite particle reinforcement on dry sliding wear of SiC/Gr/Al composites. Rare Met. Mater. Eng. **38**(11), 1894–1898 (2009)
23. S.V. Prasad, R. Asthana, Aluminum metal–matrix composites for automotive applications: tribological considerations. Tribol. Lett. **17**(3), 445–453 (2004)
24. Q. Qi, Evaluation of sliding wear behavior of graphite particle-containing magnesium alloy composites. Trans. Nonferr. Met. Soc. China **16**(5), 1135–1140 (2006)
25. M.-j. Zhang, Y.-b. Liu, X.-h. Yang, J. An, K.-s. Luo, Effect of graphite particle size on wear property of graphite and Al_2O_3 reinforced AZ91D-0.8% Ce composites. Trans. Nonferr. Met. Soc. China **18**, s273–s277 (2008)
26. M.-j. Zhang, X.-h. Yang, Y.-b. Liu, Z.-y. Cao, L.-r. Cheng, Y.-l. Pei, Effect of graphite content on wear property of graphite/Al_2O_3/Mg–9Al–1Zn–0.8Ce composites. Trans. Nonferr. Met. Soc. China **20**(2), 207–211 (2010)
27. M.-j. Zhang, Z.-y. Cao, X.-h. Yang, Y.-b. Liu, Microstructures and wear properties of graphite and Al_2O_3 reinforced AZ91D-Ce*x* composites. Trans. Nonferr. Met. Soc. China **20**, 471–475 (2010)
28. D.H. He, R. Manory, A novel electrical contact material with improved self-lubrication for railway current collectors. Wear **249**(7), 626–636 (2001)
29. H. Kato, M. Takama, Y. Iwai, K. Washida, Y. Sasaki, Wear and mechanical properties of sintered copper-tin composites containing graphite or molybdenum disulfide. Wear **255**(1–6), 573–578 (2003)
30. M. Kestursatya, J.K. Kim, P.K. Rohatgi, Wear performance of copper-graphite composite and a leaded copper alloy. Mater. Sci. Eng. A **339**(1–2), 150–158 (2003)

31. S.F. Moustafa, S.A. El-Badry, A.M. Sanad, B. Kieback, Friction and wear of copper-graphite composites made with Cu-coated and uncoated graphite powders. Wear **253**(7–8), 699–710 (2002)
32. N.K. Myshkin, V.V. Konchits, Friction and wear of metal-composite electrical contacts. Wear **158**(1–2), 119–140 (1992)
33. J. Kovácik, J. Bielek, Electrical conductivity of Cu/graphite composite material as a function of structural characteristics. Scr. Mater. **35**(2), 151–156 (1996)
34. J.M. García-Márquez, N. Antón, A. Jimenez, M. Madrid, M.A. Martinez, J.A. Bas, Viability study and mechanical characterisation of copper-graphite electrical contacts produced by adhesive joining. J. Mater. Process. Technol. **143–144**, 290–293 (2003)
35. C.H. Stoessel, J.C. Withers, C. Pan, D. Wallace, R.O. Loutfy, Improved hollow cathode magnetron deposition for producing high thermal conductivity graphite-copper composite. Surf. Coat. Technol. **76–77**(Part 2), 640–644 (1995)
36. C.J. Tu, D. Chen, Z.H. Chen, J.T. Xia, Improving the tribological behavior of graphite/Cu matrix self-lubricating composite contact strip by electroplating Zn on graphite. Tribol. Lett. **31**(2), 91–98 (2008)
37. J. Kovacik, S. Emmer, J. Bielek, L. Kelesi, Effect of composition on friction coefficient of Cu–graphite composites. Wear **265**(3–4), 417–421 (2008)
38. W. Ma, J. Lu, B. Wang, Sliding friction and wear of Cu–graphite against 2024, AZ91D and Ti_6Al_4 V at different speeds. Wear **266**(11–12), 1072–1081 (2009)
39. B. Chen, Q. Bi, J. Yang, Y. Xia, J. Hao, Tribological properties of solid lubricants (graphite, h-BN) for Cu-based P/M friction composites. Tribol. Int. **41**(12), 1145–1152 (2008)
40. K. Rajkumar, S. Aravindan, Tribological performance of microwave-heat-treated copper–graphite composites. Tribol. Lett. **37**(2), 131–139 (2009)
41. K. Rajkumar, S. Aravindan, Tribological performance of microwave sintered copper–TiC–graphite hybrid composites. Tribol. Int. **44**(4), 347–358 (2011)
42. C.S. Ramesh, R. Noor Ahmed, M.A. Mujeebu, M.Z. Abdullah, Development and performance analysis of novel cast copper-SiC-Gr hybrid composites. Mater. Des. **30**(6), 1957–1965 (2009)
43. W. Ma, J. Lu, Effect of sliding speed on surface modification and tribological behavior of copper–graphite composite. Tribol. Lett. **41**(2), 363–370 (2010)
44. W. Ma, J. Lu, Effect of surface texture on transfer layer formation and tribological behaviour of copper-graphite composite. Wear **270**(3–4), 218–229 (2011)
45. Y. Zhan, G. Zhang, The role of graphite particles in the high-temperature wear of copper hybrid composites against steel. Mater. Des. **27**(1), 79–84 (2006)
46. S. Chen, J. Wang, Y. Liu, J. Liang, C. Liu, Synthesis of new cu-based self-lubricating composites with great mechanical properties. J. Compos. Mater. **45**(1), 51–63 (2010)
47. X.C. Ma, G.Q. He, D.H. He, C.S. Chen, Z.F. Hu, Sliding wear behavior of copper-graphite composite material for use in maglev transportation system. Wear **265**(7–8), 1087–1092 (2008)
48. J.L. Johnson, J. Schreurs, High current brushes: VIII: effect of electrical load. Wear **78**(1–2), 219–232 (1982)
49. Y. Feng, J. Wang, M. Zhang, Y. Xu, The influence of pressure on the electrical tribology of carbon nanotube–silver–graphite composite. J. Mater. Sci. **42**(23), 9700–9706 (2007)
50. J. Li, D. Xiong, Tribological behavior of graphite-containing nickel-based composite as function of temperature, load and counterface. Wear **266**(1–2), 360–367 (2009)
51. J. Li, D. Xiong, Tribological properties of nickel-based self-lubricating composite at elevated temperature and counterface material selection. Wear **265**(3–4), 533–539 (2008)
52. J.-l. Li, D.-s. Xiong, W. Yi, Tribological properties of nickel-graphite composite against different counterfaces at elevated temperatures. Trans. Nonferr. Met. Soc. China **17**, 99–104 (2007)
53. Y. Wu, B. Shen, L. Liu, W. Hu, The tribological behaviour of electroless Ni–P–Gr–SiC composite. Wear **261**(2), 201–207 (2006)

Part IV
Applications

Chapter 18
Wind Turbine Tribology

Elon J. Terrell, William M. Needelman and Jonathan P. Kyle

18.1 Introduction

Wind power is of increasing interest in society due to its prospects as an environmentally friendly source of renewable energy. The use of wind turbines to extract electrical energy from wind can be dated back to the late-1800s, with the 12 kW windmill generator by Charles Brush [1], as well as the mid-1900s, with the 1250 kW Smith-Putnam wind turbine. Developments in the wind industry were encouraged by the oil crisis in 1973. Between the early developments of wind turbines and the present day, wind turbine designs have made significant developments in complexity, size, and power capacity (Fig. 18.1). Wind power is now recognized as one of the fastest growing source of energy production (Fig. 18.2), having a worldwide wind power installed capacity exceeding 120 GW [2–4]. The future prospects for wind power installation indicate that growth will continue, as the United States Department of Energy targets 20% wind-based electricity generation, i.e. over 300 GW, by 2030 [5], while TPWind predicts wind energy to account for 12–14% of the total energy production by 2020 and 25% by 2030 [6]. Additionally, China aims for 15% renewable power generation by 2020 [7, 8].

E. J. Terrell (✉) · J. P. Kyle
Mechanical Engineering Department, Columbia University,
500 West 120th Street, New York, NY 10027, USA
e-mail: eterrell@columbia.edu

J. P. Kyle
e-mail: jpk2128@columbia.edu

W. M. Needelman
Donaldson Company, Inc, 8 Hillside Court,
Huntington, NY 11743, USA
e-mail: bill.needelman@donaldson.com

M. Nosonovsky and B. Bhushan (eds.), *Green Tribology*,
Green Energy and Technology, DOI: 10.1007/978-3-642-23681-5_18,

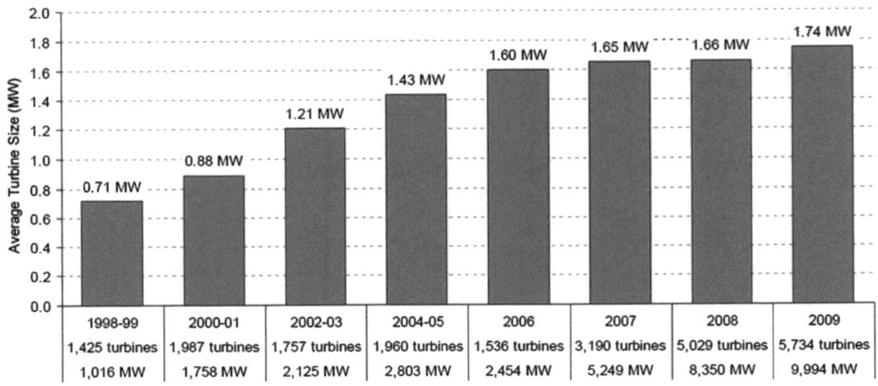

Fig. 18.1 Average nameplate power capacity of installed wind turbines in the united states between 1998 and 2009, from [9]

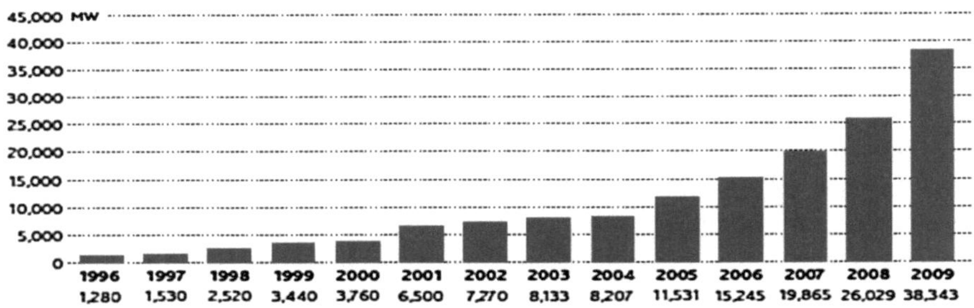

Fig. 18.2 Global annual installed wind capacity between 1996 and 2009, from [4]

Although wind turbines are being increasingly installed around the world, their power systems have a great deal of challenges related to tribology that can drastically reduce their expected lifetimes. Most wind turbines are intended for operating lifetimes of 20 years or longer; however, field reports [10] have shown that drivetrain components tend to fail much earlier than 20 years. These added maintenance and repair costs contribute significantly to the total cost of wind energy [11].

18.2 Wind Energy Siting and Maintenance Costs

Wind energy projects may be on land or offshore, and can vary in scale from small projects of one to a few turbines to large, multi-turbine projects (denoted as *utility-scale* or *wind farms*). Utility-scale projects can consist of up to hundreds of wind turbines. These turbines are normally operated by independent power producers who sell the generated power to the local utility provider [12]. Wind turbine operation and maintenance (O and M) costs, which are known to be the predominant costs that contribute to the cost of wind energy, are generally attributed to a limited number of components, including insurance, land usage, maintenance, repair, spare parts, and administration [8, 13]. For most wind turbines, maintenance and repair account for the largest share of O and M costs. These costs include the following:

- *Downtime*: The revenue lost from turbine downtime is factored into the overall cost of repair. Downtime includes the logistics time for organizing a repair crew and supplies, as well as travel time and the actual time needed to repair the affected component.
- *Labor Costs*: The cost of a service crew is factored into O and M.
- *Crane*: If major repairs or component replacements are necessary, a crane may be needed. The cost of transporting a crane (normally to a remote location) and operating it contributes significantly to O and M.
- *Materials and Consumables*: The magnitude of this cost may vary significantly depending on the component which has failed and the extent of the damage.

It must be noted that O and M costs in offshore wind turbines tend to be significantly higher than that of comparable land-based turbines because they are more difficult to access. Less frequent access for maintenance and repair can lead to large reductions in downtime costs.

18.3 Wind Turbine Theory and Designs

The term wind turbine power systems (WTPS) denotes systems that extract kinetic energy from wind and convert it into usable electrical energy. Along these lines, the size and mechanical complexity of wind turbines can range from relatively small (e.g., household wind turbines) to large, offshore wind turbines. Regardless of the size, all wind turbines incorporate airfoil blades that are pitched into and pushed by the wind, causing rotational power that is ultimately transferred to a generator.

18.3.1 Wind Energy Considerations

In the design process, a number of items about wind power must be considered, including the following:

- *Transient Wind Speed*: Wind speed can change significantly over a relatively short time; thus turbines must be able to adapt to rapidly changing loads. Turbines must also be able to protect from over speed when wind speeds are beyond operating limits.

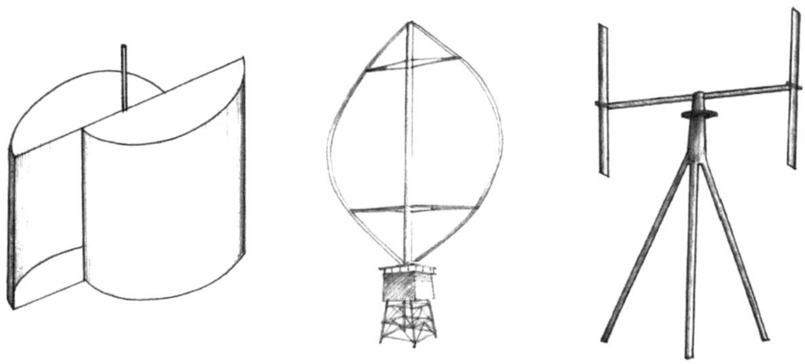

Fig. 18.3 Savonius, Darrieus (eggbeater version), and H-bar VAWTs, from [14]

- *Transient Wind Direction*: Because wind gusts can change direction quickly, wind turbines must be able to adapt to changing directions in order to function effectively.
- *Wind Shear*: Wind shear results in lower wind speeds near the ground, and larger wind speeds with increasing altitude. As a result, rotor blades transmit cyclic loads to the drivetrain.
- *Environmental Conditions*: Wind can contain dirt, dust, and water, which can accelerate damage to critical components within the turbine.
- *Environmental Temperature*: Ambient temperature can influence the viscosity of lubricating oil, which will in turn impact the operation of the entire system.

18.3.2 Wind Turbine Design and Performance

Wind turbines are generally classified into two categories: horizontal-axis wind turbines (HAWTs) and vertical-axis wind turbines (VAWTs). VAWTs incorporate a vertically oriented rotor shaft, with drivetrain components located at their base. Early VAWT designs include Savonius, Darrieus, and Giromill, or H-bar designs (Fig. 18.3). Although VAWTs have shown to have advantages over HAWTs in that they are omnidirectional and their drivetrain components easily maintained, their installation heights are limited, and their blades are prone to cyclic fatigue. Because VAWTs are relatively uncommon among modern turbine designs, they will not be discussed in this manuscript.

HAWTs incorporate drivetrains that are oriented horizontally, in a direction parallel to that of the wind. HAWTs are far more common for utility-scale applications (e.g., greater than 100 kW in capacity) than VAWTs in part due to their capability of being installed at higher altitudes, and consequently,

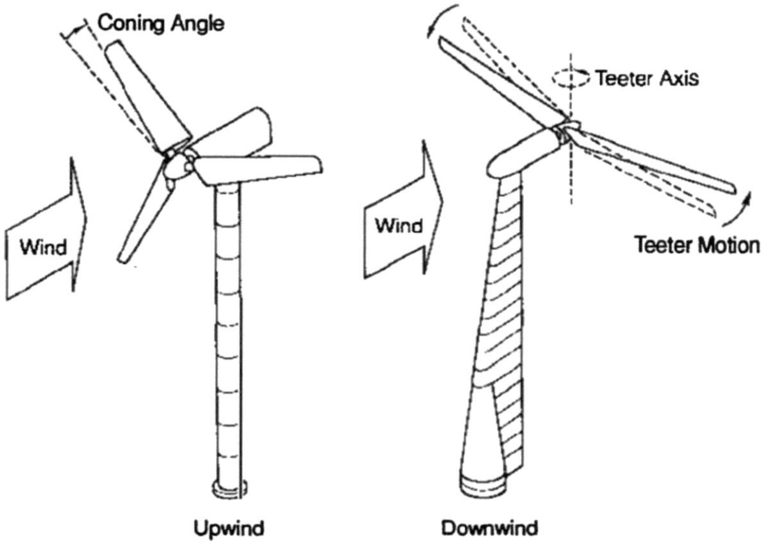

Fig. 18.4 Schematic of upwind, three-bladed HAWT, and downwind, two-bladed HAWT, from [15]

their potential to be exposed to greater wind speeds. The drivetrain components, generator, and associated systems are installed in a nacelle enclosure at the top of a tower, with the nacelle itself angled (yawed) to keep the rotor blades in the direction of the wind. Early wind turbines were designed to position the blades downstream of the support tower and be yawed passively by the wind itself (known as *downwind HAWTs*, shown in Fig. 18.4). However, the blades in downstream HAWTs are exposed to the turbulent wake that is caused by the tower, which was shown to cause fatigue failures due to cyclic loading. For this reason, almost all downwind HAWT designs have been replaced in favor of turbines that position the blades upwind of the tower (i.e., *upwind HAWTs*), as enabled by a yaw system that is actively controlled using a wind sensor and control mechanism. Because upwind HAWT designs are used almost exclusively in modern turbines, they will be the focus of this manuscript.

The power output from a wind turbine is given by the following:

$$P = \frac{1}{2} C_P \rho A U^3 \tag{18.1}$$

where ρ is the density of air, C_P is the power coefficient, A is the rotor swept area (i.e., the area of the imaginary circle formed by the blade tips), and U is the wind speed. The power coefficient C_P, which denotes the fraction of wind power that can converted into usable mechanical work, is primarily a function of the tip speed ratio (commonly denoted as λ), which is defined as the ratio of the rotor tip speed to free wind speed. The maximum theoretical power coefficient is denoted as the Betz limit, which is specified to be 0.593. In practice, lower maximum power coefficients in the range of 0.47 and below are commonly seen in utility-scale turbines, with optimal tip speed ratios between 6 and 8.

The power coefficient of a wind turbine is also dependent on the blade pitch angle, that is, the angle of attack of the blades with respect to the direction of the wind. Most modern utility-scale turbine designs use pitch angle to control the rotation of the rotor, and in doing so, fall under three main classes: (1) passive stall-controlled, (2) active stall-controlled, and (3) pitch-controlled. Passive stall control indicates that the rotor blades are designed to stall at large wind speeds, and thus do not incorporate a pitching mechanism at the blade roots.

In regard to the solidity of the rotor swept area (i.e., the total blade area divided by the swept area), it is well accepted that utility-scale turbines have three rotor blades, which corresponds to a solidity of approximately 0.0345. Having a high solidity (i.e., more than three rotor blades) results in a relatively narrow range of tip speed ratios λ at which C_P is optimum, in addition to increased production costs due to the large number of blades that must be manufactured, shipped, and installed. Meanwhile, turbines with relatively low solidity (i.e., one or two rotor blades) have been shown to experience excessive cyclic loading within their drivetrain components, and have also been shown to have less aesthetic appeal than 3-bladed turbines. For this reason, almost all utility-scale turbines have three rotor blades.

The rotor speed of the wind turbine must be limited for a number of reasons:

- The tip speed ratio has a narrow range (generally, between 6 and 8) for optimal performance, and it is prudent to maintain the rotor speed within the range for efficiency purposes.
- Extreme rotor tip speeds have been shown to cause excessive noise, because the noise emissions from rotor tips vary by the fifth power of blade speed [16]. For this reason, wind turbine designers are forced to have firmer restrictions on turbine rotational speed when the wind speed, and corresponding ambient noise levels, are relatively low. This limitation is less stringent in the case of offshore wind turbines.
- The rotor and hub must be kept within centrifugal force limits. Since centrifugal force increases with the square of rotation speed, excessive rotation speed can result in catastrophic damage to the rotor and/or bearings.

Based on Eq. 18.1, it can be easily seen that the power rating of a wind turbine is largely dependent on wind speed and rotor swept area. A wind turbine manufacturer can therefore design for increased turbine power capacity by either designing the turbine with longer rotors, or by installing the wind turbine at a location with higher wind speeds. Many wind turbine developers are thus working to install larger turbines offshore, as wind speeds tend to be larger offshore, while visual appearance is less of an issue for large, offshore wind turbines. It must be noted, however, that tip speed limitations require an increase in rotor size to be accompanied by a decrease in rotation speed.

18.3.3 Generator Requirements

The generator within the wind turbine receives rotational energy from the drive-train and converts it into electrical energy. In utility-scale turbines, utility requirements call for wind turbines to produce three-phase alternating current (AC) at a fixed frequency of 60 Hz in the United States or 50 Hz in Europe for transfer to the electrical grid. Two types of generators are common in modern turbines, namely, synchronous generators and asynchronous (otherwise known as induction) generators. Both types of generators operate by spinning a rotor within a stator, with a narrow gap, known as the air gap, separating the two. Because power is generated based upon the movement of an electromagnetic field past the windings within the stator, the frequency of the power that is generated is a function of the rotor speed.

An important parameter pertaining to generator operation is known as the synchronous speed n_s, given as follows:

$$n_s = \frac{60f}{p_P} \tag{18.2}$$

where f is the frequency of the generated AC power in Hz and p_p is the number of pole pairs within the generator. The number of pole pairs p_p is normally two, which means that if a generator is to be directly connected to the grid, it would be required to spin at a synchronous speed of $n_s = 1{,}500$ rpm (for a 50 Hz electric grid) or 1,800 rpm (for a 60 Hz grid) in order to match the frequency of generated power with that of the grid. An asynchronous generator, also known as an induction generator, operates near the synchronous speed n_s of 1,500 or 1,800 rpm that corresponds to the local grid frequency. Instead of operating directly at the synchronous speed, asynchronous generators are allowed some amount of slip, s, given as:

$$s = \frac{n_s - n_r}{n_s} \tag{18.3}$$

where n_r is the speed of the rotor. Because an increasingly negative slip causes increased generator torque while any positive slip causes the generator to behave like a motor, it is desirable to limit the magnitude of allowable slip such that the rotor speed is uniform within a tolerance of 1%. Wind turbines with asynchronous generators and direct grid connection must then operate within a relatively narrow speed range, such that they are referred to as *fixed-speed* wind turbines. Fixed-speed wind turbines often have two fixed speeds, as enabled by incorporating either two generators with different ratings and pole pairs or a single generator with two sets of windings [17]. The fixed speed system was the design used by Danish manufacturers between the 1980s and 1990s [18], and is thus considered to be the "traditional" layout. A standard squirrel-cage induction generator, which has stator windings connected to the load/excitation source and rotor windings consisting of electrically connected bars of conducting metal surrounding a soft iron core, has been the most popular choice for electrical power conversion. Because the rotor speed is relatively low (20 rpm), a multi-stage, speed-increasing gearbox is required with fixed-speed systems.

Since the late 1990s, many wind turbine manufacturers started placing increasing focus on *variable-speed* wind turbines, wherein the operating speed of the rotor and generator varies with wind speed. There are a number of advantages in using variable-speed wind turbine systems, in particular the possibilities to reduce stresses of the mechanical structure, to reduce noise, and to provide better control over the generator power [19, 20]. Variable-speed turbines typically incorporate a multi-stage gearbox and a doubly fed induction generator (DFIG). The DFIG contains windings within both the rotor and stator sections. The power that is generated within the stator is directly connected to the grid, while the induced power from the rotor windings is routed through a frequency converter to the electrical grid. Because only the power from the rotor assembly is fed through the frequency converter, the converter is typically rated as a percentage (approximately 30%) of the full generator power. It must be noted that DFIGs require a slip ring assembly to transfer power between the rotor windings and the stationary electronic components. Because the slip ring wears over time, it requires periodic maintenance and is a source of added cost [21].

Most recently, manufacturers have developed variable-speed wind turbines that incorporate synchronous, low-speed generators with no gearbox—these are known as *direct-drive* wind turbines. The generator produces variable frequency output that is proportional to the rotor speed. However, the output power from the low-speed generator is routed through a full-power frequency converter that changes the generator output power frequency to the 50 or 60 Hz that is required by the grid. Although it can be argued that direct-drive turbines improve reliability and lower cost by removing the gearbox, studies [22] have shown that the increased size, weight, and cost of a low-speed generator and full power converter (as compared to a partial converter) oftentimes compensates for the loss of the gearbox.

18.3.4 Transient Loading

It must also be noted that wind turbines, unlike other forms of electricity production, are designed to generate under a spectrum of power levels [5], and thus must contend with transient loads within their drivetrains. A commonly used means to estimate the power-producing ability and drivetrain loads is to assume that it operates under a certain wind speed distribution. The distributions that are frequently used are Weibull (Fig. 18.5) and Rayleigh distributions [23, 24]. To determine the number of cycles of a drivetrain component at a given load, it is important to determine the total number of hours per year for a particular wind speed. This is done by finding the probability of a particular wind speed for the desired distribution and multiplying it by the number of hours per year. The probability that the wind lies between two wind speeds is given by

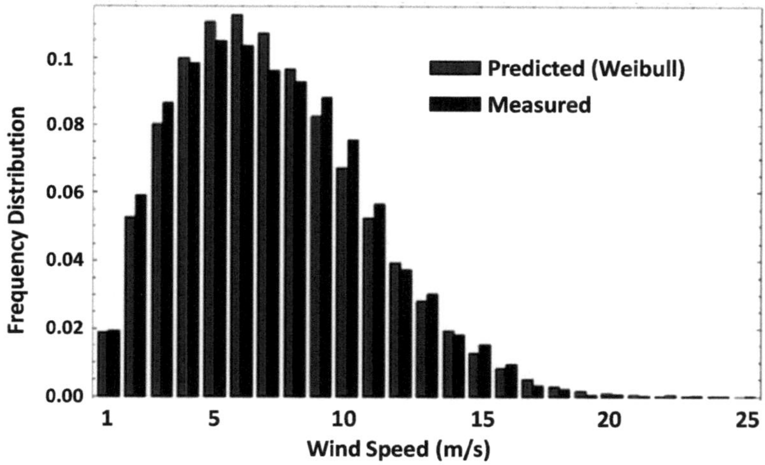

Fig. 18.5 Histogram of predicted and observed wind speeds, from [25]

$$\Pr(u + \Delta u) - \Pr(u - \Delta u) = p(u)\Delta u \tag{18.4}$$

where Pr is the probability function, and *p*(*u*) is the probability density function. For the Rayleigh distribution, the probability density function is given by

$$p(u) = \frac{\pi u}{2u_a} \exp\left[-\frac{\pi}{4}\left(\frac{u}{u_a}\right)^2\right] \tag{18.5}$$

where u_a is the average wind velocity. From this, the number of hours that a wind turbine is operated at a given speed is estimated as

$$H(u) = 8760p(u)\Delta u \tag{18.6}$$

where 8760 represents the total number of hours in a 365-day year [24]. From this analysis, designers can implement a damage criteria such as Palmgren–Miner's rule (as discussed in Sect. 18.5.2.4) to estimate the fatigue lives of components such as gears and bearings.

18.3.5 Power Control

Ideally, a wind turbine should be able to extract as much power as possible up to the rated power of the generator, then limit the extracted power to the rated level as the wind speed increases further. Modern wind turbines are able to maintain the desired amount of power, in part, by controlling the pitch of the rotor blades. Rotor pitch control can take place using one of the following methods:

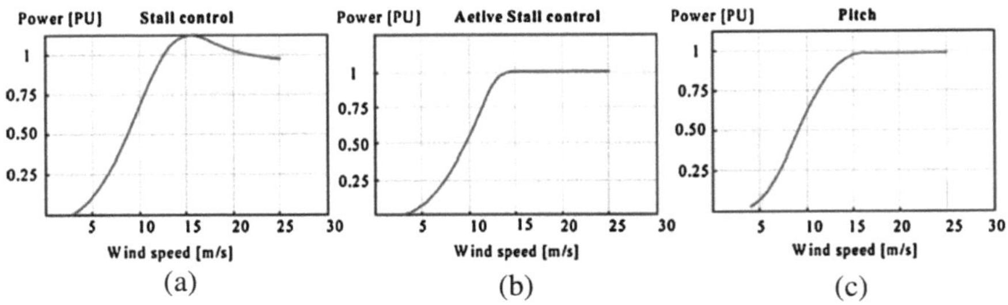

Fig. 18.6 Power characteristics of fixed-speed wind turbines under **a** stall control, **b** active stall control, and **c** pitch control, from [26]

- *Pitch-controlled*: A controller adjusts the rotor blade positions along their long axes to change their angle of attack with the wind [5]. The pitch rotation is enabled by bearings mounted in the hub of the nacelle, with the pitching mechanism activated using either hydraulic or electric stepper motor operation.
- *Passive stall-controlled*: The rotor blades are rigidly attached to the hub at a fixed angle. However, the geometry of the rotor blades are designed aerodynamically to ensure that the blades will stall (i.e., lose lift) when the wind speed exceeds a designated value. This feature serves to protect the turbine from overspeed.
- *Active stall-controlled*: The rotor blades are allowed to rotate along their long axes using a mechanism similar to that of the pitch-controlled system. In active stall-controlled systems however, the controller is programmed slightly differently such that the blades are pitched to stall when wind gusts are excessive.

Each method produces a slightly different power curve (i.e., extracted power vs. wind speed), as shown in Fig. 18.6. The generator torque is also actively controlled to maintain an optimum rotor speed, although this control mechanism can cause undesired load reversals to be transmitted through the gearbox when the wind speed is highly variable.

18.4 Drivetrain Layout

The nacelle of the wind turbine is the external housing that contains the main mechanical components of the system. As shown in Fig. 18.7, these components include the rotor shaft and bearings, a gearbox assembly (if necessary), a lubricant filtration system, a mechanical braking system, a generator, and power electronics. The gearbox transmits power from the main shaft to a high-speed shaft, which, in turn, drives the generator. The drivetrain components and nacelle cover are mounted onto a bedplate, which in turn, is positioned on top of a yaw system that is designed to actively orient the rotor into the wind.

Although a variety of wind turbine drivetrain designs are currently in use, all utility-scale designs generally fall within three main categories [27]. The first

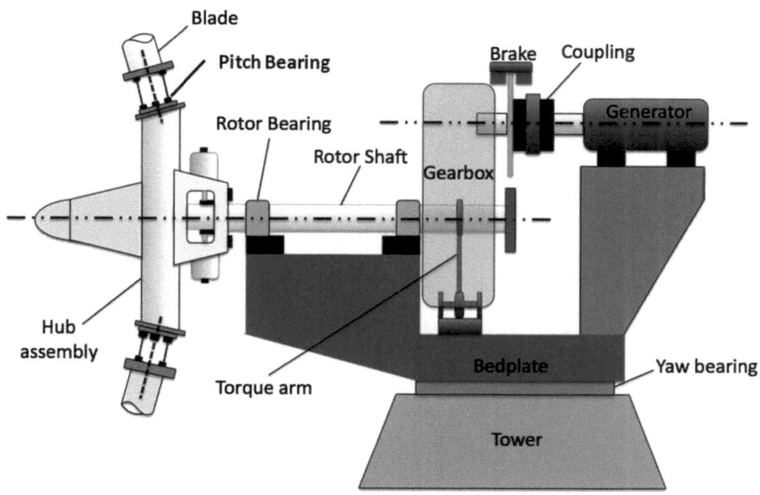

Fig. 18.7 Diagram of components in wind turbine nacelle

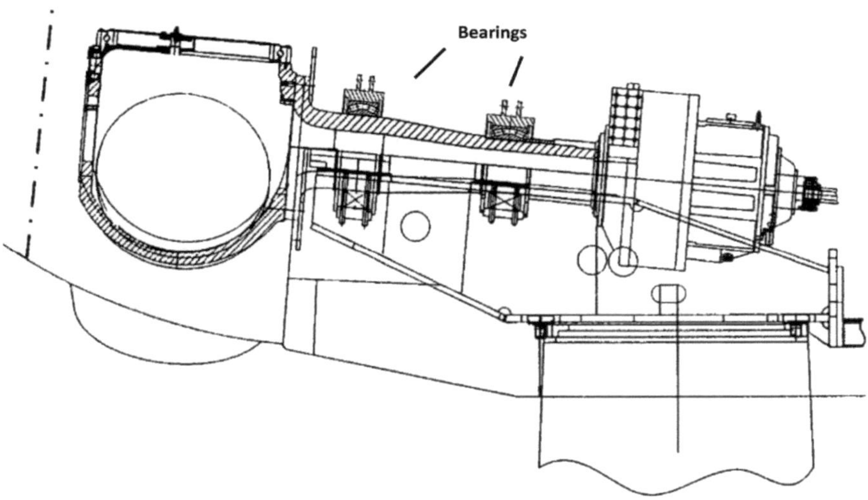

Fig. 18.8 Drivetrain layout featuring two bearings outside the gearbox supporting the main shaft, from [28]

concept (Fig. 18.8) involves the support of the main shaft using two bearings. The bearing nearest the rotor carries both axial and radial loads while the bearing near the gearbox carries radial loads. As a result, the main shaft transfers only torque into the gearbox. Because the gearbox may carry reaction torque towards the bed plate, the gearbox assembly may be fitted with torque arms to transfer the reaction torque to the bedplate. The second concept (Fig. 18.9), known as a

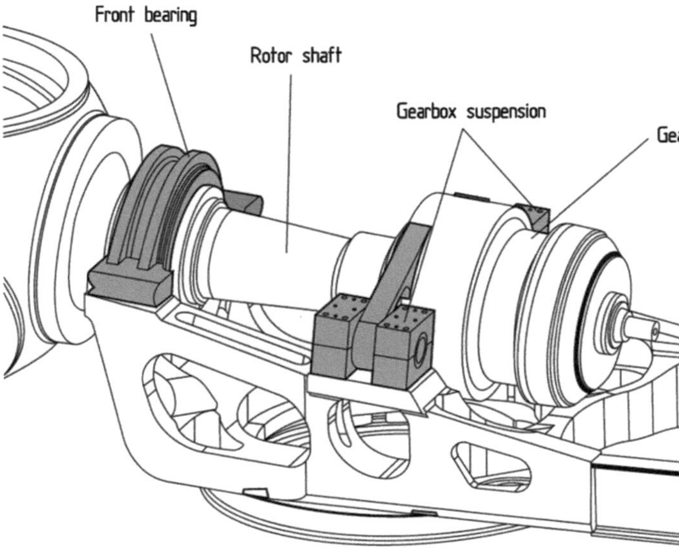

Fig. 18.9 Drivetrain layout incorporating a three-point suspension, with one rotor shaft bearing integrated into the gearbox, from [28]

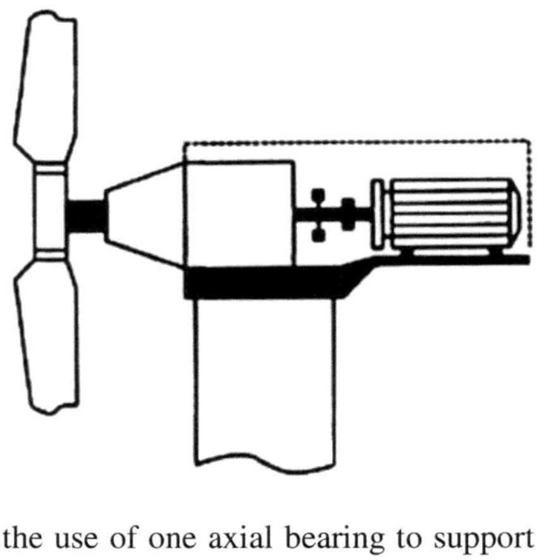

Fig. 18.10 Drivetrain layout with all main bearings integrated in the gearbox, from [27]

"three-point-suspension" design, involves the use of one axial bearing to support the main shaft near the rotor, while a radial bearing supports the opposing end of the shaft from inside the gearbox. The gearbox itself is mounted on the bed plate and is supported by two torque arms. The third category of drivetrain (Fig. 18.10) involves the use of direct integration of the gearbox into the nacelle. In this design, all loads from the rotor enter the gearbox, with all of the rotor support bearings integrated into the gearbox as well. This design can be considered advantageous for the purposes of weight reduction of the nacelle; however, incompatibilities between the gearbox and the remaining components in the nacelle can lead to early failure [27].

18.5 Wind Turbine Tribological Components and Analysis

The tribological components in a wind turbine include rotor support bearings, intermediate gearbox rotor bearings, high-speed bearings, pitch bearings in the hub, epicyclic and parallel gears in the gearbox, a mechanical brake, a roller bearing system in the yaw mechanism, and slip rings in generator (if the generator is a doubly fed induction generator).

18.5.1 Drivetrain Bearings

Rolling-element bearings are incorporated to support the rotor and rotor shaft, the gearbox shafts, and the generator input shaft, with the bearing arrangement depending heavily on the layout of drivetrain. Rotor shaft bearings support the main shaft as well as the rotor blades, operating under dynamic axial and radial loads as well as slow speeds (approximately 20–30 rpm). The rotor blades impose cyclic loads onto the main shaft, thus causing the shaft itself to bend, resulting in misalignment within the bearings. Intermediate-speed and high-speed bearings in the gearbox can also be subject to preliminary damage [29].

It is widely accepted that bearing failure is one of the major issues in wind turbine drivetrain reliability, as the bearings must contend with cyclic and transient loading as well as alignment issues. Studies by Slootweg et al. [30] and Musial et al. [29] relate the start of most drivetrain failures to faulty bearings.

Bearings can be designed to handle purely axial loads, purely radial loads, or a combination of the two. Examples of each type include [31]:

- *Bearings for pure axial load*: Four-point contact ball bearing, spherical roller thrust bearing, tapered roller bearing, and cylindrical roller thrust bearing (Fig. 18.11).
- *Bearings for pure radial load*: Cylindrical roller bearing and toroidal bearing (Fig. 18.12).
- *Bearings for combined axial and radial loads*: Spherical roller bearing, tapered roller bearing, deep-groove ball bearing, angular contact ball bearing (Fig. 18.13).

It must be noted that the locating ability of bearings is designated according to their axial load-handling abilities. A bearing that is capable of supporting axial forces in both directions is known as a *locating* bearing, while a bearing that supports only radial load is designated a *non-locating*, or floating, bearing. Finally, a bearing that supports axial loads in one direction is referred to as a *cross-locating* bearing. Various combinations of bearings can be fitted in a given location to provide complementary load-sharing abilities. For instance, a cylindrical roller bearing can be situated next to a four-point contact ball bearing, such that the bearings in tandem can support both axial and radial loads.

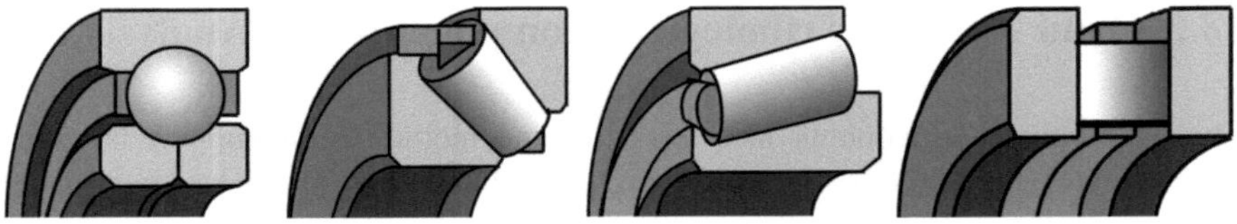

Fig. 18.11 Bearings for purely axial load **a** four-point contact ball bearing, **b** spherical roller thrust bearing, **c** tapered roller bearing, and **d** cylindrical roller thrust bearing [31]

Fig. 18.12 Bearings for purely radial load **a** cylindrical roller bearing, and **b** toroidal bearing [31]

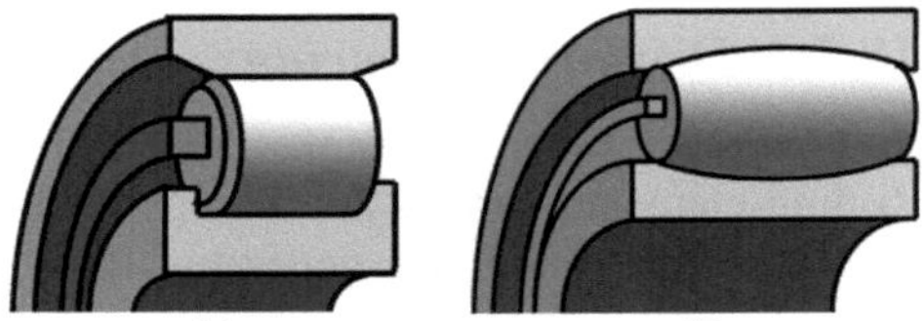

A spherical or tapered roller bearing typically supports the main shaft of most modern wind turbines upwind of the gearbox with rear, non-locating support bearings inside the gearbox [29]. Spherical bearings have the advantage of allowing the bearing's inner and outer ring to be slightly askew with each other without damaging the bearing while running. The spherical bearing has two sets of rollers, allowing support of both radial loads (across the shaft) from the weight of the rotor, shaft, etc. and large axial forces (along the shaft) resulting from wind pressure on the rotor [32]. Thus, the axial load on the main shaft is supported by the main bearing and is not transmitted to the gearbox, which is advantageous for maximizing gearbox reliability.

18.5.1.1 Lubrication Methods

The main shaft bearings are lubricated using either pressure-fed oil lubrication or grease lubrication. The pressure lubrication method uses a circulating oil delivery system that involves the use of pumps, valves, and pipes to supply oil to the bearing. This system is considered to be the most effective because the circulation system regulates the oil temperature while filtering contaminants and metallic wear particles from the oil. However, the complexity of the system and the possibility of lubricant leakage tend to make pressure lubrication impractical for the rotor bearings. For this reason, lubrication with grease is a suitable, low-maintenance alternative which does not require any delivery systems. Grease lubrication is thus used in the rotor bearings of some utility-scale wind turbines [28].

In regard to the bearings in the gearbox (i.e., intermediate-speed, or high-speed bearings), it is well accepted that bearings in the gearbox are generally lubricated using gear oil or some equivalent, which is circulated and distributed using either an oil-splash or circulating oil (i.e., pressure-fed) lubricant delivery system.

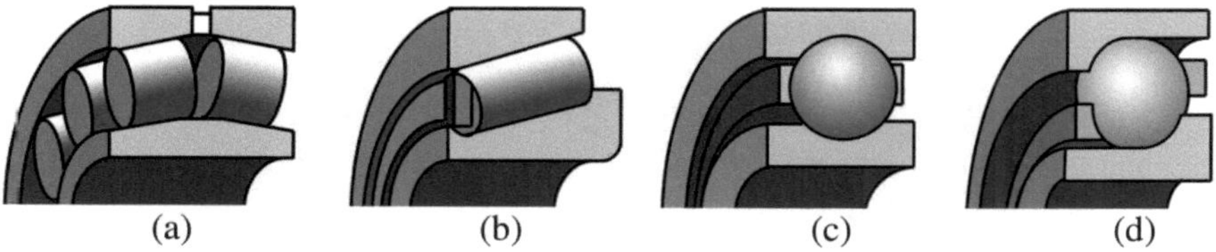

Fig. 18.13 Bearings for combined radial and axial loads **a** spherical roller bearing, **b** tapered roller bearing, **c** deep-groove ball bearing, and **d** angular contact ball bearing [31]

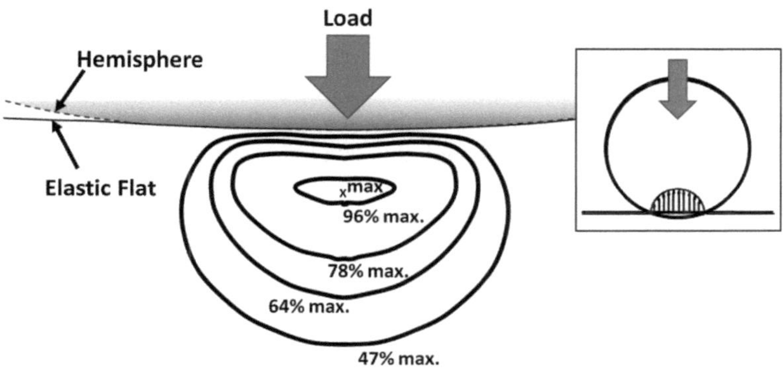

Fig. 18.14 Position of maximum stress for elastic contact between a sphere and a flat, adapted from [34]

18.5.1.2 Bearing Dynamics and Wear

The dynamics of rolling element bearings can be idealized as that of two non-conforming solids which initially touch along a point or a line. Under a compressive load in static conditions, the surfaces deform, causing a contact region of rectangular, ellipsoidal, or spherical shape depending on the geometry of the contacting bodies. Contact of this nature can be analyzed according to Hertzian contact theory [33]. Studies have shown that the contact pressure distribution is maximum in the geometrical center of the contact region, while the maximum shear stress occurs at a defined depth beneath the surface (Fig. 18.14).

As discussed by a report by Kotzalas and Doll [35], there exist a number of failure modes in wind turbine bearings. One of the most common wear modes is rolling contact fatigue, whose occurrence can often be attributed to the relatively large stresses that tend to appear below the surface of rolling elements. These subsurface stresses can produce cracks that form beneath the surface of the rolling elements or raceways, which then propagate to the surface, causing material to be removed from the surface and leaving behind small pits [36] (Fig. 18.15). Fretting wear can also occur within bearings, as wind gusts can cause the bearings to have low-amplitude motion while the system is shut down for maintenance [31].

Fig. 18.15 Micropitting in the raceway of a spherical roller bearing, from [39]

Finally, a phenomenon referred to as white-etch area flaking has also been deemed problematic by some groups. Although the detailed physics of this phenomenon is not well understood, recent studies [37, 38] have indicated that this phenomenon can be traced back to the diffusing of hydrogen from the lubricant into the steel, causing the steel to embrittle, and making the raceway surfaces susceptible to flaking wear.

18.5.1.3 Bearing Predictive Modeling

Bearing dynamics and wear are normally predicted using either statistical or numerical methods. The statistical approach involves the use of a fatigue life model (i.e., probability density function) that provides a description of the trend of failure probability, but contains empirical constants that must be filled in using experimental data. The numerical approach, on the other hand, relies on the numerical approaches to determine parameters such as the lubricant hydrodynamic pressure, lubricant film thickness, and bearing and raceway deformation.

Bearing Fatigue Life

The life theory of rolling bearings, as presented by Lundberg and Palmgren [40, 41], is a commonly used method by which the expected lifetimes of rolling bearings are predicted. According to this theory, the rating life of rolling bearings, L_{10}, in millions of revolutions, is given as:

$$L_{10} = \left(\frac{C}{P}\right)^p \tag{18.7}$$

where C is the load for which $L_{10} =$ unity, P is the actual bearing load, and p is an exponential constant that has the value of $p = 3$ for ball bearings and $p = 10/3$ for roller bearings. The term L_{10} denotes the number of millions of cycles at which 10% of bearings will begin to fail under the operating conditions given.

Lubricant Film Thickness and Lambda Ratio

Wind turbine bearings are generally lubricated using grease or high-viscosity mineral or synthetic oil to minimize surface contact. Lubrication modes are generally classified according to one of three regimes—boundary lubrication, elastohydrodynamic lubrication (EHL), and full film (i.e., hydrodynamic) lubrication. Bearing interfaces typically operate within the EHL regime due to the large applied loads. Because EHL tends to take place under high loads and/or low sliding speeds, it is characterized by the interplay between the pressure-based elastic deformation of the contacting surfaces and the viscosity and density enhancements of the lubricant [42–45]. The lubricant itself experiences a sudden rise in pressure, from ambient to over 1 GPa, within the interface, causing (under ideal conditions) separation of the solid surfaces and minimizing of surface wear. The study of EHL has been extensively investigated by a number of researchers over the past half-century, as initiated by the pioneering work of Grubin [44, 46], and Dowson and Higginson [47, 48], and collaborators.

Assuming that both surfaces are separated by a thin film of lubricant, EHL theory is governed by equations describing the lubricant pressurization, viscosity, and density enhancement, as well as an elasticity equation that describes the deformation of the gear tooth surfaces under load [49]. Assuming the contact takes place along a line (i.e., spur gear contact without edge effects), the lubricant pressure, p, is governed by the Reynolds Equation:

$$\frac{\partial}{\partial x}\left(\frac{\rho h^3}{\eta}\frac{\partial p}{\partial x}\right) = 6U\frac{\partial(\rho h)}{\partial x} + 12\frac{\partial(\rho h)}{\partial t} \tag{18.8}$$

where ρ and η are the density and viscosity, respectively, of the lubricant, t is time, and U is the sum of the surface velocities. The pressure distribution is typically subjected to zero-pressure inlet and Reynolds outlet boundary conditions. The lubricant film thickness h is given by the following:

$$h = h_0 + g_0 + \delta \tag{18.9}$$

where h_0 is the reference film thickness, g_0 is the undeflected geometric gap between the two surfaces, and δ is the composite elastic deformation of both surfaces. The geometric gap is calculated to be:

$$g_0 = \frac{x^2}{2R_{\mathrm{eq}}} \tag{18.10}$$

where $R_{\mathrm{eq}} = (1/R_A + 1/R_B)^{-1}$ is the equivalent radius of curvature as defined by Hertzian theory. The surface deflection, meanwhile, is governed by the elasticity equation for semi-infinite solids under applied load:

$$\delta(x) = -\frac{1}{\pi E'}\int_{-b}^{a} p(s)\ln|x - s|\mathrm{d}s \tag{18.11}$$

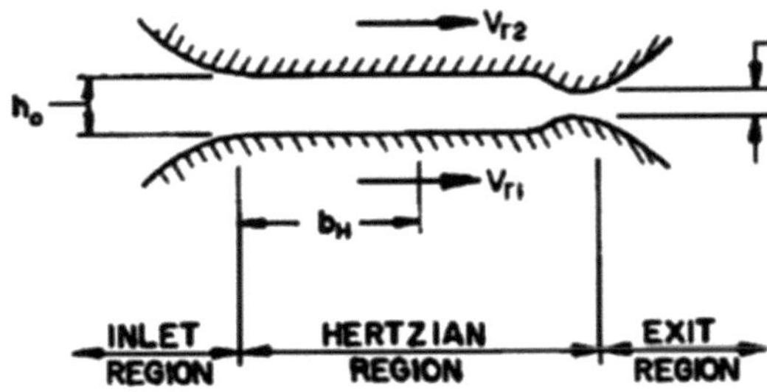

Fig. 18.16 Film thickness distribution in EHL, from [51]

where v is Poisson's ratio and $E' = 2((1 - v_A^2)/E_A + (1 - v_B^2)/E_B)^{-1}$ is the Young's modulus.

The lubricant viscosity, meanwhile, has been shown to increase with increasing pressure. A simplified model that describes this phenomenon, known as piezo-viscosity, is given by the Barus Equation, as follows:

$$\eta = \eta_0 \exp(\alpha p) \tag{18.12}$$

where η_0 is the lubricant viscosity at ambient pressure. The lubricant density, which also increases within the EHL interface, can be modeled by means of a density–pressure relationship presented by Dowson and Higginson [48]:

$$\rho = \rho_0 \frac{(1 + \gamma p)}{(1 + \lambda p)} \tag{18.13}$$

where $\gamma = 2.266 \times 10^{-9}$ Pa^{-1}, $\lambda = 1.683 \times 10^{-9}$ Pa^{-1}, and ρ_0 is the density at ambient pressure.

Equations 18.8–18.13 can be solved simultaneously using iterative or inverse numerical methods [50]. Studies have shown that the film thickness h within the EHL interface is approximately uniform, with the exception of a sudden decrease in h that occurs near the exit of the interface (Fig. 18.16).The lubricant pressure distribution, meanwhile, is found to have a similar parabolic profile as that which would be predicted for Hertzian contact (Fig. 18.17), with the exception of a pressure spike that occurs in the same location as where the film thickness suddenly drops off. Both the pressure and film thickness profiles have been analyzed extensively, and verified experimentally.

The film thickness distribution within EHL is a subject of tremendous interest because the lubricant film is the primary means by which the surface asperities of the gear teeth are separated from one another. The EHL film thickness is generally known to have a relatively flat region followed by a sharp decrease at the lubricant exit. It is thus characterized by two parameters, namely, the central film thickness h_c, and the minimum film thickness $h_{\min}$. If the combined height of the asperities on each mating gear is larger than either the central or minimum film thickness, asperity–asperity contact may occur. Such contact will inevitably lead to surface damage such as abrasive wear, micropitting, scuffing, and crack formation. To characterize the effect of film thickness separation versus asperity–asperity contact,

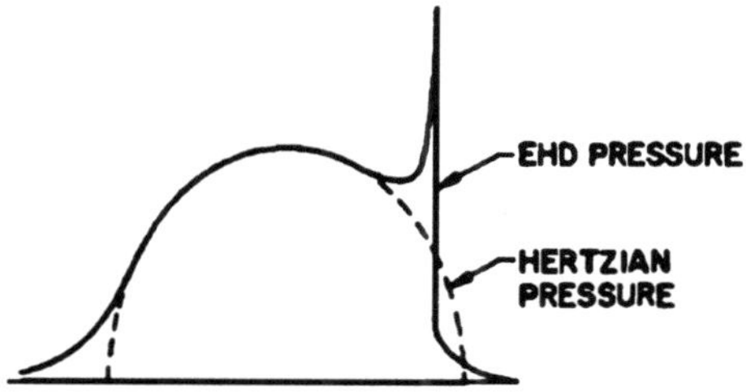

Fig. 18.17 Lubricant pressure distribution in EHL, from [51]

Table 18.1 Wear patterns for various lambda ratios (reproduced from [53])

λ value	Wear pattern
$\lambda < 1$	Surface smearing, deformation, abrasive wear
$1 \leq \lambda < 1.5$	Smoothing of rough areas, spallation
$1.5 \leq \lambda < 3$	Some smoothing of rough areas
$3 \leq \lambda < 4$	Minimal wear
$4 \leq \lambda$	Full separation by EHL film

Tallian [52] introduced a parameter known as lambda ratio λ, known as follows [53]:

$$\lambda = \frac{h_c}{\sqrt{\sigma_1^2 + \sigma_2^2}} \tag{18.14}$$

where σ_1 and σ_2 are the root-mean-square (RMS) roughness of the two contacting surfaces. Studies have shown that the lambda ratio governs the wear pattern, if any, that is experienced by the geartrain, according to Table 18.1.

It can thus be seen to be advantageous to have a lambda ratio that is as large as possible in order to minimize the possibility of gear wear. Recent efforts have thus been taken towards improving the surface finish of gears. Experimental tests by Doll and collaborators [54, 55] have shown that superfinishing (surface polishing to sub-micron roughness) causes noticeable improvement in anti-wear performance.

18.5.2 Gearbox

Utility-scale fixed-speed and variable-speed wind turbines rely upon a gearbox to increase the slow-moving, high-torque input from the rotor to the high-speed output that is required by the generator. The gearing ratio, G, for a wind turbine transmission is defined as follows:

$$G = \frac{n_{\text{generator}}}{n_{\text{rotor}}} \tag{18.15}$$

where $n_{\text{generator}}$ is the rotational speed of the generator and n_{rotor} is the rotational speed of the rotor, both in rpm. A wind turbine that incorporates a generator

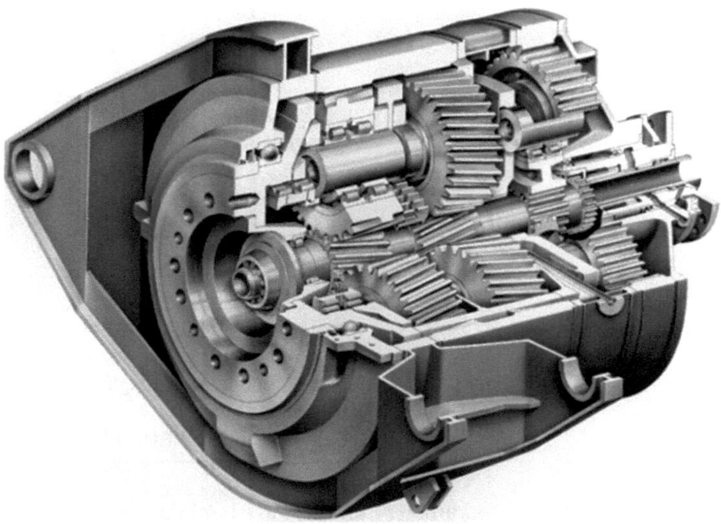

Fig. 18.18 Configuration of three-stage gear for 2–3 MW wind turbine, from [28]

operating at 1,500 rpm and rotor operating at 20 rpm would require a gearbox with gearing ratio of $G = 75$ [56]. The total gearing ratio can be considered a product of the ratio of the individual stages, i.e.,

$$G = G_1 \times G_2 \cdots \times G_N \tag{18.16}$$

where N is the number of stages in the gearbox. Modern turbines generally incorporate at least three gearing stages in the gearbox. One or more stages is generally of a planetary, or epicyclic configuration, while the remaining stages are typically of a parallel (i.e., gears with fixed, parallel axes) configuration (Fig. 18.18). Epicyclic gears incorporate multiple outer gears whose centers revolve around a single, center gear. The revolving gears are denoted as "planets," while the center gear is the "sun". An outer ring gear is located outside the planet gears [56]. For wind applications, the number of planets is normally chosen to be three. Epicyclic gears have seen significant use in wind applications due to the fact that they can transfer a higher torque density than parallel stages, which enables them to be relatively lightweight and compact as compared to parallel gears of similar load capacity. The epicyclic stage is normally designed for gear ratios up to $G_i = 7$, while parallel stages can handle gear ratios up to $G_i = 5$ [27].

Although the electrical systems in wind turbine systems are generally most prone to failure, gearbox and components tend to be the costliest to maintain and replace [10, 57]; thus gearbox reliability is considered a critical issue in wind turbine design and operation. Studies have shown that less-than-desired reliability in wind turbine gearboxes can be attributed to a number of different factors, including the following:

- Wind gusts and generator connection/disconnection from the electrical grid may cause undesired load reversals in the gearbox, which can lead to excessive contact stresses in the gear flanks [35].
- Undesired movement of the drivetrain, which can be caused by deformations within the bedplate, can cause misalignment of the gearbox with the generator shafts. This misalignment results in unexpected damage to the high-speed bearings, which leads to damage in the high-speed gears [56].
- The gearbox may be subject to particulate contamination, which may lead to surface pitting in the gears.
- Excessively high oil temperatures may result in scuffing wear in the gears [28], thus necessitating the use of oil coolers and filters in the lubricant delivery system.

18.5.2.1 Gear Geometry

The tooth profile of spur and helical gears is based on a truncated involute curve (Fig. 18.19), which is defined as a curve that connects a locus of points that are generated at the end of a taut string when it is unwound from the tangent of a base circle (known as the evolute). Litvin and Fuentes [58] involute gearing, first proposed by Euler, has many advantages in its use, including: (1) ease of manufacturability, (2) lack of transmission errors when the gear center distance is changed, and (3) the tooth-to-tooth force is applied along a constant line of action throughout the time of meshing. Through geometrical analysis, it can be found that the geometry of contacting involute gear teeth can be represented by circular discs of varying radii. For this reason, the contact theory of Hertz [33] has the ability to provide a reasonable solution for the elastic deformation, pressure distribution, and real contact area between mating teeth, although it must be noted that Hertzian analysis is based on the assumption of static, dry (nonlubricated), and frictionless conditions—none of which are experienced between moving gear teeth. A more appropriate analysis involves the combined study of lubricant flow and pressurization along with deflection of tooth surfaces, as will be discussed in Sect. 18.5.1.3.

Both spur and helical gearing geometries are common within wind turbine gearboxes. In the case of spur gears, the contact region is a straight line across the depth of the tooth, such that at any time either one or two teeth are in contact. Helical gears, however, are skewed in the axial direction, causing each tooth to appear as a segment of a helix. Because the teeth are angled with respect to the axis of rotation, the contact region is composed of a series of slanted lines, with several teeth in contact at a given time. The angled teeth engage more gradually than do spur gear teeth, causing them to run more smoothly and quietly [60]. In regard to loading, spur gears impose only radial loads on their bearings.

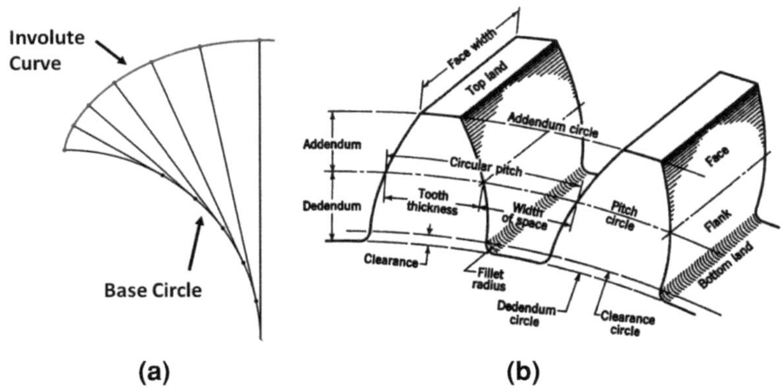

Fig. 18.19 Geometry of involute spur gears **a** involute curve generation, **b** spur gear tooth nomenclature, from [59]

Fig. 18.20 Equivalent radius of curvature for contacting gear teeth, from [62]

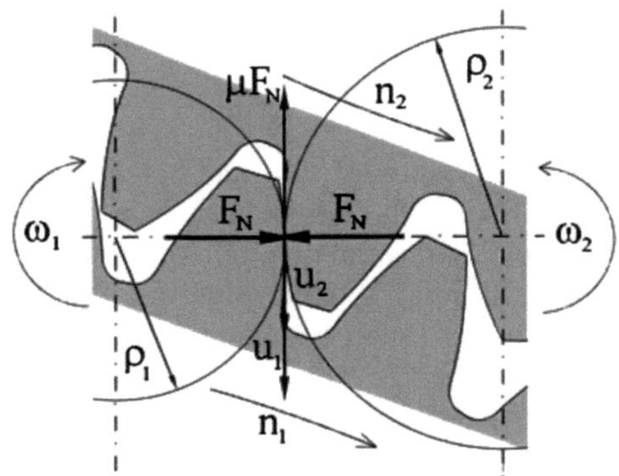

Single helical gears, meanwhile, impose both thrust and radial loads on their bearings. Double helical gears, which are side-to-side combinations of helical gears of opposing axial skewness, develop equal and opposite thrust reactions which serve to cancel out the thrust load [61].

18.5.2.2 Gear Loads and EHL Calculations

The geometry of contacting spur gear teeth is likened to that which is formed by the contact of two rotating circular discs of varying radii. The radii of curvature (Fig. 18.20) of the driving gear (denoted as *Gear A*) and the driven gear (denoted as *Gear B*) are given as:

$$R_A = r_A \sin(\psi) + S \tag{18.17a}$$

$$R_B = r_B \sin(\psi) - S \tag{18.17b}$$

where r_A and r_B are the pitch radii of *Gears A* and *B*, respectively, ψ is the contact angle, and S is the distance of the contact point from the pitch point [53].

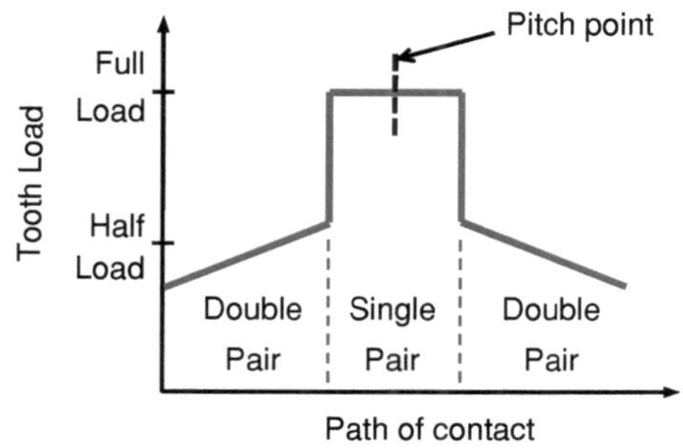

Fig. 18.21 Load sharing of gear tooth during a meshing cycle

The torque that is applied to the gear is given by [63]:

$$T = \frac{P}{\omega} \tag{18.18}$$

where P is the transmitted power, and $\omega = 2\pi N$ is the rotation speed. The normal load that is applied to each meshing tooth is then found as:

$$W_A = s_A\left(\frac{T}{R_A\cos(\psi)}\right) \tag{18.19a}$$

$$W_B = s_B\left(\frac{T}{R_B\cos(\psi)}\right) \tag{18.19b}$$

where s_A and s_B are the percentage of torques shared by the contacting teeth of *Gear A* and *Gear B*, respectively. The load sharing percentages are functions of the contact ratio m_c (a measure of the average number of teeth in contact as a given instant), as well as elastic and surface profile properties of the gear teeth [64]. The torque is shared between contacting teeth based upon their meshing positions, as it is dependent on the elastic stiffness of each tooth at the point of contact (Fig. 18.21). In the case of epicyclic gearing, studies have shown that under ideal conditions, the rotating load is divided evenly by the number of planets, which is further shared by the number of contacting teeth on each planet [65].

Using the loads W_A and W_B and the equivalent radii of curvature R_A and R_B, one can perform EHL analysis to determine the lubricant film thickness and lambda ratio between mating gear teeth in a similar manner as that done for bearings (Sect. 18.5.1.3).

18.5.2.3 Gear Dynamics and Wear

The gears in the transmission experience rolling and sliding contact along the gear tooth faces. Contacting gear teeth slide and roll against one another as they come into mesh and then transition into pure rolling contact at the pitch point. As the teeth move out of mesh, the contact mode transitions back to combined rolling and sliding.

Fig. 18.22 Gear tooth surfaces with scuffing damage, from [67]

In the case of helical gears, it must be noted that an involute helical gear can be analyzed as a set of spur gears with infinitely small width, with each of the spur gears rotated with respect to the adjacent gear to comprise the helical shape [66].

A review by Errichello [51] classifies gear failure according to the categories of overload, bending fatigue, Hertzian fatigue, wear, and scuffing. The primary failure modes in wind turbine gears are generally scuffing, pitting, and abrasive wear, as discussed in the following sections.

Scuffing

This failure mode is caused by rapid adhesion and/or welding of the asperities of contacting gear teeth, followed by the tearing of one or both surfaces as the gears teeth slide past one another (Fig. 18.22). It results in the transfer of material from one tooth surface to another. Scuffing damage can occur if the lubricant film thickness is too low, in which case the oxide layers that normally protect the gear tooth surfaces may be penetrated, and the bare metal surfaces may weld together, resulting in the tearing of the metallic junction.

Although small lubricant film thickness is one condition that is necessary for scuffing to occur, it by itself will not cause scuffing single-handedly. On the contrary, studies have shown that the metallic welding phenomenon that occurs during scuffing is preceded by localized frictional heating, which itself is the result of large contact pressures and elevated sliding velocities between the gear tooth surfaces. The critical temperature theory presented by Blok [68], a widely used criterion for predicting scuffing, predicts that scuffing will occur when the maximum contact temperature exceeds a given value. To minimize the possibility of scuffing, mineral and synthetic lubricants may include anti-scuff, or extreme pressure (EP), additives, which contain sulfur, phosphorus, or other compounds that create a protective layer on gear surfaces [31].

Pitting

This failure mode, considered to be a Hertzian failure mode, is characterized by the formation of surface or subsurface cracks, which propagate to the surface, causing small pits to remain in the surface. The formation of micron-scale pits is known as micropitting. It must be noted that when a micropitted surface is subjected to subsequent loading cycles, several micropits may grow together, leaving behind a larger pit known as a macropit. The formation of macropits generally leads to vibration and transmission error within the gearbox, as well as overall tooth failure [69]. The probability of micropitting can be reduced by keeping the lubricant film thickness (i.e., lambda ratio) high, as damage can most readily occur on gear teeth with rough surfaces or with an insufficient lubricant film. Surface treatment techniques such as carburizing and superfinishing [70] have promise for increasing pitting resistance, although the cost of these procedures may be of concern.

Abrasive Wear

This failure mode is generally caused by hard contaminant particles that entrain into the interface. Particles that are introduced between sliding interfaces abrade material off each surface creating indentations. These indents create high local stresses and are the primary cause for surface-related fatigue failures [71]. Contaminant particles can be created during the manufacturing, assembly, or run-in processes, be ingested from the environment through breathers or seals, or be internally generated. Further details on particulate contamination in wind turbine drivetrains are given in Sect. 18.6.2.

Bending Fatigue

In addition to lubricant film thickness, an important consideration in geartrain operation involves the bending strength of the gear teeth. If the bending stress extends beyond design criteria, cracks may initiate at the base of the tooth and propagate through its base, causing it to break away from the base of the gear wheel [72]. Most gears, particularly, high contact ratio spur gears, may continue to transmit load after a tooth failure. However, the loss of a gear tooth may cause increased levels of noise and vibration that may lead to subsequent damage to the bearings or other areas of the gearbox [73].

The well-known Lewis equation [59, 74], presented by Wilfred Lewis in 1892, is a classical method to estimate the bending stress in spur gear teeth. According to the Lewis equation, the bending stress σ at the tooth root is given by the following:

$$\sigma = \frac{W}{pBy}$$

where W is the transmitted load, p is the circular pitch, B is the width of the tooth face, and y is a dimensionless form factor which depends on the shape of the tooth. More recently, workers such as Dolan and Broghamer [75] and Kelley and Pedersen [76] have developed more sophisticated formulas for gear tooth bending stress based upon photoelastic experimental measurements. More recent studies have employed FEA to enable an accurate prediction of bending stresses [77, 78].

18.5.2.4 Fatigue Life Prediction

The Palmgren–Miner Linear Cumulative Fatigue Damage Theory (Miner's Rule) is used to calculate the resultant pitting or bending fatigue lives for gears that are subjected to highly varying loads [79, 80]. According to Miner's Rule, failure occurs when:

$$\frac{n_1}{N_1} + \frac{n_2}{N_2} + \cdots + \frac{n_i}{N_i} = 1$$

where n_i is the number of cycles at the ith stress level, N_i is the number of cycles to failure corresponding to the ith stress level, and n_i/N_i is the damage ratio at the ith stress level. It must be noted that a significant limitation in the Palmgren–Miner theory is that the loading sequence (i.e., the order of applied loads) is not considered.

18.5.2.5 Power Loss and Noise

Modern gearboxes produce relatively small power losses, with the main sources of power loss being attributed to tooth-flank friction. As a rule of thumb, parallel gear losses are assumed to be 2% of the full load per stage, while planetary gear losses are assumed to be 1% per stage. The frictional losses result in heat generation and noise emission [22]. Although the generated noise from the gearbox constitutes only a small fraction of the power loss, it can still be noticeable to the point where it can draw complaints from nearby residents (in the case of land-based turbines). To minimize noise, gearbox manufacturers must ensure smooth gear meshing by using high-quality design and manufacturing techniques [28]. The nacelle fairing is also acoustically insulated to prevent sound transmission into the air.

18.5.3 Pitch and Yaw Bearings

Blade pitching systems are used to control power and rotor speed, while yaw systems are used to orient the rotor perpendicular to the wind direction.

The bearings that are used for pitch and yaw systems are often discussed within a similar context, as their ranges of motion and wear modes are often similar to one another. Additionally, both pitch and yaw bearings experience similar types of loading—namely, an eccentrically applied thrust, which can be divided into a pure axial load and an overturning moment load [81]. As a result, the fatigue lives of both pitch and yaw bearings can be estimated using a single life rating predictive approach. It is expected that the static loads that are applied to pitch and yaw bearings will cause some degree of permanent deformation in the raceways and rolling elements; however, experience has shown that a deformation of 1/10,000th of the rolling element diameter is tolerable for successful operation.

18.5.3.1 Pitch Bearing

Blade pitching systems, which actively rotate blades along their longitudinal axes, are used in active-pitch and active-stall controlled wind turbine systems. The pitching motion is typically driven by hydraulic actuators [82] or electric motors. The pitch control system is critical for utility-scale turbines, as it allows the system to optimize power extraction and minimize unnecessary loads. Above all else, the pitching system is employed as a fail-safe aerodynamic brake to stop the turbine when wind speeds become excessive [28], thus making the pitching system critical for preventing catastrophic failure of the whole turbine [8]. While a pitching range of 20–25° is generally sufficient for controlling the power and speed of the rotor, aerodynamic braking necessitates that the blades have the ability to be placed into the *feathered* position (i.e., parallel to oncoming airflow), which minimizes lift, causing the turbine to brake.

The use of a pitching system necessitates that the blades be exclusively supported by rolling element bearings at their roots. The bearings themselves are subject to large static loads as well as centrifugal and cyclic bending loads when the turbine is in operation. Most modern turbines use four-point contact bearings or ball bearing slewing rings for pitch operation. Turbines with smaller rotor blades can suffice with a single-row arrangement in their pitch bearings, while larger rotor blades require double-row four-point bearings.

Pitch bearings are generally lubricated using grease with a high-viscosity base oil. Similar to drivetrain bearings, the grease in pitch bearings should maintain the thickest lubricant film possible to prevent premature damage. Because pitch bearings operate under slight oscillatory motion, fretting wear and/or false brinnelling can occur when the lubricant film is insufficient. Additionally, a report by the National Renewable Energy Laboratory (NREL) suggested the pitch bearing be periodically (i.e., once per day) rotated through a large-amplitude oscillation cycle in order to redistribute grease that has been previously displaced [81].

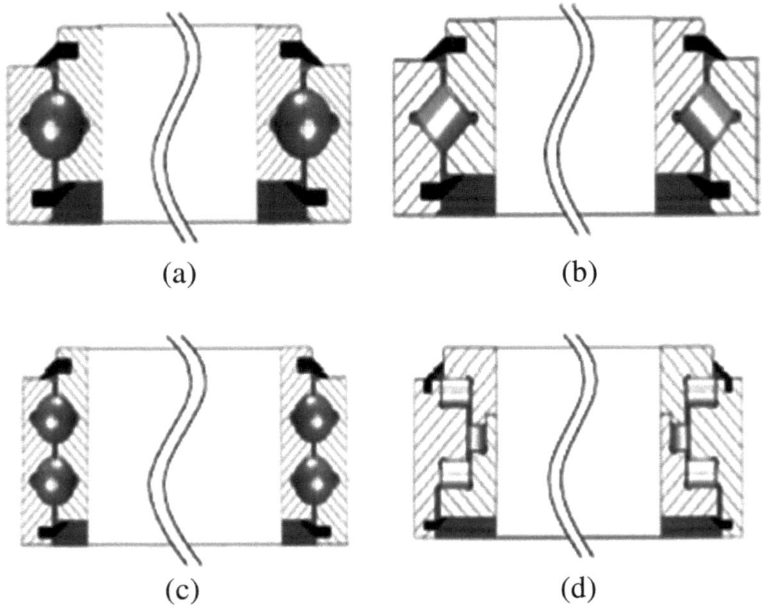

Fig. 18.23 Different types of slewing bearings **a** four-contact-point single row ball bearing, **b** crossed cylindrical roller bearing, **c** four-contact-point-double row ball bearing, and **d** three-row cylindrical roller bearing, from [83]

18.5.3.2 Yaw Bearing

The yaw mechanism rotates the nacelle about its vertical axis to keep the rotors oriented into the wind. Ideally, yaw bearings should provide smooth rotation and long service life as well as yaw damping to prevent unwanted yaw oscillations. Most turbines thus employ roller bearings or ball bearings in their yaw drives, although recent interest has been directed towards gliding bearings that use low-friction synthetic sliding elements in place of balls or rollers. Ball bearing configurations are often single-row, four-point contact ball bearings, although double-row, eight-point contact ball bearings are also used (Fig. 18.23). Double-row bearings have been shown to have the advantage of lower Hertzian stresses and increased fatigue life; however, the double-row arrangement incurs higher manufacturing costs than that of single-row bearings [81].

Similar to pitch drives, yaw drives can be equipped with either hydraulic or electric motors, although electric motors are most popular with modern, large turbines. Usually, the yaw bearings are driven using one or more electric motors in the nacelle, so the bearings contain gears in the inner or outer ring. The yawing motion is usually done only a few degrees at a time [84]. Once the nacelle is in the desired orientation, two or more yaw brakes are typically applied, acting on a brake ring located either inside the tower or the nacelle. The yaw brakes provide stabilization of the nacelle under varying wind loads, while also preventing the yaw drive from absorbing the yawing moment.

18.5.3.3 Fatigue Life Prediction for Pitch and Yaw Bearings

Because pitch and yaw bearings exhibit oscillatory motion and do not spin continuously, the life estimation theory of Lundberg and Palmgren (Eq. 18.7), which predicts lifetime in terms of number of rolling cycles to probable failure, is not directly applicable. However, studies have shown that the Lundberg–almgren estimation can be modified to apply to oscillating bearings based upon the critical angle of oscillation θ_{crit}, which is calculated as:

$$\theta_{\text{crit}} = \begin{cases} \dfrac{720^\circ}{Z(1-\gamma)} & \text{(outer raceway)} \\ \dfrac{720^\circ}{Z(1+\gamma)} & \text{(inner raceway)} \end{cases} \tag{18.20}$$

where Z is the total number of rolling elements (whether loaded or unloaded) and γ is the contact angle. If the oscillation angle (denoted as θ) is less than θ_{crit}, each rolling element has its own discrete stressed volume within the raceway and the modified axial load rating is given by:

$$C_{a,\text{osc}} = \begin{cases} C_a \left(\dfrac{180^\circ}{\theta}\right)^{3/10} Z^{0.033} & \text{(ball bearings)} \\ C_a \left(\dfrac{180^\circ}{\theta}\right)^{2/9} Z^{0.028} & \text{(roller bearings)} \end{cases} \tag{18.21}$$

where C_a is the actual axial load. However, if θ is greater than θ_{crit}, the contact stresses of the individual rolling elements overlap, and the modified axial load rating is given by:

$$C_{a,\text{osc}} = C_a \left(\frac{180^\circ}{\theta}\right)^{1/p} \tag{18.22}$$

where $p = 3$ for ball bearings and $p = 4$ for roller bearings. If the speed of pitch and yaw bearings is denoted in terms of oscillations per minute, the fatigue life, expressed in terms of millions of oscillations, is calculated as [36, 81]:

$$L_{10} = \left(\frac{C_{a,\text{osc}}}{P_{\text{ea}}}\right)^3 \text{ for thrust ball bearings, or} \tag{18.23a}$$

$$L_{10} = \left(\frac{C_{a,\text{osc}}}{P_{\text{ea}}}\right)^{10/3} \text{ for thrust roller bearings,} \qquad (18.23b))$$

where P_{ea} is the dynamic equivalent axial load.

18.5.4 Mechanical Brake

Modern wind turbines incorporate a mechanical brake on the high-speed section of the drivetrain, normally between the gearbox and the generator. This mechanical brake, which almost always is in the form of a disk brake, is primarily responsible for locking down the rotor during shutdown such that servicing and repair work can take place. Beyond functioning as a parking brake, some turbines employ the mechanical brake as a secondary braking system to complement aerodynamic braking during operation. On larger wind turbines, the mechanical brake is situated on the high-speed side of the gearbox to minimize the size and weight of the brake disk. Because of its position on the high-speed shaft, the use of the mechanical brake can be detrimental towards gearbox reliability. At standstill, braking loads oftentimes contend with forces from wind turbulence, causing small oscillatory movements of the gear teeth. These motions can result in fretting wear in the gear teeth [28].

18.6 Contamination Effects

Gear oil recirculates through wind turbine gearboxes, lubricating mating surfaces and removing heat. Unfortunately, contaminant particles suspended in the gear oil, as well as water in the parts per million (ppm) range, can significantly degrade the performance and reliability of gearboxes.

Sources of contaminant particles include: cutting and grinding swarf built-in from manufacturing, ingression of airborne abrasives through vents and mechanical seals, and internally generated wear debris and metal oxide corrosion products. To quantify particulate contamination levels, the International Organization for Standardization (ISO) created cleanliness codes that serve as a universal standard for measuring and reporting contamination levels in fluids. Based upon a milliliter sample of lubricant, ISO codes are defined according to the nomenclature $\alpha/\beta/\gamma$, where α, β, and γ denote the number of particles greater than 4, 6, and 14 μm in size, respectively, per milliliter of lubricant. It can be easily seen that the lower the code, the cleaner the lubricant, and thus the aim for wind turbine filtration and monitoring is to keep the code as low as possible. In modern turbines, well-filtered gearbox lubricants may have a code of 16/14/11 or below.

In addition to particulate contaminants, small amounts of water dissolve in lubricating oils and hydraulic fluids. The maximum amount of dissolved water (the saturation level) is typically 300–500 ppm, depending on base stock and additives. Water contamination in excess of the saturation level is termed *free water*, which settles to low points of the system. Emulsified free water—droplets on the order of 1 μm—remains suspended in the oil and gives it a hazy to milky appearance depending on the amount present. Sources of water include humidity ingressing

Table 18.2 Hard particle contamination problems in wind turbine gearboxes

Problem	Summary
Surface-initiated fatigue spalling	3-body wear in rolling contacts. Hard ductile particles dent surfaces, followed by crack propagation leading to fatigue spalling, a.k.a. pitting, cratering of surfaces
Abrasive wear	3-body wear in sliding contacts. Hard particles plough through and cut away surface material, leading to loss of clearance and rough surfaces accompanied by high friction
Accelerated oil oxidation	Catalytic surfaces of fresh metal wear debris accelerate oil oxidation, lead to acidity, oil thickening, and sticky fouling deposits

Table 18.3 Water contamination problems in wind turbine gearboxes

Problem	Summary
Corrosion	Due to free water, especially if acids present from oil degradation and/or microbial growth. Ion currents in aqueous solution. Leads to pitting, leakage, and breakage
Foaming	Due to free water. Leads to air blockage within the oil distribution system, and spillage
Loss of oil film	Due to free water. Water in contact zone cannot support load, allowing opposing surfaces to contact. Results in adhesive wear, high friction, and seizure
Additive drop-out	Due to free and dissolved water. Depletion of hydrophilic additives. Also breaking colloidal suspensions of additives. Leads to loss of additives and fouling of parts
Microbial growth	Due to free water. Colonization of oil by bacteria and/or fungi. Results in: acids, fouling slimes; health issues
Surface-initiated Fatigue spalling	Due to dissolved water carried by gear oil to the tips of propagating cracks. Surfaces of fresh cracks are highly reactive, dissociating water molecules into O_2 and H_2. H_2 migrates into and weakens steel by hydrogen embrittlement. Cracks spread faster, reducing life of rolling elements and resulting in fatigue spalling, pits, and craters
Accelerated oil oxidation	Due to dissolved water, especially if metal wear debris present. Increases rate of oil oxidation by up to 2 orders of magnitude. Leads to acidity, oil thickening, fouling deposits
Hydrolysis	Due to dissolved water. Decomposition of ester-based additives. Leads to loss of additives, formation of acids, and sometime fouling gels

through vents and mechanical seals as well as liquid water incurred during transportation and storage.

18.6.1 Contamination-Based Tribological Problems

Problems caused by particle and water contamination in wind turbine gearboxes are listed in Tables 18.2 and 18.3. Synopses of these problems are provided in the following sections.

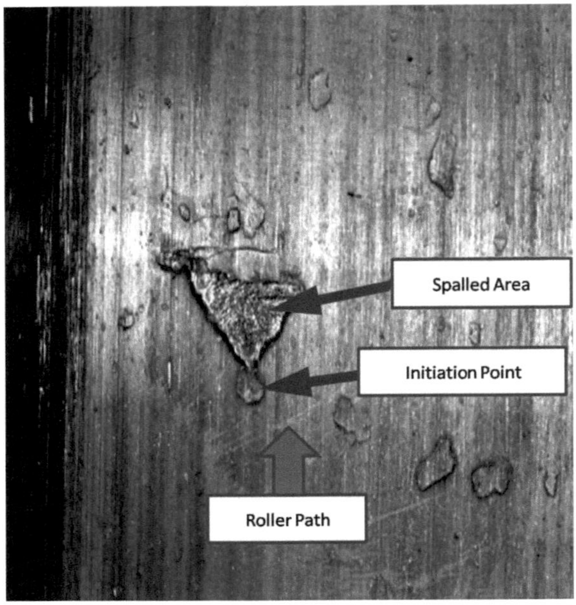

Fig. 18.24 New spall formed on inner raceway of a wind turbine roller bearing. With continued operation the spall would rapidly grow and peel off the entire raceway surface. (From [89])

18.6.1.1 Surface-Initiated Fatigue Spalling

The dynamic clearances between moving parts are on the order of 1 μm, which is the same dimension as many hard contaminant particles. Hard particles trapped in rolling contacts indent raceway surfaces creating raised metal shoulders that act as stress risers [85–88]. The pressure increase around a dent can surpass the normal Hertzian pressure. During subsequent rolling cycles, tangential surface stress plus lubricant forced into the cracks cause the cracks to propagate beneath the surface. Cracks eventually undermined raceway surfaces, creating a fatigue spall (Fig. 18.24) in a similar manner as that described in Sect. 18.5.1.2. Copious quantities of wear-hardened steel are thus released as wear debris, which can lead to secondary damage if not quickly removed. As indicated by Fig. 18.25, hard ductile steel particles (including gear wear debris) are the most damaging. However, large quantities of any hard particle will seriously degrade the bearing life. This makes the wind turbine gearbox a severe application for bearing survival if contamination is not properly managed.

As cracks propagate within the steel, chemically reactive metal is exposed at newly opened crack tips. Dissolved water carried by the lubricant to these crack tips dissociates into its component chemical elements, hydrogen and oxygen gas (H_2 and O_2). The oxygen reacts with both lubricant and steel surfaces. Because hydrogen is the smallest possible molecule, it easily diffuses into grain boundaries where it weakens the steel by hydrogen embrittlement. This accelerates crack propagation and shortens the time to spall formation and component failure.

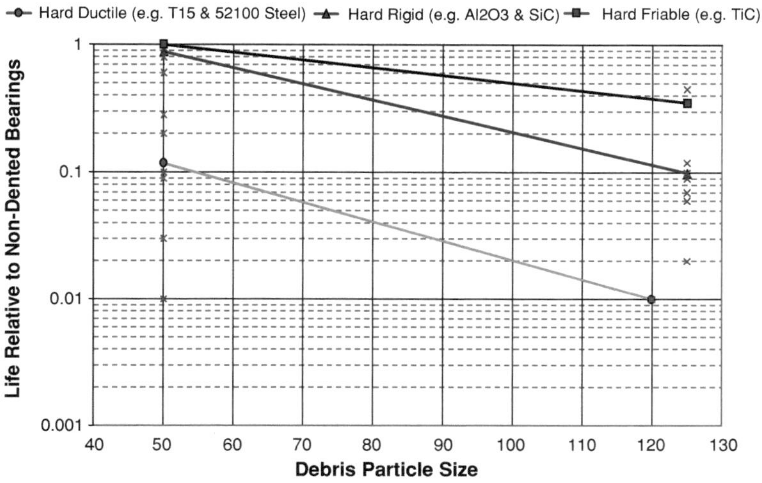

Fig. 18.25 Hard ductile particles, such as gear tooth wear debris, produce the greatest decrement to bearing life. More fragile particles tend to shatter in contact zone and produce less damage per particle. However, damage also increases with concentration for any type of hard particle contaminant [90]

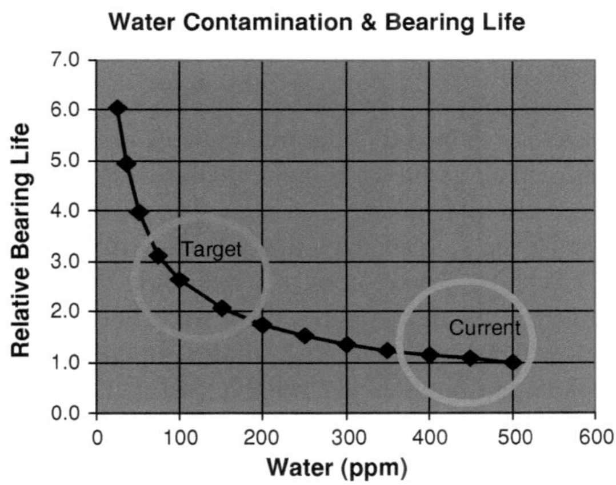

Fig. 18.26 Dissolved water and bearing life. Saturation level of water in oil is 500 ppm. Bearings operating in oil with *lower* levels of dissolved water have longer life. 'Current' refers to typical water contamination levels in wind turbine gearboxes; 'Target' refers to 125 ppm water level achieved through improved contamination control. (Original data from [91], and revised in [92])

Figure 18.26 shows decreasing bearing life as the concentration of dissolved water increases.

18.6.1.2 Abrasive Wear

Hard particles in sliding contacts (such as between gear teeth and in journal bearings) plough through and cut away component surfaces, resulting in frictional losses, elevated oil temperature, loss of fit and tolerance, and reduction in power transmission efficiency. The major offenders are abrasives used in manufacturing and gritty airborne minerals such as silica dust.

Table 18.4 Effects of metals and water on oil oxidation. Adapted from [93]

Condition	Hours (h)	TAN increase
Clean and dry	3,500	0.17
Iron	3,500	0.65
Copper	3,500	0.79
Water	3,00	0.80
Water + Iron	500	8.1
Water + Copper	40	11.2

18.6.1.3 Oil Oxidation

Because of difficulty, expense, and safety issues, wind turbine operators strive to minimize the frequency of gear oil changes. The enemy is oil degradation, which includes loss of additives, excessive accumulations of fine particles, and especially oil oxidation. This well-documented oxidation mechanism starts when oxygen atoms are incorporated into oil molecules producing chemically reactive free radicals, which in turn initiate chain reactions involving thousands of oil molecules and producing acids and polymeric compounds. The acids promote corrosion. The polymers are gummy substances that: (1) thicken the oil, fostering filter bypassing and lubricant starvation during cold starts, (2) foul passages and flow controls, another source of lubricant starvation, and (3) coat and thermally insulate heat exchange surfaces, fostering excessive heat build-up and elevated temperatures during operation.

Antioxidants (free-radical scavengers) are incorporated into gear oil additive packages to intercept the free radicals and terminate the chain reaction. When the antioxidant additives become depleted, oxidation takes over, and the degraded oil requires replacement. A strategy for extending oil life and oil change intervals is to reduce the rate of depletion of antioxidant additives. This is accomplished by: (1) keeping the oil cool, (2) inhibiting the generation of metal wear debris and sequestering fresh debris away from hotter zones, and (3) keeping water levels below 100–200 ppm. Regarding temperature, the rate of oil oxidation tends to follow the simple, yet reliable, estimation for oil temperature: The rate of a chemical reaction approximately doubles for every 10°C increase in temperature. Maintaining gear oil operating temperature below ~50°C will alleviate most oxidation issues due to hot oil.

The surfaces of fresh metal wear particles are catalytic, promoting the formation of free radicals and accelerating oil oxidation. As summarized in Table 18.4, one classic study found fresh metal surfaces accelerated oxidative oil degradation by 6–8 times, as measured by increasing *Total Acid Number* (TAN) values. Per unit surface area, copper was the worst offender (such as from tubing and bronze bushings). However, the greater amounts of iron-containing wear debris found in gearboxes are likely the major offenders for wind turbines. In the same study, water by itself was found to accelerate oil oxidation as rapidly as copper. However, water together with either copper or iron accelerated the rate of oil oxidation by a factor of 300.

Fig. 18.27 Water-induced additive drop-out fouling a gearbox thermostat, rendering it inoperative (COT-Puritech, Inc.)

18.6.1.4 Hydrolysis

In addition to oxidation, wind turbine gear oils can degrade through the direct hydrolysis of ester-based additives. These types of additives are synthesized by reacting alcohols and acids to produce the required esters, along with water as a by-product. In operating systems, dissolved water drives the reaction in reverse, decomposing esters back into alcohols (mostly innocuous) and acids (quite harmful), along with depletion of the ester-based additives. Acids produced by hydrolysis promote corrosion and can also react with metals to produce fouling gels.

18.6.1.5 Additive Drop-Out

Another form of oil degradation is additive drop-out. Some gear oil additives have strong affinity for water (hydrophilic polar additive molecules). They become unavailable by congregating in and around water droplets. Furthermore, high concentrations of dissolved water can break colloidal suspensions of finely divided powders sometimes used as anti-wear additives, resulting in dumping of massive amounts of material. Not only are these additives inactivated by water, but as illustrated in Fig. 18.27, additive drop-outs can completely foul components and make them inoperable.

18.6.1.6 Corrosion

Galvanic corrosion requires current of ions in aqueous solution, and therefore the presence of free water. Even a thin film of water suffices. NaCl salt accelerates the corrosion of metals, making offshore and coastal marine environments especially sensitive to this wear mode. Corrosion results in pitting, leakage, weakening, and breaking of parts, and release of abrasive particles into the oil (such as iron oxide, better known as rust).

18.6.1.7 Microbial Growth

To grow and multiply microbes need three requirements: moderate temperatures, food, and free water. Many strains of bacteria and molds will metabolize gear oil. If free water is present along with temperatures ranging from 15 to 52°C (60–125 F), these will thrive. Consequences include accumulation of acids (promoting corrosion since free water is present), and the formation of biological slimes that foul flow passages and moving parts. Microbial colonization of lubricants is also associated with fetid odors, asthma, and skin allergies.

18.6.2 Contamination Control

Strategies for ameliorating the harmful effects of oil contamination have been developed over the past 40 years in the construction, mining, agriculture, and aerospace industries, and some of the knowledge gained is currently being applied to the wind turbine industry. Some of the practices that are currently being recommended for the manufacture, assembly, and maintenance of wind turbines are listed in Table 18.5.

Additionally, an appropriately designed oil filtration system is critical for minimizing the possibility of early gear or bearing failure. Most wind turbines incorporate one or more filtration systems to remove debris and contaminants from the lubricant. Inline filters, which are placed within the same circulation line as the delivery system, are now standard on most current turbines. To keep the gear oil cool, a heat exchanger is used prior to returning the filtered oil back to the gearbox, typically keeping the maximum oil temperature below 70°C [94]. Some turbines are also designed with an offline or bypass filtration system. These systems include a separate pump that circulates oil through an independent filter. Figure 18.28 depicts a typical contamination control system, complete with a water absorption filter. Offline particle filters are needed only if online filters are insufficient to maintain lubricant cleanliness.

18.6.2.1 Rating Particle Filters

The international standard method for rating lube and hydraulic filters is the Multi-Pass Test [95]. As illustrated in Fig. 18.29, a slurry of silica particles is continuously fed into the test circuit. As particles flow into the filter under test, some are captured, and the remainder continues to recirculate. Throughout the

Table 18.5 Summary of recommended contamination control practices for wind turbines

Area	Description
Design	Drain oil from gears directly back into the sump, bypassing bearing packages. This allows filters to remove gear wear debris upstream of bearings. Use dynamic seals with minimal breathing
Materials	Avoid brittle materials that can shed hard particles. Avoid copper containing alloys contacting lubricants
Component and subsystem cleanliness	Remove manufacturing swarf, grinding/polishing compounds, machine chips, airborne dirt. Avoids early damage leading to premature failures. See ISO 10949
Roll-off cleanliness	Systems and replacement parts should be shipped in cleaner condition than operating systems
Transportation	Ship in sealed packaging to prevent ingression of airborne dust and water from rain and splashing
On-site storage	Seal and store systems and replacement parts in protected enclosures
Prevent contaminant ingression	Install regenerable breather dryers and/or pressured dry air blankets
Rapid contaminant removal	Install 5 μm full-flow (inline) particle filters
Off-line contaminant removal	Install water removing cartridges offline; use dry air blanket systems as needed. If 5 μm full-flow filters not available, supplement with 3 μm offline filters

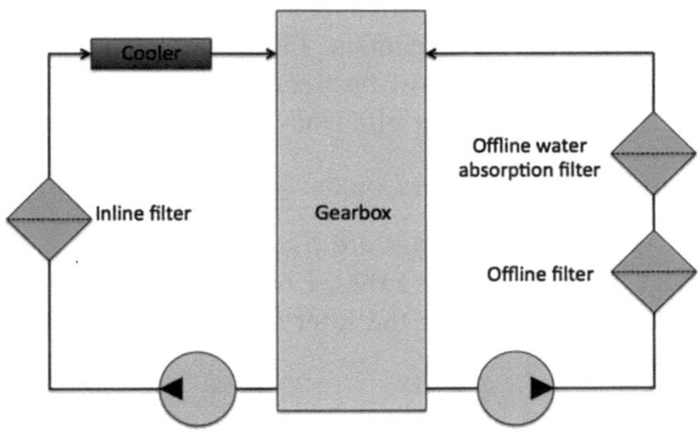

Fig. 18.28 Flowchart of a contamination control system

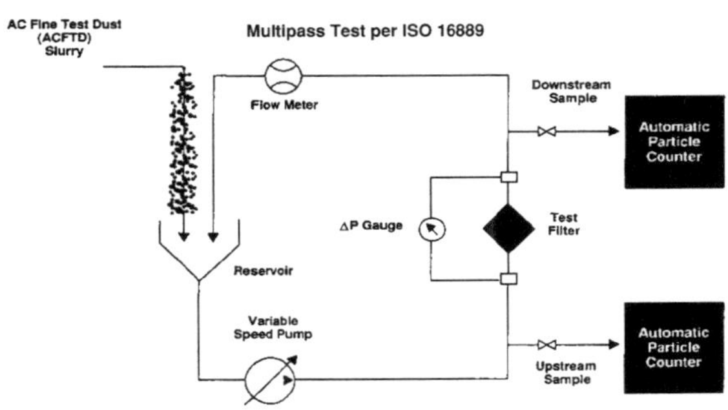

Fig. 18.29 Schematic of the multi-pass test for rating filter efficiency. A fine powder of contaminant, ISO medium test dust, is continuously fed into the system. Upstream and downstream particle counts taken during the test are used to calculated the beta value ($\beta_{X(C)}$), as described in the text

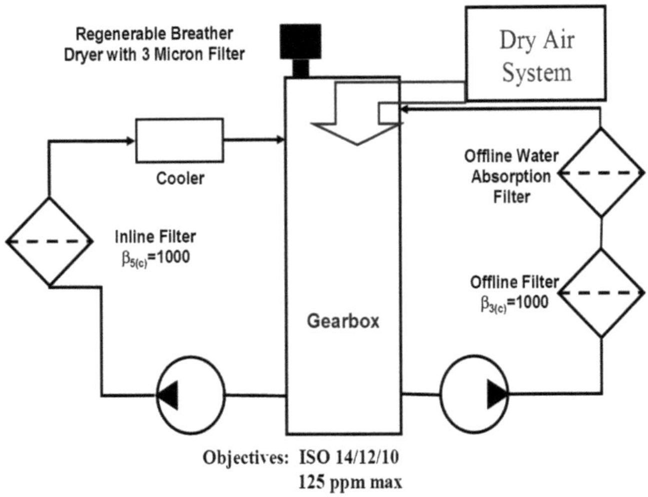

Fig. 18.30 Detailed diagram of recommended filtration cycle for ingression prevention and removal. Main lubricant circuit on *left*, with 5 μm ($\beta_{5(C)} \geq 1{,}000$) installed. Side-loop (offline) circuit shown with optional 3 μm ($\beta_{3(C)} \geq 1{,}000$) particle filter and water absorbing cartridge. Also shown is optional dry air system for removing water from gear oil [96]

test, the number of upstream and downstream particles is measured with electronic automatic particle counters. Filter efficiency for particle size X (in microns) is reported as the ratio of upstream to downstream counts recorded during the test. This measure of filter efficiency is termed the 'beta ratio', $\beta_{X(C)}$:

$$\beta_{X(C)} = (\text{Upstream Counts } \geq X\mu\text{m}) \div (\text{Downstream Counts } \geq X\mu\text{m}) \quad (18.24)$$

Modern filter ratings are $\beta_{X(C)} \geq 200$ and $\beta_{X(C)} \geq 1000$. As an example, for a filter rated at $\beta_{5(C)} = 1{,}000$, 1 out of every 1,000 particles equal to or greater than 5 μm passed through the filter during the test, while the other 999 particles were captured by the filter.

18.6.2.2 Recommended Oil Particle Filters

A detailed model of a wind turbine contamination control system is shown in Fig. 18.30. When available, full-flow filters rated at $\beta_{5(C)} \geq 1{,}000$ are recommended. As shown in Fig. 18.31, bearing life can be extended 2–5 times simply by installing 5 μm full-flow gear oil filters in place of the older, less-efficient 10–25 μm filters found in some wind turbine gearboxes. In addition, wear debris released by gear teeth, bushings, and some dynamic seals are diminished in clean oil maintained by 5 μm filters, resulting in less abrasive wear. Full-flow filters also capture catalytic fresh metallic wear debris before passing into hot sections of the gearbox, reducing the rate of oil oxidation and extending oil life.

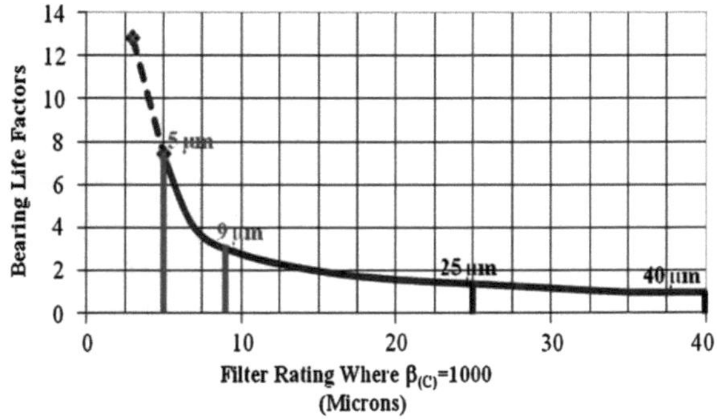

Fig. 18.31 Bearing life factors as a function of filter ratings (Original life factors in [97], revised to reflect changes in filter rating method by [98].)

A difficulty with older types of high-efficiency full-flow filters is excessive flow restriction (i.e., large pressure drop across the filter). This can be especially demanding when filtering high viscosity wind turbine gear oils. However, this problem has been solved with the introduction of a new generation of filters using bi-component fiber technology [99]. As illustrated in Fig. 18.32, replacing resin binders that film over and occlude pores with fibers that reinforce the porous structure results in filter media that is stronger, thinner, more open, and less flow restricting. As summarized in Table 18.6, new-generation 5 μm full-flow filters ($\beta_{5(C)} \geq$ 1,000) installed into standard housings have maintained wind turbine gear oils consistently cleaner than the best conventional wind turbine gearbox filters, and have service lives greater than 12 months.

When 5 μm full-flow filters ($\beta_{5(C)} \geq 1000$) are not available for a particular model gearbox, then 10 μm or coarser filters are installed. In these cases high efficiency offline (a.k.a. side-loop) filters may be considered to supplement the coarser full-flow filter. It is recommended that offline filters have a rating of 3 or 5 μm ($\beta_{3(C)}$ or $\beta_{5(C)} \geq$ 1,000), depending on the recommendation of the oil suppliers. Offline filter circuits require their own pump and source of power, and typically are kept on continuously. Because they function at very low rates of flow, typically 1/20 or less of full-flow, offline filters do not remove particles fast enough to protect against recent ingressions of hard particles. They can help maintain oil cleanliness by slowly removing the particles 5 μm and smaller that coarser full-flow filters cannot.

18.6.2.3 Recommended Air Breathers

To prevent ingression of abrasive airborne particles, vents should be equipped with air breathers capable of removing particles as small as 1 μm. Because vents also provide pathways for humidity to pass into gearboxes, these breathers typically combine an air particle filter and a humidity remover, as described below.

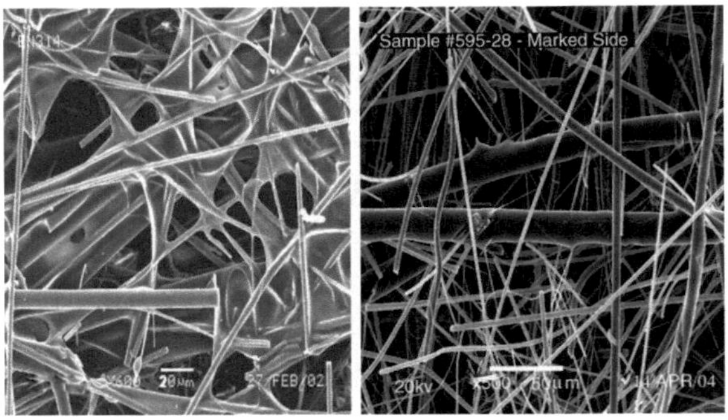

Fig. 18.32 Scanning electron photomicrographs of two types of 5 μm filter media. In conventional media on *left*, flow channels occluded by binder resin filming *over* microglass fibers. On *right*, unobstructed flow passages in bi-component media held together by larger binder fibers to which microglass fibers adhere, resulting in *lower* flow restriction (ΔP) and greater dirt holding capacity

Table 18.6 Polyalphaolefin-based gear oils from three different manufacturers. Filter rating X based on $\beta_{X(C)} \geq 1000$, per ISO 16899 [99]

Gear oil	Filter type (μm)	ISO averages
A	10	19/16/13
	5	17/15/12
B	10	21/17/13
	5	19/16/12
C	10	21/18/13
	5	19/16/11

18.6.3 Water Control

Component life can be significantly extended by keeping gear oil dry. For example, as shown in Fig. 18.31, compared to typical water contamination levels of 400–500 ppm found in many wind turbine gearboxes, maintaining gear oil near ~125 ppm increases bearing life two to three times.

Two approaches are normally used for minimizing water contamination in operating wind turbines:

- *Minimize humidity ingression through vents*: Ingression is minimized by installing vent breathers that prevent humidity from passing into the gearbox. Older types of desiccant breathers using silica gel are effective. However, these types of units have low water holding capacity, resulting in short service lives on the order of weeks to days in humid environments. This problem is avoided by installing regenerable breather driers on vents. These units have unlimited service life with respect to water. As illustrated in Fig. 18.33, humidity is removed from incoming air during the inhalation phase by thin-film adsorption; then during exhalation the unit regenerates by releasing and returning the previously captured water back into the atmosphere.

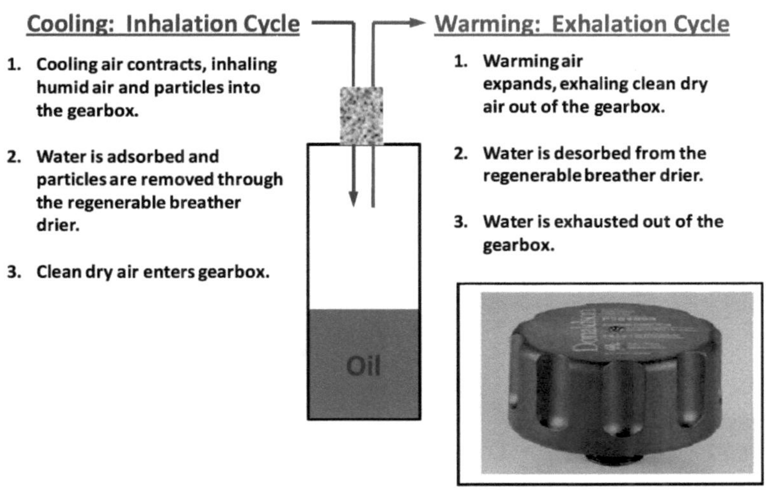

Fig. 18.33 The regenerable breather dryer (*inset*) contains a thin film of adsorbent distributed *over* a large porous surface area, as well as a fine air particle filter. During inhalation water vapor and airborne particles are removed and prevented from entering the system. During start-up air warms inside the reservoir, reducing its relative humidity. During exhalation water vapor retained by the thin film desorbs back into the warmed air and passes back to the environment [100]

Fig. 18.34 At time zero emulsified water injected into oil to 11,000 ppm (1.1%) free water concentration. Oil then recirculated through a single water absorbing cartridge at a turnover rate of 1 every 6 min. All free water removed in 12 turnovers. Saturation level of dissolve water in this oil = 200 ppm. (Turnover rate = fluid volume/flow rate)

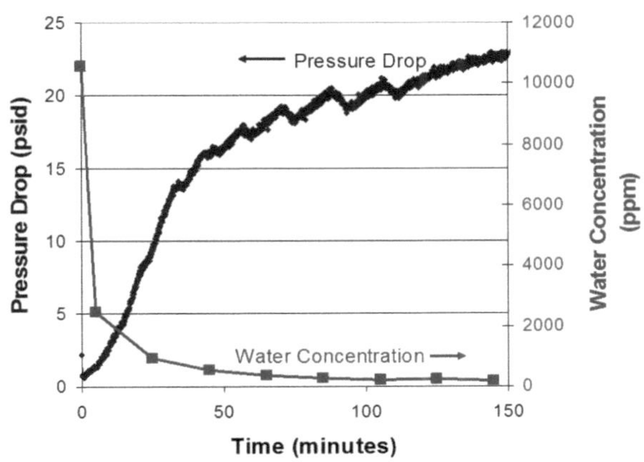

Water may also enter wind turbine gearboxes during maintenance activities or through external seals. A recommended practice for removing this water is to install a water absorbing cartridge in the off-line position. Externally these units appear similar to particle filters, but inside have quite different constructions. Some are made from porous cellulose-based materials with limited rates of removal and water holding capacity. Preferred units incorporate super-absorbent polymers (SAP) with 95% water removal efficiency that rapidly take out free water and also a portion of the dissolved water, as shown in Fig. 18.34.

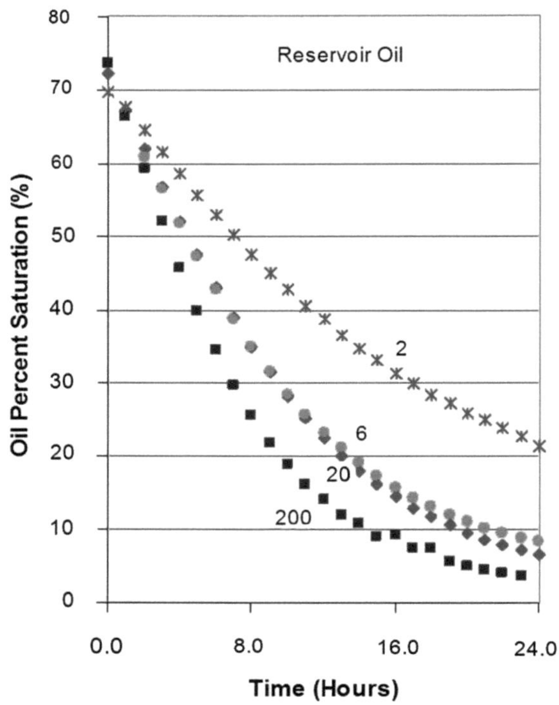

Fig. 18.35 Drying oil with a dry air blanket. Low flow of very dry air (−40°C dew point) sweeps *over* surface of oil in reservoir, bringing oil to less than 20% saturation (<100 ppm for most gear oils). Values on graph indicate the number of headspace turnovers per hour [100]

Finally, a proven technology known as *dry air blankets* has recently been adapted to wind turbines for water contamination removal. In this system, compressed air passes through a pressure swing absorber producing very dry air (−40°C dew point). Flow is arranged so that a blanket of the dry air sweeps over the oil surface in the reservoir. Gear oil is continuously dehumidified by contacting the dry air, as shown in Fig. 18.35. Slight pressurization of the system also inhibits ingression of humid air. Thus dry air blankets minimize water ingression and simultaneously remove water from gearboxes.

18.7 Condition Monitoring

Wind turbine condition monitoring systems allow early warnings of mechanical and electrical faults, which allow the operator to plan system repair before severe failure takes place [101]. Because early damage to components can oftentimes be detected while the turbine is still operational, repair work can be planned well in advance. This is considered to be of importance particularly in regard to offshore wind turbines, where bad weather conditions can prevent repair actions for several weeks [3]. Typical wind turbine sensors include those for measuring the following: (1) rotor position and speed, (2) generator position and speed, (3) power output,

(4) tower vibrations and deflections, (5) blade deformations, and (6) controller inputs such as pitch and yaw settings [102].

Condition monitoring can be performed in one of two ways, namely:

- *Offline*: Not to be confused with offline filtration as described in Sect. 18.6.2, this method involves periodic inspections that require the turbine to be shut down, and may also require the attention of an operator. This method may also incorporate the use of diagnostic equipment.
- *Online*: This method involves the use of automated sensory equipment that monitor the components in the turbine continuously during operation. The system may report raw measurements (stress, deflection, etc.), or may also incorporate processing systems for data analysis.

Condition monitoring techniques that pertain to the tribological condition of the system, namely vibration analysis and oil analysis, are described in the following sections.

18.7.1 Vibration Analysis

Vibration sensing is one of the most well-known methods for monitoring the condition of gears and bearings. The sensors that are commonly used depend on the frequency range that is to be measured, according to the following:

- Position transducers for monitoring low-frequency vibrations
- Velocity transducers for monitoring middle-frequency vibrations
- Accelerometers for monitoring high-frequency vibrations
- Spectral emitted energy (SEE) sensors for very high-frequency vibrations

Additionally, specialized offline acoustic monitoring can be used to determine the presence of cracks or defects in wind turbine components. Monitoring of this sort involves the attachment of acoustic sensors directly onto the component using glue with low attenuation, which then monitor the acoustic frequency of the component. The monitoring can take place either passively (i.e., the excitation is generated by the component itself), or actively (i.e., the excitation is externally applied).

18.7.2 Oil Analysis

Oil analysis can provide information about the health of gears and bearings, as well as detect the presence of particulate or water contamination. For turbines that are monitored offline, experience has shown that oil analysis should be performed after every 4,000 h of operation. After taking a sample of oil from the gearbox, analyses that are commonly performed include:

- *Oxidation*: By analyzing the visual appearance of the used lubricant, one can determine whether it is oxidized. Ideally, the lubricant should appear clear and bright; lubricant-darkened colors may indicate oxidation. Also, the odor of lubricant can be analyzed, in that a sour, pungent, or "burnt" smell indicates that oxidation has taken place. Finally, a measure of the lubricant viscosity may allow for the determination of oxidation, as an increase in viscosity over that of fresh oil can be caused by oxidative effects.
- *Water Content*: Water contamination can be determined by a *crackle* test (i.e., placing the oil on a hot plate to determine the degree of bubble formation), or by more sophisticated methods, such as distillation, infrared, and Karl Fischer analysis.
- *Particulate Contamination*: Similar to the oxidation test, the visual appearance of lubricant (i.e., darkened colors), as well as an increase in the lubricant viscosity, may also indicate particulate contamination. More sophisticated methods such as spectrochemical analysis and automatic particle counting can provide significant details as to the size and composition of contaminants.

Although offline oil analysis is frequently used, an increasing number of wind turbine systems are being equipped with online oil monitoring systems which provide continuous monitoring of a small fraction of the lubricant. These systems may rely on optical, electromagnetic, or other physical principles to sense the degree of particle contamination. It must also be noted that a specialized class of lubricant monitoring, denoted as inline monitoring systems, involves the monitoring of all of the circulating fluid [57].

18.8 Closure

Wind turbine designs have evolved significantly since their inception and all indications point to continued growth in both installation rate and technological development. Wind turbines incorporate a number of tribological systems, including low-speed, intermediate-speed, and high-speed drivetrain gears and bearings, oscillatory pitch and yaw bearings, and a mechanical brake. As increasingly powerful wind turbines have been designed over the years, more attention has been paid towards solving the tribological challenges in wind turbine systems, although many challenges remain. In particular, advanced filtration technology is of upmost importance, as clean lubricant is crucial towards maximizing the lifetime of critical drivetrain components. It is also important to have a detailed understanding of the loads that are being applied to wind turbine tribological systems. Having this knowledge will aid in the selection of the most appropriate tribological component for each section of the turbine, and will also allow for reasonable estimations of the failure lifetime of each component. Finally, condition monitoring is important for monitoring the health of tribological components.

Wind turbine manufacturers are trending towards developing wind turbines of increasingly larger size and power capacity. This is especially the case for offshore wind turbine power plants where the wind loads are complex and highly variable. To ensure the reliability of extra-large-scale turbines, it is important to have robust designs of the tribological systems as well as sufficient filtration and online condition monitoring to make the system less reliant on routine checkups and maintenance.

References

1. B.C. Babu, K.B. Mohanty, Doubly-fed induction generator for variable speed wind energy conversion systems–modeling and simulation. Int. J. Comput. Electr. Eng. **2**(1), 141–147 (2010)
2. H. Chandler (ed.), *Wind Energy–The Facts, in European Wind Energy Association* (2003)
3. Y. Amirat, M.E.H. Benbouzid, B. Bensaker, R. Wamkeue, *Condition Monitoring and ault Diagnosis in Wind Energy Conversion Systems: A Review in Electric Machines and Drives Conference. IEMDC '07, IEEE International*, 2007
4. Global Wind 2009 Report, Global Wind Energy Council, 2010
5. 20% Wind Energy by 2030, U.S. Department of Energy Technical Report, 2008
6. Strategic research agenda: market deployment strategy from 2008 to 2030, European Wind Energy Technology Platform, 2008
7. Mid and long range plan for renewable energy development, Chinese Committee for National Development and Reform, 2007
8. B. Lu, Y. Li, X. Wu, Z. Yang, in *A review of recent advances in wind turbine condition monitoring and fault diagnosis* (Power Electronics and Machines in Wind Applications, PEMWA, IEEE, 2009), pp. 1–7
9. 2009 wind technologies market report, U.S. Department of Energy, 2010
10. J. Ribrant, L. Bertling, Survey of failures in wind power systems with focus on Swedish wind power plants during 1997–2005. IEEE Power Engineering Society General Meeting, 2007
11. E.J. Terrell, W.M. Needelman, J.K. Kyle, Current and future tribological challenges in wind turbine power systems, in *STLE/ASME international joint tribology conference*, ASME IJTC2009-15220, 2009
12. Wind Energy Siting Handbook, American Wind Energy Association, 2008
13. L. Mumper, Wind turbine technology turns on bearings and condition monitoring. Utilities Manager, 2006
14. M. Islam, D.S.K. Ting, A. Fartaj, Aerodynamic models for Darrieus-type straight-bladed vertical axis wind turbines. Renew. Sustain. Energy Rev. **12**(4), 1087–1109 (2008)
15. A.C. Hansen, C.P. Butterfield, Aerodynamics of horizontal-axis wind turbines. Annu. Rev. Fluid Mech. **25**(1), 115–149 (1993)
16. S. Oerlemans, P. Sijtsma, B. Méndez López, Location and quantification of noise sources on a wind turbine. J. Sound Vib. **299**(4–5), 869–883 (2007)
17. A.D. Hansen, L.H. Hansen, Wind turbine concept market penetration over 10 years (1995–2004). Wind Energy **10**, 81–97 (2007)
18. L.H. Hansen, L. Helle, F. Blaabjerg, E. Ritchie, S. Munk-Nielsen, H. Binder, P. Soerensen, B. Bak-Jensen, Conceptual survey of generators and power electronics for wind turbines. Riso National Lab Technical Report, 2001
19. T. Burton, D. Sharpe, N. Jenkins, E. Bossanyi, *Wind Energy* (Wiley, NY, 2001)
20. A. Petersson, *Analysis, modeling and control of doubly-fed induction generators for wind turbines*, in *Department of Energy and Environment*, Chalmers University of Technology, 2005
21. H. Li, Z. Chen, Overview of different wind generator systems and their comparisons. Renew. Power Gener. IET **2**(2), 123–138 (2008)
22. H. Polinder, F.F.A. van der Pijl, G.J. de Vilder, P.J. Tavner, Comparison of direct-drive and geared generator concepts for wind turbines. Energy conversion. IEEE Trans **21**(3), 725–733 (2006)

23. G.L. Johnson, *Wind Energy Systems* (Prentice-Hall, Englewood Cliffs, 1985)
24. D.S. Zinger, E. Muljadi, Annualized wind energy improvement using variable speeds. IEEE Trans. Indus. Appl. **33**(6), 1444–1447 (1997)
25. S.A. Akdag, Ö. Güler, Comparison of wind turbine power curve Models, in *International Renewable Energy Congress*, Sousse, Tunisia, 2010
26. C. Zhe, J.M. Guerrero, F. Blaabjerg, A review of the state of the art of power electronics for wind turbines. Power Electron. IEEE Trans. **24**(8), 1859–1875 (2009)
27. J. Peeters, *Simulation of Drive Train Loads in a Wind Turbine* (Katholieke Universiteit Leuven, Lueven, 2006)
28. E. Hau, *Wind Turbines* (Springer-Verlag, Berlin, 2006)
29. W. Musial, S. Butterfield, B. McNiff, Improving wind turbine gearbox reliability, in *Proceedings of the European Wind Energy Conference*, Milan, Italy, 2007
30. H. Slootweg, E. De Vries, Inside wind turbines-fixed vs variable speed. Renew. Energy World **6**(1), 3041 (2003)
31. American Gear Manufacturer's Association, Standard for design and specification of gearboxes for wind turbines, ANSI/AGMA/AWEA 6006-A03, 2004
32. H. Stiesdal, *The Wind Turbine Components and Operation* (Bonus-Info, Denmark, 1999)
33. K.L. Johnson, *Contact Mechanics* (Cambridge University Press, Cambridge, 1985)
34. D.H. Buckley, *Surface Effects in Adhesion, Friction, Wear, and Lubrication* (Elsevier, Amsterdam, 1981)
35. M.N. Kotzalas, G.L. Doll, Tribological advancements for reliable wind turbine performance. Philos. Trans. Royal Soc. A: Math. Phys. Eng. Sci. **368**(1929), 4829–4850 (2010)
36. T.A. Harris, M.N. Kotzalas, *Essential Concepts of Bearing Technology* (Taylor and Francis, Boca Raton, 2007)
37. K. Iso, A. Yokouchi, H. Takemura, Research work for clarifying the mechanism of white structure flaking and extending the life of bearings. in SAE Technical Paper 2005-01-1868, Society of Automotive Engineers, 2005
38. M. Kohara, T. Kawamura, M. Egami, Study on mechanism of hydrogen generation from lubricants. Tribol. Trans. **46**, 5360 (2006)
39. A.V. Olver, The mechanism of rolling contact fatigue: an update. J. Eng. Tribol. **219**(5), 313–330 (2005)
40. G. Lundberg, A.Z. Palmgren, Dynamic capacity of rolling bearings. Proc. R. Swed. Acad. Eng. Sci. **196**, 50 (1947)
41. E.V. Zaretsky, Palmgren revisited–a basis for bearing life prediction. Lubr. Eng. **54**, 1823 (1998)
42. F.T. Barwell, Report on papers in Session 3 (lubrication), in *Proceedings of the International Conference on Gearing* (1958), pp. 23–25
43. D. Dowson, G.R. Higginson, *Elastohydrodyanmic Lubrication*, 2nd edn. (1977)
44. A.N. Grubin, *Investigation of the Contact of Machine Components* (Central Scientific Research Institute for Technology and Mechanics, Moscow, 1949), Book 30
45. R.D. Britton, C.D. Elcoate, M.P. Alanou, H.P. Evans, R.W. Snidle, Effect of surface finish on gear tooth friction. J. Tribol. **122**(1), 354–360 (2000)
46. A.N. Grubin, *Fundamentals of the hydrodynamic theory of lubrication of heavily loaded cylindrical surfaces*. Investigation of the Contact Machine Components, No. 30, 115–166 (1949)
47. D. Dowson, G.R. Higginson, A numerical solution to the elasto-hydrodynamic problem. J. Mech. Eng. Sci. **1**(1), 6–15 (1959)
48. D. Dowson, G.R. Higginson, *Elastohydrodynamic Lubrication* (Pergamon Press, Oxford, 1977)
49. S. Li, A. Kahraman, Prediction of spur gear mechanical power losses using a transient elastohydrodynamic lubrication model. Tribol. Trans. **53**, 554–563 (2010)
50. R. Gohar, *Elastohydrodynamics* (Wiley, New York, 1988)
51. R. Errichello, Friction, lubrication, and wear of gears, in *ASM Handbook Friction, Lubrication, and Wear Technology*, vol. 18, ed. by P. Blau (ASM International, Materials Park, Ohio, 1992), pp. 535–545

52. T.E. Tallian, Simplified contact fatigue life prediction model–part I: review of published models. J. Tribol. **114**(2), 207–213 (1992)
53. G. Stachowiak, A.W. Batchelor, *Engineering Tribology* (Elsevier, Oxford, 2005)
54. G.L. Doll, B.K. Osborn, Engineering surfaces of precision steel components. Proc. Annu. Tech. Conf. Soc. Vac. Coaters **44**, 78–84 (2001)
55. G.L. Doll, C.R. Ribaudo, R.D. Evans, Engineered surfaces for steel rolling element bearings and gears. Mater. Sci. Technol. **2**, 367–374 (2004)
56. A. Ragheb, M. Ragheb, Wind turbine gearbox technologies, in *International Nuclear and Renewable Energy Conference* (*INREC10*) (Amman, Jordan, 2010)
57. R.W. Hyers, J.G. Mcgowan, K.L. Sullivan, J.F. Manwell, B.C. Syrett, Condition monitoring and prognosis of utility scale wind turbines. Energy Mater.: Mater. Sci. Eng. Energy Syst. **1**(3), 187–203 (2006)
58. F.L. Litvin, A. Fuentes, *Gear Geometry Applied Theory* (Cambridge University Press, Cambridge, 2004)
59. J.E. higley, *Mechanical Engineering Design* (McGraw-Hill, New York, 1963)
60. P. Lynwander, *Gear Drive Systems: Design and Application* (Marcel Dekker, NY, 1983)
61. F.L. Litvin, A. Fuentes, I. Gonzalez-Perez, L. Carvenali, K. Kawasaki, R.F. Handschuh, Modified involute helical gears: computerized design simulation of meshing and stress analysis. Comput. Methods Appl. Mech. Eng. **192**(33–34), 3619–3655 (2003)
62. J. Kleemola, A. Lehtovaara, Experimental simulation of gear contact along the line of action. Tribol. Int. **42**, 1453–1459 (2009)
63. G.M. Maitra, *Handbook of Gear Design* (Tata McGraw-Hill, New Delhi, 1997)
64. A.H. Elkholy, Tooth load sharing in high contact ratio spur gears. J. Mech. Trans. Autom. Des. **107**(1), 11–16 (1985)
65. S. Avinash, Application of a system level model to study the planetary load sharing behavior. J. Mech. Des. **127**(3), 469–476 (2005)
66. C.R.M. Rao, G. Muthuveerappan, Finite element modelling and stress analysis of helical gear teeth. Comput. Struct. **49**(6), 1095–1106 (1993)
67. B.R. Hohn, K. Michaelis, Influence of oil temperature on gear failures. Tribol. Int. **37**(2), 103–109 (2004)
68. H. Blok, Les temperatures de surface dans les conditions de graissage sons pression extreme, in *Second World Petroleum Congress*, Paris, 1937
69. E.E. Shipley, *Failure analysis of coarse-pitch, hardened, and ground gears*. Paper No. P229.26, (American Gear Manufacturers Association, Alexandria, 1982), pp. 1–24
70. S. Tanaka, *Appreciable increases in surface durability of gear pairs with mirror-like finish*, Paper No. 84-DET-223, (American Society of Mechanical Engineers, Alexandria, 1984), pp. 1–8
71. X. Ai, Effect of debris contamination on the fatigue life of roller bearings. Proc. Inst. Mech. Eng. Part J: J. Eng. Tribol. **215**(6), 563–575 (2001)
72. I. Allison, E. Hearn, A new look at the bending strength of gear teeth. Exp. Mech. **20**(7), 217–225 (1980)
73. J. Hiremagalur, B. Ravani, Effect of backup ratio on root stresses in spur gear design. Mech. Based Des. Struct. Mach.: Int. J. **32**(4), 423–440 (2004)
74. W. Lewis, Investigation of the strength of gear teeth, in *Proceedings of Engineers Club*(1892), pp. 16–23
75. T.J. Dolan, E.L. Broghamer, *A photoelastic study of stresses in gear tooth fillets*. University of Illinois Bulletin, vol 355 (1942)
76. B.W. Kelley, R. Pedersen, The beam strength of modern gear tooth design. Transactions of the S.A.E., 1957
77. X.Q. Peng, L. Geng, W. Liyan, G.R. Liu, K.Y. Lam, A stochastic finite element method for fatigue reliability analysis of gear teeth subjected to bending. Comput. Mech. **21**(3), 253–261 (1998)
78. J.D. Andrews, A finite element analysis of bending stresses induced in external and internal involute spur gears. J. Strain Anal. Eng. Des. **26**(3), 153–163 (1991)

79. M.A. Miner, Cumulative damage in fatigue. J. Appl. Mech. **67**, A159–A164 (1945)
80. A.Z. Palmgren, Die Lebensdauer von Kugelagern. 2 ver Deutsch Ing **68**, 339–341 (1924)
81. T.A. Harris, J.H. Rumbarger, C.P. Butterfield, Wind turbine design guideline DG03: yaw and pitch rolling bearing life, NREL Technical Report No. NREL/TP-500-42362, 2009
82. T. Senjyu, R. Sakamoto, N. Urasaki, T. Funabashi, H. Fujita, H. Sekine, Output power leveling of wind turbine generator for all operating regions by pitch angle control. IEEE Trans. Energy Convers. **21**(2), 467–475 (2006)
83. J. Aguirrebeitia, R. Aviles, I.F.d. Bustos, M. Abasolo, Calculation of general static load-carrying capacity for the design of four-contact-point slewing bearings. J. Mech. Des. **132**(6), P064501 (2010)
84. W.J. Bartz, Tribological aspects of wind power plants, in *Proceedings of the World Tribology Congress III*, Washington, D.C., USA, 2005
85. J.C. Enthoven, H.A. Spikes, Visual observation of the process of scuffing, in *Lubricants and Lubrication, Proceedings of the 21st Leeds-Lyon Symposium on Tribology* (1995), pp. 487–494
86. G.K. Nikas, R.S. Sayles, E. Ioannides, Effects of debris particles in sliding/rolling elastohydrodynamic contacts. J. Eng. Tribol. **212**(5), 333–343 (1998)
87. G. Xu, F. Sadeghi, J.D. Cogdell, Debris denting effects on elastrohydrodynamic lubricated contacts. Trans. ASME J. Tribol. **119**, 579–587 (1997)
88. X. Ai, H.S. Cheng, The influence of moving dent on point EHL contacts. Tribol. Trans. **37**(2), 323–335 (1994)
89. S. Butterfield, R. Errichello, B. McNiff, Wind turbine gearbox issues and lubrication, in *IJTC2008-71361*, 2008
90. M.N. Kotzalas, W.M. Needelman, D.R. Lucas, G.L. LaVallee, Improving wind turbine gearbox life, in *AWEA Windpower Conference*, Houston, 2008
91. R.E. Cantley, The effect of water in lubricating oil on bearing fatigue life. ASLE Trans. **20**(3), 244–248 (1977)
92. E.V. Zaretsky (ed.), *Life Factors for Rolling Bearings* (Society of Tribologists and Lubrication Engineers, Park Ridge, 1992)
93. E. Abner, in *Handbook of Lubrication*, ed. by E.R. Booser, Lubricant deterioration in service (CRC Press, Boca Raton, 1983)
94. W.M. Needelman, M.A. Barris, and G.L. LaVallee, Contamination Control for Wind Turbine Gearboxes. Power Engineering
95. Hydraulic fluid power—Filters—Multi-pass method for evaluating filtration performance of a filter element. ISO 16889. 2008: International Organization for Standardization
96. W.M. Needelman, M.A. Barris and G.L. LaVallee, Contamination control for wind turbine gearboxes. Power Engineering, 2009
97. W.M. Needelman, E.V. Zaretsky, New equations show oil filtration effect on bearing life. Power Transm. Des. **33**(8), 65–68 (1991)
98. W.M. Needelman, E.V. Zaretsky, Recalibrated equations for determining effect of oil filtration on rolling bearing life, in *STLE Annual Meeting*, Orlando (2009)
99. W.M. Needelman, M.A. Barris, G.L. LaVallee, Reducing cost of operation in harsh conditions with new generation filters, in *International Fluid Power Exposition*, Las Vegas (2011)
100. G.L. LaVallee, W.M. Needelman, Dry air blankets: An effective and economical method for eradicating water contamination. Part I: principles of operation, in *STLE Annual Meeting*, Las Vegas (2010)
101. C.S. Gray, S.J. Watson, Physics of failure approach to wind turbine condition based maintenance. Wind Energy **13**(5), 395–405 (2010)
102. Z. Hameed, Y.S. Hong, Y.M. Cho, S.H. Ahn, C.K. Song, Condition moitoring and fault detection of wind turbines and related algorithms: a review. Renew. Sustain. Energy Rev. **13**(1), 1–39 (2009)

Chapter 19
Ecological Aspects of Water Desalination Improving Surface Properties of Reverse Osmosis Membranes

Tyler G. Hurd, Saman Beyhaghi and Michael Nosonovsky

Abstract The lack of fresh clean water is an economical and ecological problem which affects half of humanity. More than 97.5% of all water on the Earth is seawater, so the ability to harvest even a small fraction as fresh water would have a huge impact on water scarcity. Reverse osmosis (RO) is currently the main technique of seawater desalination. During RO, salt water under pressure exceeding the fluids osmotic pressure is forced through a semipermeable membrane. RO requires significant energy inputs and affects the environment due the greenhouse gas emissions (usually associated with an external power source), the output of brine with high salt concentration, and other negative effects. Improving the efficiency and environmental impact of RO plants involves several challenges, some of which are related to surface science and tribology. This involves mimicking water filtration by cell membranes, as well as creating biomimetic antifouling coatings on membranes. We present a comprehensive review of RO and other desalination techniques and suggest how a composite material can improve permeability and antifouling properties of RO membranes.

19.1 Introduction

According to the 2010 UN-Water Global Annual Assessment of Sanitation and Drinking-Water [1], around 1.1 billion people in developing countries have inadequate access to water, while 2.6 billion lack basic sanitation. Close to half of

T. G. Hurd · S. Beyhaghi · M. Nosonovsky (✉)
Department of Mechanical Engineering, College of Engineering and Applied Science, University of Wisconsin-Milwaukee, Milwaukee, WI 53211, USA
e-mail: nosonovs@uwm.edu

M. Nosonovsky and B. Bhushan (eds.), *Green Tribology*,
Green Energy and Technology, DOI: 10.1007/978-3-642-23681-5_19,

all people in developing countries at any given time suffer from health problems caused by water and sanitation deficits. As a result of unsafe drinking-water, inadequate sanitation or poor hygiene, 11.9 million children die annually, which accounted for 18% of health-related deaths in children in 2008 [2]. In addition to health related issues, the lack of fresh, safe water affects many more equally important facets of human life. These include food production, economic growth, ecosystem balance, pharmaceutical applications, energy cost, industrial progression and education, to name a few. With dwindling water supplies we find it necessary to develop more efficient and robust methods of providing clean water.

On average, seawater has a salt concentration of about 3.5% (35 g/L or 35,000 parts per million), i.e. approximately 35 g of dissolved salt molecules per one liter of seawater, predominantly Na^+ and Cl^- ions. The process that removes some amount of salt and other minerals from seawater is called *desalination*. The salinity of typical seawater ranges from 3 to 5%, while saline aquifers with salt concentration of 0.05–3% are considered brackish (typical for situation when fresh water meets seas, such as estuaries). Water with salt concentrations of >5% is called brine, which is the byproduct of desalination. Due to the large difference in salt concentration when compared to fresh, potable water (3.5–4.5% vs. <0.1%, respectively), desalination of the seawater is required for consumption and many other applications.

Arid countries lacking simple access to water or developing countries for example, require the ability to effectively sanitize and reuse water that may have been used for irrigation or other purposes. On the other hand, countries with access to seawater require effective, energy-efficient desalination techniques. As the global population continues to rise and is expected to pass 9 billion over the next three decades [3], the demand for new progressive water filtration technologies will push researchers toward developing more efficient and affordable methods to provide clean water to industrialize and developing countries alike.

Between 1990 and 2008, the International Desalination Association (IDA) reported that the total global desalination capacity grew almost fourfold, from 14.8 million m^3 per day up to over 52 million m^3 per day. In the past 3 years the contracted global capacity has increased 90%. The growth in Cumulative global desalination capacity since 1967 is illustrated in Fig. 19.1. Desalination plants continue to grow around the world, with Saudi Arabia, the US, Kuwait, China, and Japan having the leading positions (Fig. 19.2). Notably large desalination plants also exist in Aruba, Australia, Cyprus, Gibraltar, Israel, Maldives, the UK, and many Middle Eastern countries. Current strategies for water desalination and filtration range from mimicking the natural hydrological water cycle, to single-walled carbon nanotube membrane filtration. The main methods of desalination however include Reverse Osmosis (RO), multi-stage flash (MSF), multi-effect distillation (MED), and electrodialysis (ED). The trend of each methods contribution to the global desalination capacity is shown in Fig. 19.3. The largest desalination plant in the world is the Jebel Ali Plant located in the United Arab Emirates. This massive plant uses MSF desalination and is capable of producing 300 million m^3 of water per year. Comparatively, the largest desalination plant in

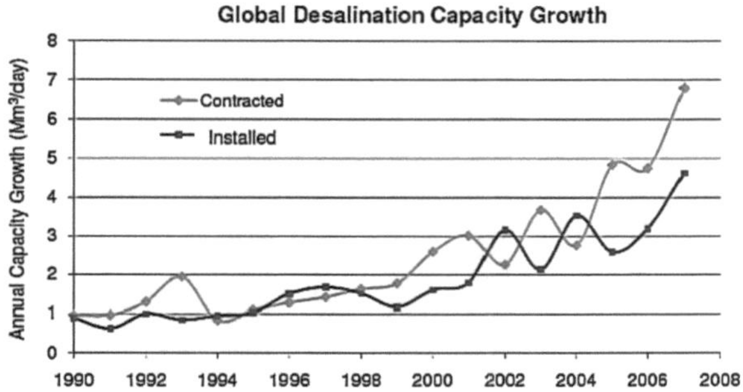

Fig. 19.1 Cumulative global desalination capacity growth since 1967 [4]

the US, located in Tampa Bay, FL, is capable of desalinating around 35 million m^3 of water per year. In 2008, 13,080 desalination plants produced more than 12 billion gallons of water a day [7]. Small-scale desalination plants are also often used onboard ships and submarines, and were among the first recorded devices used for desalination.

Despite recent progress, desalination remains an expensive and energy-intensive process, involving pumping seawater to the desalination plant, pretreatment of the intake water, disposal of the concentrate (brine), and process maintenance [8]. Estimated costs of pumping from sea intake to the desalination plant depend on height and distance from the water source. The total cost and increased environmental concerns have limited the widespread adoption of desalination technologies. Currently, a state-of-the-art RO system uses as little as 2.2 kWh per 1,000 L of potable water solely in the RO process. Total energy usage of a large scale plant however, including the aforementioned electrical plus modest conveyance energy needs, can be on the order of 5 kWh per 1,000 L. The cost breakdown of a typical RO and a typical thermal desalination plant is shown in Fig. 19.4. Comparing this to typical energy consumption per capita (ranging from 3.2 kWh in China to 30 kWh in the USA) shows that supplying even 50 L a day of potable water per capita costs 0.25 kWh, which constitutes a small fraction of daily energy usage [8]. Speaking practically, the cost of desalinated water can be estimated currently at about $0.5 per 1,000 L which sounds reasonably low, however, it is still cheaper in many areas to either pump the water out of the ground or import it.

From a thermodynamic approach, the energy required to desalinate a given amount of seawater can be calculated as that required to overcome osmotic pressure. Removing of small amount of fresh water from a large amount of seawater (low recovery limit) requires the minimum amount of energy, which is an estimated 0.7 kWh per 1,000 L. Approaching the theoretical thermodynamic minimum however is impractical because it would require massive facilities with high capital costs [8].

Several environmental concerns are associated with desalination and must also be addressed. First, the intake of seawater may have a negative impact on local marine life, destroying plankton, fish eggs, etc. Second, the brine, which is often returned to the ocean, constitutes environmental hazards and should therefore be

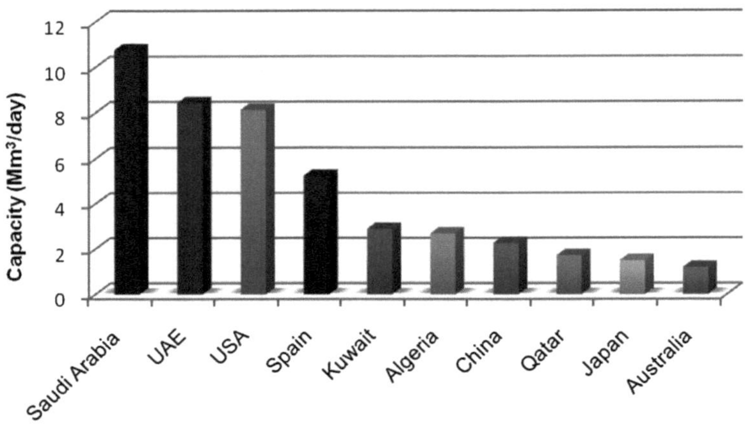

Fig. 19.2 Leading countries in desalination capacity [4]

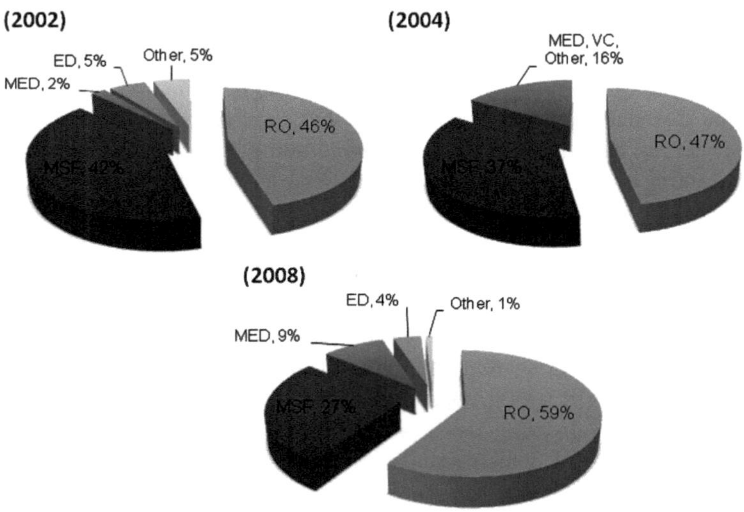

Fig. 19.3 Trends in global desalination capacity by reverse osmosis (*RO*), multi-stage flash (*MSF*), multi-effect distillation (*MED*), electrodialysis (*ED*), and other methods [4–6]

properly disposed off. Third, desalination requires considerable energy inputs, thus air pollution and greenhouse gas emissions from its energy source are significant concerns as well. Some of the advantages and disadvantages of the RO desalination method is outlined in Table 19.1.

In efforts to reduce energy consumption and maintenance costs, more specific surface science issues have been related to RO desalination technology. In particular,

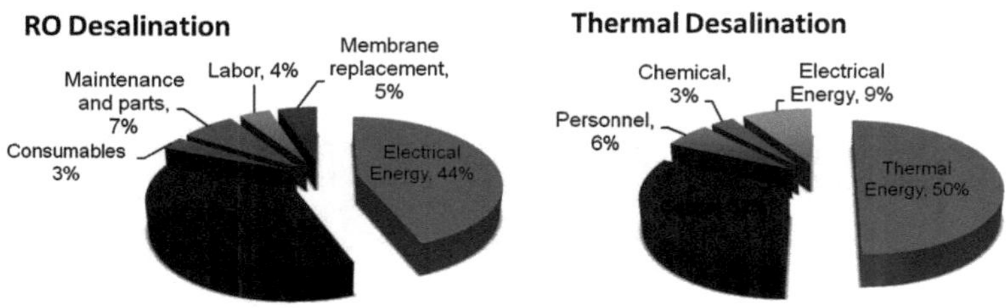

Fig. 19.4 Cost breakdown comparison of RO and thermal desalination

Table 19.1 Pros and cons of reverse osmosis

Advantages	Disadvantages
• Desalinating clean, treated feed water is a relatively simple process	• Advanced pre-treatment, microfiltration and nanofiltration steps are required for clean feed water supply to RO filter in order to minimize fouling
• Installation can be simple, portable, and inexpensive	• Current methods of manufacturing efficient RO membranes are expensive and time consuming
• Unlimited, reliable source of seawater	• Many countries suffering from fresh water deficits do not have adequate access to seawater
• Progression in research promises membranes with high flux, selectivity, and resistance to wear	• Current RO membranes require frequent cleaning due to fouling, and replacing due to failure caused by the resulting high feed pressure
• In two decades, the energy required to produce 1 m^3 of fresh water from RO desalination has been reduced by over 60% and continues to drop	• It is still cheaper in many regions to either pump fresh water from the ground or import it
• RO process removes a very high degree of contaminants, minerals, and sulfates producing exceptionally pure water	• The concentrated brine solution produced as a byproduct can be hazardous to the environment and difficult to dispose of

biofouling or biofilming (i.e. contamination of the surface with living microorganisms such as bacteria, fungi, plankton, and organic contaminants) is a factor that reduces the lifetime of the membranes, increases osmotic pressure, and causes contamination of the output fresh water. To reduce biofouling, self-cleaning anti-fouling membranes using the principle of the lotus-effect (underwater superoleophobicity) have been suggested. The use of self-assembled, superhydrophobic polymeric membranes for RO desalination [9] has also recently been proposed. Biomimetic membranes based on an array of aligned carbon nanotubes, mimicking biological protein channels by functionalizing the open tips of the nanotubes with a synthetic peptide, are another area of research [10]. In addition, attempts to mimic the natural water cycle for water desalination using solar energy for water

Fig. 19.5 Ship desalination device using distillation. Kura-Hulanda Museum, Willemstad, Curacao, Netherlands Antilles (photo by M. Nosonovsky 2005)

evaporation have been considerably more popular in the past due to the processes relatively of simple nature. This is especially relevant because the demand for desalination is high in dry desert regions where solar energy is abundant.

In this chapter, we will address some of the currently utilized desalination techniques and discuss the energy consumption, cost, and feasibility of each. Our focus however, will remain on the research and development of advanced membrane technologies with self-cleaning properties for use in RO water desalination. Furthermore, we will concentrate on the application of biomimetics to RO membranes to achieve self-cleaning (antifouling) properties such as the Lotus effect, manipulating wetting properties and studying the subsequent effect on hydrophobicity and oleophobicity, as well as mimicking the simple hydrological cycle for desalination.

19.2 Methods of Desalination

Seawater desalination has been used for centuries. The earliest known method, water distillation, which was first recorded in the First century AD when alchemists in Hellenistic Egypt and Greece described it, particularly Alexander of Aphrodisias (cir. 200 AD) [11]. During the Age of Discovery, when long marine expeditions to the Americas and other parts of the world were launched, the need for fresh distilled water was particularly important for the ships. Thus, Dutch ships used for the colonization of the Caribbean in the sixteenth–seventeenth centuries often had desalination devices using the distillation method (Fig. 19.5). Later, more advanced desalination methods were developed, involving membrane processes, freezing, ion exchange, etc.

In this section, an overview of some of the more commonly used desalination techniques is presented. Note that selection of each of these methods is dependent on several factors such as available resources, climate, capitol, etc. Desalination methods can be divided into two major categories; *thermal methods* and *membrane-based methods*. The former are always associated with phase change phenomena. Salt-water partially vaporizes after receiving sufficient energy from a heat source, and can be subsequently condensed and collected as purified water. This heat could be provided from a number of sources, including but not limited to fossil fuel, solar radiation [12–16], and geothermal energy [17, 18]. MSF, MED, Vapor Compression (VC), and Natural Vacuum Distillation (NVD) are common methods of thermal desalination, and shall be briefly reviewed in this section.

The second category of desalination is the membrane-based process, in which the highly concentrated saline solution is forced through a membrane filter. While water molecules and equally small minerals may pass, other contaminants will be removed by the membrane. Selectivity of the membrane is based on various parameters such as particle size (as in RO) and particle charge (as in ED). These desalination techniques are also briefly reviewed in this section. The MSF and RO collectively account for more than 80% of the world's desalination capacity [19]. The MSF process represents more than 93% of the water production by thermal methods, while RO process provides more than 88% of the membrane-based water purification [17, 20].

19.2.1 Thermal Desalination

19.2.1.1 Multi-Stage Flash Desalination

Multi-Stage Flash (MSF) distillation is a process that distills seawater by flashing a portion of the water into steam in multiple stages of what are essentially countercurrent heat exchangers. This process, patented by Silver in 1957, became quite popular shortly after its introducion [21]. Many commercial MSF plants have been constructed and utilized worldwide over the past four decades, particularly in the Persian Gulf [22]. The worldwide contracted capacity of all MSF desalination plants was 0.027 million m^3 per day between 1950 and 1959 and gradually increased until it reached 17 million m^3 per day in 2004 [4].

In an MSF plant, seawater is flash-heated into steam in a series of vacuum-pressure chambers, with successively decreasing temperature and pressure. Current MSF plants generally have 10–30 stages, where the temperature gradient is around 2°C per stage [23]. Note that in such low-pressure chambers, water boils at a relatively low temperature, which means much less energy is required for evaporation. The schematic of this process is shown in Fig. 19.6 and the procedure is summarized as follows:

1. The cold saline water is pumped from the sea into the condensing coils, which pass through the top region of the vacuum chambers. This will preheat the seawater before entering the heat exchanger and, more importantly, result in condensation of the flashed steam in the vacuum chambers producing purified water.

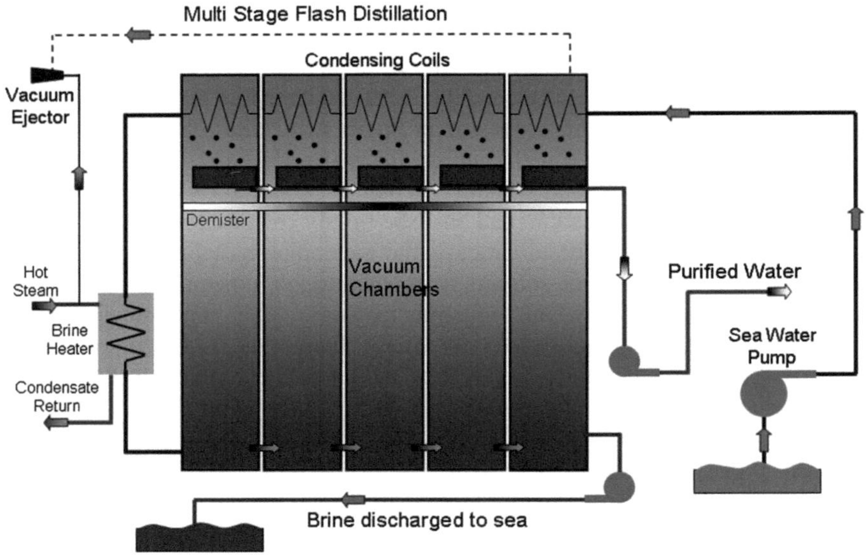

Fig. 19.6 Schematic of an MSF desalination process

2. The warm seawater passes through an indirect heat exchanger and is heated to 90–110°C. This temperature will be close to the saturation temperature at the maximum pressure of the system. (Note seawater boils at a temperature slightly greater than fresh water)
3. The hot saline then enters the first vacuum chamber through an orifice resulting in a slight pressure reduction. Because the water was near saturation temperature corresponding to a higher pressure, some portion of the water will flash into steam when introduced to the lower pressure chamber.
4. Steam then travels upward in the chamber toward the cold coils. Upon contact with the coils, the steam condenses to form pure, fresh water. The formed condensate then drips into a distillation tray for collection.
5. The more concentrated brine will remain at the bottom of the chamber, moving toward the next chamber where the water will again have a chance to vaporize.
6. After passing through multiple vacuum chambers, the highly concentrated brine will either be recirculated or returned to the sea.
7. Fresh water collected in the distillation tray will be pumped out of the chambers for use.

Note that the MSF method requires an external supply of steam for heating up the brine and creating the vacuum condition. A total fresh-water recovery of 25–50% for high temperature plants (with recirculation) is commonly achieved [19].

19.2.1.2 Multi-Effect Distillation

Similar to MSF, multi-effect distillation (MED) involves a series of stages of successively decreasing temperature and pressure. From a thermodynamic standpoint, it is more efficient than MSF, however, performance of MED processes are directly proportional to the number of effects used. MED plants typically use an external steam supply at a temperature of around 70°C. Small capacity MED plants were used since the early 1900s for water desalination onboard ships, whereas the first land-based MED plant was constructed in the 1930s in Saudi Arabia [22]. The first large capacity MED plant for seawater desalination, with a capacity of 3.5 million gallons per day and a vertical tube evaporator, was built in Kazakhstan in 1964. Construction of similar plants followed in 1966–1970 in the same country [22]. The number of MED plants with such large capacities has grown since then. A review of MED plants installed in the twentieth century is offered in [24].

In MED, thermal energy is absorbed by the seawater in the first chamber (or 'effect') resulting in partial vaporization (Fig. 19.7). The steam generated will then be used in the following effect to heat the more concentrated brine [12]. Note the next stage is at a lower temperature and pressure allowing thermodynamic feasibility of this heat transfer process. MED typically has a recovery rate between 40 and 65%, however, due to difficulties in maintenance and operation, use of such plants has declined.

The MED process can be summarized as follows:

1. Feed water is pumped through the first chamber or 'effect'. Meanwhile, the superheated steam is supplied from an external source and flows through a tube in the same chamber (indirect heat exchanger). The water is sprinkled on the evaporation tubes resulting in partially condensed steam, which then leaves the system and returns to the boiler.
2. Some portion of the feed water evaporates after receiving heat from the steam.
3. The newly generated vapor is directed to the next chamber by flowing through another tube. In the next stage, the steam serves as the heat source for evaporation.
4. The solute-concentrated water enters the second 'effect'.
5. The condensed vapor in the tubes merges with the main condensate pipe, eventually collecting in the form of fresh water.
6. This process is repeated in subsequent effects.
7. The highly concentrated brine is drained from the bottom of the last effect.
8. Vapor resulting from the last stage passes through an external heat exchanger and is collected in the form of condensed water.

In comparison with the MSF desalination method, MED has proven to have a greater potential for designs with high performance ratios [26] and demonstrate greater corrosion resistance [27].

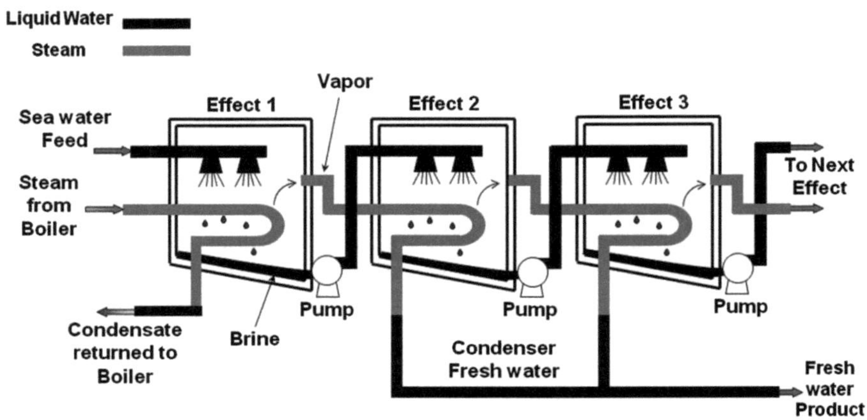

Fig. 19.7 Schematic of an MED plant [25]

19.2.1.3 Mechanical Vapor Compression

Vapor compression systems are divided into two main categories: mechanical vapor compression (MVC) and thermal vapor compression (TVC) systems. In MVC and TVC plants, vapor is produced from the saline solution then compressed thermally, or mechanically, to generate additional production. VC units were first used in naval vessels powered by diesel engines during WWII [22]. A number of MVC-based desalination plants are installed worldwide shortly after, producing fresh water for industrial and municipal purposes. These plants however, have the disadvantage of restricted capacity due to scale limitations for large size vapor compressors [28]. Another imperative issue is compressor maintenance due to the high possibility of corrosion, consequently many MVC plants must operate at relatively low temperatures in order to minimize the formation of scaling and corrosion of materials.

The MVC distillation process is mainly used for small-scale desalination units. The evaporation heat is supplied by compression of vapor rather than direct heat transfer from the steam produced in the boiler. The compressor creates a partial vacuum in the chamber, compresses the vapor, then condenses it in a bundle of tubes in the chamber. This mechanical compressor is usually electrically driven. Note that MVC cycles are commonly combined with MED processes in order to achieve better performance [29] (Fig. 19.8).

The MVC process of desalination can be summarized as following:

1. The seawater (saline feed) is pumped through a pre-heater. As we shall see later, the heat is provided by further condensation of a partially condensed vapor leaving the evaporation chamber.
2. The saline feed is further heated by passing through another heat exchanger where it receives energy from the relatively hot discharged brine.

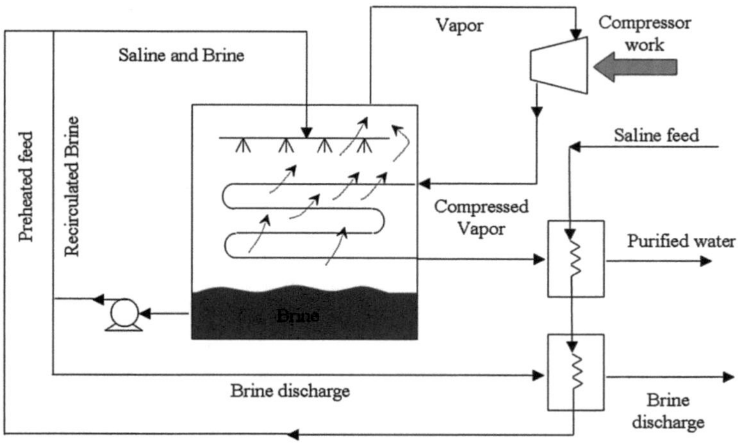

Fig. 19.8 Schematic of a mechanical vapor compression (MVC) process

3. The water then enters the chamber and is sprinkled on the outside of heated tubes containing high-pressure steam.
4. As some of the water boils, the vapor is drawn by a vacuum line. It is then compressed after passing through a mechanical compressor [30].
5. Some portion of the brine collected in the bottom of the chamber is recirculated and merged to the preheated feed line, while the rest is directed to the saline pre-heater and finally discharged (brine level in the evaporation chamber must be monitored carefully).
6. The vapor leaving the compressor flows through a bundle of tubes and acts as a heat source for the incoming seawater as noted in (1.).
7. Steam passes through a heat exchanger pre-heating the cold seawater. As a result, the steam condenses and is collected as fresh water.

19.2.1.4 Thermal Vapor Compression

As mentioned above, MVC systems mechanically compress vapor (which is later used for evaporation of saline), whereas TVC systems thermally compress vapor utilizing a steam jet compressor. The main problems associated with the TVC process are [31]:

1. Vapor generated in the chamber may contain some solute molecules which lead to corrosion of compressor blades and other components when drawn into the compressor.

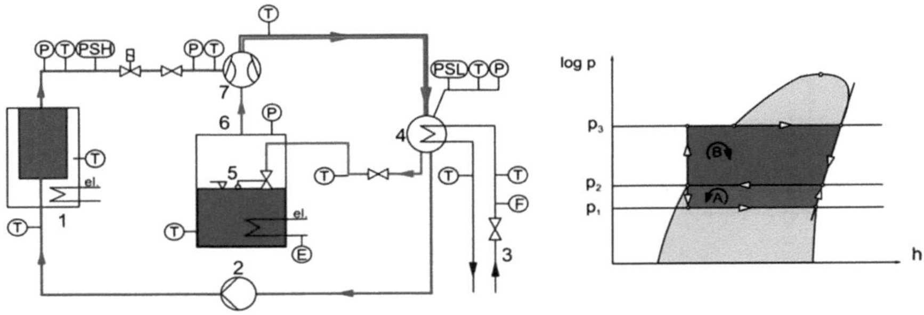

Fig. 19.9 (*Left*) A schematic diagram of a Thermal Vapor Compression system [32], with the following notations: *1* steam generator, *2* pump, *3* cooling water connections, *4* condenser, *5* float valve, *6* evaporator, *7* steam jet compressor, *T* temperature, *P* pressure, *PSL* & *PSH* pressure switch, *F* flow rate, *E* voltage and amperage. (*Right*) The log-scale pressure-enthalpy diagram of the two involving thermodynamic cycles; *A* refrigeration cycle, *B* steam cycle, P_1 pressure in the evaporator, P_2 pressure in the condenser, and P_3 pressure in the steam generator

2. Due to the restricted capacity of the compressor, there are limitations on TVC plant size. Thermal vapor systems are designed for projects where steam is sufficiently available.

A schematic diagram of a steam jet compressor typically used in TVC systems along with the pressure-enthalpy diagram of the two thermodynamic cycles is illustrated in Fig. 19.9 [32].

19.2.1.5 Natural Vacuum Distillation

The majority of currently used water desalination methods require a large amount of energy input in the form of heat or electricity. Use of renewable energy sources such as wind, solar, geothermal and biomasses will result in a cleaner environment and sustainable energy production. Different desalination methods can take advantage of a number of these renewable energy sources. Among these methods is a relatively new procedure which aims to mimic the natural hydrological cycle of water, known as natural vacuum distillation [33–35]. This system is composed of a large inverted U-shaped tube, which holds 10.33 m of water vertically, consequently creating vacuum. Note, due to its density, thermodynamic properties and surrounding atmospheric pressure, seawater cannot elevate in a vacuum higher than ~10.33 m before evaporating. In order to create this vacuum, natural means (gravity, changing tides, and atmospheric pressures) can be used. As a result of the vacuum condition, water can evaporate at a reduced temperature. Heat can be provided by solar radiation, which was sufficiently available in the country where one of the first studies of this method was performed [16]. The pressure drop as a result of the natural vacuum can reach as low as 4 kPa (the National Institute of

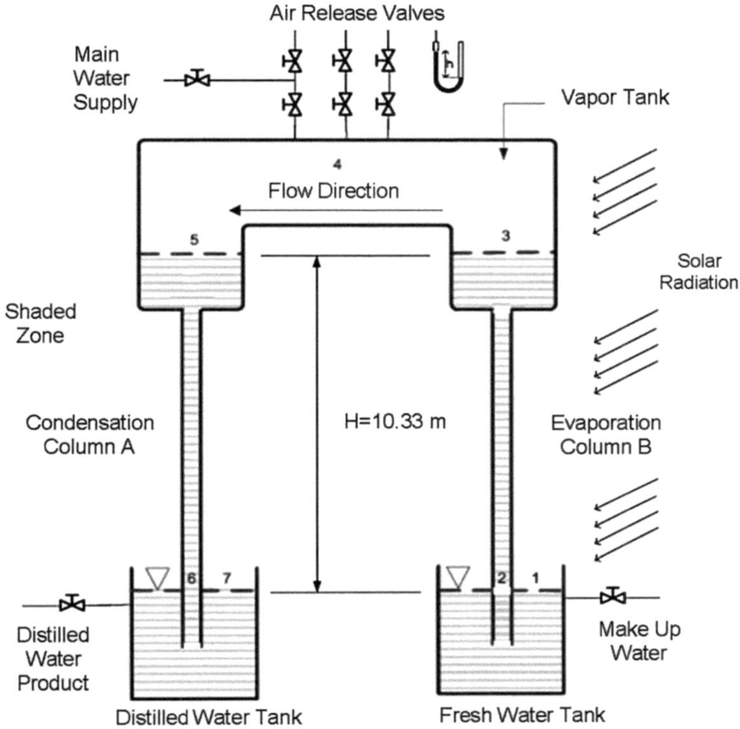

Fig. 19.10 Schematic of a natural vacuum distillation system [16]

Standards and Technology (NIST) regards 101.325 kPa, or 1 atm, standard atmospheric pressure) [36]. A schematic diagram of the NVD process is shown in Fig. 19.10:

This natural evaporation system consists of an evaporation column, vapor tank, and condensation column. A fan and a vacuum pump can be also added to the system to improve efficiency. Additionally, an electric heat battery for an increased evaporation rate and a coolant battery to aid condensation could be employed. The procedure of the NVD method is as follows:

1. A water column (Fig. 19.10: right) is formed due to the vacuum and is heated by solar radiation.
2. Vacuum condition holds elevated water levels within the vapor tank.
3. Due to relatively low pressure and raised temperature, some water evaporates at the surface and flows toward the condensation column (left, often aided by a fan).
4. Distilled water is then collected in the condensation column and transported to the outlet.

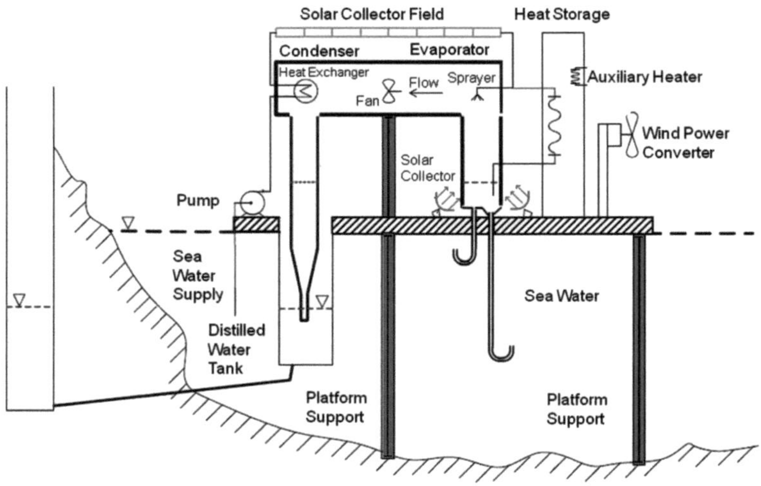

Fig. 19.11 A more detailed representation of the NVD system used by [16]

In these systems, the vacuum condition in the upper tank gives rise to a much lower pressure and therefore boiling temperature. Water elevated at such levels ($\sim$10.33 m) evaporates at a saturation temperature around 55°C. The schematic diagram of the marine NVD system used in [16] (Fig. 19.11).

19.2.2 Membrane-Based Processes

19.2.2.1 Reverse Osmosis

Reverse Osmosis (RO) is a process which forces seawater through a semi-permeable membrane (usually made of a polymeric material) under high pressure allowing water molecules to pass, while salt and other contaminants are rejected forming a concentrated brine solution [37]. This process requires high amounts of electricity to drive the pump which pressurizes the saline solution. The amount of pressure required depends on a number of factors such as the solute concentration, the solutions molecular structure (number of particles generated after ionization of each molecule, see *van't Hoff's factor*), membrane flux and the operating temperature. This pressure (the minimum required for desalination) is referred to as Osmotic pressure, and can be as large as 7,000 kPa for seawater desalination. More details of this process are provided in the Sect. 19.3.

19.2.2.2 Electrodialysis

Electrodialysis (ED) is another membrane-based process, which reduces the salinity of the feed water by repelling the electrically charged particles (salt ions) regardless of their size. Unlike RO, this technology is limited to the less concentrated brackish feed water [19]. Electrodialysis utilizes DC current to extract Na^+ and Cl^- ions from water through selective membranes (located at the two oppositely charged electrodes), while rejecting water molecules (Fig. 19.12). In an actual ED process, a large number of alternating cation and anion membranes are stacked together and separated by flow spacers to allow the passage of water. In other words, cation-permeable and anion-permeable membranes separate the two electrodes, and the resulting gap between these membranes will have a lower salt concentration [38].

Since the energy required for salt ionization is proportional to the salt concentration of the water, this method is most efficient when salinity levels are below 0.6% [23]. On the other hand, since the electrical conductivity of pure water is quite low, the energy needed for ionization will be very high for low concentration feed water. Therefore, this method is not economically feasible for solute concentration of less than 0.04%.

19.2.3 Other Methods of Desalination

In addition to the thermal and membrane methods of desalination, there are also other methods that have been recently suggested. Among these include freezing desalination and various types of renewable energy-based desalination, such as solar desalination and geothermal desalination [19, 23, 39, 40].

When seawater freezes, salts initially tend to freeze with the water. However, due to mass transport in the semisolid layer the salinity throughout the layer decreases with time. As a result, glacial and sea ice have low salt concentrations. The freezing desalination was used for centuries by inhabitants of Central Asia and Siberia to obtain water for their cattle. Freezing desalination has certain advantages over the thermal methods, such as lower corrosion rates of steel pipes, however, this method cannot be considered for large scale reproduction due to its lack of economical and thermodynamic feasibility.

Solar desalination in the form of low-tech solar stills was suggested more than 50 years ago. Currently, more sophisticated methods are used including the solar humidification-dehumidification (HDH) process, which mimics the natural water cycle on a shorter time frame by evaporating and condensing water in order to separate it from other substances. Geothermal desalination was suggested in the 1990s. The method requires less maintenance than the RO and uses the eco-friendly energy of geothermal heat. Prototypes of the device have been built and their feasibility has been demonstrated.

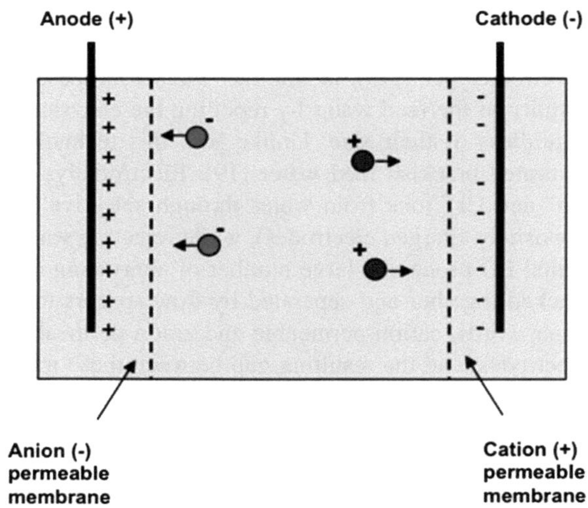

Fig. 19.12 Schematic of an electrodialysis desalination system

19.3 Physico-Chemical Principles of Reverse Osmosis

Osmosis is defined as the transport of a solvent through a semipermeable membrane into a solution of higher solute concentration. This transport tends to equalize the chemical potentials of the solute on either side of the membrane over time (Fig. 19.13). A semipermeable membrane allows solvent molecules (such as H_2O) to pass through the membrane, while larger solute molecules (such as NaCl) or ions cannot penetrate through. As a result, when a tube is separated into two sections by a semipermeable membrane isolating solute molecules in one section, water molecules will equilibrate (eventually having the same partial pressure on both sides of the membrane), whereas the solute molecules or ions will create excessive partial pressure upon the membrane. This is the osmotic pressure required to maintain equilibrium with no movement of the solvent.

In naturally occurring osmosis, the solvent is generally water and the semipermeable membrane is a membrane of a living cell. This diffusion of molecules is a form of passive transport, in which no applied energy is required. Passive transport occurs due to the molecular internal energy and concentration. This is very important in many biological processes where water penetrates into living cells through cell membranes and in the transport of various substances through blood cells.

When separated by a membrane, a solution with a higher chemical potential will permeate molecules through the membrane to a solution of lower potential due to the difference in energy. Because an RO membrane is semipermeable allowing only the smaller solvent molecules to diffuse across to equilibrate, the resulting transfer of solvent molecules will yield one solution more massive.

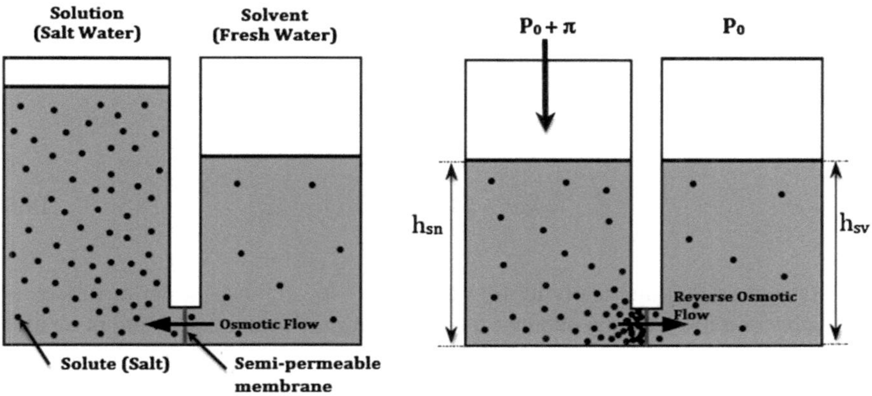

Fig. 19.13 Comparison of natural osmosis and reverse osmosis under pressure $P_0 + \pi$, with chemical potentials μ_{sn} and μ_{sv} for the solution and solvent, respectively

This difference in mass creates excess hydrostatic pressure, known as osmotic pressure, and can be estimated via the Morse [41] equation

$$\pi = iRTc \tag{19.1}$$

where π is the osmotic pressure, $R = 8.31\ \mathrm{JK^{-1}mol^{-1}}$ is the universal gas constant, T is absolute temperature (K), and c is the molar concentration of the solute (mol $\mathrm{m^{-3}}$). The variable i, known as the van't Hoff Factor [42], is defined as the degree of dissociation or association of solute particles in a solution. This is the ratio of the number of moles of solute particles to the number of moles of solute dissolved in a solution.

For the derivation of the Morse Equation, we can start with Gibbs free energy per unit mass $g(T, P, n_{st}, n_{sv})$ as a function of the number of moles of solvent (n_{sv}) and solute (n_{st}), so that its differential is given as

$$\mathrm{d}g = \bar{V}\mathrm{d}P - S\mathrm{d}t + \mu_{st}\mathrm{d}n_{st} + \mu_{sv}\mathrm{d}n_{sv} \tag{19.2}$$

where is the molar volume and S is specific entropy. The chemical potential of the solution $\mu_{sn}(P, X_{st})$ is a function of the solutions concentration of solute molecules X_{st}, and pressure P. In order to achieve equilibrium and prevent the osmotic flow of the solvent, the pressure of the solution should be increased from P_0 to $P_0 + \pi$ so that the chemical potentials at both sides of the membrane are equal.

$$\mu_{sn}(P + \pi, X_{st}) = \mu_0(P) \tag{19.3}$$

where μ_0 is the chemical potential of the solvent (H_2O) The chemical potentials can be written explicitly as

$$\mu_{sn}(P+\pi, X_{st}) = \mu_0(P) + \int_{P}^{P+\pi} \bar{V}_{sv}\mathrm{d}P + RT\int_{0}^{X_{st}} dX_{st}/(1-X_{st}) \tag{19.4}$$

where $\bar{V}_{sv}$ is the molar volume of the solvent. Therefore,

$$\int_{P_0}^{P_0+\pi} \bar{V}_{sv}\mathrm{d}P + RT\int_{0}^{X_{st}} dX_{st}/(1-X_{st}) \tag{19.5}$$

After integrating and applying $\ln(1-X_{st}) \approx -X_{st} = -n_{st}/n_{sv}$ for a sufficiently small concentration X_{st} one obtains an equation for the osmotic pressure of an ideal solution

$$\pi = RTn_{st}/V = RTc \tag{19.6}$$

where $\bar{V}_{sv}n_{sv} = V$, $n_{sv}/V = c$, and c represents the molar concentration of solute molecules in the solution. Note that this equation is very similar to the ideal gas law. Because ionic solutions, such as NaCl solute ions in water, are not ideal we must approximate the osmotic pressure by taking into consideration the dissociation of the solute molecules into ions and the corresponding loss in chemical energy. For this type of irreversible process, van't Hoff suggests a factor i, which represents every ion given off by one molecule of solute when dissolved in a solvent. With this, we conclude the derivation of Morse's Law for osmotic pressure $\pi = iRTc$ (Eq. 19.1) [43]. Examples of simple i calculations include the dissolution of salt into water,

$$NaCl_{(s)} \xrightarrow{H_2O} Na^{+}_{(aq)} + Cl^{-}_{(aq)} \tag{19.7}$$

yielding 2 ions ($i = 2$), or dissolving calcium chloride into water,

$$CaCl_{2(s)} \xrightarrow{H_2O} Ca^{2+}_{(aq)} + 2Cl^{-}_{(aq)} \tag{19.8}$$

yielding 3 ions ($i = 3$). Because a solution's critical temperatures is dependent on its concentration, we can experimentally determine van't Hoff's factor of a given solute, solvent, and concentration of solute. This is calculated as the difference in measured freezing temperature of a solution divided by the difference in its calculated freezing temperature:

$$i = \Delta T_F\,(\text{measured}) \;/\; \Delta T_F\,(\text{calculated}) \tag{19.9}$$

This can also be determined using respective boiling points. A common application of this is the use of calcium chloride as a road salt rather than sodium chloride because it has a greater i and when dissolved in water, the resulting solution has a lower freezing temperature [42].

The phenomenon of osmosis in semipermeable membranes was discovered in 1748 by the French physicist, Jean-AntioneNollet [44]. Though pressure-driven

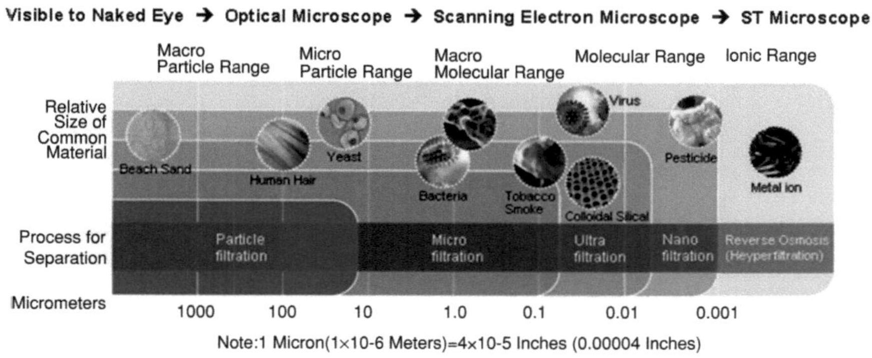

Fig. 19.14 Contaminant removal by filtration stage [45]

membrane desalination has been around for nearly 100 years [44], only in recent decades have researchers successfully integrated RO and selective surface science into membrane desalination. Most RO membrane desalination processes involve force-feeding seawater through four stages of filtration: micro, ultra, nano, and lastly, reverse osmosis filtration (Fig. 19.14). Once the feed solution enters the RO membrane, it should be sufficiently free of contaminant particles, with the exception of Na^+ and Cl^- ions, and other nano-scale contaminants.

The standard commercial RO filter contains a thin film composite (TFC) membrane spiral wound about a perforated tube and sealed in a pressure vessel. A permeate material is sandwiched between the TFC membrane, which in turn lies between two feed channel spacers (Fig. 19.15). The feed channel spacer allows the feed solution (seawater) to pass through the filter, while creating turbulence to aid in the filtration process. The TFC membrane allows the passage of water molecules which are then absorbed into the permeate material, or membrane. The permeate material is wrapped around a perforated tube at the center of the filter, in which the pure water enters and flows freely.

To many, RO membrane desalination has not been regarded as a viable solution to fresh water issues due to the immense amount of pressure required to overcome osmotic pressure and produce fresh water at a reasonable rate. The theoretical minimum amount of energy (for an ideal process) required for recoveries of 25, 50, and 75% (percent fresh water from sea water) is calculated at 0.81, 0.97 and 1.29 kW h m^{-3}, respectively [46]. Though the specific energy required to produce one cubic meter of potable water via RO desalination has been reduced from over 10 kW h m^{-3} in the 1980s to less than 4 kW h m^{-3}, there is much progress to be made in membrane technology to further reduce energy consumption. The Affordable Desalination Collaboration (ADC) has recently published comparable results from a scale seawater reverse osmosis (SWRO) plant at the US Navy's Seawater Desalination Test Facility in Port Hueneme, California in March of 2006 [47].

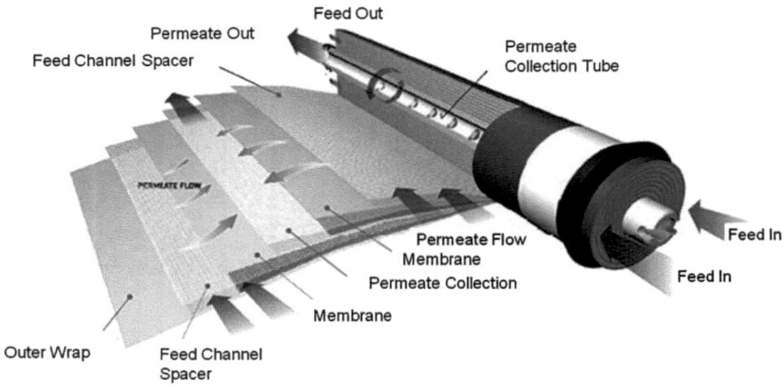

Fig. 19.15 Spiral wound RO filter [45]

Under ideal conditions (new membranes, no fouling, and low flux rates), ADC demonstrated an exceptionally low specific energy usage in seawater desalination at 1.58 kW h m^{-3} with a 42% recovery rate, which is relatively close to the theoretical minimum [47]. In real cases however, this theoretical minimum is not possible to due inherent loses in energy of the irreversible thermodynamic process, as maintained by the 2nd Law of Thermodynamics. In the case of seawater desalination, membrane pore diameters must be small enough to exclude ions such as Na^+ and Cl^-, which significantly increases enthalpic energy requirements. Build up of such ions and other fouling matter at the pore wall is one of the biggest factors in energy consumption. As flux rate increases, the build-up of rejected solute particles locally increases. In other words, increased flux rates result in a higher solute concentration at the membrane wall, which is a critical factor in energy or osmotic pressure requirements. From a diffusion membrane model, the flux of water across a membrane can be calculated as

$$J_w = \frac{D_{\text{wm}} C_{\text{wm}} V_w}{RT\delta} (\Delta P - \pi X_F - \pi X_P) \tag{19.10}$$

where D_{wm} is the water diffusivity in the membrane, C_{wm} is the water concentration in the membrane, V_w is the partial molar volume of water, R is the gas constant, T is absolute temperature and δ is the thickness of the membrane. A is known as the water permeability constant, and can be written as $A \equiv D_{\text{wm}} C_{\text{wm}} V_w /$ RTδ. Substituting into (19.10) we have an equation for the flux of water across a membrane as a function of A, change in pressure across the membrane ΔP, and osmotic pressure π as a function of solute concentration at the feed side of the membrane X_F, and concentration at the permeate side X_P.

$$J_w = A(\Delta P - \pi X_F - \pi X_P) \tag{19.11}$$

From (19.11) it is clear that both flux and pressure are functions of the solutions concentration. One method to reduce solute concentration locally at the membrane wall and improve flux is Electrodialysis (see Sect. 19.2.2.2). Active transport, which occurs naturally in living cells, can also significantly reduce energy requirements. The use of major intrinsic proteins (MIP) embedded in biomimetic membranes as a form of actively transporting molecules (either water or ions) is another area of research which could prove valuable to RO desalination [48].

Countless studies have concluded there is an innate trade-off between selectivity and permeability, which can be determined using variations on kinetic theory, such as the Hagen-Poiseuille equation and the Knudsen model for sub-micron size channels [49]. The issue is, for a given semipermeable membrane, the reduction in pore size increases filtration or selectivity, but results in a decrease in permeate flux through the membrane. Research on mass transport through small double walled carbon nanotubes (DWNT) serving as membrane pores has exhibited flux rates that exceed continuum hydrodynamic models by more than three orders of magnitude [49]. This increase in flux can be attributed to the DWNT smooth, hydrophobic walls which allow DWNT models high levels of selectivity as well as flux, however, these nanotubes are prone to wear and fouling. Despite their potential, the manufacturing of such membranes is currently very difficult and expensive, so developing new inexpensive and efficient means to create DWNT membranes is another challenge to address.

Current research promises membranes exhibiting relatively high degrees of selectivity and flux, however, the most complex problem remains with reducing biofouling of the membrane. The development of new RO membranes with sufficient selectivity and permeability, along with advanced anti-fouling properties will give rise to more reliable, accessible water filtration, and desalination methods worldwide. In the past decade, there has been a large increase in the use of RO membranes in water desalination due progressing technology, while less controllable thermal methods such as MSF, MED, and VC have declined.

19.4 Surface Roughness and Reinforcement Effect on Permeability and Fouling

As current methods of RO water desalination requires a considerable amount of energy and maintenance cost, the emphasis of this research will be to reduce energy consumption by maximizing membrane permeability, while maintaining a desirably high level of salt rejection (retention). Equally important is optimizing the antifouling (self-cleaning) properties of the membranes. This can be achieved, for example, by manufacturing membranes with a high density of small pores (~ 10 nm), as well as by introducing nanoparticle reinforcement to the filter material allowing for modification of wetting properties, and to facilitate self-cleaning of the membranes.

Researchers from the Sandia National Lab suggested the use of biomimetic membranes for desalination. Most living cells contain specialized "water-channels" in the cell membrane which transport water and exclude salt with very high efficiency. Sandia aims to model and understand the function of the natural water channel and build a synthetic membrane which mimics the natural function [50].

Researchers at Lawrence Livermore National Laboratory have created a membrane made of carbon nanotubes (CNT) and silicon that may offer, among many possible applications, a less expensive desalination application. Billions of CNTs act as the pores in the membrane allowing the solvent to rapidly flow through, while the tiny pore size can block larger molecules of the solute. The membrane was created by filling the gaps between aligned carbon nanotubes with a ceramic matrix material. The pores are so small that only six water molecules could fit across their diameter [51, 49].

Bormashenko et al. [9] reported a method of manufacturing free-standing, micrometer-scale honeycomb polyetherimide films. Films are manufactured with a dip-coating technique under water-assisted self-assembly. They showed that the addition of poly(organosilane/siloxane)s and poly(ethylene glycol) allows the formation of regular honeycomb patterns. The films demonstrated the high thermal stability inherent for polyetherimide. The wetting properties of films are reported and the presence of nanopores was revealed with SEM imaging of the films. The makeup of the films allows their use as asymmetric membranes for reverse osmosis and ultrafiltration.

19.4.1 Theoretical Model

As mentioned above, one of the major challenges in the use of reverse osmosis for desalination is the accumulation of minerals and living micro-organisms on the front surface and inside the pores of the membrane, and consequently the formation of biofilms. As a result, the fluid channels will partially clog and therefore a larger pressure is required in order to maintain the water flow rate through the membrane.

One of the novel antifouling mechanisms investigated is altering the wetting properties of the membrane surface. The extent of the wetting of a solid surface by a liquid is determined by the respective contact angle, which in turn is dependent on the interfacial energies of the three involved surfaces (i.e., liquid–air, liquid–solid, and solid–air).

Generally, contact angles below 90° indicate relatively large wetting of the solid by the liquid (*hydrophilic* for water), whereas formation of droplets with contact angles greater than 90° means smaller wetting (or weaker liquid–solid attachment). In the latter case, the solid surface can be considered as a good repellant of the liquid, and is called *hydrophobic* when water is used as liquid. The surfaces which show contact angles above 150° and have a small contact angle hysteresis (smaller than 5°) are called superhydrophobic, and usually have suitable self-cleaning properties. Self-cleaning surfaces find interesting applications, such as in industrial paints, textiles, solar panels, etc. [52–54].

In RO processes, the goal is to control the wetting of the membrane surface by contaminants existing in the water. Since the organic contaminants (oily compounds) generally have very low surface energies, the anti-fouling process should be aimed for removal of these compounds (worst case scenario). Therefore, creating self-cleaning surfaces for underwater antifouling is associated with manufacturing *superoleophobic* surfaces (rather than superhydrophobic surfaces). As a result, the trio of water–solid–air is replaced with oil–solid–water. In this section, we use two well-known wetting models (Wenzel and Cassie–Baxter) in order to study the possibility of creating superoleophobic surfaces by adding reinforcement particles to the main matrix of the polymeric membranes. It is suggested that the reinforcement particles are inserted in the bulk of the main polymeric matrix. This way, even if the particles at the surface are washed away due to wear, a new surface with a very similar roughness pattern comes into contact with water and therefore, the membrane will retain its self-cleaning properties.

For a composite interface built of two materials with different structural properties with the fractional areas of f_1 and f_2 (so that $f_1 + f_2 = 1$), the composite surface energy can be expressed as a weighted average of the corresponding surface energies. The contact angle is then given by the Cassie equation [55]

$$\cos\theta = f_1 \cos\theta_1 + f_2 \cos\theta_2 \tag{19.12}$$

where θ_1 and θ_2 are the contact angles of the fractions. If a composite material has a matrix and reinforcement with the volume fractions of f_m and f_r (so that $f_m + f_r = 1$) forming a rough surface the contact angle is then given by

$$\cos\theta = R_{\mathrm{fm}}(1 - f_r)\cos\theta_m + R_{\mathrm{fr}} f_r \cos\theta_r \tag{19.13}$$

where θ_m and θ_r are the contact angles for the matrix and reinforcement materials, and R_{fm} and R_{fr} are corresponding roughness factors [56]. The roughness factor of each of surface (reinforcement and matrix) is defined as the ratio of the actual surface area A_{act} to the projected area, A_{proj}. If we assume the composite surface to be made of a perfectly smooth matrix reinforced with spherical particles, the roughness factor of each surface can be easily estimated as follows:

- *Matrix*: neglecting the roughness asperities in the main matrix, the actual surface area is the same as the projected area, and therefore $R_{\mathrm{fm}} = 1$.
- *Reinforcement*: spherical particles with different radii are placed randomly in the surface, where a portion of each particle is above the surface. This configuration is schematically shown in Fig. 19.16.

For each spherical asperity of the surface, one can obtain the actual and projected areas based on the depth of the particle's penetration in the surface. As shown in Fig. 19.16, different particles can intersect the surface at different depths H $(-R < H < R)$. The corresponding actual surface area can be found as

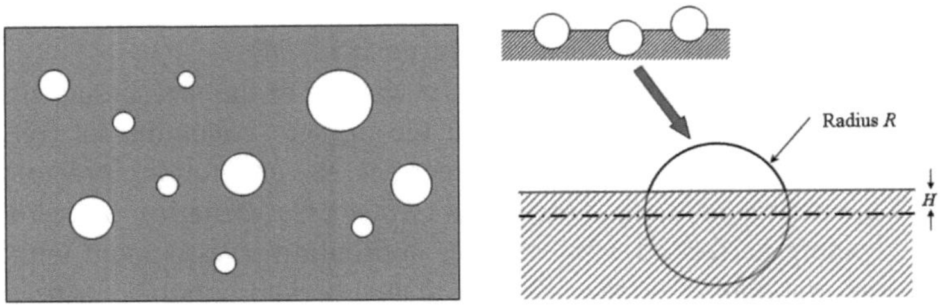

Fig. 19.16 Schematic of a matrix reinforced with spherical particles, (*left*) top view, and (*right*) front view (cross section)

$$A_{\text{act}} = \int\limits_{\sin^{-1}(H/R)}^{\pi/2} 2\pi R\cos\theta\, R\mathrm{d}\theta = 2\pi R^2\left(1 - \frac{H}{R}\right) \tag{19.14}$$

and the corresponding projected area will be

$$A_{\text{proj}} = \pi\left(R^2 - H^2\right) \tag{19.15}$$

The roughness factor of this spherical asperity will be therefore found as

$$R_{\text{fr}} = \frac{A_{\text{act}}}{A_{\text{proj}}} = \frac{2\pi R^2\left(1 - \frac{H}{R}\right)}{\pi(R^2 - H^2)} = \frac{2}{1 + H/R} \tag{19.16}$$

The roughness factor of each spherical particle is dependent on the particle radius R and depth H. In order to estimate the impact of the reinforcement particles on the overall contact angle, the roughness factor is estimated using the average particle radius, particle size distribution, and average penetration depth. A constant value for the roughness factor can be determined using micro or nano-size particles with sufficiently narrow size distribution. Note it is more accurate to consider a normal distribution for H, as it will vary between $-R$ and R.

Solving (19.13) for the reinforcement fraction yields the volume of the reinforcement fraction as a function of the desired contact angle θ

$$f_r = \frac{\cos\theta - R_{\text{fm}}\cos\theta_m}{R_{\text{fr}}\cos\theta_r - R_{\text{fm}}\cos\theta_m} \tag{19.17}$$

We can further assume $H = 0$ for all the spherical particles (surface intersects with the mid-plane of all of the spherical particles) and therefore, $R_{f,r} = 2$. Moreover, assuming $R_{\text{fm}} = 1$ (no roughness except from the reinforcement particles), and $\theta = 180°$ (the theoretical superhydrophobic limit) yields

$$f_r = \frac{-1 - \cos\theta_m}{2\cos\theta_r - \cos\theta_m} \tag{19.18}$$

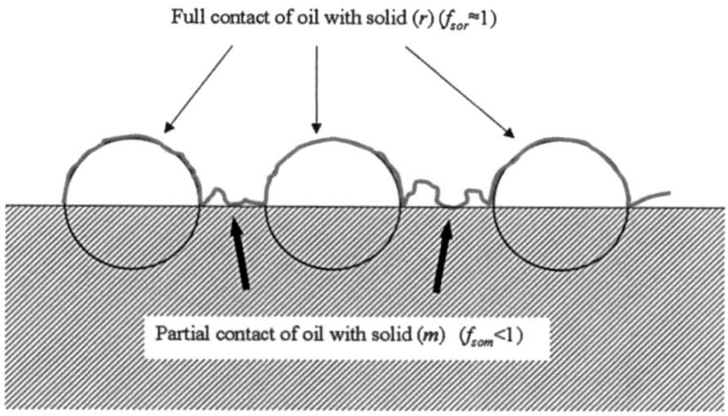

Fig. 19.17 Contact between the liquid (oil) and the solid surfaces: matrix and reinforcements

which has a solution ($f_r < 1$) if $\theta_r > 120°$. Thus it is difficult to produce a composite interface by using only the reinforcement roughness [55]. If the liquid (here: oil) forms partial contact with the solid surfaces of both reinforcement and the matrix (composite or Cassie–Baxter) with the fractional solid liquid contact areas f_{Som} and f_{Sor} such that $f_{\mathrm{sor}} + f_{\mathrm{swr}} = 1$ and $f_{\mathrm{som}} + f_{\mathrm{swm}} = 1$ (Fig. 19.17), the contact angle can be found as

$$\cos\theta = R_{\mathrm{fm}}(1 - f_r)\cos\theta_m f_{\mathrm{som}} + R_{\mathrm{fr}} f_r \cos\theta_r f_{\mathrm{sor}} - 1 + f_r f_{\mathrm{sor}} + (1 - f_r) f_{\mathrm{som}} \tag{19.19}$$

where the subscripts *s*, *o*, and *w* represent solid, oil, and water, respectively.

Solving (19.19) for the reinforcement fraction yields the volume of the reinforcement fraction as a function of the desired contact angle θ [55].

$$f_r = \frac{1 + \cos\theta - R_{\mathrm{fm}}\cos\theta_m f_{\mathrm{som}} - f_{\mathrm{som}}}{f_{\mathrm{sor}} + R_{\mathrm{fr}}\cos\theta_r f_{\mathrm{sor}} - R_{\mathrm{fm}}\cos\theta_m f_{\mathrm{som}} - f_{\mathrm{som}}} \tag{19.20}$$

Estimation of f_r in the current form might be very difficult. Making the assumptions of $R_{fr} = 2$, $R_{fm} = 1$, $\theta = 180°$, and further assuming that oil completely touches the reinforcement surfaces can simplify (19.20) to

$$f_r = \frac{-\cos\theta_m f_{\mathrm{som}} - f_{\mathrm{som}}}{1 + 2\cos\theta_r - \cos\theta_m f_{\mathrm{som}} - f_{\mathrm{som}}} \tag{19.21}$$

For a given matrix, reinforcement, and contact angles θ_r and θ_m, one can plot the reinforcement volume fraction as a function of the f_{som}. Alternatively, the variation of the overall contact angle θ with f_r can be plotted for different values of f_{som}. Two such plots are shown in Fig. 19.18. In Fig. 19.18a, the main matrix is oleophobic ($\theta_m = 120°$, as obtained through an underwater experiment); while in Fig. 19.18b, the matrix is oleophilic ($\theta_m = 45°$). The contact angle of the reinforcement particles is considered to be 140°. For each case, the contact angle is plotted as a function of the volumetric fraction of the reinforcement for four different values of f_{som}. Comparing the results reveals that for composite materials with hydrophobic matrices, smaller amounts of reinforcement particles are needed

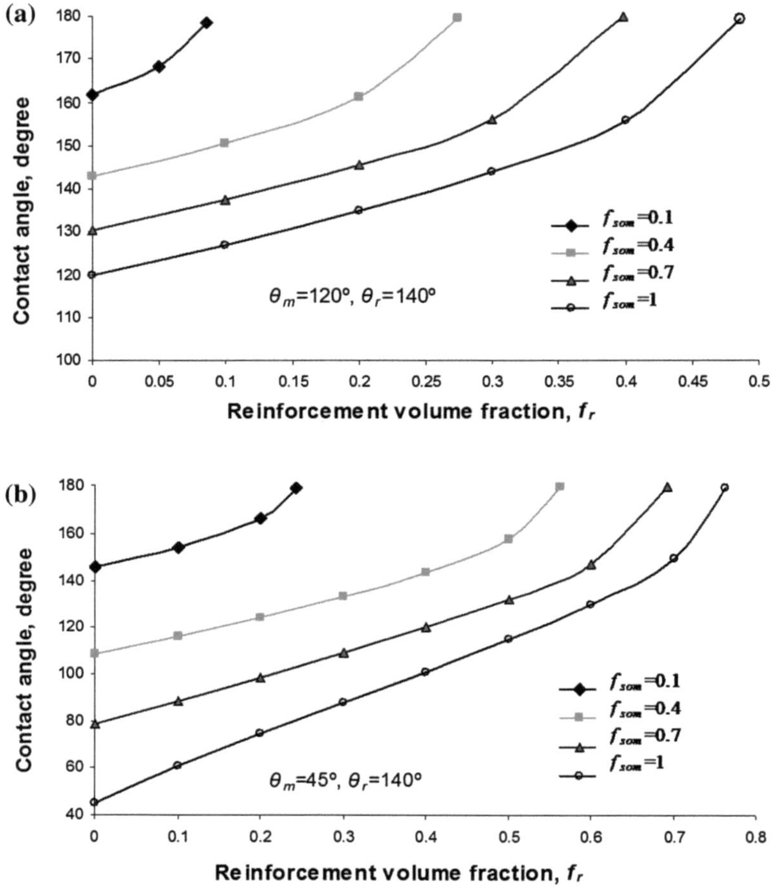

Fig. 19.18 The underwater contact angle of oil with a composite surface as a function of the reinforcement volume fraction (Cassie–Baxter regime) for different values of f_{som}. **a** Oleophobic matrix, **b** Oleophilic matrix

to create a superoleophobic surface. It is also desired to observe the effect of reinforcement volume fraction on the extent of wetting of the main matrix f_{som}. This will indicate whether using the Cassie–Baxter model is a valid assumption. This relationship can be shown as

$$f_{\text{som}} = \frac{\cos\theta - R_{\text{fr}} f_r \cos\theta_r f_{\text{sor}} + 1 - f_r f_{\text{sor}}}{(1 - f_r)(1 + R_{\text{fm}} \cos\theta_m)} \tag{19.22}$$

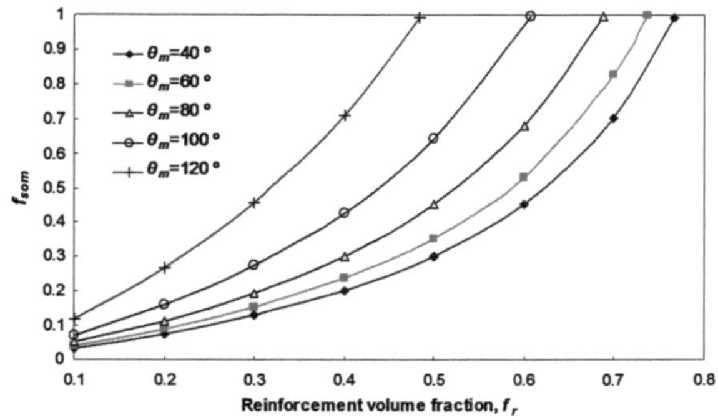

Fig. 19.19 The variation of f_{som} with reinforcement volume fraction f_r for different matrix contact angles

Results are shown in Fig. 19.19 for the composite surfaces with a wide range of matrix contact angles (from 40° to 120°). It is observed that for more hydrophobic matrices, the liquid droplet fully touches the matrix (f_{som} becomes equal to unity) at a lower reinforcement volume fraction f_r.

If an oil droplet is placed on a solid surface in water, the contact angle of an oil droplet in water, θ_{OW}, is given by Young's equation:

$$\cos\theta_{OW} = \frac{\gamma_{OA}\cos\theta_O - \gamma_{WA}\cos\theta_W}{\gamma_{OW}} \tag{19.23}$$

where θ_O and θ_W are contact angles of oil and water with the solid surface, γ_{OA}, γ_{WA}, and γ_{OW} are interfacial energies for oil–air, water–air, and oil–water interfaces [57]. As a consequence, a superhydrophobic (in air) surface can become oleophobic when immersed in water, under certain circumstances, which are summarized in Table 19.2. An oleophobic surface repels organic liquids and thus prevents organic contaminants from accumulation and decrease the adhesion of bacteria, thus facilitating antifouling properties.

Similarly to the superhydrophobic surfaces, besides the homogeneous solid–oil interface (Wenzel state), a composite solid–oil–water interface (Cassie–Baxter or Cassie state) with water pockets trapped between the solid and the oil droplet can exist. The contact angle is then given by

$$\cos\theta_{OW} = \frac{\gamma_{OA}\cos\theta_O - \gamma_{WA}\cos\theta_W}{\gamma_{OW}} f_{SO}R_f - 1 + f_{SO} \tag{19.24}$$

where $0 < f_{SO} < 1$ is the fractional solid–oil contact area and R_f is the roughness factor. The rules of the Cassie–Wenzel wetting regime transition are the same as in the case of superhydrophobic surfaces, as it was discussed in the preceding chapters. A more complex four-phase solid–oil–water-air system can form if both water and air bubbles are present at the composite interface. It is expected that with time air will dissolve so that $f_{SA} \to 0$.

Table 19.2 Oleophobic and oleophilic interfaces (based on [57])

Interface					
Solid–air–water	Hydrophobic $\gamma_{SA} > \gamma_{SW}$			Hydrophobic $\gamma_{SA} < \gamma_{SW}$	
Solid–air–oil	Oleophilic if $\gamma_{SA} > \gamma_{SO}$	Oleophobic if $\gamma_{SA} < \gamma_{SO}$		Oleophilic if $\gamma_{SA} > \gamma_{SO}$	Oleophobic if $\gamma_{SA} < \gamma_{SO}$
Solid–water–oil	oleophilic	Oleophilic if $\gamma_{OA}\cos\theta_0 < \gamma_{WA}\cos\theta_W$	Oleophobic if $\gamma_{OA}\cos\theta_0 > \gamma_{WA}\cos\theta_W$	Oleophilic if $\gamma_{OA}\cos\theta_0 < \gamma_{WA}\cos\theta_W$	Oleophobic if $\gamma_{OA}\cos\theta_0 < \gamma_{WA}\cos\theta_W$

Fig. 19.20 GE RO filter

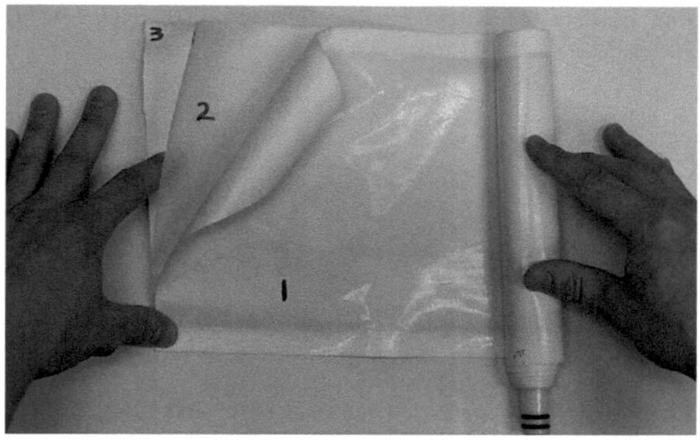

Fig. 19.21 GE RO filter unrolled displaying two 2 identical RO membranes (labeled 1 and 3) and the permeate material labeled (19.2)

19.4.2 Experimental Study

To study current RO filter wetting properties, specifications, and how to improve them, we tested the GE RO Membrane model TFM-75 (Fig. 19.20). This standard RO filter contains a thin film composite membrane rated for using desalinating water with a salinity of 0.5 g/L (brackish water), and can produce 75 gallons a day under 65 psi, with an average salt rejection of 96%. Using a Ramè-Hart goniometer we measure the contact angles of the inner and outer surfaces of both the membrane and the permeate material (which also acts as a feed spacer in this application) (Fig. 19.21). Contact angles from both the water–surface–air (Figs. 19.22, 19.23) and oil–surface–water (Fig. 19.24) regimes were measured for each material's surface.

The membrane's outer surface was found to be more hydrophilic than the inner (36° and 58°, respectively), yet this outer surface was less oleophilic than the inner (45.2° compared to 36.7°). This relationship is quite intuitive considering the function of these two surfaces. The outer surface of this TFC membrane will be in contact with the brackish, contaminated water and therefore will require a more oil-resistant surface structure designed to repel bio-contaminants and prevent bio-filming. Furthermore, it would be desirable for this surface to easily allow water to permeate through, which is suggested by the large degree of wetting, or small contact angle, the water droplet forms.

Once the water permeates through the membrane it comes in contact with two surfaces; the inner membrane surface and the outer permeate material surface. As stated, the inner surface of the membrane is more hydrophobic, however,

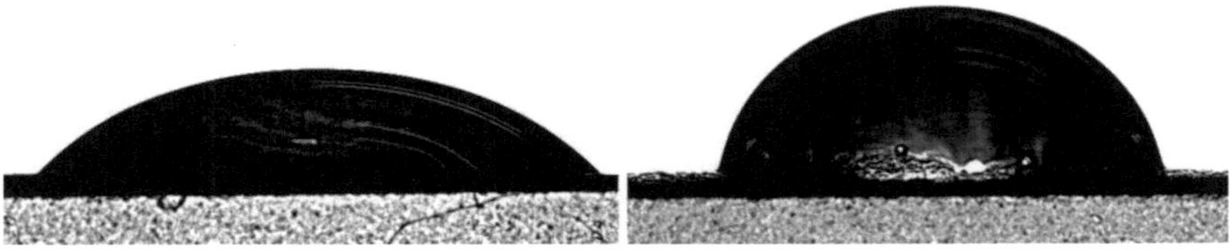

Fig. 19.22 RO membrane outer (*left*) and inner (*right*) surfaces having a water contact angle of 36° and 58°

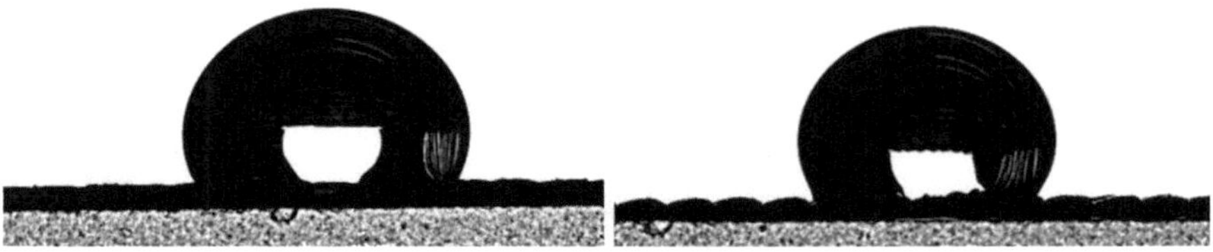

Fig. 19.23 Filter permeate material outer and inner surfaces having a water contact angle of 114.7° and 123°

Fig. 19.24 RO membrane outer and inner surfaces having an oil contact angle of 45.2° and 36.7°

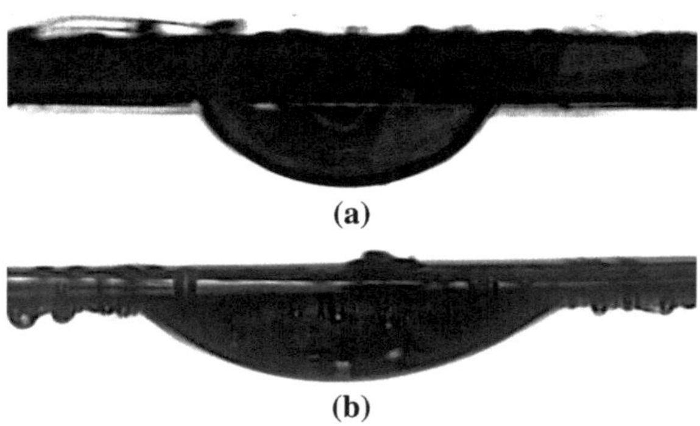

the interesting bit lies in the function of the permeate material. When looking closely at the inner surface structure of the permeate material we can clearly see small, consistent humps. This roughness allows for the surface to be nearly superhydrophobic (Fig. 19.23). More importantly, when spiral-wound in the filter this inner surface plays two important roles:

1. The surface will be compressed upon bending, effectively reducing the distance between and increasing the height of the asperities, or "humps". Based on the Cassie–Baxter Equation it can be assumed that the change in fractional areas due to this compression will result in a higher contact angle, which could likely reach the superhydrophic limit of 150°.

2. Because the inner surface's roughness will create gaps between itself and the membrane and because the outer surface is relatively smooth and more hydrophilic, the flowing water will naturally tend to the inner surface of the permeate material.

Because the filtered water has to flow such a long distance within the permeate material to reach the center, this materials hydrophobicity is a major factor in the filter's performance. The facilitated flow will allow the fresh, filtered water to travel along a superhydrophobic surface and greatly increase the filter's flux and efficiency.

Along with the importance of water permeability, it is also important to create outer membrane surfaces with high degrees of oleophobicity and selectivity. By incorporating an appropriate composite matrix to these polymers we can modify their surface structures and facilitate desired wetting properties greatly improving the efficiency of such RO filters.

19.5 Conclusions

Reverse Osmosis is currently one of the prevalent techniques of desalination and its global capacity continues to grow. The main concern with RO is that it requires significant energy inputs which affect the environment due to greenhouse gas emissions, and the output of brine with high salt concentrations. Improving efficiency and reducing the environmental hazards of RO plants involves several challenges, some of which are related to surface science and tribology. In particular, the field of biomimetic membranes and surface modification to reduce fouling are related to the Green Tribology. By mimicking water filtration of cell membranes and utilizing biomimetic antifouling coatings of membranes, we can improve efficiency and ecological impact of RO desalination.

Acknowledgements The authors acknowledge the support of the University of Wisconsin-Milwaukee (UWM) Research Growth Initiative (RGI) NSF I/UCRC for Water Equipment and Policy, and UWM Research Foundation Bradley Catalyst grants and the UWM SURF program. Authors are also thankful to Prof. Pradip K. Rohatgi from the UWM Center for Composite materials for filter samples.

References

1. Anonymous, GLAAS 2010: UN-Water global annual assessment of sanitation and drinking-water (2010a), http://whqlibdoc.who.int/publications/2010/9789241599351_eng.pdf. Accessed 16 May 2011
2. Anonymous, The 3rd United Nations world water development report: water in a changing world (2009), http://www.unesco.org/water/wwap/wwdr/wwdr3/tableofcontents.shtml. Accessed 16 May 2011

3. Anonymous, United Nation news center (2011a), http://www.un.org/apps/news/story.asp?NewsID=38253&Cr=Population. Accessed 16 May 2011
4. K. Wangnick, IDA Worldwide Desalting Plants Inventory. Report No. 18. Wangnick Consulting, Gnarrenburg (2004)
5. K. Wangnick, 2002 IDA Worldwide Desalting Plants Inventory. Wangnick Consulting (for the International Desalination Association) Gnarrenburg (2002)
6. K. Kranhold, Water, water everywhere. The Wall Street Journal. 17 January 2008 (2008), http://online.wsj.com/article/SB120053698876396483.html. Accessed 10 May 2011
7. Anonymous, 21st GWI/International Desalination Association, Worldwide Desalting Plant Inventory: Global Market Snapshot (2008), http://www.idadesal.org/PDF/ida%20desalination%20snapshot_october%202008.pdf. Accessed 25 May 2011
8. M.A. Shannon, P.W. Bohn, M. Elimelech, J.G. Georgiadis, B.J. Marinas, A.M. Mayes, Science and technology for water purification in the coming decades. Nature **452**, 301–310 (2008)
9. E. Bormashenko, A. Schechter, O. Stanevsky, T. Stein, S. Balter, A. Musin, Y. Bormashenko, R. Pogreb, Z. Barkay, D. Aurbach, Free-standing, thermostable, micrometer-scale honeycomb polymer films and their properties. Macromol. Mater. Eng. **293**, 872–877 (2008)
10. P. Nednoor, V.G. Gavalas, N. Chopra, B.J. Hinds, L.G. Bachas, Carbon nanotube based biomimetic membranes: mimicking protein channels regulated by phosphorylation. J. Mater. Chem. **17**, 1755–1757 (2007)
11. R.J. Forbes, *A Short History of the Art of Distillation: From the Beginnings up to the Death of Cellier Blumenthal* (Brill, Leiden, 1970)
12. S. Kalogirou, Survey of solar desalination systems and system selection. Energy **22**, 69–81 (1997)
13. G. Fiorenza, V.K. Sharma, G. Braccio, Techno-economic evaluation of a solar powered water desalination plant. Energy Convers. Manag. **44**, 2217–2240 (2003)
14. C. Koroneos, A. Dompros, G. Roumbas, Renewable energy driven desalination systems modeling. J Cleaner Prod. **15**, 449–464 (2007)
15. H.M. Qiblawey, F. Banat, Solar thermal desalination technologies. Desalination **220**, 633–644 (2008)
16. T. Ayhan, H. Al Madani, Feasibility study of renewable energy powered seawater desalination technology using natural vacuum technique. Renew. Energy **35**, 506–514 (2010)
17. L. Garcia-Rodriguez, Renewable energy applications in desalination: state of the art. Sol. Energy **75**, 381–393 (2003)
18. A. Subramani, M. Badruzzaman, J. Oppenheimer, J.G. Jacangelo, Energy minimization strategies and renewable energy utilization for desalination: a review. Water Res. **45**, 1907–1920 (2011)
19. M.A. Eltawil, Z. Zhengming, L. Yuan, A review of renewable energy technologies integrated with desalination systems. Renew. Sustain. Energy Rev. **13**, 2245–2262 (2009)
20. L.R. Evans, J.E. Miller, *Sweeping Gas Membrane Desalination Using Commercial Hydrophobic Hollow Fiber Membranes*. SAND 2002-0138 (Sandia National Laboratories, 2002)
21. R.S. Silver, Multi-stage flash distillation. The first 10 years. Desalination **9**, 3–17 (1971)
22. K.V. Reddy, N. Ghaffour, Overview of the cost of desalinated water and costing methodologies. Desalination **205**, 340–353 (2007)
23. S.A. Kalogirou, Seawater desalination using renewable energy sources. Prog. Energy Combust. Sci. **31**, 242–281 (2005)
24. M. Al-Shammiri, M. Safari, Multi-effect distillation plants: state of the art. Desalination **126**, 45–59 (1999)
25. T. Dabbagh, P. Sadler, A. Al Saquabi, Desalination: An Emergent Option, in *Water in the Arab World: Perspectives and Prognoses*, 1st edn., ed. by P. Rogers, P. Lydon (Harvard University Press, Cambridge, 1994)
26. N.M. Wade, Technical and economic evaluation of distillation and reverse osmosis desalination processes. Desalination **93**, 343–363 (1993)

27. F.M. Mubeen, Workshop identifies options for nuclear desalination. Int. Desalin. Water Reuse **11**, 15–19 (2001)
28. Anonymous, *IAEA Introduction of Nuclear Desalination; a Guidebook. Technical Report Series no. 400* (International Atomic Energy Agency, Vienna, 2000)
29. C. Fernandez-Lopez, A. Viedma, R. Herrero, A.S. Kaiser, Seawater integrated desalination plant without brine discharge and powered by renewable energy systems. Desalination **235**, 179–198 (2009)
30. C. Mustacchi, V. Cena, *Solar Water Distillation, Technology for solar energy utilisation* (New York, 1978)
31. R.M. Morris, W.T Hanbury, Renewable energy and desalination—a review. in *Proceedings of the new technologies for the use of renewable energy sources in water desalination*, 1991
32. Anonymous, Steam Jet Compressor in Refrigeration Engineering, G.U.N.T. Gerätebau GmbH (2010b), http://www.gunt.de/networks/gunt/sites/s1/mmcontent/produktbilder/06135200/Datenblatt/0613200%202.pdf. Accessed 17 May 2011
33. A. Midilli, Waste water distillation via natural vacuum technique. Dissertation, Karadeniz Technical University, 1997
34. A. Midilli, T. Ayhan, Natural vacuum distillation technique—part I: theory and basics. Int. J. Energy Res. **28**, 355–371 (2004)
35. A. Midilli, T. Ayhan, Natural vacuum distillation technique—part II: Experimental investigation. Int. J. Energy Res. **28**, 373–389 (2004)
36. S. Al-Kharabsheh, D.Y. Goswami, Analysis of an innovative water desalination system using low-grade solar heat. Desalination **156**, 323–332 (2003)
37. R. Oldach, Matching renewable energy with desalination plants (Muscat, Sultanate of Oman: The Middle East Desalination Research Center, MEDRC). MEDRC Series of R&D Reports, MEDRC Project: 97-AS-006a (2001)
38. L.H. Shaffer, M.S. Mintz, Electrodialysis, in *Principles of Desalination, Part A*, 2nd edn., ed. by K.S. Spiegler, A.D.K. Laird (Academic Press, New York, 1980)
39. E. Barbier, Geothermal energy technology and current status: an overview. Renew. Sustain. Energy Rev. **6**, 3–65 (2002)
40. K. Bourouni, M.T. Chaibi, T. Tadrist, Water desalination by humidification and dehumidification of air: state of the art. Desalination **137**, 167–176 (2001)
41. H.N. Morse, Ueber eine neue Darstellungsmethode der Acetylamidophenole. Berichte der deutschen chemischen Gesellschaft **11**, 232–233 (1878). doi:10.1002/cber.18780110151
42. J.H. vant Hoff, The Role of Osmotic Pressure in the Analogy Between Solutions and Gases, in *The Modern Theory of Solution*, 1st edn., ed. by H.C. Jones (Harper & Brothers Publishers, London, 1899)
43. W.R. Salzman, Colligative properties (2004), http://www.chem.arizona.edu/~salzmanr/480a/480ants/colprop/colprop.html. Accessed 13 May 2011
44. J. Glater, The early history of reverse osmosis membrane development. Desalination **117**, 297–309 (1998)
45. Anonymous, Construction of COWAY R/O membrane filter (2011b), http://www.coway-usa.com/technology/01_WaterFiltrationSystem_03.html. Accessed 17 May 2011
46. K.S. Spiegler, Y.M. El-Sayed, The energetics of desalination processes. Desalination **134**, 109–128 (2001)
47. T.F. Seacord, S.D. Cooker, J. MacHarg, Affordable desalination collaboration 2005 results. Int. Desalin. Water Reuse Q. **16**, 1–10 (2006)
48. C.H. Nielsen, Major Intrinsic Proteins in Biomimetic Membranes, in *MIPs and Their Role in the Exchange of Metalloids*, 1st edn., ed. by T.P. Jahn, G.P. Bienert (Landes Bioscience, Austin, 2009)
49. J.K. Holt, H.G. Park, Y. Wang, M. Stadermann, A.B. Artyukhin, C.P. Grigoropoulos, A. Noy, O. Bakajin, Fast mass transport through sub-2-nanometer carbon nanotubes. Science **312**, 1034–1037 (2006)

50. Anonymous, Sandia National Laboratory: Research and Development, Membrane Technologies (2011c), http://www.sandia.gov/water/desal/research-dev/membrane-tech.html. Accessed 15 April 2011
51. Anonymous, Lawrence Livermore National Laboratory, Nanotube Membranes Offer Possibility of Cheaper Desalination (2006), https://www.llnl.gov/news/newsreleases/2006/NR-06-05-06.html. Accessed 15 Apr 2011
52. M. Nosonovsky, B. Bhushan, Superhydrophobic surfaces and emerging applications: non-adhesion, energy, green engineering. Curr. Opin. Colloid Interface Sci. **14**, 270–280 (2009)
53. K. Teshima, H. Sugimura, Y. Inoue, O. Takai, A. Takano, Transparent ultra water repellent poly(ethylene terephthalate) substrates fabricated by oxygen plasma treatment and subsequent hydrophobic coating. Appl. Surf. Sci. **244**, 619–622 (2005)
54. H. Yabu, M. Shimomura, Single-step fabrication of transparent superhydrophobic porous polymer films. Chem. Mater. **17**, 5231–5234 (2005)
55. M. Nosonovsky, V. Hejazi, A.E, Nyong, P.K. Rohatgi, *Metal Matrix Composites for Sustainable Lotus-Effect Surfaces* (submitted to *Langmuir*, 2011)
56. M. Nosonovsky, B. Bhushan, *Multiscale Dissipative Mechanisms and Hierarchial Surfaces: Friction Superhydrophobicity and Biomimetics* (Springer, Berlin, 2008)
57. Y.C. Jung, B. Bhushan, Wetting behavior of water and oil droplets in three-phase interfaces for hydrophobicity/philicity and oleophobicity/philicity. Langmuir **25**, 14165–14173 (2009)

Chapter 20
Biorefining: A Green Tribological Perspective

P. L. de Vaal, L. F. Barker, E. du Plessis and D. Crous

Abstract As a developing country, South Africa is in a unique position with regard to establishing, maintaining and expanding infrastructure to ensure compliance with international trends with respect to *environmental regulations*, while at the same time establishing the means to provide access to affordable energy to all its citizens to share the potential of its resources. In many respects, tribology plays an important role in saving of energy as well as ensuring that requirements with regard to protecting the environment are complied with. Green tribology can rightly be regarded as an approach which is timely and which has an impact on many activities like electricity generation, production of synthetic fuels and lubricants, mining operations and protection of the environment and its resources. Focusing on the interface between Tribology and *Biorefining*, several interesting possibilities open up. With the constant rise in the price of oil, alternatives to crude oil as primary energy source and as basic feedstock for fuels and chemicals are becoming more and more attainable. In this chapter an overview is provided of the above, from a South African perspective. A number of case-study

P. L. de Vaal (✉) · D. Crous
Department of Chemical Engineering,
University of Pretoria,
Pretoria, South Africa
e-mail: pdv@up.ac.za

L. F. Barker
Eskom Generation Business Engineering,
Johannesburg, South Africa
e-mail: leslie.barker@eskom.co.za

E. du Plessis
Producut Lubrication Technologies (Pty) Ltd,
Post Box 2309 Silverton, 0127, South Africa
e-mail: producut@icon.co.za

M. Nosonovsky and B. Bhushan (eds.), *Green Tribology*,
Green Energy and Technology, DOI: 10.1007/978-3-642-23681-5_20,

examples are given which indicate that a "green" approach in finding engineering solutions to tribological problems which could have a far-reaching impact on the environment. Three examples are used, namely how proper selection of tailor-made *lubricants* could decrease *energy usage* in gear-driven systems. The focus here is on the power industry, where coal-based power plants are the only economically feasible solution to the increasing demand for electricity in a developing economy with virtually no crude oil reserves. The success atttained in this endeavour should stimulate similar projects in the mining sector of the country. In the second instance, ingenious application of tribology with respect to application of specialised lubricants from a renewable source, namely plant oils, can decrease cost of lubrication and, in addition, can resolve difficult issues with regard to disposal of contaminated waste in *metal cutting* operations, indicating the value of a "green tribology approach". Thirdly, combining the concept of biorefineries, tribology and the ability to synthesise products to suit specific requirements, including formulation of lubricants and fuels, can lead to substantially improved products, impacting in a positive way on the environment.

20.1 Introduction

South Africa's economy is heavily dependent on coal. According to the World Energy Council (WEC) South African coal resources were estimated to be approximately 34 billion short tons at the end of 2007, accounting for 95% of African coal reserves and 4% of world reserves. Coal provided an estimated 72% share of the country's total primary energy supply in 2007 and accounts for approximately 85% of electricity generation capacity. Coal is also a major feedstock for the country's synthetic fuel industry.

It is understandable that coal should and will play a significant role in the South African industry. In addition, South Africa's petrochemical industry is largely based on conversion of coal to chemicals, via Sasol's *Fischer–Tropsch* process, which makes it unique in the world.

In 2008, South Africa produced 195,000 barrels per day (bbl/d) of oil, of which about 160,000 bbl/d was synthetic liquids processed from coal and natural gas. South African oil consumption is estimated to be 575,000 bbl/d, of which, approximately 380,000 bbl/d is imported (67% of consumption). South Africa has a highly developed synthetic fuels (synfuel) industry, creating liquid fuels from coal and natural gas. State company PetroSA and petrochemicals giant Sasol are the major players. PetroSA manages the world's largest commercial natural gas-to-liquids plant at Mossel Bay in the Western Cape, with a capacity of 45,000 bbl/d. Sasol, an oil, gas and chemical company, operates the world's only commercial coal-to-liquids synthetic fuels facility. According to the Department of Minerals and Energy, approximately 36% of liquid fuel demand in South Africa is met by synthetic fuels.

The concept of biorefining has been tested and tried in many ways—one of the main reasons holding back full exploitation of the concept is the economical feasibility of such processing when it has to compete with processes using crude oil. From an environmental perspective, the concept of a biorefinery satisfies many of the potential problems posed by conventional refineries—providing a means to convert renewable resources into chemicals of a widely-varying nature and only limited by the ingenious application of technology.

With rapidly dwindling resources of crude oil, utilising renewable feedstocks forms the key to the successful application of biorefining: If renewable feedstocks could be used, many of the problems related to diminishing resources would be solved. It is however important to relate the scale of petroleum- or coal-based refineries to the current potential of refineries using renewable feedstocks. Initially only in special cases, based on the market-demand for specialised chemicals, would a venture in this direction be economically feasible.

Processing alternatives are largely determined by technical and by economical viabiliity. It is always interesting to focus on benefits of existing technology, and how that can be extended or enhanced by new technology.

In the case of Sasol's *coal-to-lquid (CTL)* FT-based operations in South Africa, one of the benefits of producing liquid products starting from synthesis gas, is that there is potentially more control over the process of synthesising chemical compounds.

In addition, *gas-to-liquid (GTL)-technology*, also making use of the FT-process, enables the same benefit, which leads to production of petrol, diesel, jet fuel as well as polyalphaolefins, the base material for the production of a variety of lubricants.

Due to South Africa's unique position with regard to generation of electricity via primarily coal and due to current difficulties of generating sufficient electricity, many production plants in the country, now focus strongly on utilisation of excess, or unused energy sources, which are available in a variety of forms, depending on the type of operation, e.g. paper and pulp, smelters in the minerals processing industry, the agricultural sector, etc.

The term "green fuels" can be interpreted in a variety of ways: In the first instance it could point towards the origin of the fuels, i.e. not based on depletable resources, like crude oil, but rather based on a renewable resource, of an origin which could include a variety of vegetable oils, including rapeseed, soybean, sunflower seed, castor oil etc., but also several other possibilities, including sugar cane, cellulose and cellulose waste and a variety of algae, and then applying any of several processing routes, including fermentation, gasification and further modification into virtually any hydrocarbon-based commodity for which the demand can be identified.

An alternative interpretation of green fuels could be to focus on the effect that the products of combustion may have on the environment. To this end, several exciting possibilities are under investigation.

Current *fuel specifications* were developed around crude-oil-based products and engine technology of a bygone era. It took many years to ensure that well-controlled product ranges can be produced internationally with reasonable comfort and ease.

With stricter demands being placed on the effect that fuels and combustion gases have on the environment, several properties of fuels and lubricants need to be determined based on their desired behaviour in a much more demanding environment. This is in a process of giving rise to a brand new set of required product specifications.

With regard to fuel, South Africa has a lot of experience in the production of synthetic fuels via the Sasol-route. In the case of diesel fuel, the Sasol fuels are essentially sulphur-free. The lubricity of coal-derived diesel has to be improved by suitable additives. Active research in this field is ongoing locally. Successful development of synthetic jet fuel is another achievement of late.

Supplementing diesel by biodiesel, a renewable resource which is known to have good lubricity properties is an alternative approach that has to be investigated. Contrary to the approach taken in other parts of the world, considerable attention has to be given to the idea of using a feedstock that could have been used for human consumption, like maize and sunflower, which are the two crops cultivated locally with the best potential given the climatic conditions before this is converted and added to supplement fuel supplies.

20.2 Green Tribology with Applications to Biorefineries

Biorefining is the concept of processing renewable biomass feedstock into chemical products and energy. Unlike current commercial bioprocesses that mainly focus on producing a single product and treating everything else obtained as a by-product, in a biorefinery all the available feedstock should be utilised, gaining as much value out of it as possible. The *biorefinery* concept is very much like a modern day fossil fuel refinery, producing a broad range of products while using as much of the feedstock as possible. The main difference is however the fact that the feedstock of a biorefinery would be a renewable resource, while the products would be hydrocarbon-based, similar to a conventional crude-oil-based refinery [1].

The Sasol Coal-to-Liquid (CTL) plants in South Africa use the Fisher-Tropsch (FT) route to produce liquid hydrocarbons. To this end, *fuels*, namely gasoline, diesel and jet fuel (aviation turbine fuel) are produced, complying with international specifications, but with the added advantage that some of these products have improved performance characteristics, reducing the unwanted compounds in exhaust gases. Also, the potential of synthesising liquid lubricants via selective manufacturing, enables prodution of synthetic lubricants, which have tailor-made properties.

Biorefining as a concept is not new and references to this date back to the nineteenth century. As an example the German chemist, Mitscherlich, reported in 1878 how sugar, a waste product of the sulphite pulping process, can be fermented and how glue can be made from the waste liquor of a pulping plant [1]. The reason

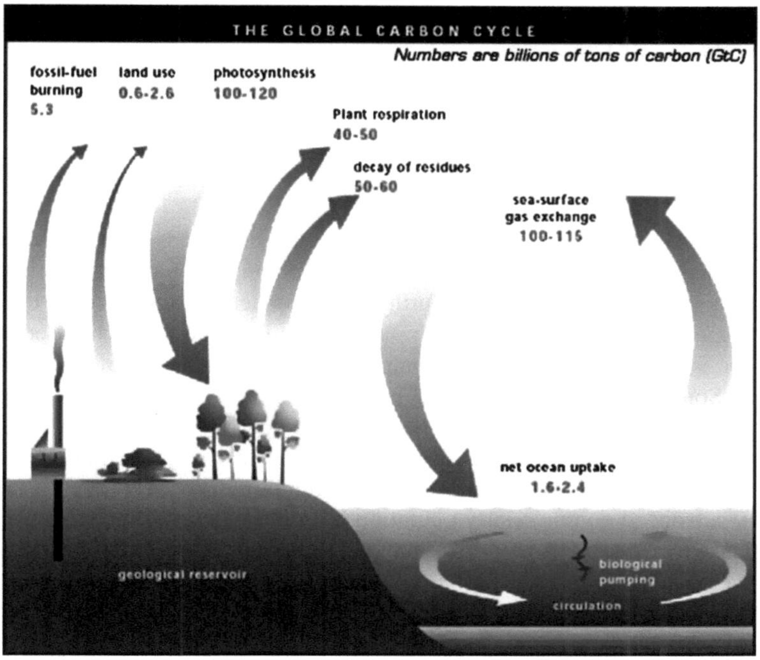

Fig. 20.1 Representation of the carbon cycle [4]

so much attention has turned to it lately is because of increasing environmental awareness and the realisation that fossil fuel supplies will run out.

When the word biorefining is heard, the product most commonly thought of is *biofuels*. The largest current producer of bio-ethanol is Brazil, producing a total of 27.5 billion litres of bio-ethanol during the 2008/2009 harvests [2]. The bio-ethanol is used extensively in their local industries, as a transport fuel and is exported around the world. The acceptance of ethanol as a transport fuel in Brazil has been extremely successful to the point where some manufacturers have made engines specifically for use with the ethanol blends available in Brazil. The total number of "flex-fuel" cars sold in 2008 accounted for 91% of all car sales in Brazil during 2008 [3].

While the use of fossil fuels creates an increase of carbon in the world's carbon cycle (Fig. 20.1) biorefining has the potential of being carbon neutral. Their products can also have direct green house gas emission reduction effects, as in the case of bio-ethanol where it has been shown that using ethanol in transport fuels reduce carbon emissions, even when used in a mixture (mixed with fossil fuels), by as much as 19% [3]. Bio-degradable bio-plastics, another potential product, have the obvious benefit to reduce plastic wastes. Some plants might be used to directly reduce CO_2 emissions, as in the case of algae which use photosynthesis to turn CO_2, sunlight and water into usable products. Current "wastes" that are rich in carbon and biomass all have the possibility of becoming feedstock, furthering the contribution of biorefineries.

20.2.1 Classification

According to Van Ree and Annevelink [5], the current bio-based industries (maize mills, pulp– and ethanol plants) are not seen as complete biorefineries. Even so they are a good starting point for further development as they use methods for separating and processing *biomass* similar to the ones expected to be used in new biorefineries but still require more complete use of their feed materials and side streams. Some of these current day industries are so similar to the expected biorefinery concept that research has been done to upgrade them to biorefineries by adding some additional processes that makes further use of their side streams, for instance heat and *power generation* using the lignin by-products [6, 7]. Currently the most advanced commercially operating biorefinery is owned and operated by Borregaard in Norway [8]. It uses lingo-cellulosic feedstock (wood) to create speciality cellulose, lignin, vanillin (by modifying lignin) and sugars.

The classification of biorefineries depends on their preferred feedstock and the main process routes involved. These classifications are as follows:

- Two platform biorefineries
- Lingo-cellulosic biorefineries
- Whole crop biorefineries
- Green biorefineries
- Aquatic/Marine biorefineries
- Thermo-chemical biorefineries

20.2.1.1 Two Platform Biorefinery

The "two platform" biorefinery can accommodate essentially any biomass type with high carbohydrate content. In this concept the feedstock is separated into a sugar platform chemical and the residues thereof (Fig. 20.2). This is obtained through enzymatic–, acid– or catalytic hydrolysis to saccharify the biomass.

The sugar side of the biorefinery has an immense range of possible products with varied options of processes that can be used to get to those products. Screening of the products showed that the best candidates are high-value products, as mass produced, low-value commodity products have low production costs using conventional production processes.

For the syngas platform the best products (short term) are hydrogen and methanol as production of alcohols, aldehydes, mixed alcohols and *Fischer–Tropsch* products are not competitive with their fossil fuel derived counterparts

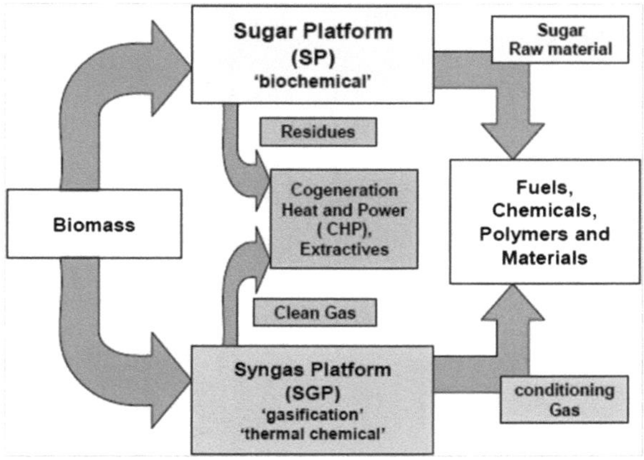

Fig. 20.2 Two-platform biorefinery [1]

at the current oil price [9]. Compared with coal, to create the same amount of fuel from wood would require almost 4 times the weight in wood at best, taking into account that natural wood is almost 50% (w/w) water and has a carbon content of around 48% [10], while coal has 80–99%. For fermentation only 20% of oven dried wood is usable carbon that can be turned into ethanol fuel [11].

20.2.1.2 Ligno-Cellulosic Feedstock Biorefinery

Ligno-cellulosic feedstock (LCF) biorefineries are good candidates for being one of the most promising types of biorefineries because of their favourable feedstock materials and product possibilities [1]. Their feedstock can be any kind of plant based biomass, from forestry and papermaking waste to agricultural waste like maize stover and straw. The reason for the LCF's wide range of feedstock usage is that it separates the feedstock into its main components namely cellulose, hemi-cellulose and lignin (Fig. 20.3). This allows for specialised reactions tailored to each of the components that can then produce a range of products from innovative bio-based chemicals to even traditional *petrochemical* based products.

While the array of products that can be produced is extensive, the state of lignin processing as of 2007 is not very efficient with respect to effective usage of lignin [12] as lignin is mostly used "as is" for fuel or as a binding agent in various applications. Processing of lignin could yield a lot of high-value aromatics that could significantly boost the profitability of its use in a biorefinery but the techniques to do so still need considerable development.

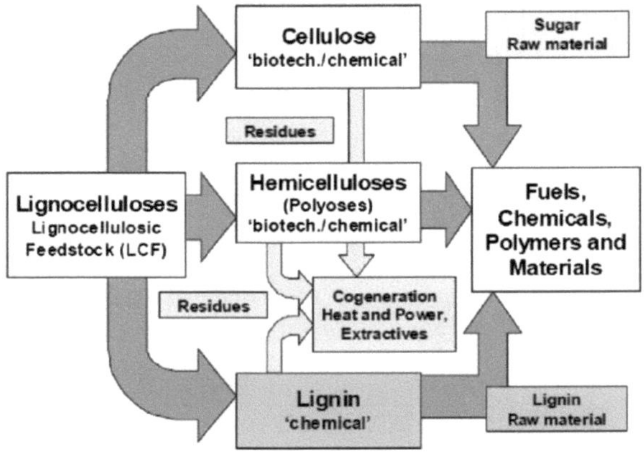

Fig. 20.3 Ligno-cellulosic feedstock biorefinery [1]

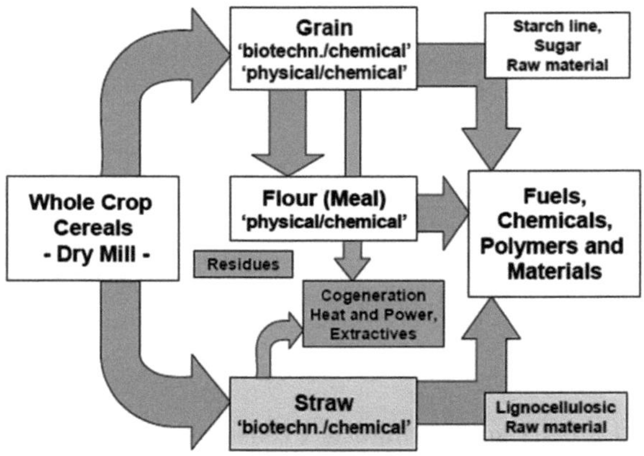

Fig. 20.4 Whole crop biorefinery overview (dry milling) [1]

20.2.1.3 Whole Crop Biorefinery

As the name suggests a whole crop biorefinery uses an entire harvested crop (rye, wheat or maize) along with all parts that would typically be seen as waste. For this reason the most common first step is to separate the “waste” straw (leaves, nodes, ears and chaff) from the grains (90:10 weight ratio, respectively) mechanically by dry milling it [1]. Each of the components can then be processed in a different platform (Fig. 20.4).

The straw side can be separated in the same way as with an LCF *biorefinery* and processed accordingly. Another option for the straw is to convert it to syngas using pyrolysis. The grain can be processed to a variety of products.

An alternative to dry milling is wet milling, where the grain is swelled and pressed to release high-value oils (Fig. 20.5). The advantage of wet milling is that the production of the natural structures (starch, cellulose, oil and protein) are kept high while the resulting processes used are common [1]. The products produced

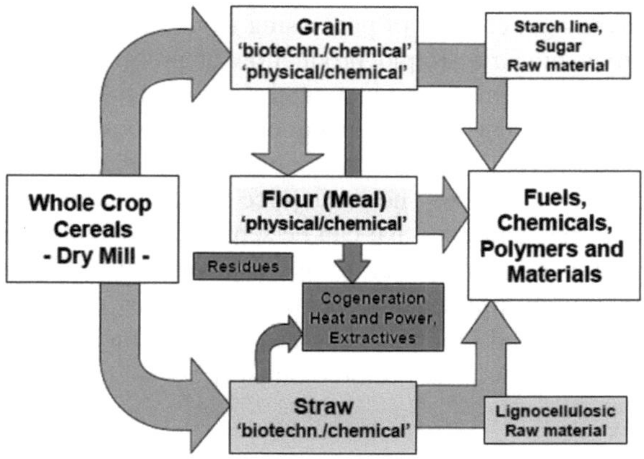

Fig. 20.5 Whole crop biorefinery (wet milling) [1]

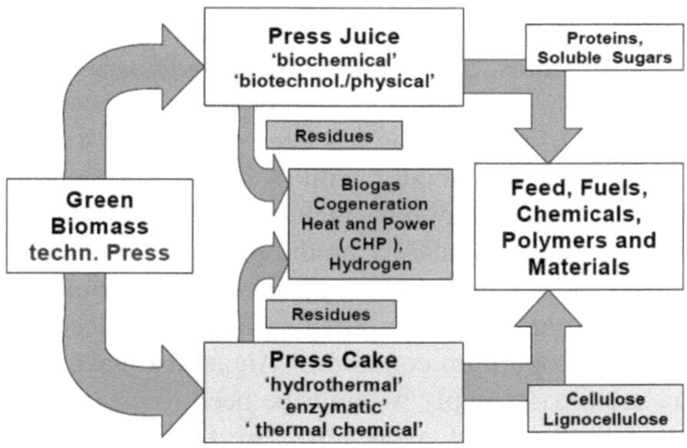

Fig. 20.6 Green biorefinery [1]

from wet milling are some high-value pharmaceutical cosmetic chemicals though the feedstock costs are quite high when compared to others. A major advantage is that this technology has been in use with maize as a feedstock since 1992 yielding primarily fuel alcohol, high-fructose syrup, starch and dextrose, amongst other products.

20.2.1.4 Green Biorefinery

A green biorefinery uses "green" biomass (grasses, lucerne, clover and immature cereals) in a process called green fractionation to produce two main platforms, a press juice platform containing mostly liquids and a press cake platform with the remaining solids (Fig. 20.6) [1]. Green fractionation is wet fractionation of green biomass and allows for the co-production of food and non-food products, a great benefit considering that a lot of the biomass used in various biorefineries could cut into their use as a food resource. The scalability of a green biorefinery is also a big advantage, capable of processing from a few tons per hour to 100 tons per hour of *green biomass.* Rapid production or preservation methods will have to be used with the primary products and the feedstock both of which various preservation methods have various effects on the composition of the materials.

Numerous products can be obtained from either the press juice, also green juice, or the press cake. The press juice contains large quantities of proteins, lactic acid, vitamins, and dyes that can be separated out and processed further. The press cake mainly contains crude drugs, pigments and other organic compounds that are commonly pressed into pellets for various uses but containing high amounts of cellulose and lignin that can be processed further, however it is not as accessible as that found in a LCF biorefinery [1].

20.2.1.5 Aquatic/Marine Biorefinery

Aquatic/marine biorefineries (AMB) use either macro-algae (*seaweeds*) or micro-algae (diatoms, green- and golden *algae* as well as blue-green algae, cyanobacteria) either as a feedstock and/or for the processing of CO_2. The CO_2 reduction potential of micro-algae has been known for some time and seaweeds have been exploited commercially with 8 Mt of wet biomass being produced in 2003. Products from seaweeds include food, feed, chemicals, cosmetic- and pharmaceutical products mainly produced in Asian countries. On average, seaweeds are more proficient at converting solar energy to biomass than land-based plants owing to their simpler cellular structure and easier access to CO_2, sunlight and nutrients given the optimum conditions. Algae also have the benefit of increasing a country's biomass supply without the need to clear land. [13].

Current uses of micro-algae are focused around supplemental and feed uses. Proposed micro-algae biorefineries fall in two main categories, in the one the algae are harvested and used as a biomass source (similar to what is currently done) while another option is to use modified algae that produce a specific product like ethanol that does not need harvesting. Algae can be grown in open pond systems or in enclosed photobioreactors that are either transparent or use light fibres or tubes to carry sunlight to the algae. It has also been suggested to use heterotrophic algae, which use the carbon as an energy source as well as a carbon source, in conventional fermentors. On the one side open pond systems are very cheap while the photobioreactors require a higher capital cost, however, the photobioreactors are much smaller, eliminate contamination and allow for precise control over conditions meaning better efficiencies.

Production of *bio-diesel* from micro-algae has been investigated but it was found that competing prices from other bio-diesel sources were just too low (less than €1/kg) [13]. It has also been shown using the best assumed growth rates, most favourable selling prices and the most optimistic green house abatement that it will still be infeasible to produce only fuels from the micro-algae though this could change with increased fossil fuel prices.

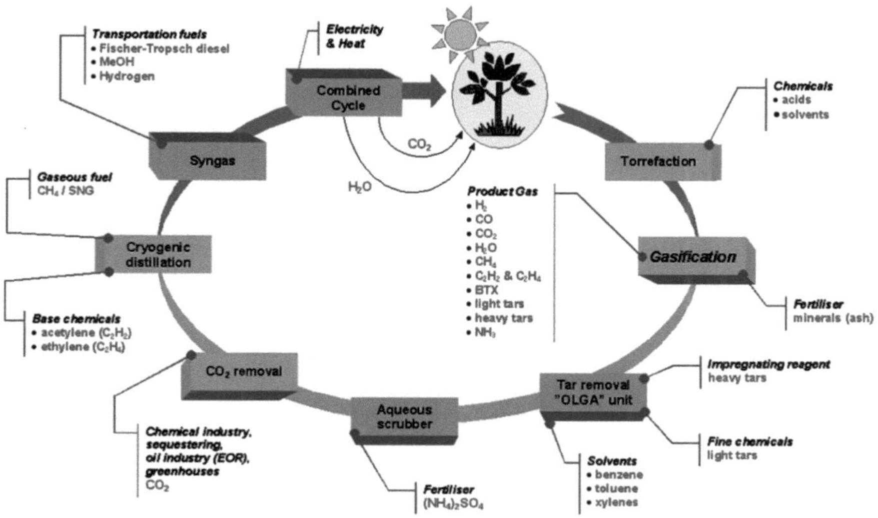

Fig. 20.7 Thermo-chemical biorefinery [5]

Other proposed uses for aquatic biomass include electricity generation through co-firing, biomethane production through fermentation and liquid fuel generation through pyrolysis (bio-oil). But producing algae solely for these purposes would still not make it economically viable. More promise has been shown in its ability to capture CO_2, possibly directly from the exhausts of industries and power plants, however screening is required for algae that can handle high CO_2 saturation levels. In the same way algae can be used to process waste water, removing excess nutrients, xenobiotics and heavy metals. High-value products that can be obtained from algae include methane, iodine, algin, mannitol, polymers, high-value oils, colourants, nanomaterials, new pharmaceuticals and cosmetics [13].

20.2.1.6 Thermo-Chemical Biorefinery

A thermo-chemical biorefinery (TCB) relies on a series of different processes such as pyrolysis, torrefaction (a milder pyrolysis performed on biomass) and gasification to process biomass into different products (Fig. 20.7). Even though all these processes work on more or less the same principles (elevated temperatures and pressure) each produces a different range of the products that can be taken off between the processes (Fig. 20.8) in what is also called the staged catalytic biorefinery.

An alternative option to having a TCB as a stand alone biorefinery is to incorporate it into a current petrochemical plant by using any biomass or biomass derived intermediates in the proposed TCB processes to supplement the use of petrochemical feedstock. Referred to as "green upgrading" this could be the first commercial use of the TCB [5].

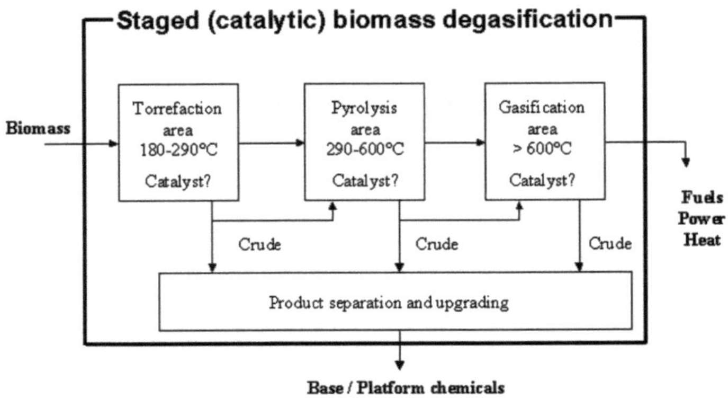

Fig. 20.8 Staged catalytic biorefinery [5]

20.2.2 Processes and Products

An overview of the different categories of *biorefineries* with respect to the feed materials, processes and status is given in Table 20.1. Once the feedstock has been fractioned into their platform chemicals/materials they are ready for further processing. The most common non-mechanical types of processes that are used can be categorised as biochemical or thermo-chemical processes.

The biochemical processes are a collection of processes derived from natural processes using enzymes, bacteria and yeasts. The possible products from biochemical processes are enormous and are mostly products from sugars, with the prime example being ethanol production. There are other high-value products as well, such as xylitol and furfural. Xylitol is a sugar alcohol made from the sugar xylose, which can only be found in plants, that has commercial application as a non-nutritive sweetener. Furfural is a sugar derived building block with a growing market where a wide range of new uses are being investigated [9].

Thermo-chemical processes, as the name suggests, require elevated temperatures and so more energy. The products from these processes are very much like those obtained from petrochemical feed-stocks such as fuels. Thermo-chemical processes include pyrolysis (no air), gasification (partial air), liquefaction (no air, high pressure), Fischer–Tropsch and combustion (excess air).

Depending on the residence time and temperature, *pyrolysis* can produce charcoal, char, bio-oil, chemicals and gas. The bio-oils obtained from pyrolysis are made up of a mixture of water, guaiacols, catecols, syringols, vanallins, furancaboxaldehydes, isoeugenol, pyrones, acetic acid, formic acid, other carboxylic acids, sugars and phenolics [14].

Gasification is similar to pyrolysis except that some oxygen and air is added to allow for partial oxidation of the products. Apart from chemical products the gas from *gasification* is an efficient source to generate electricity from, either by producing steam or directly in a gas turbine or engine, the alternative being

Table 20.1 Overview of biorefinery platforms [5]

Concept	Type of feedstock	Predominant technology	Development phase
Green Biorefineries (GBR)	wt biomass: green grasses and green crops, such as lucerne and clover	Pretreatment, pressing, fractionation, separation, digestion	Pilot plant (and R&D)
WholeCrop Biorefineries (WCBR)	whole crop (including straw) cereals such as rye, wheat and maize	Dry or wet milling, biochemical conversion	Pilot plant (and Demo)
Ligno-Cellulosic Feedstock Biorefineries (LCFBR)	Lingo-cellulosic-rich biomass: e.g. straw, chaff, reed, miscanthus, wood	Pretreatment, chemical and enzymatic hydrolysis, fermentation, separation	R&D/Pilot plant (EC), Demo (US)
Two Platform Concept Biorefineries (TPCBR)	All types of biomass	Combination of sugar platform (biochemical conversion) and syngas platform (thermo-chemical conversion)	Pilot plant
Thermo-chemical Biorefineries (TCB	All types of biomass	Thermo-chemical conversion: torrefaction, pyrolysis, gasification, HTU, product separation, catalytic synthesis	Pilot plant (R&D and Demo)
Marine Biorefineries (MBR)	Aquatic biomass: micro algae and macro-algae (seaweed)	Cell disruption, product extraction and separation	R&D (and Pilot plant)

combustion of the biomass to produce steam. Synthetic gas (syngas), a mixture of hydrogen and carbon monoxide, is a very flexible platform chemical. Depending on the ratio of these and sometimes the addition of water many different hydrocarbons (liquids, gases and solids (waxes) can be produced. The *Fischer–Tropsch* process uses syngas to produce a wide range of products, all of which are currently produced in the petrochemical industry.

20.2.3 Biomass

Feedstock to a biorefinery should be sustainable (renewable) meaning that unlike fossil fuels there should not come a time when it inevitably runs out. For this reason the prime feedstock for biorefineries is plant material. This biomass can be harvested or collected and then regrown, converting CO_2 in the air into sugars, oils and other useful chemicals, thereby staying carbon "neutral". Most of the suggested feed-stocks are already being used in various processes, e.g. woody biomass for pulp manufacture, sugarcane and maize for ethanol production and algae for their various uses. The leap to second and third-generation *biorefineries* (Table 20.2) will be to fully utilise the non-food feedstock potential and fine tune the processes involved. A third-generation demonstration biorefinery has been

Table 20.2 Classification of biorefineries based on their generation technologies showing their feedstocks and products [14]

Generation	Feedstocks	Examples
First-generation	Sugar, Starch, vegetable oils or animal fats	Bioalcohols, vegetable oil, biodiesel, biosyngas, biogas
Second-generation	Non-food crops, wheat straw, maize, wood, solid waste and energy crop	Bioalcohols, bio-oil, bio-DMF, biohydrogen, bio-Fishcer-Tropsch diesel
Third-generation	Algea	Vegetable oil, biodiesel
Fourth-generation	Vegetable oil, biodiesel	Biogasoline

commissioned but large integrated third-generation biorefineries are not expected to become established until 2020 [14].

Apart from water, the main components of plant matter are cellulose, hemicellulose, starches, sugars, lignin and extractives. These components are present in all plant biomass and vary in composition depending on the exact plant, which is the reason for all the different types of biorefineries.

Hemicellulose and *cellulose* are sugar polymers that make up the cell walls of plants. Cellulose has a rigid, crystalline structure made of β bonded (1–4) glucose molecules in a straight chain that is extremely resistant to hydrolysis. Hemicellulose, a branched polymer, is made up of different sugar monomers, among them are xylose (predominantly), galactose, mannose and arabinose. While not as strong as cellulose, hemicellulose is easier to hydrolyse and has a wider range of sugars to offer.

Lignin is a complex polymer with no set structure or pattern that provides bonding between different layers in a plant. It is made up of various aromatics which, if they can be hydrolysed without being destroyed could be valuable products. The technologies required to do so are still not fully developed and requires considerable research before they could be used in industry. Lignin as a whole is currently used as a bonding agent, compressed into pellets as an energy source, or directly burned to provide heat and power.

Extractives generally make up a small part of a plant and consist of fats, waxes, resins, photosterols and non-volatile hydrocarbons [Paperonweb]. Because they make up such a small part of the overall biomass they are largely ignored.

20.2.4 Biorefining in South Africa

Biorefining was first investigated in South Africa as early as 1970 by the CSIR when alternate sources of petroleum-based fuels were of great interest [15], but was abandoned after 1994 as an active area of research. With recent growing international focus on biorefining and utilization of biomass research in this field has seen a steady growth in South Africa.

As far as biomass production goes South Africa is in a very good position to be a competitor in *biorefining* with its range of climates for diverse agriculture and its established forestry industries. In 2005 South Africa was ranked 9th in the world for its maize (primary source of bio-ethanol production in the USA) production and 12th for its sugarcane (primary source of bio-ethanol in Brazil) production [16].

The forestry industry in South Africa is well established since the government started a plantation after World War 1 and the expansion by the two leading paper manufacturers (Sappi and Mondi) [17] in 1980. Currently South Africa has 1,2 million hectares of timber plantation producing a total of 18,5 million tons of wood used in a variety of industries, but primarily the production of paper and pulp, which has a lot of similarities with biorefining and would find adaptation to the biorefining industry fairly easy with their current operations in the biomass processing field.

While both the agricultural and forestry industries are very good sources for biorefineries South Africa also has a few non-conventional sources for biomass. One of these is the "Working for Water" incentive, started in 2005, where alien plant material is removed to increase water flow throughout the country. It has been estimated that the plant material removed up to date under this incentive could supply the entire paper and pulp industry for a year (almost 11 million tons) and could be used as an alternative biomass source for biorefineries, at least until the "Working for Water" incentive comes to an end [18].

The South African plants developed by Sasol utilising Fischer–Tropsch technology (High-Temperature Fischer–Tropsch, or HTFT), produce a product spectrum of products, including gasoline, diesel and aviation *fuel*, comlying with international standards, but with the added benefit that they contain virtually no sulphur, low levels of nitrogen-containing compounds, poly-aromatics, olefins, olefins and oxygenates, resulting in more environmentally friendly products than those produced via crude-oil refining [19].

Although the Sasol plants cannot be regarded as biorefineries, they are the largest plants in the world and have been producing almost 50% of the country's requirements in terms of fuel. The operational experience of using this technology can, when applied to biorefineries, can contribute to satifying a range of products which have to comply with ever-stricter *environmental* requirements.

20.2.5 Concluding Remarks

An overview of the current state of biorefineries and the technologies they need was given. While there are a lot of options not only in the types of biorefineries available, that will mostly depend on the feedstocks and products required, but also the current state of their development. While some of the processes suggested for these plants are in common use, such as hydrolysis, others are still in their research phase, like the suggested algae plants. The project aim of being able to select different processes should therefore take the level of development of each process into account and be able to discern between these development stages.

20.3 Reducing Power Consumption in Helical Gear Drives Using Synthetic Lubricants: A Case Study

The use of *energy* efficient gear oils to reduce power consumption has been tested and shown to produce significant and measurable results. In a program that included a laboratory trial and an extensive in-service trial energy reduction of at least 2.2% have been demonstrated in helical gearboxes.

The potential impact of these savings if expanded to South African Industry could save in excess of 200 MW. This saving could have significant impact on the need to install additional generating capacity in the future. This case study shows a detailed methodology for the laboratory and in-service testing of *lubricants* that may reduce energy consumption.

Energy consumption is an increasingly important topic globally. Increasing electricity prices driven by demand and supply constraints have forced companies globally to initiate projects and measures that will reduce specific power consumption. In South Africa, this problem has been highlighted by recent supply constraints in which power cuts have been experienced. The availability of electricity will remain under pressure for many years as the country puts in place measures to increase installed capacity and replace or maintain existing capacity. This situation has lead to a need for increased electricity costs to partly fund the new capacity requirements.

The sustainable reduction in *power consumption* by all is one part of an overall strategy to reduce the overall need for more power. All areas of consumption need to be analysed and every small incremental saving can contribute to a reduced new capacity requirement.

20.3.1 Eskom's Energy Efficiency Drive

Eskom have embarked on an external and internal energy efficiency awareness campaign and various programs have been implemented to encourage savings. In some cases, as with the solar water heating and energy efficient motor projects, financial incentives in the form of rebates or discounts have been put in place to encourage take up of new energy efficient technologies.

Internally Eskom has a savings initiative which aims to save 1 billion KWh of electricity. Most of the initiatives to date have focused on savings from electrical applications such as energy efficient lighting and thermal insulation in households. More recently this program has been extended to look at process efficiencies and auxiliary or internal power consumption.

This paper reports the results of the testing of various synthetic gear oils in helical gear drives. The project was initiated to confirm claims by numerous vendors as to the energy reduction benefits of synthetic gear oils. In addition it was felt that independently verified data would encourage not only the use of this technology in-house but also encourage general industry and other large lubricant users to implement the findings.

Lubricant specifications generally do not have any requirements for energy efficiency. As such there are relatively few oils designed specifically with energy saving as a primary objective. An exception would be in the automotive arena where energy efficiency is a component of various specifications. This requirement is driven by the need to reduce fuel consumption, which in turn is driven by environmental requirements, i.e., the need to reduce exhaust emissions. Even with this in mind, not many users purchase these oils with the intent to reduce fuel consumption. No such requirements are in place to drive energy consumption in the industrial arena; however some marketers of synthetic and high performance oils have been promoting this as part of the benefits of their oils for some time. Most synthetic oils are marketed on the basis of extending oil change intervals and reduction in wear. Due to high differential cost versus standard mineral oils, synthetic oils have also been confined to extreme operating conditions, mainly high temperatures applications and extremely low temperature applications. Synthetic oils, for example, have been used for many years in compressor applications and are now a requirement in many modern engine designs. Nonetheless the increasing availability and use of *synthetic oils* has brought with it the possibility of reducing power consumption via the use of these oils.

In Eskom, a study was undertaken to identify the areas where energy reduction could be implemented on a continuous and sustainable basis. The amount of savings also needed to be measurable to have a long term significant impact. In addition, the type of application selected should be generic enough as to be applicable across the business but also applicable to South African Industry in general.

The main types of lubricant used in Eskom are turbine oils, hydraulic oils and gear oils. These three oil types account for 90% of Eskom's lubricant consumption in the Generation Group.

From an analysis of lubricant usage and lubricant applications in Eskom, gearing applications were identified as a potential candidate. Literature surveys revealed little guidance as to the extent of savings actually achievable. Most data was produced by *lubricant* vendors in laboratory tests and the few in-service tests described concentrated mainly on worm gearing or single applications. Based on the known mechanical efficiencies of helical gearboxes, a saving of about 0.5–0.75% per gear stage was considered possible.

The air cooled condenser at Matimba was identified as an ideal candidate site to test oils in gearboxes due to the large population of gearboxes in the condenser. Matimba's air-cooled condenser (ACC) has 288 *gearboxes*. These gearboxes operate continuously and form part of the drive train that powers large 9 m diameter cooling fans. The fans are used to provide cooling air to the condenser radiator banks where steam from the turbines is condensed before returning to the boiler.

20.3.2 Oil Selection

Numerous oil vendors, representing a wide range of lubricant technologies are represented in the South African market. Eskom wanted to identify which technologies, if any, could provide significant and provable savings. A number of oil types were identified for inclusion in a laboratory screening test. The main constraints in oils selection were the budget allowed for testing and the specific application constraints in service. A range of oils representing various synthetic types and grades were identified. Semi-synthetic oil was chosen as well as some high performance mineral oils. The reference oil chosen for the laboratory trial was an SAE 50 diesel engine oil. The engine oil was included as a reference oil as most of the gearboxes were using this oil in service. The engine oil was used on the recommendation of the contracted oil vendor due to high operating temperatures experienced in-service. This oil was changed to a standard extreme pressure gear oil for the baseline in-service trial.

It must be noted that specific additive chemistries for test oils were not identified in detail but rather the general base oil and additive types. The performance standards to which each oil complies were identified via data submitted by various oil suppliers in the form of technical data sheets and material safety data sheets. No aftermarket additives were considered as the use of these products is generally prohibited in Eskom.

The costs of synthetic oils are always a concern but this was not used as a criterion in selecting oils. No attempt was made to entertain all the oils on the market but rather to select a range of technologies from various vendors. No sample donations were allowed and all lubricants were purchased via the normal channels.

20.3.3 Laboratory Testing

In order to identify potential candidate oils and to verify some of the oil company marketing claims, 19 oils of different *viscosities* and oil types were selected for a laboratory trial. Testing was performed on a modified *FZG machine* at the Tribology Laboratory of the Department of Chemical Engineering at the University of Pretoria [20]. The modified FZG method was initially developed by this laboratory as a part of an earlier Eskom project to test open gear lubricants [21, 22].

The standard FZG machine (Figs. 20.1, 20.2) was modified to allow for constant temperature operation [23]. The modification was intended to measure the wear performance of lubricants at a constant temperature through increasing loads. The modification allows cooling water flow rate to be adjusted according to load and lubricant tested. This modification allows for the measurement of energy usage via the requirement for cooling water. A heat balance was developed to

Table 20.3 Lubricants for laboratory testing

ID. No.	Type	ISO viscosity grade
LUB 1	Synthetic (PAO ester)	460
LUB 2	Synthetic (PAO ester)	320
LUB 3	Synthetic (PAO)	320
LUB 4	Mineral oil (Semi –synthetic)	220
LUB 5	Mineral oil (Diesel engine oil)	SAE 50 (ISO VG 220)
LUB 6	Mineral oil	460
LUB 7	Mineral oil	220
LUB 8	Synthetic (PAO)	460
LUB 9	Synthetic (PAO)	220
LUB 10	Synthetic	460
LUB 11	Synthetic	320
LUB 12	Synthetic	220
LUB 13	Mineral	460
LUB 14	Mineral	220
LUB 15	Mineral oil (plus moly additive)	220
LUB 16	Synthetic (PAO)	220
LUB 17	Synthetic (PAO)	460
LUB 18	Synthetic (PAO)	220
LUB 19	Synthetic (PAO ESTER Blend)	220

demonstrate that heat generated via friction can be equated to energy consumption. It was also found in the laboratory trial that the measurement of temperature or heat generated by friction is a better and more reliable indicator of energy savings potential [24]. This is especially true when savings are low and measurements of savings are affected by fluctuating electrical currents. FZG *gears* are intentionally "poorly" designed in order to produce a result in the laboratory test [25]. The result is less-than-perfect gear meshing and fluctuating electrical currents. Table 20.3 is a list of oil types and grades included in the laboratory test.

The test was designed to compare various technologies and oil grades against the traditional lubricant selection parameters of load, speed and temperature. Each oil was put through the same sequence of tests. Each stage of the test was defined by a specific load stage, a specific set temperature and a specific speed. At the end of each stage, one parameter was changed and the test continued. A total of 5 stages constituted a test for each oil. Each test sequence lasted 3½ h. The total cooling water usage required to calculate the energy removed was recorded for each lubricant (Fig. 20.9).

Due to the short duration of the test the mass loss was too low for all lubricants to be considered significant. In addition, the power measurements recorded could not conclusively differentiate between lubricants. The heat removed via cooling water however produced significant results (Fig. 20.10).

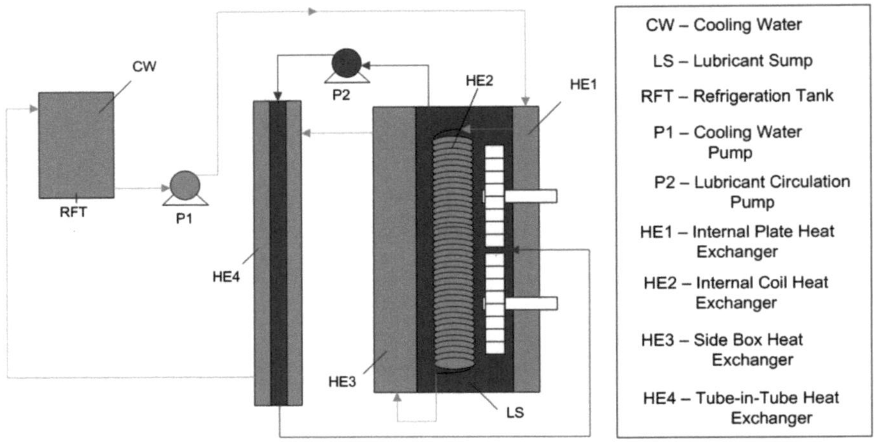

Fig. 20.9 The modified FZG layout showing cooling circuit and oil circuit [5]

Fig. 20.10 Standard FZG apparatus [25]

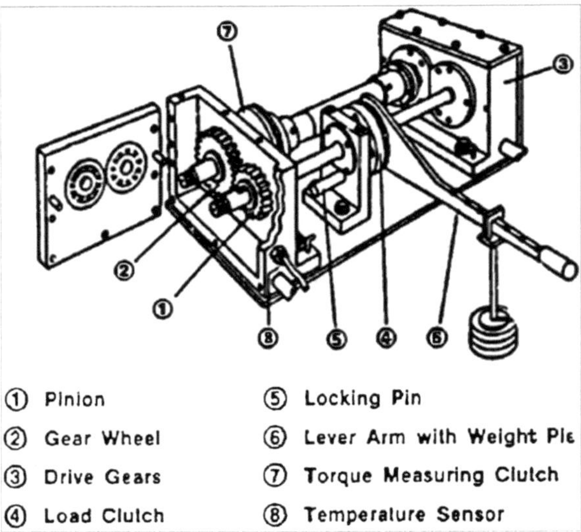

20.3.4 Laboratory Results

The graph shows that the heat removed differed significantly between the different lubricants. What was surprising was that not all *synthetic oils* performed well despite vendor claims. Nonetheless, in general the synthetic oils showed a lower energy requirement than the mineral oils. Some synthetic oils of the polyglycol type performed well but were excluded from the in-service trial for compatibility

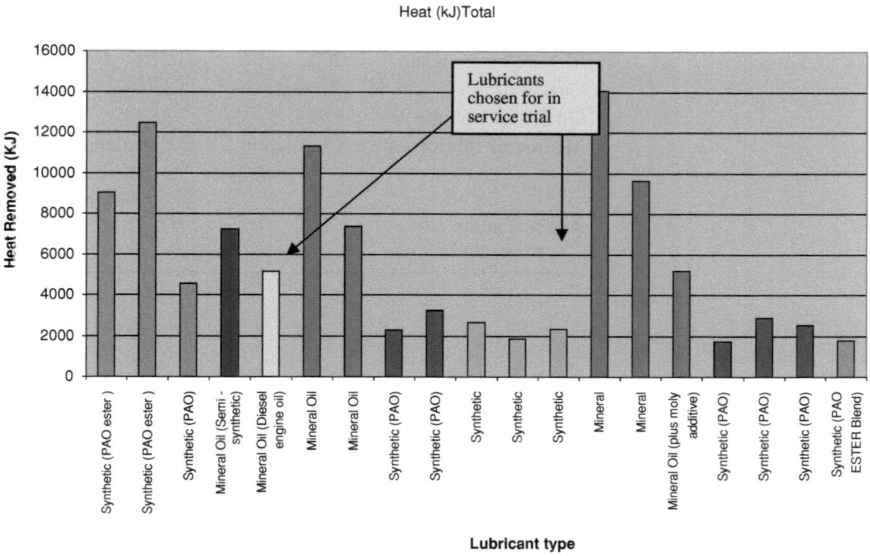

Fig. 20.11 Heat removed during laboratory testing

reasons. Three of the best performing polyalphaolefin type gear oils were selected for an in-service trial (Fig. 20.11).

20.3.5 In-Service Trial

The in-service trial of selected PAO gear oils was performed at Matimba Power station from June 2010 to December 2010. Three different oil brands were selected, with two different viscosity grades, an ISO VG 220 and an ISO VG 320.

Thirteen gearboxes, from two different gearbox manufacturers, were selected for the trial. Three older design *gearboxes* from manufacturer A and 10 new gearboxes from manufacturer B. The basic gearbox data is given in the table. Gearbox type A was originally installed during the 1980s and has been in service since then. Most of these gearboxes have been refurbished more than once and may be replaced in future by a more modern and efficient design. In order to ensure the trial was not compromised the 3 A type gearboxes were fully refurbished before the trial started (Table 20.4).

The 10 new gearboxes, of the B design, were brand new gearboxes of the same basic design intended for Eskom's two new coal-fired power stations, namely Medupi and Kusile *Power Stations*. These gearboxes are much lighter in weight and have a significantly lower oil volume compared with the original gearbox. Gearbox A has a 160 L oil sump whilst Gearbox B has a 35 L oil sump. Both gearboxes transmit the same power, approximately 220 KW.

Table 20.4 Test gearbox data

	Gearbox A	Gearbox B
Motor Power (KW)	270	270
Gear Ratio	11.84:1	11.76:1
Oil Sump Volume	160 l	35 l
Method of lubrication	Circulation, no cooling	Circulation with in-line filtration and no cooling
Average sump temperature	70°C	95°C

The basic test methodology was simply to test all gearboxes on standard extreme pressure gear oils and to measure the current drawn as well as the bulk oil and ambient temperatures. The data would be gathered after a minimum run-in period of about 1 month. After baselines were verified, the oil was to be changed to one of the identified synthetic oils. After a period of one month to allow for any changes and to comply with external measurement and verification criteria, the same data would be collected again.

Data collected included measurement of the current drawn by each motor as well as the measurement of bulk oil temperature and ambient air temperature. PT100 probes were installed in all test gearboxes and a data logger was attached to collect ambient air and bulk oil temperature readings at regular intervals. As with the *laboratory trial*, it was expected that the synthetic oils would show a significant, measurable drop in sump temperature. What was of more importance was the rise in oil temperature over ambient temperature.

Electrical current is measured continuously by the *power station* and recorded on the plant database. It is possible to obtain historical data for many years of operation as well as the average ambient temperature on each day. The ambient temperature is logged by the plant operators as the temperature has a direct influence on the volume and quality of cooling air available to the condenser. The power drawn by the fan drops as the temperature increases, and increases as the temperature drops. Essentially the power requirement is a function of air density.

An analysis of the current drawn on a number of *gearboxes* before the test showed that current could fluctuate as much as 5% either way, giving a power fluctuation of about 10%. This caused initial concern as the anticipated saving before the test began was expected to be in the order of 1–1.5%. There was some doubt as to whether an average current drop of 1.5% would be detectable in the scatter of measurements. Nonetheless, it was expected that due to smoothing of data, the oil temperature measurements would be a better and more reliable indicator of *energy savings*.

In addition to the temperature and current measurements, vibration baselines and oil samples were taken of the synthetic oils at the start of the trial. These gearboxes will be monitored using vibration analysis and oil analysis on a regular basis until an oil change is required. This is done in order to detect premature gear wear and the life of the synthetic oils in service at the operating temperature. This data will be used to determine planned oil change intervals. The oil change interval is also required to finalise the life cycle costs of these oils. To date neither oil nor vibration data show any significant or noticeable changes.

All test boxes were installed on different dates over a 3 month period. The main constraint being the availability of maintenance personnel to install gearboxes. Due to production constraints it was not possible to stop any particular generating unit to install 13 test gearboxes. As mentioned earlier, the station has 288 gearboxes installed in the air cooled condenser, and there is a regular failure rate due to the age of the station. As gearboxes "fail" or are removed from service for maintenance, they were replaced with test gearboxes. Every attempt was made to try and locate these boxes in the same general area but this was not possible. The result is that the running-in period of these boxes ranged from less than one month to as much as 3 months. This was considered insignificant based on the normal life of such a gearbox, i.e., greater than 100,000 operating hours.

All test boxes were filled with standard ISO VG 320 extreme pressure gear oil, typically meeting DIN 51517 Part 3 CLP requirements. The oil grade used was suitable for both gearbox designs. In fact both gearbox designs can operate on an ISO VG 220 oil. The intention of using the higher grade was to cater for the normally high operating temperatures in Gearbox A. Gearbox type A was found to fail the required thermal rating for the application but has remained in-service for many years with a reduced oil change interval.

All oil vendors participating in the trial were allowed to share data relevant to their oils being tested. In order to maintain objectivity and transparency, Eskom did not accept any oil donations and all samples tested were purchased through normal purchasing channels.

The current oil supplier decided to employ the services of an external M&V (measurement and verification) consultant to measure and verify the savings recorded during the trial. Eskom have established a process whereby *energy savings* need to be externally verified in order to be accredited. Although this was initially not a requirement for this trial, it was decided to allow the external consultant full access to all data. The consultant was trained and accredited by Eskom's demand side management process. The requirements imposed by the consultant were that data for baseline purposes needed to be gathered for at least one month prior to the oil change. In addition, it was interpreted that the gearboxes would have to operate for a minimum of one month before the new data was recorded. An additional requirement, again not part of this test, is that data has to be verified annually for a 5 year period after the change.

In the case of this trial, the intention was to establish whether energy could be saved by correctly choosing and using synthetic gear oils and to verify the extent of this saving. This objective was met as is discussed.

20.3.6 Power Savings

Data gathered during the trial was captured in weekly intervals due to limitations of the data recorders. There was no limitation on data available for current drawn; however occasional glitches resulted in corrupted data from time to time. The result is that to ensure repeatable and consistent results, data was gathered more than once over a six month period.

The graphs and tables show the average calculated power consumption for each gearbox for specific weeks during the trial; from the graphs it is evident that the power consumption using mineral oils was consistently higher than that of the synthetic oils. This data is actual data and has not been corrected for any ambient temperature influences. On average, a power consumption drop in the order of 4.64% was noted. Table 20.5 is a summary of *power consumption* averaged for all the data recording periods.

The power consumption shown in graphs (Figs. 20.12, 20.13) is the average for each gearbox for a specific week of recorded data. The weeks reported here coincide with the same weeks in which *temperature* data was logged. The external verification was based purely on electrical current data and did not coincide with the temperature data collection periods.

From the graphical results it is clear that power was reduced with the introduction of *synthetic oils*. The average weekly differences can be attributed mainly to differences in ambient conditions. No changes to the plant configuration were allowed when replacing the mineral oil with synthetic oils. The differences between the various gearboxes can be attributed to the location of the gearbox in the condenser. In this case air flows differ depending on the specific installed location within the condenser.

The average *power savings* documented by the external M&V verifier was 3.8%. From this perspective the results are in agreement with each other. It must be noted that the external verification used an average of one month's data for the base line and one month's data for the power consumption after the oil change. The periods chosen by the external consultant differed slightly from the results presented here, however after comparison of results it was agreed that the basic methodology used to analyse and interpret the data was consistent between the two parties.

In order to determine the clean or net energy savings it is necessary to take into account changes in efficiency of the motor and gearbox drive system due to the drop in power required. In effect the motor has a specific power factor versus load profile. Data on the specific motors showed that the motor average power factor may be affected by the change in required or absorbed power. In essence the motor becomes slightly less efficient resulting in higher power consumption. To allow for this, on advice from Eskom in-house electrical engineers, a power factor of 0.86 was used for all calculations using mineral oil. For synthetic oils, due to the drop in absorbed power, a power factor of 0.85 was used in power calculations. On average this would adjust the gross power saving by about 1%.

Table 20.5 Average power consumption for mineral and synthetic oils

Gearbox number	Manufacturer	Oil	ISO VG	Average power (kW) (mineral)	Average power (kW) (synthetic)	Ave min–ave syn	% diff
3-22-20	Gearbox A	Synthetic 220	220	245.38	234.91	10.47	4.27
3-22-50	Gearbox A	Synthetic 220	220	223.48	213.69	9.79	4.38
3-23-20	Gearbox A	Synthetic 320	320	220.35	210.07	10.28	4.67
3-23-60	Gearbox A	Synthetic 220	220	216.60	203.36	13.24	6.11
3-24-20	Gearbox A	Synthetic 220	220	228.04	217.05	10.99	4.82
3-24-40	Gearbox A	Synthetic 220	220	234.78	224.53	10.25	4.37
3-31-10	Gearbox A	Synthetic 320	320	224.69	214.23	10.47	4.66
3-31-20	Gearbox A	Synthetic 320	320	210.91	201.12	9.79	4.64
3-32-60	Gearbox A	Synthetic 320	320	210.30	203.28	7.02	3.34
1-22-50	Gearbox B	Synthetic 220	220	251.77	237.04	14.73	5.85
2-22-30	Gearbox B	Synthetic 320	320	204.13	195.81	8.32	4.07
3-31-40	Gearbox A	Synthetic 320	320	203.41	194.77	8.64	4.25
3-33-20	Gearbox B	Synthetic 220	220	228.83	217.58	11.25	4.92
							4.64

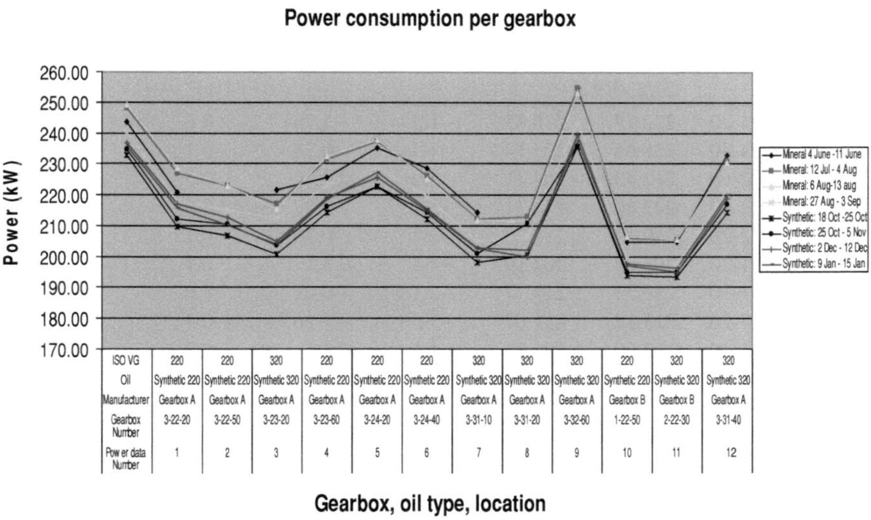

Fig. 20.12 Gross power consumption

In order to compensate for ambient *temperature* differences a factor of 0.3237% reduction was applied for every degree increase in average ambient temperature. This figure is derived from the change in air density per degree at an altitude of 1,400 m above sea level. This correction is in line with the air cooled condenser OEM design philosophy. In the case where ambient temperature dropped on average, the savings were increased accordingly. Where possible, data was

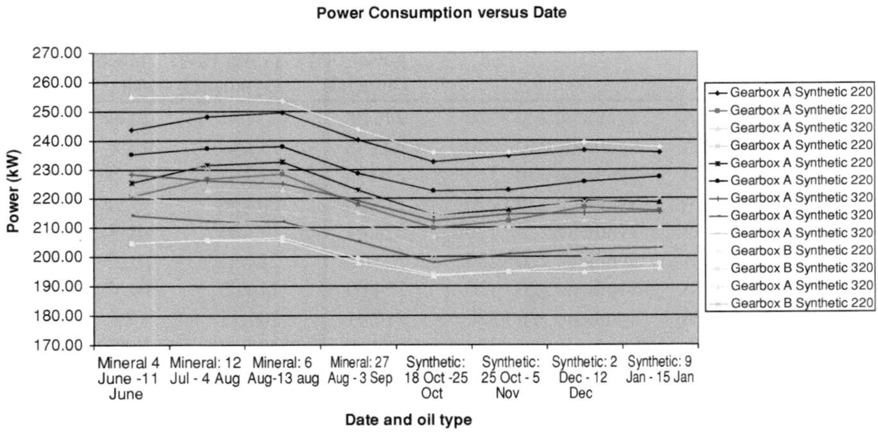

Fig. 20.13 Gross power consumption

Table 20.6 Corrected power consumption for end August and beginning December 2010

Oil gearbox number	Gearbox number	Power mineral oil 27/8 -3/9	Power synthetic oil	% drop in power	Change in ambient temp	Correction for change in ambient (%)	Nett Power Saving (kW) (%)
7	3-24-20	222.77	218.65	−1.8	−1.78	−0.6	2.4
8	3-24-40	228.66	225.14	−1.5	−0.38	−0.1	1.7
6	3-23-60	212.40	204.21	−3.9	0.91	0.3	3.6
4	3-22-50	243.80	238.71	−2.1	−0.07	0.0	2.1
3	3-22-20	240.19	236.20	−1.7	−0.36	−0.1	1.8
10	3-31-20	205.08	202.19	−1.4	0.37	0.1	1.3
9	3-31-10	219.33	214.77	−2.1	−0.01	0.0	2.1
5	3-23-20	215.12	211.87	−1.5	−1.26	−0.4	1.9
11	3-31-40	197.80	194.56	−1.6	0.49	0.2	1.5
12	3-32-60	206.64	199.33	−3.5	−0.16	−0.1	3.6
13	3-33-20	221.73	218.51	−1.5	−0.55	−0.2	1.6
1	1-22-50	243.80	235.62	−3.4		0.0	3.4
2	2-22-30	199.39	196.42	−1.5	−0.05	0.0	1.5

selected where ambient temperature would have had little or no effect on the resultant drop in power consumption. This approach is different to the formal M&V process in which only gross or actual power measurements are used. The results here are therefore conservative but realistic and in line with what were expected before the trial began.

Table 20.6 shows results for the last week of August 2010 and the first week of December 2010. These two weeks were chosen after carefully studying load and data from all gearboxes. The data for these weeks were not influenced significantly by load changes and data recorded was by and large free from anomalies.

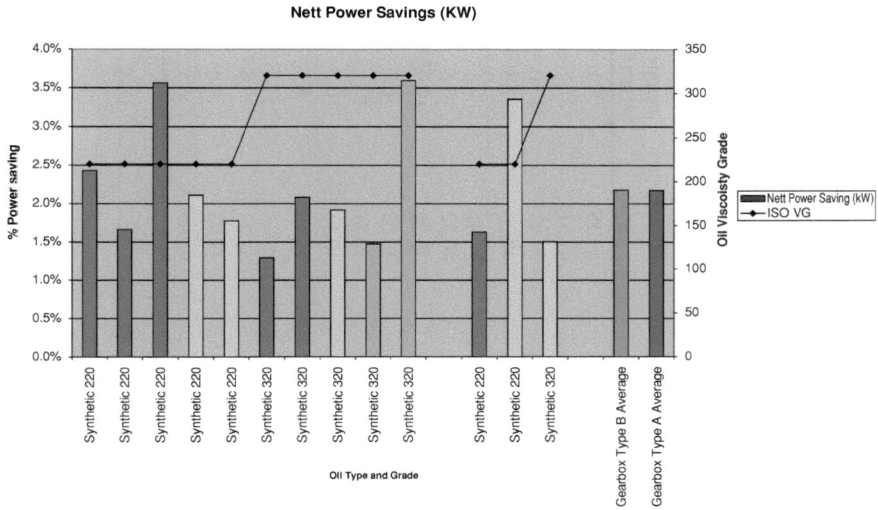

Fig. 20.14 Net power savings after temperature Correction

All *gearboxes* ran steadily at a relatively constant load for the entire duration of the trial period. In addition ambient temperatures were not significantly different.

From Table 20.6 and Fig. 20.14, the average nett power saving per *gearbox* can be calculated to be 2.2%. This is below the gross saving of 3.8% verified externally and below the average gross power drop of 4.64% using the actual current drawn or absorbed power figures. The nett saving is however in-line with the 1.5% savings predicted before the test program began.

20.3.7 Temperature Drop

In addition to measuring current, the bulk oil temperature and ambient temperature were recorded. From the results in Table 20.7 it is clear that the rise in temperature above ambient is significantly lower using synthetic gear oils. On average there is a 6°C drop in operating temperature. It appears that the percentage drop in oil temperature is roughly 3 times the drop in *power consumption*. There is no correction for this value as it can be seen from the graphical data that the operating temperature tracks the change in ambient temperature. This confirms that temperature can be a more reliable indicator of energy savings in applications where the percentage current savings are low.

Two different *grades of oil* were used in the trial, an ISO VG 220 and an ISO VG 320. From Fig. 20.15 it is clear that the lower viscosity grade provides additional power savings. This illustrates the need to ensure correct viscosity grade selection up front. Due to the higher viscosity index associated with the synthetic

Table 20.7 Bulk oil, ambient and oil temperature rise

Gearbox number	Oil	Bulk oil temp (mineral oil) °C	Ambient temp (mineral oil) °C	Oil temp rise (mineral oil) °C	Bulk oil temp (synthetic oil) °C	Ambient temp (synthetic oil) °C	Oil temp rise (synthetic oil) °C	% drop in bulk oil temperature rise (%)
3-24-20	Synthetic 220	78.83	29.13	49.70	70.73	27.35	43.38	9.1
3-24-40	Synthetic 220	76.48	29.15	47.33	70.28	28.77	41.51	11.5
3-23-60	Synthetic 220	75.16	30.20	44.96	69.08	31.11	37.97	17.6
3-22-50	Synthetic 220	71.83	28.42	43.41	70.58	28.35	42.23	2.5
3-22-20	Synthetic 220	75.39	28.23	47.16	70.37	27.88	42.49	9.1
3-31-20	Synthetic 320	71.34	26.98	44.37	68.17	27.35	40.82	8.8
3-31-10	Synthetic 320	74.77	28.55	46.23	68.79	28.54	40.25	12.9
3-23-20	Synthetic 320	72.80	29.13	43.66	67.37	27.88	39.50	6.7
3-31-40	Synthetic 320	71.93	27.77	44.16	68.47	28.26	40.21	10.1
3-32-60	Synthetic 320	74.82	29.39	45.43	69.66	29.23	40.43	10.7
3-33-20	Synthetic 220	94.00	28.36	65.64	89.50	27.81	61.69	5.2
1-22-50	Synthetic 220	92.03	26.85	65.18				
2-22-30	Synthetic 320	88.54	26.91	61.63	88.10	26.86	61.24	0.6
	Gearbox Type B Average	74.51	28.70	45.64	69.35	28.47	41.03	9.6
	Gearbox Type A Average	91.52	27.37	64.15	88.80	27.34	61.46	4.1

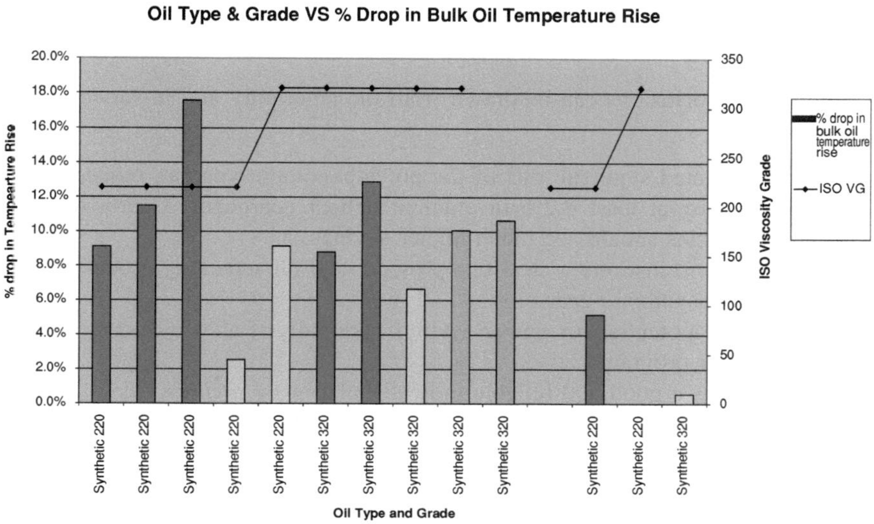

Fig. 20.15 Percentage drop in temperature using synthetic oils

oils and the lower operating temperature it should be possible to further reduce the oil grade and achieve higher savings without compromising gear protection.

As is clear from the results presented above, there is both a power saving and a temperature drop. The temperature drop should also equate to longer oil life and future reduced oil consumption in these gearboxes.

20.3.8 Cost Savings

The potential energy savings at Matimba have been calculated at 1,425 kW if all 288 gearboxes are converted to the *synthetic oil*. This equates roughly to 11,500,000 KWh per annum. In Eskom, the conservative potential savings could be as high as 20 MW or 140,000,000 MWh if all major gearboxes were converted to a suitable synthetic oil. One of the major considerations is that the use of these oils would require a significant change in mindset and maintenance practises to allow the oil to achieve its full life without the need for an early oil change.

In the South African market Eskom's oil usage is estimated at less than 0.5%, whilst the *mining industry* alone accounts for at least 15–20%. This makes the potential for energy savings potential a factor of 30 times higher in the mining industry. It is difficult to estimate all sectors but potential savings in gearing applications alone could exceed 200 MW and be as high as 600 MW. It is not likely that all these applications would be converted but even a low percentage uptake would have a significant impact in reducing power consumption consistently [26].

20.3.9 Conclusions

A number of conclusions can be drawn from the laboratory and in-service trials, they are listed:

- Correctly selected synthetic oils of the polyalphaolefin type can reduce power consumption by at least 2.2% in efficient helical gearboxes. Gearboxes with lower efficiencies should see much higher savings.
- It is advisable to institute a clear test protocol and oil screening process before implementing synthetic gear oils.
- There is huge potential for power savings nationally if all major users implement a similar program.
- Temperature is a good and reliable indicator of *energy savings*.
- Other application types should be explored and tested for energy savings.

20.4 Metal-Working in South Africa: A Green Tribology Perspective

South Africa is a country richly endowed with minerals and coal energy and with a modern industrial production base. This enables us to produce base metals such as steel, stainless steel, yellow metals (copper, brass, bronze etc.) and aluminum, ultimately ending up in manufactured final consumer or industrial products. From the virgin metal to the final product, diverse *metal-working* operations are performed which invariably involve the use of either neat or water soluble oils applied in copious quantities with often negative consequences to the environment. Traditionally, the lubricant was a mineral oil and the disposal of spent neat oil is by incineration in a conventional boiler. Disposal of spent water soluble oil however, is more of a problem because of smaller volumes and longer distances between source of oil and facility to separate the oil and water. Hence the use of evaporation ponds to dispose of such waste oil. Unfortunately, the easier but environmentally unacceptable way is to dispose of such waste by discharging it in an open drain or municipal sewer system. With the rapid urbanisation of our population and South Africa being a water scarce country, such disposal of neat or soluble oil has severe pollution potential for our rivers, dams and underground water resources.

The South African manufacturing industry, like elsewhere in the world, is confronted with the negative impact of spilled *lubricant* on the production floor influencing workers' health (skin disorders such as dermatitis, tramp oil mist etc.) and the cost of keeping production equipment and the production floor free from excess oil with detrimental consequences.

The cleaning of parts from the conventional neat/soluble oil manufacturing plant covered with liberal quantities of oil, results in higher cost of cleaning and requires costly cleaning machines, again impacts on the environment as spent washing fluid must be recycled or disposed of. Another cost factor is tool life, because the cost of tooling is playing an increasingly more important part of production cost as more sophisticated tools are continually being introduced with modern coatings. The change of tooling not only implies a direct cost, but also the cost of down time for tool replacement and the loss of production when running in the tool in ramping up to full production rate.

Being part of a highly competitive global production environment, South African industry is continuously under pressure to keep costs down. As the cost of lubricants, tooling and associated in the European field of automotive parts production approaches 20% [27], more and more attention will be given to cost saving lubrication techniques, including the cost of energy.

20.4.1 Introducing Minimum Quantity Lubrication

Seen against this background, a South African company, Producut Lubrication Technologies started with the introduction of *Minimum Quantity Lubrication (MQL)* in the early 1990s with a "home brewed" approach to address the realities of metal and part production in the African environment where realities are often remote from First World operations but not escaping the competition of a modern global manufacturing village.

The application on the technique starts with primary metal production (e.g. at smelters, steel rolling mills etc.) through semi-products (rolling mills, extrusion plants, pipe mills etc.) ending up as pipes, machined parts, fabricated products, produced via milling, turning, stamping, forging, pressing etc. in our homes, our buildings, our mines, our automobile plants, defence production plants, electrical industry etc. In almost all these metal-working operations, lubrication of some cost is required.

The introduction of the MQL technique was prompted to address the negative consequences of conventional lubrication by using extremely small quantities of lubricant, the cleaning up of the production environment and eliminating spent fluid disposal, reducing the impact on production personnel etc. Costs come down by using far less lubricant, having a better tool life, reduced part cleaning costs and no spent oil disposal costs. *Micro-dosing* also facilitates production logistics in view of small volumes to be handled. Energy savings are achieved by eliminating fluid circulation/filtering equipment and by keeping tools sharper for longer times. In short, the dramatic reduction of the lubricant carbon footprint and environmentally friendly production floor and of course, bottom line cost savings, are the real drivers for introducing MQL in the market place. Our case study (below), aims to practically illustrate these claims.

20.4.2 Use of Plant Oils

A further milestone in our journey to make a contribution to *green metal-working* [28], came with the introduction of plant oils and oils derived from plant (renewable) feed stock, such as synthetic esters. We started our MQL with conventional mineral oils but soon realised its negative qualities. In MQL, the lubricant is applied in an open environment, stressing the need for environmental–friendly qualities of a lubricant. We then changed to plant oils/synthetic esters to enjoy the benefits of almost zero toxicity, safety (high flash points), high smoke points, good biodegradability, wide temperature application ranges and of course good lubricating properties. This was verified by film strength determination in an Optimol 1 SRV® tribometer simulating MQL continuous flow on lubricant on the test piece. Furthermore, costs of plant oils compares favourably with that of mineral oil and is readily available from our soil. However, we had to contend with the negative features of plant oils i.e. its poor resistance to oxidation stability and polymerisation to create rubber-like deposits. Thanks to modern anti-oxidant packages, it is possible to inhibit these negative features substantially. When severe metal-working is performed in the boundary lubrication regime and high rates of metal removal are required, we blend in Extreme Pressure and Anti-Wear enhancing additives, meeting the US FDA's clearance for formulating FDA approved lubricants.

20.4.3 The Case Study

In the early 1990s [29] a trial was performed at an *aluminium smelter* on a continuous horizontal slab casting line, which was fitted with a circular saw ("old saw") and lubricated by soluble oil being sprayed liberally onto the blade. The production floor and environment was pretty wet and housekeeping was difficult and occasional incidences of skin disorders (*dermatitis*) were reported by operating personnel. Accurate data on operating parameters on the saw was recorded.

Then an identical new saw ("new saw") was fitted onto this line but this time with an MQL lube system supplying neat synthetic oil. Again accurate data on saw and lubricant was collected. Table 20.8 shows the cutting performance of both saws, cutting 99.7% aluminium cast ingot material, 490 × 1675 mm, using a TCT blade 1870 mm diameter with 18 mm cutting kerf. The cutting parameters (feed and speed) pertaining to the sawing operation was in essence the same for both saws.

The outcome of this trial bears out the virtue of a "green tribology" operation i.e.

- The volume of neat lubricant used was almost 600 times lower when using the MQL technique, drastic reduction in carbon foot print
- The cost implications of lubricant usage (71:1) speaks for itself, which makes the pay-back period for the MQL equipment installed on the saw months rather than years, making it the financial/production man's dream with a normal pay-back period of 3–5 years.

Table 20.8 Cutting performance of two saws

Parameter monitored	New saw (MQL lube)	Old saw (Soluble lube)	Ratio new:Old
Duration of trial	12 months	12 months	1:1
Tons sawn (Metric Tons)	43148	43148	1:1
Lubricant consumed (Litres)	110	65100	1:592
Cost of Lubrication (p.a.)	**$628**	**$44,880**	1:71,4
Ave Blade Changes (p.a.)	18	69	1:3,8
Blade repair cost (p.a.)	**$1,323**	**$14,857**	1:11.2
Lube Disposal cost (p.a.)	Nil	**$2,800**	–
Total Cost	**$1,951**	**$62,537**	1:32

Note: The above amounts refer to US dollars

- The significant reduction in blade changes (1:4) agrees with our findings in cutting operations. In our mind this can be attributed to the cutting oil with good lubricity, high film strength and anti-wear properties keeping the cutting tip sharp over longer periods.
- A last but significant advantage of *MQL* is that saw chips (18 mm kerf) can in the case of MQL be recycled by simply compressing chips into a block for downstream re-melting. In the case of soluble oil cutting, it was necessary to dry the chips before recycling could take place and is normally sold at scrap prices as drying costs were too high.

20.4.4 Practical Applications of MQL

20.4.4.1 The Aluminium Value Chain

Aluminium processing starts at the smelter producing molten aluminium which then goes to the cast house where it is cast in either slab or T-Bar in a horizontal casting machine or in a vertical cast into poles (Fig. 20.1). Our involvement started with mould *lubrication* of the casting of T-Bar material where lubrication plays an important role when just solidified T-Bar is moved from the mould and where the mould release properties of a lubricant are required to allow a smooth flow of aluminium in the cast. A break in the cast results in considerable cost, hence the necessity for reliable dosing equipment and lubricant being plant oil (Fig. 20.16).

Our real challenge came when we introduced MQL on a saw cutting cast ingot or slab, as described in the above case study. Again lubrication of the sawing operation was of major importance as the saw moved with the cast ingot from the melting furnace. A jam in the cut would result in aborting the cast and considerable cost of re-starting the cast cycle. We shortly afterwards introduced the MQL cutting on smaller circular saws at the cast house where pole-like solid aluminium castings is cut into logs for downstream use by extruders of aluminium profiles.

Fig. 20.16 Inside the T-Bar cast house (Courtesy: Hertwich Engineering)

Fig. 20.17 Scalping Alu cast ingot

At the aluminium rolling plant, cast ingot is rolled down to plate sheet and foil. The process starts with a cast ingot either from virgin material or from scrap, and the first step is to remove the hard aluminium oxide layer from the top, the bottom and the sides of the cast ingot. The operation is very severe as cuts of up to 18 mm deep are taken and the rate of metal removing is very high. When we got involved, the scalper in question used copious quantities of neat oil and it impacted on the production environment in that excess oil was not only a menace to the production floor, but it also caused air pollution when oil-soaked chips were burnt off in the re-melt furnace. Furnace linings also suffered as a result of this excess oil. We managed to "dry" out this operation and today *scalping* is performed with the MQL system and the problems pertaining to a clean environment and a recycling of chips without excessive cutting oil have been solved (Fig. 20.17).

On the rolling mills, MQL pays an important role as absolute minimal quantities of lubricant are required for the lubrication of wheel cutters without

Fig. 20.18 Cutting stacked Alu plate (Courtesy: Hulamin Ltd and Alu-Cut Inc)

Fig. 20.19 Edge milling of cast Alu plate on casting machine

depositing oil on plate etc. causing downstream stains resulting in rejected material.

When it comes to cutting the rolled aluminium with guillotines or with circular saws (Fig. 20.18), MQL plays its role of keeping the *cutting* edges sharp, to give a good finish to the cut, avoid the build up of aluminium on the tools whilst avoiding soiled products ready for packing to the end user. In modern day plate production, the plate casting technique is used where molten aluminium is directly cast into plate form in the casting machine and the plate exiting from the machine is trimmed on its edges with controlled MQL lubrication (Fig. 20.19). This eliminates the risk of *stains* on the material, protects the cutters, makes housekeeping around the machine easy and allows off-cut material to be recycled without drying.

Concerning the recycling of aluminum scrap, either from own production or from scrap metal vendors, the material has to be cut into small strips in a shredding machine and then into small blocks, ready for use in the re-melt operation. Without

Fig. 20.20 MQL in alloy wheel turning

lubrication, severe pick up of aluminium on the cutting tips of these two machines will occur, disrupting the production process.

In the re-melt of aluminium, an important feature of our additive package in the plant oil/ester lubricant is the absence of any metals to avoid metal building up in the molten re-melts, hence the importance of ash-less additives in the lubricant.

We introduced MQL to lubricate a set of large bearings on a rolling mill where contamination of the oil from the bearings soiled the rolled stock in the mill with dire financial consequences. We use air as a carrier and MQL as the source of minute quantities of stain free neat *lubricant* to keep the bearings properly lubricated, continuously replacing the lubricant inside the bearing

Now we reach the downstream processing of aluminium to create extrusions at the aluminium extruder company. MQL is imminently suitable to cut aluminum logs (diameter up to 213 mm) into smaller billets which will be converted into extrusions in the press The fact that the cut billet is free from moisture and almost oil free, facilitates the heating thereof before entering the chamber of the press. Cutting of extruded profiles into lengths at the end of the extrusion cycle is another excellent application of MQL, as cutting of profiles requires sharp cutting tools and very little lubricant with low staining properties, remaining on the cut profile. When profiles require heat treatment, the manual removal of excess oil to avoid lube stains on heat treated items is costly and time consuming. With MQL, we succeeded in avoiding this scenario.

A major project which we tackled was the replacement of soluble oil cutting with MQL in an alloy wheel manufacturing facility (Fig. 20.20). The cast wheel with 9–12% silicon content is highly abrasive and chips have the tendency to weld onto the *cutting tool*. Cast wheels are first cut to size in a turning operation and then stud holes are drilled in the casting. The rate of metal-working is very high and we had to contend with the problem of stringy chips interfering with the spray of the lubricant in the MQL mode onto the working surface. We employed a

Fig. 20.21 Special spray gun used in alloy wheel turning

special spray gun that was able to spray lubricant in accurately controlled drops onto the cutting face in a remote fashion (Fig. 20.21).

A major change was the introduction of MQL in high speed stud hole drilling operations. The rate of metal removal was high, the aluminium highly abrasive and removal of chips critical. We were however fortunate to supply lubricant in MQL mode through the centre of the drill and later contended with a faster drilling cycle by adding smaller quantities of water together with the MQL lubricant. The elimination of copious quantities of soluble oil as lubricant had a significant impact on the air quality in the plant, especially as hot humid conditions gave rise to rapid bacterial growth in the soluble oil. The ultimate benefit was the elimination of the frequent disposals of spent lubricant at considerable cost, and having an improved finish and better tool life as bonuses of the MQL venture.

The use of MQL *lubricated drilling*, tapping, milling etc. operations of *aluminium* materials in diverse engineering applications, is well established and in use in our local manufacturing industry.

20.4.4.2 Yellow Metals

Virgin copper is produced by electrolytic refining with *copper* cathode material leaving the refinery. When converting copper into flat semi's (sheet, plate etc.) or when producing profiles or tubes, MQL starts to play its role. The same applies to converting copper scrap into extruded or cast billet material. Cutting of hot cast billet material in a casting machine with MQL, proved to be a cost efficient way of changing cast ingot into logs. The elimination of water in a molten metal environment added to safety and good housekeeping. Like in the aluminium industry, cutting of cast logs into billets for extrusion operations (e.g. producing copper pipes, brass profiles etc.) was performed with MQL. Likewise, the cutting of small billets and the creation of small copper/brass products like valves, taps etc. by hot forging, is successfully performed by applying micro-quantities of lubricant, often replacing graphite in oil as lubricant, the latter producing unwanted carbon monoxide. When it comes to producing small delicate parts for use in electrical and defense industries, MQL proved its value in drilling, tapping and blanking operations to giving better tool life, improving the quality of the finished product and saving on part cleaning costs.

In coin production, chips from scalping of ingot (containing high percentage of nickel) can be economically recycled because minute residual lubricant from MQL making re-melt feasible whilst water wet chips requires costly drying before re-melting. Tool life also benefited from small drops of lubricant being applied in a continuous mode.

Where aluminium is a relative soft material with a low melting point, copper is a much harder material with higher proof stress and hence we had to change our *chemistry* of our lubricant to accommodate harsher conditions to perform the plastic deformation of the chip in the cutting zone. In particular, smoke point of the oil had to be carefully evaluated to avoid MQL contribution in a negative way to the health in the production environment.

20.4.4.3 MQL in Steel Working Operations

Steel finds wide application in our manufacturing industry ranging from automobile parts, construction, building, defense, building etc. At present, the use of MQL in *steel working* processes is limited to working of light products because of the limited cooling available when using MQL. High temperatures in the cutting zone [30] caused by the high proof stress of the steel being worked, coupled to the low heat dissipation characteristics of steel, mitigates against the use of MQL in working of solid steels. MQL however is very useful in cutting steel tube in tube mills as conventional water soluble fluid (mill fluid) enters the pipe often causing downstream spilling of fluid in handling/processing operations. In blanking and pressing of steel parts, MQL with plant oils is gaining ground as a substitute for conventional heavy chlorinated paraffins which are environmentally unacceptable substances [31]. However to cope with these stringent demands placed on plant oils when working steel, sulphurized esters and increased EP/AW additives were blended into the plant base oil, This lubricant formulation has been in use for more than a decade with much success.

Robotic drilling and cutting of a steel parts using the MQL technique is very successful as the dispensing nozzle(s) moves freely with the robotic arm as there are no circulation of fluid or spillage of excess oil on the parts/robot.

20.4.4.4 Stainless Steel and MQL

A very successful product range being produced in South Africa, is catalytic converters and exhaust systems for automobiles produced from stainless steel. As the platinum, the chrome and nickel all comes from our soil, cutting and pressing of parts for catalytic converters are widely performed using the MQL technique.

We often use remotely positioned *spray* guns to shoot accurately controlled minute drops of lubricant into the press thereby eliminating the severe oversupply of lubricant and resulting negative consequences to the environment and the quality of the pressing. In multi-stage tooling presses, lubricant very often does not reach critical parts of the product or tooling i.e. lubricant starvation occurs, again causing reduced tool life and/or unacceptable quality of pressings. The remotely placed gun, offers a solution when the press opens for short time duration. In general, stainless tube cutting has benefited significantly by using MQL, cutting down cost in tooling and lubricant usage.

The problem of high temperatures in the cutting zone, as with steel, again limits the application of the MQL technique. However we are encouraged by the work of Autret et al. [32] who evaluated MQL vs. flood cooling in working of hardened bearing grade steel showing the technical feasibility of MQL. This was confirmed by Beno et al. [33–36], who have shown that it is possible to work difficult-to-machine metals such as Ti-6Al4 V, Waspalloy, Inconel 718 and Greek Ascaloy, using the MQL technique.

20.4.5 The Future

As local legislation pertaining to the *environment* and in particular to protect our scarce water resources, becomes more stringent, more attention must be given to how we can widen the scope of MQL application. Two areas are being investigated:

- First, we are studying the concurrent application of minute drops of water with the lubricant to use the latent heat of evaporation to bring down the temperature of the steel/stainless steel metal removal/forming operation, without ending up with a wet production environment. The work by Wang et al. [37] and Joubert [38] give good indications that this approach can be applied in practice.
- Second, we must continue our search for better lubricants without sacrificing our commitment to the environment. In this regard, the use of nano-sized solid particles dispensed in the plant oil [39] is an area of hope in our potential quest for a cleaner and sustainable environment.

References

1. B. Kamm, P. Gruber, M. Kamm, *Biorefineries—Industrial Processes and Products, Status Quo and Future Directions*, vol. 1 (Wiley-VCH, Weinheim, 2006)
2. UNICA, Brazilian Sugarcane Industry Association, "2008/2009 Harvest" (2009), http://english.unica.com.br/dadosCotacao/estatistica/, 01 Dec 2010
3. J. Martines-Filho, H.L. Burnquist, C.E.F. Vian, Bioenergy and the rise of sugarcane-based ethanol in Brazil. Choices **21**(2), 91–96 (2006)

4. Carbon Recycling, The carbon cycle (2010), http://www.carbonrecycling.co.uk/http://assets/images/carbon_cycle.jpg , 28 Nov 2010
5. R. Van Ree, B. Annevelink, Status Report Biorefinery 2007, Report for SenterNovem and the Dutch Ministry of Agriculture, Nature and Food Quality (2007)
6. M. Marinova, E. Mateos-Espejel, N. Jemaa, J. Paris, Addressing the increased energy demand of a Kraft mill biorefinery: the hemicellulose extraction case. Chem. Eng Res. Des. **87**, 1269–1275 (2009)
7. S. Consonni, R.E. Katofsky, E.D. Larson, A gasification-based biorefinery for the pulp and paper industry. Chem. Eng. Res. Des. **87**, 1293–1317 (2009)
8. The world's most advanced biorefinery, http://www.borregaard.com/eway/default.aspx, 15 Dec 2010
9. T. Werpy, G. Peterson (eds.), Top Value Added Chemicals From Biomass, Volume I: Results of Screening for Potential Candidates from Sugars and Synthesis Gas, Report prepared by members of NREL and PNNL (2004)
10. S.H. Lamlom, R.A. Savidge, A reassessment of carbon content in wood: variation within and between 41 North American species. Biomass Bioenergy **24**(4), 381–388 (2003)
11. J.I. Zerbe, Liquid fuels from wood—ethanol, methanol and diesel. World Resour. Rev. **3**(4), 406–414 (1991)
12. J.E. Holladay, J.J. Bozel, J.F. White, D. Johnson, Top Value Added Chemicals From Biomass, Volume II – Results of Screening for Potential Candidates from Biorefinery Lignin, Report prepared by members of NREL, PNNL and University of Tennessee (2007)
13. D. Bowles (ed.), Micro- and macro-algae: Utility for industrial applications, Outputs from the EPOBIO project (2007)
14. A. Demirbas, *BiorefineriesBiorefineries: For Biomass Upgrading Facilities* (Springer-Verlag London limited, 2010)
15. L.R. Lynd, H. von Blottnitz, B. Tait, J. de Boer, I.S. Pretorius, K. Rumbold, W.H. van Zyl, Converting plant biomass to fuels and commodity chemicals in South Africa: a third chapter? S. Afr. J. Sci. **99**, 499–507 (2003)
16. Major Food And Agricultural Commodities And Producers; Food and Agricultural Organisation of the United Nations http://www.fao.org/es/ess/top/commodity.html?lang=en&item=156&year=2005, 06 Dec 2010
17. R.M. Byrnes (ed.), South Africa: A Country Study. Washington: GPO for the Library of Congress, (1996), http://countrystudies.us/south-africa/67.htm, 06 Dec 2010
18. Working for Water: A South African Sustainability Case, http://www.unep.org/training/programmes, 04 Dec 2010
19. B.I. Kamara, J. Coetzee, Overview of high-temperature Fischer-Tropsch gasoline and diesel quality. Eng. Fuels **23**, 2242–2247 (2009)
20. P.L. de Vaal, H. Guidi, Eskom Internal Research Report, A Test Method to Measure Heat Loss Through Friction Using the FZG Machine (2009)
21. H. Guidi, P.L. de Vaal, L.F. Barker, Lubricant performance comparison using frictional heat characteristics on a laboratory test rig; South African Chemical Engineering Conference (SACEC), Cape Town, 20–23 Sept. 2009
22. P.L. De Vaal, L.F. Barker, Quantifying friction-reducing properties of gear oils at constant temperature with a modified FZG test; Nordtrib 2010 Conference, Lulea, 8–11 June 2010
23. P.L. de Vaal, B.M. Meister, A constant temperature FZG test for open gear lubricants; Proceedings of the 3rd World Tribology Congress, Washington 12–15, Sept 2005
24. H. Guidi-Figueroa, P.L. de Vaal, The effect of operating temperature on open gear lubricant performance; Proceedings of the 9th SAIT International Tribology Conference, "Tribology 2008", University of Pretoria Conference Centre; Pretoria, 2–4 April 2008, (ISBN: 978-0-620-38082-9)
25. DIN 51354, Testing of lubricants, Mechanical testing of gear oils in the FZG rig test machine (1970)
26. P.L. de Vaal, H. Guidi-Figueroa, L.F. Barker, Synthetic Lubricants do reduce friction–but what about their Environmental Impact? SAIT Seminar on Lubricants and the Environment, South African Institute of Tribology, 6 Oct 2009

27. F. Klocke, G. Eisenblätter, Dry Cutting. Annals of the CIRP **46**(2), 519–526 (1997)
28. P. Jost, Green Tribology. A footprint where economics and environment meet, Address to the 4th World Tribology Congress, Kyoto (2009)
29. E. Du Plessis, Green Tribology. Environmentally friendly micro-dosing of lubricants in Aluminium metal-working operations–a case in point, *SAIT International Symposium*, Pretoria 2011
30. E.M. Trent, *Metal Cutting* (Butterworths, London, 1977)
31. N. Bay, A. Azushima, P. Groche, I. Ishibashi, M. Merklein, M. Morishita, T. Nakamura, S. Schmid, M. Yoshida, Environmentally benign tribo-systems for metal forming. CIRP Ann.–Manuf. Technol. **59**, P771–P776 (2010)
32. R. Autret, S.Y. Liang, Minimum Quantity Lubrication in finish hard turning, Proceedings of the HNICEM International Conference, March 2003
33. T. Beno, M. Isaksson, L. Pejryd, Investigation of Minimal Quantity Cooling Lubrication in Turning of Inconel 718. ICTPM, Yokohama, pp. 281–286 (2007)
34. T. Beno, M. Isaksson, L. Pejryd, Investigation of Minimal Quantity Lubrication in Turning of Waspaloy. CIRP-LCE, Tokyo, pp. 196–202 (2007)
35. T. Beno, M. Isaksson, L. Pejryd, Machining of Greek Ascaloy with the use of Minimal Quantity cutting fluid, CIRP-LCE 2008, Sydney (2007, to be published)
36. T. Beno, M. Isaksson, L. Pejryd, Investigations of the effects of minimal quantity cutting fluid in turning of Titanium 6–4, 9th International Tribology Conference of the South African Institute of Tribology (2007)
37. S. Wang, A. Clarens, Modelling metalworking fluid penetration in the cutting zone to understand EAL, ASME/STLE International Joint Tribology Conference, Oct 2010
38. H.J. Joubert, High speed machining of aluminium alloys using Minimal Quantity Lubrication Methods, B Ind Eng Thesis, Stellenbosch University, Oct 2006
39. K. Mistry, A. Erdemir, W. Zhang, A. Malshe, Tribochemical interaction of Boron and MoS2 based nano-colloidal lubrication under severe tribological conditions, ASME/STLE International Joint Tribology Conference, Oct 2010

Chapter 21
Tribological and Mechanical Design Considerations for Wave Energy Collecting Devices

Andrew Fronek, Michael Nosonovsky, Ben Barger and Ilya Avdeev

Abstract Creating new sources of renewable and sustainable energy is a critical problem in the world today and will continue to be so in the future. A major sector of renewable energy is the hydro energy, and in particular, harvesting wave energy of seas and lakes. Various types of designs have been suggested for marine energy sources. These new applications bring new problems for mechanical engineers and tribologists. Operating in marine conditions is difficult due to the biofouling, water contamination, particulate contamination, and other tribological issues. Efficiency and optimization are major issues in the energy generation industry are. In this chapter, we review offshore marine and lake wave energy systems and discuss in more detail wave energy collector (WEC) that is being developed and optimized at the University of Wisconsin-Milwaukee (UWM).

21.1 Introduction

A priority of many countries is to find reliable and sustainable energy sources to combat increasing energy demands, the threat of climate change, and dwindling fuel reserves [17]. Electricity is the primary source of consumed energy in the United States and across the globe, so developing an ecologically clean method of conversion of mechanical energy into electrical energy, in particular renewable and alternative methods of energy harvesting, is now a very important issue for the current technology market. The interest in promoting the use of renewable energy in electricity

A. Fronek · M. Nosonovsky (✉) · B. Barger · I. Avdeev
Department of Mechanical Engineering, College of Engineering and Applied Science, University of Wisconsin-Milwaukee, Milwaukee, WI 53211, USA
e-mail: nosonovs@uwm.edu

M. Nosonovsky and B. Bhushan (eds.), *Green Tribology*,
Green Energy and Technology, DOI: 10.1007/978-3-642-23681-5_21,

production has existed for a number of years and even decades in the United States; however, the market share of non-hydro renewable energy sources of electricity production has remained at about 2% over the past decade [11]. Furthermore, the United States currently relies heavily on non-renewable fossil fuels for most of its energy needs. The share of net electricity generation in the United States produced from renewable sources, including pre-established hydropower, declined from nearly 12% in 1990 to about 9% in 2002, indicating an increase in fossil fuel-based electricity energy generation during that period [15].

There are many economic, environmental, and social reasons to consider alternative sources of energy. Current electricity production systems usually meet the energy demands. However, other critical issues, conserving finite natural resources, maintaining stable and reasonable electricity costs, and protecting the environment and human health, are potentially affected by the non-renewable energy. As a consequence, new policies are emerging to address the issues of using new technologies, upgrading efficiency and developing new sources of electricity [9]. At this time, the electricity market does not yet fully recognize the possibility that current electricity generation methods may impose a heavy cost on the economy, the environment, as well as society as a whole through greenhouse gas emissions and other costs associated with the long-term damage to the environment [8]. However, the change is happening and increasing public attention is paid to green and sustainable energy technology.

Wind, hydro, biomass, wave, tidal, stream, and solar energies constitute emerging methods of renewable energy production [8]. These methods serve as alternatives to the current usage of non-renewable fossil fuel, and they can be utilized to satisfy energy demands while protecting the environment [15]. It is expected that, within decades, many nations will be forced into establishing significant alternative energy technologies because of economic and geopolitical constraints on non-renewable resources [14]. The current and future prices of CO_2 emissions is another major incentive to invest in these emerging technologies. A new generation of green technology provides a significant reduction to the various economic, political, and environmental pressures incurred by CO_2 emissions [9].

Among alternative and environment-friendly sources of energy, marine energy technologies constitute a major resource to be tapped in the future to cut down energy costs and emissions [17]. Marine energy systems are sustainable and provide an energy source that does not cause significant environmental harm. Harvesting the energy of waves and currents present one of the least impactful opportunities to produce large-scale amounts of electricity without hampering the use of land for other, non-associated ventures. Tribological issues, including friction, wear, and lubrication, play a very significant role in the design of these energy devices.

21.2 Marine Energy Sources

In this section, we briefly review various types of marine energy devices.

21.2.1 Tidal Turbines

Marine currents and waves can be used for reliable energy extraction. Many design approaches have been suggested and studied for marine current energy conversion, although most are based on an underwater horizontal axis tidal turbine (HATT). Alternative concepts include the marine current turbine [5], as well as vertical axis and oscillatory foil machines.

The HATT devices extract kinetic energy from the tidal flow through moving a lifting surface (blade or foil) connected to an electrical power generator. As tidal turbine developments lag those of wind turbines by about 15 years, considerable experience and knowledge is being transferred from wind turbines [3] as well as from the understanding of ship propellers [4]. Wood et al. [17] discussed the similarity and differences between the use modern wind turbines and HATT devices. Medium-sized wind turbines are operating routinely at similar power and/or torque levels to the generic tidal device, so certain major components or subsystems (e.g. gearbox) could be translated into tidal current service with little or no modification. However, tidal turbines operate in very different conditions. For example, tidal current velocities are predictable over long timescales and are not subject to random excursions over these values, whereas wind velocity is correlated to a series of stochastic processes, including random gusts. As a result, the tidal turbine needs lower structural margins on rotor blades. Conversely, the tidal turbine working environment is corrosive, with suspended solids leading to the possibility of erosive damage over the lifetime of the device. Another important consideration for tidal turbines is the presence of marine life [17].

The field of tribology of Wind turbines has been developing for a number of years [7], and several approaches can be borrowed from that field. Despite that, Wood et al. [17] pointed out that there are many differences between the wind and tidal turbines. Typically, there are only two tidal directions, which are usually close to 180° apart. Topography constraints due to a finite depth, channel width, as well as the presence of the free surface constrain the tidal volume flow rate, furthermore, the vapor pressure of water limits the rotational speed of the blade before significant cavitation effects cause a rapid increase in blade drag and loss of lift. Centripetal loads dominate in the long and slender blades of wind turbines which restrict bending. In water turbines, however, bending loads dominate. For a 1 MW tidal turbine the thrust is about 100 t, which is twice that of similarly rated wind turbine [17].

21.2.2 Water Current Energy Extractors

A different type of device is intended to collect the energy of water currents. This can be a horizontal-axis turbine or vertical-axis turbine. Among the vertical-axis turbines, the notorious example is the Gorlov helical turbine (GHT),

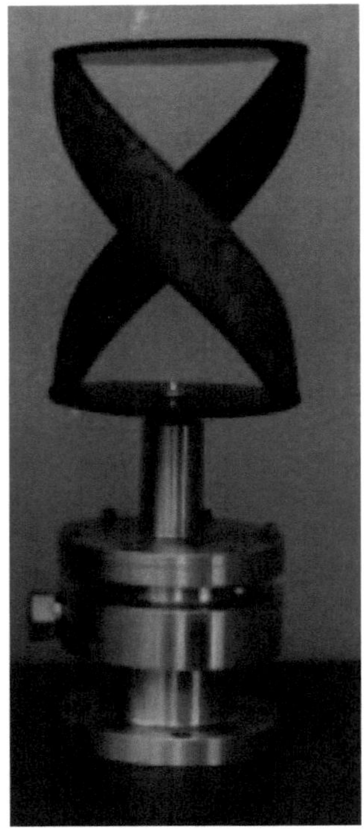

Fig. 21.1 Double-helix Gorlov turbine for underwater installation [6]

which was invented in the 1990s by Prof. A. M. Gorlov from Northeastern university in Boston (Fig. 21.1). It uses helical blades designed in such a manner that one component of force acting from water current upon the blade creates torque in the same direction, independent of the direction of the current [6]. This makes the GHT a convenient tool to operate in the conditions where water current direction can change, including rivers, lakes, and seas. The GHT is an example of damless hydropower technology, which is considered safer and more environment-friendly than traditional methods. It is claimed that a GHT spins slowly enough that fish can see it soon enough to swim around it [13].

21.2.3 Wave Energy Collectors

Another type of hydropower devices is the wave energy collector (WEC). The momentum of the movement of water bodies is a significant resource for energy harvesting. Ocean waves present a form of consistent energy by winds passing over open water. It has been demonstrated that capturing wave energy in

Table 21.1 Classification of wave energy collectors

Device classification	Description	Orientation
Attenuators	Long multi-segment devices	Devices placed parallel to wave travel
Overtopping devices	Large reservoirs filled then released, gravity then pulls water to harness energy	Large reservoirs can be mobilized to any convenient location
Point absorbers	Vertical oscillating motion converted to rotational motion	Floating structure anchored to ground
Terminators	Generally piston pressured air driven turbines	Onshore or near-shore devices placed perpendicular to wave travel direction

offshore locations is possible [2]. In general energy harvesting applications, the power generation system consists of a turbine, generator, and power electric module [10]. Many designs using this type of system have matured into the commercial prototype testing stage (Table 21.1).

There are four major types of wave energy collectors: Terminators, Attenuators, Point Absorbers, and Overtopping Devices. Terminators extend perpendicular to the direction of wave travel. These devices are usually located onshore or near shore. Attenuators are long multi-segmented floating structures which are oriented parallel to the vector of the wave travel. Point absorbers are pin point devices which have a small horizontal dimension in comparison to their vertical dimension. Point absorbers use the rise and fall of the wave height at a single point for the purpose of harvesting the wave energy. Overtopping Devices use water reservoirs that are filled by waves attaining levels above the average surrounding ocean. The reservoirs are then released and the gravity force causes a water pressure. The exiting water velocity is used to generate electricity using turbines or different conversion methods [2].

An example of an already commercially viable point absorbing wave energy converting device is the Aegir Dynamo. The vertical motion of the buoyancy component is transferred to the rotational motion to power the motor by the Aegir Dynamo [1]. Generally, water movement is used to displace an electromagnetic coil to generate electricity through a turbine or other mechanism, but other systems use mechanical vibration to apply strain energy to a piezoelectric material [12]. An example of a hydrodynamic energy conversion device utilizing piezoelectric technology is the Energy Harvesting Eel. The Eel uses piezoelectric polymers that convert mechanical flow energy of water into electrical power. The Eel uses the concept of traveling vortices behind a bluff body to put strain on the piezoelectric components of the device. The motion of the device is similar to that of a natural eel moving through water. These devices will be fairly inexpensive and are easy to fit to a specific system for which each Eel will be scaled to generate from milliwatts to many watts. The overall electrical output is a function of the size of the Eel and the flow velocity of the specific environment in which the device is to be implemented [16].

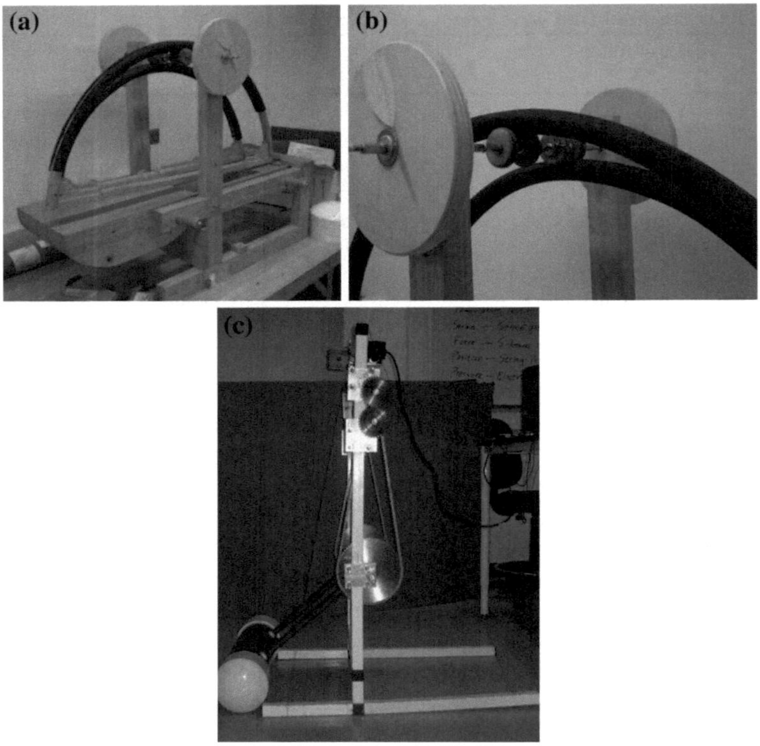

Fig. 21.2 The UWM coastline Wave energy conversion machine (**a**) the earlier model with a wooden frame (**b**) detail, showing the bushing and the rubber gearing (**c**) newer model with the metal frame

21.3 UW-Milwaukee Wave Energy Conversion Device

In this section we will discuss a wave energy collection device developed at the UWM, The machine is a coastline application wave energy machine, which uses a buoy and a rocker arm to change the linear, vertical motion of the wave into mechanical rotational energy. Two models of the machine were built (Fig. 21.2). The earlier model was developed in 2010 by Dr. Patel from Milwaukee (Fig. 21.2a). It used wood frame and bicycle bushing-based ratcheting mechanism, which guaranteed that the output shaft always rotates in the same direction. A simple friction-based rubber bushing was provided (Fig. 21.2b). The second more advanced model with a steel frame was designed in 2011 by Ben Barger and Dr. Ilya Avdeev, at Advanced manufacturing laboratory of the UWM (Fig. 21.2b).

21.3.1 Experimental Study of WEC Device Efficiency

The basic concept of Dr. Patel's initial wave energy machine is a buoy that floats on top of the wave as it passes underneath. The buoy is set up in a 'teeter-totter' fashion, having the buoyant part on the leading edge of the pivot. This buoy has large arches attached over the platform, which rotate above and below an overhead shaft, creating a rotating motion. The linearization of the crankshaft movement is attained through a series of clutch bearing installed in dowels on the shaft. These clutch bearings allow the upward, as well as the downward motion of the wave to create a single directional spin on the crankshaft. This is necessary as an oscillating motion would be difficult to convert to electrical energy. This method, however, is very effective at producing dc current, as all energy put in will translate to losses and electrical energy out, as opposed to having a missed power stroke in either the upward or downward direction.

The buoy itself is only about a foot wide, which makes it great as a proof-of-concept machine. Had the buoy been much wider, it would then require a much wider wave to produce an effective power supply to justify such a large machine. A final comparison will be made as to the efficiencies of this initial machine and the efficiencies of a newly prototyped machine.

The main buoy, shown on the left, rises and falls with the waves causing the series of gears to increase rotations/second to the generator, shown attached to top of the frame. The wave energy machine itself has enough stability and area to hold the variety of components we will be using to test the machine. The device uses the simple harmonic motion of waves and the buoyancy force provided by the water to drive a crankshaft, after the conversion of the vertical motion to the rotational motion necessary for the motor to generate electricity. The buoyancy force of the water coupled with the gravitational force on the block to maintain a consistent torque on the crankshaft. This rotational energy is then put through a series of gears to increase RPM, as well as three clutch bearings to create a linear rotational motion at the electrical generator. The generator at the top then transmits voltage and amperage to a load.

There is a torque applied about the base axis, causing rotation of the semicircular device components. There is a mechanical advantage between the driving semicircular components and the drive shaft. This step up in velocity is proportional to both the base and pinion radii. Unidirectional rotational motion is assured due to the ratcheting mechanisms in the crankshaft. Each of the two semicircular components is attached to a separate ratchet, which are oriented opposite to one another, as seen on the right. This type of orientation causes one ratchet to be engaged on the rise of the device, the other engaged on the fall of the waves. The device is to be anchored on a fixed base of some sort, an example being a post in the ground or a rock outcropping. The overall goal of this specific device is to provide constant electrical power to an onshore system. For the purposes of the analysis performed, the system remains ambiguous, as the analysis simply provides general design considerations for a vast array of possible system setups.

The target goal for the efficiency of wave energy harvesting devices is around 35%. The current model is clearly below this. Having a high efficiency would

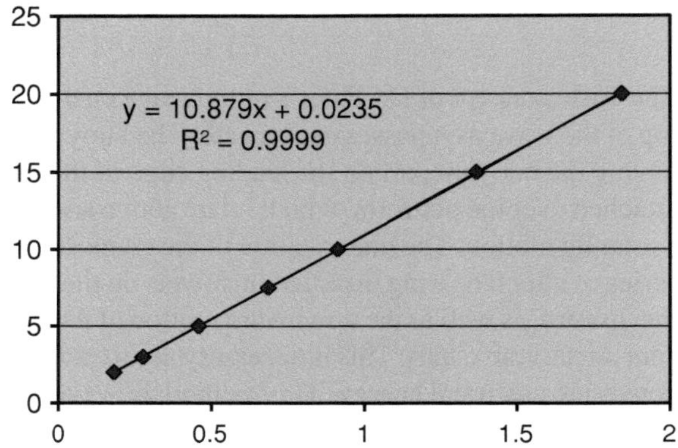

Fig. 21.3 Calibration curve used for the voltage divider system

make the machine more competitive with other energy sectors such as solar and wind which have relative efficiencies of 20 and 32%, respectively. In order to investigate and improve the efficiency, a number of experiments were conducted. A potentiometer was used to determine displacement of the buoy as it rises and falls. The rotation of the potentiometer results in a change in resistance. Output voltage was calibrated and converted to an angle of rotation. This angle will then be used to determine the height of the buoy, thus, its potential energy.

A voltage divider circuit was used to determine the output of the generator on the wave energy machine, since the voltage allowed for the data acquisition unit was only in the range of ±10 V. The divider consists of a 1 KW and 100 W resistor in series, with the 1 kW being the leading resistor. The voltage of the machine is then taken after the 1 kW resistor and put to the data acquisition unit, resulting in roughly one-tenth the power to the system. These voltages were then calibrated to determine actual power generated by the wave energy machine. The data was analyzed using data acquisition units and Labview software. The general idea is to convert the calibrated outputs from the potentiometer and voltage divider to energy units (Joules) to determine the overall efficiency of the system. The graphs generated in Labview will then be postprocessed in Excel to attain the Joule Ratio Desired. The potentiometer as well as the voltage divider had to be calibrated. The calibration curves were done in a similar fashion, and the curve for the voltage divider is depicted in Fig. 21.3. The results were accepted as accurate as the R-squared value was extremely close to 1. It was determined through experimentation that the accuracy of both units was in the 95% range.

The results of this experiment can be seen below in a series of figures. Each set of data was portrayed in a graph as to show all systems and conversions working harmoniously. After acquiring the data in Labview, the chart data was exported to Excel for further processing. The data for the buoy was output in Joules, so the difference in initial Joules to final Joules had to be taken to determine actual energy input into the buoy. This graphical data is displayed in Fig. 21.4. The wattage of the generator output had to be modified significantly. The power in Watts was averaged over the time of output, and then multiplied by the length of

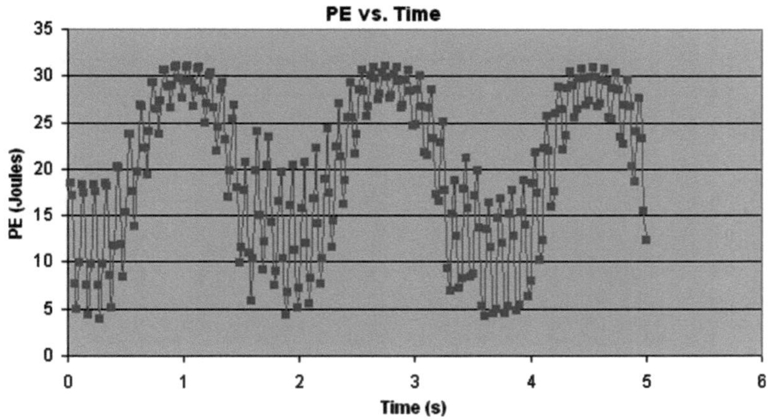

Fig. 21.4 Potential energy of the buoy versus time

time of the averages. This results in converting the Watts to Joules, allowing for a comparison of input and output resulting in efficiency of the system. The Excel graph for the Watts produced by the generator is depicted in Fig. 21.5. It could also be seen that there was significantly less power in the output than in the input. After analyzing a significant number of trials, the final efficiency of the machine was found at 2.5%. The initial theoretical prediction was around 2%.

21.3.2 Optimizing the Dynamics of the UW-Milwaukee Wave Energy Conversion Device

The major controllable device parameters are the radius of the base, radius of the pinion on the crankshaft, volume of the buoyancy block, mass of the buoyancy block, and desired power output for the system that is being powered. These are the key parameters to be altered in a more optimized design. In order for the device to be more effective in a pinpointed practical application, the desired power output is a very necessary design criterion for optimal power output efficiency. Congruency of the power factor with the system being powered is of great importance to guarantee high efficiency [10].

A diagram of the WEC device is presented in Fig. 21.6. Using the numerical analysis program MATLAB, the derived equations of motion of the device can be visualized and optimized. In this model of the UWM WEC, the power output, angular velocity of the driveshaft, and torque applied on the driveshaft can all be analyzed at different points in time as functions of the base radius and pinion radius. With the known critical parameters of the motion of the waves, the points in time of interest are easily pinpointed and analyzed. With this analysis, some interesting points become prevalent. The figure on the right is a MATLAB output

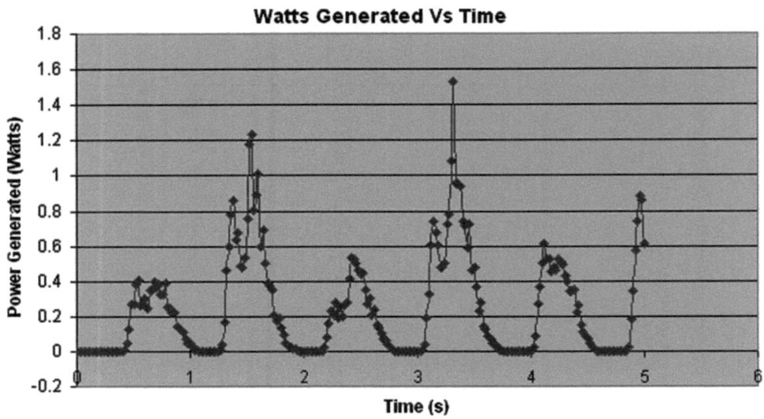

Fig. 21.5 Power generated by the machine versus time

of the current wave energy machine design, with the variables of pinion radius and base radius plotted for overall power output. Agreeing with the conservation of energy, the overall power output is only a function of the base radius, which affects the input torque to the system. This is due to the step up in the angular velocity being exactly proportional to the step down in applied torque. The kinematics of the model break down when the base radius value is exceeded by the amplitude of the wave. This makes sense because when the wave height exceeds the height of the device, the wave would dampen the flywheel and the crankshaft causing the motion of the driving components of device to be greatly slowed.

A fairly simple concept for providing a constant applied torque to the crankshaft is equalizing the vertical sums of forces which provide the torque for the rising and falling situations. Working off the assumption that the UWM WEC is symmetric except for the buoyancy block, the calculation of the relationship between the mass and volume of the block can be easily determined. During the rising portion of the motion of the device, the sum of forces in the vertical axis is the buoyancy force less the gravitational force. In the falling portion of the motion, the only significant vertical force is the gravitational force on the buoyancy device. Setting sum of forces for each situation equal to one another, it is found that the buoyancy force is to be twice that of the gravitational force for equal torque in the rise and fall phases of the device.

A major tribological factor affecting the efficiency of the transmission components of the UWM WEC is the potential slip between the crankshaft pinions and the semicircular drivers. The current design has a rubber like material to increase the friction between the contact surfaces, as seen in the figure on the right. There is significant slip that can be associated with this configuration, especially when operating in marine conditions when the rubber like material becomes wetted. In a more efficient design, a chain system could be implemented in order to reduce loss in the transmission. With a chain or pulley design, deterioration of the chains and chain lubricant is a clear concern which must be considered during the design process.

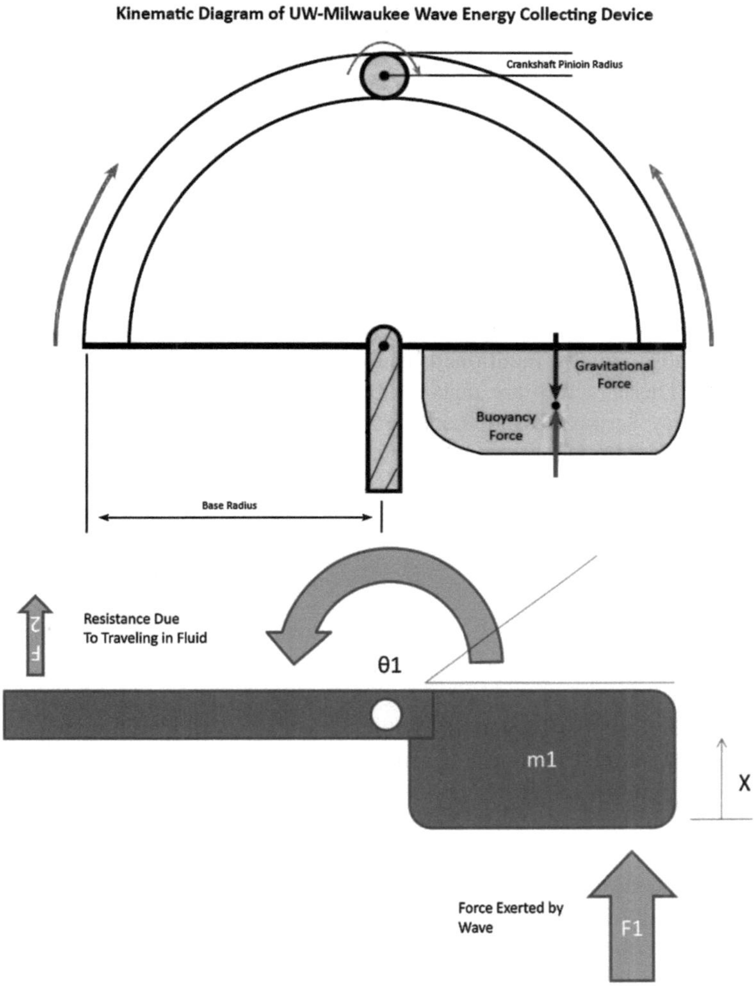

Fig. 21.6 Schematic diagram of the WEC device (model 1)

21.4 Tribological Challenges of Wave Energy Conversion Devices

The main challenge of wave energy machine design is the ability to create a device that is efficient and able to withstand the harsh conditions that is present in both fresh and salt water. Since these machines have many parts involving solid surfaces in relative motion, tribological considerations play a significant role. In addition to pure tribological considerations, such as lubrication, friction, and wear, in water energy collection system there are additional factors related to the corrosive effects of salt water, fouling growth, and abrasive suspended particles. Seals, welds, bearing surfaces, and electrical insulation materials are some of the locations most important to the wave energy converter which are threatened by biofouling. Solid particles, for example, sand, can infiltrate the system and become trapped between contacting surfaces. These hard particles are a source of abrasive wear and increased friction, both damaging the system components and reducing efficiency. Ice build up on the machine can cause the balance of the system to be interrupted and the efficiency to be greatly affected. This misbalancing can cause unknown and strong forces on device components including very high torques throughout the entire device, as well as lubricant degradation. Salt water contamination and contamination from the atmosphere are also factors which, in addition to those already mentioned, can possibly reduce the working life of the device [17]. These factors are unavoidable when working in marine conditions. Therefore, very serious design considerations are necessary to make a device economically viable. Having a consistently working device is possibly the most important monetary factor to determine if the device is worth investing in.

21.5 Conclusions

Currently, alternative sources of energy are at the forefront of the energy industry. Being the most widely used source of energy, electricity generation is an enormous industry. Dwindling fossil fuels are less viable because of environmental and economic reasons, while generating electricity from natural sources, such as wave energy, is becoming a focused goal of research and implementation throughout the world. Renewable sources of electricity have been decreasing as a proportion of overall energy generated in the United States over the past decades. New policies are beginning to encourage development of new renewable energy technology helping to conserve natural resources and making the electricity generation industry less dependent on non-renewable energy as well as protecting the environment. A number of tribological constraints will be involved into the development of these devices. The Wave Energy Collection Device which is being developed at UWM is an example of such technology to help satisfy these future energy goals.

References

1. A. Al-Havaibeh, D. Su, J. McCague, A. Knight, An innovative approach for energy generation from waves. GCReeder **51**(8), 1664–1668 (2010)
2. Anonymous, Technology White Paper on Wave Energy. Minerals Management Service, Renewable Energy and Alternate Use Program, U.S. Department of the Interior, 2006, http://ocsenergy.anl.gov
3. T. Burton, D. Sharpe, N. Jenkins, E. Bossanyi, *Wind Energy Handbook* (Wiley Blackwell, New Jersey, 2001)

4. J.S. Carlton, *Marine Propellers and Propulsion* (Butterworth-Heinemann, London, 2007)
5. P.L. Fraenkel, Power from marine turbines. Proc. Inst. Mech. Eng. A **216**, 1–14 (2002)
6. A.N. Gorban', A.M. Gorlov, V.M. Silantyev, Limits of the turbine efficiency for free fluid flow. ASME J. Energy Res. **123**, 311–317 (2001)
7. M.N. Kotzalas, G.L. Doll, Tribological advancements for reliable wind turbine performance. Phil. Trans Royal. Soc. A. **368**, 4829–4850 (2010)
8. B. Kuri, F. Li, Allocation of emission allowances to effectively reduce emissions in electricity generation. Power and Energy Society General Meeting, PES '09. IEEE (2009), pp. 1–8
9. X.I. Li, C.W. Yu, Impacts of emission trading on carbon, electricity and renewable markets: a review, IEEE Power and Energy Society General Meeting (2010), pp. 1–7
10. S.E. Lyshevski, High-power density miniscale power generation and energy harvesting systems. Energy Convers. Manage. **52**(1), 46–52 (2011)
11. F.C. Menz, Green electricity policies in the United States: case study. Energy Policy **33**(18), 2398–2410 (2005)
12. M. Pereyma, Overview of the modern state of the vibration energy harvesting devices. Perspective Technologies and Methods in MEMS Design. MEMSTECH (2007), pp. 107–112
13. R. Petkewich, Technology solutions: creating electricity with undammed hydropower. Environ. Sci. Technol. **38**, 55A–56A (2004)
14. R. Ramakumar, W.L. Hughes, Renewable energy sources and rural development in developing countries. IEEE Trans. Edu. **E-24**(3), 242–251 (1981)
15. M. Stefanovich, Wave energy and public opinion in the state of Oregon, USA.. OCEANS MTS/IEEE Biloxi—Marine technology for our future: global and local challenges (2009), pp. 1–10
16. G.W. Taylor, J.R. Burns, S.M. Kammann, W.B. Powers, T.R. Welsh, The energy harvesting Eel: a small subsurface ocean/river power generator. IEEE J. Oceanic Eng. **26**(4), 539–547 (2001)
17. R.J.K. Wood et al., Tribological design constraints of marine renewable energy systems. Phyl. Trans. R. Soc. A **368**, 4807–4827 (2010)

Editor's Biography

Dr. Bharat Bhushan received an M.S. in Mechanical Engineering from the Massachusetts Institute of Technology in 1971, an M.S. in Mechanics and a Ph.D. in Mechanical Engineering from the University of Colorado at Boulder in 1973 and 1976, respectively, an MBA from Rensselaer Polytechnic Institute at Troy, NY in 1980, Doctor Technicae from the University of Trondheim at Trondheim, Norway in 1990, a Doctor of Technical Sciences from the Warsaw University of Technology at Warsaw, Poland in 1996, and Doctor Honouris Causa from the National Academy of Sciences at Gomel, Belarus in 2000. He is a registered professional engineer. He is presently an Ohio Eminent Scholar and The Howard D. Winbigler Professor in the College of Engineering, and the Director of the Nanoprobe Laboratory for Bio- & Nanotechnology and Biomimetics (NLB^2) at the Ohio State University, Columbus, Ohio. His research interests include fundamental studies with a focus on scanning probe techniques in the interdisciplinary areas of bio/nanotribology, bio/nanomechanics and bio/nanomaterials characterization, and applications to bio/nanotechnology and biomimetics. He is an internationally recognized expert on bio/nanotribology and bio/nanomechanics using scanning probe microscopy, and is one of the most prolific authors. He is considered by some as a pioneer of the tribology and mechanics of magnetic storage devices. He has authored 7 scientific books, more than 90 handbook chapters, more than 700 scientific papers (h-index—52+; ISI Highly Cited in Materials Science, since 2007), and more than 60 technical reports, edited more than 50 books, and holds 17 US and foreign

M. Nosonovsky and B. Bhushan (eds.), *Green Tribology*,
Green Energy and Technology, DOI: 10.1007/978-3-642-23681-5,

patents. He is co-editor of Springer NanoScience and Technology Series and co-editor of Microsystem Technologies. He has given more than 400 invited presentations on six continents and more than 160 keynote/plenary addresses at major international conferences.

Dr. Bhushan is an accomplished organizer. He organized the first symposium on Tribology and Mechanics of Magnetic Storage Systems in 1984 and the first international symposium on Advances in Information Storage Systems in 1990, both of which are now held annually. He is the founder of an ASME Information Storage and Processing Systems Division founded in 1993 and served as the founding chair during 1993–1998. His biography has been listed in over two dozen Who's Who books including Who's Who in the World and has received more than two dozen awards for his contributions to science and technology from professional societies, industry, and US government agencies. He is also the recipient of various international fellowships including the Alexander von Humboldt Research Prize for Senior Scientists, Max Planck Foundation Research Award for Outstanding Foreign Scientists, and the Fulbright Senior Scholar Award. He is a foreign member of the International Academy of Engineering (Russia), Byelorussian Academy of Engineering and Technology and the Academy of Triboengineering of Ukraine, an honorary member of the Society of Tribologists of Belarus, a fellow of ASME, IEEE, STLE, and the New York Academy of Sciences, and a member of ASEE, Sigma Xi, and Tau Beta Pi.

Dr. Bhushan has previously worked for the R&D Division of Mechanical Technology Inc., Latham, NY; the Technology Services Division of SKF Industries Inc., King of Prussia, PA; the General Products Division Laboratory of IBM Corporation, Tucson, AZ; and the Almaden Research Center of IBM Corporation, San Jose, CA. He has held visiting professor appointments at University of California at Berkeley, University of Cambridge, UK, Technical University Vienna, Austria, University of Paris, Orsay, ETH Zurich, and EPFL Lausanne.

Michael Nosonovsky is an Assistant Professor at University of Wisconsin–Milwaukee. He got his M.Sc. from St. Petersburg Polytechnic University (Russia) and Ph.D. from Northeastern University (Boston). After that he worked at Ohio State University and National Institute of Standards and Technology. His interests include biomimetic surfaces, capillary effects, nanotribology, and friction-induced self-organization. Among his research findings are:

- Multiscale roughness can stabilize a composite superhydro/oleophobic interface, which is used for new omniphobic surfaces.

- Friction-induced vibrations and instabilities of various types can lead to self-organized structures (e.g., protective in situ tribofilms), and the phenomenon is investigated by non-equilibrium thermodynamics. The transient running-in process can be treated as self-organization (since surfaces adjust to each other) in accordance with the minimum entropy production theorem. Multiscale thermodynamic models of self-healing materials bridged the gap between the theories of self-organization and practical efforts in synthesizing self-healing materials.
- Empirical laws of friction and wear (irreversible energy dissipation and material deterioration) are consequences of the 2nd law of thermodynamics, and thus friction is a fundamental physical phenomenon rather than a collection of various unrelated mechanisms.
- Pressure in condensed water capillary bridges between nanoasperities is deeply negative (tensile stress), possibly showing the world record of negative pressure in water (-150 MPa).

Index

A
Abrasive, 447
Accumulation, 155
Active transport, 551
Acyloxylation, 311
Adams-matrins instabilities (ami), 59, 62
Additives, 114, 267
Adhesion, 174–176, 178, 203–205, 210, 211, 213
Adhesion force, 203
Adhesion hysteresis, 27, 29
Adhesive tapes, 11
Adhesive wear, 460
Adhesive-driven, 55
Aerospace, 456
Afm, 57
Air gaps, 160
Algae, 574
Al-si-graphite, 451
Alumina, 168
Aluminium smelter, 596
Aluminium, 601
Aluminum alloy, 155
Aluminum fiber, 439
Aluminum-graphite, 450
Amontons-coulomb, 73
Amorphous, 153
Amorphous polymers, 206, 208
Animal-fat, 17
Anodic aluminum oxide (AAO), 188, 206
Antifouling, 149, 551
Antifriction materials, 440
Antifriction, 339, 448
Anti-icing, 12
Antimony doped tin oxide (ATO), 194
Anti-wear, 306
Aramid fibers, 431
Arithmetic average, 169
Asbestos, 330, 431
Aspect ratio, 165, 188, 201
Aspect ratios, 188, 190, 192
Atomic force microscopy (AFM), 176, 213
Attractive force, 176, 177
Auger electrons, 212
Automobile, 456
Automotive, 274, 430
Avalanche dynamics, 61
Avalanche-like, 43
Avalanches, 62
Az91d, 461

B
Backscattered electrons, 212
Bacteria, 117
Base oil, 266
Bearing, 612
Being locally attuned, 119
Bénard cell-like, 69, 71
Binder, 430
Biodegradability, 120, 287, 289, 322
Biodegradable, 266
Biodegradable lubricants, 291, 302, 323
Biodegradable lubrication, 8
Bio-diesel, 574
Biofilming, 535
Biofouling, 551
Biofuels, 569
Biological tribosurfaces, 115
Biology, 109

M. Nosonovsky and B. Bhushan (eds.), *Green Tribology*,
Green Energy and Technology, DOI: 10.1007/978-3-642-23681-5,

B (*cont.*)
Biomass, 570
Biomimetic, 173, 214, 244, 251
Biomimetic materials, 9
Biomimetic superhydrophobic, 194
Biomimetics, 109, 536
Biomimicry, 223–225, 262
Biomimicry guild, 105, 116
Biomineralization, 117
Biorefineries, 576
Biorefineries, 577
Biorefinery, 568, 572
Biorefining, 565, 579
Bio-waste, 337
Boric acid, 269, 339
Boric oxide, 270
Boron nitride, 462
Bottom ash, 435
Brackish, 545
Brake disk, 431
Brake lining, 434
Brake materials, 430
Brake, 330
Braking system, 430
Breaking-in, 114
Brine, 561
Buoy, 612
Buoyancy, 613

C
Canola oil, 278
Capillary force, 10, 175, 176, 178, 190
Carbon fiber, 330, 434
Carbon fibers, 197, 201
Carbon nanotube, 532
Carbon nanotubes, 201, 202
Cashew, 339
Cassie to Wenzel transition, 191
Cassie–Baxter model, 160, 181, 202, 203
Cassie–baxter, 553
Cassie–wenzel wetting regime transition, 155
Casting, 448
Causal knowledge, 109
Cell adhesions, 80, 81, 92, 96
Cellulose, 578
Cenosphere, 432
Characteristic X-rays, 212, 213
Characterization, 211, 213
Chemical bath deposition, 206
Chemical composition, 168
Chemical modifications, 302
Chemical potentials, 546
Chemical treatments, 331
Chemical vapor deposition (CVD), 192
Chemistry, 602
Climate change, 607
Closed tribosystems, 117
Closely packed arrays, 199
Cloud point, 320
Coal, 431
Coal-to-lquid (ctl), 567
Coating, 151
Coatings, 448
Coaxial electrospinning, 187
Coconut oil, 293
Coefficient of friction, 293, 294, 306
Coefficient, 305
Coir fiber, 342
Collagen, 117
Collapse of societies, 106
Com, 343
Complex, 115
Composite against, 461
Composite materials, 446
Composite, 151, 200, 201, 553
Composites, 202, 214
Condensation, 540
Conductivity, 244, 258
Contact angle goniometer, 211, 212
Contact angle hysteresis, 150, 186, 189, 204, 210
Contact angle, 150, 179–181, 552
Contact area, 176, 190, 223, 233, 234, 243, 248, 249, 251, 257, 261
Contact mechanics, 178
Contact pressure, 450
Contact strip, 460
Contaminants, 150
Cooperative relationships, 119
Copper base alloy, 168
Copper, 275, 601
Copper-graphite, 459
Corn oil, 298
Corrosion, 290, 321, 540
Corrosive, 166
Cotton, 197, 198, 201, 330, 335
Cottonseed oil, 298
Counterpart, 450
Crankshaft, 613, 616
Crystallization, 203, 206–208
Crevices, 157
Cross-sectional area, 152
Current densities, 459
Current density, 450
Cutting edges, 599
Cutting tool, 600

D
Damping characteristic, 56
Darkling beetle, 14
Deformation-driven friction, 55
Degradation, 72–74
Degree of unsaturation, 304
De-icing, 14
Delamination, 458
Density, 448
Dermatitis, 596
Desalination, 540
Destabilization, 43, 49–52, 54, 59, 74
Deterioration, 73, 151
Development, 118
Diatoms, 113
Diffuse, 546
Dimerisation, 309, 311
Dimethylformamide (DMF), 186
Dip-coating method, 201
Dlvo theory, 176
Double bonds, 292, 303, 304, 311, 319
Double layer, 175, 176
Drinking water, 19
Drum, 431
Dry adhesion, 174
Dry sliding, 452
Dynamic characteristics, 370

E
Edible vegetable oils, 293
E-factor, 7
Electrical conductivity, 459
Electrical contacts, 470
Electrical current, 460
Electrical sliding, 459
Electrodialysis, 532
Electrospinning, 182, 187, 188
Electrospinning, 182, 186
Electrostatic force, 10
Elemental analysis, 213
Energy dissipation, 73
Energy efficiency, 108
Energy harvesting, 607, 611
Energy savings, 586, 587, 594
Energy usage, 566
Energy, 105, 580
Enthalpy, 542
Entropy, 47, 547
Environment, 118, 450, 603
Environmental regulations, 565
Environmental, 579
Environmentally friendly lubricants, 302
Environmentally friendly, 271, 330, 446
Epoxidation, 303, 304
Epoxy resins, 203
Equilibrium value, 154
Erosion wear, 469, 472
Erosion, 460
Erosive wear, 166
Eskom, 580
Etchant, 151
Evaporation, 543
Experimental details, 379
Ex-situ, 68

F
Fatty acids, 304, 310, 312, 316
Feedback loop, 49, 66, 119
Fiber modification, 333
Fiber orientation, 333
Fibers, 165
Fillers, 430
Finishing, 114, 115
Fire points, 320
Fischer-tropsch, 566, 570, 577
Flash, 320
Fly ash, 431
Form, 117
Free fatty acid value, 319
Friction and wear, 55
Friction coefficient, 430
Friction, 174, 210, 267, 330, 446
Frictional additives/modifiers, 430
Frictional sliding, 47, 49, 61
Friedel–crafts acylation, 312
Friedel–crafts alkylation, 311
Fuel specifications, 567
Fuel, 579
Fuels, 568
Function, 117
Fzg machine, 582

G
Galling, 450
Gas constan, 547
Gasification, 576
Gas-to-liquid (gtl)-technology, 567
Gearbox, 492, 591
Gearboxes, 581, 585, 586, 591
Gears, 583
Gecko, 28
Gecko-effect, 174, 214
Generalists, 112
Geothermal energy, 18
Geothermal, 537

G (*cont.*)
Gibbs, 547
Glass fiber, 330
Glass fiber, 431
Global challenges, 105
Global challenges, 121
Goniometer, 159
Grades of oil, 591
Graphite film, 449
Graphite particle size, 455
Graphite, 159, 339, 438, 447
Green biomass, 574
Green chemistry, 19
Green composites, 339
Green engineering, 19
Green liquid lubricants, 267
Green lubricants, 267
Green metal-working, 595
Green nanotribology, 105
Green tribology, 19, 173, 174, 214
Greenhouse gas, 561
Groundnut oil, 299

H
Hardness, 434
Helical turbine, 609
Hemispherical, 163
Heterogeneity, 149
Hierarchical structure, 201, 208
Hierarchical structures, 191, 193, 203
Hierarchical, 115
Hierarchy, 118
High-temperature, 447, 466
Hip joint, 121
Holographic lithography, 210
Homogeneous, 170
Honeycomb polyetherimide films, 552
Hot embossing, 193
Humidification-dehumidification, 545
Humidity, 240
Hybrid composite, 446
Hydroformylation, 306
Hydrogenation, 307
Hydrological water cycle, 532
Hydrophilic cotton, 201
Hydrophilic, 79, 82, 90, 92, 93, 97, 99, 179–181, 196, 201, 209, 552
Hydrophobic, 79, 82, 90, 92, 97, 99, 151, 196, 200, 201, 203, 204, 209, 552
Hydrophobic poly (ethylene terephthalate) PET, 198
Hydrostatic pressure, 547

I
Ice friction, 223
Ice, 13
Information, 117
Ingredients, 339
Initial consolidation, 114
Injection molding, 189
In-situ tribofilms, 68
In-situ, 62
Instability mechanism, 59
Interface film growth, 63
Interfacial energies, 155
International desalination association, 532
Introduction, 349
Iodine value, 319
Irreversible, 41, 49, 55

J
Johnson–Kendall–Roberts (JKR), 175, 178

K
Kinematic viscosity, 319

L
Laboratory trial, 586
Lamellar solids, 268
Layer-by-layer (LBL), 196
Layered, 447
Leverage of cyclic processes, 121
Liquid lubricants, 266
Liquid-like layer, 227
Liquid-likelayer, 223, 228, 230–234, 236
Lithography, 193
Local resources, 117
Lotus effect, 10, 26, 151
Lotus leaf , 194, 203, 204
Lotus leaves, 193, 201
Lotus, 114
Lotus-effect, 174, 535
Lubricant, 266, 446, 581, 594, 600, 618
Lubricants, 319, 322, 566, 580
Lubricated drilling, 601
Lubrication, 597

M
Maglev transportation system, 467
Magnesium-graphite, 456
Magnets, 117
Manufacturing, 274
Marine energy, 608

Material selection, 114
Materials, 450
Matrix, 151, 552
Mechanical mixed layer, 477
Mechanical wear, 471
Mechanicalvapor compression, 540
Mems, 13
Meniscus, 176
Metal cutting, 566
Metal matrix composites (MMC), 149, 432, 446
Metal-working, 594
Methyltrichlorosilane (MTS), 206
Micro-dosing, 595
Micromechanically optimized, 121
Micron-size, 280
Micropatterns, 191, 200
Micropillar, 192, 203
Microscale, 43, 54, 69, 71
Microstructured surface, 12, 13, 151
Micro-topography, 151
Microwave-heat-treated, 463
Millennium project, 105
Mining industry, 593
Modelling and analysis using computational fluid dynamics, 378
Molybdenum disulfide, 266, 447
Montmorillonite, 311
Morphology, 182, 186, 196, 207, 208, 211, 212
Morse, 547
Multi-effect distillation, 532
Multi-functional, 115
Multifunctionality, 118
Multi-stage flash, 532
Mustard oil, 299

N
Nanoagents, 108
Nanocomposite materials, 15
Nanocomposite, 187, 193
Nanocomposites, 202, 203, 213
Nanodiversity, 120
Nanofabrication, 188
Nanofibers, 188–190
Nanohair, 190
Nanoimprint lithography (NIL), 193
Nanoparticle, 551
Nanopatterns, 188, 189, 192, 193
Nanopillars, 190, 191
Nanoprocesses, 108
Nanoscale, 200, 278
Nanostructures, 79–81, 83, 90, 92, 96–99
Nanosurfaces, 108
Nanotechnology, 121
Nanotopography, 81–83
Nanotribology, 15
Nanotubes, 190, 201, 202, 551
Natural composites, 338
Natural fibers, 330, 430
Natural lubrication, 8
Natural vacuum distillation, 537
Negative mold, 194, 195
Nickel mold, 194
Nickel-graphite, 473
Nile roses, 338
Non-deterioration effect, 42, 52
Non-edible oils, 293
Non-equilibrium process, 47
Non-equilibrium thermodynamics, 42
Non-solvent, 207–209
Normal force, 223, 227, 236–238, 241, 242, 244, 257
Normal load, 453
Nylon, 441

O
Oil filters, 520
Oil-based lubricants, 121
Oleophobicity, 154
Oligomerisation, 309
Onsager, 72
Oxidation stability, 322
Oxidation wear, 458
Oxygen plasma etching, 199

P
Particle size, 458
Particles, 152
Particulate contaminants, 512
Pattern formation, 52, 63, 64
Permeability, 550
Petal effect, 27
Petrochemical, 571
Phase separation, 207, 208
Phenolic composites, 339
Phenolic matrix, 434
Phenolic resin, 332
Pin-on-disk experiment, 62
Pin-on-disk, 278
Pitch bearings, 495
Plasma treatment, 206, 209
Pollution, 115
Poly (e-caprolactone) (PCL), 187
Poly (furfuryl alcohol) (PFA), 202

P (*cont.*)
Poly (L-lactic acid) (PLLA)
superhydrophobic, 209
Poly (vinyl chloride) (PVC), 187, 208
Polyamid6 (PA6), 187
Polydimethylsiloxane (PDMS), 181
Polyelectrolytes, 196
Polyester, 333
Polyethylene (PE), 189, 198
Polyimide, 196
Polymer composites, 337
Polymer-matrix composites, 432
Polypropylene (PP), 189, 190, 207
Polystyrene (PS), 181, 187, 190, 199, 203, 209
Polytetrafluoroethylene (PTFE), 180
Polyurethane (PU), 200
Polyvinylidene fluoride (PVDF), 198, 206
Pour point, 293, 294, 320
Powder lubricants, 17, 268
Powder metallurgy, 448
Power consumption, 580, 588, 591
Power generation, 570
Power savings, 588
Power stations, 585
Precipitation, 206–209
Precipitator, 432
Prigogine, 42
Prigogine's, 54
Profile entropy, 55
Profilometer, 169
Protrusions, 150
Pulley, 616
Pulsed electron deposition (PED), 210
Pyrolysis, 576

R
Radio frequency (RF), 210
Rayleigh waves, 59
Reaction-diffusion, 43
Recycling, 117, 432
References, 390
Reinforcement particles, 553
Reinforcement volume fraction, 171
Reinforcement, 149
Relative humidity, 223, 240, 241
Renewable energy, 542, 607, 618
Renewable power generation, 483
Renewable sources of energy, 18
Repair, 118
Replication, 190, 193, 194
Repulsive force, 177
Resilience, 121
Responsive, 115
Responsiveness, 119
Results and discussion, 382
Retention, 551
Reusing, 117
Reverse osmosis, 561
Review of literature, 353
Reynolds equation, 361
Rice straw, 339
Rise husk, 339
Root-mean-square (RMS), 155
Rosa hybrid tea, 32
Rose petal effect, 29
Rose petals, 204
Roughness factor, 27
Roughness, 155, 180, 181, 186–188,
189–193, 195–198, 200, 201,
203, 204, 206–210, 214, 226,
228, 235, 243, 246, 249, 250,
251, 257, 260, 261, 282, 553
Running-in, 41, 55, 74

S
Saline, 532
Salinity, 532
Sand fish, 15
Sandpile model, 61
Saturated fatty acids, 292
Scalping, 598
Scanning electron microscope (SEM), 168
Science and technology, 105
Seaweeds, 574
Secondary electrons, 212, 213
Secondary structures, 42, 52, 62, 68
Seed oil palm, 337
Seizure, 448
Selectivity, 537
Self-assembly, 190, 196, 203, 204
Self-cleaning surfaces, 16
Self-cleaning, 10, 26, 42, 152, 174, 206,
207, 551
Self-healing, 42, 55, 70, 74
Self-lubric, 43
Self-lubricating, 42, 52
Self-lubrication, 43, 44, 54, 55, 70, 74, 446
Self-organization, 42–44, 52
Self-organized criticality (soc), 43
Self-repairing, 118
Self-replenishing lubrication, 17
Semipermeable, 546
Shannon entropy, 55, 57, 62
Shark-skin, 174
Silane treatment, 333
Silica nanoparticles, 193, 195–198

Silica, 117
Silicon carbide, 446
Silver-graphite, 470
Simultaneous spinning, 187
Single molecules, 119
Sisal, 331
Sliding angle, 196, 206, 207, 209
Sliding angles, 190
Sliding velocity, 450
Snails, 114
Soc, 61
Social behavior, 119
Soft materials, 114
Solar energy, 535
Sol-gel technique, 197
Solid lubricants, 266, 447
Solute, 546
Solution-casting method, 190
Solvation force, 177
Solvent, 121, 546
South africa, 578
Spectroscopic techniques, 211, 212
Spider, 9
Spin coating, 199–201
Spray-coating technique, 203
Spray, 603
Sputter deposition, 210
Stability analysis, 376
Stability criteria, 373
Stains, 599
Steady state characteristics, 362
Steel working, 602
Stick-slip, 59
Stiffness and damping coefficients, 371
Structure, 447
Structure-function relationships, 117
Submicron-sized, 283
Subsurface deformation, 466
Sugarcane, 333
Summary and conclusions, 389
Sun flower, 338
Sunflower oil, 294
Superhydrophilic, 195
Superhydrophobic, 181, 182, 186, 187, 189, 190, 192, 193, 195, 196, 198–204, 206–210, 552
Superhydrophobic cotton, 197, 198
Superhydrophobic Poly(methyl methacrylate) (PMMA), 209
Superhydrophobic rose petals, 204
Superhydrophobicity, 10, 26, 150, 174
Superoleophobic, 553
Surface chemistry, 186, 196, 202, 211
Surface energy, 174, 186, 187, 191, 192, 196, 198, 199, 206, 209
Surface roughening, 56, 57
Surface structure, 223, 225, 231, 243, 246, 248, 249
Surface texture, 466
Sustainability, 118
Sustainable nanotribology, 105
Sustainable solutions, 121
Synergetics, 42
Synthetic oil, 593
Synthetic oils, 581, 584

T

Tapping mode, 214
Temperature, 214, 223, 227–231, 233, 236–238, 242, 245, 246, 254, 255–261, 450, 466, 588
Tetrahydrofuran, 187
Theory, 361
Thermal conductivity, 257–259, 223, 238, 244, 245, 251, 257, 261, 288, 446
Thermal expansion, 447
Thermal stability, 290, 321
Thermal vapor compression, 540
Thermoelastic instabilities, 62
Thin film composite, 549
Third body, 69
Tidal power turbines, 18
Tidal turbine, 609
Tilting angle, 204
Tilting plate, 153
Tokay gecko, 12
Topography, 190, 196, 199, 202, 203, 206, 209, 211, 212, 214
Transient analysis, 374
Transient loading, 490
Transmission electron microscopy (TEM), 212
Triacylglycerol, 268
Triacylgylcerols, 292
Tribofatigu, 53
Tribological properties, 294
Tribological, 293
Tribology, 561
Triglycerides, 307

U

Unburn carbon, 437
Unsaturated molecule, 303
UV-curable resin, 191, 193

V
Van der waals force, 10
Van der waals forces, 174, 176, 178, 190
Van't hoff factor, 547
Vaporization, 539
Vegetable oil-based lubricants, 303
Vegetable-oil, 17
Vegetable-oils, 266
Velocity, 223, 227, 229, 236–240, 245, 247, 248, 251, 255–257
Viscosities, 582
Viscosity index, 319

W
Water contact angle, 179–181, 186, 187
Water contact angles, 199
Water contamination, 513
Water current, 609
Water lubrication, 17
Water soluble, 114
Water, 121
Water-based chemistry, 120
Water-based lubricants, 121
Wave energy collector, 607
Wave energy collectors, 611
Wave energy conversion device, 615, 618
Wave energy, 18
Wear maps, 456
Wear particles, 17
Wear preventive characteristics, 322
Wear rate, 332
Wear rates, 151
Wear resistance, 281, 441, 447
Wear, 174, 267, 430, 551
Wear-resistant, 163
Wenzel model, 157
Wenzel, 553
Wenzel's model, 180
Wettability, 151, 187, 189, 191, 196, 200–202, 204, 211, 244, 249–251
Wetting, 174, 180, 182, 186, 192, 198, 199, 202–204, 210, 211, 214, 223, 551
Wheat, 330
Why water lubrication?, 352
Wind power turbines, 18
Wind power, 5, 483
Winter sport, 223
Winter sports, 224, 235
Woven glass, 337

Y
Yaw system, 492
Young equation, 26
Young's equation, 154, 179

MIX
Papier aus verantwortungsvollen Quellen
Paper from responsible sources
FSC® C105338

If you have any concerns about our products,
you can contact us on
ProductSafety@springernature.com

In case Publisher is established outside the EU,
the EU authorized representative is:
Springer Nature Customer Service Center GmbH
Europaplatz 3, 69115 Heidelberg, Germany

Printed by Libri Plureos GmbH
in Hamburg, Germany